AF335664

....

Cystic Fibrosis in the 21st Century

Progress in
Respiratory Research

Vol. 34

Series Editor   *Chris T. Bolliger*, Cape Town

Basel · Freiburg · Paris · London · New York ·
Bangalore · Bangkok · Singapore · Tokyo · Sydney

Cystic Fibrosis in the 21st Century

<table>
<tr><td>Volume Editors</td><td>

Andrew Bush, London
Eric W.F.W. Alton, London
Jane C. Davies, London
Uta Griesenbach, London
Adam Jaffe, London

</td></tr>
</table>

83 figures, 15 in color, and 47 tables, 2006

Basel · Freiburg · Paris · London · New York ·
Bangalore · Bangkok · Singapore · Tokyo · Sydney

Prof. Andrew Bush
Department of Pediatric Respiratory Medicine
Royal Brompton Hospital
Sydney Street
London SW3 6NP, UK

Library of Congress Cataloging-in-Publication Data

Cystic fibrosis in the 21st century / volume editor, Andrew Bush … [et al.].
      p. ; cm. – (Progress in respiratory research, ISSN 1422-2140 ; v. 34)
    Includes bibliographical references and indexes.
    ISBN 3-8055-7960-8 (hard cover : alk. paper)
    1. Cystic fibrosis.   2. Cystic fibrosis–Genetic aspects.   I. Bush, Andrew,
  1954 Apr. 24–.   II. Title: Cystic fibrosis in the twenty-first century.
    III. Series.
    [DNLM: 1. Cystic Fibrosis–complications.   2. Cystic Fibrosis–genetics.
  3. Cystic Fibrosis Transmembrane Conductance Regulator–genetics.
  4. Cystic Fibrosis Transmembrane Conductance Regulator–metabolism.
  5. Lung–physiopathology.   WI 820 C9978 2006]
    RC858.C95C972 2006
    616.3′72–dc22

2005023395

Contents

Contents

Foreword

Most of the volumes of the book series *Progress in Respiratory Research* deal with subjects relating to adult pulmonology. To address the needs of the pediatric community the most recent volume, No. 33 in the series, covered a long overdue topic, namely *Paediatric Pulmonary Function Testing*, and was launched just a couple of months ago. When planning the next volume I thought it would be an appropriate and logical sequel to have an update on cystic fibrosis, a topic which interests both pediatricians as well as adult pulmonologists.

True to the vision of our series, the volume should not be yet another textbook, but rather a state-of-the-art overview of the most recent advances in the field. As usual I asked one of the leaders in the area to be the volume editor. When I proposed this to Andi Bush, he enthusiastically accepted but then came back saying he would like to do this book together with four other volume editors. Although I agreed I was quite worried that with so many 'cooks', the book would never materialize. How wrong I was! Not only did the five editors manage to share the work without any problems, they also made sure that the authors delivered their papers on time, and that their contributions represented truly cutting-edge research. This is – among other things – illustrated in the larger number of most recent references of 2004 and even 2005, which has always been my aim for the series.

The usual speed and quality of the publisher, S. Karger AG, Basel, guaranteed that the book was printed within the shortest possible production time.

Looking at the many different aspects of cystic fibrosis covered, the final result is a magnificent book, which will appeal to many more specialists than just pediatric and adult pulmonologists.

All that remains for me to say is well done and a big thank you to the authors, editors, and all people involved at the publisher! The many readers of this volume, No. 34 in the *Progress in Respiratory Research* series, will appreciate its quality.

C.T. Bolliger
Cape Town

Preface

Less than 70 years ago, cystic fibrosis (CF) was a disease that was uniformly fatal in the first year of life, and that could only be differentiated from other gastro-intestinal diseases at autopsy. Over the years, advances such as the development of the sweat test enabled greater diagnostic accuracy and the beginnings of understanding of at least some aspects of pathophysiology.

The real explosion in the knowledge of fundamental airway biology and CFTR function came with the identification of the CFTR gene in 1989. Since then, the tools of molecular and cellular biology, transgenic animals and modern physiology, combined with big strides in modern, multidisciplinary care, have challenged virtually every previously held concept of the disease. The diagnosis is no longer a matter of a positive sweat test as the gold standard: atypical forms of the disease are being recognized. Far from dying in babyhood, patients are increasingly surviving into old age. Treatment goals have moved from the sole (laudable) focus of dealing with symptoms, towards the development of genotype-specific, molecular therapy. Even the pathophysiology is being challenged; the sweat test, which has stood us in such good stead diagnostically, may have deceived us into thinking that all manifestations of CF are related to chloride transport.

There are numerous large and excellent standard textbooks on CF; what is the need for yet another tome? We believe that the rapid advances in CF have reached the point where keeping abreast of research in scientific and clinical areas has become a major challenge for the individual. There is a need for a concise and up-to-date summary of the current knowledge in all the various areas in which the study of CF is being pushed forward.

In this volume, we feel fortunate to have been able to bring together the finest scientists and clinicians to present a state of the art in their respective fields. They have assumed a basic knowledge of the subject; this is not intended to be a comprehensive text book of CF, and the reader will not find extensive reviews of valuable, but older work. The authors have been tasked to write brief chapters, citing mainly only recent literature, and to make their subject accessible to workers in the field of CF from other disciplines. The aim is that the reader will by the end of the volume be up to date in all of the key areas in this rapidly expanding field. We are very grateful to our authors for the enthusiasm and skill with which they have tackled their tasks. We have certainly learned a huge amount from editing this book, and with due modesty, believe that as a result of their efforts, this volume will be of interest to anyone working, or intending to work, in any area of CF.

Andrew Bush
Eric W.F.W. Alton
Jane C. Davies
Uta Griesenbach
Adam Jaffe

The Basics

Bush A, Alton EWFW, Davies JC, Griesenbach U, Jaffe A (eds): Cystic Fibrosis in the 21st Century.
Prog Respir Res. Basel, Karger, 2006, vol 34, pp 2–10

The CFTR Gene: Structure, Mutations and Specific Therapeutic Approaches

Malka Nissim-Rafinia   Liat Linde   Batsheva Kerem

Department of Genetics, Life Sciences Institute, Hebrew University, Jerusalem, Israel

Abstract

Fifteen years ago the gene responsible for cystic fibrosis (CF), the most common severe autosomal recessive disorder among Caucasians, was identified. In this chapter we describe the cloning of the cystic fibrosis transmembrane conductance regulator (CFTR) gene, the spectrum of the CFTR mutations and classification of the mutations by their mechanisms of CFTR dysfunction. Last but not least, we summarize the contribution of all these data to the development of mutation-specific therapy.

The CFTR Gene Structure

Cystic fibrosis (CF) is an autosomal recessive lethal disease affecting 1 in 2,500 newborns among Caucasians (though rare among Orientals, 1:90,000) [1, 2]. The disease was described first by Anderson [3] in 1938 as 'cystic fibrosis of the pancreas', to point out the destruction of the pancreatic exocrine function. In 1953 Di Sant'Agnese et al. [4] demonstrated that excessive salt loss occurs in the sweat of CF patients. This finding led to the use of sweat electrolytes measurements as a diagnostic tool. The major clinical characteristics of CF are pancreatic insufficiency and progressive lung disease, caused by thick and dehydrated airway mucus frequently infected with *Pseudomonas* and *Staphylococcus*, leading to respiratory failure and CF mortality. In addition, most males are infertile, due to congenital bilateral absence of the vas deferens. Other CF characteristics include bile duct obstruction, reduced fertility in females, high sweat chloride, intestinal obstruction, nasal polyp formation, chronic sinusitis, liver disease and diabetes [1, 2, 5].

Fifty years following the first description of the CF disease the biochemical basis was still unknown. In 1983 it was first shown that sweat duct cells derived from CF patients lack chloride efflux [6]. However, this information was not sufficient for the identification of the defective protein in CF patients. Hence, a positional cloning approach was undertaken to identify the gene responsible for CF. In 1985 identification of polymorphic markers in close proximity to the disease mapped the gene to chromosome 7 [7–9]. The cloning was performed using chromosome walking, which enables the identification of overlapping cloned DNA fragments, and chromosome jumping, which enables to skip over uncloned DNA segments and hence allows additional walking start points. In parallel, linkage disequilibrium analysis was performed to determine the walking and jumping direction and to indicate the proximity of the isolated clones to the target gene [10]. Identification of potential genes, along the cloned region, was performed by comparing DNA sequences among different organisms, based on the assumption that coding sequences are conserved during evolution [11]. Furthermore, the tissue expression pattern of each putative CF gene was correlated to the pathology of CF [12]. This led finally to the identification of the cystic fibrosis transmembrane conductance regulator (CFTR) gene (fig. 1). The gene comprises 27 coding exons, spanning over 250 kb

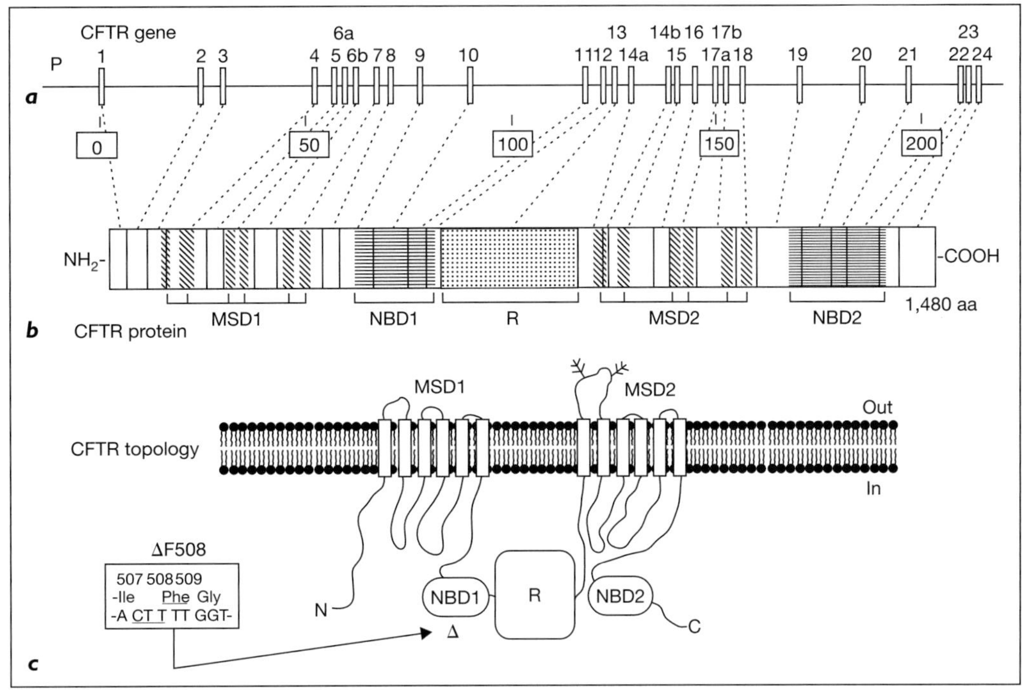

Fig. 1. Schematic diagram of the CFTR gene. *a* Structure of the CFTR gene consisting of promoter region (P) and 27 exons. *b* CFTR polypeptide with predicted domains (highlighted). *c* Topology of the CFTR protein relative to the cytoplasmic membrane and position of the most common mutation, ΔF508. Box: Deletion of 3 nucleotides, CTT (underlined), and subsequent loss of phenylalanine 508 (underlined) [adapted from 5].

on chromosome 7q31.2, and the transcript is 6.5 kb. Sequence comparison between alleles from patients and their parents revealed the major CF mutation, a 3-bp deletion, causing a loss of phenylalanine at position 508 of the protein, designated ΔF508 [10]. Most of the chromosomes carrying the ΔF508 mutation share the same haplotype (a series of alleles found at linked loci on a single chromosome), which is rare in the normal population, indicating that the ΔF508 mutation occurred only once [10, 13].

The protein encoded by the CFTR gene is a chloride (Cl^-) channel in the apical membrane of exocrine epithelial cells [12] (fig. 1). It comprises 1,480 amino acids with a molecular weight of ~170 kDa. The protein comprised five domains: two membrane-spanning domains (MSD1 and MSD2), each composed of six transmembrane segments (TM1 to TM12) that form the channel, two nucleotide-binding domains (NBD1 and NBD2), capable of ATP hydrolysis, and a regulatory domain (R), which contains numerous phosphorylation sites [12, 14]. This protein structure indicates that the CFTR is part of the ATP-binding cassette (ABC) transporter proteins. Consistent with its structure it was found that the phosphorylation of sites in the R domain by protein kinase A, regulated by cyclic adenosine monophosphate (cAMP), and the hydrolysis of ATP by the NBDs are essential for activating the chloride channel [1, 15–19]. In addition to the CFTR function as a chloride channel, it appears to have an effect on a growing number of proteins. The CFTR modifies the function and properties of other ion transporters including chloride, sodium and potassium channels and the Cl^-/HCO_3^- exchanger. Moreover, it has an effect on water permeability, ATP transport, and mucus secretion [reviewed in 20, 21] (for more detail see chapters 4–7).

The high incidence of CF carriers can be explained by a possible protective effect in these individuals compared with healthy individuals. Several works showed indeed that there is a heterozygote advantage for carriers of CFTR mutations. For example, *Salmonella typhi*, which is the etiologic agent of typhoid fever, was found to bind the CFTR. This binding mediates translocation of this pathogen into the gastrointestinal submucosa. Thus, decreased CFTR levels in the gastrointestinal epithelium of mice heterozygous for a CFTR mutation resulted in an increased resistance to typhoid fever, in comparison to wild-type mice [22]. It was also shown that mice carrying a CFTR mutation have an increased resistance to cholera toxin [23]. Similarly, heterozygous mice were also found to be more resistant to *Pseudomonas aeruginosa* keratitis than wild-type mice [24].

Spectrum of CFTR Mutations

Over 1,300 sequence variations (mutations which are involved in disease expression and polymorphisms which have no effect on the phenotype) have been identified so far along the entire CFTR gene [25]. ΔF508 is found in ~70% of the CF chromosomes worldwide; however, its frequency varies greatly among different ethnic groups, between 100% in the isolated Faroe Islands of Denmark to 18% in Tunisia. In Europe there is a clear decreasing gradient in the frequency of ΔF508 from northeast to southwest. All the other mutations are mostly rare and only 11 were found in more than 100 patients [26, 27]. Several of the rare mutations, however, appear with high incidence in isolated populations, e.g. the Q359K/T360K among Georgian Jews and M1101K among the Hutterite (table 1). Furthermore, populations in geographical proximity may share their mutations (see mutations 3120 + 1 G→A and N1303K, table 1).

As can be seen in table 2, 48.7% of the mutations are missense, 19.5% are frameshifts caused by small insertions or deletions, 15.7% are splicing and 12.9% are nonsense mutations. The remaining (3.2%) affect other sequence variations, like in-frame insertions or deletions and mutations in the promoter [25]. Although hot spots for mutations along the CFTR gene were not found, several CFTR amino acids or even specific CFTR nucleotides show higher probability for mutations (e.g. amino acids R117, R347, I506, S549 and nucleotides 460 and 1058) [25]. Furthermore, the density of mutations is higher in the first half of the protein (particularly in MSD1 and NBD1), while very few occur in the R domain (table 2) suggesting a different role for each domain. Recently it was shown that heterodimerization of the two CFTR NBDs exhibited 2- to 3-fold enhancement in ATPase activity relative to homodimerization of each NBD [28], which indicates a separate role for the two NBDs. Such a separate role was shown for another ABC protein, MRP1, in which ATP binding affinity and hydrolysis differ between the two NBDs [29].

Classification of CFTR Mutations

The different CFTR mutations can be divided into five major classes according to their effect on CFTR function (fig. 2).

Table 1. CFTR mutations with high incidence in specific populations

Mutation	Frequency in specific populations[a], %	Frequency in the general population[b], %	Ref. No.
Q359K/T360K	Georgian Jews, 88		73
M1101K	Hutterite Brethren, 69		74
S549K	United Arab Emirates, 61.5		74
W1282X	Ashkenazi Jews, 48	1.2	73
	Tunisian Jews, 17		
	Israeli Arabs, 10.6		
405+1 G→A	Tunisian Jews, 48		73
	Libyan Jews, 18		
3120+1 G→A	Bantu, Africa, 46.4		74, 75
	South African, 17.4		
	African American, USA, 13.9		
	African American, Africa, 12.2		
	Saudi Arabia, 10		
N1303K	Egyptian Jews, 33	1.3	73, 74
	Israeli Arabs, 21		
	Algeria, 20		
	Lebanon, 10		
G85E	Turkish Jews, 30		73
1898+5 G→T	Taiwan, 30		74
394delTT	Finland, 28.8		74
	Estonia, 13.3		
621+1 G→T	Saguenay Lac-Saint-Jean, Canada, 24.3	0.7	74
	Northern Greece, 12.1		
Y122X	Reunion Island, 24		74
3905insT	Amish, Mennonite, 16.7		74
	Switzerland, 9.8		
Y569D	Pakistani, UK, 15.4		74
T338I	Italy, Sardinia, 15.1		77
1548delG	Saudi Arabia, 15		75
R553X	Switzerland, 14	0.7	74
3120+1 kb del8.6 kb	Israeli Arabs, 13		76
I1234V	Saudi Arabia, 13		75
R347P	Turkish population, Bulgaria, 11.7	0.2	77
Q98X	Pakistani, UK, 11.5		74
G542X	South Spain, 11.4	2.4	77
711+1 G→T	Algeria, 10		74
4010del4	Lebanon, 10		74
R1162X	Northeast Italy, 9.8	0.3	74
1525-1 G→A	Pakistani, UK, 9.6		74

[a] Mutations were included only if their frequency in a specific population was at least 10%, excluding ΔF508. The mutations are listed in decreasing order of their frequency (in case of more than one population, the frequency was listed according to the highest).

[b] The frequency in the general population was listed only if it reached >0.1%, based on CF mutation database [25].

Table 2. Distribution of sequence variation (mutations and polymorphisms) along the CFTR gene

a Mutation distribution

	Pro	MSD1	ExLs1	InLs1	NBD1	R	MSD2	ExLs2	InLs2	NBD2	Other[a]	Total
Missense		68	18	47	93	47	46	23	59	59	91	551 (48.7)
Frameshift (PTC)		20	5	18	27	33	15	6	22	26	49	221 (19.5)
Splicing		12	0	25	22	5	13	6	18	23	54	178 (15.7)
Nonsense (PTC)		13	4	7	17	26	12	2	10	15	40	146 (12.9)
In-frame in/del		2	1	4	4	2	1	0	5	2	7	28 (2.5)
Noncoding	8											8 (0.7)
Total mutations	8	115	28	101	163	113	87	37	114	125	241	1,132

b Mutation density

	Pro	MSD1	Exl1	Inl1	NBD1	R	MSD2	Exl2	Inl2	NBD2	Other[a]	Total
Total mutations, n	8	115	28	101	163	113	87	37	114	125	241	1,132
Size (aa)		129	20	121	152	242	127	50	126	146	267	1,480
Density[b]		0.89	1.4	0.83	1.07	0.47	0.69	0.74	0.9	0.86	0.9	0.76

c Variation distribution

	Pro	MSD1	Exl1	Inl1	NBD1	R	MSD2	Exl2	Inl2	NBD2	Other[a]	Total
Total mutations, n	8	115	28	101	163	113	87	37	114	125	241	1,132
Polymorphism	6	16	1	14	23	12	14	10	19	19	62	196
Total variations	14	131	29	115	186	125	101	47	133	144	303	1,328

The data is based on the CF mutation database [25]. Figures in parentheses represent percentage. Pro = Promoter; MSD = membrane spanning domain; Exls = extracellular loops within the MSD; Inls = intracellular loops within the MSD; NBD = nucleotide binding domain; R = regulator; in/del = insertion/deletion; aa = amino acid.
[a] Mutations in the intracellular domains.
[b] Number of mutations per domain size.

Class I: Defective Protein Synthesis

Class I includes mutations which lead to the disruption of the CFTR protein synthesis. The mutations in this class include nonsense and frameshifts, which lead to the creation of premature termination codons (PTCs) (>30%, table 2). As can be seen in table 1, PTCs (W1282X, G542X, etc.) are among the more frequent mutations in the population. PTCs were known to result in truncated proteins; however, it is now apparent that they have additional effects on transcripts carrying these mutations. PTCs can dramatically decrease the half-lives of mutant mRNAs by the nonsense-mediated mRNA decay pathway, as well as alter the pattern of pre-mRNA splicing. Therefore, such mutations are expected to produce little or no protein. Indeed, genotype-phenotype studies revealed that CFTR PTCs are associated with a severe form of the disease [30, 31] (see also chapter 8).

A specific therapy for PTCs has been suggested, aiming to read through the nonsense codon, allowing synthesis of full-length proteins. Aminoglycoside antibiotics, in addition to their antimicrobial activity, can inefficiently interact with the A site of eukaryotic rRNA, leading to alteration in RNA conformation, which reduces the accuracy between codon-anticodon pairing. This can lead to read-through of the PTCs by binding of any tRNA to the nonsense codon, thereby permitting protein translation to continue to the normal end of the transcript [32–35]. Since the normal termination of eukaryote genes consists of several termination codons, the aminoglycosides are not expected to affect the normal termination. In addition, in cases where even low levels of physiologically functional proteins are sufficient to restore the function, aminoglycosides might be suitable for treatment.

Several in vitro studies demonstrated that aminoglycosides can read through PTCs in the CFTR gene, and lead to

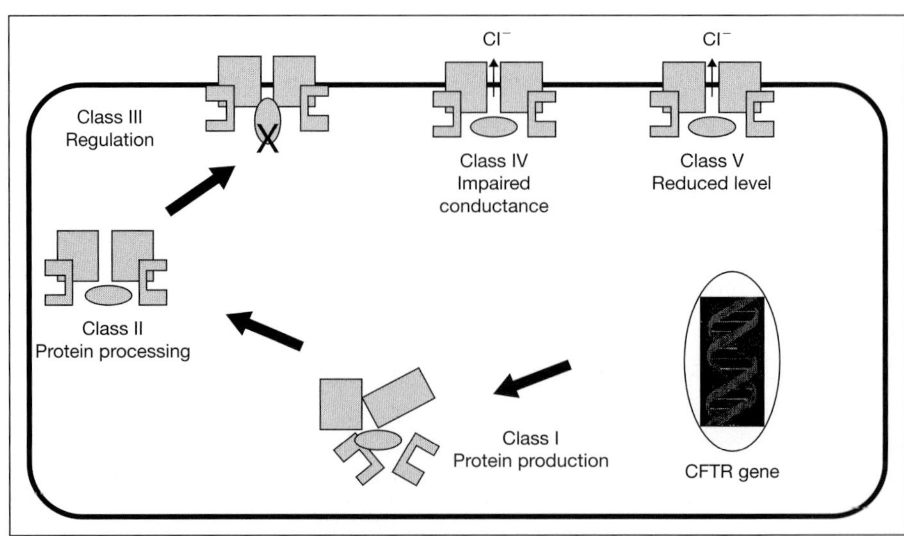

Fig. 2. Classes of CFTR mutation.

functional full-length CFTR proteins [36, 37]. Ex vivo exposure of airway cells from CF patients carrying nonsense mutations led to the identification of surface-localized CFTR in a dose-dependent fashion [38]. Clinical studies provided evidence that the aminoglycoside gentamicin can read through PTCs in vivo. A pilot study in 9 patients with CF carrying at least one nonsense mutation demonstrated a significant correction of the basic electrophysiological abnormalities characteristics of CF, using the application of gentamicin drops to the nasal epithelium [39]. In most patients the main effect of gentamicin was activation of transmembrane chloride transport that approached the normal range. An additional clinical study in which systemic gentamicin was administrated also showed correction of the CFTR abnormalities [38]. Recently, in a double-blind, placebo-controlled, crossover study we have demonstrated the expression of full-length CFTR proteins and restoration of CFTR function following topical application of gentamicin to the nasal epithelium of 19 CF patients carrying the W1282X mutation [40]. Complete normalization of the electrophysiological abnormalities was found in 21% of the patients and in 68% there was restoration of either chloride or sodium transport. Furthermore, a significant increase in peripheral and surface staining for full-length CFTR proteins was observed in the nasal epithelial cells of the patients following the treatment [40]. Together, these results suggest that gentamicin treatment can read through PTCs. It is important to note that studies of other genetic diseases also showed that aminoglycoside have a potential to read through PTCs, and restore the function of defective proteins encoded by nonsense alleles, both ex vivo and in vivo [41–46].

Class II: Defective Protein Processing
Class II mutations are associated with defective protein processing. Upon completion of the CFTR protein translation, the normal protein undergoes a series of processes in the endoplasmic reticulum (ER) and the Golgi apparatus. This includes glycosylation and folding that enable the protein trafficking to the apical cell membrane (see also chapter 3). Class II mutations cause impairment of this process, which leads to degradation of the abnormally processed protein. The major mutation, ΔF508, results in the synthesis of a CFTR protein that is unable to correctly fold into its appropriate tertiary conformation. Consequently, this protein is retained in the ER and abnormally degraded (>99 vs. 75% in normal proteins). In addition, more recently it was found that ΔF508 CFTR proteins when they reach the plasma membrane undergo abnormal exocytosis and recycling into the membrane. However, most of the exocyted ΔF508 proteins will be marked for degradation and will not be recycled into the membrane [47]. This process further reduces the level of the defective protein in the membrane.

In vitro studies of the ΔF508-CFTR protein demonstrated that this mutant polypeptide can function as a cAMP-dependent chloride channel once it reaches the cell membrane, suggesting that a therapy aimed at correcting protein folding and trafficking might partially correct the CFTR defect. A number of different chaperones within the lumen of the ER and in the cytosol can stabilize the

Nissim-Rafinia/Linde/Kerem

misfolded structures and promote ΔF508 CFTR trafficking. Among the molecular chaperones are the 70-kDa heat shock proteins, Hsp70 and Hsc70, and sodium-4-phenylbutyrate, a histone deacetylase inhibitor that downregulates Hsc70 [48, 49] and upregulates Hsp70 [50]. Another molecular chaperone is calnexin, a calcium-binding transmembrane protein chaperone that assists newly synthesized proteins to fold into a normal structure in the ER. The mutant CFTR undergoes a prolonged specific association with calnexin and with Hsp70. Further information regarding pharmacological approaches can be found in chapter 28.

Class III and IV: Defective Protein Regulation and Altered Conductance

Phosphorylation and dephosphorylation of the CFTR is considered the major pathway by which the chloride channel activity is physiologically regulated. In addition, the normal gating cycle of CFTR (both opening and closing) requires ATP binding and hydrolysis, at the two NBDs. Class III includes mutations that lead to the production of proteins (e.g. G551D and Y569D), which reach the plasma membrane; however, their regulation is defective and, thus, they cannot be activated by ATP or cAMP. Class IV mutations are associated with altered conductance (e.g. R347P, R117H and D1152H) such that the rate of chloride transport is reduced. Thus, mutations in both class III and IV lead to CFTR proteins that can be produced, processed, transported and inserted into the apical membrane, but display a defective conductance. Investigators have searched for exogenous compounds that are potential therapeutic activators of class III and IV mutant protein (including flavonoids, like genistein and NS-004 and xanthine derivatives, like CPX and IBMX). Genistein was shown to increase open probabilities of phosphorylated channels by binding directly to one or both of the NBDs without raising cAMP concentration and without affecting either protein kinases or protein phosphatases. IBMX seems, on the other hand, to affect CFTR through combined effects of raising cAMP levels and blocking protein phosphatases. Further information regarding pharmacological approaches can be found in chapter 28.

Class V: Reduced CFTR Level

Class V mutations lead to the production of normal proteins, however at reduced levels. This class includes promoter mutations that reduce transcription and amino acid substitutions that cause inefficient protein maturation. Yet, most of the mutations are splicing mutations, which affect the normal splicing of the pre-mRNA and thus reduce the levels of correctly spliced mRNA, by partial exon skipping

or inclusion of intronic sequences. The alteration in the splicing pattern is caused by disrupting or generating intronic splicing motifs, required for exon recognition. These mutations account for >5% of CFTR mutations and include mutations which are relatively frequent in the general population (such as 3849+10 kb C→T, the 12th most common mutation) and/or in specific populations (1898+5 G→T, 3120+1 kb del8.6 kb, see table 1). In addition, there are mutations and polymorphisms that disrupt exonic splicing motifs, which also affect the splicing pattern. Class V splicing mutations (e.g. 3849+10 kb C→T, 3272-26 A→G, IVS8-5T, D565G and G576A) can lead to variable levels of correctly spliced transcripts among different patients and among different organs of the same patient [51, 52, reviewed in 53]. These levels were found to inversely correlate with the variable disease expression, such that lower levels of correctly spliced transcripts are associated with a severe disease, while higher levels are associated with milder disease [54, 55].

Splicing is regulated through the interaction of a complex repertoire of splicing factors with various splicing motifs [reviewed in 56]. Differences in the levels of functional splicing factors were found among different tissues, which have been suggested to regulate the level of alternatively spliced transcripts. Initially, the effect of overexpression of splicing factors on the level of correctly spliced CFTR transcripts was studied in minigenes carrying mutations, which lead to partial skipping of exons 9, 12, and the 5′ end of exon 13 and the 3849+10 kb C→T mutation, which results in partial inclusion of an 84-bp sequence from intron 19. Most (10/11) of the minigenes were modulated by splicing factors. Higher levels of correctly spliced transcripts were generated by several of these factors: Htraα and E4-ORF3 promoted exon 13 and 9 inclusion, respectively, and hnRNP A1 and E4-0RF6 promoted skipping over the cryptic 84-bp exon [52, 57–60]. Subsequently, we showed that Htra2-β1 and SC35 increased the level of correctly spliced mRNA transcribed from an endogenous CFTR allele carrying the 3849+10 kb C→T mutation [60]. Importantly, this increase activated the CFTR channel and restored its function. Overexpression of other splicing factors had no effect on the transcript level and did not restore the CFTR function.

Therapeutic approaches for this class aim to increase the level of correctly spliced transcripts and upregulation of CFTR expression. One such approach is using antisense oligonucleotides designed to inhibit cryptic splicing. Antisense oligonucleotides for the 84-bp exon, cotransfected with CFTR cDNA carrying the 3849+10 kb C→T mutation, resulted in a decrease at the level of aberrant

CFTR transcripts containing the 84-bp 'exon' [61]. Recently, a similar approach was taken for the SMN2 (survival motor neuron) and BRCA1 (breast cancer) genes. Chimeric antisense oligonucleotides comprising two parts were designed, one complementary to the aberrantly spliced exon providing exon specificity, and the other containing binding motifs for recruitment of splicing factors to the mutation site [62, 63]. An increase in the binding of splicing factors by such oligonucleotides resulted in an increased level of correctly spliced transcripts.

Another approach is the identification of small molecules, which may lead to an increase in the level of correctly spliced transcripts. We have recently shown that administration of sodium butyrate, a histone deacetylase inhibitor which upregulates the expression of splicing factors [64, 65], led to a decrease in the level of aberrant CFTR transcripts containing the 84-bp 'exon'. Importantly, this decrease resulted in activation of the CFTR channel and restoration of the CFTR function [60]. Several other small molecules were shown to increase the level of correctly spliced mRNA transcribed from other genes, including aclarubicin, sodium vanadate and valproic acid in SMN2 and EGCG [(−)-epigallocatechin gallate] and kinetin in IKAP (IkB kinase complex-associated protein) [65–69]. These molecules among others might be appropriate for CFTR therapy.

CFTR Polymorphisms

As mentioned above, DNA sequence polymorphisms are defined as sequence variations, which do not lead to disease expression. Yet, several polymorphisms in the CFTR gene were shown to modify disease severity. For example, the number of TG repeats in IVS8 correlates with the level of exon 9 skipping, and therefore with the disease severity [70–72]. Similarly, R668C was shown to affect exon 12 skipping [52]. In addition, M470V was shown to have an effect on the CFTR channel activity and correlate with disease severity [70]. It should be mentioned that CFTR polymorphisms were also found in patients with CF-related diseases, with no other mutations in the CFTR. Thus, sequence variations that were defined as CF polymorphisms (noncausing CF) can be defined as mutations causing CF-related diseases.

Future Prospects

Identifying all the mutations in the CFTR gene which are involved in typical CF and CF-related diseases and developing simple and inexpensive methods for screening a large number of mutations will enable early genetic diagnosis of all the patients before the development of symptoms. Additional studies aiming to better understand the effect of different mutations on CFTR function will broaden our understanding.

Further studies aiming to investigate the potential modulation of the CF phenotype by polymorphisms in the CFTR sequence and/or by other genes will enable us to learn more about the genetic complexity of the disease. Furthermore, development of mutation-specific therapies using high-throughput screening for small molecules will lead to pharmacotherapy targeting the basic CFTR defect.

References

1 Collins FS: Cystic fibrosis: Molecular biology and therapeutic implications. Science 1992; 256:774–779.

2 Welsh MJ, Tsui LC, Boat TF, Beaudet AL: Cystic fibrosis; in Scriver CL, Beaudet AL, Sly WS, Valle D (eds): The Metabolic Basis of Inherited Disease. New York, McGraw-Hill, 1995, pp 3799–3876.

3 Anderson DH: Cystic fibrosis of the pancreas and its relation to celiac disease: A clinical and pathological study. Am J Dis Child 1938;56: 344.

4 Di Sant'Agnese PA, Darling RC, Perera GA, Shea E: Abnormal electrolyte composition of sweat in cystic fibrosis of the pancreas; clinical significance and relationship to the disease. Pediatrics 1953;12:549–563.

5 Zielenski J: Genotype and phenotype in cystic fibrosis. Respiration 2000;67:117–133.

6 Quinton PM: Chloride impermeability in cystic fibrosis. Nature 1983;301:421–422.

7 Tsui LC, Buchwald M, Barker D, Braman JC, Knowlton R, Schumm JW, Eiberg H, Mohr J, Kennedy D, Plavsic N: Cystic fibrosis locus defined by a genetically linked polymorphic DNA marker. Science 1985;230:1054–1057.

8 White R, Woodward S, Leppert M, O'Connell P, Hoff M, Herbst J, Lalouel JM, Dean M, Vande-Woude G: A closely linked genetic marker for cystic fibrosis. Nature 1985;318:382–384.

9 Wainwright BJ, Scambler PJ, Schmidtke J, Watson EA, Law HY, Farrall M, Cooke HJ, Eiberg H, Williamson R: Localization of cystic fibrosis locus to human chromosome 7cen-q22. Nature 1985;318:384–385.

10 Kerem B, Rommens JM, Buchanan JA, Markiewicz D, Cox TK, Chakravarti A, Buchwald M, Tsui LC: Identification of the cystic fibrosis gene: Genetic analysis. Science 1989;245:1073–1080.

11 Rommens JM, Iannuzzi MC, Kerem B, Drumm ML, Melmer G, Dean M, Rozmahel R, Cole JL, Kennedy D, Hidaka N: Identification of the cystic fibrosis gene: Chromosome walking and jumping. Science 1989; 245:1059–1065.

12 Riordan JR, Rommens JM, Kerem B, Alon N, Rozmahel R, Grzelczak Z, Zielenski J, Lok S, Plavsic N, Chou JL: Identification of the cystic fibrosis gene: Cloning and characterization of complementary DNA. Science 1989;245: 1066–1073.

13 Morral N, Bertranpetit J, Estivill X, Nunes V, Casals T, Gimenez J, Reis A, Varon-Mateeva R, Macek MJ, Kalaydjieva L: The origin of the major cystic fibrosis mutation (delta F508) in European populations. Nat Genet 1994;7: 169–175.

14 Sheppard DN, Welsh MJ: Structure and function of the CFTR chloride channel. Physiol Rev 1999;79:S23–S45.

15 McIntosh I, Cutting GR: Cystic fibrosis transmembrane conductance regulator and the etiology and pathogenesis of cystic fibrosis. FASEB J 1992;6:2775–2782.

16 Welsh MJ, Anderson MP, Rich DP, Berger HA, Denning GM, Ostdgaard LS, Sheppard DN, Cheng SH, Gregory RJ, Smith AE: Cystic fibrosis transmembrane conductance regulator: A chloride channel with novel regulation. Neuron 1992;8:821–829.

17 Gabriel SE, Clarke LL, Boucher RC, Stutts MJ: CFTR and outward rectifying chloride channels are distinct proteins with a regulatory relationship. Nature 1993;363:263–268.

18 Frizzel RA: Functions of the cystic fibrosis transmembrane conductance regulator protein. Am J Respir Crit Care Med 1995;151: S54–S58.

19 Li C, Ramjeesingh M, Wang W, Garami E, Hewryk M, Lee D, Rommens JM, Galley K, Bear CE: ATPase activity of the cystic fibrosis transmembrane conductance regulator. J Biol Chem 1996;271:28463–28468.

20 Kunzelmann K, Schreiber R, Nitschke R, Mall M: Control of epithelial Na$^+$ conductance by the cystic fibrosis transmembrane conductance regulator. Pflügers Arch 2000;440:193–201.

21 Greger R, Schreiber R, Mall M, Wissner A, Hopf A, Briel M, Bleich M, Warth R, Kunzelmann K: Cystic fibrosis and CFTR. Pflügers Arch 2001;443(suppl 1):S3–S7.

22 Pier GB, Grout M, Zaidi T, Meluleni G, Mueschenborn SS, Banting G, Ratcliff R, Evans MJ, Colledge WH: *Salmonella typhi* uses CFTR to enter intestinal epithelial cells. Nature 1998;393:79–82.

23 Gabriel SE, Brigman KN, Koller BH, Boucher RC, Stutts MJ: Cystic fibrosis heterozygote resistance to cholera toxin in the cystic fibrosis mouse model. Science 1994;266:107–109.

24 Pier GB: Role of the cystic fibrosis transmembrane conductance regulator in innate immunity to *Pseudomonas aeruginosa* infections. Proc Natl Acad Sci USA 2000;97:8822–8828.

25 The Cystic Fibrosis Genetic Analysis Consortium, CFGAC. http://www.genet.sickkids.on.ca/cftr/

26 Population variation of common cystic fibrosis mutations. The Cystic Fibrosis Genetic Analysis Consortium. Hum Mutat 1994;4:167–177.

27 Zielenski J, Tsui LC: Cystic fibrosis: Genotypic and phenotypic variations. Annu Rev Genet 1995;29:777–807.

28 Kidd JF, Ramjeesingh M, Stratford F, Huan LJ, Bear CE: A heteromeric complex of the two nucleotide binding domains of cystic fibrosis transmembrane conductance regulator (CFTR) mediates ATPase activity. J Biol Chem 2004;279:41664–41669.

29 Hou YX, Riordan JR, Chang XB: ATP binding, not hydrolysis, at the first nucleotide-binding domain of multidrug resistance-associated protein MRP1 enhances ADP.Vi trapping at the second domain. J Biol Chem 2003;278: 3599–3605.

30 Shoshani T, Augarten A, Gazit E, Bashan N, Yahav Y, Rivlin Y, Tal A, Seret H, Yaar L, Kerem E: Association of a nonsense mutation (W1282X), the most common mutation in the Ashkenazi Jewish cystic fibrosis patients in Israel, with presentation of severe disease. Am J Hum Genet 1992;50:222–228.

31 Correlation between genotype and phenotype in patients with cystic fibrosis. The Cystic Fibrosis Genotype-Phenotype Consortium. N Engl J Med 1993;329:1308–1313.

32 Davies J, Gorini L, Davies BD: Misreading of RNA codewords induced by aminoglycoside antibiotics. Mol Pharmacol 1965;1:93–106.

33 Singh A, Ursic D, Davies J: Phenotypic suppression and misreading *Saccharomyces cerevisiae*. Nature 1979;277:146–148.

34 Palmer E, Wilhelm JM, Sherman F: Phenotypic suppression of nonsense mutants in yeast by aminoglycoside antibiotics. Nature 1979;277:148–150.

35 Martin R, Mogg AE, Heywood LA, Nitschke L, Burke JF: Aminoglycoside suppression at UAG, UAA and UGA codons in *Escherichia coli* and human tissue culture cells. Mol Gen Genet 1989;217:411–418.

36 Howard M, Frizzell RA, Bedwell DM: Aminoglycoside antibiotics restore CFTR function by overcoming premature stop mutations. Nat Med 1996;2:467–469.

37 Bedwell DM, Kaenjak A, Benos DJ, Bebok Z, Bubien JK, Hong J, Tousson A, Clancy JP, Sorscher EJ: Suppression of a CFTR premature stop mutation in a bronchial epithelial cell line. Nat Med 1997;3:1280–1284.

38 Clancy JP, Bebok Z, Ruiz F, King C, Jones J, Walker L, Greer H, Hong J, Wing L, Macaluso M, Lyrene R, Sorscher EJ, Bedwell DM: Evidence that systemic gentamicin suppresses premature stop mutations in patients with cystic fibrosis. Am J Respir Crit Care Med 2001;163:1683–1692.

39 Wilschanski M, Famini C, Blau H, Rivlin J, Augarten A, Avital A, Kerem B, Kerem E: A pilot study of the effect of gentamicin on nasal potential difference measurements in cystic fibrosis patients carrying stop mutations. Am J Respir Crit Care Med 2000;161:860–865.

40 Wilschanski M, Yahav Y, Yaacov Y, Blau H, Bentur L, Rivlin J, Aviram M, Bdolah-Abram T, Bebok Z, Shushi L, Kerem B, Kerem E: Gentamicin-induced correction of CFTR function in patients with cystic fibrosis and CFTR stop mutations. N Engl J Med 2003; 349:1433–1441.

41 Barton-Davis ER, Cordier L, Shoturma DI, Leland SE, Sweeney HL: Aminoglycoside antibiotics restore dystrophin function to skeletal muscles of mdx mice. J Clin Invest 1999;104:375–381.

42 Keeling KM, Brooks DA, Hopwood JJ, Li P, Thompson JN, Bedwell DM: Gentamicin-mediated suppression of Hurler syndrome stop mutations restores a low level of alpha-L-iduronidase activity and reduces lysosomal glycosaminoglycan accumulation. Hum Mol Genet 2001;10:291–299.

43 Helip-Wooley A, Park MA, Lemons RM, Thoene JG: Expression of CTNS alleles: Subcellular localization and aminoglycoside correction in vitro. Mol Genet Metab 2002;75: 128–133.

44 Sleat DE, Sohar I, Gin RM, Lobel P: Aminoglycoside-mediated suppression of nonsense mutations in late infantile neuronal ceroid lipofuscinosis. Eur J Paediatr Neurol 2001;5(suppl A):57–62.

45 Schulz A, Sangkuhl K, Lennert T, Wigger M, Price DA, Nuuja A, Gruters A, Schultz G, Schoneberg T: Aminoglycoside pretreatment partially restores the function of truncated V(2) vasopressin receptors found in patients with nephrogenic diabetes insipidus. J Clin Endocrinol Metab 2002;87:5247–5257.

46 Lai CH, Chun HH, Nahas SA, Mitui M, Gamo KM, Du L, Gatti RA: Correction of ATM gene function by aminoglycoside-induced readthrough of premature termination codons. Proc Natl Acad Sci USA 2004;101:15676–15681.

47 Sharma M, Pampinella F, Nemes C, Benharouga M, So J, Du K, Bache KG, Papsin B, Zerangue N, Stenmark H, Lukacs GL: Misfolding diverts CFTR from recycling to degradation: Quality control at early endosomes. J Cell Biol 2004;164:923–933.

48 Rubenstein RC, Zeitlin PL: Sodium 4-phenylbutyrate downregulates Hsc70: Implications for intracellular trafficking of DeltaF508-CFTR. Am J Physiol Cell Physiol 2000;278: C259–C267.

49 Rubenstein RC, Lyons BM: Sodium 4-phenylbutyrate downregulates HSC70 expression by facilitating mRNA degradation. Am J Physiol Lung Cell Mol Physiol 2001;281:L43–L51.

50 Choo-Kang LR, Zeitlin PL: Induction of HSP70 promotes DeltaF508 CFTR trafficking. Am J Physiol Lung Cell Mol Physiol 2001;281:L58–L68.

51 Ramalho AS, Beck S, Meyer M, Penque D, Cutting GR, Amaral MD: Five percent of normal cystic fibrosis transmembrane conductance regulator mRNA ameliorates the severity of pulmonary disease in cystic fibrosis. Am J Respir Cell Mol Biol 2002;27:619–627.

52 Pagani F, Stuani C, Tzetis M, Kanavakis E, Efthymiadou A, Doudounakis S, Casals T, Baralle FE: New type of disease causing mutations: The example of the composite exonic regulatory elements of splicing in CFTR exon 12. Hum Mol Genet 2003;12: 1111–1120.

53 Nissim-Rafinia M, Kerem B: Splicing regulation as a potential genetic modifier. Trends Genet 2002;18:123–127.

54 Augarten A, Kerem BS, Yahav Y, Noiman S, Rivlin Y, Tal A, Blau H, Ben-Tur L, Szeinberg A, Kerem E: Mild cystic fibrosis and normal or borderline sweat test in patients with the 3849+10 kb C→T mutation. Lancet 1993; 342:25–26.

55 Kerem E, Rave-Harel N, Augarten A, Madgar I, Nissim-Rafinia M, Yahav Y, Goshen R, Bentur L, Rivlin J, Aviram M, Genem A, Chiba-Falek O, Kraemer MR, Simon A, Branski D, Kerem B: A cystic fibrosis transmembrane conductance regulator splice variant with partial penetrance associated with variable cystic fibrosis presentations. Am J Respir Crit Care Med 1997;155:1914–1920.

56 Black DL: Mechanisms of alternative premessenger RNA splicing. Annu Rev Biochem 2003;72:291–336.

57 Nissim-Rafinia M, Chiba-Falek O, Sharon G, Boss A, Kerem B: Cellular and viral splicing factors can modify the splicing pattern of CFTR transcripts carrying splicing mutations. Hum Mol Genet 2000;9:1771–1778.

58 Pagani F, Buratti E, Stuani C, Romano M, Zuccato E, Niksic M, Giglio L, Faraguna D,

Baralle FE: Splicing factors induce cystic fibrosis transmembrane regulator exon 9 skipping through a nonevolutionary conserved intronic element. J Biol Chem 2000;275: 21041–21047.

59 Aznarez I, Chan EM, Zielenski J, Blencowe BJ, Tsui LC: Characterization of disease-associated mutations affecting an exonic splicing enhancer and two cryptic splice sites in exon 13 of the cystic fibrosis transmembrane conductance regulator gene. Hum Mol Genet 2003;12:2031–2040.

60 Nissim-Rafinia M, Aviram M, Randell SH, Shushi L, Ozeri E, Chiba-Falek O, Eidelman O, Pollard HB, Yankaskas JR, Kerem B: Restoration of the cystic fibrosis transmembrane conductance regulator function by splicing modulation. EMBO Rep 2004;5: 1071–1077.

61 Friedman K, Kole J, Cohn JA, Knowles MR, Silverman LM, Kole R: Correction of aberrant splicing of the cystic fibrosis transmembrane conductance regulator (CFTR) gene by antisense oligonucleotides. J Biol Chem 1999; 274:36193–36199.

62 Skordis LA, Dunckley MG, Yue B, Eperon IC, Muntoni F: Bifunctional antisense oligonucleotides provide a *trans*-acting splicing enhancer that stimulates SMN2 gene expression in patient fibroblasts. Proc Natl Acad Sci USA 2003;100:4114–4119.

63 Cartegni L, Krainer AR: Correction of disease-associated exon skipping by synthetic exon-specific activators. Nat Struct Biol 2003; 10:120–125.

64 Chang JG, Hsieh-Li HM, Jong YJ, Wang NM, Tsai CH, Li H: Treatment of spinal muscular atrophy by sodium butyrate. Proc Natl Acad Sci USA 2001;98:9808–9813.

65 Brichta L, Hofmann Y, Hahnen E, Siebzehnrubl FA, Raschke H, Blumcke I, Eyupoglu IY, Wirth B: Valproic acid increases the SMN2 protein level: A well-known drug as a potential therapy for spinal muscular atrophy. Hum Mol Genet 2003;12:2481–2489.

66 Andreassi C, Jarecki J, Zhou J, Coovert DD, Monani UR, Chen X, Whitney M, Pollok B, Zhang M, Androphy E, Burghes AH: Aclarubicin treatment restores SMN levels to cells derived from type I spinal muscular atrophy patients. Hum Mol Genet 2001;10:2841–2849.

67 Zhang ML, Lorson CL, Androphy EJ, Zhou J: An in vivo reporter system for measuring increased inclusion of exon 7 in SMN2 mRNA: Potential therapy of SMA. Gene Ther 2001;8:1532–1538.

68 Anderson SL, Qiu J, Rubin BY: EGCG corrects aberrant splicing of IKAP mRNA in cells from patients with familial dysautonomia. Biochem Biophys Res Commun 2003; 310:627–633.

69 Slaugenhaupt SA, Mull J, Leyne M, Cuajungco MP, Gill SP, Hims MM, Quintero F, Axelrod FB, Gusella JF: Rescue of a human mRNA splicing defect by the plant cytokinin kinetin. Hum Mol Genet 2004;13:429–436.

70 Cuppens H, Lin W, Jaspers M, Costes B, Teng H, Vankeerberghen A, Jorissen M, Droogmans G, Reynaert I, Goossens M, Nilius B, Cassiman JJ: Polyvariant mutant cystic fibrosis transmembrane conductance regulator genes. The polymorphic (Tg)m locus explains the partial penetrance of the T5 polymorphism as a disease mutation. J Clin Invest 1998;101: 487–496.

71 Niksic M, Romano M, Buratti E, Pagani F, Baralle FE: Functional analysis of *cis*-acting elements regulating the alternative splicing of human CFTR exon 9. Hum Mol Genet 1999; 8:2339–2349.

72 Groman JD, Hefferon TW, Casals T, Bassas L, Estivill X, Des Georges M, Guittard C, Koudova M, Fallin MD, Nemeth K, Fekete G, Kadasi L, Friedman K, Schwarz M, Bombieri C, Pignatti PF, Kanavakis E, Tzetis M, Schwartz M, Novelli G, D'Apice MR, Sobczynska-Tomaszewska A, Bal J, Stuhrmann M, Macek M Jr, Claustres M, Cutting GR: Variation in a repeat sequence determines whether a common variant of the cystic fibrosis transmembrane conductance regulator gene is pathogenic or benign. Am J Hum Genet 2004;74:176–179.

73 Kerem E, Kalman YM, Yahav Y, Shoshani T, Abeliovich D, Szeinberg A, Rivlin J, Blau H, Tal A, Ben-Tur L: Highly variable incidence of cystic fibrosis and different mutation distribution among different Jewish ethnic groups in Israel. Hum Genet 1995;96: 193–197.

74 Bobadilla JL, Macek M Jr, Fine JP, Farrell PM: Cystic fibrosis: A worldwide analysis of CFTR mutations – correlation with incidence data and application to screening. Hum Mutat 2002;19:575–606.

75 Banjar H, Kambouris M, Meyer BF, al-Mehaidib A, Mogarri I: Geographic distribution of cystic fibrosis transmembrane regulator gene mutations in Saudi Arabia. Ann Trop Paediatr 1999;19:69–73.

76 Laufer-Cahana A, Lerer I, Sagi M, Rachmilewitz-Minei T, Zamir C, Rivlin JR, Abeliovich D: Cystic fibrosis mutations in Israeli Arab patients. Hum Mutat 1999;14: 543.

77 The Molecular Genetic Epidemiology of Cystic fibrosis. Report of a joint meeting of WHO/ECFTN/ICF(M)A/ACFS, Genoa, 2002.

Prof. Batsheva Kerem
Department of Genetics
The Life Sciences Institute
The Hebrew University of Jerusalem
IL-91904 Jerusalem (Israel)
Tel. +972 2 658 5689
Fax +972 2 658 4810
E-Mail kerem@cc.huji.ac.il

Bush A, Alton EWFW, Davies JC, Griesenbach U, Jaffe A (eds): Cystic Fibrosis in the 21st Century.
Prog Respir Res. Basel, Karger, 2006, vol 34, pp 11–20

Exquisite and Multilevel Regulation of *CFTR* Expression

Ann E.O. Trezise

School of Biomedical Science, University of Queensland, Brisbane, Australia

Abstract

Despite the enormous interest in cystic fibrosis (CF) and *CFTR*, with thousands of researchers worldwide working on this gene and the disease, there has not been a comprehensive review of the distribution and regulation of *CFTR* expression. This is surprising as knowledge of the sites of expression and the signals that modulate *CFTR* expression can provide critical insights into the pathogenesis of CF and the functions of *CFTR*. Also, understanding the regulation of endogenous *CFTR* expression is fundamental to the development of appropriate and specific gene therapy for CF and the possibility of therapeutic manipulation of endogenous *CFTR* expression. In this chapter I address the sites and signals that modulate *CFTR* expression in vivo and in vitro and then go on to examine our current understanding of the mechanisms that regulate *CFTR* expression. More than anything else, the work reviewed in this chapter shows that *CFTR* is not a housekeeping gene (a gene expressed at uniformly low levels in all cells), as once thought, but is subject to extensive and exquisite regulation in response to a variety of signals in space and in time.

Tissue-Specific and Developmental Regulation of in vivo *CFTR* Expression

Tissue-Specific Expression: Epithelial Tissues

The predominant site of *CFTR* expression is many of the epithelial surfaces throughout the body and for the most part these sites of expression are conserved across mammals and other vertebrates. Most of the epithelial sites of *CFTR* expression correspond well with the sites of cystic fibrosis (CF) disease: the submucosal glands and airway surface epithelium [1], the pancreatic ductal epithelium [2], the epithelium of the crypts of Lieberkuhn throughout the gastrointestinal tract [2], the epithelium of sweat glands [3], the epithelium of the developing genital ducts, adult epididymis and vas deferens [4, 5], the cervical and uterine epithelium [2, 5], the ductal and acinar epithelium of the salivary glands [2, 6], and the epithelial lining of the intrahepatic bile ducts and gall bladder [7].

However, there are some exceptions and not all the epithelial sites of *CFTR* expression correspond with known sites of CF disease. These sites of *CFTR* expression include: the kidney collecting duct epithelium [8] and the epithelium of the Brunner's glands, the submucosal glands of the duodenum [6].

Table 1 summarizes the known sites of *CFTR* expression in vivo, gives a broad indication of the relative levels of *CFTR* expression, where known, and provides references to some of the key articles describing *CFTR* expression. I have also indicated whether *CFTR* expression is regulated at a particular site and, where known, identified the molecular signal(s) resulting in altered *CFTR* expression.

Tissue-Specific Expression: Non-Epithelial Cells and Tissues

While *CFTR* is primarily thought of as a gene specifically expressed in epithelial cells, there are increasing numbers of reports describing *CFTR* expression in non-epithelial tissues (see table 1, fig. 1). These include ventricular cardiomyocytes [23], neuronal expression in the brain

Table 1. In vivo *CFTR* expression

Site of expression	Cell types and regulation of expression	Level[a]	Ref. No.
Gastrointestinal tract			
Salivary glands	Duct epithelial cells of all salivary glands	+++	2
	Also, mucous acini of submaxillary gland and submandibular acinar cells	++	6
Stomach	Diffuse expression throughout the mucosa	+	6
	Specific expression in the epithelium of the pyloric glands	+++	
Pancreas	Low, but detectable expression in ductal epithelium in rodents	+	2
	Comparatively high expression in human duct epithelium beginning during mid-trimester fetal development	++++	4
	Pancreatic acinar cells	+	
Liver and gall bladder	Specific expression in the apical membrane of gall bladder epithelial cells and intrahepatic biliary epithelial cells	+++	7
	Increased transcription during liver regeneration		22
Sweat gland	Reabsorptive duct (higher) and secretory coil (lower)	+++	3
Intestine	Brunner's gland acinar and duct cells are the site of highest expression in vivo	+++++	6
	High expression in intestinal crypt epithelial cells, with two perpendicular gradients of expression: intestinal crypt (high) to villus (low), and duodenum (high) to colon (low)	++++	2
Respiratory tract			
Nose	Nasal respiratory epithelium at very low level	+	1
Lung	Low level diffuse expression throughout the surface epithelium and lamina propria of bronchi and bronchioles	+	2, 1
	Submucosal glands of the larger airways are the site of highest expression in the adult respiratory tract	+++	4
	Highly expressed in fetal airway epithelium as early as 13 weeks' gestation	++++	16
	Expression is increased by combined oestrogen and progesterone		21
Urogenital tract			
Kidney	Cortical collecting duct epithelial cells	+++	8
Female reproductive tract	Luminal and glandular uterine epithelium and cervical epithelium	++++	2
	Expression is regulated during the oestrous cycle being maximal at pro-oestrus		5
	Moderate expression in the uterine epithelium of third trimester human fetus		21
	Expression is increased by oestrogen		
Male reproductive tract	In rodents, high levels in post-meiotic male germ cells	++++	5
	Low expression in Sertoli cells in rodents and humans	+	
	High expression in the rodent and human initial segment and caput of the epididymis and medium expression in the vas deferens	++++	
	Low expression throughout human embryonic male genital ducts (18 weeks)	++	4
Other epithelial sites			
Thyroid	Moderate expression in isolated thyroid follicular epithelial cells		41
Early embryo	Expression in 2-, 4- and 8-cell embryos, morulae and the inner cell mass and polar trophectoderm of human blastocysts		20
	Expression also detected in xenopus oocytes and blastula		
Non-epithelial sites			
Heart	Predominantly expressed in ventricular cardiomyocytes	++	9
	Two perpendicular expression gradients: epicardial (higher) to endocardial (lower), and apical (higher) to basal (lower) regions of the left ventricle in rabbit heart		42
	Some species differences exist; expression in human heart still controversial; developmental and pathological regulation		
Brain	Neurons of the hypothalamus, thalamus and amygdaloid nuclei, medial pre-optic area and cortex in rodents	++	10
	Decreased expression in human hypothalamic neurons in Alzheimer's disease		17
Endothelial cells	Human bovine and rabbit corneal endothelium and mouse aorta endothelium	+++	11
	Human endothelial cells from umbilical vein and lung microvasculature		12
Smooth muscle	Measured functionally in rat aortic smooth muscle cells		13
Lymphocytes	Low level expression in freshly isolated blood lymphocytes, neutrophils, monocytes and alveolar macrophages	+	14

[a] + indicates low expression and +++++ indicates highest expression. If blank then relative expression level could not be assessed.

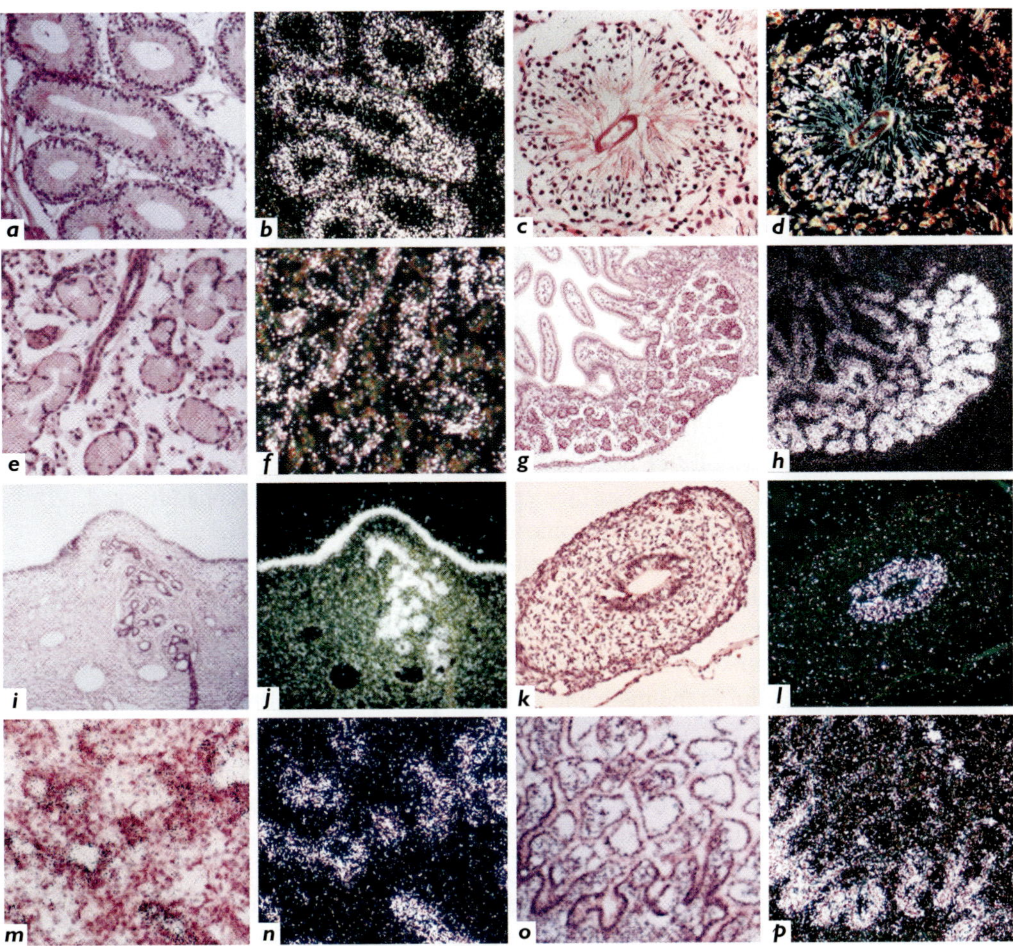

Fig. 1. In vivo sites of *CFTR* expression. Expression of CFTR mRNA was detected by RNA in situ hybridization. For each tissue two panels are shown, one with bright-field illumination to present cellular histology and the other with dark-field illumination, which reveals the silver grains (white signal) that overlie the cells expressing *CFTR*. The *CFTR*-expressing tissues shown here are: initial segment of the mouse epididymis (*a*, *b*), haploid spermatids in rat testis (*c*, *d*), ductal epithelium and mucous acini of mouse submaxillary salivary gland (*e*, *f*), mouse Brunner's glands, the site of highest in vivo *CFTR* expression, and duodenal epithelium (*g*, *h*), rat uterine surface and glandular epithelium at pro-oestrus (*i*, *j*), mouse umbilical endothelium (*k*, *l*), human fetal airway epithelium (*m*, *n*), and human fetal small intestine showing *CFTR* expression in the developing crypt epithelium and in rare villous epithelial cells (red arrowheads) (*o*, *p*). *a–f*, *o*, *p* ×100. *g–j* ×40. *k–n* ×200.

[10], corneal [11] and vascular endothelial cells [12], aortic smooth muscle cells [13] and lymphocytes [14]. In general, *CFTR* expression is lower in non-epithelial cells than in epithelial cells. Also, non-epithelial sites of *CFTR* expression have not been identified as major sites of CF disease, which raises the question of the role of *CFTR* in these cells. I would speculate that in these sites loss of *CFTR* expression might be compensated for by the expression of other gene(s). This may also be the case for epithelial sites of *CFTR* expression unaffected in CF, such as kidney collecting duct epithelium. This hypothesis is supported by the observation that while the heart is not a site of primary CF disease in most cases, there exists a distinct subgroup of CF patients (≈2%) that display a severe cardiac phenotype of extensive myocardial necrosis with scarring fibrosis that leads to sudden unexpected fatal cardiac arrest in CF infants usually less than 24 months of age [15]. The basis for this myocardial necrosis and fibrosis phenotype is

thought to be genetic, due to familial concordance, and points to an important role for *CFTR* in ventricular cardiomyocytes as part of an integrated network of complementary functions.

Developmental, Inducible and Coordinate Regulation of CFTR Expression

As well as distinct patterns of tissue and cell-specific expression, *CFTR* is also developmentally regulated [4, 16] (chapter 7), altered in pathological conditions [9, 17], responds to a variety of chemical and physical stimuli and is coordinately regulated with the expression of the multidrug resistance gene (*MDR1*) [18, 19].

The most well-known site of developmental regulation of *CFTR* expression is the airway surface epithelium, with relatively high expression during embryonic and fetal development and then a marked decrease in expression at birth (see fig. 1). This switch in expression occurs as the airway switches from a secretory to an absorptive epithelium [4, 16]. In addition to the airway, *CFTR* expression is subject to developmental regulation in the male and female reproductive tracts and the heart. *CFTR* has also been detected in the very early embryo, being expressed as early as the 2-cell stage, expression continuing in all cells through to the development of the blastula and then becoming restricted to the inner cell mass and polar trophectoderm [20].

CFTR expression in vivo is also modulated by a variety of chemical, physical and pathological stimuli. Most notably *CFTR* is regulated by the female sex hormones oestrogen and progesterone [21 and references therein], is significantly increased in regenerating liver [22], is also increased in many tissues by the peptide hormone guanylin [23] and is subject to complex regulation in the heart during the development of ventricular hypertrophy [9, 24]. Table 1 summarizes the in vivo sites of *CFTR* expression and the signals that modulate *CFTR* expression. Figure 1 shows the cellular localization of CFTR mRNA in a selection of tissues.

Constitutive and Regulated *CFTR* Expression in vitro

CFTR expression has been characterized in a large number of transformed cell lines, mostly of human origin but also in some cell lines originating from rodents and other vertebrate species. The analysis of *CFTR* expression in vitro provides some unique opportunities to investigate the stimuli that modulate *CFTR* expression and the underlying transcriptional and post-transcriptional mechanisms. The

established cell lines used in CF research have recently been reviewed [25]. Table 2 lists some cell lines known to endogenously express *CFTR*. Where *CFTR* expression is modulated the stimuli and the mechanism of regulation, where known, are indicated.

On the whole, the cellular and tissue origin of the transformed cells that endogenously express *CFTR* reflects the cell-specific expression of *CFTR* in vivo, as would be expected. A variety of second messenger, cytokine, hormonal and other stimuli have been shown to modulate *CFTR* expression. Many of these signals act through post-transcriptional mechanisms that ultimately regulate CFTR mRNA stability. However, the molecular mechanisms underlying these changes in CFTR mRNA stability are largely unaddressed.

Mechanisms Regulating *CFTR* Expression

Transcriptional Regulation

Precise control of gene expression is fundamentally important in health and disease. Morphology, histology and physiology of every cell are controlled by the precise regulation of expression of the subset of genes specific for that cell and tissue. Deranged gene expression is also a major disease process and for these reasons it is important to understand how *CFTR* expression is regulated. *CFTR* expression is primarily controlled by regulating *CFTR* gene transcription. Beyond this, the ultimate level of CFTR protein expressed can be modulated by alternative splicing of CFTR mRNA, by varying the efficiency of translation and stability of CFTR mRNA (see section Post-Transcriptional Regulation below), and by post-translational regulation of CFTR protein trafficking and stability (see chapter 3).

Control of gene transcription is mediated by the promoter of each gene. The gene promoter is made up of DNA sequences, usually located immediately upstream of the transcription start site, that form binding sites for the RNA polymerase transcription complex. In addition, the promoter includes DNA sequence element binding sites for a variety of transcription factor protein complexes that enhance or suppress the basal rate of gene transcription, by RNA polymerase, in response to a variety of temporal, spatial, physiological and pathological signals. These enhancer and suppressor DNA elements may be located upstream, downstream or within the intronic sequences of *CFTR* and can be tens or even hundreds of kilobases from the protein coding region of the gene. By identifying the DNA sequence elements and transcription factors that control *CFTR* expression we will gain important insights into the

Table 2. In vitro *CFTR* expression

Cell type	Cell line and regulation of expression	Level[a]	Ref. No.
Human epithelial cells			
Intestine	T84: cAMP, acting through PKA, is essential for basal expression	++++	26,
	Interferon-γ decreases CFTR mRNA stability		35,
	Caco2: Differentiation increases CFTR mRNA stability, but decreases protein	+++	43,
	HT29: Tumour necrosis factor-α decreases CFTR mRNA stability but no change in transcription	+++	36,
	Protein kinase C activation decreases *CFTR* transcription		30,
	PKC inhibition increases *CFTR* transcription		
	Differentiation increases CFTR mRNA stability, but decreases protein		
	Interferon-γ decreases CFTR mRNA stability		44,
	Induction of *MDR1* expression decreases *CFTR* expression		18
	HT29-18: Low glucose-induced differentiation increases CFTR mRNA		
Cervical	HeLa	++	26
Lymphoblastoid	B3.1.0	++	26
Pancreatic duct	PANC-1	+	26
	Capan-1: CFTR mRNA increases with days in culture	++	25
Fetal tracheal	56FHTE-8o-	++	25
Airway	Calu-3: Interleukin-1β increases *CFTR* transcription 16HBE	++	25
Mouse epithelial cells			
Kidney	M-1: Mouse cortical collecting duct cell line		25
Intestine	CMT-93: Mouse rectal adenocarcinoma cells	+++	45
Rat epithelial cells			
Intestinal	IEC-18: Derived from small intestinal crypts		
Uterine	UIT 1.16: Oestrogen is necessary for *CFTR* expression		25
Non-epithelial cells			
Human blood	U-937: histiocytic lymphoma cells	++	14
cells	K-562: erythroleukaemia cells	++	
Neurons	GT1-7: GnRH-expressing GT1-7 hypothalamic neurons express CFTR		46

^a + indicates low expression and +++++ indicates highest expression. If blank then relative expression level could not be assessed.

physiological role and regulation of *CFTR* in healthy tissues, mechanisms of CF disease and this may in the future allow us to manipulate *CFTR* expression and alleviate CF symptoms (see below).

Early work aimed at understanding the promoter and transcriptional regulation of *CFTR* had some success in identifying the DNA elements responsible for *CFTR* expression in vitro, in cell lines such as T84, HT29, Caco-2 and some pancreatic cell lines [26]. However, identifying the DNA elements and understanding the mechanisms that direct *CFTR* expression in vivo proved more elusive. To identify the in vivo *CFTR* promoter, the DNA thought to encode the promoter elements is linked to a reporter gene, such as β-galactosidase or luciferase, and this promoter-reporter gene construct is inserted into the genome of a transgenic mouse. If the DNA construct contains the complete *CFTR* promoter then the reporter gene will be expressed in all the same cells as the endogenous *CFTR* gene. However, early promoter-reporter gene transgenic mice did not show any significant reporter gene expression.

The first transgenic mouse to recapitulate many, but not all, of the sites of endogenous *CFTR* expression was produced by Huxley's group [6] in 1997. Rather than the traditional promoter-reporter gene construct, this mouse harboured a yeast artificial chromosome containing the complete human *CFTR* locus, including 50 kb of DNA upstream of *CFTR* exon 1. This human *CFTR* (*hCFTR*) transgene recapitulated endogenous mouse *CFTR* expression qualitatively and quantitatively at many sites and with sufficient accuracy to restore CFTR function in a CF mouse model. The expression of the transgene did vary from endogenous expression at some sites; for example, the *hCFTR* transgene did not show increased expression in fetal lung, and while the transgene was appropriately expressed in the intestinal crypt epithelium, it was not expressed in the Brunner's glands of the small intestine. Given the highly conserved nature of transcriptional regulatory mechanisms, with many human transcription factors able to function in the cells of other mammals, insects and even yeast, the most likely explanation for the lack of *hCFTR* transgene expression in fetal lung and Brunner's glands is that the required DNA regulatory elements were not contained within the transgene construct. But it is also possible that the human regulatory sequences are not recognized by the mouse transcription factors. The major insight provided by this work came with the realization of the previously unappreciated magnitude of the complexity of in vivo *CFTR* regulation.

More recently this approach, of using a *hCFTR* transgene, has identified a DNA element in *hCFTR* intron 1 that is responsible for *CFTR* expression in the intestinal crypt epithelium. Deletion of this element results in loss of *hCFTR* transgene expression specifically in the intestinal crypt epithelium. It is likely that this *hCFTR* intron 1 element, previously identified as a DNase I hypersensitive site, functions through binding hepatic nuclear factor 1 alpha (HNF1α) [27].

Additional DNA elements that have been shown to be important for either maintaining or inducing *CFTR* expression in vivo or in vitro include: (1) a nuclear factor (NF) κB element at $-1,040$ bp that responds to interleukin-1β signalling via binding of NF-κB and increases *CFTR* expression in Calu-3 cells [28], (2) an inverted CCAAT (Y-box) element at $+2$ bp that, when bound by the CCAAT displacement protein/cut homologue (CDP/cut), inhibits *CFTR* transcription through histone deacetylation [29], and (3) adjacent to the inverted Y-box is a functional cAMP response element (CRE) located at $+14$ bp that, when bound by CRE-binding protein (CREB), confers cAMP-dependent protein kinase (PKA)-dependent activation on basal *CFTR* transcription and thereby mediates increased *CFTR* expression in response to cAMP [30].

Figure 2 shows a diagram of the genomic structure of the region surrounding *CFTR* exon 1 indicating the relative positions of these regulatory elements. Numbering is relative to the first transcription start site ($+1$) identified, which is located 134 bp upstream of the translation initiation codon in *CFTR* exon 1. This is consistent with the CF mutation database (http://www.genet.sickkids.on.ca/cftr/).

There is also evidence that incompletely understood, cell-specific interactions occur between the protein complexes bound to the adjacent inverted Y-box and CRE elements that further modulate *CFTR* expression. Mutational analysis shows that both elements are required for basal

CFTR expression [29, 30]. Conservation across large evolutionary distances is often indicative of important function. As such, it is perhaps significant that the inverted Y-box, the CRE sequence and the intron 1 HNFα DNA elements are amongst the few non-protein coding sequences conserved in the pufferfish *Fugu rubripes CFTR* gene [31]. The most recent common ancestor of the pufferfish and humans lived over 400 million years ago. Therefore, these *CFTR* regulatory elements have been maintained throughout 800 million years of independent evolution.

Currently the CF mutation database lists 20 DNA sequence variations in the 5′-untranslated region (5′UTR) of *CFTR* exon 1 and the *CFTR* promoter (up to 1,060 bp upstream of the translation initiation codon in *CFTR* exon 1). A number of these sequence variations have been listed as regulatory or promoter mutations either because no other coding sequence mutation was identified, or because they coincide with a known or putative regulatory element. The most thoroughly investigated sequence variation of relevance to *CFTR* expression is T to A at -102. Rather than destroying a control element, this change creates a Yin Yang 1 (YY1) box and leads to increased *CFTR* expression. When the -102T>A mutation is present as a complex allele with the severe S549R(T>G) mutation, this results in a much milder CF disease phenotype than produced by the S549R(T>G) mutation alone [32]. These findings establish an important principle, that is that increasing the expression of processing defect mutations such as S549R can allow enough CFTR protein to reach the plasma membrane for restoration of partial function and the relief of CF disease symptoms (see chapter 3 for full discussion of CFTR protein processing defects).

Investigations into 'the *CFTR* promoter' to date have concentrated on the DNA sequences immediately upstream of *CFTR* exon 1. However, multiple transcription start sites and the presence of at least six alternative, *CFTR* first exons [9, 26, 33, 34] complicate analysis of the transcriptional control of *CFTR* and again highlight the complexity of *CFTR* regulation. Alternative *CFTR* first exons, which splice directly to exon 2, were first characterized by Koh et al. [26] in 1993 and similar alternative first exons were found in the sheep *CFTR* gene [34]. In both cases these alternative exons were located within 1 kb of *CFTR* exon 1, they did not contain an alternative translation start codon to substitute for the AUG start codon present in *CFTR* exon 1 and the functional significance of these CFTR transcript isoforms remains undetermined. In contrast, cardiac-specific alternative first *CFTR* exons, that do include an alternative translation initiation codon, have been identified and account for over 90% of the CFTR transcripts expressed in

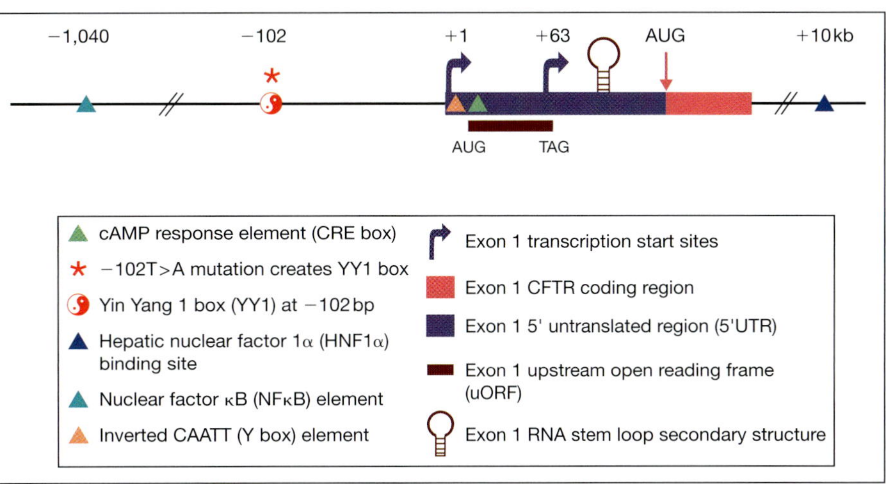

Fig. 2. Conserved regulatory elements in and around human *CFTR* exon 1. Bent arrows indicate the two major sites of transcription initiation in human *CFTR* exon 1. Similar transcription start sites are used in other species where this has been measured (mouse and rabbit). Transcripts initiating at these sites account for the great majority of *CFTR* expression in epithelial cells. Symbols indicate the relative positions of evolutionarily conserved regulatory elements that have been confirmed to functionally regulate *CFTR* expression at either the transcriptional or post-transcriptional level.

the rabbit heart. These cardiac-specific exons are located approximately 10 kb upstream of the traditional *CFTR* exon 1 and highly conserved sequences are present in the human genome [29]. It is likely that a cardiac-specific *CFTR* promoter is responsible for driving *CFTR* transcription from these distant alternative first exons, although this remains to be investigated. Clearly, *CFTR* expression is very complex and involves multiple, tissue-specific transcription start sites, alternative first exons and alternatively spliced transcripts. However, most *CFTR* expression in epithelial cells initiates at one of two major transcription start sites in exon 1 (see fig. 2).

The complexity, diversity and our comparatively sophisticated understanding of the spatial, temporal and inducible patterns of *CFTR* expression provide a stark contrast to our understanding of the DNA elements and mechanisms that control *CFTR* expression in vivo. It is clear that there are many aspects of *CFTR* expression that are not explained by the known transcription control elements. As is evident from the work to date, it is a long, hard road to understanding the mechanisms controlling *CFTR* expression in vivo, but it is a road worth travelling. While gene therapy will not be the panacea for CF, the future development of appropriately regulated CFTR gene therapy vectors will rely on mechanistic knowledge of the regulation of *CFTR* expression. This knowledge will also be crucial for the development of therapies aimed at artificially increasing endogenous *CFTR* expression to restore sufficient function to alleviate some of the symptoms of CF.

Post-Transcriptional Regulation

Many of the signals that modulate *CFTR* expression, by as much as 3- to 4-fold, have been reported to act via undefined post-transcriptional mechanism(s) [22, 35, 36]. As the −102T>A mutation produces only a small increase in *CFTR* expression (approximately 2-fold), but this is sufficient to partially compensate for the S549R processing defect and moderate CF symptoms, it is worthwhile investigating the post-transcriptional regulation of *CFTR*. This may provide an alternative therapeutic avenue and we will gain insight into the physiological signals and mechanisms that regulate *CFTR* expression.

Recently, we have gained some understanding of one mechanism of post-transcriptional *CFTR* regulation. This mechanism has been shown in yeast to coordinately regulate translation efficiency and mRNA stability [37] and is based on the presence of one or more upstream open reading frames (uORFs) and predicted RNA secondary structures in the CFTR mRNA 5′UTR. Translation of the uORF reduces the number of ribosomes available to translate the main coding region because many ribosomes are released from the mRNA at the translation stop codon of the uORF. Release of ribosomes at the uORF stop codon depends on the presence of RNA secondary structure and the G/C content of the sequence immediately downstream of the stop codon, and on the presence of mostly undefined *trans*-acting protein factors. Reduced density of ribosomes translating the main coding region makes the mRNA more vulnerable to degradation and results in an overall decrease in mRNA stability [37].

One or more uORFs and RNA secondary structures are encoded in the 5′UTRs of the *CFTR* cardiac- and testis-specific alternative first exons and in *CFTR* exon 1, the predominant CFTR mRNA isoform expressed in epithelial cells. Similar elements are present in exon 1 of human, chimp, rabbit and mouse *CFTR* [9]. Analysis of *CFTR* transcription initiation across species and in multiple tissues identifies two major regions of transcription initiation in *CFTR* exon 1 [9, 26, 33, 34]. One group of transcription start sites occur at and near 69 bp upstream of the translation initiation codon of the CFTR coding region. The second group of start sites are located around 132 bp upstream of the CFTR AUG codon. Transcription start site selection will determine whether the uORF is included in the expressed mRNA (see fig. 2). Reporter gene assays have been used to show that the uORF and RNA secondary structure encoded in the rabbit CFTR exon 1 5′UTR are functional. Transcripts that initiate at the –69 bp start site do not encode an uORF and produce more than twice as much protein as transcripts that initiate at the –132 bp site and do include the uORF in their 5′UTR [9]. Also, the uORFs and RNA secondary structures encoded in the 5′UTRs of different CFTR mRNA isoforms regulate differential CFTR transcript stabilities, which contributes to regulated *CFTR* expression in the heart [24]. This work identifies a completely new mechanism of *CFTR* regulation and highlights the caution that must be used in extrapolating from CFTR mRNA levels to functional channel activity.

The exon 1 uORF is conserved in mammalian, marsupial and amphibian *CFTR* genes. Pufferfish and other bony fish also encode an uORF in the 5′UTR of *CFTR* exon 1, but it is located closer to the CFTR protein coding sequence, in a similar position to a second uORF in exon 1 of mouse *CFTR* (Trezise et al., unpubl. data). The evolutionary conservation of the *CFTR* exon 1 uORF indicates that this is an important mechanism of post-transcriptional regulation of *CFTR*.

Perspectives and Future Directions

The benefits of studying the sites and regulation of *CFTR* expression are well illustrated by the investigations into the regulation of *CFTR* expression by female sex hormones. Oestrogen and progesterone regulation of *CFTR* is not just restricted to the uterine epithelium, where it was first identified [5], but is also evident in airway epithelia [21]. Both CFTR and ENaC (the epithelial sodium channel) mRNA levels are greater in the lungs of adult female rats compared to males. Also, combined administration of both oestrogen and progesterone increased the expression of both CFTR and ENaC mRNAs in the lung. ENaC has an important role in the fluid-absorptive capacity of the airway epithelium and higher *ENaC* expression in females results in increased water absorption in the lungs [21]. Excessive ENaC activity and increased water absorption is already a major pathogenic mechanism in the airways of CF patients [38]. The increased airway expression of *ENaC* in women will only exacerbate this effect and may provide the molecular basis for the documented 9-year difference in the median survival age for women with CF (27.8 years) compared to male CF patients (36.7 years [39]). Another study found no significant difference in lung function between male and female CF patients [40], but this study did not separate pre- and post-pubescent groups, an important factor in the hormonal regulation of *CFTR* and *ENaC* expression. Overall, the analysis of *CFTR* expression and the identification of the molecular signals that regulate *CFTR* expression have provided significant insights into the pathogenesis of CF.

Our understanding of the molecular basis of CF will continue to grow as data from microarray gene expression profiling (see also chapter 14) and the ENCODE Project are beginning to be widely disseminated. These data can be accessed via a number of different web sites (http://www.genome.ucsc.edu/encode/encode.html; http://www.genome.ucsc.edu/; http://www.ensembl.org/; http://www.ncbi.nlm.nih.gov/geo/). Microarray expression profiling is providing insights into how loss of *CFTR* expression and/or function in CF impacts on the expression profile of cells and tissues. The ENCODE (ENCyclopedia Of DNA Elements: see http://www.genome.gov/10005107) Project aims to identify all the functional elements in the human genome sequence. *CFTR* is one of the first regions of the human genome to be targeted for full functional annotation. As more data are generated by these projects it is expected that we will gain a better understanding of some of the features of CF that are not easily explained by the loss of a membrane chloride channel. Understanding and documenting the sites, mechanisms and regulation of *CFTR* expression is not simply a matter of stamp collecting. It is through this process that we can gain critical insights into the pathogenesis of CF, develop hypotheses about the mechanisms of disease progression and with this we can move forward to develop novel, logical strategies to combat these pathogenic mechanisms.

Acknowledgements

I would like to thank all the people in the CF community that I have had the privilege to work with over the years. In particular I would like to thank Manuel Buchwald, Bill Colledge, Chris Higgins, Alan Cuthbert and Sir Martin Evans. I would also like to thank Alison Jones for assistance and Bill Colledge for critical reading of the manuscript. Over the years my work on CF has been supported by the

Canadian Cystic Fibrosis Foundation, the North American Cystic Fibrosis Foundation, the Beit Memorial Trust for Medical Research, the UK Cystic Fibrosis Research Trust, the Wellcome Trust, the British Heart Foundation and the Australian Research Council.

References

1 Engelhardt JF, Yankaskas JR, Ernst SA, Yang Y, Marino CR, Boucher RC, Cohn JA, Wilson JM: Submucosal glands are the predominant site of CFTR expression in the human bronchus. Nat Genet 1992;2:240–248.

2 Trezise AE, Buchwald M: In vivo cell-specific expression of the cystic fibrosis transmembrane conductance regulator. Nature 1991;353:434–437.

3 Kartner N, Augustinas O, Jensen TJ, Naismith AL, Riordan JR: Mislocalization of delta F508 CFTR in cystic fibrosis sweat gland. Nat Genet 1992;1:321–327.

4 Trezise AE, Chambers JA, Wardle CJ, Gould S, Harris A: Expression of the cystic fibrosis gene in human foetal tissues. Hum Mol Genet 1993;2:213–218.

5 Trezise AE, Linder CC, Grieger D, Thompson EW, Meunier H, Griswold MD, Buchwald M: CFTR expression is regulated during both the cycle of the seminiferous epithelium and the oestrous cycle of rodents. Nat Genet 1993;3: 157–164.

6 Manson AL, Trezise AE, MacVinish LJ, Kasschau KD, Birchall N, Episkopou V, Vassaux G, Evans MJ, Colledge WH, Cuthbert AW, Huxley C: Complementation of null CF mice with a human CFTR YAC transgene. EMBO J 1997;16:4238–4249.

7 Yang Y, Raper SE, Cohn JA, Engelhardt JF, Wilson JM: An approach for treating the hepatobiliary disease of cystic fibrosis by somatic gene transfer. Proc Natl Acad Sci USA 1993; 90:4601–4605.

8 Todd-Turla KM, Rusvai E, Naray-Fejes-Toth A, Fejes-Toth G: CFTR expression in cortical collecting duct cells. Am J Physiol 1996;270: F237–F244.

9 Davies WL, Vandenberg JI, Sayeed RA, Trezise AE: Cardiac expression of the cystic fibrosis transmembrane conductance regulator involves novel exon 1 usage to produce a unique amino-terminal protein. J Biol Chem 2004;279:15877–15887.

10 Mulberg AE, Weyler RT, Altschuler SM, Hyde TM: Cystic fibrosis transmembrane conductance regulator expression in human hypothalamus. Neuroreport 1998;9:141–144.

11 Sun XC, McCutheon C, Bertram P, Xie Q, Bonanno JA: Studies on the expression of mRNA for anion transport related proteins in corneal endothelial cells. Curr Eye Res 2001; 22:1–7.

12 Tousson A, Van Tine BA, Naren AP, Shaw GM, Schwiebert LM: Characterization of CFTR expression and chloride channel activity in human endothelia. Am J Physiol 1998; 275:C1555–C1564.

13 Robert R, Thoreau V, Norez C, Cantereau A, Kitzis A, Mettey Y, Rogier C, Becq F: Regulation of the cystic fibrosis transmembrane conductance regulator channel by beta-adrenergic agonists and vasoactive intestinal peptide in rat smooth muscle cells and its role in vasorelaxation. J Biol Chem 2004;M279: 21160–21168.

14 Yoshimura K, Nakamura H, Trapnell BC, Chu CS, Dalemans W, Pavirani A, Lecocq JP, Crystal RG: Expression of the cystic fibrosis transmembrane conductance regulator gene in cells of non-epithelial origin. Nucleic Acids Res 1991;19:5417–5423.

15 Zebrak J, Skuza B, Pogorzelski A, Ligarska R, Kopytko E, Pawlik J, Rutkiewicz E, Witt M: Partial CFTR genotyping and characterisation of cystic fibrosis patients with myocardial fibrosis and necrosis. Clin Genet 2000;57: 56–60.

16 McCray PB Jr, Reenstra WW, Louie E, Johnson J, Bettencourt JD, Bastacky J: Expression of CFTR and presence of cAMP-mediated fluid secretion in human fetal lung. Am J Physiol 1992;262:L472–L481.

17 Lahousse SA, Stopa EG, Mulberg AE, de la Monte SM: Reduced expression of the cystic fibrosis transmembrane conductance regulator gene in the hypothalamus of patients with Alzheimer's disease. J Alzheimers Dis 2003;5: 455–462.

18 Breuer W, Slotki IN, Ausiello DA, Cabantchik IZ: Induction of multidrug resistance down-regulates the expression of CFTR in colon epithelial cells. Am J Physiol 1993;265: C1711–C1715.

19 Trezise AE, Ratcliff R, Hawkins TE, Evans MJ, Freeman TC, Romano PR, Higgins CF, Colledge WH: Co-ordinate regulation of the cystic fibrosis and multidrug resistance genes in cystic fibrosis knockout mice. Hum Mol Genet 1997;6:527–537.

20 Ben-Chetrit A, Antenos M, Jurisicova A, Pasyk EA, Chitayat D, Foskett JK, Casper RF: Expression of cystic fibrosis transmembrane conductance regulator during early human embryo development. Mol Hum Reprod 2002;8:758–764.

21 Sweezey N, Tchepichev S, Gagnon S, Fertuck K, O'Brodovich H: Female gender hormones regulate mRNA levels and function of the rat lung epithelial Na channel. Am J Physiol 1998;274:C379–C386.

22 Tran-Paterson R, Davin D, Krauss RD, Rado TA, Miller DM: Expression and regulation of the cystic fibrosis gene during rat liver regeneration. Am J Physiol 1992;263(1 Pt 1): C55–C60.

23 Kulaksiz H, Schlenker T, Rost D, Stiehl A, Volkmann M, Lehnert T, Cetin Y, Stremmel W: Guanylin regulates chloride secretion in the human gallbladder via the bile fluid. Gastroenterology 2004;126:732–740.

24 Davies WL, Vandenberg JI, Sayeed RA, Trezise AE: Post-transcriptional regulation of the cystic fibrosis gene in cardiac development and hypertrophy. Biochem Biophys Res Commun 2004;319:410–418.

25 Gruenert DC, Willems M, Cassiman JJ, Frizzell RA: Established cell lines used in cystic fibrosis research. J Cyst Fibros 2004;3:191–196.

26 Koh J, Sferra TJ, Collins FS: Characterization of the cystic fibrosis transmembrane conductance regulator promoter region: Chromatin context and tissue-specificity. J Biol Chem 1993;268:15912–15921.

27 Rowntree RK, Vassaux G, McDowell TL, Howe S, McGuigan A, Phylactides M, Huxley C, Harris A: An element in intron 1 of the CFTR gene augments intestinal expression in vivo. Hum Mol Genet 2001;10:1455–1464.

28 Brouillard F, Bouthier M, Leclerc T, Clement A, Baudouin-Legros M, Edelman A: NF-kappa B mediates up-regulation of CFTR gene expression in Calu-3 cells by interleukin-1beta. J Biol Chem 2001;276:9486–9491.

29 Li S, Moy L, Pittman N, Shue G, Aufiero B, Neufeld EJ, LeLeiko NS, Walsh MJ: Transcriptional repression of the cystic fibrosis transmembrane conductance regulator gene, mediated by CCAAT displacement protein/cut homolog, is associated with histone deacetylation. J Biol Chem 1999;274: 7803–7815.

30 Matthews RP, McKnight GS: Characterization of the cAMP response element of the cystic fibrosis transmembrane conductance regulator gene promoter. J Biol Chem 1996;271: 31869–31877.

31 Davidson H, Taylor MS, Doherty A, Boyd AC, Porteous DJ: Genomic sequence analysis of *Fugu rubripes* CFTR and flanking genes in a 60 kb region conserving synteny with 800 kb of human chromosome 7. Genome Res 2000; 10:1194–1203.

32 Romey MC, Pallares-Ruiz N, Mange A, Mettling C, Peytavi R, Demaille J, Claustres M: A naturally occurring sequence variation that creates a YY1 element is associated with increased cystic fibrosis transmembrane conductance regulator gene expression. J Biol Chem 2000;275:3561–3567.

33 White NL, Higgins CF, Trezise AE: Tissue-specific in vivo transcription start sites of the human and murine cystic fibrosis genes. Hum Mol Genet 1998;7:363–369.

34 Mouchel N, Broackes-Carter F, Harris A: Alternative 5′ exons of the CFTR gene show developmental regulation. Hum Mol Genet 2003;12:759–769.

35 Nakamura H, Yoshimura K, Bajocchi G, Trapnell BC, Pavirani A, Crystal RG: Tumor necrosis factor modulation of expression of the cystic fibrosis transmembrane conductance regulator gene. FEBS Lett 1992;314: 366–370.

36 Sood R, Bear C, Auerbach W, Reyes E, Jensen T, Kartner N, Riordan JR, Buchwald M: Regulation of CFTR expression and function during differentiation of intestinal epithelial cells. EMBO J 1992;11:2487–2494.

37 Vilela C, Ramirez CV, Linz B, Rodrigues-Pousada C, McCarthy JE: Post-termination ribosome interactions with the 5′UTR modulate yeast mRNA stability. EMBO J 1999;18: 3139–3152.

38 Mall M, Grubb BR, Harkema JR, O'Neal WK, Boucher RC: Increased airway epithelial Na$^+$ absorption produces cystic fibrosis-like lung disease in mice. Nat Med 2004;10: 487–493.

39 Corey M, Farewell V: Determinants of mortality from cystic fibrosis in Canada, 1970–1989. Am J Epidemiol 1996;143:1007–1017.

40 Verma N, Bush A, Buchdahl R: It there really a gender gap in cystic fibrosis (CF)? Pediatr Pulmonol Suppl 2004;27:326.

41 Devuyst O, Golstein PE, Sanches MV, Piontek K, Wilson PD, Guggino WB, Dumont JE, Beauwens R: Expression of CFTR in human and bovine thyroid epithelium. Am J Physiol 1997;272:C1299–C1308.

42 Nagel G, Hwang TC, Nastiuk KL, Nairn AC, Gadsby DC: The protein kinase A-regulated cardiac Cl$^-$ channel resembles the cystic fibrosis transmembrane conductance regulator. Nature 1992;360:81–84.

43 Bargon J, Trapnell BC, Yoshimura K, Dalemans W, Pavirani A, Lecocq JP, Crystal RG: Expression of the cystic fibrosis transmembrane conductance regulator gene can be regulated by protein kinase C. J Biol Chem 1992;267:16056–16060.

44 Besancon F, Przewlocki G, Baro I, Hongre AS, Escande D, Edelman A: Interferon-gamma downregulates CFTR gene expression in epithelial cells. Am J Physiol 1994;267: C1398–C1404.

45 Huertas D, Howe S, McGuigan A, Huxley C: Expression of the human CFTR gene from episomal oriP-EBNA1-YACs in mouse cells. Hum Mol Genet 2000;9:617–629.

46 Weyler RT, Yurko-Mauro KA, Rubenstein R, Kollen WJ, Reenstra W, Altschuler SM, Egan M, Mulberg AE: CFTR is functionally active in GnRH-expressing GT1–7 hypothalamic neurons. Am J Physiol 1999;277:C563–C571.

Ann E.O. Trezise
School of Biomedical Science
University of Queensland
Brisbane, Qld. 4072 (Australia)
Tel. +617 3365 2715, Fax +617 3365 1299
E-Mail ann.trezise@uq.edu.au

Bush A, Alton EWFW, Davies JC, Griesenbach U, Jaffe A (eds): Cystic Fibrosis in the 21st Century.
Prog Respir Res. Basel, Karger, 2006, vol 34, pp 21–28

Intracellular Processing of CFTR

Hervé Barriere   Gergely L. Lukacs

Hospital for Sick Children Research Institute, Program in Cell and Lung Biology and
Department of Laboratory Medicine and Pathobiology, University of Toronto, Toronto, Canada

Abstract

The cystic fibrosis transmembrane conductance regulator
(CFTR or ABCC7), a cAMP-regulated plasma membrane chloride channel, belongs to the ATP-binding cassette (ABC)
transporter superfamily. Cystic fibrosis (CF) mutations could
be grouped into three major categories based on their cellu-
lar/biochemical phenotype: impaired CFTR production, function and cell surface stability, or a combination of these. Each
of these mechanisms can be attributed to a variety of molecular defects. Since the majority of the mutations, including the
most common deletion phenylalanine 508 (ΔF508 CFTR),
result in premature degradation of the channel at the endoplasmic reticulum, significant efforts have been devoted to
relocate the mutant, but functional CFTR to the cell surface in
an attempt to alleviate the CF phenotype. To increase the rescue efficiency of the trafficking defects and unravel the cellular
and molecular basis of CF, it is imperative to understand the
principles that govern the biogenesis, trafficking, and degradation of CFTR. Here we provide a brief overview of recent
progress regarding the cellular processing of CFTR and some
highlights of its perturbations caused by clinically relevant
mutations.

Introduction

Deletion of a single residue, phenylalanine 508 [ΔF508
cystic fibrosis transmembrane conductance regulator
(CFTR)], in the first nucleotide binding domain (NBD1) is

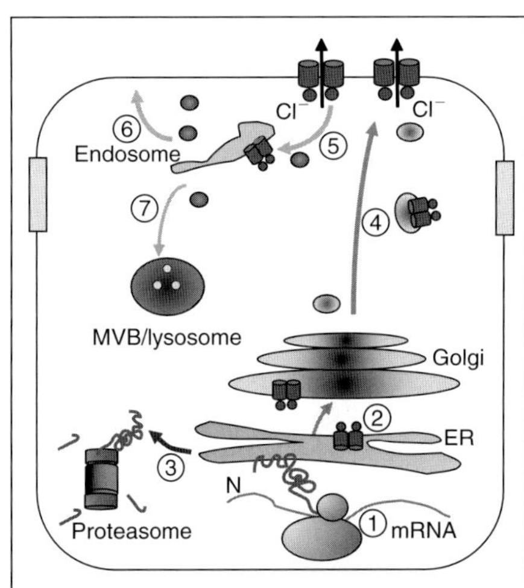

Fig. 1. Schematic illustration of cellular and biochemical pathways
involved in the biogenesis, plasma membrane targeting and degradation of CFTR. 1 = Cotranslational folding; 2 = posttranslational
folding at the ER; 3 = proteasome-dependent degradation; 4 =
biosynthetic processing and vesicular transport; 5 = endocytosis;
6 = recycling; 7 = degradation.

responsible for 66% of the cystic fibrosis (CF) cases and
could be identified in 90% of CF alleles [1]. Most of the
newly synthesized ΔF508 CFTR degrades via the ubiquitin-proteasome pathway at the endoplasmic reticulum (ER)
(fig. 1) [2, 3], likely due to its temperature-sensitive folding
defect [4, 5]. The majority of mutations, including the

ΔF508 CFTR, lead to defective trafficking with partially preserved channel activity. A better understanding of the molecular basis of the wild-type and mutant CFTR processing will aid the design of more efficient pharmacological interventions to alleviate the misprocessing of CF mutations [6].

Domain Assembly of CFTR during Biogenesis at the ER

The primary quality control mechanism of membrane protcins is associated with the ER [7]. Membrane proteins that are unable to fold or assemble are recognized by the aid of molecular chaperones and targeted for proteolysis by the ER-associated degradation (ERAD) to prevent the accumulation of toxic, aggregation-prone polypeptides in the cell [7, 8]. In nonpolarized, heterologous overexpression systems the wild-type (wt) CFTR folds inefficiently. No more than 20–40% of the nascent chain attain its folded conformation [9, 10], while the remaining molecules are targeted for ERAD, partially via the ubiquitin-proteasome proteolytic pathway with similar kinetics than the ΔF508 CFTR [2, 3].

CFTR contains five domains in two symmetrical halves, each comprising a membrane-spanning domain (MSD1 and MSD2), six transmembrane segments and a nucleotide binding domain (NBD1 and NBD2) with ATP hydrolytic capacity [1]. The two halves are connected by the fifth, regulatory (R) domain [1], which in coordination with the NBDs regulates the channel gating [11] (see chapter 4). This complex multidomain assembly is thought to be responsible for the limited folding efficiency of wt CFTR. CFTR appears to have a significantly higher folding efficiency (60–100%) in epithelia, expressing the channel endogenously [10, 12]. The initial steps of biosynthetic processing, including the recognition, tagging and selective degradation of nonnative conformers of the wt and mutant CFTR, overlap at the ER and are summarized below.

Reconstitution of CFTR biogenesis using in vitro transcription-translation assays and truncated molecules revealed that the transmembrane segments are cotranslationally inserted into the ER and undergo N-linked glycosylation [13, 14]. Although efficient plasma membrane delivery of the N-terminal half channel (consisting of the MSD1, NBD1 and the R domains) requires its coexpression with C-terminal half (MSD2 and NBD2), indicating cooperative domain folding/assembly, protein kinase A (PKA)-regulated chloride current could be observed in the presence of CFTR lacking its NBD2 [15–17]. This observation suggests that the truncated CFTR is able to fold independently of NBD2. Dissociation of heat shock protein 40 (Hsp40) and Hsp70 from the completely translated N-terminal half CFTR molecules also supports the notion that the MSD2 and NBD2 are dispensable for the N-terminal half CFTR to attains its near native conformation in vitro [18].

We applied limited proteolysis followed by immunoblot analysis to probe the conformational changes of the NBD1 and NBD2 in the nascent CFTR chain [5]. These studies provided direct evidence for the vectorial folding of the cytosolic domains of CFTR [19]. While the protease resistance of NBD1 was attained cotranslationally, the folding of NBD2, similar to the full-length CFTR assembly, was completed only posttranslationally [19]. Systematic mutagenesis also revealed that productive folding of NBD2 requires hydrophobic side chain interactions of the 508 residues [19]. Whether the contact surface of the F508 residue is required for the native tertiary structure of the transmembrane segments via direct interactions with cytosolic loops of the MSDs, or it facilitates the folding transition of the NBD2 directly, remains to be established. The former possibility is favored by interactions of F508 homologue residues with the cytoplasmic loops of the MSDs in the crystal structure of the vitamin B_{12} importer (BtuCD) and the lipid flippase MsbA of *Escherichia coli* [20, 21]. Perturbation of the MSD topology was indeed proposed in ΔF508 CFTR [22].

The Role of Molecular Chaperones in the Biogenesis of CFTR

During the co- and posttranslational folding, CFTR binds to several cytosolic (Hsc70, Hsp40 and Hsp90) and ER resident (calnexin) molecular chaperones as well as ubiquitin ligating (E3) enzyme [8, 23–25]. Complex formation with chaperones and co-chaperones [human DnaJ homologue 1/2 (Hdj1/2); carboxyl terminus of Hsc70-interacting protein (CHIP) and Hsp70 co-chaperone heat shock protein-binding protein 1 (HspBP1)] not only protects the channel from aggregation, but facilitates folding, as well as the degradation of nonnative conformers [26–28]. While in vitro data suggest that Hsp70 and Hdj-2 promote the folding of NBD1 and the N-terminal half of CFTR, in vivo overexpression of Hsp70 in cooperation with the Hsp40 homologous (Ydj1p/Hlj1p) augmented the ERAD of CFTR in yeast [18, 26, 29, 30]. Induction of Hsp70 and simultaneous downregulation of Hsc70, on the other hand, promoted the folding of ΔF508 CFTR [31, 32], whereas overexpression of Hsp70 had no discernable effect on the stability of the wt and mutant CFTR in mammalian cells [33]. Cell type-specific differences may explain the discordant effect of chaperones on the mutant CFTR. Intriguingly, coexpression of the cysteine string protein (Csp), a co-chaperone of Hsc70, blocked the conversion of immature CFTR to mature form by stabilizing the core-glycosylated channel [34]. The physical interaction of this Hsc70-binding protein with immature CFTR suggests that alterations of Csp activity may represent another level of modulation of CFTR biogenesis at the ER [34].

Inhibition of Hsp90 activity by geldanamycin blocked channel biogenesis and accelerated the degradation of the nascent chain in heterologous expression systems, but had the opposite effect in vitro [25, 35]. Inactivation of the yeast Hsp90 homologous (HSC82 and HSP82) also destabilized CFTR, supporting the role of Hsp90 in the posttranslational folding of CFTR [26]. The physiological relevance of calnexin association with the folding intermediate of wt and ΔF508 CFTR remains incompletely understood. Calnexin remains associated with the mutant until its degradation by ERAD, but dissociates from folded wt CFTR prior to its exit from the ER [24]. In light of the undisrupted biogenesis of the CFTR lacking N-linked oligosaccharides, it is plausible to assume that calnexin has no significant role in CFTR biogenesis [25]. This conclusion is in line with the report that the degradation kinetics of CFTR in calnexin-deficient yeast remains unaltered [27].

Hsc70 and Hsp90 are not only involved in the folding, but also in the degradation of misfolded wt and mutant CFTR. Hsc70/Hsp90 facilitate the recruitment of CHIP, a soluble U-box containing E3 ubiquitin ligase, to the nascent CFTR chain, a prerequisite for CHIP-mediated ubiquitination of the nonnative CFTR [27]. The human orthologues of Ubc5/Ubc6, ubiquitin-conjugating enzymes (E2), appear to cooperate with CHIP in covalent attachments of ubiquitin (Ub) to the partially folded wt and mutant CFTR [27, 36], a prerequisite for the targeting of nonnative CFTR for proteasome-mediated degradation at the ER. Besides direct binding of the poly-Ub chain of CFTR to the 26S proteasome, recruitment of the co-chaperone BAG-1 that harbors a ubiquitin-like domain also facilitates the recruitment of proteasome to the partially folded CFTR [37]. Intriguingly, the co-chaperone HspBP1 attenuates the ubiquitin ligase activity of CHIP, causing enhanced CFTR maturation [28]. The latter mechanism offers an explanation for the variable folding efficiency of CFTR in nonpolarized and polarized cells with different HspBP1 expression levels. In addition to CHIP, an N-glycan conformation specific E3 ubiquitin ligase complex (Skp1-Cullin1-Fbs-Roc1; SCF^{Fbs1}) [38] was shown to participate in the ubiquitination of CFTR, in the cytoplasm [39, 40]. Retrotranslocation of ubiquitinated CFTR from the ER into the cytoplasm is assisted by the cytosolic trimeric AAA (ATPase associated with a variety of cellular activities) protein complex. Inhibition of the ATPase activity of the complex significantly delayed the ERAD of misfolded CFTR [40, 41].

Folded CFTR, similar to other cargo molecules, is presumably concentrated and transported out of the ER via a common cytosolic budding machinery [42]. The coordinated action of different transport signals may regulate the ultimate fate of CFTR in the ER. Misfolding of CFTR, in addition to provoke the ubiquitination of the channel, may expose an additional ER retention signal. Multiple short peptide motifs (RXR; where X represents any amino acid residue interposed between two arginine residues) were identified in CFTR, contributing to the retrieval of misfolded conformers from the *cis*-Golgi to the ER compartment via transport vesicles coated with coat protein complex I (COPI vesicles) [43]. According to an alternative and perhaps complementary model, the exit code consists of two acidic amino acids that are exposed in the native NBD1 [42]. The export signal assists the recruitment of the channel via the Sec23/24 cargo selection complex into transport vesicle harboring the coat protein complex II (COPII) [42], and ferries the folded channel to the Golgi compartment. Although alanine replacement of the di-acidic signal (563YKDAD567) resulted in complete ER retention of CFTR, presumably by interfering with the cargo selection process [42], whether the mutant channel preserved the native domain structure or the mutations caused other secondary effects remains to be seen.

The complexity of the protein-protein interactions required for the normal folding and vesicular transport of CFTR from the ER to the distal compartments of the biosynthetic pathway provides a plausible explanation for the sensitivity of these processes to structural disturbances in CFTR caused by mutation.

CFTR Targeting to the Cell Surface

Two alternate trafficking pathways have been described for the biosynthetic processing of native CFTR, following its export via the COPII-mediated pathway from the ER [44]. The classical pathway of core-glycosylated CFTR, detected in HeLa and HEK293 cells, consists of its progressive transfer via the *cis*-, medial and *trans*-Golgi compartments, concomitant with its complex-N-glycan modification (fig. 1) [44]. The nonclassical processing route involves the direct vesicular transfer of CFTR between the ER and *trans*-Golgi network/endosomal compartments followed by syntaxin 13-dependent recycling through the Golgi compartment and is, therefore, insensitive to dominant negative Arf1, Rab1a/Rab2 and syntaxin 5 [44]. The nonclassical biosynthetic route is consistent with the negligible accumulation of CFTR in the Golgi compartment of BHK and CHO cells [44], while its presence in polarized epithelia remains to be established.

The post-ER biosynthetic processing of CFTR is modulated by several protein-protein interactions. Among these, the high affinity association of multivalent PDZ domain-containing molecules with the conserved C-terminal peptide ([1476]DTRL) of CFTR has been studied most extensively [45, 46]. The PDZ (PSD-95/Dlg/ZO-1) domain-containing CFTR-associated ligand (CAL) downregulates CFTR expression by impeding its biosynthetic processing at the Golgi compartment [46]. NHERF (Na$^+$/H$^+$ exchanger regulatory factor), the first protein to be reported to bind the C-terminus of CFTR [47, 48], contains two class I PDZ domains (D1 and D2) and a C-terminal ezrin/radixin/moezin (ERM)-binding domain. The PDZ domains bind with nanomolar affinity to the [1477]DTRL motif [47]. The ERM-binding domain tethers the complex via ezrin to cytoskeletal elements in a phosphorylation-dependent manner and to the catalytic and regulatory subunit of PKA [48, 49]. NHERF may not only activate CFTR by allosteric mechanisms and phosphorylation [49, 50], but also contribute to the apical expression of CFTR in epithelia [51–54]. Deletion of the last 26 amino acids, including the DTRL motif of CFTR, provoked the accumulation of the channel at the lateral membrane in Madin-Darby canine kidney II (MDCKII) and bronchial epithelia [51, 53] and caused almost a complete (>97%) loss of the PKA-activated CFTR channel function [53]. These observations, however, are inconsistent with the clinical phenotype of the Δ26 CFTR. Individuals harboring the Δ26 CFTR on one allele and a nonsense mutation on the second allele have only moderately elevated sweat chloride concentration without obvious pancreatic and pulmonary phenotype [55]. Furthermore, disruption of complex formation between CFTR and NHERF by deleting the NHERF binding motif or masking the DTRL motif by epitope-tag attachment had no significant effect on the apical localization, metabolic stability and PKA-activated chloride conductance of CFTR in polarized respiratory, pancreatic, intestinal and kidney epithelia [56, 57]. These results strongly suggest that polarized expression of CFTR at the apical membrane domain involves sorting signals other than the C-terminal 26 amino acid residues and the PDZ-binding motif. Although a recent report proposed that tropomyosin isoforms 5a and 5b can regulate the insertion/retention of CFTR at the apical plasma membrane of intestinal epithelia the cellular and biochemical mechanisms that account for the polarized expression of CFTR at the apical plasma membrane domain remain to be elucidated [58].

Dynamics of CFTR Endocytosis and Recycling

The constitutive internalization of CFTR has been well established in heterologous and endogenous expression systems (fig. 1) [59]. Clathrin-dependent endocytosis of CFTR relies on tyrosine-based and di-leucine-based motifs in the C-terminal tail of CFTR [60–62]. Recruitment of CFTR to clathrin-coated pits and/or subsequent budding of clathrin-coated vesicles are facilitated by CFTR complex formation with myosin VI (an actin-dependent, minus end-directed mechanoenzyme), the myosin VI adaptor protein Disabled 2 (Dab2), and clathrin [54]. Interfering with CFTR binding to the μ2 subunit of the AP2 adaptor complex, which is essential in the recognition of Tyr-based endocytic signals, as well as overexpression of dominant negative dynamine and Rab5, prevented the channel internalization [62–65]. Obliterating the binding of myosin VI to CFTR or disrupting the actin cytoskeleton also inhibited the channel internalization [54].

Following internalization, CFTR recycles by default back to the cell surface with high efficiency. Recycling is absolutely necessary to maintain the long residence time of CFTR ($t_{1/2}$ ~14–18 h) at the cell surface in the presence of constitutive internalization (5–8%/min) [54, 66, 67]. Interfering with the recycling of CFTR by overexpressing dominant negative Rab11 and Rme-1 indeed resulted in the accumulation of CFTR at early and recycling endosomes, respectively, with the concomitant depletion of CFTR at the plasma membrane [65, 66].

The internalization and recycling dynamics of CFTR offer multiple regulatory mechanisms for altering the cell surface density of CFTR. cAMP-dependent protein kinase (PKA) stimulates CFTR translocation to the plasma membrane in a cell type-dependent manner [59, 68]. Arginine vasotocin, a hormone that increases the cellular cAMP level, mobilized CFTR from an intracellular compartment to the apical plasma membrane in the amphibian kidney (A6) epithelia [69]. Forskolin, a PKA activator, also led to the accumulation of CFTR at the apical membrane of MDCKII epithelia [70]. In contrast, translocation could not be observed upon PKA activation in MDCKI epithelia, a model of renal distal tubule and collecting duct [71].

However, the possibility that the extent of apical transloca-tion of CFTR was below the detection limit cannot be precluded.

In native tissues, cAMP elicits significant CFTR translocation from the subapical endosomal pool to the apical plasma membrane. Reduction of the subapical CFTR pool was demonstrated by immunofluorescence microscopy in shark rectal gland upon PKA activation [72]. Similar results were obtained using biochemical and morphological techniques of perfused rat intestinal loops upon stimulation with vasoactive intestinal peptide, leading to a 3-fold increase in CFTR density at the brush border [73]. These observations, collectively, support the physiological significance of cAMP-induced translocation of CFTR in certain epithelia and the notion that phosphorylation not only increases the open probability, but also the channel density at the apical membrane. Inhibition of the constitutive internalization of CFTR upon PKA activation is a plausible mechanism accounting for the redistribution of the channel from the endosomal compartment to the cell surface [74, 75].

Mutations Causing Defective Trafficking of CFTR

CF mutations could impair the channel production, function and stability or a combination of these, based on a variety of molecular defects (fig. 1) [76]. For example, CFTR protein production could be compromised at the transcriptional, splicing or at the posttranslational level (see chapter 1) [76]. CFTR stability at the plasma membrane may be attenuated by accelerated endocytosis, as well as impaired recycling. In the following paragraphs three types of trafficking defects, caused by ER retention, accelerated internalization and defective recycling, will be discussed.

Posttranslational Folding Defect of the ΔF508 CFTR
The paradigmatic ΔF508 mutation is widely believed to disrupt the folding of the ΔNBD1 domain and therefore recognized by the ER quality control and eliminated by the ubiquitin-proteasome pathway with similar mechanism and kinetics as the incompletely folded fraction of wt CFTR [6]. Recent progress, however, suggests that the F508 residue has a minor role in the folding of the NBD1. The ΔF508 mutation has a modest effect on the X-ray crystal structure of the recombinant human wt NBD1 [42a] and the protease susceptibility of the NBD1 in the full-length CFTR [19]. In contrast, ΔF508, similar to 508 amino acid substitutions that prevented the CFTR processing, substantially compromised the conformation of the NBD2 [19]. Perturbation of cytosolic domain-domain assembly likely affects the packing of the transmembrane helices as well [22] and serves as a recognition signal for the ubiquitination machinery. In addition, exposure of RXR ER retention motifs [43] and masking of a di-acidic export signal [77] may contribute to the ER retention of the ubiquitinated channel. The relative contribution of these signals to the ER retention of misfolded CFTR will have to be defined in future studies.

Accelerated Internalization of N287YCFTR
The N287Y mutation, which resides in the second intracellular loop, causes mild CF in the background of ΔF508 as the second allele [78]. While the biosynthetic processing efficiency and single channel characteristics of the N287Y CFTR were identical to its wt counterpart, the mutation increased the internalization rate of the channel 2-fold [78]. Accelerated endocytosis resulted in 50% reduction of the steady-state cell surface expression of the channel, providing a plausible explanation for the mild clinical phenotype of the patients at the cellular level. This example also demonstrates that missense mutations may lead to impaired trafficking without imposing a folding defect on CFTR at the ER [78].

Misfolding of Mature CFTR Leads to Recycling Defect
It is known that conformationally unstable plasma membrane proteins, including the rescued ΔF508 CFTR (rΔF508) that has been redirected to the plasma membrane by incubating the cells at reduced temperature and the C-terminally truncated (Δ70) CFTR, are rapidly degraded [79]. Severely impaired recycling account for the ~10-fold faster turnover rate of the rΔF508 and Δ70 CFTR and the 10-fold lower steady-state expression of Δ70 CFTR at the cell surface [67]. Importantly, no difference was detected in the endocytosis rates of the mutant and wt CFTR [67]. The recycling defect could be reverted to some extent by the overexpression of rab11, a small GTP-binding protein that is critical in directing recycling endosome fusion to the cell surface, using proteasome inhibition and inactivating the ubiquitination machinery [65, 67]. Preferential ubiquitination of the misfolded mutant seems to play a pivotal role in the rapid degradation of the conformationally defective ΔF508 and Δ70 CFTR from post-Golgi compartments. This hypothesis is supported by the observation that (1) the mutant CFTR displayed nearly 20-fold increased ubiquitination relative to its wt counterpart, (2) in-frame fusion of a single ubiquitin to the wt CFTR mimicked the recycling defect of the mutants, and (3) the recycling defect of the chimera was completely reverted upon disrupting the fused Ub recognition by components of the Ub-dependent

endosomal sorting machinery (e.g. Hrs and STAM) [67]. In light of the preserved biosynthetic maturation of the $\Delta 70$ CFTR, these observations imply that impaired recycling accounts for the severe clinical phenotype of $\Delta 70$ CFTR [80]. These and other observations support an emerging paradigm for the role of peripheral quality control mechanism that rapidly eliminates misfolded membrane proteins, including mutant CFTR, from the cell surface. The peripheral quality control of CFTR entails the redirection of misfolded channels from the constitutive recycling pathway towards the lysosomal degradation pathway in cooperation with the Ub-dependent endosomal sorting machinery, in order to protect the cell from the accumulation of nonnative channels at the periphery [67].

Conclusion

A large number of studies not only verified that CFTR processing follows the fundamental principles of membrane protein translocation between compartments, but also

revealed new layers of complexity in the protein quality control mechanisms. The synchronized actions of multiple quality control check points ensure that nonnative CFTR molecules are rapidly eliminated from both the ER and post-Golgi compartments. In light of these considerations, identifying pharmacological chaperones that can rescue the folding defect of mutant channels seems to be the most efficient way to restore normal trafficking and alleviate the phenotypic consequences of CFTR misprocessing.

Acknowledgement

The work in the laboratory of G. Lukacs was supported by grants from CIHR, CCFF, CF USA, NIH, Ontario Thoracic Society and The Hospital for Sick Children Foundation.

References

1 Riordan JR, Rommens JM, Kerem B, Alon N, Rozmahel R, Grzelczak Z, Zielenski J, Lok S, Plavsic N, Chou JL, et al: Identification of the cystic fibrosis gene: Cloning and characterization of complementary DNA. Science 1989; 245:1066–1073.

2 Jensen TJ, Loo MA, Pind S, Williams DB, Goldberg AL, Riordan JR: Multiple proteolytic systems, including the proteasome, contribute to CFTR processing. Cell 1995;83: 129–135.

3 Kopito RR, Ward CL, Omura S: Degradation of CFTR by the ubiquitin-proteasome pathway. Cell 1995;83:121–127.

4 Denning GM, Anderson MP, Amara JF, Marshall J, Smith AE, Welsh MJ: Processing of mutant cystic fibrosis transmembrane conductance regulator is temperature-sensitive. Nature 1992;358:761–764.

5 Zhang F, Kartner N, Lukacs GL: Limited proteolysis as a probe for arrested conformational maturation of ΔF508 CFTR. Nat Struct Biol 1998;5:180–183.

6 Gelman MS, Kopito RR: Rescuing protein conformation: Prospects for pharmacological therapy in cystic fibrosis. J Clin Invest 2002; 110:1591–1597.

7 Ellgaard L, Helenius A: Quality control in the endoplasmic reticulum. Nat Rev Mol Cell Biol 2003;4:181–189.

8 Ahner A, Brodsky JL: Checkpoints in ER-associated degradation: Excuse me, which way to the proteasome? Trends Cell Biol 2004;14:474–478.

9 Kopito RR, Ward CL: Intracellular turnover of cystic fibrosis transmembrane conductance regulator. J Biol Chem 1994;269: 25710–25718.

10 Lukacs GL, Mohamed A, Kartner N, Chang XB, Riordan JR, Grinstein S: Conformational maturation of CFTR but not its mutant counterpart (delta F508) occurs in the endoplasmic reticulum and requires ATP. EMBO J 1994;13: 6076–6086.

11 Sheppard DN, Welsh MJ: Structure and function of the CFTR chloride channel. Physiol Rev 1999;79:S23–S45.

12 Varga K, Jurkuvenaite A, Wakefield J, Hong JS, Guimbellot JS, Venglarik CJ, Niraj A, Mazur M, Sorscher EJ, Collawn JF, et al: Efficient intracellular processing of the endogenous cystic fibrosis transmembrane conductance regulator in epithelial cell lines. J Biol Chem 2004;279:22578–22584.

13 Cheng SH, Gregory RJ, Marshall J, Paul S, Souza DW, White GA, O'Riordan CR, Smith AE: Defective intracellular transport and processing of CFTR is the molecular basis of most cystic fibrosis. Cell 1990;63:827–834.

14 Xiong X, Chong E, Skach WR: Evidence that endoplasmic reticulum (ER)-associated degradation of CFTR is linked to retrograde translocation from the ER membrane. J Biol Chem 1999;274:2616–2624.

15 Sheppard DN, Ostedgaard LS, Rich DP, Welsh MJ: The amino terminal portion of CFTR forms a regulated chloride channel. Cell 1994; 76:1091–1098.

16 Chan KW, Csanady L, Nairn AC, Gadsby DC: Deletion analysis of CFTR channel R domain using severed molecules. Biophys J 1999;76: A405.

17 Akabas MH: Cystic fibrosis transmembrane conductance regulator: Structure and function of an epithelial chloride channel. J Biol Chem 2000;275:3729–3732.

18 Meacham GC, Lu Z, King S, Sorscher E, Tousson A, Cyr DM: The Hdj-2/Hsc 70 chaperone pair facilitates early steps in CFTR biogenesis. EMBO J 1999;18:1492–1505.

19 Du K, Sharma M, Lukacs GL: The ΔF508 cystic fibrosis mutation impairs domain-domain interactions and arrests posttranslational folding of CFTR. Nat Struct Mol Biol 2005;12:17–25.

20 Locher KP, Lee AT, Rees DC: The *E. coli* BtuCD structure: A framework for ABC transporter architecture and mechanism. Science 2002;296:1091–1098.

21 Chang G: Structure of MsbA from *Vibrio cholera*: A multidrug resistance ABC transporter homolog in a closed conformation. J Mol Biol 2003;330:419–430.

22 Chen EY, Bartlett MC, Loo TW, Clarke DM: The DeltaF508 mutation disrupts packing of the transmembrane segments of the cystic fibrosis transmembrane conductance regulator. J Biol Chem 2004;279:39620–39627.

23 Brodsky JL: Chaperoning the maturation of the cystic fibrosis transmembrane conductance regulator. Am J Physiol Lung Cell Mol Physiol 2001;28:L39–L42.

24 Pind S, Riordan JR, Williams DB: Participation of the endoplasmic reticulum chaperone calnexin (p88, IP90) in the biogenesis of the cystic fibrosis transmembrane conductance regulator. J Biol Chem 1994;269: 12784–12788.

25 Loo MA, Jensen TJ, Cui L, Hou J, Chang XB, Riordan JR: Perturbation of Hsp90 interaction with nascent CFTR prevents its maturation and accelerates its degradation by the proteasome. EMBO J 1998;17:6879–6887.

26 Youker RT, Walsh P, Beilharz T, Lithgow T, Brodsky JL: Distinct roles for the Hsp40 and Hsp90 molecular chaperones during cystic fibrosis transmembrane conductance regulator degradation in yeast. Mol Biol Cell 2004;15: 4787–4797.

27 Meacham GC, Patterson C, Zhang W, Younger JM, Cyr DM: The Hsc70 co-chaperone CHIP targets immature CFTR for proteasomal degradation. Nat Cell Biol 2001;3:100–105.

28 Alberti S, Bohse K, Arndt V, Schmitz A, Hohfeld J: The cochaperone HspBP1 inhibits the CHIP ubiquitin ligase and stimulates the maturation of the cystic fibrosis transmembrane conductance regulator. Mol Biol Cell 2004;15:4003–4010.

29 Strickland E, Qu BH, Millen L, Thomas PJ: The molecular chaperone Hsc70 assists the in vitro folding of the N-terminal nucleotide-binding domain of the CFTR. J Biol Chem 1997;272:25421–25424.

30 Zhang Y, Nijbroek G, Sullivan ML, McCracken AA, Watkins SC, Michaelis S, Brodsky JL: Hsp70 molecular chaperone facilitates endoplasmic reticulum-associated protein degradation of cystic fibrosis transmembrane conductance regulator in yeast. Mol Biol Cell 2001;12:1303–1314.

31 Rubenstein RC, Zeitlin PL: Sodium 4-phenylbutyrate downregulates Hsc70: Implications for intracellular trafficking of ΔF508-CFTR. Am J Physiol 2000;278:C259–C267.

32 Choo-Kang LR, Zeitlin PL: Induction of HSP70 promotes DeltaF508 CFTR trafficking. Am J Physiol 2001;281:L58–L68.

33 Farinha CM, Nogueira P, Mendes F, Penque D, Amaral MD: The human DnaJ homologue (Hdj)-1/heat-shock protein (Hsp) 40 co-chaperone is required for the in vivo stabilization of the CFTR by Hsp70. Biochem J 2002;366: 797–806.

34 Zhang H, Peters KW, Sun F, Marino CR, Lang J, Burgoyne RD, Frizzell RA: Cysteine string protein interacts with and modulates the maturation of the cystic fibrosis transmembrane conductance regulator. J Biol Chem 2002;277: 28948–28958.

35 Ji HL, Chalfant ML, Jovov B, Lockhart JP, Parker SB, Fuller CM, Stanton BA, Benos DJ: The cytosolic termini of the beta- and gamma-ENaC subunits are involved in the functional interactions between cystic fibrosis transmembrane conductance regulator and epithelial sodium channel. J Biol Chem 2000;275: 27947–27956.

36 Lenk U, Yu H, Walter J, Gelman MS, Hartmann E, Kopito RR, Sommer T: A role for mammalian Ubc6 homologues in ER-associated protein degradation. J Cell Sci 2002; 115:3007–3014.

37 Luders J, Demand J, Hohfeld J: The ubiquitin-related BAG-1 provides a link between the molecular chaperones Hsc70/Hsp70 and the proteasome. J Biol Chem 2000;275: 4613–4617.

38 Yoshida Y, Chiba T, Tokunaga F, Kawasaki H, Iwai K, Suzuki T, Ito Y, Matsuoka K, Yoshida M, Tanaka K, et al: E3 ubiquitin ligase that recognizes sugar chains. Nature 2002;418: 438–442.

39 Bebok Z, Mazzochi C, King SA, Hong JS, Sorscher EJ: The mechanism underlying cystic fibrosis transmembrane conductance regulator transport from the endoplasmic reticulum to the proteasome includes Sec61beta and a cytosolic, deglycosylated intermediary. J Biol Chem 1998;273: 29873–29878.

40 Gnann A, Riordan JR, Wolf DH: Cystic fibrosis transmembrane conductance regulator degradation depends on the lectins Htm1p/EDEM and the Cdc48 protein complex in yeast. Mol Biol Cell 2004;15:4125–4435.

41 Dalal S, Rosser MF, Cyr DM, Hanson PI: Distinct roles for the AAA ATPases NSF and p97 in the secretory pathway. Mol Biol Cell 2004;15:637–648.

42 Wang X, Matteson J, An Y, Moyer B, Yoo JS, Bannykh S, Wilson IA, Riordan JR, Balch WE: COPII-dependent export of cystic fibrosis transmembrane conductance regulator from the ER uses a di-acidic exit code. J Cell Biol 2004;167:65–74.

42a Lewis HA, Zhao X, Wang C, Sauder JM, Rooney I, Noland BW, Lorimer D, Kearins MC, Conners K, Condon B, et al: Impact of the delta F508 mutation in NBD1 of human CFTR on domain folding and structure. J Biol Chem 2005;280:1346–1353.

43 Chang XB, Cui L, Hou YX, Jensen TJ, Aleksandov AA, Mengos A, Riordan JR: Removal of multiple arginine-framed trafficking signals overcomes misprocessing of ΔF508 CFTR present in most patients with cystic fibrosis. Mol Cell 1999;4:137–142.

44 Yoo JS, Moyer BD, Bannykh S, Yoo HM, Riordan JR, Balch WE: Non-conventional trafficking of the cystic fibrosis transmembrane conductance regulator through the early secretory pathway. J Biol Chem 2002; 277: 11401–11409.

45 Medintz I, Wang X, Hradek T, Michels CA: A PEST-like sequence in the N-terminal cytoplasmic domain of *Saccharomyces* maltose permease is required for glucose-induced proteolysis and rapid inactivation of transport activity. Biochemistry 2000;39:4518–4526.

46 Cheng J, Moyer BD, Milewski M, Loffing J, Ikeda M, Mickle JE, Cutting GR, Li M, Stanton BA, Guggino WB: A Golgi-associated PDZ domain protein modulates cystic fibrosis transmembrane regulator plasma membrane. J Biol Chem 2002;277:3520–3529.

47 Hall RA, Premont RT, Chow CW, Blitzer JT, Pitcher JA, Claing A, Stoffel RH, Barak LS, Shenolikar S, Weinman EJ, et al: The beta2-adrenergic receptor interacts with the Na⁺/H⁺-exchanger regulatory factor to control Na⁺/H⁺ exchange. Nature 1998;392:626–630.

48 Short DB, Trotter KW, Reczek D, Kreda SM, Bretscher A, Boucher RC, Stutts MJ, Milgram SL: An apical PDZ protein anchors the cystic fibrosis transmembrane conductance regulator to the cytoskeleton. J Biol Chem 1998;273: 19797–19801.

49 Sun F, Hug MJ, Bradbury NA, Frizzell RA: Protein kinase A associates with cystic fibrosis transmembrane conductance regulator via an interaction with ezrin. J Biol Chem 2000; 275:14360–14366.

50 Raghuram V, Mak DD, Foskett JK: Regulation of cystic fibrosis transmembrane conductance regulator single-channel gating by bivalent PDZ-domain-mediated interaction. Proc Natl Acad Sci USA 2001;98:1300–1305.

51 Moyer BD, Denton J, Karlson KH, Reynolds D, Wang S, Mickle JE, Milewski M, Cutting GR, Guggino WB, Li M, et al: A PDZ-interacting domain in CFTR is an apical membrane polarization signal. J Clin Invest 1999;104: 1353–1361.

52 Milewski MI, Mickle JE, Forrest JK, Stafford DM, Moyer BD, Cheng J, Guggino WB, Stanton BA, Cutting GR: A PDZ-binding motif is essential but not sufficient to localize the C terminus of CFTR to the apical membrane. J Cell Sci 2001;114:719–726.

53 Moyer BD, Duhaime M, Shaw C, Denton J, Reynolds D, Karlson KH, Pfeiffer J, Wang S, Mickle JE, Milewski M, et al: The PDZ-interacting domain of cystic fibrosis transmembrane conductance regulator is required for functional expression in the apical plasma membrane. J Biol Chem 2000;275: 27069–27074.

54 Swiatecka-Urban A, Duhaime M, Coutermarsh B, Karlson KH, Collawn J, Milewski M, Cutting GR, Guggino WB, Langford G, Stanton BA: PDZ domain interaction controls the endocytic recycling of the cystic fibrosis transmembrane conductance regulator. J Biol Chem 2002;277: 40099–40105.

55 Mickle JE, Macek M, Fulmer-Smentek SB, Egan MM, Schwiebert E, Guggino WB, Moss R, Cutting GR: A mutation in the cystic fibrosis transmembrane conductance regulator gene associated with elevated sweat chloride concentrations in the absence of cystic fibrosis. Hum Mol Genet 1998;7:729–735.

56 Ostedgaard LS, Randak C, Rokhlina T, Karp P, Vermeer D, Ashbourne EKJ, Welsh MJ: Effects of C-terminal deletions on cystic fibrosis transmembrane conductance regulator function in cystic fibrosis airway epithelia. Proc Natl Acad Sci USA 2003;100:1937–1942.

57 Benharouga M, Sharma M, So J, Haardt M, Drzymala L, Popov M, Schwapach B, Grinstein S, Du K, Lukacs GL: The role of the C terminus and Na⁺/H⁺ exchanger regulatory factor in the functional expression of cystic fibrosis transmembrane conductance regulator in nonpolarized cells and epithelia. J Biol Chem 2003;278:22079–22089.

58 Dalby-Payne JR, O'Loughlin EV, Gunning P: Polarization of specific tropomyosin isoforms in gastrointestinal epithelial cells and their impact on CFTR at the apical surface. Mol Biol Cell 2003;14:4365–4375.

59 Bertrand CA, Frizzell RA: The role of regulated CFTR trafficking in epithelial secretion. Am J Physiol Cell Physiol 2003;285:C1–C18.

60 Prince LS, Peter K, Hatton SR, Zaliauskiene L, Cotlin LF, Clancy JP, Marchase RB, Collawn JF: Efficient endocytosis of the CFTR requires a tyrosine-based signal. J Biol Chem 1999;274:3602–3609.

61 Hu W, Howard M, Lukacs GL: Multiple endocytic signals in the C-terminal tail of the cystic fibrosis transmembrane conductance regulator. Biochem J 2001;354:561–572.

62 Weixel KM, Bradbury NA: The carboxyl terminus of the cystic fibrosis transmembrane conductance regulator binds to AP-2 clathrin adaptors. J Biol Chem 2000;275:3655–3660.

63 Weixel KM, Bradbury NA: Mu 2 binding directs the cystic fibrosis transmembrane conductance regulator to the clathrin-mediated endocytic. J Biol Chem 2001;276:46251–46259.

64 Cheng J, Wang H, Guggino WB: Modulation of mature cystic fibrosis transmembrane regulator protein by the PDZ domain protein CAL. J Biol Chem 2004;279:1892–1898.

65 Gentzsch M, Chang XB, Cui L, Wu Y, Ozols VV, Choudhury A, Pagano RE, Riordan JR: Endocytic trafficking routes of wild type and DeltaF508 cystic fibrosis transmembrane conductance regulator. Mol Biol Cell 2004;15:2684–2696.

66 Picciano JA, Ameen N, Grant BD, Bradbury NA: Rme-1 regulates the recycling of the cystic fibrosis transmembrane conductance regulator. Am J Physiol 2003;285: C1009–C1018.

67 Sharma M, Pampinella F, Nemes C, Benharouga M, So J, Du K, Bache KG, Papsin B, Zerangue N, Stenmark H, et al: Misfolding diverts CFTR from recycling to degradation: Quality control at early endosomes. J Cell Biol 2004;164:923–933.

68 Bradbury NA: Intracellular CFTR: Localization and function. Physiol Rev 1999;79:S175–S191.

69 Morris RG, Tousson A, Benos DJ, Schafer JA: Microtubule disruption inhibits AVT-stimulated Cl$^-$ secretion but not Na$^+$ reabsorption in A6 cells. Am J Physiol 1998;274:F300–F314.

70 Howard M, Jiang X, Stolz DB, Hill WG, Johnson JA, Watkins SC, Frizzell RA, Bruton CM, Robbins PD, Weisz OA: Forskolin-induced apical membrane insertion of virally expressed, epitope-tagged CFTR in polarized MDCK cells. Am J Physiol 2000;279: C375–C382.

71 Moyer BD, Loffing J, Schwiebert EM, Loffing-Cueni D, Halpin PA, Karlson KH, Ismailov II, Guggino WB, Langford GM, Stanton BA: Membrane trafficking of the cystic fibrosis gene product, CFTR, tagged with green fluorescent protein in Madin-Darby canine kidney cells. J Biol Chem 1998;273: 21759–21768.

72 Lehrich RW, Aller SG, Webster P, Marino CR, Forrest JN Jr: Vasoactive intestinal peptide, forskolin, and genistein increase apical CFTR trafficking in the rectal gland of the spiny dogfish, *Squalus acanthias*. J Clin Invest 1998; 101:737–745.

73 Ameen NA, Marino C, Salas PJ: cAMP-dependent exocytosis and vesicle traffic regulate CFTR and fluid transport in rat jejunum in vivo. Am J Physiol 2003;284:C429–C438.

74 Prince LS, Workman RB, Marchase RB Jr: Rapid endocytosis of the cystic fibrosis transmembrane conductance regulator chloride channel. Proc Natl Acad Sci USA 1994;91: 5192–5196.

75 Lukacs GL, Segal G, Kartner N, Grinstein S, Zhang F: Constitutive internalization of CFTR occurs via clathrin-dependent endocytosis and is regulated by protein phosphorylation. Biochem J 1997;328:353–361.

76 Zielenski J: Genotype and phenotype in cystic fibrosis. Respiration 2000;67:117–133.

77 Wang X, Matteson J, An Y, Moyer B, Yoo JS, Bannykh S, Wilson IA, Riordan JR, Balch WE: COPII-dependent export of cystic fibrosis transmembrane conductance regulator from the ER uses a di-acidic exit code. J Cell Biol 2004;167:65–74.

78 Silvis MA, Picciano JA, Bertrand CA, Weixel KM, Bridges RJ, Bradbury NA: A mutation in the cystic fibrosis transmembrane conductance regulator generates a novel internalization sequence and enhances endocytic rates. J Biol Chem 2003;278:11554–11560.

79 Krebs MP, Noorwez SM, Malhotra R, Kaushal S: Quality control of integral membrane proteins. Trends Biochem Sci 2004;29:648–655.

80 Haardt M, Benharouga M, Lechardeur D, Kartner N, Lukacs GL: C-Terminal truncations destabilize the cystic fibrosis transmembrane conductance regulator without impairing its biogenesis. A novel class of mutation. J Biol Chem 1999;274: 21873–21877.

Gergely L. Lukacs
Hospital for Sick Children Research Institute
Program in Cell Biology, 555 University Av.
Toronto, Ont. M5G 1X8 (Canada)
Tel. +1 416 813 5125, Fax +1 416 813 5771
E-Mail glukacs@sickkids.ca

Bush A, Alton EWFW, Davies JC, Griesenbach U, Jaffe A (eds): Cystic Fibrosis in the 21st Century.
Prog Respir Res. Basel, Karger, 2006, vol 34, pp 29–37

Structure of the Cystic Fibrosis Transmembrane Conductance Regulator

Fiona L.L. Stratford   Christine E. Bear

Programme in Structural Biology and Biochemistry, Research Institute, Hospital for Sick Children, and
Departments of Physiology and Biochemistry, University of Toronto, Toronto, Canada

Abstract

CFTR is a distinctive member of the ATP-binding cassette
(ABC) superfamily of membrane transport proteins. As
expected, on the basis of this designation, CFTR is a multidomain membrane protein which utilizes cellular ATP to regulate
the flux of its substrate through the membrane. Furthermore,
as for related family members, the conformation of the membrane-spanning domains (MSDs) is regulated by nucleotide
interactions with both cytosolic nucleotide-binding domains
(NBD1 and NBD2). CFTR is distinct in that it functions as a
channel and its MSDs form a pore through which chloride
ions can be conducted. A further distinction relates to the
presence of a unique regulatory domain, the 'R' domain, which
contributes to the regulation of channel opening and closing
when phosphorylated. To date, mutagenesis studies have
guided much of our understanding of the molecular basis of
the chloride permeation pathway and its regulation by the
NBDs and the 'R' domain. However, recent X-ray crystallization studies of prokaryotic ABC proteins and the crystal
structure of CFTR-NBD1 provide the first high resolution
models of the molecular basis for ATP binding and permit further modelling of the domain-domain interactions essential
for activity of CFTR.

CFTR is a multidomain membrane protein possessing
an internal duplication like other related members of the
ATP-binding cassette (ABC) superfamily of transport proteins [1, 2]. The CFTR protein is composed of two halves,

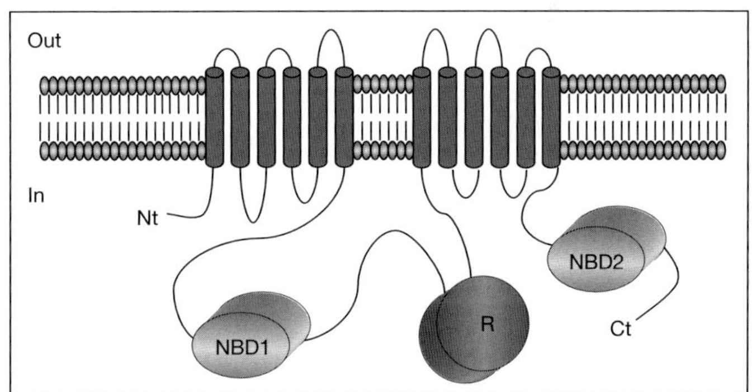

Fig. 1. The putative secondary structure of CFTR, highlighting
the two MSDs, two NBDs (NBD1 + 2) and the regulatory (R)
domain.

each possessing a membrane-spanning domain (MSD),
comprised of several helical membrane segments (probably six) with intervening extracellular and intracellular
loops followed by a cytosolic nucleotide-binding domain
(NBD) (fig. 1). In the first half of the molecule, the putative NBD (NBD1) is followed by a large cytoplasmic
domain called the 'R' domain, which contains multiple serine residues that are phosphorylated by protein kinase A
and protein kinase C to regulate CFTR channel function [3,
4]. The 'R' domain connects the first and the second half of
the molecule, with the second half of the molecule comprised of the second MSD (MSD2) and the second NBD
(NBD2).

CFTR is the only member of the ABC superfamily that functions as a chloride channel. The regulation of CFTR channel function exhibits several intriguing properties. For example, channel activity by CFTR exhibits obligatory regulation by phosphorylation of its large cytosolic regulatory 'R' domain, a structure that is unique to CFTR in the ABC family. CFTR distinguished itself as a chloride channel in that it also possesses intrinsic ATP-binding and ATPase activities which are involved in opening and closing of the gate at the mouth of the channel. However, the molecular mechanism underlying ATP-dependent channel gating in CFTR remains an area of active investigation. In the following review, we will discuss the current biochemical and structural findings which guide our understanding of the mechanism of action of CFTR.

provide further evidence that these purified proteins are monomeric [11, 12].

However, biochemical studies of CFTR localized exclusively at the cell surface suggest that it may exist as a dimer in the plasma membrane. CFTR-containing complexes, the size of dimers, can be cross-linked at the surface of mammalian cells using cross-linkers with short spacer arms of approximately 11 Å, consistent with the close contact of these proteins [7]. Li et al. [13] recently reported that CFTR dimers could be cross-linked along with signalling molecules in the apical membrane of Calu-3 epithelial cells. Therefore, these findings suggest that while dimeric association of CFTR molecules is not necessary for channel function, it may have an impact on trafficking to and/or interaction with other proteins at the cell surface.

Quaternary Structure of CFTR: Monomer or Dimer?

There has been considerable interest regarding the quaternary structure of CFTR which mediates its chloride channel activity. Biochemical studies of cellular and purified CFTR suggest that CFTR is monomeric. Differentially tagged versions of CFTR fail to show co-immunoprecipitation from whole cell homogenates in a variety of detergents, arguing that in studies representative of the total cellular pool of CFTR, there is little detectable association between CFTR molecules [5, 6]. Gel filtration studies of detergent solubilized CFTR derived from whole cell homogenates showed that CFTR migrated primarily as a monomer [7]. In biochemical studies, we found that monomers of CFTR can be purified and functionally reconstituted, and were capable of mediating chloride electrodiffusion. Furthermore, detailed single channel electrophysiological studies revealed that purified monomeric CFTR can mediate regulated chloride channel and ATPase activity (a function discussed in depth in subsequent paragraphs) [8].

Recent structural determinations of prokaryotic and eukaryotic ABC proteins support the idea that the minimal functional unit of CFTR is monomeric. X-ray crystal structures of prokaryotic ABC proteins; including the lipid A transporter, MsbA [9] and the vitamin B_{12} transporter, BtuCD [10] reveal multi-subunit proteins comprised of two MSDs and two NBDs. These essential components of the functional proteins are contained within a single CFTR polypeptide. Finally, low resolution structures of CFTR and P-glycoprotein generated by electron crystallography

Molecular Basis for Anion Permeation Through CFTR

Chloride Ion Permeation
The molecular basis for chloride permeation through CFTR remains poorly understood. On the basis of the structures of prokaryotic ABC proteins [9, 10], one would predict that the translocation path for chloride ions would be formed at the interface between the two MSDs (MSD1 and MSD2) of CFTR (each comprised of six putative helical segments). The sixth putative transmembrane (TM6) helix within MSD1 has been studied most extensively through mutagenesis and/or by monitoring the accessibility of substituted cysteine residues along its length using sulphydryl reagents [14, 15]. Multiple residues appear to contribute to anion binding in TM6, including amino acids 334–344. Studies by Gong et al. [15] and Smith et al. [16] highlight the potential importance of the positively charged residues: arginine at position 334 and lysine, at position 335, in forming an external vestibule which attracts anions toward the selectivity filter. Gong and Linsdell [15, 17] have performed elegant studies that suggest that the anion selectivity filter may be confined to a relatively narrow segment within the pore (including residues: F337–S341 of TM6) K95, a positively charged residue, lines the relatively wide inner vestibule of the CFTR pore and likely attracts chloride ions (and possibly other organic anions into the pore) [18]. Mutations in comparable regions of TM5 and TM2–4 led to minor or negligible effects on the above biophysical properties arguing that TM1, TM6 and to a minor degree, TM5 contribute to the chloride permeation path, whereas TM2–4 likely exert a supportive role [19].

Stratford/Bear

The contribution of MSD2 to anion conduction through CFTR is somewhat less clear. Recently, we found that purified and reconstituted MSD2 can dimerize to reconstitute a chloride-selective pore [20], arguing that MSD2, as well as MSD1, possesses structural features important for chloride conduction. In fact in site-directed mutagenesis studies, Gupta et al. [21] reported that mutations in TM12 (T1134, M1137, S1141) caused minor effects on the anion selectivity and unitary conductance of the channel. Similarly, mutations at S1118 in TM11 and T1134 in TM12 altered the affinity and/or voltage dependence of channel blockade by: DPC or NPPB [22]. In summary, studies to date suggest that the anion conductance path is comprised of multiple segments interspersed throughout the putative membrane domain.

Permeation of Other Physiologically Significant Anions

Patch clamp studies have revealed that CFTR can directly mediate the conduction of bicarbonate ion although the relative permeability of bicarbonate ion is significantly less than that of chloride ion ($P_{HCO_3}^-/P_{Cl^-} \sim 0.1$) [23, 24]. Studies by Reddy and Quinton [25] and subsequently by Shcheynikov et al. [26] support the concept that the pore of CFTR may exist in two forms, one form which may be bicarbonate permeant and the other, relatively chloride permeant. However, the two groups favour different regulatory mechanisms whereby selectivity may be altered. The evidence reported by Reddy and Quinton from studies using perfused human sweat ducts, support a model wherein the relative chloride and the bicarbonate-conducting functions of CFTR can be differentially regulated by the intracellular amino acid, glutamate and nucleotides. On the other hand, in *Xenopus* oocyte expression studies, Shcheynikov et al. determined that external chloride ion concentrations played a pivotal role in modulating the relative $Cl^-: HCO_3^-$ permeability, wherein a reduction of extracellular chloride ion concentration to 20–30 mM led to an increase in the relative HCO_3^- permeability. To summarize, although the mechanism is unclear at present, the anion selectivity of the CFTR pore appears to be regulated, possibly reflecting long range conformational changes induced by anion binding to internal and/or external binding sites.

Linsdell and Hanrahan [27] were the first to show that glutathione (and other large organic anions) could be conducted through membranes expressing CFTR although their permeability relative to that of chloride ions is very low, i.e. $P_{GSH}/P_{Cl} \sim 0.08$. Furthermore, these authors reported that glutathione conductance exhibited a distinctive dependence on nucleotide interaction. In our recent work, we also determined that the nucleotide requirements for glutathione conductance through purified CFTR were different from those required for chloride conductance [28]. These findings complement those discussed for bicarbonate ion above and suggest that the structure of the conductance pore through CFTR may be regulated by distinct ligand interactions.

Molecular Basis for Regulation by Phosphorylation

It is well known that phosphorylation of the 'R' domain of CFTR by PKA is absolutely required for nucleotide-dependent channel activity of CFTR [29]. There are ten dibasic consensus sites for PKA phosphorylation on the 'R' domain and it has been proposed that upon phosphorylation, they act in concert to mediate changes in the conformation and/or activity of other domains, as no single phosphoserine residue appears to be critical for function. However, the mechanisms underlying phosphorylation-dependent regulation remains unclear.

Electrophysiological studies suggest that the functional 'R' domain likely extends from a residue located somewhere between position 634–673 to residue 835 of CFTR [30, 31]. Deletion mutants of the 'R' domain and the functional analysis of isolated peptides derived from the 'R' domain reveal that there may be structures responsible for inhibition or stimulation within the 'R' domain. Deletion of residues 760–783 led to constitutive, phosphorylation-independent activity of CFTR, suggesting that this region may have an inhibitory effect on CFTR channel gating [32]. On the other hand, studies by Winter and Welsh [31] showed that addition of a PKA-phosphorylated 'R' domain peptide, corresponding to residues 708–831 caused activation of the CFTR channel (missing this region), suggesting that this large peptide includes regions which can engage in stimulatory interactions. As this large 'R' domain peptide enhanced nucleotide-dependent gating to the channel open state, primarily by enhancing the nucleotide affinity of the activity, it was suggested that it interacts with the NBDs and modifies them. Currently, however, it remains unclear whether the 'R' domain may also interact with other regions of the protein, i.e. the MSDs to mediate or modify channel gating transitions. Analyses by circular dichroism and NMR showed that this same peptide is largely unstructured, prompting speculation that the 'R' domain likely interacts with multiple regions of the protein simultaneously.

Molecular Basis for Regulation by Nucleotides

Functional Regulation of CFTR by Nucleotides:
Biochemical Studies of the ATPase Activity of CFTR
Purified, reconstituted full length CFTR exhibits a low
level of intrinsic ATPase activity [33]. To date, experimental data suggest that the ATPase activity of CFTR (as for
other ABC proteins) is mediated by an intramolecular complex between NBD1 and NBD2 [34]. For example, recombinant NBD1 and NBD2 proteins of CFTR functionally
interact to generate significantly greater ATPase activity
than either individual domain [34]. In the context of the full
length CFTR protein, a mutation in either the ABC signature motif of NBD1 (G551D) or in the Walker A:ATP binding consensus motif of NBD2 (K1250A) completely
abrogates ATPase activity [33, 35], pointing to the functional significance of NBD heterodimerization and the
importance of these particular motifs.

Functional Regulation of CFTR by Nucleotides:
Electrophysiological Studies of the Role of Nucleotides
in Channel Gating
CFTR channel opening and closing, i.e. gating, has been
linked to nucleotide binding and ATPase activity by one or
both of its NBDs [36]. There have been many studies of
nucleotide-dependent gating of CFTR [35, 37–40] and certain general concepts emerge from such studies. First,
opening of the CFTR channel is likely regulated by
nucleotide binding and/or events secondary to ATP binding. It has been suggested that channel closure may be
linked (directly or indirectly) to ATPase activity at one or
more sites in CFTR. Mutations that decrease ATPase activity, such as mutation of the Walker A lysine residue in
NBD2 (K1250A) and mutations which disrupt the
Walker B motif of NBD2; D1370N and E1371S, lead to
prolonged channel open times with a significant decrease
in the rate of channel closure [35, 36]. These studies argue
that there is a relationship between the ATPase activity and
the rate of channel closing.

Vergani et al. [36] have incorporated the biochemical
and electrophysiological studies described above to generate a model for the molecular basis underlying channel gating. The model proposes that nucleotide interaction with
two sites, one containing the Walker A of NBD1 and the
other, the Walker A of NBD2, leads to effective interaction
of the two NBD domains and conformational changes associated with chloride channel gating to the open state.
Subsequent hydrolysis of nucleotide at one or both sites
returns the protein to its ground level and the channel gate
resumes the closed conductance state. This model resembles the 'ATP Switch' model recently proposed by Higgins
and Linton [42] for ABC proteins in general, wherein the
power stroke associated with transporter function and substrate translocation is conferred by nucleotide binding and
induction of a 'closed' nucleotide dimer. Subsequent
nucleotide hydrolysis 'opens' the nucleotide dimer and
results in restoration of the basal conformation of the membrane domains. The above model has yet to be rigorously
tested for CFTR.

Structural Studies: NBDs of Prokaryotic ABC Proteins
Crystal structures of prokaryotic proteins have provided
useful tools to model the structural basis for CFTR NBD
interactions [43–45]. These NBDs share a common three-
dimensional fold and nucleotide-binding site architecture.
The NBD is divided roughly into two sub-domains, termed
lobes I and II [46]. The NBDs contain several conserved
motifs which are involved in ATP binding and hydrolysis.
These are the Walker A, Walker B, ABC signature motif
and the Q loop and H (or Switch) loop (named, respectively, for glutamine and histidine residues) [10]. The N terminus of the α-helix in lobe I contains the Walker A motif.
Lobe II is mostly α-helical and contains the ABC signature
motif. Both lobes are connected by a shared β-sheet that
forms an interlobe interface and contains the Walker B
motif [43].

Several studies have shown that ABC NBDs work in
homodimeric or heterodimeric pairs and that dimeric
assembly is essential for ABC ATPase function [47, 48].
The crystal structure of the ATPase domain of Rad50 in
complex with ATP [43] was the first to reveal a head to tail
arrangement of the two NBD domains. In this structure, two
functional active sites were formed at the dimer interface by
juxtaposition of the ABC signature motif of one monomer
with the Walker A and B motifs of the other monomer.
Biochemical and crystal structure models [10, 44, 47] suggest that this relative orientation of the dimer interface is
conserved amongst several ABC proteins (fig. 2a).

It is predicted that the two NBDs of CFTR also interact
in a head-tail orientation (fig. 2b). However, unlike the
prokaryotic NBDs, the NBDs of CFTR are structurally
asymmetrical (fig. 2c) and hence interaction of CFTR
NBDs may confer one rather than two catalytic sites. NBD1
possesses the canonical ABC signature motif but the
Walker A motif is modified, i.e. S/T substituted for T/S
[49]. It also lacks conservation of two residues implicated
in interaction with the hydrolytic water, a Walker B glutamate [48] and a histidine residue [44]. NBD2 possesses
sequence conservation in canonical Walker A, B and H
motifs, but lacks conservation within the ABC signature

motif. Molecular modelling of the NBDs of CFTR, using prokaryotic NBD dimers as a template, predicts that site 1 comprises a 'conventional' catalytic site (comprising the Walker A, B motifs of NBD2 and the ABC signature motif of NBD1). Site 2, comprising the Walker A, B of NBD1 with the Walker C of NBD2 forms a 'non-conventional' site [49].

Structural Studies of CFTR-NBD1

Lewis et al. [50, 51] recently determined high-resolution crystal structures for monomeric mouse (m) and human (h) CFTR NBD1 (fig. 3, 4). Overall, the core tertiary structures for both CFTR NBD1s were similar to other ABC transporters NBDs, however they differed from typical ABC domains in having two major insertional regions. mNBD1 contains an insertion of two short α-helices (H1b and H1c) separated by a flexible linker region between β-strands S1 and S2 (fig. 3). This insertion in mNBD1 results in an altered binding geometry for the base and ribose. Interestingly, in the hNBD1 structure [51], this insertion, is reorientated such that the canonical base stacking interaction is restored, suggested that this region may be somewhat flexible. Furthermore, the C-termini of mNBD1 and hNBD1 include a long α-helix (H9B) that is not present in other ABC-NBD domain structures. The authors suggest that this region may correspond to a neighbouring, but distinct domain, the 'R' domain, as it contains one (Ser 660) of multiple phosphorylation sites [50]. This extension is also flexible as its' orientation in the mNBD and hNBD structures is displaced by 180°.

The crystal structures of mNBD1 and hNBD1 revealed the nucleotide-binding site of this domain. As in other ABC NBDs, the following residues formed hydrogen bonds with the phosphates and/or coordinated the Mg^{2+} ion: Lys464 and Thr465 (Walker A), Asp572 (Walker B) and Gln493 (Q loop). As displayed in figure 4a, b, the regions involved in nucleotide interaction line up at the surface of the domain. However, there were no structural changes induced by nucleotide binding in NBD1, possibly reflecting the absence of NBD2, its functional partner in ATP hydrolysis.

Lewis et al. [50] then tested the prediction that the catalytic site(s) formed by the CFTR-NBDs requires interaction between NBD1 and NBD2 as a labile nucleotide 'sandwich' by creating a model of an NBD1-NBD2 heterodimer, docking the mNBD1 structure and a homology model of NBD2 onto the only available dimeric structure of the E171Q mutant MJ0796 homodimer [44]. However, in this heterodimeric model, there are severe main chain steric clashes between NBD1 and NBD2 [50]. Specifically, the C-terminal α-helix H9b of mNBD1 is thrust directly into

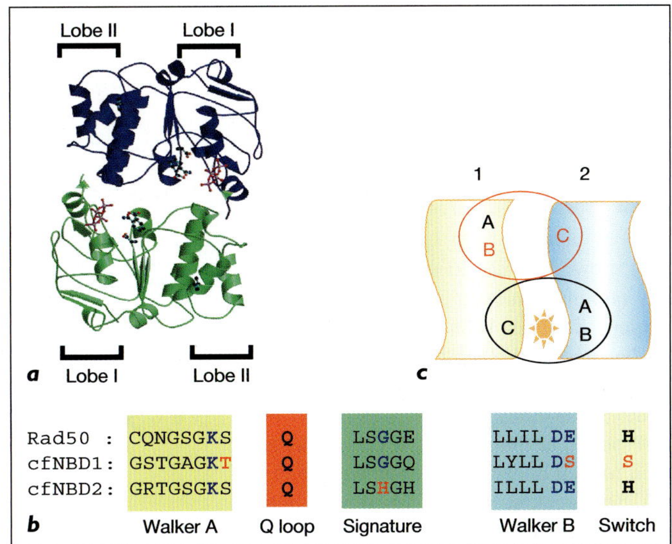

Fig. 2. NBDs – primary and tertiary structure. *a* The BtuD dimer solved by Locher et al. [10] and rendered by G.D. Smith using Molscript. Each monomer possesses two lobes which interact in a head to tail orientation within the dimeric complex. Pyrophosphate is 'sandwiched' within the two NBDs at two distinct sites. *b* Alignment of the key nucleotide binding motifs in Rad50 with CFTR-NBD1 and CFTR-NBD2. Modified from Locher et al. [10]. *c* Cartoon of the putative interaction between the two NBDs of CFTR in a head-to-tail orientation. The circled star indicates the site (site 1) at which the canonical nucleotide binding motifs A, B and signature are conserved with the rest of the family. The empty circle indicates the site (site 2) at which the motifs are degenerate.

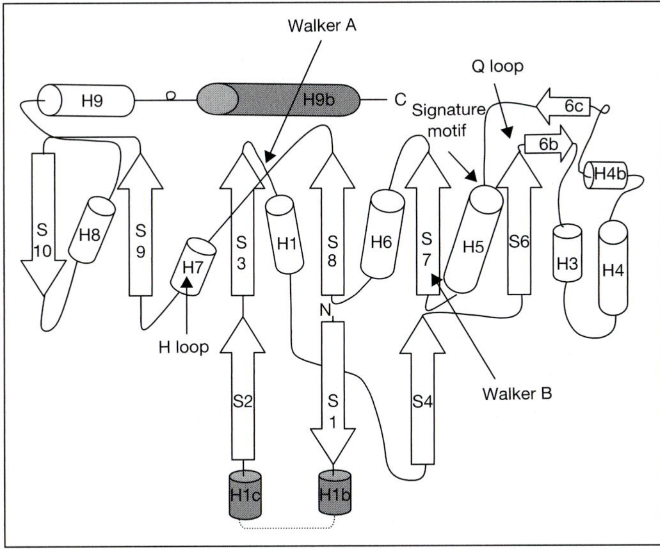

Fig. 3. Topology diagram of mNBD1. Regions of mNBD1 that are different from previous ABC structures are shown in grey. Modified from Lewis et al. [50].

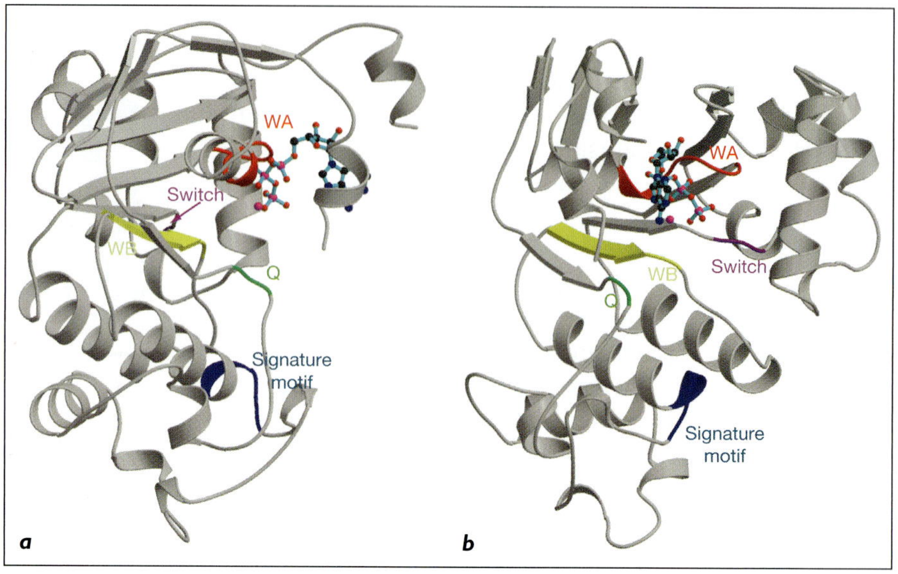

Fig. 4. Two aspects of a ribbon diagram of mNBD1 adapted from Lewis et al. [50]. *a* ATP shown in ball and stick representation. Key region interacting with ATP: the Walker A motif (red) has been shown. The location of degenerate Walker B (yellow) motif of NBD1 is also shown. The signature motif (blue) is spatially separated from the one ATP bound to this protein. *b* The model has been rotated by 90° in order to show the linear arrangement of all of these motifs Models were generated by Lewis et al. [50] and rendered by G.D. Smith using Molscript.

the opposing NBD2. The steric hindrance conferred by this region is apparent in the model rendered in figure 4a. These authors suggest there are several ways this conflict may be resolved, possibly through a conformational change in NBD1 upon interdomain interaction or through displacement of clashing segments upon phosphorylation [50].

Structural Studies of Intact Prokaryotic ABC Proteins by Three-Dimensional X-Ray Crystallography

Recently, high-resolution crystal structures have been reported for three intact prokaryotic ABC proteins, the lipid flippase MsbA proteins [from *Vibrio cholera* (VC-MsbA) and *Escherichia coli* (EC-MsbA)] and the vitamin B_{12} transporter (BtuCD), [9, 10, 52]. These structures reveal three structural features critical for ABC transporter function: the translocation pathway in the membrane, the association of the cytosolic NBDs and the transmission interface, or connection between the membrane incorporated and cytosolic domains.

In all three protein structures, the translocation pathway is located at the interface of the two MSDs. However, the MSDs of BtuCD appear to interact in a tighter complex than those of VC-MsbA. For both BtuCD and VC-MsbA, the MSDs and the NBDs are connected via a helical loop or a 'U'-like structure extending from the membrane or intra-cytosolic domains, respectively. The NBD structures of these proteins share the conserved architecture of two sub-domains, lobe I containing the Walker A and lobe II con-

taining the conserved signature motif sequence. In BtuCD, two ATP binding/hydrolysis sites are formed in the presence of cyclotetravanadate molecules (mimicking the α- and β-phosphates of ATP). The cyclotetravanadate molecules are sandwiched at the interface of two NBD subunits arranged in a head to tail orientation as found in other ABC transporters [10]. The NBDs of VC-MsbA are generally oriented relative to one another in manner similar to the BtuCD structure [9]. However, in the absence of nucleotide or analogues/mimetics, their interaction is relatively open, postulated to represent the catalytically inactive conformation.

Locher et al. [10] showed that helical segments (L1-L2) extend from the membrane domains of BtuCD to provide a docking site for the NBDs (fig. 5). The L1-L2 helical segments are bisected by a glycine residue, which confers a bend critical in creating the interface. Locher et al. [10] aligned this loop region in BtuD with regions in MSD1 (residues 248–261) and MSD2 (1056–1069) of CFTR, which are the sites of many disease-causing mutations [53, 54]. Interestingly, the phenylalanine residue (F508) which is deleted in most cases of cystic fibrosis, aligns with a residue in the NBD:BtuD located at the L1, L2-NBD interface [10]. The crystal structure of human NBD1 lacking F508 was recently published by Lewis et al. [51]. Although deletion of F508 caused only minimal perturbations in the structure or folding of the domain, it did lead to a change in the local topography on the surface of the NBD in the same region modelled to interact with the MSD. These findings

     Stratford/Bear

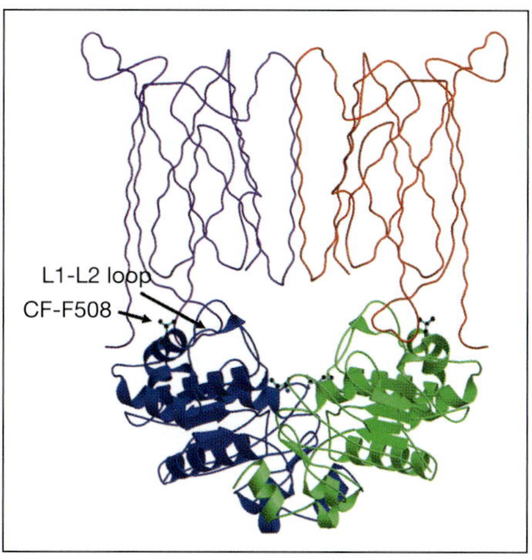

Fig. 5. Model of BtuCD structure generated by Locher et al. [10] and rendered by G.D. Smith using Molscript. Backbone of membrane-spanning helices (BtuC) drawn in purple and red. Ribbon diagrams of NBDs (BtuD) drawn in blue and green. The location of the L1-L2 loop in the MSD, forming the interaction between the MSDs and the NBDs has been shown. The putative position of F508 (deleted in most cases of cystic fibrosis) in the NBD is rendered as a ball and stick model (arrow). Figure rendered by D.G. Smith.

corroborate biochemical studies of the intact mutant protein that suggest the molecular basis for defective folding of ΔF508-CFTR lies in the perturbation of domain-domain interactions throughout the full length molecule [55–57].

Structural Studies of Intact Eukaryotic ABC Proteins by Electron Crystallography

Recently, several low resolution structures of full length eukaryotic ABC transporters have been determined [11, 12, 58, 59]. The first three-dimensional structure of an ABC protein (at 8 Å resolution) was determined for P-glycoprotein by Rosenberg et al. [11] allowing detection of the orientation of helices within the membrane domain of the intact transporter as well as the relative orientation of the NBDs. As predicted, 12 cylindrical densities, oriented in two symmetrical bundles (corresponding to MSD1 and MSD2) were observed and these densities could be modeled as α-helices within and extending beyond the membrane. The helical extensions connect with the NBDs and likely relate to the intracytoplasmic domains described for VC-MsbA by Chang [9]. The resolution was not sufficient to permit identification of the particular helices. Similarly, the regions corresponding to the NBDs lack sufficient resolution to permit definition of β-strands or helices. However, the MJ0796 NBD structure with ATP bound [44] and the BtuCD NBD structure [10] could be superimposed on the NBDs of P-glycoprotein, suggesting that the NBDs of P-glycoprotein may associate in an identical manner as these prokaryotic NBDs. Most recently, Rosenberg et al. [12] determined a low resolution structure of CFTR by electron crystallography of negatively stained two-dimensional crystals. The overall features of the CFTR structure resembled that of P-glycoprotein [11]. The two-dimensional crystal studies of CFTR in the presence of the non-hydrolyzable ATP analogue: MgAMP-PNP, reported two quite distinct conformations which the authors ascribe to the nucleotide-bound and unbound states. These inspiring advances in structural studies of ABC proteins and CFTR in particular promise a future in which we will eventually have high resolution models with which to determine the molecular mechanisms of activity.

Conclusions

The tremendous recent progress in structural determination of ABC protein domains and intact ABC proteins has stimulated excitement and optimism that we may be able to generate meaningful models of the molecular mechanisms underlying CFTR function soon. Within the next few years we hope to understand the molecular basis for ATPase activity by the NBDs of CFTR, the nature of the interaction of 'R' domain with the NBDs and in the longer term, the structural basis for the regulation of the membrane-spanning, pore-forming region of CFTR by these cytoplasmic domains. Once these models are in place, understanding the consequences of cystic fibrosis disease-causing mutations on CFTR structure and function and development of structure-based therapeutics will become realistic goals.

Acknowledgements

The structural models rendered in this paper were prepared by G. David Smith (Emeritus Scientist, Programme in Structural Biology and Biochemistry, Research Institute, Hospital for Sick Children). The work performed in the author's laboratory was possible through the generous support by the Canadian Cystic Fibrosis Foundation.

References

1 Riordan JR, Rommens JM, Kerem B, Alon N, Rozmahel R, Grzelczak Z, Zielenski J, Lok S, Plavsic N, Chou JL, et al: Identification of the cystic fibrosis gene: Cloning and characterization of complementary DNA Science 1989; 245:1066–1073.
2 Higgins CF, Linton KJ: Structural biology: The xyz of ABC transporters. Science 2001; 293:1782–1784.
3 Seibert FS, Loo TW, Clarke DM, Riordan JR: Cystic fibrosis: Channel catalytic and folding properties of the CFTR protein. J Bioenerg Biomembr 1997;29:429–442.
4 Chappe V, Hinkson DA, Howell LD, Evagelidis A, Liao J, Chang XB, Riordan JR, Hanrahan JW: Stimulatory and inhibitory protein kinase C consensus sequences regulate the cystic fibrosis transmembrane conductance regulator. Proc Natl Acad Sci USA 2004; 101:390–395.
5 Chen JH, Chang XB, Aleksandrov AA, Riordan JR: CFTR is a monomer: Biochemical and functional evidence. J Membr Biol 2002;188:55–71.
6 Marshall J, Fang S, Ostedgaard LS, O'Riordan CR, Ferrara D, Amara JF, Hoppe HT, Scheule RK, Welsh MJ, Smith AE, et al: Stoichiometry of recombinant cystic fibrosis transmembrane conductance regulator in epithelial cells and its functional reconstitution into cells in vitro. J Biol Chem 1994;269:2987–2995.
7 Ramjeesingh M, Kidd JF, Huan LJ, Wang Y, Bear CE: Dimeric cystic fibrosis transmembrane conductance regulator exists in the plasma membrane. Biochem J 2003;374(pt 3): 793–797
8 Ramjeesingh M, Li C, Kogan I, Wang Y, Huan LJ, Bear CE: A monomer is the minimum functional unit required for channel and ATPase activity of the cystic fibrosis transmembrane conductance regulator. Biochemistry 2001;40:10700–10706.
9 Chang G: Structure of MsbA from *Vibrio cholera:* A multidrug resistance ABC transporter homolog in a closed conformation. J Mol Biol 2003;330:419–430.
10 Locher KP, Lee AT, Rees DC: The *E. coli* BtuCD structure: A framework for ABC transporter architecture and mechanism. Science 2002;296:1091–1098.
11 Rosenberg MF, Callaghan R, Modok S, Higgins CF, Ford RC: Three-dimensional structure of P-glycoprotein: The transmembrane regions adopt an asymmetric configuration in the nucleotide-bound state J Biol Chem 2005;280:2857–2862.
12 Rosenberg MF, Kamis AB, Aleksandrow LA, Ford RC, Riordan JR: Purification and crystallization of the cystic fibrosis transmembrane conductance regulator (CFTR). J Biol Chem 2004;279:39051–39057.
13 Li C, Roy K, Dandridge K, Naren AP: Molecular assembly of cystic fibrosis transmembrane conductance regulator in plasma membrane. J Biol Chem 2004;279: 24673–24684.
14 Akabas MH: Probing CFTR channel structure and function using the substituted-cysteine-accessibility method. Methods Mol Med 2002;70:159–174.
15 Gong X, Linsdell P: Molecular determinants and role of an anion binding site in the external mouth of the CFTR chloride channel pore. J Physiol 2003;549(pt 2):387–397.
16 Smith SS, Liu X, Zhang ZR, Sun F, Kriewall TE, McCarty NA, Dawson DC: CFTR: Covalent and noncovalent modification suggests a role for fixed charges in anion conduction. J Gen Physiol 2001;118:407–431.
17 Gong X, Linsdell P: Maximization of the rate of chloride conduction in the CFTR channel pore by ion-ion interactions. Arch Biochem Biophys 2004;426:78–82.
18 Linsdell P: Location of a common inhibitor binding site in the cytoplasmic vestibule of the CFTR chloride channel pore. J Biol Chem epub January 2005.
19 Ge N, Muise CN, Gong X, Linsdell P: Direct comparison of the functional roles played by different membrane spanning regions in the cystic fibrosis transmembrane conductance regulator chloride channel pore. J Biol Chem 2004;279:55283–55289.
20 Ramjeesingh M, Ugwu F, Li C, Dhani S, Huan LJ, Wang Y, Bear CE: Stable dimeric assembly of the second membrane-spanning domain of CFTR reconstitutes a chloride-selective domain. Biochem J 2003;375(pt 3):633–641.
21 Gupta J, Evagelidis A, Hanrahan JW, Linsdell P: Asymmetric structure of the cystic fibrosis transmembrane conductance regulator chloride channel pore suggested by mutagenesis of the twelfth transmembrane region. Biochemistry 2001;40:6620–6627.
22 Zhang ZR, Zeltwanger S, McCarty NA: Direct comparison of NPPB and DPC as probes of CFTR expressed in *Xenopus* oocytes. J Membr Biol 2000;175:35–52.
23 Linsdell P, Tabcharani JA, Rommens JM, Hou YX, Chang XB, Tsui LC, Riordan JR, Hanrahan JW: Permeability of wild-type and mutant cystic fibrosis transmembrane conductance regulator chloride channels to polyatomic anions. J Gen Physiol 1997;110: 355–364.
24 Illek B, Yankaskas JR, Machen TE: cAMP and genistein stimulate HCO_3^- conductance through CFTR in human airway epithelia. Am J Physiol 1997;272:L752–L761.
25 Reddy MM, Quinton PM: Control of dynamic CFTR selectivity by glutamate and ATP in epithelial cells. Nature 2003;423:756–760.
26 Shcheynikov N, Kim KH, Kim KM, Dorwart MR, Ko SB, Goto H, Naruse S, Thomas PJ, Muallem S: Dynamic control of cystic fibrosis transmembrane conductance regulator Cl^-/HCO_3^- selectivity by external Cl^-. J Biol Chem 2004;279:21857–21865.
27 Linsdell P, Hanrahan JW: Glutathione permeability of CFTR. Am J Physiol 1998;275(1 pt 1):C323–C326.
28 Kogan I, Ramjeesingh M, Li C, Kidd JF, Wang Y, Leslie EM, Cole SP, Bear CE: CFTR directly mediates nucleotide-regulated glutathione flux. EMBO J 2003;22:1981–1989.
29 Hanrahan JW, Tabcharani JA, Becq F, Mathews CJ, Augustinas O, Jensen TJ, Chang XB, Riordan JR: Function and dysfunction of the CFTR chloride channel. Soc Gen Physiol 1995;Ser 50:125–137.
30 Chan KW, Csanady L, Seto-Young D, Nairn AC, Gadsby DC: Severed molecules functionally define the boundaries of the cystic fibrosis transmembrane conductance regulator's NH_2-terminal nucleotide binding domain. J Gen Physiol 2000;116:163–180.
31 Winter MC, Welsh MJ: Stimulation of CFTR activity by its phosphorylated R domain. Nature 1997;389:294–296.
32 Xie J, Adams LM, Zhao J, Gerken TA, Davis PB, Ma J: A short segment of the R domain of cystic fibrosis transmembrane conductance regulator contains channel stimulatory and inhibitory activities that are separable by sequence modification. J Biol Chem 2002; 277:23019–23027.
33 Li C, Ramjeesingh M, Wang W, Garami E, Hewryk M, Lee D, Rommens JM, Galley K, Bear CE: ATPase activity of the cystic fibrosis transmembrane conductance regulator. J Biol Chem 1996; 271:28463–28468.
34 Kidd JF, Ramjeesingh M, Stratford F, Huan LJ, Bear CE: A heteromeric complex of the two nucleotide binding domains of cystic fibrosis transmembrane conductance regulator (CFTR) mediates ATPase activity. J Biol Chem 2004;279:41664–41669.
35 Ramjeesingh M, Li C, Garami E, Huan LJ, Galley K, Wang Y, Bear CE: Walker mutations reveal loose relationship between catalytic and channel-gating activities of purified CFTR (cystic fibrosis transmembrane conductance regulator). Biochemistry 1999;38: 1463–1468.
36 Vergani P, Nairn AC, Gadsby DC: On the mechanism of MgATP-dependent gating of CFTR Cl^- channels. J Gen Physiol 2003;121: 17–36.
37 Gunderson KL, Kopito RR: Conformational states of CFTR associated with channel gating: The role ATP binding and hydrolysis. Cell 1995;82:231–239.
38 Aleksandrov AA, Chang X, Aleksandrov L, Riordan JR: The non-hydrolytic pathway of cystic fibrosis transmembrane conductance regulator ion channel gating. J Physiol 2000; 528(pt 2):259–265.
39 Schultz BD, Venglarik CJ, Bridges RJ, Frizzell RA: Regulation of CFTR Cl^- channel gating by ADP and ATP analogues. J Gen Physiol 1995;105:329–361.
40 Berger AL, Ikuma M, Welsh MJ: Normal gating of CFTR requires ATP binding to both nucleotide-binding domains and hydrolysis at the second nucleotide-binding domain. Proc Natl Acad Sci USA 2005;102:455–460.
41 Ikuma M, Welsh MJ: Regulation of CFTR Cl^- channel gating by ATP binding and hydrolysis. Proc Natl Acad Sci USA 2000;97:8675–8680.
42 Higgins CF, Linton KJ: The ATP switch model for ABC transporters. Nat Struct Mol Biol 2004;11:918–926.
43 Hopfner KP, Karcher A, Shin DS, Craig L, Arthur LM, Carney JP, Tainer JA: Structural biology of Rad50 ATPase: ATP-driven conformational control in DNA double-strand break repair and the ABC-ATPase superfamily. Cell 2000;101:789–800.

44 Smith PC, Karpowich N, Millen L, Moody JE, Rosen J, Thomas PJ, Hunt JF: ATP binding to the motor domain from an ABC transporter drives formation of a nucleotide sandwich dimer. Mol Cell 2002;10:139–149.

45 Chen J, Lu G, Lin J, Davidson AL, Quiocho FA: A tweezers-like motion of the ATP-binding cassette dimer in an ABC transport cycle. Mol Cell 2003;12:651–661.

46 Hung LW, Wang IX, Nikaido K, Liu PQ, Ames GF, Kim SH: Crystal structure of the ATP-binding subunit of an ABC transporter. Nature 1998;396:703–707.

47 Fetsch EE, Davidson AL: Vanadate-catalyzed photocleavage of the signature motif of an ATP-binding cassette (ABC) transporter. Proc Natl Acad Sci USA 2002;99:9685–9690.

48 Moody JE, Millen L, Binns D, Hunt JF, Thomas PJ: Cooperative ATP-dependent association of the nucleotide binding cassettes during the catalytic cycle of ATP-binding cassette transporters. J Biol Chem 2002;277:21111–21114.

49 Callebaut I, Eudes R, Mornon JP, Lehn P: Nucleotide-binding domains of human cystic fibrosis transmembrane conductance regulator: Detailed sequence analysis and three-dimensional modeling of the heterodimer. Cell Mol Life Sci 2004;61:230–242.

50 Lewis HA, Buchanan SG, Burley SK, Conners K, Dickey M, Dorwart M, Fowler R, Gao X, Guggino WB, Hendrickson WA, Hunt JF, Kearins MC, Lorimer D, Maloney PC, Post KW, Rajashankar KR, Rutter ME, Sauder JM, Shriver S, Thibodeau PH, Thomas PJ, Zhang M, Zhao X, Emtage S: Structure of nucleotide-binding domain 1 of the cystic fibrosis transmembrane conductance regulator. EMBO J 2004;23:282–293.

51 Lewis HA, Zhao X, Wang C, Sauder JM, Rooney I, Noland BW, Lorimer D, Kearins MC, Conners K, Condon B, Maloney PC, Guggino WB, Hunt JF, Emtage S: Impact of the delta F508 mutation in NBD1 of human CFTR on domain folding and structure. J Biol Chem 2005;280:1346–1353.

52 Chang G, Roth CB: Structure of MsbA from *E. coli:* A homolog of the multidrug resistance ATP binding cassette (ABC) transporters. Science 2001;293:1793–1800.

53 Cotten JF, Ostedgaard LS, Carson MR, Welsh MJ: Effect of cystic fibrosis-associated mutations in the fourth intracellular loop of cystic fibrosis transmembrane conductance regulator. J Biol Chem 1996;271:21279–21284.

54 Seibert FS, Linsdell P, Loo TW, Hanrahan JW, Clarke DM, Riordan JR: Disease-associated mutations in the fourth cytoplasmic loop of cystic fibrosis transmembrane conductance regulator compromise biosynthetic processing and chloride channel activity. J Biol Chem 1996;271:15139–15145.

55 Cheng SH, Gregory RJ, Marshall J, Paul S, Souza DW, White GA, O'Riordan CR, Smith AE: Defective intracellular transport and processing of CFTR is the molecular basis of most cystic fibrosis. Cell 1990;63:827–834.

56 Thibodeau PH, Brautigam CA, Machius M, Thomas PJ: Side chain and backbone contributions of Phe508 to CFTR folding. Nat Struct Mol Biol 2005;12:10–16.

57 Du K, Sharma M, Lukacs GL: The deltaF508 cystic fibrosis mutation impairs domain-domain interactions and arrests post-translational folding of CFTR. Nat Struct Mol Biol 2005;12:17–25.

58 Lee JY, Urbatsch IL, Senior AE, Wilkens S: Projection structure of P-glycoprotein by electron microscopy evidence for a closed conformation of the nucleotide binding domains. J Biol Chem 2002;277:40125–40131.

59 Velarde G, Ford RC, Rosenberg MF, Powis SJ: Three-dimensional structure of transporter associated with antigen processing (TAP) obtained by single particle image analysis. J Biol Chem 2001;276:46054–46063.

Christine E. Bear
Programme in Structural Biology and
Biochemistry
Research Institute
Hospital for Sick Children
555 University Avenue
Toronto (Canada)
E-Mail bear@sickkids.on.ca

Bush A, Alton EWFW, Davies JC, Griesenbach U, Jaffe A (eds): Cystic Fibrosis in the 21st Century.
Prog Respir Res. Basel, Karger, 2006, vol 34, pp 38–44

Function of CFTR Protein: Ion Transport

Jeng-Haur Chen   Zhiwei Cai   Hongyu Li   David N. Sheppard

Department of Physiology, University of Bristol, School of Medical Sciences, Bristol, UK

Abstract

The cystic fibrosis transmembrane conductance regulator (CFTR) is a unique member of the ATP-binding cassette (ABC) transporter superfamily that plays a critical role in fluid and electrolyte transport across epithelial tissues. CFTR is composed of two membrane-spanning domain (MSD)-nucleotide-binding domain (NBD) motifs linked by a unique regulatory (R) domain. The MSDs assemble to form a transmembrane pore with deep intracellular and shallow extracellular vestibules that funnel anions towards a selectivity filter, which determines the permeation properties of CFTR. Anion flow through the CFTR pore is powered by cycles of ATP binding and hydrolysis at two ATP-binding sites. Stable ATP binding occurs at one ATP-binding site (site 1), whereas rapid ATP turnover occurs at the other (site 2). These ATP-binding sites are located at the interface of the two NBDs, which are themselves organized as a head-to-tail dimer. The R domain contains multiple consensus phosphorylation sites on the surface of an unstructured domain. Phosphorylation of the R domain stimulates CFTR function by enhancing ATP-dependent channel gating at the NBDs. Thus, CFTR is an anion channel with exquisite regulation. Malfunction of CFTR in cystic fibrosis has profound consequences for transepithelial ion transport.

Introduction

Cystic fibrosis transmembrane conductance regulator (CFTR) is a unique member of the ATP-binding cassette (ABC) transporter superfamily that forms an anion channel with exquisite regulation [1]. CFTR is principally expressed in the apical membrane of epithelia where it provides a pathway for Cl^- and HCO_3^- movement and controls the rate of fluid flow through (1) its role as an anion channel and (2) regulating the function of ion channels and transporters in epithelial cells [1]. Thus, CFTR plays a fundamental role in transepithelial fluid and electrolyte transport. In sweat duct epithelia, CFTR drives the reabsorption of salt, while in intestinal, pancreatic and respiratory airway epithelia CFTR powers the secretion of Cl^- and HCO_3^-. The importance of CFTR is dramatically highlighted by the consequences of CFTR malfunction in CF and related disorders [1].

CFTR has several distinguishing characteristics [for a review, see 2]. First, CFTR has a small single-channel conductance (6–10 pS). Second, the current-voltage (I-V) relationship of CFTR is linear. Third, the anion permeability sequence of CFTR is $Br^- > Cl^- > I^- > F^-$. Fourth, CFTR shows time- and voltage-independent gating behaviour. Fifth, the activity of CFTR is regulated by cAMP-dependent phosphorylation and intracellular nucleotides. These hallmarks are conferred on CFTR by the function of the different domains from which CFTR is assembled: the two membrane-spanning domains (MSDs) that are each composed of six transmembrane segments, the two nucleotide-binding domains (NBDs) that each contain motifs, which interact with ATP and the unique R domain that contains multiple consensus phosphorylation sites and many charged amino acids (fig. 1). Here we discuss the relationship between CFTR structure and function.

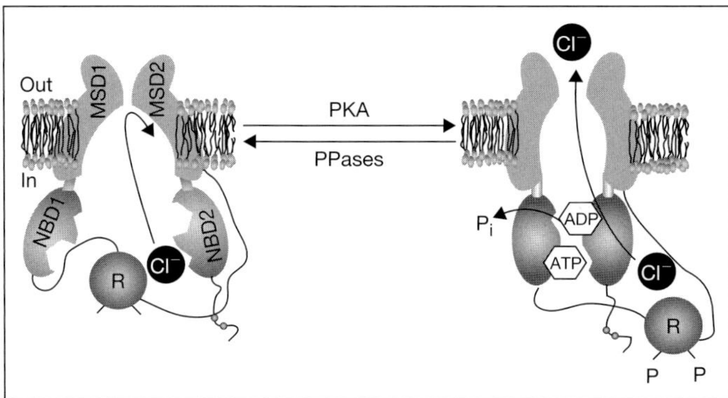

Fig. 1. The CFTR Cl⁻ channel is regulated by phosphorylation and intracellular ATP. This simplified model shows a CFTR Cl⁻ channel under quiescent and activated conditions. P = Phosphorylation of the R domain; P_i = inorganic phosphate; PKA = cAMP-dependent protein kinase; PPase = protein phosphatase. In and out denote the intra- and extracellular sides of the membrane, respectively. See text for further information.

Anion Flow: The MSDs

Architecture of CFTR Pore

In most ABC transporters the MSDs assemble to form a translocation pathway that shuttles substrates across the cell membrane. In contrast, the MSDs of CFTR form an anion-selective pore through which anions stream across the cell membrane driven by the transmembrane electrochemical gradient [2]. Excitingly, Rosenberg et al. [3] recently determined the three-dimensional structure of CFTR by electron crystallography (fig. 2). Their data suggest several important conclusions: (1) the quaternary structure of CFTR is monomeric [for a review, see 4], (2) CFTR exists in different conformations with either single- or double-barreled central cavities [3] and (3) the CFTR pore has a deep wide intracellular cavity, but a shallow extracellular cavity [3]. This image of the CFTR pore shows notable similarity to that predicted by functional studies.

Anion permeation studies indicate that the narrowest part of the CFTR pore is ~0.53–0.60 nm in diameter [e.g. 5], widening under certain circumstances to a diameter of ~1.3 nm [6]. On the intra- and extracellular sides of this constriction, the pore enlarges (fig. 1). The voltage dependence of channel block by large organic anions suggests that the CFTR pore contains a wide intracellular vestibule that funnels blocking anions deep into the pore where they bind, occlude the pore and block Cl⁻ permeation [for a review,

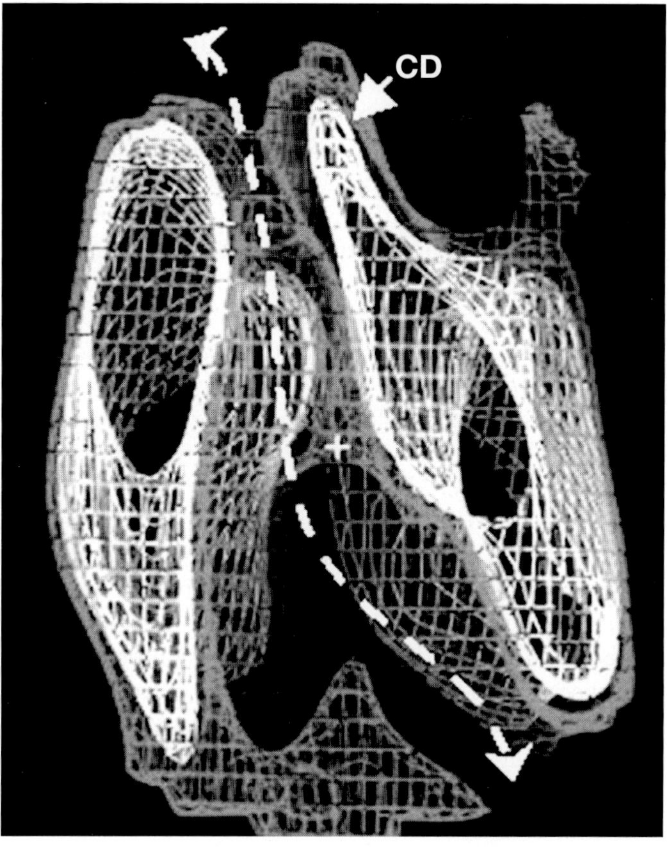

Fig. 2. Structure of CFTR. The image shows a cross section through the three-dimensional structure of CFTR determined by electron crystallography. The dashed line shows the channel lumen, through which anions pass from the intra- (bottom) to the extracellular (top) sides of the membrane. CD = Central domain, formed from the MSDs. The scale bar is 5 nm [reproduced with permission from 3].

see 7]. Based on the inability of open-channel blockers to reach their binding sites when added to the extracellular solution and the short length of extracellular loops 3 and 6, McCarty [8] proposed that CFTR has a small extracellular vestibule. Consistent with this idea, CFTR exhibits an asymmetric permeability to large organic anions [6] and current flow through the CFTR pore weakly inwardly rectifies [9].

The number, identity and organization of the transmembrane segments that line the CFTR pore are unknown at the present time. In the ABC transporter P-glycoprotein (P-gp), five α-helices from each MSD exhibit pseudo-2-fold symmetry with extensive interfaces along the length of long α-helices [10]. Moreover, functional studies [e.g. 11] suggest that the translocation pathway of P-gp is lined by transmembrane segment 6 (M6) and M12 [10]. Given the similarities in the crystal structures of CFTR and P-gp [3], M6 and M12 might be predicted to be important determinants of the pore properties of the CFTR Cl^- channel. Numerous studies have demonstrated that amino acid residues in M6 line the CFTR pore (see below). However, the role of M12 and other transmembrane segments in determining CFTR's pore properties is much less certain [for discussion, see 2].

Identification of Anion Binding Sites and Location of the Selectivity Filter

M6 plays a crucial role in determining the pore properties of CFTR [2, 8, 12]. Within M6, the residues arginine (R) 334, lysine (K) 335, phenylalanine (F) 337, threonine (T) 338, serine (S) 341, isoleucine (I) 344 and possibly R352 contribute to Cl^--binding sites. The narrowest part of the CFTR pore, the location of the selectivity filter, likely occurs in the region of F337 and T338 because mutation of these residues altered dramatically the anion permeability sequence of CFTR, whereas mutation of other residues in M6 had less marked effects on anion permeation [e.g. 12]. These data suggest that three Cl^--binding sites (S341, I344 and R352) might be located in a spacious intracellular vestibule, whereas two Cl^--binding sites (R334 and K335) might be located in a more confined extracellular vestibule. An important caveat of this model is that present knowledge of the mechanism of anion permeation by the CFTR Cl^- channel is incomplete and it is quite likely that residues in other transmembrane segments contribute to Cl^--binding sites.

Mechanism of Conduction and Permeation

CFTR is a multi-ion pore capable of holding multiple anions simultaneously [for review, see 2, 13]. This hallmark of the CFTR pore is a key determinant of the channel's conduction properties because repulsive interactions between anions inside the pore ensure rapid anion permeation [14]. Of note, some CF mutants [e.g. R334 (W) tryptophan] likely cause CFTR malfunction by disrupting anion–anion interactions within the CFTR pore [14].

As discussed by Liu et al. [15], anion flow through the CFTR pore is determined by (1) anion permeability, the ease with which anions enter the CFTR pore and (2) anion binding, the tightness of the interaction between anions and the CFTR pore. Studies of polyatomic anions of known dimensions [e.g. 16] demonstrate that the anion permeability sequence of CFTR follows a lyotropic sequence. This suggests that anion permeation is determined by the hydration energy of anions with large, weakly hydrated anions

being most permeant [16]. Similarly, anion binding exhibits a lyotropic sequence with large anions binding tighter to the CFTR pore. This tight binding of large anions within the CFTR pore (e.g. $Au(CN)_2^-$ [12]) explains why these anions block Cl^- permeation. Models with a series of binding sites (or wells) separated by energy barriers have been developed to describe the anion permeation properties of the CFTR Cl^- channel. While highly speculative, these models simulate accurately many of the characteristics of the CFTR pore [13].

Regulation of CFTR Channel Gating by Intracellular ATP: The NBDs

The ATP Dependence of CFTR Channel Gating: Role of Binding and Hydrolysis

In ABC transporters, the NBDs are the site of ATP hydrolysis. In most ABC transporters, the energy released during the hydrolysis of ATP is used to actively transport substrates across the cell membrane. However, for the reasons described above, it was unclear why CFTR should have two domains that might hydrolyze ATP. Nevertheless, studies of channel regulation by ATP indicated that non-hydrolyzable ATP analogues and Mg^{2+}-free ATP were unable to support channel activity [for review, see 2, 13, 17]. The data suggested that ATP hydrolysis is a prerequisite for channel function.

Site-directed mutation of conserved residues in the NBDs suggest that ATP hydrolysis at the NBDs regulates channel gating. Mutation of the conserved Walker A lysine residues in the NBDs produced specific alterations in channel gating without altering the pore properties of CFTR. Initial studies were interpreted to suggest that ATP hydrolysis at NBD1 initiates channel activity, while ATP hydrolysis at NBD2 terminates activity [for a review, see 2]. However, subsequent studies have advocated different models to explain how the NBDs control channel gating. For example, using kinetic analyses of single-channel recordings, Zeltwanger et al. [18] developed an irreversible gating scheme to describe CFTR channel gating. While electrophysiology has been invaluable for the acquisition of knowledge and understanding of CFTR channel gating, biochemical approaches have proved instrumental to the success of recent studies about the NBDs.

The Functional Unit of the NBDs Is a Head-to-Tail Dimer

Crystal structures of several NBDs (e.g. NBD1 of murine CFTR [19]) and a few complete ABC transporters

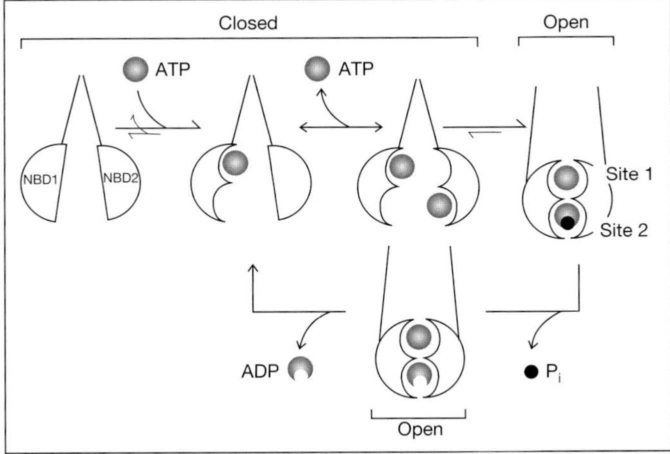

Fig. 3. ATP-dependent NBD dimerization drives channel opening. Simplified scheme linking the interaction of ATP with ATP-binding sites located at the interface of the two NBDs with NBD dimerization and channel opening. See text for explanation [modified from 23 by copyright permission of The Rockefeller University Press].

(e.g., BtuCD, the vitamin B_{12} transporter of *Escherichia coli* [20]) have recently emerged. These data suggest strongly that the two NBDs of CFTR function as a head-to-tail dimer with the ATP-binding sites located at the interface of the two subunits (fig. 1). The data suggest that one ATP-binding site is formed by the Walker A and B motifs of NBD1 and the LSGGQ motif of NBD2 (termed site 1), while the other is formed by the Walker A and B motifs of NBD2 and the LSGGQ motif of NBD1 (termed site 2 [19]). Consistent with this model of the NBDs, Kidd et al. [21] demonstrated that the association of NBD1 and NBD2 is required for optimal ATPase activity by CFTR.

Mechanism of Action of the NBDs

To study ATP binding to the NBDs, several groups have employed photoaffinity labelling. Using this approach, Aleksandrov et al. [22] demonstrated that NBD1 is a site of stable nucleotide interaction, whereas NBD2 is a site of rapid nucleotide turnover. Based on these and other data, Vergani et al. [23] proposed a model to describe CFTR channel gating (fig. 3). In this model, MgATP binding to both ATP-binding sites drives a conformational change in the NBDs leading to the formation of an NBD dimer. During this period, the channel remains closed. Channel opening requires the formation of a prehydrolysis complex at site 2. Hydrolysis of this ATP drives a conformation

change leading to closure of the channel pore and separation of the two NBDs. However, provided that ATP remains stably bound at site 1, cycles of NBD association and disassociation and, hence, channel opening and closing are driven by ATP binding and hydrolysis at site 2 [23] (fig. 3). Based on this model of CFTR channel gating, Riordan [4] described CFTR as a hydrolyzable-ligand-gated channel with intracellular ATP acting as the ligand.

Phosphorylation-Dependent Regulation of CFTR: The R Domain

Boundaries and Characteristics of the R Domain

Phosphorylation of the R domain is a crucial determinant of CFTR function [2, 13, 17]. The R domain was originally defined as exon 13 (amino acids (aa) 590–830), which links the two MSD-NBD motifs of CFTR. However, structural [e.g. 19] and functional [e.g. 24] studies suggest that the N-terminus of the R domain lies between glutamine (Q) 634 and I708, while the C-terminus is close to aspartate (D) 836.

The R domain contains multiple consensus phosphorylation sites for PKA and protein kinase C (PKC) [for review, see 2, 13, 17]. It is also rich in charged amino acid residues. Most of the basic residues are associated with consensus phosphorylation sites, but the acidic residues tend to be clustered in two regions (aa 725–733 and aa 817–838). Outside of these regions, there is little sequence similarity across species. Consistent with this idea, Ostedgaard et al. [25] demonstrated that a recombinant R domain (aa 708–831) was unstructured with the vast majority of the protein being random coil. Of note, phosphorylation was without marked effect on R domain structure [25].

Phosphorylation of the R Domain

PKA is the most important kinase responsible for regulating CFTR. However, other kinases phosphorylate the R domain and stimulate the channel including PKC, the type II isotype of cGMP-dependent protein kinase and tyrosine kinases [2, 17]. Dephosphorylation of the R domain is cell-type specific [2, 17].

The most important phosphorylation sites for CFTR regulation by PKA are S660, S700, S737, S795 and S813. Cyclic AMP agonists phosphorylate these sites in vivo [e.g. 26]. Moreover, the largest decrement in CFTR activity occurred with the simultaneous mutation of S660, S737, S795 and S813 [e.g. 27]. The data argue that PKA stimulation of CFTR is redundant with no individual phosphoserine residue being essential. They also suggest that there are functional interactions between different phosphoserines. However, it remains unclear how individual phosphoserines control channel activity. For example, while most phosphoserines are stimulatory, S737 and S768 [e.g. 28] are inhibitory. Perhaps individual phosphoserines have differential sensitivity to protein kinases, phosphatases and other regulatory molecules.

Evidence for the importance of sequential phosphorylation events comes from studies of PKC. By itself, PKC only weakly stimulates human CFTR (see below). However, PKC greatly potentiates the onset and magnitude of channel activation when PKA is subsequently applied [29]. Using a CFTR mutant (termed 9CA) that annuls nine PKC consensus sequences in NBD1 and the R domain, Chappe et al. [30] demonstrated that the effects of PKC on CFTR function are a direct consequence of PKC-dependent phosphorylation of CFTR. Using 9CA as a backbone, Chappe et al. [31] investigated how individual PKC consensus sequences control CFTR function. Several conclusions can be drawn from their work: (1) residues T582 and T604 in NBD1 and S686 in the R domain are crucial for CFTR activation by PKA either by itself or following PKC pretreatment, (2) S686 is essential and S641 contributes to CFTR activation by PKC and (3) S641 and T682 likely inhibit CFTR activation by PKA and PKC, respectfully [31]. Interestingly, the identification of T682 as an inhibitory site helps to explain the marked species-dependent differences in the effects of PKC on CFTR function. In amphibian CFTRs, which lack the inhibitory PKC site T682, PKC and PKA activate CFTR equipotently [32]. However, amphibian CFTRs also possess a potent stimulatory PKC site at T665, which is absent in human CFTR [32]. Based on these data, future studies should elucidate how PKC controls CFTR function.

Mechanism of Action of the R Domain

The R domain was originally proposed to regulate CFTR by keeping the channel closed at rest. In this model, the unphosphorylated R domain acts as an inhibitor maintaining the CFTR pore in a closed state; channel inhibition is relieved by either phosphorylation or deletion of the R domain [2]. However, the activity of phosphorylated CFTR greatly exceeded that of CFTR variants, which did not require phosphorylation to open in the presence of ATP [e.g. 33, but see 24]. These data suggest that the phosphorylated R domain might stimulate the activity of CFTR. Consistent with this idea, phosphorylation augments ATP binding and hydrolysis by the NBDs [34, 35] to accelerate the rate of channel opening [34]. Thus, the R domain does not function solely as an 'on-off' switch, but rather as a 'tethered' enzyme stimulating the interaction of ATP with the NBDs. Based on the model proposed by Ostedgaard et al. [25], multiple phosphoserines situated on the outside of an unstructured R domain would interact with different sites on the NBDs to stimulate CFTR function.

CFTR: A Chloride Channel with Complex Regulation

The data reviewed in this chapter highlight the complexity of the specific domains from which CFTR is assembled. While current knowledge remains far from complete, our understanding is growing such that we can begin to understand how the function of individual domains determines the physiological role of CFTR. A major challenge now is to learn how the different domains interact. For example, how the R domain stimulates the interaction of ATP with the NBDs [34].

A particularly fascinating question is how the interaction of ATP with the NBDs drives conformation changes in the MSDs that cause the pore to open and close. Because the movement of ions through ion channels is passive, driven by the electrochemical gradient across the membrane, it is perplexing that the energy of ATP hydrolysis is a prerequisite for CFTR channel gating. A novel answer to this enigma has been proposed by Randak and Welsh [36]. These authors demonstrated that CFTR has adenylate kinase activity and that this activity regulates channel gating. Because the energy cost of adenylate kinase activity is very low compared with that of ATP hydrolysis, Randak and Welsh [36] proposed that adenylate kinase and not ATP hydrolysis might control CFTR channel gating at physiological concentrations of nucleotides. But how does work performed by the NBDs gate the CFTR pore? Because F508 is located on the surface of NBD1 [19] and its deletion disrupts CFTR channel gating [37], the region around F508 is likely involved in coupling the NBD engine to channel opening and closing. Similarly, the effects of CF mutations in the intracellular loops on the function of CFTR [2] suggest that these loops transduce the activity of the NBDs to the CFTR pore.

A novel feature of the CFTR pore is that anion selectivity is 'dynamic', with the pore switching between conformations permeable to Cl^- and large anions (e.g. HCO_3^- and glutathione) in the presence of different factors including glutamate [38], non-hydrolyzable ATP analogues [39] and external Cl^- ions [40]. The switch of the CFTR pore from Cl^- to HCO_3^- selectivity plays a valuable role in HCO_3^- secretion (but see [41] for reciprocal regulation of

     Chen/Cai/Li/Sheppard

SLC26 Cl^--HCO_3^- exchangers by CFTR in HCO_3^- secretion). Of note, in airway epithelia CFTR mediates the transport of glutathione, a valuable antioxidant, following a change in its selectivity [39]. Elucidating how dynamic selectivity is achieved will have significant implications for the role of CFTR in transepithelial ion transport and its malfunction in CF.

In conclusion, knowledge and understanding of CFTR structure and function are crucial for understanding the physiological role of CFTR, its malfunction in CF and related diseases and the development of rational new approaches to therapy.

Acknowledgements

We thank Prof. D.C. Gadsby and Dr. R.C. Ford for permission to reproduce their work and our laboratory colleagues for stimulating discussions. J.-H. Chen is the recipient of a scholarship from the University of Bristol and an ORS award from Universities UK. The CF Trust supported the preparation of this review.

Appendix: Glossary

Current-voltage (I-V) relationship: Obtained by plotting the amount of current that flows through a single ion channel by the applied voltage over a range of voltages (e.g. -80 to $+60\,mV$). If the ion concentrations on either side of the membrane are similar and current flow through the channel obeys Ohm's law, the single channel I-V relationship is linear and the slope of the line is the single-channel conductance.

Electrochemical gradient: Electrical and chemical forces that drive ion movement across a permeable membrane.

Gating behaviour: A characteristic property of ion channels that describes the pattern of transitions between the closed (non-conducting) conformation and the open (conducting) conformation.

LSGGQ motif: Highly conserved sequences found in ABC transporters. This motif contributes to the site where ATP is bound and hydrolyzed.

Lyotropic sequence: A rank order of ion selectivity where large ions that weakly bind water (and are most easily dehydrated) are selected first and small ions that strongly immobilize water (and are most difficult to dehydrate) are selected last.

Multi-ion pore: An ion channel that contains more than one ion at a time.

Permeability: A measure of the ease with which ions cross the membrane.

Polyatomic anion: An anion composed of more than one chemical group, e.g. gluconate, $HOCH_2[CH(OH)]_4COO^-$.

Rectification: A term used to describe the I-V relationship of an ion channel that is non-linear as a result of asymmetries in the ion concentration on either side of the membrane or the architecture of the channel pore. Rectification is inward when current flow at voltages negative to the reversal potential exceeds that at voltages positive to the reversal potential. In the converse situation rectification is outward.

Reversal potential: Voltage at which no current flows through an ion channel.

Selectivity: The ability of an ion channel to discriminate between different ions allowing some types of ions to flow through the channel and others not.

Selectivity filter: The site that determines which ions pass through an ion channel.

Single-channel conductance: A measure of the current that flows through an open channel in response to a given electrochemical driving force [units: picosiemens (pS)].

Walker A and B motifs: Highly conserved sequences found in ATP- and GTP-binding proteins. The Walker A lysine (K) interacts with either the α- or γ-phosphate of ATP during ATP hydrolysis. The Walker B aspartate (D) interacts with Mg^{2+} during the binding of ATP.

References

1   Welsh MJ, Ramsey BW, Accurso F, Cutting GR: Cystic fibrosis; in Scriver CR, Beaudet AL, Sly WS, Valle D (eds): The Metabolic and Molecular Basis of Inherited Disease. New York, McGraw-Hill, 2001, pp 5121–5188.

2   Sheppard DN, Welsh MJ: Structure and function of the cystic fibrosis transmembrane conductance regulator chloride channel. Physiol Rev 1999;79:S23–S45.

3   Rosenberg MF, Kamis BF, Aleksandrov LA, Ford RC, Riordan JR: Purification and crystallization of the cystic fibrosis transmembrane conductance regulator (CFTR). J Biol Chem 2004;279:39051–39057.

4   Riordan JR: Assembly of functional CFTR chloride channels. Annu Rev Physiol 2005;67: 701–718.

5   Cheung M, Akabas MH: Identification of cystic fibrosis transmembrane conductance regulator channel-lining residues in and flanking the M6 membrane-spanning segment. Biophys J 1996;70:2688–2695.

6   Linsdell P, Hanrahan JW: Adenosine triphosphate-dependent asymmetry of anion permeation in the cystic fibrosis transmembrane conductance regulator chloride channel. J Gen Physiol 1998;111:601–614.

7   Hwang T-C, Sheppard DN: Molecular pharmacology of the CFTR Cl^- channel. Trends Pharmacol Sci 1999;20:448–453.

8   McCarty NA: Permeation through the CFTR chloride channel. J Exp Biol 2000;203: 1947–1962.

9   Cai Z, Scott-Ward TS, Sheppard DN: Voltage-dependent gating of the cystic fibrosis transmembrane conductance regulator Cl^- channel. J Gen Physiol 2003;122:605–620.

10   Rosenberg MF, Callaghan R, Szabolcs M, Higgins CF, Ford RC: 3-D structure of P-glycoprotein: The transmembrane regions adopt an asymmetric configuration in the nucleotide-bound state. J Biol Chem 2005;280: 2857–2862.

11   Loo TW, Bartlett MC, Clarke DM: Simultaneous binding of two different drugs in the binding pocket of the human multidrug resistance P-glycoprotein. J Biol Chem 2003;278: 39706–39710.

12   Gong X, Burbridge SM, Cowley EA, Linsdell P: Molecular determinants of $Au(CN)_2^-$ binding and permeability within the cystic fibrosis transmembrane conductance regulator Cl^- channel pore. J Physiol 2002;540:39–47.

13   Hanrahan JW, Gentzsch M, Riordan JR: The cystic fibrosis transmembrane conductance regulator (ABCC7); in Holland IB, Cole SPC, Higgins CF, Kuchler K (eds): ABC Proteins: From Bacteria to Man. New York, Elsevier Science, 2002, pp 589–618.

14 Gong X, Linsdell P: Maximization of the rate of chloride conduction in the CFTR channel pore by ion-ion interactions. Arch Biochem Biophys 2004;426:78–82.

15 Liu X, Smith SS, Dawson DC: CFTR: What's it like inside the pore? J Exp Zoolog A Comp Exp Biol 2003;300:69–75.

16 Linsdell P, Tabcharani JA, Rommens JM, Hou Y-X, Chang X-B, Tsui L-C, Riordan JR, Hanrahan JW: Permeability of wild-type and mutant cystic fibrosis transmembrane conductance regulator chloride channels to polyatomic anions. J Gen Physiol 1997;110:355–364.

17 Gadsby DC, Nairn AC: Control of cystic fibrosis transmembrane conductance regulator channel gating by phosphorylation and nucleotide hydrolysis. Physiol Rev 1999;79:S77–S107.

18 Zeltwanger S, Wang F, Wang G-T, Gillis KD, Hwang T-C: Gating of cystic fibrosis transmembrane conductance regulator chloride channels by adenosine triphosphate hydrolysis: Quantitative analysis of a cyclic gating scheme. J Gen Physiol 1999;113:541–554.

19 Lewis HA, Buchanan SG, Burley SK, Conners K, Dickey M, Dorwart M, Fowler R, Gao X, Guggino WB, Hendrickson WA, Hunt JF, Kearins MC, Lorimer D, Maloney PC, Post KW, Rajashankar KR, Rutter ME, Sauder JM, Shriver S, Thibodeau PH, Thomas PJ, Zhang M, Zhao X, Emtage S: Structure of nucleotide-binding domain 1 of the cystic fibrosis transmembrane conductance regulator. EMBO J 2004;23:282–293.

20 Locher KP, Lee AT, Rees DC: The *E. coli* BtuCD structure: A framework for ABC transporter architecture and mechanism. Science 2002;296:1091–1098.

21 Kidd JF, Ramjeesingh M, Stratford F, Huan L-J, Bear CE: A heteromeric complex of the two nucleotide binding domains of cystic fibrosis transmembrane conductance regulator (CFTR) mediates ATPase activity. J Biol Chem 2004;279:41664–41669.

22 Aleksandrov L, Aleksandrov AA, Chang X-B, Riordan JR: The first nucleotide binding domain of cystic fibrosis transmembrane conductance regulator is a site of stable nucleotide interaction, whereas the second is a site of rapid turnover. J Biol Chem 2002;277:15419–15425.

23 Vergani P, Nairn AC, Gadsby DC: On the mechanism of MgATP-dependent gating of CFTR Cl⁻ channels. J Gen Physiol 2003;120:17–36.

24 Csanády L, Chan KW, Seto-Young D, Kopsco DC, Nairn AC, Gadsby DC: Severed channels probe regulation of gating of cystic fibrosis transmembrane conductance regulator by its cytoplasmic domains. J Gen Physiol 2000;116:477–500.

25 Ostedgaard LS, Baldursson O, Vermeer DW, Welsh MJ, Robertson ADA: Functional R domain from cystic fibrosis transmembrane conductance regulator is predominantly unstructured in solution. Proc Natl Acad Sci USA 2000;97:5657–5662.

26 Cheng SH, Rich DP, Marshall J, Gregory RJ, Welsh MJ, Smith AE: Phosphorylation of the R domain by cAMP-dependent protein kinase regulates the CFTR chloride channel. Cell 1991;66:1027–1036.

27 Rich DP, Berger HA, Cheng SH, Travis SM, Saxena M, Smith AE, Welsh MJ: Regulation of the cystic fibrosis transmembrane conductance regulator Cl⁻ channel by negative charge in the R domain. J Biol Chem 1993;268:20259–20267.

28 Wilkinson DJ, Strong TV, Mansoura MK, Wood DL, Smith SS, Collins FS, Dawson DC: CFTR activation: Additive effects of stimulatory and inhibitory phosphorylation sites in the R domain. Am J Physiol 1997;273:L127–L133.

29 Tabcharani JA, Chang X-B, Riordan JR, Hanrahan JW: Phosphorylation-regulated Cl⁻ channel in CHO cells stably expressing the cystic fibrosis gene. Nature 1991;352:628–631.

30 Chappe V, Hinkson DA, Zhu T, Chang X-B, Riordan JR, Hanrahan JW: Phosphorylation of protein kinase C sites in NBD1 and the R domain control CFTR channel activation by PKA. J Physiol 2003;548:39–52.

31 Chappe V, Hinkson DA, Howell LD, Evagelidis A, Liao J, Chang X-B, Riordan JR, Hanrahan JW: Stimulatory and inhibitory protein kinase C consensus sequences regulate the cystic fibrosis transmembrane conductance regulator. Proc Natl Acad Sci USA 2004;101:390–395.

32 Button B, Reuss L, Altenberg GA: PKC-mediated stimulation of amphibian CFTR depends on a single phosphorylation consensus site. Insertion of this site confers PKC sensitivity to human CFTR. J Gen Physiol 2001;117:457–467.

33 Rich DP, Berger HA, Cheng SH, Travis SM, Saxena M, Smith AE, Welsh MJ: Regulation of the cystic fibrosis transmembrane conductance regulator Cl⁻ channel by negative charge in the R domain. J Biol Chem 1993;268:20259–20267.

34 Winter MC, Welsh MJ: Stimulation of CFTR activity by its phosphorylated R domain. Nature 1997;389:294–296.

35 Li C, Ramjeesingh M, Wang W, Garami E, Hewryk M, Lee D, Rommens JM, Galley K, Bear CE: ATPase activity of the cystic fibrosis transmembrane conductance regulator. J Biol Chem 1996;271:28463–28468.

36 Randak C, Welsh MJ: An intrinsic adenylate kinase activity regulates gating of the ABC transporter CFTR. Cell 2003;115:837–850.

37 Dalemans W, Barbry P, Champigny G, Jallat S, Dott K, Dreyer D, Crystal RG, Pavirani A, Lecocq J-P, Lazdunski M: Altered chloride ion channel kinetics associated with the ΔF508 cystic fibrosis mutation. Nature 1991;354:526–528.

38 Reddy MM, Quinton PM: Control of dynamic CFTR selectivity by glutamate and ATP in epithelial cells. Nature 2003;423:756–760.

39 Kogan I, Ramjeesingh M, Li C, Kidd JF, Wang Y, Leslie EM, Cole SPC, Bear CE: CFTR directly mediates nucleotide-regulated glutathione flux. EMBO J 2003;22:1981–1989.

40 Shcheynikov N, Kim KH, Kim K-M, Dorwart MR, Ko SBH, Goto H, Naruse S, Thomas PJ, Muallem S: Dynamic control of cystic fibrosis transmembrane conductance regulator Cl⁻/HCO₃⁻ selectivity by external Cl⁻. J Biol Chem 2004;279:21857–21865.

41 Ko SB, Zeng W, Dorwart MR, Luo X, Kim KH, Millen L, Goto H, Naruse S, Soyombo A, Thomas PJ, Muallem S: Gating of CFTR by the STAS domain of SLC26 transporters. Nat Cell Biol 2004;6:343–350.

D.N. Sheppard, PhD
Department of Physiology
University of Bristol
School of Medical Sciences
University Walk, Bristol BS8 1TD (UK)
Tel. +44 117 928 8992
Fax +44 117 928 8923
E-Mail D.N.Sheppard@bristol.ac.uk

Bush A, Alton EWFW, Davies JC, Griesenbach U, Jaffe A (eds): Cystic Fibrosis in the 21st Century.
Prog Respir Res. Basel, Karger, 2006, vol 34, pp 45–53

Cystic Fibrosis

Function of CFTR Protein: Regulatory Functions

R.D. Coakley   M.J. Stutts

Cystic Fibrosis and Pulmonary Research and Treatment Center, University of North Carolina,
Chapel Hill, N.C., USA

Abstract

The pathophysiology of cystic fibrosis (CF) is complex and not neatly ascribed to loss of the Cl^- channel function of CFTR. Instead, cellular processes seemingly unrelated to Cl^- conductance, such as Na^+ absorption in airways, are abnormal in CF patients. This circumstance gives rise to the concept that CFTR has multiple functions, including the capability to regulate the function of other proteins. Despite much research, little molecular detail for the proposed regulatory functions of CFTR has emerged. This chapter reviews the major regulatory functions proposed for CFTR, and evaluates evidence for their mechanistic bases.

Introduction and Perspective

To evaluate critically the 'regulatory functions of CFTR', it is important to review the origin of this concept and to consider its present scope. The notion that CFTR regulates the functions of other proteins grew from the understanding of cystic fibrosis (CF) pathophysiology that existed when the CF gene was discovered in 1989 [1]. At that time, ion transport in multiple epithelial organs affected in patients with CF was known to be abnormal. Net Na^+ and Cl^- reabsorption was clearly deficient in CF sweat ducts, whereas pancreatic juice was known to be low in HCO_3^- [2, 3]. In CF airways, Cl^- permeability was decreased and Na^+ absorption was increased [4, 5]. These seemingly pleiotropic defects in the composition of secretions and underlying epithelial ion transport processes in CF set the stage for the discovery of the CF gene. Recognition that the newly found gene belonged to the regulatory ABC transporter superfamily fitted well with the perception that the CF gene product possessed regulatory functions [1]. Its discoverers judiciously gave it the name 'cystic fibrosis transmembrane conductance regulator' (CFTR). Subsequently, CFTR was shown to be a cAMP-regulated Cl^- channel, and this function was found to be uniformly absent in epithelia affected in CF. The molecular basis for every imaginable aspect of the channel function of CFTR Cl^- has been worked out in great detail [6–9]. In contrast, despite a long list of potential regulatory targets (table 1), detailed molecular mechanisms for any of the many proposed regulatory functions of CFTR have been slow in developing. All the while, new examples of CFTR regulation of other processes continue to be reported [10, 11]. Discovering molecular bases for any of the reported CFTR regulatory actions would help substantiate the regulatory role of CFTR in CF pathophysiology, allow a more rational evaluation of CFTR regulatory functions as potential therapeutic targets, and may be useful in defining surrogate endpoints for novel therapies. This chapter accepts a broad definition of 'regulatory functions of CFTR' to include any protein whose activity is altered by CFTR, but focuses on potential direct and indirect molecular mechanisms that underlie these phenomena.

Table 1. Reported examples of CFTR regulatory functions

Protein/process reported to be regulated	Protein function	Gene	Pathophysiologic role	Proposed Mechanisms of regulation
Aquaporin [11]	H_2O channel	Aq3	Disrupted H_2O permeability	Unknown
ATP release	Conduct nucleotides	Unknown	Disrupted paracrine signaling	Unknown
Outward Rectifying Cl^- Channels (ORCC)	cAMP regulated Cl^- conductance	Unknown	Reduced Cl^- conductance	Unknown
CaCC [12–14]	Ca^{2+} regulated Cl^- conductance	Unknown	Alternative Cl^- conductance	Unknown
HCO_3^- secretion	HCO_3^- conductance/ exchange	SLC26	Diminished alkalinity of secretions	Direct, Indirect
NHE3	Na^+/H^+ exchange	NHE3	Hyper Na^+ absorption	Scaffolding, indirect
ROMK	K^+ channel	KCNJ1	Renal K^+ recycling	Direct or scaffolding
TRPV4 [10]	Swelling-induced Ca^{2+} entry	TRPV4	Defective regulatory volume decrease	Unknown
Cytokine expression [15]	Inflammatory mediators	Multiple	Lung disease	Unknown
ENaC	Na^+ channel	α,β,γ ENaC	Dehydrated ASL	Direct, indirect

Molecular Mechanisms of CFTR Regulatory Functions

Regulation Through Direct Physical Interaction

The most straightforward influence of one protein on the function of another is exerted through direct physical interaction. Biology is replete with examples of protein-protein interactions that have clear-cut functional consequences, ranging from multimeric ion channels with separate regulatory and conducting subunits to enzymes with distinct regulatory or anchoring subunits [16, 17]. Two models of CFTR regulation exerted through physical interaction with other proteins have received serious scrutiny. The first is based on the membership of CFTR in the adenosine-binding cassette (ABC) transport superfamily [1], and the second comes from ability of CFTR to bind proteins containing PDZ domains (protein-protein interaction sites named for *p*ostsynaptic density the Drosophila septate junction protein *d*iscs-large, and the epithelial tight junction protein *Z*O-1) [18].

Among ABC transporters, the outstanding example of direct regulatory interaction is the tight binding between the sulfonyl urea receptor (SUR1) and inward rectifying K channels (K6.02) [19]. These two proteins co-immunopurify and this stable association conveys regulation of K_{ATP} channels by ATP and by sulfonylurea compounds [20]. Not surprisingly, the possibility that CFTR regulates other ion channels in the way that SUR regulates inward rectifying K^+ channels engendered considerable interest. Indeed, two examples of functional interactions of CFTR with other channels bear obvious similarities with the role of SUR in regulation of K_{ATP} channels. The most analogous observation is the role of CFTR in forming the ATP-sensitive K^+ channel ROMK [21]. CFTR appears to mediate sensitivity of this K^+ channel to both cytosolic ATP and to sulfonylurea compounds. However, the biochemical basis of this regulation does not appear to involve formation of a stable complex by direct interaction of the two proteins. Instead, the functional and physical association of CFTR and ROMK is strengthened by their common binding to a scaffolding protein (see below) [22].

Interestingly, there is another well-documented occurrence of CFTR regulation of a second channel, with observations that are functionally similar the SUR-K_{ATP}. Prior to

　Coakley/Stutts

the discovery of CFTR, multiple laboratories observed a distinctive Cl^- channel that was activated by depolarization and rectified outward current [23]. Notably, this outward rectifying chloride channel (ORCC) was stimulated by cAMP in cells derived from normals, but not in cells derived from CF patients [23, 24]. This made ORCC the leading candidate for the CF gene product up until CFTR was discovered and established itself to be a Cl^--channel of very different physical properties. Subsequently, the ORCC fell from favor as it became clear that this channel was most abundant in nonepithelial cells and arguably not present in the apical membrane of epithelia. Lack of a molecular identity for ORCC stymied conventional approaches to learn the basis of the functional linkage between CFTR and ORCC. However, Gabriel et al. [25] found that the ORCC in excised patches of normal mouse nasal epithelial cells were immediately active upon excision into a bath containing ATP and PKA. In stark contrast, in patches from CFTR$-/-$ mouse nasal epithelial cells ORCC never appeared in this maneuver, and could only be demonstrated to be present in the patch following prolonged strong voltage depolarization. This unequivocal demonstration that CFTR conveys PKA sensitivity onto ORCC, coupled with the recent demonstration that ORCC become sensitive to the CFTR inhibitor gliblenclamide when the two channels are co-expressed points to a genuine regulatory function of CFTR [26]. Although candidates for the molecular identity of ORCC have been debated [27, 28], the true identity of this channel, its functions and mode of interaction with CFTR remain unsettled.

Amiloride-sensitive Na^+ absorption is elevated 2- to 3-fold in the conducting airways of CF patients [29]. Na^+ absorption at other sites, such as the kidney or sweat duct, does not appear to be affected in CF [30]. However, numerous observations have confirmed the abnormality in CF airways and increased Na^+ absorption appears to be a strong component of CF pathophysiology in the lung [31, 32]. Therefore, the role of CFTR in determining the rate of Na^+ absorption across airway epithelia is of critical importance in understanding the pathophysiology of CF lung disease. Na^+ absorption in airways and other epithelia is mediated by the amiloride-sensitive epithelial Na^+ channel (ENaC), a heteromultimer comprised of distinct but homologous α, β- and γ-subunits [33]. Accordingly, it is inferred that CFTR exerts a 'brake' on ENaC activity, specifically in airway epithelia. The possibility that CFTR inhibits ENaC by direct interaction has been tested with varying results. Kunzelmann et al. [34] studied co-expression of ENaC with wild-type CFTR, G551D CFTR and CFTR fragments in *Xenopus* oocytes. They observed inhibition of ENaC upon stimulation of wild-type but not G551D, and speculated that a portion of the cytosolic region of CFTR between transmembrane domains 6 and 7 (encompassing nucleotide binding fold 1 and the regulatory domain) interacted with ENaC. They concluded, based on yeast 2-hybrid analysis, that CFTR$_{351-830}$ interacted with cytosolic C-terminus of α-ENaC [34]. In contrast, Ji et al. [35], also working in *Xenopus* oocytes, reported co-immunoprecipitation of CFTR and EnaC, and CFTR-mediated inhibition of ENaC that required the cytosolic termini of β- and γ-ENaC. Despite these reports, direct interaction of CFTR with ENaC is not firmly established. The validity of CFTR inhibition of ENaC in oocytes has been questioned on technical grounds by Nagel et al., who showed that expression of any Cl^- conductance in oocytes decreased the electrical driving force for Na^+ entry [36]. Reconciliation of the differing biochemical bases for CFTR-ENaC interactions and the applicability of these results from oocytes to mammalian airway epithelia require studies using more detailed approaches in a different expression system. The possibilities of less direct modes of CFTR-ENaC interaction are considered below.

Facilitated Interaction

The extreme C-terminal amino acid sequence of CFTR is highly conserved across species and forms a cognate binding motif for interaction with PDZ domain containing proteins [18]. PDZ proteins are important organizers of receptors, ion transporters and regulatory elements at the apical membrane of epithelia [37]. The interactions of CFTR with the PDZ proteins NHERF2 (Na^+/H^+ exchange regulatory factor 2), CAL (CFTR-associated ligand) and others have been reported to affect its own function and regulation [38–42]. It is additionally tempting to speculate that CFTR-PDZ interactions sustain close physical association of CFTR with other proteins that bind PDZ domains. This mechanism could enable regulatory interactions of CFTR with other proteins that would be difficult to detect biochemically. Indeed, CFTR-ROMK physical interactions and functional consequences were strengthened when both channels were co-expressed in oocytes with NHERF1 or NHERF2 [22].

Perhaps the most compelling evidence for PDZ-mediated CFTR regulatory functions is from protein-protein interactions detected between CFTR and the SLC26T family of bicarbonate/chloride exchangers. Defective CFTR-dependent bicarbonate secretion has been demonstrated in several CF epithelia. These include the airway [43, 44],

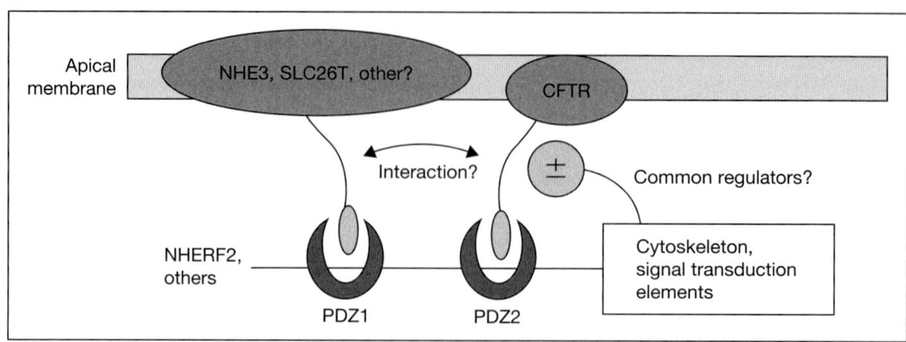

Fig. 1. Model of CFTR-PDZ interactions that potentially accounts for CFTR effects on the activity of other transporters. CFTR, NHE3 and SLC26T family members have been shown to interact with NHERF2. Conceptually, NHERF2 binding keeps CFTR in close physical proximity to other transporters that is required for regulatory interaction. Additionally, co-localization could allow for coordinated regulation of activity by highly compartmentalized signal transduction elements.

pancreas [45, 46], duodenum [47], salivary gland [48] and urogenital organs [11, 49]. The mechanism underlying this phenomenon is uncertain. However, the highly variable HCO_3^- secretory capacity of different tissues makes it unlikely that the intrinsic HCO_3^- conductance of CFTR is the sole explanation (as demonstrated in heterologous expression systems [50, 51]). Instead it seems necessary to invoke regulatory interactions between CFTR and other HCO_3^- transport pathways.

Such an interaction has been proposed between CFTR and members of the SLC26T family of HCO_3^-/Cl^- exchangers, which exhibit tissue-specific expression patterns [50, 52–54]. In HEK293 cells, Ko et al. [55] demonstrated that SLC26A6 and SLC26T DRA expression could potently activate CFTR, increasing the channels open probability by up to 6-fold. Importantly, this functional interaction was shown to be dependent upon PDZ domain interactions and binding of the conserved STAS domain of SLC26T family proteins to the R domain of CFTR. These authors also described reciprocal activation of electrogenic members of the SLC26T family (DRA, SLC26A6 and pendrin) by CFTR [50]. Although these earlier studies did not investigate a direct dependence of these effects on protein-protein interactions, it is tempting to speculate that a similar mechanism is implicated. In this manner, in different tissues, CFTR could differentially regulate HCO_3^- secretion by interacting with specific isoforms of the SCL26T family, thus facilitating secretion of a HCO_3^--rich solution into the pancreatic lumen and less alkaline solutions into the lumina of other organs.

The Na^+/H^+ exchanger NHE3 is another key apical membrane transporter that is potentially regulated through a PDZ-mediated association with CFTR. NHE3 mediates electroneutral Na^+ absorption in the kidney and intestine and is normally inhibited by cAMP [56]. NHERF was first identified as a soluble protein factor required for cAMP regulation of NHE3 [57]. Subsequently, NHERF1 and NHERF2 were shown to interact with NHE3 via their PDZ domains [58]. As CFTR is also required for cAMP-mediated inhibition of NHE3 [59] and binds NHERF2 [18], it is possible that PDZ-organized complexes containing CFTR and NHE3 are central to the regulation of NHE3 [42, 60]. A generalized scheme for CFTR effects on other proteins through common interactions with PDZ domain proteins is presented in figure 1.

Indirect Mechanisms

Most CF patients have effectively null expression of CFTR. CFTR is a Cl^- channel, and not withstanding its potential as a regulator of other proteins, the absence of CFTR from cells where it is normally expressed can be predicted to impact many cellular processes. Models that explain how loss of CFTR could indirectly affect the function of other ion transporters include loss of CFTR transport of regulatory substances, loss of Cl^- conductance and altered genomic expression.

Loss of Paracrine/Autocrine Function

One of the most studied models of an indirect regulatory role of CFTR is that of a paracrine/autocrine function of CFTR in mediating transport of bioactive solutes that can modify the surface milieu and/or signal through receptors on the epithelial surface, analogous to other ABC transport proteins. Careful comparison of CFTR to other ABC transporters suggests it is unlikely that CFTR actively transports a regulatory solute [61, 62]. However, given the function of CFTR as a passive ion channel, its role as a conduit in down-hill transfer of several compounds has been examined. Most controversial is regulated release of cytosolic ATP [63]. Regulated ATP release is undoubtedly an important autocrine/paracrine phenomenon, and its mediation by CFTR would indeed explain the propensity of CFTR to regulate other proteins [64]. Some investigators concluded that CFTR itself conducted ATP [65], others came to the position that while not conducting ATP itself, CFTR was regulating its release through another protein [66]. Still others find no participation of CFTR in ATP release at all [67, 68]. Critically, repeated and careful comparisons of ATP levels in airway surface liquid between normal and CF patients detect no differences [69].

A role for CFTR in supplying glutathione to the cell surface has been suggested based on the small, but measurable, conductance of glutathione through CFTR anion channels [70]. Glutathione levels in airway surface liquid of CF patients are diminished, leading to the suggestion glutathione permeability of CF airways epithelia is lost due to CF mutations [71].

Altered Electrochemical Potential

CFTR is a Cl^- conductance and its presence in a membrane contributes to the transmembrane electrochemical potential that affects the net activity of all other processes that carry out membrane transport of charged species. It is most important to consider how Na^+ and HCO_3^- are affected by the absence of CFTR Cl^- conductance. Because of the importance of airway disease in CF morbidity, the cause of excessive Na^+ absorption has been examined in detail. Microelectrode analyses unambiguously identified amiloride-sensitive apical membrane Na^+ conductance as the rate-limiting step in Na^+ absorption and revealed a 2- to 3-fold increased apical membrane amiloride-sensitive conductance in CF [72]. From the same analyses, it was observed that the absence of CFTR-mediated Cl^- conductance in the apical membrane of airway epithelia depolarized the apical membrane, which reduces the electrical potential favoring Na^+ entry into the cell [73]. Thus, the primary data do not favor a role for a CFTR-determined electrochemical potential as an indirect mechanism by which loss of CFTR Cl^- conductance enhances Na^+ transport via ENaC.

Reciprocal regulatory interactions between CFTR and members of the SLC26T family of HCO_3^-/Cl^- exchangers, via protein-protein interactions, have been discussed above. However, CFTR may regulate HCO_3^- secretion by indirect mechanisms. A model has proposed an indirect mechanism by which CFTR can regulate the activity of anion exchangers (AE), where CFTR facilitates AE function by supplying the required chloride ions to the apical surface for exchange with cytosolic HCO_3^-. Although Ko et al. [50] did not detect regulatory effects of CFTR on the SLC4 family of

AEs (AE 1–4), Clarke et al. [74], studying the CFTR-knockout mouse, characterized a cAMP-stimulated HCO_3^- secretion across the duodenum, involving both electrogenic secretion via a CFTR HCO_3^- conductance as well as electroneutral secretion via a CFTR-dependent Cl^-/HCO_3^- exchange which was sensitive to $4',4'$ -diisothiocyanatostilbene-2,$2'$ disulfonic acid (DIDS). However, these observations were not reproduced by Spiegel et al. [75] in studies of rat and rabbit proximal duodena mounted in Ussing chambers. In these studies, DIDS, as well as removal of lumenal Cl^-, did not affect cAMP-activated HCO_3^- secretion, arguing against a role for CFTR in activating Cl^-/HCO_3^- exchange. Underscoring the complexity of the regulatory role of CFTR in HCO_3^- secretion, studies of cultured, polarized, airway epithelial cells revealed no evidence of a DIDS-sensitive electroneutral AE process. Instead, they suggested that HCO_3^- efflux in this tissue was sensitive instead to the CFTR blocker, diphenylamine-2-carboxylate (DPC) [76], suggesting a dominant role for direct conductance through CFTR itself.

One of the earliest attempts to explain how defective CFTR could affect so many seemingly disparate ion transport processes was a model in which CFTR Cl^- conductance was required in intracellular organellar membranes for proton ATPases to generate intravesicular acidity [77]. Due to the strong pH optima of numerous enzymes, including those needed for proper glycosylation, it was hypothesized that the proteins responsible for many transport activities would be ineffectively modified during biosynthesis. Although direct evidence against this hypothesis has been presented [78], it continues to draw interest as an attractive explanation for the seemingly endless capability of CFTR to affect functions of other proteins [79].

Genomic Changes in Expression Induced by CF

Cells routinely respond to environmental stimuli, hormonal treatment and other perturbations by changing gene expression. Pathways that link such stimuli to gene expression are intensively studied. Thus, it is not a surprise that CF mutations which effectively eliminate CFTR function from cells, result in significantly reduced or increased expression of numerous genes [80]. Altered gene expression in response to loss of CFTR will cause 'CFTR-dependent' changes in function. Whether this constitutes true 'regulation' by CFTR is a semantic issue, but this mechanism could apply to some regulatory functions that have been attributed to CFTR.

Evidence of direct interaction between CFTR and ENaC or an indirect interaction due to electrochemical coupling is not presently sufficient to account for the increased Na^+

permeability directly measured in the apical membrane of CF airway epithelia. Increased expression of one or more ENaC subunits in CF airway epithelial cells was tested and eliminated as a basis for this observation [81]. Observations that amiloride-sensitive Na^+ absorption was not only increased in CF airways, but was abnormally stimulated by cAMP, point to a defect in ENaC regulation in the absence of CFTR [82]. Although this defect has yet to be identified with certainty, much has been learned about normal regulation of ENaC. ENaC is a complexly regulated channel that is essential for salt and water homeostasis carried out by the kidney, intestine and sweat ducts [83]. In such salt-conserving tissues ENaC is regulated by mineralocorticoids and vasopressin in response to changes in total body volume. In other tissues, such as the airways, the regulation of ENaC is tailored to control the volume of liquid on the epithelial surface. ENaC in airways is not regulated by mineralocorticoids, but by pathways responsive to local stimuli, such as P2Y2 receptors [84]. Nonetheless, all ENaC regulatory pathways converge on common mechanisms to control ENaC number, conductance or open probability [85]. One of the proximal pathways regulating ENaC is its ubiquitination by the ubiquitin ligase Nedd4–2 [86]. Recent data suggest this interaction is affected by serum-glucocorticoid-kinase 1 (sgk1) and protein kinase A, components of signaling pathways likely to be involved in sensing local conditions [87]. Because other mechanisms of CFTR regulation of ENaC have not been conclusively established, pathways important for ENaC regulation could be affected at the genomic level by null function of CFTR in CF cells. For example, sgk1 expression has been found by in situ hybridization to be increased in CF airways, providing a logical explanation for the abnormal Na^+ absorption measured at that site [88].

CFTR may affect bicarbonate secretion by regulating gene transcription. Activation of SLC26T exchangers by CFTR was observed in studies of CFPAC-1 cells (cultured pancreatic duct epithelial cells with physiological features similar to CF) where the activity of SCL26 DRA function was enhanced 2-fold by transfection with functional CFTR [89]. In these studies, Northern hybridization studies indicated the induction of downregulated-in-adenoma (DRA) in cells expressing functional CFTR. Gawenis et al. [90] studied CFTR-knockout mice and postulated that chronically affected activity of one transporter (i.e. CFTR) may alter expression of functionally linked transport process (i.e. NHE3).

Early observations in human and murine nasal epithelia, in vivo, revealed a larger Ca^{2+}-activated Cl^- conductance in CF than in normal subjects [91, 92]. By inference, these data support an inhibitory role of CFTR on the Ca^{2+}-sensitive Cl^- conductance in the apical membrane of airway epithelial cells. Elucidation of the molecular details of this observation has been hampered because the molecular identity for the Ca^{2+}-regulated Cl^- channel (CaCC) remains unknown, making direct testing of the effect of CFTR on CaCC expression impossible.

Conclusions

The co-expression of functional CFTR protein with ENaC, SLC26a, ORCC, ROMK and perhaps other proteins alters the function of these channels and transporters. Nonetheless, the available evidence does not identify a global regulatory function of CFTR that combines these observations. Specifically, none of the cited examples of CFTR regulation support a model in which CFTR and the regulated protein join together in a stable binary complex. A more promising model rests on PDZ domain proteins binding to CFTR and other proteins, to position CFTR close to regulatory complexes and other transporters or channels. In this model, physical interaction of the cytoplasmic domains of CFTR with other proteins could exert a 'regulatory' function. However, it must be pointed out that the molecular events that effect such regulation, assuming a model that allows for sustained association, are not even hypothetically known. There remains little doubt that CFTR influences the activity of other channels and transporters indirectly, but the molecular mechanisms responsible are still not known well. An obvious indirect mechanism, the transport by CFTR of soluble regulators that affect other functions in a paracrine/autocrine process, has been controversial. In particular, CFTR-mediated ATP release from cells has been hotly debated. Whereas CFTR appears by the most exacting analyses not to conduct ATP, some proponents of the role of CFTR in ATP export from cells now view this as yet another example of an unidentified protein (a putative ATP release protein) being regulated by CFTR. Substances that do permeate the CFTR anion channel include glutathione [93]. Loss of glutathione permeability may influence redox balance on the surface of epithelia, and secondarily affect multiple events important in CF pathogenesis [71]. Confusion exists regarding the electrochemical coupling between CFTR Cl^- conductance and other processes that operate via electrochemical gradients. As an example, an active CFTR Cl^- conductance will contribute to the resting potential of the cell and will conduct Cl^- so as to supply substrate to anion exchangers. This coupling could be considered regulatory in the sense that

　　Coakley/Stutts

CFTR Cl⁻ conductance is itself regulated. Finally, CFTR null function due to defective processing undoubtedly changes the expression of a significant number of genes [80, 94]. This may be the main mechanism by which CFTR appears to influence many other cellular functions, but it is clearly not regulation in the sense the discoverers of the CF gene had in mind when they added 'regulator' to 'cystic fibrosis transmembrane conductance'.

References

1 Riordan JR, Kerem B, Alon N, Rozmahel R, Grzelczak Z, Zielenski J, Lok S, Plavsic N, Chou JL et al: Identification of the cystic fibrosis gene: Cloning and characterization of complementary DNA. Science 1989;245: 1066–1073.

2 Davidson GP, Kirubakaran CP, Ratcliffe G, Cooper DM, Robb TA: Abnormal pancreatic electrolyte secretion in cystic fibrosis: Reliability as a diagnostic marker. Acta Paediatr Scand 1986;75:145–150.

3 Bijman J, Quinton PM: Influence of abnormal Cl⁻ impermeability on sweating in cystic fibrosis. Am J Physiol 1984;247(1 pt 1): C3–C9.

4 Knowles MR, Stutts MJ, Yankaskas JR, Gatzy JT, Boucher RC Jr: Abnormal respiratory epithelial ion transport in cystic fibrosis. Clin Chest Med 1986;7:285–297.

5 Knowles M, Gatzy J, Boucher R: Relative ion permeability of normal and cystic fibrosis nasal epithelium. J Clin Invest 1983;71: 1410–1417.

6 Chang XB, Hou YX, Jensen TJ, Riordan JR: Mapping of cystic fibrosis transmembrane conductance regulator membrane topology by glycosylation site insertion. J Biol Chem 1994;269:18572–18575.

7 Hanrahan JW, Tabcharani JA, Becq F, Mathews CJ, Augustinas O, Jensen TJ, Chang XB, Riordan JR: Function and dysfunction of the CFTR chloride channel. Soc Gen Physiol Ser1995;50:125–137.

8 Aleksandrov AA, Riordan JR: Regulation of CFTR ion channel gating by MgATP. FEBS Lett 1998;431:97–101.

9 Sheppard DN, Welsh MJ: Structure and function of the CFTR chloride channel. Physiol Rev 1999;79:23–45.

10 Arniges M, Vazquez E, Fernandez-Fernandez JM, Valverde MA: Swelling-activated Ca²⁺ entry via TRPV4 channel is defective in cystic fibrosis airway epithelia. J Biol Chem 2004; 279:54062–54068.

11 Cheung KH, Leung CT, Leung GP, Wong PY: Synergistic effects of cystic fibrosis transmembrane conductance regulator and aquaporin-9 in the rat epididymis. Biol Reprod 2003;68:1505–1510.

12 Wei L, Vankeerberghen A, Cuppens H, Cassiman JJ, Droogmans G, Nilius B: The C-terminal part of the R-domain, but not the PDZ binding motif, of CFTR is involved in interaction with Ca²⁺-activated Cl⁻ channels. Pflügers Arch 2001;442:280–285.

13 Wei L, Vankeerberghen A, Cuppens H, Eggermont J, Cassiman JJ, Droogmans G, Nilius B: Interaction between calcium-activated chloride channels and the cystic fibrosis transmembrane conductance regulator. Pflügers Arch 1999:438:635–641.

14 Tarran R, Loewen ME, Paradiso AM, Olsen JC, Gray MA, Argent BE, Boucher RC, Gabriel SE: Regulation of murine airway surface liquid volume by CFTR and Ca²⁺-activated Cl⁻ conductances. J Gen Physiol 2002; 120:407–418.

15 Stecenko AA, King G, Torii K, Breyer RM, Dworski R, Blackwell TS, Christman JW, Brigham KL: Dysregulated cytokine production in human cystic fibrosis bronchial epithelial cells. Inflammation 2001;25:145–155.

16 Lagrutta AA, Bond CT, Xia XM, Pessia M, Tucker S, Adelman JP: Inward rectifier potassium channels: Cloning, expression and structure-function studies. Jpn Heart J 1996;37: 651–660.

17 Liu X, Songu-Mize E: Effect of Na⁺ on Na+,K+-ATPase alpha-subunit expression and Na⁺-pump activity in aortic smooth muscle cells. Eur J Pharmacol 1998;351:113–119.

18 Short DB, Trotter KW, Reczek D, Kreda SM, Bretscher A, Boucher RC, Stutts MJ, Milgram SL: An apical PDZ protein anchors the cystic fibrosis transmembrane conductance regulator to the cytoskeleton. J Biol Chem 1998;273: 19797–19801.

19 Bryan J, Vila-Carriles WH, Zhao G, Babenko AP, Aguilar-Bryan L: Toward linking structure with function in ATP-sensitive K⁺ channels. Diabetes 2004;53(suppl 3):S104–S112.

20 Shyng SL, Nichols CG: Octameric stoichiometry of the KATP channel complex. J Gen Physiol 1997;110:655–664.

21 McNicholas CM, Nason MW Jr, Guggino WB, Schwiebert EM, Hebert SC, Giebisch G, Egan ME: A functional CFTR-NBF1 is required for ROMK2-CFTR interaction. Am J Physiol 1997;273(5 pt 2):F843–F848.

22 Yoo D, Flagg TP, Olsen O, Raghuram V, Foskett JK, Welling PA: Assembly and trafficking of a multiprotein ROMK (Kir 1.1) channel complex by PDZ interactions. J Biol Chem 2004;279:6863–6873.

23 Yankaskas JR, Stutts MJ, Cotton CU, Knowles MR, Gatzy JT, Boucher RC: Cystic fibrosis airway epithelial cells in primary culture: Disease-specific ion transport abnormalities. Prog Clin Biol Res 1987;254:139–149.

24 Li M, McCann JD, Liedtke CM, Nairn AC, Greengard P, Welsh MJ: Cyclic AMP-dependent protein kinase opens chloride channels in normal but not cystic fibrosis airway epithelium. Nature 1988:331:358–360.

25 Gabriel SE, Clarke LL, Boucher RC, Stutts MJ: CFTR and outward rectifying chloride channels are distinct proteins with a regulatory relationship. Nature 1993;363:263–268.

26 Julien M, Verrier B, Cerutti M, Chappe V, Gola M, Devauchelle G, Becq F: Cystic fibrosis transmembrane conductance regulator (CFTR) confers glibenclamide sensitivity to outwardly rectifying chloride channel (ORCC) in Hi-5 insect cells. J Membr Biol 1999;168:229–239.

27 Ogura T, Furukawa T, Toyozaki T, Yamada K, Zheng YJ, Katayama Y, Nakaya H, Inagaki N: ClC-3B, a novel ClC-3 splicing variant that interacts with EBP50 and facilitates expression of CFTR-regulated ORCC. FASEB J 2002;16:863–865.

28 Gentzsch M, Cui L, Mengos A, Chang XB, Chen JH, Riordan JR: The PDZ-binding chloride channel ClC-3B localizes to the Golgi and associates with cystic fibrosis transmembrane conductance regulator-interacting PDZ proteins. J Biol Chem 2003;278:6440–6449.

29 Boucher RC, Stutts MJ, Knowles MR, Cantley L, Gatzy JT: Na⁺ transport in cystic fibrosis respiratory epithelia: Abnormal basal rate and response to adenylate cyclase activation. J Clin Invest 1986;78:1245–1252.

30 Reddy M, Quinton P: Functional interaction of CFTR and ENaC in sweat glands. Pflügers Arch 2003;445:499–503.

31 Matsui H, Grubb BR, Tarran R, Randell SH, Gatzy JT, Davis CW, Boucher RC: Evidence for periciliary liquid layer depletion, not abnormal ion composition, in the pathogenesis of cystic fibrosis airways disease. Cell 1998;95:1005–1015.

32 Mall M, Grubb BR, Harkema JR, O'Neal WK, Boucher RC: Increased airway epithelial Na⁺ absorption produces cystic fibrosis-like lung disease in mice. Nat Med 2004;10: 487–493.

33 Canessa CM, Schild L, Buell G, Thorens B, Gautschi I, Horisberger JD, Rossier BC: Amiloride-sensitive epithelial Na⁺ channel is made of three homologous subunits. Nature 1994;367:463–467.

34 Kunzelmann K, Kiser GL, Schreiber R, Riordan JR: Inhibition of epithelial Na⁺ currents by intracellular domains of the cystic fibrosis transmembrane conductance regulator. FEBS Lett 1997;400:341–344.

35 Ji HL, Chalfant ML, Jovov B, Lockhart JP, Parker SB, Fuller CM, Stanton BA, Benos DJ: The cytosolic termini of the beta- and gamma-ENaC subunits are involved in the functional interactions between cystic fibrosis transmembrane conductance regulator and epithelial sodium channel. J Biol Chem 2000;275: 27947–27956.

36 Nagel G, Szellas T, Riordan JR, Friedrich T, Hartung K: Non-specific activation of the epithelial sodium channel by the CFTR chloride channel. EMBO Rep 2001;2:249–254.

37 Sheng M, Sala C: PDZ domains and the organization of supramolecular complexes. Annu Rev Neurosci 2001;24:1–29.

38 Raghuram V, Mak DD, Foskett JK: Regulation of cystic fibrosis transmembrane conductance regulator single-channel gating by bivalent PDZ-domain-mediated interaction. Proc Natl Acad Sci USA 2001;98:1300–1305.

39 Hung AY, Sheng M: PDZ domains: Structural modules for protein complex assembly. J Biol Chem 2002;277:5699–5702.

40 Sun F, Hug MJ, Lewarchik CM, Yun CH, Bradbury NA, Frizzell RA: E3KARP mediates the association of ezrin and protein kinase a with the cystic fibrosis transmembrane conductance regulator in airway cells. J Biol Chem 2000;275:29539–29546.

41 Wang S, Yue H, Derin RB, Guggino WB, Li M: Accessory protein facilitated CFTR-CFTR interaction, a molecular mechanism to potentiate the chloride channel activity. Cell 2000;103:169–179.

42 Ahn W, Kim KH, Lee JA, Kim JY, Choi JY, Moe OW, Milgram SL, Muallem S, Lee MG: Regulatory interaction between the cystic fibrosis transmembrane conductance regulator and HCO_3^- salvage mechanisms in model systems and the mouse pancreatic duct. J Biol Chem 2001;276:17236–17243.

43 Coakley RD, Grubb BR, Paradiso AM, Gatzy JT, Johnson LG, Kreda SM, O'Neal WK, Boucher RC: Abnormal surface liquid pH regulation by cultured cystic fibrosis bronchial epithelium. Proc Natl Acad Sci USA 2003;100:16083–16088.

44 Smith JJ, Welsh MJ: cAMP stimulates bicarbonate secretion across normal, but not cystic fibrosis airway epithelia. J Clin Invest 1992;8:1148–1153.

45 Johansen PG, Anderson CM, Hadorn B: Cystic fibrosis of the pancreas. A generalised disturbance of water and electrolyte movement in exocrine tissues. Lancet 1968;1:455–456.

46 Kopelman H, Corey M, Gaskin K, Durie, Weizman Z, Forstner G: Impaired chloride secretion, as well as bicarbonate secretion, underlies the fluid secretory defect in the cystic fibrosis pancreas. Gastroenterology 1988;95:349–355.

47 Hogan DL, Crombie DL, Isenberg JI, Svendsen P, Schaffalitzky de Muckadell OB, Ainsworth MA: CFTR mediates cAMP- and Ca^{2+}-activated duodenal epithelial HCO_3^- secretion. Am J Physiol Gastrointest Liver Physiol 1997;272:G872–G878.

48 Lee MG, Choi JY, Luo X, Strickland E, Thomas PJ, Muallem S: Cystic fibrosis transmembrane conductance regulator regulates luminal Cl^-/HCO_3^- exchange in mouse submandibular and pancreatic ducts. J Biol Chem 1999;274:14670–14677.

49 Wang XF, Zhou CX, Shi QX, Yuan YY, Yu MK, Ajonuma LC, Ho LS, Lo PS, Tsang LL, Liu Y, Lam SY, Chan LN, Zhao WC, Chung YW, Chan HC: Involvement of CFTR in uterine bicarbonate secretion and the fertilizing capacity of sperm. Nat Cell Biol 2003;5:902–906.

50 Ko SB, Shcheynikov N, Choi JY, Luo X, Ishibashi K, Thomas PJ, Kim JY, Kim KH, Lee MG, Naruse S, Muallem S: A molecular mechanism for aberrant CFTR-dependent HCO_3^+ transport in cystic fibrosis. EMBO J 2002;21:5662–5672.

51 Linsdell P, Tabcharani JA, Rommens JM, Hou YX, Chang XB, Tsui LC, Riordan JR, Hanrahan JW: Permeability of wild-type and mutant cystic fibrosis transmembrane conductance regulator chloride channels to polyatomic anions. J Gen Physiol 1997;110:355–364.

52 Mount DB, Romero MF: The SLC26 gene family of multifunctional anion exchangers. Pflügers Arch 2004;447:710–721.

53 Everett LA, Glaser B, Beck JC, Idol JR, Buchs A, Heyman M, Adawi F, Hazani E, Nassir E, Baxevanis AD, Sheffield VC, Green ED: Pendred syndrome is caused by mutations in a putative sulphate transporter gene (PDS). Nat Genet 1997;17:411–422.

54 Lohi H, Kujala M, Makela S, Lehtonen E, Kestila M, Saarialho-Kere U, Markovich D, Kere J: Functional characterization of three novel tissue-specific anion exchangers SLC26A7, -A8, and -A9. J Biol Chem 2002;277:14246–14254.

55 Ko SB, Zeng W, Dorwart MR, Luo X, Kim KH, Millen L, Goto H, Naruse S, Soyombo A, Thomas PJ, Muallem S: Gating of CFTR by the STAS domain of SLC26 transporters. Nat Cell Biol, 2004;6:343–350.

56 Donowitz M, Welsh MJ: Ca^{2+} and cyclic AMP in regulation of intestinal Na, K, and Cl transport. Annu Rev Physiol 1986;48:135–150.

57 Ko SB, Zeng W, Dorwart MR, Luo X, Kim KH, Millen L, Goto H, Naruse S, Soyombo A, Thomas PJ, Muallem S: cAMP-mediated inhibition of the epithelial brush border Na^+/H^+ exchanger, NHE3, requires an associated regulatory protein. Proc Natl Acad Sci USA 1997;94:3010–3015.

58 Weinman EJ, Steplock D, Tate K, Hall RA, Spurney RF, Shenolikar S: Structure-function of recombinant Na/H exchanger regulatory factor (NHE-RF). J Clin Invest 1998;101:2199–2206.

59 Clarke LL, Harline MC: CFTR is required for cAMP inhibition of intestinal Na^+ absorption in a cystic fibrosis mouse model. Am J Physiol Gastrointest Liver Physiol 1996;270:G259–G267.

60 Weinman EJ, Minkoff C, Shenolikar S: Signal complex regulation of renal transport proteins: NHERF and regulation of NHE3 by PKA. Am J Physiol Renal Physiol 2000;279:F393–F399.

61 Klein I, Sarkadi B, Varadi A: An inventory of the human ABC proteins. Biochim Biophys Acta 1999;1461:237–262.

62 Homolya L, Varadi A, Sarkadi B: Multidrug resistance-associated proteins: Export pumps for conjugates with glutathione, glucuronate or sulfate. Biofactors 2003;17:103–114.

63 Cantiello H: Electrodiffusional ATP movement through CFTR and other ABC transporters. Pflügers Arch 2001;443:S22–S27.

64 Schwiebert EM, Benos DJ, Fuller CM: Cystic fibrosis: A multiple exocrinopathy caused by dysfunctions in a multifunctional transport protein. Am J Med 1998;104:576–590.

65 Prat AG, Reisin IL, Ausiello DA, Cantiello HF: Cellular ATP release by the cystic fibrosis transmembrane conductance regulator. Am J Physiol Cell Physiol 1996;270:C538–C545.

66 Braunstein GM, Roman RM, Clancy JP, Kudlow BA, Taylor AL, Shylonsky VG, Jovov B, Peter K, Jilling T, Ismailov II, Benos DJ, Schwiebert LM, Fitz JG, Schwiebert EM: Cystic fibrosis transmembrane conductance regulator facilitates ATP release by stimulating a separate ATP release channel for autocrine control of cell volume regulation. J Biol Chem 2001;2769:6621–6630.

67 Grygorczyk R, Hanrahan JW: CFTR-independent ATP release from epithelial cells triggered by mechanical stimuli. Am J Physiol 1997;272(3 pt 1):C1058–C1066.

68 Watt WC, Lazarowski ER, Boucher RC: Cystic fibrosis transmembrane regulator-independent release of ATP. Its implications for the regulation of P2Y2 receptors in airway epithelia. J Biol Chem 1998;273:14053–14058.

69 Donaldson SH, Lazarowski ER, Picher M, Knowles MR, Stutts MJ, Boucher RC: Basal nucleotide levels, release, and metabolism in normal and cystic fibrosis airways. Mol Med 2000;6:969–982.

70 Kogan I, Ramjeesingh M, Li C, Kidd JF, Wang Y, Leslie EM, Cole SP, Bear CE: CFTR directly mediates nucleotide-regulated glutathione flux. EMBO J 2003;22:1981–1989.

71 Hudson VM: Rethinking cystic fibrosis pathology: The critical role of abnormal reduced glutathione (GSH) transport caused by CFTR mutation. Free Radic Biol Med, 2001;30:1440–1461.

72 Willumsen NJ, Boucher RC: Transcellular sodium transport in cultured cystic fibrosis human nasal epithelium. Am J Physiol 1991;261(2 pt 1):C332–C341.

73 Willumsen NJ, Boucher RC: Sodium transport and intracellular sodium activity in cultured human nasal epithelium. Am J Physiol 1991;261(2 pt 1):C319–C331.

74 Clarke LL, Harline MC: Dual role of CFTR in cAMP-stimulated HCO_3^- secretion across murine duodenum. Am J Physiol Gastrointest Liver Physiol 1998; 274:G718–G726.

75 Spiegel S, Phillipper M, Rossmann H, Riederer B, Gregor M, Seidler U: Independence of apical Cl^-/HCO_3^- exchange and anion conductance in duodenal HCO_3^- secretion. Am J Physiol Gastrointest Liver Physiol 2003;285:G887–G897.

76 Paradiso AM, Coakley RD, Boucher RC: Polarized distribution of HCO_3^- transport in human normal and cystic fibrosis nasal epithelia. J Physiol 2003;548(pt 1):203–218.

77 Barasch J, Kiss B, Prince A, Saiman L, Gruenert D, al-Awqati Q: Defective acidification of intracellular organelles in cystic fibrosis. Nature 1991;352:70–73.

78 Seksek O, Biwersi J, Verkman AS: Evidence against defective trans-Golgi acidification in cystic fibrosis. J Biol Chem 1996;271:15542–15548.

79 Poschet J, Perkett E, Deretic V: Hyperacidification in cystic fibrosis: links with lung disease and new prospects for treatment. Trends Mol Med 2002;8:512–519.

80 Xu Y, Clark JC, Aronow BJ, Dey CR, Liu C, Wooldridge JL, Whitsett JA: Transcriptional

Coakley/Stutts

adaptation to cystic fibrosis transmembrane conductance regulator deficiency. J Biol Chem 2003;278:7674–7682.

81 Burch LH, Talbot CR, Knowles MR, Canessa CM, Rossier BC, Boucher RC: Relative expression of the human epithelial Na$^+$ channel subunits in normal and cystic fibrosis airways. Am J Physiol 1995;269(2 pt 1): C511–C518.

82 Boucher RC, Stutts MJ, Knowles MR, Cantley L, Gatzy JT: Na$^+$ transport in cystic fibrosis respiratory epithelia. Abnormal basal rate and response to adenylate cyclase activation. J Clin Invest 1986;78:1245–1252.

83 Schild L: The epithelial sodium channel: From molecule to disease. Rev Physiol Biochem Pharmacol 2004;151:93–107.

84 Kunzelmann K, Bachhuber T, Regeer R, Markovich D, Sun J, Schreiber R: Purinergic inhibition of the epithelial Na$^+$ transport via hydrolysis of PIP2. FASEB J 2005;19: 142–143.

85 Rossier BC: Hormonal regulation of the epithelial sodium channel ENaC: N or P_o? J Gen Physiol 2002;120:67–70.

86 Verrey F, Loffing J, Zecevic M, Heitzmann D, Staub O: SGK1: aldosterone-induced relay of Na+ transport regulation in distal kidney nephron cells. Cell Physiol Biochem 2003;13: 21–28.

87 Thomas CP, Campbell JR, Wright PJ, Husted RF: cAMP stimulated Na$^+$ transport in H441 distal lung epithelial cells: Role of PKA, phosphatidylinositol 3-kinase, and sgk1. Am J Physiol Lung Cell Mol Physiol 2004;287: L843–L851.

88 Wagner CA, Ott M, Klingel K, Beck S, Melzig J, Friedrich B, Wild KN, Broer S, Moschen I, Albers A, Waldegger S, Tummler B, Egan ME, Geibel JP, Kandolf R, Lang F: Effects of the serine/threonine kinase SGK1 on the epithelial Na$^+$ Channel (ENaC) and CFTR: Implications for cystic fibrosis. Cell Physiol Biochem 2001;11:209–218.

89 Greeley T, Shumaker H, Wang Z, Schweinfest CW, Soleimani M: Downregulated in adenoma and putative anion transporter are regulated by CFTR in cultured pancreatic duct cells. Am J Physiol Gastrointest Liver Physiol 2001;281:G1301–G1308.

90 Gawenis LR, Hut H, Bot AG, Shull GE, de Jonge HR, Stien X, Miller ML, Clarke LL: Electroneutral sodium absorption and electrogenic anion secretion across murine small intestine are regulated in parallel. Am J Physiol Gastrointest Liver Physiol 2004;287: G1140–G1149.

91 Knowles MR, Clarke LL, Boucher RC: Activation by extracellular nucleotides of chloride secretion in the airway epithelia of patients with cystic fibrosis. N Engl J Med 1991;325:533–538.

92 Gabriel SE, Makhlina M, Martsen E, Thomas EJ, Lethem MI, Boucher RC: Permeabilization via the P2X7 purinoreceptor reveals the presence of a Ca^{2+}-activated Cl^- conductance in the apical membrane of murine tracheal epithelial cells. J Biol Chem 2000;275: 35028–35033.

93 Linsdell P, Hanrahan JW: Glutathione permeability of CFTR. Am J Physiol 1998;275(1 pt 1):C323–C326.

94 Virella-Lowell I, Herlihy JD, Liu B, Lopez C, Cruz P, Muller C, Baker HV, Flotte TR: Effects of CFTR, interleukin-10, and *Pseudomonas aeruginosa* on gene expression profiles in a CF bronchial epithelial cell line. Mol Ther 2004;10:562–573.

M. Jackson Stutts
6023 Thurston Bowles Building
University of North Carolina
Chapel Hill, NC 27599 (USA)
Tel. +1 919 966 7056, Fax +1 919 966 5178
E-Mail hoopster@med.unc.edu

Bush A, Alton EWFW, Davies JC, Griesenbach U, Jaffe A (eds): Cystic Fibrosis in the 21st Century.
Prog Respir Res. Basel, Karger, 2006, vol 34, pp 54–60

Function of CFTR Protein: Developmental Role

Deborah Gill[a]   Janet E. Larson[b]

[a]Gene Medicine Group, Nuffield Department of Clinical Laboratory Sciences, University of Oxford,
John Radcliffe Hospital, Oxford, UK;
[b]Neonatology, Stony Brook University Health Sciences Center, Stony Brook, N.Y., USA

Abstract

In the lung, CFTR is largely expressed in undifferentiated multipotent stem cells of the first- and second-trimester airway epithelium. As differentiation proceeds, the expression of CFTR dissipates and the adult lung expresses a small fraction of what was expressed in the fetus. There is growing evidence that infants with CF have lungs that are functionally and structurally abnormal prior to the appearance of clinical infection. This chapter examines the expression and role of CFTR in developmental cascades in the lung.

Introduction

Cystic fibrosis (CF) is a progressive disease in which the lung is traditionally thought to be normal at birth and injured by recurrent infection. However, there is increasing evidence that the lung may be functionally and structurally abnormal prior to the appearance of clinical infection. CFTR is highly expressed in fetal tissues and this chapter will examine the role of CFTR in regulatory cascades during development.

CFTR Expression during Development

CFTR is a member of the ABC (ATP-binding cassette) family of transporters and the regulation of stem cell biology by these proteins is an important new field of study [1].

Several other members of the ABC family also show developmentally regulated expression patterns [2, 3]. In most organs investigated, the pattern of CFTR expression is similar in tissues from the adult and the developing human fetus. The exception is the lung, where comparatively high fetal CFTR expression eventually disappears at birth [4]. Quantification of CFTR expression in fetal sheep lung demonstrated that CFTR mRNA expression correlated with specific developmental stages. High levels were observed at the start of the second trimester which later declined throughout gestation until birth [5]. Analysis of CFTR mRNA expression in human fetal, newborn and infant tissues [6, 7] showed significant levels of CFTR mRNA in the primordial epithelium of the developing lung throughout gestation (table 1). As lung development progressed, CFTR expression decreased in cells of the future alveolar spaces and was gradually restricted to the epithelium of the small airways. After birth the observed pattern of expression changed dramatically with reduced levels of CFTR mRNA throughout the airways and for the first time, expression in the submucosal glands was observed. Interestingly, CFTR expression in premature newborns showed a CFTR expression pattern that was intermediate between third-trimester fetal and newborn infants. Thus, initially CFTR is primarily expressed in the undifferentiated multipotent stem cells of the first- and second-trimester pulmonary epithelium and as differentiation proceeds, the expression of CFTR dissipates. Together these results suggest a role for CFTR during lung development and imply that CF babies may be born with functionally immature epithelia.

Table 1. Summary of the pattern of expression of CFTR in the developing human lung

Epithelium	T1 (n = 4)	T2 (n = 5)	T3 (n = 4)	P (n = 2)	N (n = 5)
Distal	+++	+++	+	+	
Small airways	+++	+++	+++	++	++
Trachea and large bronchi	+++ d	+++ p	+++ i		
Submucosal glands					++

Expression of CFTR mRNA was detected by in situ hybridization in human respiratory tissues. T1 = First trimester; T2 = second trimester; T3 = third trimester; P = premature newborns; N = newborns and infants; +++ = 15- to 20-fold over background; ++ = 8- to 12-fold over background; + = 3- to 6-fold over background; d = diffuse; i = individual cells; p = patchy [reproduced with permission from 7].

Regulation of CFTR during Lung Development

Lung development involves a complex sequence of cell proliferation, branching morphogenesis, alveolarization, and cell lineage differentiation. Although there is considerable interspecies variation in timing, the pattern of lung development is similar in all mammals. It is during this complex period of organogenesis that CFTR is expressed predominantly in undifferentiated epithelial cells that serve as progenitors for the airways and the subsequent alveolar epithelium. Thus, expression of CFTR, or the lack thereof, has the potential to affect all of these cell types in the mature organ.

During development the CFTR-expressing epithelium in the lung is exposed to, and interacts with, the amniotic fluid and its constituents, including epidermal growth factor, insulin-like growth factor-1, fibroblast growth factor, transforming growth factor-α, hepatocyte growth factor and α-fetoprotein [8–10], as well as a host of cytokines that increase during parturition such as interleukin-1β and tumour necrosis factor-α [11]. The effects of these cytokines on the CFTR-expressing epithelium are largely unknown.

The regulation of CFTR during lung development is relatively unstudied, although two factors were shown to affect CFTR expression during this critical time. During early lung development, the expression of CFTR in the proximal developing airways is downregulated by retinoic acid thought to be responsible for restricting lung branch-ing morphogenesis [12]. In addition to its regulation during airway development, CFTR has been shown to be upregulated by Foxa2, one of the forkhead transcription factors thought to play a role in alveolarization [13, 14]. Therefore, CFTR expression appears to be controlled by factors involved in both early airway development and in the later differentiation of the gas-exchanging part of the lung.

Paradoxes in CF Disease

As a pleiotropic disease, CF is a group of seemingly unrelated phenotypes controlled by a single gene. The precise role of CFTR in the airway, and the mechanism for its direct participation in the disease pathology, remain unclear. CFTR controls cellular functions such as mediation of vesicular trafficking, modulation of intracellular pH, regulation of glycosylation, and increase of extracellular ATP. In its role as an epithelial chloride channel, CFTR affects the disease phenotype with numerous contradictions. First, in vitro protein interactions of CFTR do not correlate with those observed in vivo. The epithelial sodium channel (ENaC) is downregulated by CFTR in cell culture [15, 16], but sodium absorption is variable in CF patient organs despite the expression of ENaC in all of these tissues [17, 18]. The calcium-regulated chloride channel is downregulated by CFTR in tissue culture [19, 20] but CF patients show both intact airway and defective intestinal activity of this same channel [21]. The outwardly rectifying chloride channel, which is upregulated by CFTR in vitro and has decreased activity in CF tissues [18, 21, 22], is not co-expressed with CFTR [23, 24].

Second, the expression pattern of CFTR mRNA often shows little correlation to location of disease pathology. For example, the CF airway goblet cell secretes defective mucus, but the normal goblet cell has undetectable levels of CFTR [25]. Although there is moderate expression of CFTR in the adult intestine and pulmonary submucosal glands, on average only two copies of CFTR mRNA are present in each lung epithelial cell [25].

Third, patients can manifest the CF phenotype without aberrant CFTR gene expression. Groman et al. [26] described 74 patients with CF pathology of which 45 had either only one or no detectable mutation in the CFTR gene. Half of these patients had normal chloride sweat tests yet had CF pathology in one or more organs. Patients have also been found with CFTR mutations associated with elevated sweat chloride concentrations in the absence of the CF phenotype [27]. Many of the phenotypic changes in CF probably occur via indirect actions of CFTR. Several paradoxes seen in CF that cannot be explained in terms of direct protein interactions are better explained as the downstream results of regulatory pathways disrupted by lack of CFTR expression. The question remains whether it is possible that the downstream effects are observed in daughter cells many years after their progenitors expressed CFTR.

CFTR Gene Transfer in Utero

When in utero gene transfer technology was developed to circumvent the problems of inflammation seen with adenoviral-mediated gene therapy [28] it was discovered that the in utero transfer of CFTR to normal rat fetuses resulted in phenotypic changes in the lungs of neonates. At the time of gene transfer, the targeted epithelial cells were undifferentiated somatic stem cells and administration of the CFTR to this epithelium using an adenovirus vector resulted in persistent phenotypic changes in cells, although the transgene was expressed only transiently. These data provided the first insight that transient overexpression of CFTR in the fetus could alter the differentiation of lung epithelial stem cells. The permanent functional changes in the in utero CFTR-treated rats included an enhanced resistance to pulmonary bacterial infection 3 months after birth [29]. Gene transfer of CFTR into normal fetal mice increased pulmonary epithelial cell proliferation, and lung weight/body weight ratios when compared to those treated with control vector. DNA content also confirmed that the cell number was increased [30]. Thus, one unexpected outcome from CFTR overexpression in utero was that the proliferating fetal lungs had increased numbers of differentiated cells [30].

During the late stages of lung morphogenesis, the cuboidal pre-type II epithelial cells of the peripheral lung differentiate into cuboidal type II and squamous type I cells. Perinatal maturation of type II cells is associated with marked ultrastructural and biochemical changes that include reduced glycogen content, enhanced surfactant protein and lipid synthesis, and increased numbers of lamellar bodies. A mature type II cell contains an increased volume proportion of lamellar bodies and mitochondria and a decreased volume proportion of glycogen [31, 32]. In the control (treated with an adenovirus encoding the B-galactosidase gene under the control of a CMV promoter) lungs, 22% of the cells were undifferentiated with large pools of glycogen in their cytoplasm. In the CFTR-treated animals, there were no glycogen-containing cells in the parenchyma (p < 0.001) [30]. Moreover, morphometric analysis revealed that the volume proportion of lamellar bodies

increased from 5% in the control mice to 9.5% in the in utero CFTR-treated animals (p = 0.008). Type II cells are thought to be the precursors of the terminally differentiated type I cells of the epithelium. Nuclear counts showed an increase (from 10.1 to 20.1%) in the proportion of type I cells in the in utero CFTR-treated animals, while the proportion of type II cells remained the same within the two groups. These data suggested that CFTR played a role in the proliferation and differentiation of the lung epithelium [30].

Overexpression of CFTR in fetal primates also resulted in accelerated differentiation of the lung. Treatment with recombinant adenovirus expressing CFTR at 110 days' gestation (equivalent to 10–15 weeks' gestation in human lung development) and examination at 140 days showed increased alveolarization and thinning of alveolar walls characteristic of lungs from a more mature fetus [33] (fig. 1). In control treated fetuses, alveolar type II cells exhibited the level of differentiation expected for their developmental stage, with large euchromatic nuclei surrounded by a cytoplasm filled with glycogen and only a few mitochondria (fig. 1c). In contrast, CFTR-treated animals exhibited a more mature pulmonary epithelium, with highly differentiated alveolar type II cells identifiable by their large multilamellar bodies and secreted surfactant (fig. 1d) [33]. Thus, transient overexpression of CFTR in the developing lung leads to significant changes in the differentiation of cell types in the lungs.

Recently, a transient in utero knockout technology was used to address the developmental role of CFTR in the rat lung. Rat fetuses transiently treated with antisense *cftr* in utero developed pathology that replicated aspects of the human CF phenotype. Lung fibrosis, chronic inflammation, reactive airway disease, and expression of CF antigen (MRP8/14), a marker for CF in human patients, was documented [34].

Early Lung Changes in the Human CF Infant

While it is clear that CF is progressive, there is increasing evidence that the lung is structurally and functionally abnormal prior to the appearance of clinical infection. Studies have shown changes in the newborn CF lung, including early pulmonary function differences, and inflammation and infection in presymptomatic infants.

Changes in Pulmonary Function Tests
Pulmonary function testing in CF infants has revealed early functional changes in clinically asymptomatic infants.

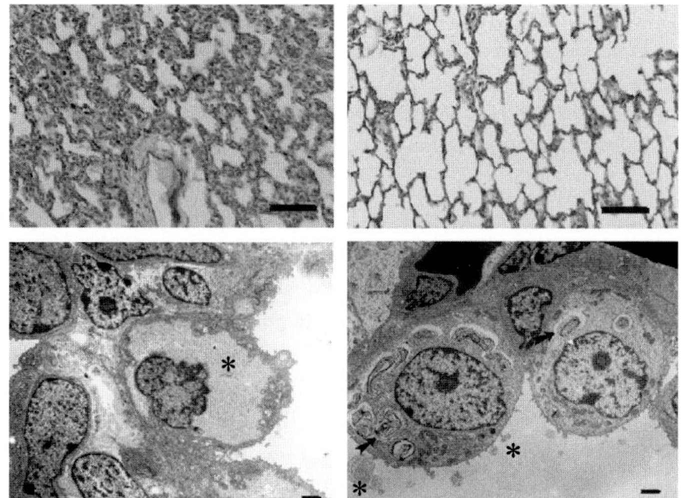

Fig. 1. Altered primate fetal lung structure following cftr gene transfer in utero. Fetal primate lungs were treated in utero at 110 days of gestation with recombinant adenovirus expressing either luciferase (**a, c**) or luciferase and cftr (**b, d**), and examined at 140 days of gestation by light (**a, b**; 100-nm reference bar) or electron (**c, d**; 1-μm reference bar) microscopy. In control fetuses, alveolar type II cells exhibited differentiation expected for their developmental stage (**c**), with large euchromatic nuclei surrounded by a cytoplasm filled with glycogen (indicated by *) and only a few mitochondria. In contrast, *cftr*-treated animals exhibited highly differentiated alveolar type II cells identifiable by large multilamellar bodies (indicated by arrows) and secreted surfactant (indicated by *; **d**) [reprinted from 33, with permission from Elsevier].

Evidence of hyperinflation is present in asymptomatic infants with CF [35, 36] and is thought to be one of the earliest manifestations [37]. In addition, asymptomatic CF infants demonstrate a decrease in airflow as measured by V_{maxFRC} or forced expiratory flow at functional residual capacity, a measure of peripheral airway function [36, 38–40]. Another study comparing healthy CF infants to controls demonstrated differences in airway function based on the measure of $FEF_{50\%}$, thought to reflect early airway obstruction in these infants [41]. Recent studies using the raised-volume rapid thoracoabdominal compression technique have reported diminished airway function in infants soon after diagnosis with CF [42]. Many of these infants had no history of prior clinical infection [42, 43], although this was not excluded by bronchoalveolar lavage.

Airway reactivity has also been assessed in CF infants, both by response to a bronchodilator and to bronchial challenge [36]. Newly diagnosed asymptomatic infants have decreased total airway resistance following treatment with a bronchodilator and chest physiotherapy [44]. CF infants

also had an increase in bronchial hyperresponsiveness as measured by V_{maxFRC} [39, 45]. In one study, infants homozygous for the ΔF508 deletion, but not other infants, responded to albuterol with a decrease in respiratory system resistance, suggesting that infants with asymptomatic CF homozygous for the ΔF508 deletion have early evidence of airway obstruction [46]. These studies of airway function suggest that there are physiological abnormalities in airway function in asymptomatic infants with CF [36].

Early Infection

The extensive genetic screening programs have identified early infection in asymptomatic CF infants. Forty percent of CF infants have lower respiratory infection within the first 6 months of age [47]. More interesting, and more contentious, is the observation that airway inflammation is present in infants with CF at 4 weeks of age, before they have clinically apparent disease [48]. CF infants have been shown to have IL-8 concentrations and neutrophil infiltrate that is independent of the presence of CF pathogens [48–50]. Many have theorized that the CF lung is pro-inflammatory in the absence of CFTR. Lung grafts from human fetal CF airways grafted into SCID mice develop progressive intraluminal inflammation that leads to the destruction of lung parenchyma in the absence of infection [51, 52].

Early Airway Changes

Recently, high-resolution computed tomography imaging has demonstrated that infants with CF have more dilated airways with thicker walls [53]. This study, performed in minimally symptomatic infants, demonstrated structural changes in both smaller and larger airways that worsened with age. These findings suggest that these early structural airway changes lead to bronchiectasis and that chronic infection may be a secondary contributor to airway remodelling.

The Disrupted Development of the Lung

The in utero CFTR experiments suggested that the lack of CFTR during the fetal period could disrupt normal lung development. Recently, the use of sophisticated techniques to evaluate rodent pulmonary function has been able to identify a readily quantifiable lung phenotype in the *cftr* knockout mouse. In addition, the heterozygote *cftr+/−* mouse had a distinguishable pulmonary function phenotype from that observed in either the homozygous normal or mutant mice [54]. These data are consistent with CFTR-dependent, physiological changes in the structure and function of the lung.

This concept is not unprecedented, as many laboratories have documented the lack of differentiation in other CF organs. While the CF knockout mouse intestine has the general appearance of being normal, it does not have all of the histological signs of colonic differentiation [55]. In the pancreas, CFTR is thought to be a marker of human duct cell development and differentiation [56], and early studies of the human pancreas suggested that there was a lack of maturation of exocrine tissue associated with CF [57].

Terminally differentiated cells of the trachea, the ciliated cells [58], have been shown to be abnormal in CF humans in several studies. Ultrastructural examination of 17- to 21-week-gestation human tracheal epithelial cells revealed convoluted tight junctions, abnormal cilia, aberrant mitochondria, and Golgi [59]. Fetuses examined following prenatal diagnosis of CF at 19–23 weeks' gestation had tracheal epithelial atrophy and the absence of cilia in most tracheal epithelial cells [60].

In addition, pulmonary neuroendocrine cells, which occur in high numbers during lung development and decrease with differentiation [61], have been shown to be elevated in CF [62, 63]. These observations suggest that lack of differentiation and persistence of a fetal pattern also occurs in the CF lung.

Conclusions

It remains to be seen how many of the observed paradoxes in CF can be attributed to CFTR's role in developmental regulation. While recent experiments demonstrate that increasing airway sodium absorption initiates a CF-like lung disease [64] the full pathophysiology of the disease has not been fully replicated. There are many examples in CF that may have to be explained by the developmental component of the disease. For instance, sulfation of lung glycoprotein is characteristic of the fetal lung and sialylation increases during lung differentiation. The CF airway goblet cell secretes defective glycoconjugates with altered sulfation and sialylation, yet non-CF goblet cells have undetectable levels of CFTR mRNA after birth [25]. Although the goblet cell may not express CFTR in a mature lung, its undifferentiated airway progenitor did. Thus, the CF goblet cell may remain in a poorly differentiated state throughout life. Other seemingly unrelated observations, such as the fact that CF patients have decreased surfactant-associated protein A, can also be explained by this paradigm. Surfactant-associated protein A is a functional differentiation marker in the parenchymal type II cell. The type II cell is a descendent of lung epithelial progenitors that expressed CFTR during differentiation. In addition, a mutation in another gene coordinately expressed with CFTR in the developmental cascade would explain the findings of Groman et al. [26]. Disruption of selected genes in this developmental cascade would have the potential to result in a CF phenotype without two mutated copies of CFTR.

It appears that there are early structural and functional changes in the infant CF lung reflecting disrupted development during the fetal period. CFTR is proposed to function as part of a developmental cascade for cells in the lung, intestine, pancreas, and other secretory organs. Disruption of this pathway could occur from mutations in either CFTR or in other genes required in this developmental cascade. The lack of CFTR function during development would lead to incomplete differentiation of epithelial cells and loss of function. In addition, the failure of cell differentiation would lead to constitutive expression of cytokines that normally function during development as agents of differentiation. Once the infant is in the extrauterine environment, these same cytokines could assume a pro-inflammatory role, leading to chronic inflammation and fibrosis.

References

1 Bunting KD: ABC transporters as phenotypic markers and functional regulators of stem cells. Stem Cells 2002;20:11–20.

2 Chaudhary PM, Roninson IB: Expression and activity of P-glycoprotein, a multidrug efflux pump, in human hematopoietic stem cells. Cell 1991;66:85–94.

3 Zhou S, Schuetz JD, Bunting KD, Colapietro AM, Sampath J, Morris JJ, Lagutina I, Grosveld GC, Osawa M, Nakauchi H, et al: The ABC transporter Bcrp1/ABCG2 is expressed in a wide variety of stem cells and is a molecular determinant of the side-population phenotype. Nat Med 2001;7:1028–1034.

4 Trezise AE, Chambers JA, Wardle CJ, Gould S, Harris A: Expression of the cystic fibrosis gene in human foetal tissues. Hum Mol Genet 1993;2:213–218.

5 Broackes-Carter FC, Mouchel N, Gill D, Hyde S, Bassett J, Harris A: Temporal regulation of CFTR expression during ovine lung development: Implications for CF gene therapy. Hum Mol Genet 2002;11:125–131.

6 Tizzano EF, Chitayat D, Buchwald M: Cell-specific localization of CFTR mRNA shows developmentally regulated expression in human fetal tissues. Hum Mol Genet 1993;2:219–224.

7 Tizzano EF, O'Brodovich H, Chitayat D, Benichou JC, Buchwald M: Regional expression of CFTR in developing human respiratory tissues. Am J Respir Cell Mol Biol 1994;10:355–362.

8 Hirai C, Ichiba H, Saito M, Shintaku H, Yamano T, Kusuda S: Trophic effect of multi-

ple growth factors in amniotic fluid or human milk on cultured human fetal small intestinal cells. J Pediatr Gastroenterol Nutr 2002;34: 524–528.

9 Keel BA, Eddy KB, Cho S, May JV: Human alpha-fetoprotein purified from amniotic fluid enhances growth factor-mediated cell proliferation in vitro. Mol Reprod Dev 1991;30: 112–118.

10 Okamura M, Kurauchi O, Itakura A, Morikawa S, Suganuma N, Mizutani S, Tomoda Y: Hepatocyte growth factor in human amniotic fluid promotes the migration of fetal small intestinal epithelial cells. Am J Obstet Gynecol 1998;178(1 Pt 1):175–179.

11 Keelan JA, Blumenstein M, Helliwell RJ, Sato TA, Marvin KW, Mitchell MD: Cytokines, prostaglandins and parturition – a review. Placenta 2003;24(suppl A):S33–S46.

12 Chazaud C, Dolle P, Rossant J, Mollard R: Retinoic acid signaling regulates murine bronchial tubule formation. Mech Dev 2003; 120:691–700.

13 Levinson-Dushnik M, Benvenisty N: Involvement of hepatocyte nuclear factor 3 in endoderm differentiation of embryonic stem cells. Mol Cell Biol 1997;17:3817–3822.

14 Wan H, Kaestner KH, Ang SL, Ikegami M, Finkelman FD, Stahlman MT, Fulkerson PC, Rothenberg ME, Whitsett JA: Foxa2 regulates alveolarization and goblet cell hyperplasia. Development 2004;131:953–964.

15 Chabot H, Vives MF, Dagenais A, Grygorczyk C, Berthiaume Y, Grygorczyk R: Downregulation of epithelial sodium channel (ENaC) by CFTR co-expressed in Xenopus oocytes is independent of Cl⁻ conductance. J Membr Biol 1999;169:175–188.

16 Mall M, Bleich M, Kuehr J, Brandis M, Greger R, Kunzelmann K: CFTR-mediated inhibition of epithelial Na⁺ conductance in human colon is defective in cystic fibrosis. Am J Physiol 1999;277:G709–G716.

17 Beck S, Kuhr J, Schutz VV, Seydewitz HH, Brandis M, Greger R, Kunzelmann K: Lack of correlation between CFTR expression, CFTR Cl⁻ currents, amiloride-sensitive Na⁺ conductance, and cystic fibrosis phenotype. Pediatr Pulmonol 1999;27:251–259.

18 Kunzelmann K: The cystic fibrosis transmembrane conductance regulator and its function in epithelial transport. Rev Physiol Biochem Pharmacol 1999;137:1–70.

19 Kunzelmann K, Mall M, Briel M, Hipper A, Nitschke R, Ricken S, Greger R: The cystic fibrosis transmembrane conductance regulator attenuates the endogenous Ca²⁺ activated Cl⁻ conductance of Xenopus oocytes. Pflügers Arch 1997;435:178–181.

20 Wei L, Vankeerberghen A, Cuppens H, Eggermont J, Cassiman JJ, Droogmans G, Nilius B: Interaction between calcium-activated chloride channels and the cystic fibrosis transmembrane conductance regulator. Pflügers Arch 1999;438:635–641.

21 Anderson MP, Welsh MJ: Calcium and cAMP activate different chloride channels in the apical membrane of normal and cystic fibrosis epithelia. Proc Natl Acad Sci USA 1991;88: 6003–6007.

22 Jovov B, Ismailov II, Berdiev BK, Fuller CM, Sorscher EJ, Dedman JR, Kaetzel MA, Benos DJ: Interaction between cystic fibrosis transmembrane conductance regulator and outwardly rectified chloride channels. J Biol Chem 1995;270:29194–29200.

23 Ward CL, Krouse ME, Gruenert DC, Kopito RR, Wine JJ: Cystic fibrosis gene expression is not correlated with rectifying Cl⁻ channels. Proc Natl Acad Sci USA 1991;88: 5277–5281.

24 Wine JJ, Brayden DJ, Hagiwara G, Krouse ME, Law TC, Muller UJ, Solc CK, Ward CL, Widdicombe JH, Xia Y: Cystic fibrosis, the CFTR, and rectifying Cl⁻ channels. Adv Exp Med Biol 1991;290:253–272.

25 Jiang Q, Engelhardt JF: Cellular heterogeneity of CFTR expression and function in the lung: Implications for gene therapy of cystic fibrosis. Eur J Hum Genet 1998;6:12–31.

26 Groman JD, Meyer ME, Wilmott RW, Zeitlin PL, Cutting GR: Variant cystic fibrosis phenotypes in the absence of CFTR mutations. N Engl J Med 2002;347:401–407.

27 Mickle JE, Macek M Jr, Fulmer-Smentek SB, Egan MM, Schwiebert E, Guggino W, Moss R, Cutting GR: A mutation in the cystic fibrosis transmembrane conductance regulator gene associated with elevated sweat chloride concentrations in the absence of cystic fibrosis. Hum Mol Genet 1998;7:729–735.

28 Sekhon HS, Larson JE: In utero gene transfer into the pulmonary epithelium. Nat Med 1995;1:1201–1203.

29 Morrow SL, Larson JE, Nelson S, Sekhon HS, Ren T, Cohen JC: Modification of development by the CFTR gene in utero. Mol Genet Metab 1998;65:203–212.

30 Larson JE, Delcarpio JB, Farberman MM, Morrow SL, Cohen JC: CFTR modulates lung secretory cell proliferation and differentiation. Am J Physiol Lung Cell Mol Physiol 2000; 279:L333–L341.

31 Snyder JM, Magliato SA: An ultrastructural, morphometric analysis of rabbit fetal lung type II cell differentiation in vivo. Anat Rec 1991;229:73–85.

32 Williams MC, Dobbs LG: Expression of cell-specific markers for alveolar epithelium in fetal rat lung. Am J Respir Cell Mol Biol 1990;2:533–542.

33 Larson JE, Morrow SL, Delcarpio JB, Bohm RP, Ratterree MS, Blanchard JL, Cohen JC: Gene transfer into the fetal primate: Evidence for the secretion of transgene product. Mol Ther 2000;2:631–639.

34 Cohen JC, Larson JE: Pathophysiologic consequences following inhibition of a CFTR-dependent developmental cascade in the lung. BMC Dev Biol 2005;5:2.

35 Phelan PD, Gracey M, Williams HE, Anderson CM: Ventilatory function in infants with cystic fibrosis. Physiological assessment of halation therapy. Arch Dis Child 1969;44: 393–400.

36 Sharp JK: Monitoring early inflammation in CF. Infant pulmonary function testing. Clin Rev Allergy Immunol 2002;23:59–76.

37 Gappa M, Ranganathan SC, Stocks J: Lung function testing in infants with cystic fibrosis: Lessons from the past and future directions. Pediatr Pulmonol 2001;32:228–245.

38 Tepper RS, Hiatt PW, Eigen H, Smith J: Total respiratory system compliance in asymptomatic infants with cystic fibrosis. Am Rev Respir Dis 1987;135:1075–1079.

39 Ackerman V, Montogomery G, Eigen H, Tepper RS: Assessment of airway responsiveness in infants with cystic fibrosis. Am Rev Respir Dis 1991;144:344–346.

40 Beardsmore CS, Bar-Yishay E, Maayan C, Yahav Y, Katznelson D, Godfrey S: Lung function in infants with cystic fibrosis. Thorax 1988;43:545–551.

41 Davis S, Jones M, Kisling J, Howard J, Tepper RS: Comparison of normal infants and infants with cystic fibrosis using forced expiratory flows breathing air and heliox. Pediatr Pulmonol 2001;31:17–23.

42 Ranganathan SC, Dezateux C, Bush A, Carr SB, Castle RA, Madge S, Price J, Stroobant J, Wade A, Wallis C, et al: Airway function in infants newly diagnosed with cystic fibrosis. Lancet 2001;358:1964–1965.

43 Ranganathan SC, Stocks J, Dezateux C, Bush A, Wade A, Carr S, Castle R, Dinwiddie R, Hoo AF, Lum S, et al: The evolution of airway function in early childhood following clinical diagnosis of cystic fibrosis. Am J Respir Crit Care Med 2004;169:928–933.

44 Hardy KA, Wolfson MR, Schidlow DV, Shaffer TH: Mechanics and energetics of breathing in newly diagnosed infants with cystic fibrosis: Effect of combined bronchodilator and chest physical therapy. Pediatr Pulmonol 1989;6:103–108.

45 Stick SM, Turner DJ, LeSouef PN: Lung function and bronchial challenges in infants: Repeatability of histamine and comparison with methacholine challenges. Pediatr Pulmonol 1993;16:177–183.

46 Mohon RT, Wagener JS, Abman SH, Seltzer WK, Accurso FJ: Relationship of genotype to early pulmonary function in infants with cystic fibrosis identified through neonatal screening. J Pediatr 1993;122:550–555.

47 Armstrong DS, Grimwood K, Carlin JB, Carzino R, Gutierrez JP, Hull J, Olinsky A, Phelan EM, Robertson CF, Phelan PD: Lower airway inflammation in infants and young children with cystic fibrosis. Am J Respir Crit Care Med 1997;156:1197–1204.

48 Khan TZ, Wagener JS, Bost T, Martinez J, Accurso FJ, Riches DW: Early pulmonary inflammation in infants with cystic fibrosis. Am J Respir Crit Care Med 1995;151: 1075–1082.

49 Rosenfeld M, Gibson RL, McNamara S, Emerson J, Burns JL, Castile R, Hiatt P, McCoy K, Wilson CB, Inglis A, et al: Early pulmonary infection, inflammation, and clinical outcomes in infants with cystic fibrosis. Pediatr Pulmonol 2001;32:356–366.

50 Balough K, McCubbin M, Weinberger M, Smits W, Ahrens R, Fick R: The relationship between infection and inflammation in the early stages of lung disease from cystic fibrosis. Pediatr Pulmonol 1995;20:63–70.

51 Tirouvanziam R, de Bentzmann S, Hubeau C, Hinnrasky J, Jacquot J, Peault B, Puchelle E:

Inflammation and infection in naive human cystic fibrosis airway grafts. Am J Respir Cell Mol Biol 2000;23:121–127.

52 Tirouvanziam R, Khazaal I, Peault B: Primary inflammation in human cystic fibrosis small airways. Am J Physiol Lung Cell Mol Physiol 2002;283:L445–L451.

53 Long FR, Williams RS, Castile RG: Structural airway abnormalities in infants and young children with cystic fibrosis. J Pediatr 2004; 144:154–161.

54 Cohen JC, Lundblad LK, Bates JH, Levitzky M, Larson JE: The 'Goldilocks effect' in cystic fibrosis: Identification of a lung phenotype in the cftr knockout and heterozygous mouse. BMC Genet 2004;5:21.

55 Hinojosa-Kurtzberg AM, Johansson ME, Madsen CS, Hansson GC, Gendler SJ: Novel MUC1 splice variants contribute to mucin overexpression in CFTR-deficient mice. Am J Physiol Gastrointest Liver Physiol 2003;284: G853–G862.

56 Hyde K, Reid CJ, Tebbutt SJ, Weide L, Hollingsworth MA, Harris A: The cystic fibrosis transmembrane conductance regulator as a marker of human pancreatic duct development. Gastroenterology 1997;113: 914–919.

57 Imrie JR, Fagan DG, Sturgess JM: Quantitative evaluation of the development of the exocrine pancreas in cystic fibrosis and control infants. Am J Pathol 1979;95:697–707.

58 Chang LY, Wu R, Nettesheim P: Morphological changes in rat tracheal cells during the adaptive and early growth phase in primary cell culture. J Cell Sci 1985;74: 283–301.

59 Gosden CM, Gosden JR: Fetal abnormalities in cystic fibrosis suggest a deficiency in proteolysis of cholecystokinin. Lancet 1984;ii: 541–546.

60 Ornoy A, Arnon J, Katznelson D, Granat M, Caspi B, Chemke J: Pathological confirmation of cystic fibrosis in the fetus following prenatal diagnosis. Am J Med Genet 1987;28: 935–947.

61 Warburton D, Wuenschell C, Flores-Delgado G, Anderson K: Commitment and differentiation of lung cell lineages. Biochem Cell Biol 1998;76:971–995.

62 Johnson DE, Wobken JD, Landrum BG: Changes in bombesin, calcitonin, and serotonin immunoreactive pulmonary neuroendocrine cells in cystic fibrosis and after prolonged mechanical ventilation. Am Rev Respir Dis 1988;137:123–131.

63 Dovey M, Wisseman CL, Roggli VL, Roomans GM, Shelburne JD, Spock A: Ultrastructural morphology of the lung in cystic fibrosis. J Submicrosc Cytol Pathol 1989; 21:521–534.

64 Mall M, Grubb BR, Harkema JR, O'Neal WK, Boucher RC: Increased airway epithelial Na$^+$ absorption produces cystic fibrosis-like lung disease in mice. Nat Med 2004;10: 487–493.

Dr. Janet E. Larson
Neonatology
T11060
Stony Brook University Health Sciences Center
Stony Brook, NY 11794-8111(USA)
E-Mail: janet.larson@stonybrook.edu

Bush A, Alton EWFW, Davies JC, Griesenbach U, Jaffe A (eds): Cystic Fibrosis in the 21st Century.
Prog Respir Res. Basel, Karger, 2006, vol 34, pp 61–68

Genotype-Phenotype Correlations in Cystic Fibrosis

Ruslan Dorfman   Julian Zielenski

The Hospital for Sick Children, Toronto, Canada

Abstract

The primary defect in cystic fibrosis (CF) is dysfunction of cystic fibrosis transmembrane conductance regulator (CFTR) caused by mutations in the CFTR gene. In order to better understand the effect of CFTR mutations on disease severity and progression, genotype-phenotype (G-P) correlation studies are being conducted. These studies reveal a complex relationship between the CF phenotype and underlying CFTR mutations, with variable correlation in different organs. Mutant alleles that completely abolish CFTR expression and function produce a classic CF phenotype, whereas genotypes associated with at least one CFTR variant with residual expression or function tend to protect against the most severe consequences, especially pancreatic insufficiency, meconium ileus, liver disease, diabetes and to a lesser extent severe lung disease. The level of penetrance and expressivity (severity) varies among clinical manifestations in CF. For example, penetrance and expressivity of exocrine pancreatic insufficiency is almost 100% for most of the severe CFTR genotypes. Some other clinical traits are not fully penetrant (meconium ileus, liver disease or diabetes) and may require the involvement of other factors such as CF modifiers and/or environment. Finally, although lung disease is almost fully penetrant late in advanced stages of the disease, its expressivity is the most variable of all clinical manifestations in CF due to more complex gene-environment interactions involving both primary and secondary genetic factors as well as the environment. Extending G-P studies beyond CF revealed a role of CFTR variation in the pathogenesis of other diseases with similarities to CF. In the long term, the G-P correlation studies will have more practical implications to diagnosis and prognosis of CF and provide foundation for improved management and therapy in CF.

Defining Genotype and Phenotype in Cystic Fibrosis

Delineation of genotype and phenotype in cystic fibrosis (CF) is the key to understanding the relationship between them, and in more practical terms to what extent a CF clinical phenotype can be derived from an underlying cystic fibrosis transmembrane conductance regulator (CFTR) genotype. The scope of CFTR genotype and CF phenotype definitions has evolved over the years and reflects advances in clinical and genetic knowledge of CF.

CFTR Genotype Causing CF

Due to an autosomal recessive mode of inheritance, CF is caused by coinheritance of two disease-causing mutant alleles of the CFTR gene residing on the maternal and paternal chromosome 7s (with the exception of uniparental disomy). Typically a CFTR mutant allele harbors one well-defined DNA alteration affecting the gene's expression or function or both; occasionally, however, a CFTR allele may consist of more than one distinct change. The latter is termed a complex allele [1]. A continuing effort to identify CFTR mutations in various CF populations initiated 15 years

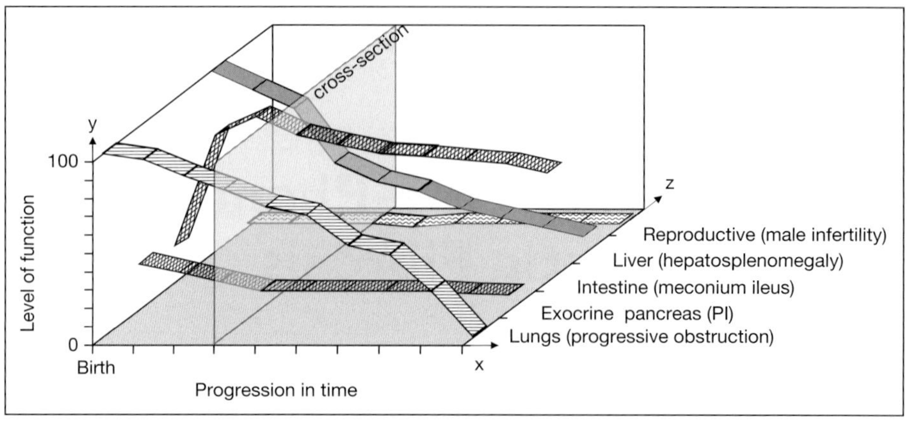

Fig. 1. Schematic representation of a hypothetical clinical phenotype in CF. The phenotype is described as a configuration of manifestations, their severity (level of function) and progression in time. CF phenotype at a given time point is evident in the cross section (y-z axes). The CF phenotype defined by x-y axes yields a longitudinal description (e.g. decline of pulmonary function). PI = Pancreatic insufficiency.

ago led to the identification of approximately 1,200 established or putative disease-causing mutations [2]. These mutations have been classified on the basis of their molecular mechanisms. Generally five major classes of CFTR mutations are recognized (for details see chapter 1). Three of them, class I (defective synthesis), class II (defective processing and maturation) and class III (defective regulation), tend to completely abolish CFTR expression and/or function. Mutations in the two remaining classes, class IV (defective conductance) and class V (reduced function/synthesis), produce CFTR variants with residual expression and/or function.

Due to the widespread prevalence (~70% worldwide) of the ΔF508 mutation (3-base pair nucleotide deletion leading to loss of phenylalanine at position 508) among Caucasian CF chromosomes, the most common CFTR genotype is ΔF508/ΔF508 [3]. As the prevalence of ΔF508 in various CF populations ranges between 20 and 90% [4] the proportion of patients homozygous for ΔF508 may vary substantially. The remaining CFTR genotypes frequently consist of compound heterozygotes with ΔF508 and rarely of two non-ΔF508 alleles. The net consequence of various genotypes ranges from complete lack of CFTR expression and function to preserving some level of both. The large number of genotypes formed by combinations of mutations representing different classes, and the consequences of their molecular and functional mechanisms, are considered the primary determinant of genetic heterogeneity in CF populations.

In summary, a CF-associated genotype is defined typically by two CFTR alleles affecting CFTR function/expression and contributing to development of the disease.

CF Phenotype
The CF clinical phenotype is complex and variable, due to the involvement of multiple organs as well as the broad spectrum of disease severity [5]. Moreover, the nosologic scope of CF has evolved over the years [6]. CF was described initially as a disease of the pancreas [7] and only later as a generalized disease with multiple manifestations including respiratory symptoms [8]. Broadening of the clinical spectrum of CF can be ascribed to several factors including extension of CF diagnosis to milder, nonclassical and atypical forms (see chapter 9 for more detail) as well as recognition of new CF complications (e.g. diabetes, osteoporosis) due to the better current prognosis. In essence, CF can be defined as a cross-sectional phenotype described by current disease manifestations and their severity or as a longitudinal phenotype in the context of three variables: configuration of clinical manifestations, their severity and time progression (fig. 1). The latter provides important information about the change of the function in time for quantitative clinical parameters (e.g. decline of the pulmonary function). In summary, the CF phenotype consists of all clinical manifestations present at a given point of time (cross-sectional definition) or through specific time durations (longitudinal definition).

Genotype-Phenotype Studies at Various Phenotypic Levels

The CF disease phenotype observed at each time point is a result of complex etiopathogenesis originating from (but not limited to) the presence of two mutant alleles affecting CFTR-related cellular functions. Understanding the development of a particular phenotype is a daunting task and requires insight into processes on various phenotypic levels: molecular, cellular, functional and clinical. This area of research is targeted by the genotype-phenotype (G-P) correlation studies [references in 2]. Their main

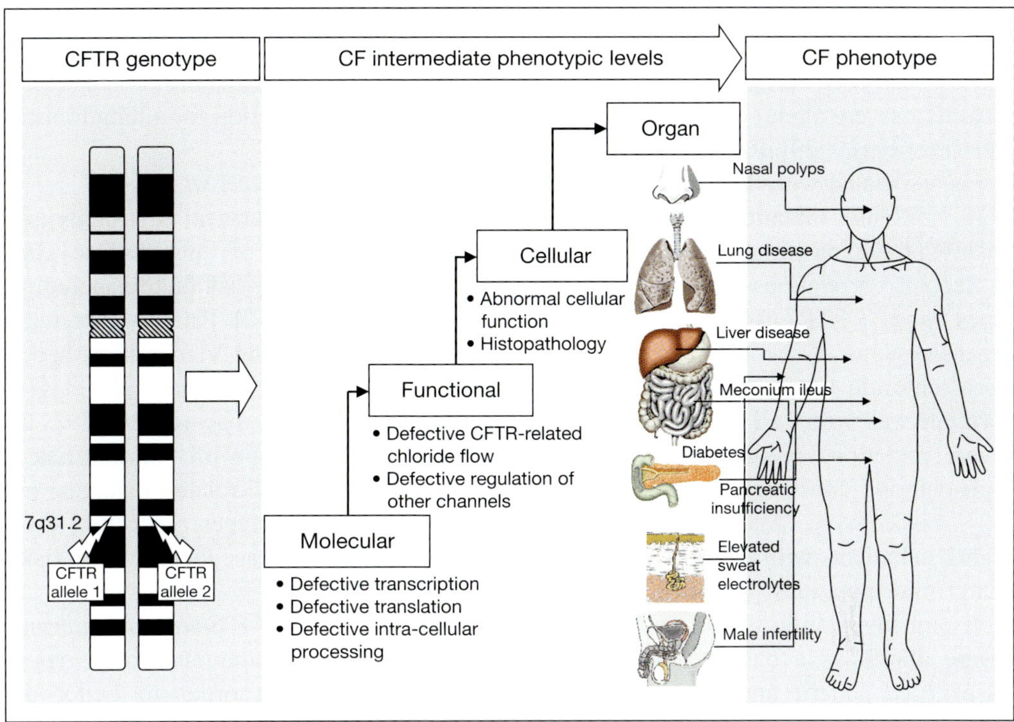

Fig. 2. CFTR genotype and intermediate phenotypic levels leading to a CF clinical (disease) phenotype. The figure is highly schematic and depicts a particular phenotype. It should be noted that CF phenotypes vary considerably in terms of presence of specific manifestations as well as their severity. Also classification of intermediate phenotypes is often oversimplified because of difficulty distinguishing between molecular and cellular processes.

objective is to establish to what extent CF phenotype results from underlying CFTR mutations. Most of the G-P studies have been conducted at the clinical level. These are admittedly very important studies with practical significance, however they are of limited value when it comes to understanding processes and mechanisms underlying the development of particular disease phenotypes. This is because the final clinical CF phenotype is a result of many interrelated processes occurring in the network of genetic, environmental and stochastic interactions. For this reason G-P studies investigating molecular, cellular and functional phenotypes (intermediate phenotypic levels) are of great importance for better understanding and prediction of clinical outcomes (fig. 2).

Clinical Phenotype

Due to the multiorgan involvement in CF, G-P correlation studies analyze the relationship between CFTR genotypes and specific clinical phenotypes (e.g. pancreatic status, lung function, liver disease). These studies typically use cross-sectional phenotypic data and to a lesser extent longitudinal data (decline of function). There are two general approaches to G-P correlation studies: first, evaluation of trends in CF patients carrying specific CFTR mutations [2], and second, analysis of genotypes in patients with a common clinical profile (e.g. age of diagnosis, pancreatic status, severe lung disease, liver disease or diabetes) [9–16].

G-P studies have led to several important observations. First, in studies where specific mutations were analyzed in CF patients in homozygous or heterozygous state (typically in *trans* with ΔF508) phenotypic effects were in good agreement with mutation classification (class I–V). Generally, genotypes consisting of two severe alleles (class I–III mutations in various combinations) were associated with more severe CF phenotype (pancreatic insufficiency, liver disease, diabetes, more advanced lung disease) than genotypes with at least one mild mutation (class IV or V) [17, 18]. Further, dominant phenotypic effects have been documented for mild CFTR alleles occurring in combinations with severe alleles [11, 12]. There were however some phe-notypic differences within the same classes. For example patients carrying various class I nonsense mutations (in *trans* to a severe allele) present with a lung disease of variable severity [19, 20]. It is not clear what molecular mechanisms associated with the processing of transcripts derived from nonsense mutations may be involved and consequently producing variable phenotypes in airways.

Second, correlations between associated clinical phenotypes and CFTR genotypes vary between affected organs/systems. Generally, exocrine pancreatic function, sweat chloride levels, and male reproductive tract status correlate well with CFTR genotype, where presence of two severe mutations predicts pancreatic insufficiency [12], high chloride concentration [17, 21] and male infertility [22], respectively. Also malnutrition was significantly higher in patients with a severe genotype than in those carrying mild mutations [23]. In addition, several less common clinical manifestations require the presence of two severe alleles as a baseline condition along with other unspecified genetic and environmental factors. Some of these complications are meconium ileus, CF liver disease and CF-related diabetes. Conversely, CF patients manifesting with pancreatitis tend to carry a genotype consisting of at least one mild allele [16]. Due to the importance of lung disease in CF a lot of focus has been on its correlation with CFTR genotype [10, 15, 17, 18, 24–29]. Some degree of correlation between CFTR genotype (and corresponding mutation classes) and CF lung disease has been documented [15, 18, 24, 28, 29]. Patients carrying mild CFTR mutations tend to present with a less severe lung disease (e.g. better pulmonary function, lower rate of *Pseudomonas aeruginosa* colonization) than patients carrying severe CFTR alleles. However, there is considerable individual variation of lung phenotype in all genotype groups. Overall, the severity of lung disease and its progression are highly variable and only partially correlated with CFTR genotype.

Third, several specific CFTR mutations are associated with higher than expected phenotypic variation (including pancreatic status) [30–33]. Such variation can be partially explained by the impact of a second site mutation (complex alleles) [31, 32]. Another possibility is modulation by poly-variant haplotypes, i.e. configurations of common subclinical DNA sequence variants (polymorphisms) in the CFTR gene (see chapter 1) [34, 35]. Examples are a combination of TG repeat variants with variable length of the poly-thymidyl tract in intron 8 or the specific common nucleotide variant in exon 10, 1540A/G (M470V).

Fourth, for CF phenotypes, such as pulmonary function, that can be monitored as a quantitative measure, longitudi-nal studies [27, 36, 37] are more powerful than cross-sectional studies in predicting the course of lung disease [37] and assisting critical therapeutic decisions such as patient selection for a lung transplant.

Survival

Survival is strongly affected by the nature and progression of lung disease [38], and has been correlated with many risk factors including CFTR genotype [15, 25, 26, 36, 39, 40]. Patients carrying milder alleles consisting of class IV and V mutations had significantly better survival than those with severe alleles (class I–III) largely due to a milder lung disease [26, 39]. In addition to lung deterioration being a life-limiting risk factor, other complications (such as CF-related diabetes) more frequent in patients carrying genotypes associated with class I–III mutations may also increase the mortality risk [41, 42].

G-P Studies at Intermediate Levels: Molecular and Functional Phenotypes

In order to better understand CF pathogenesis, an increasing number of G-P studies attempt to establish links between CFTR genotype, intermediate phenotypes at molecular and functional levels and CF clinical status (fig. 2). The molecular phenotype is usually assessed by CFTR mRNA transcript analysis in cells or tissues (e.g. airways, nasal epithelium) obtained from patients carrying specific CFTR mutations [43–47]. CFTR mRNA transcript levels are being correlated with severity of selected functional or clinical parameters. As an example, a good correlation has been established between the splicing efficiency of 3849 + 10 kb C→T splice mutation and measures of severity of lung disease [44, 45]. The mRNA levels were variable between patients but well-correlated with their pulmonary function. mRNA levels also varied substantially between organs of affected human fetuses [45], which could partially justify variable pathophysiology in specific organs even in very early stages of the disease.

One important aspect of G-P studies at the molecular phenotype is the concept of a minimal level of gene expression required to sustain CFTR-related cell functions and to protect against severe disease complications characteristic of the classic form of CF. According to a recent study [48], just below 5% of the normal level of wild-type CFTR mRNA could be sufficient to prevent development of severe lung disease. In addition, assessment of CFTR expression in relevant tissues [43, 45, 46] led to an important observation that levels of normal transcripts may be grossly perturbed based on the occurrence of aberrant transcripts related to specific splice mutations that were found

            Dorfman/Zielenski

to vary considerably between tissues and correlate with the extent of their involvement in the disease [43, 45, 46]. According to these studies, males carrying one severe mutation and the mRNA splice variant with a short polypyrimidine tract in intron 8 (IVS8-5T) have higher levels of truncated (CFTR delta exon 9) transcripts in the epididymal epithelia than in nasal epithelia. This observation might explain how males carrying genotypes with this variant develop infertility due to congenital absence of the vas deferens but not CF-like pulmonary disease [43, 46].

Expression of a variety of common nonsense and frameshift mutations were also evaluated [20]. The consequences vary from mRNA reduction to exon skipping to intact expression and may explain their various phenotypic effects in different organs (e.g. pancreas vs. airways) [49]. Other studies of the molecular phenotype strongly suggest that the phenotypic penetrance of mutations can be affected by common intragenic variants [34, 35].

Several hundred putative CFTR missense mutations were reported worldwide [2] and the vast majority of them require complex studies on the protein and functional levels to establish their precise mechanisms leading to disease. In several studies the expression, intracellular processing, stability and chloride channel activity of disease variants with known phenotypic consequences were evaluated [50–54]. The majority of these variants corresponded to known mild CF or congenital bilateral absence of the vas deferens-causing alleles. For many of them expression, maturation or chloride channel-related activities were only partially affected. The residual CFTR function associated with these mutants could readily explain their mild phenotypic effect.

G-P studies correlating CFTR mutations with functional phenotypes focus on the electrophysiology (e.g. chloride efflux; transepithelial conductance; nasal potential difference, NPD) of CFTR-expressing epithelia (nasal or rectal epithelial cells). Interestingly, some of these studies using nasal epithelial cells demonstrate poor correlation between specific CFTR genotypes and cAMP-mediated chloride channel activity. In contrast, channel activity does correlate well with disease severity represented by lung-related phenotypic variables (spirometry, bacterial colonization) or composite clinical score [55–57]. This effect may be also age-dependent since the correlation between CFTR chloride activity and clinical severity was shown in adult patients carrying severe genotypes (mostly homozygous for ΔF508) [56, 57] but not in young patients with comparable genotypes [58]. However, correspondence between cAMP-activated chloride channel and phenotype severity was not always observed suggesting the impact of other factors [55]. An alternative possibility of pulmonary function being related to sodium hyperabsorption rather than a defect in chloride secretion has been recently demonstrated by Fajac et al. [59]. The study correlated NPD measurements with pancreatic status or respiratory function in patients carrying various CFTR genotypes. According to this study pancreatic status was clearly related to NPD under conditions corresponding to chloride secretion, while pulmonary function was related only to sodium transport. This is in agreement with the recently implicated role of epithelial sodium channel (ENaC) in development of lung disease in transgenic mice overexpressing β subunit of ENaC [60].

Another functional approach has been to analyze rectal cAMP-mediated chloride secretion as a measure of CFTR function in patients [61–63]. Analysis of the chloride secretion in cohort of patients carrying various CFTR genotypes led to the conclusion that their age of diagnosis, pancreatic status and underlying genotypes correlated well with the residual intestinal chloride secretion [61]. In patients homozygous for ΔF508 with variable CF presentation the severity of the disease (as measured by composite pulmonary/growth phenotype) was related to chloride secretion [63]. A recent study showed that the residual chloride secretion in colon correlates with milder CFTR mutations and better clinical presentation (later disease onset, pancreatic sufficiency and less severe lung disease) [62].

Another functional approach has been to analyze rectal cAMP-mediated chloride secretion as a measure of CFTR function in patients [61, 62, 64]. In patients homozygous for ΔF508 the severity of clinical manifestations correlated well with chloride secretion [61, 64].

Although most of the studies focus on evaluating the relationship between CFTR function as a chloride channel and CF phenotype, the clinical severity may also be a consequence of impairment of CFTR regulatory function(s). To test this hypothesis, the effect of specific mutations on CFTR as a chloride channel and regulator of outwardly rectifying chloride channel was evaluated [65, 66]. Interestingly, mutations with milder phenotypic effect not only retained residual CFTR-related chloride function but also regulatory function towards outwardly rectifying chloride channel while the severe mutations in the NBD1 region led to loss of regulatory capacity. More studies are required to map phenotypic impact of various mutations to their functional consequences.

Overall, the studies of correlation between CFTR genotype and CF phenotype (including intermediate phenotypes) have significantly advanced our knowledge and understanding of how and to what extent CFTR mutations contribute to severity of CF. They show that the relationship is often more complex than anticipated from Mendelian inheritance and requires the contribution of factors other than the primary genotype of CFTR.

Significance and Impact of the G-P Studies in CF

CFTR Genotype as the Primary Determinant of CF

G-P studies have led to a better understanding of the contribution of CFTR gene mutations to the variability of clinical phenotype in CF. Although correlation between specific CFTR alleles and severity of CF symptoms is prominent, there is also dependence of the organ context. The correlation ranges from good (for pancreatic disease, sweat chloride, male infertility) to poor (for lung disease, liver disease). In other words, the contribution of CFTR genotype as the primary determinant of CF phenotype conforms to a more Mendelian type of inheritance for some of the clinical traits and to a more complex, non-Mendelian one for the others. Some of the traits in the latter group appear to have reduced penetrance (e.g. meconium ileus, liver disease or diabetes) and variable expressivity (e.g. lung disease) manifesting with a broad spectrum of severity. Also, the direct impact of the CFTR genotype, as a disease trigger, seems to be largely limited to the initial phase of CF with other factors strongly influencing the later phases especially in terms of lung disease [67]. In fact, severity of lung disease, especially in the more advanced clinical phase, seems to be more closely associated with other genetic and environmental factors [68].

Secondary Genetic Modulation in CF – Genetic Modifiers

As anticipated by many G-P studies there is a growing body of evidence that secondary genetic factors (modifiers) play a role in modulating severity of CF (see chapter 10). Modifier genes may affect severity of typical CF traits (e.g. lung disease) or development of less common clinical complications (e.g. meconium ileus) [5]. The involvement of modifiers is expected to be minimal in traits strongly correlating with CFTR genotype. The opposite can be true for CF phenotypes with limited G-P correlations.

It would be anticipated that genotypes of primary CF determinant (CFTR) combined with genotypes of secondary CF determinant (CF modifiers) will provide higher prognostic and diagnostic value for CF than CFTR genotypes alone.

Dissecting Genetic and Environmental Aspects of CF

Although frequently acknowledged as a factor contributing to CF variation (particularly lung function) the environmental impact has not been sufficiently investigated. This impact is multifactorial and includes, among others, exposure to pollutants, and airborne pathogens of variable virulence as well as nutritional and socioeconomic status. In one study of patients homozygous for the $\Delta F508$ mutation heavy exposure to tobacco smoke was significantly associated with lower Shwachman scores and poorer pulmonary function [69]. Gene-environment interactions slowly come into focus with better understanding of the contribution of primary and secondary genetic factors (CFTR genotype and modifiers) to CF. A longitudinal study of lung function in California school children has shown that current levels of air pollution lead to clinically significant deficits in lung function [70], and that variants of a common genetic locus are associated with poor lung growth [71]. In addition, long-term effects of *in utero* exposure to smoking in a genetically susceptible group of children has been associated with altered lung function [72]. Such large scale population studies will provide candidate modifiers and gene-environment hypotheses for future G-P correlation studies in CF.

Impact on CF Diagnosis, Management and Treatment

Recent advances in the area of genetics of CF have or are expected to have an important impact on various aspects of CF patient care including diagnosis, management and treatment. Defining the prognostic value of specific CFTR genotypes will allow not only for making more firm diagnosis of CF (especially in cases with uncertain diagnosis) but also for applying improved and optimized CF management and treatment based on the predicted course of the disease. Dissecting the impact of various genetic, molecular and environmental factors contributing to CF will permit a more targeted approach to CF treatment including new therapies (gene and protein based) as well as better control over potentially adverse environmental factors.

Acknowledgements

We would like to express our appreciation to Dr. P.R. Durie, Dr. M. Corey and Dr. J. Rommens for their critical reading of the manuscript. We would also like to acknowledge support from the Canadian Cystic Fibrosis Foundation and the Genome Canada through the Ontario Genomics Institute.

References

1 Zielenski J, Tsui LC: Cystic fibrosis: Genotypic and phenotypic variations. Annu Rev Genet 1995;29:777–807.
2 Consortium CFGA: Cystic Fibrosis Mutation Database. http://www.genet.sickkids.on.ca/cftr/.
3 Tsui LC: Mutations and sequence variations detected in the cystic fibrosis transmembrane conductance regulator (CFTR) gene: A report from the Cystic Fibrosis Genetic Analysis Consortium. Hum Mutat 1992;1:197–203.
4 Bobadilla JL, Macek M Jr, Fine JP, Farrell PM: Cystic fibrosis: A worldwide analysis of CFTR mutations – correlation with incidence data and application to screening. Hum Mutat 2002;19:575–606.
5 Zielenski J: Genotype and phenotype in cystic fibrosis. Respiration 2000;67:117–133.
6 Busch R: On the history of cystic fibrosis. Nord Medicinhist Arsb 1991;95–98.
7 Andersen D: Cystic fibrosis of the pancreas and its relation to celiac disease: A clinical and pathological study. Am J Dis Child 1938;56:344–399.
8 Farber S: Pancreatic insufficiency and the celiac syndrome. N Engl J Med 1943;229:653–682.
9 Gilljam M, Ellis L, Corey M, Zielenski J, Durie P, Tullis DE: Clinical manifestations of cystic fibrosis among patients with diagnosis in adulthood. Chest 2004;126:1215–1224.
10 Kraemer R, Birrer P, Liechti-Gallati S: Genotype-phenotype association in infants with cystic fibrosis at the time of diagnosis. Pediatr Res 1998;44:920–926.
11 Ahmed N, Corey M, Forstner G, Zielenski J, Tsui LC, Ellis L, Tullis E, Durie P: Molecular consequences of cystic fibrosis transmembrane regulator (CFTR) gene mutations in the exocrine pancreas. Gut 2003;52:1159–1164.
12 Kristidis P, Bozon D, Corey M, Markiewicz D, Rommens J, Tsui LC, Durie P: Genetic determination of exocrine pancreatic function in cystic fibrosis. Am J Hum Genet 1992;50:1178–1184.
13 Wilschanski M, Rivlin J, Cohen S, Augarten A, Blau H, Aviram M, Bentur L, Springer C, Vila Y, Branski D, Kerem B, Kerem E: Clinical and genetic risk factors for cystic fibrosis-related liver disease. Pediatrics 1999;103:52–57.
14 Cotellessa M, Minicucci L, Diana MC, Prigione F, Di Febbraro L, Gagliardini R, Manca A, Battistini F, Taccetti G, Magazzu G, Padoan R, Pizzamiglio G, Raia V, Iapichino L, Cardella F, Grinzich G, Lucidi V, Tuccio G, Bignamini E, Salvatore D, Lorini R: Phenotype/genotype correlation and cystic fibrosis related diabetes mellitus (Italian Multicenter Study). J Pediatr Endocrinol Metab 2000;13:1087–1093.
15 Schaedel C, de Monestrol I, Hjelte L, Johannesson M, Kornfalt R, Lindblad A, Strandvik B, Wahlgren L, Holmberg L: Predictors of deterioration of lung function in cystic fibrosis. Pediatr Pulmonol 2002;33:483–491.
16 Durno C, Corey M, Zielenski J, Tullis E, Tsui LC, Durie P: Genotype and phenotype corre-

lations in patients with cystic fibrosis and pancreatitis. Gastroenterology 2002;123:1857–1864.
17 Consortium TCFG-P: Correlation between genotype and phenotype in patients with cystic fibrosis. The Cystic Fibrosis Genotype-Phenotype Consortium. N Engl J Med 1993;329:1308–1313.
18 Koch C, Cuppens H, Rainisio M, Madessani U, Harms H, Hodson M, Mastella G, Navarro J, Strandvik B, McKenzie S: European Epidemiologic Registry of Cystic Fibrosis (ERCF): Comparison of major disease manifestations between patients with different classes of mutations. Pediatr Pulmonol 2001;31:1–12.
19 Rolfini R, Cabrini G: Nonsense mutation R1162X of the cystic fibrosis transmembrane conductance regulator gene does not reduce messenger RNA expression in nasal epithelial tissue. J Clin Invest 1993;92:2683–2687.
20 Will K, Dork T, Stuhrmann M, von der Hardt H, Ellemunter H, Tummler B, Schmidtke J: Transcript analysis of CFTR nonsense mutations in lymphocytes and nasal epithelial cells from cystic fibrosis patients. Hum Mutat 1995;5:210–220.
21 Wilschanski M, Zielenski J, Markiewicz D, Tsui LC, Corey M, Levison H, Durie PR: Correlation of sweat chloride concentration with classes of the cystic fibrosis transmembrane conductance regulator gene mutations. J Pediatr 1995;127:705–710.
22 Patrizio P, Zielenski J: Congenital absence of the vas deferens: A mild form of cystic fibrosis. Mol Med Today 1996;2:24–31.
23 Dray X, Kanaan R, Bienvenu T, Desmazes-Dufeu N, Dusser D, Marteau P, Hubert D: Malnutrition in adults with cystic fibrosis. Eur J Clin Nutr 2005;59:152–154.
24 Hubert D, Bienvenu T, Desmazes-Dufeu N, Fajac I, Lacronique J, Matran R, Kaplan JC, Dusser DJ: Genotype-phenotype relationships in a cohort of adult cystic fibrosis patients. Eur Respir J 1996;9:2207–2214.
25 Johansen HK, Nir M, Hoiby N, Koch C, Schwartz M: Severity of cystic fibrosis in patients homozygous and heterozygous for delta F508 mutation. Lancet 1991;337:631–634.
26 Lai HJ, Cheng Y, Cho H, Kosorok MR, Farrell PM: Association between initial disease presentation, lung disease outcomes, and survival in patients with cystic fibrosis. Am J Epidemiol 2004;159:537–546.
27 Loubieres Y, Grenet D, Simon-Bouy B, Medioni J, Landais P, Ferec C, Stern M: Association between genetically determined pancreatic status and lung disease in adult cystic fibrosis patients. Chest 2002;121:73–80.
28 Rosenecker J: Relations between the frequency of the DeltaF 508 mutation and the course of pulmonary disease in cystic fibrosis patients infected with *Pseudomonas aeruginosa*. Eur J Med Res 2000;5:356–359.
29 Kerem E, Corey M, Kerem BS, Rommens J, Markiewicz D, Levison H, Tsui LC, Durie P: The relation between genotype and phenotype in cystic fibrosis – analysis of the most com-

mon mutation (delta F508). N Engl J Med 1990;323:1517–1522.
30 Estivill X, Ortigosa L, Perez-Frias J, Dapena J, Ferrer J, Pena L, Pena L, Llevadot R, Gimenez J, Nunes V, et al: Clinical characteristics of 16 cystic fibrosis patients with the missense mutation R334W, a pancreatic insufficiency mutation with variable age of onset and interfamilial clinical differences. Hum Genet 1995;95:331–336.
31 Kiesewetter S, Macek M Jr, Davis C, Curristin SM, Chu CS, Graham C, Shrimpton AE, Cashman SM, Tsui LC, Mickle J, et al: A mutation in CFTR produces different phenotypes depending on chromosomal background. Nat Genet 1993;5:274–278.
32 Monaghan KG, Highsmith WE, Amos J, Pratt VM, Roa B, Friez M, Pike-Buchanan LL, Buyse IM, Redman JB, Strom CM, Young AL, Sun W: Genotype-phenotype correlation and frequency of the 3199del6 cystic fibrosis mutation among I148T carriers: Results from a collaborative study. Genet Med 2004;6:421–425.
33 Vazquez C, Antinolo G, Casals T, Dapena J, Elorz J, Seculi JL, Sirvent J, Cabanas R, Soler C, Estivill X: Thirteen cystic fibrosis patients, 12 compound heterozygous and one homozygous for the missense mutation G85E: A pancreatic sufficiency/insufficiency mutation with variable clinical presentation. J Med Genet 1996;33:820–822.
34 Cuppens H, Lin W, Jaspers M, Costes B, Teng H, Vankeerberghen A, Jorissen M, Droogmans G, Reynaert I, Goossens M, Nilius B, Cassiman JJ: Polyvariant mutant cystic fibrosis transmembrane conductance regulator genes. The polymorphic (Tg)m locus explains the partial penetrance of the T5 polymorphism as a disease mutation. J Clin Invest 1998;101:487–496.
35 Steiner B, Truninger K, Sanz J, Schaller A, Gallati S: The role of common single-nucleotide polymorphisms on exon 9 and exon 12 skipping in nonmutated CFTR alleles. Hum Mutat 2004;24:120–129.
36 Corey M, Edwards L, Levison H, Knowles M: Longitudinal analysis of pulmonary function decline in patients with cystic fibrosis. J Pediatr 1997;131:809–814.
37 Rosenbluth DB, Wilson K, Ferkol T, Schuster DP: Lung function decline in cystic fibrosis patients and timing for lung transplantation referral. Chest 2004;126:412–419.
38 Pilewski JM, Frizzell RA: Role of CFTR in airway disease. Physiol Rev 1999;79:S215–S255.
39 McKone EF, Emerson SS, Edwards KL, Aitken ML: Effect of genotype on phenotype and mortality in cystic fibrosis: A retrospective cohort study. Lancet 2003;361:1671–1676.
40 O'Connor GT, Quinton HB, Kahn R, Robichaud P, Maddock J, Lever T, Detzer M, Brooks JG: Case-mix adjustment for evaluation of mortality in cystic fibrosis. Pediatr Pulmonol 2002;33:99–105.

Genotype-Phenotype Correlations in CF

41 Lanng S: Glucose intolerance in cystic fibrosis patients. Paediatr Respir Rev 2001;2: 253–259.

42 Koch C, Rainisio M, Madessani U, Harms HK, Hodson ME, Mastella G, McKenzie SG, Navarro J, Strandvik B: Presence of cystic fibrosis-related diabetes mellitus is tightly linked to poor lung function in patients with cystic fibrosis: Data from the European Epidemiologic Registry of Cystic Fibrosis. Pediatr Pulmonol 2001;32:343–350.

43 Mak V, Jarvi KA, Zielenski J, Durie P, Tsui LC: Higher proportion of intact exon 9 CFTR mRNA in nasal epithelium compared with vas deferens. Hum Mol Genet 1997;6:2099–2107.

44 Chiba-Falek O, Kerem E, Shoshani T, Aviram M, Augarten A, Bentur L, Tal A, Tullis E, Rahat A, Kerem B: The molecular basis of disease variability among cystic fibrosis patients carrying the 3849+10 kb C→T mutation. Genomics 1998;53:276–283.

45 Chiba-Falek O, Parad RB, Kerem E, Kerem B: Variable levels of normal RNA in different fetal organs carrying a cystic fibrosis transmembrane conductance regulator splicing mutation. Am J Respir Crit Care Med 1999; 159:1998–2002.

46 Teng H, Jorissen M, Van Poppel H, Legius E, Cassiman JJ, Cuppens H: Increased proportion of exon 9 alternatively spliced CFTR transcripts in vas deferens compared with nasal epithelial cells. Hum Mol Genet 1997;6: 85–90.

47 Rave-Harel N, Kerem E, Nissim-Rafinia M, Madjar I, Goshen R, Augarten A, Rahat A, Hurwitz A, Darvasi A, Kerem B: The molecular basis of partial penetrance of splicing mutations in cystic fibrosis. Am J Hum Genet 1997;60:87–94.

48 Ramalho AS, Beck S, Meyer M, Penque D, Cutting GR, Amaral MD: Five percent of normal cystic fibrosis transmembrane conductance regulator mRNA ameliorates the severity of pulmonary disease in cystic fibrosis. Am J Respir Cell Mol Biol 2002;27: 619–627.

49 Gasparini P, Borgo G, Mastella G, Bonizzato A, Dognini M, Pignatti PF: Nine cystic fibrosis patients homozygous for the CFTR nonsense mutation R1162X have mild or moderate lung disease. J Med Genet 1992;29: 558–562.

50 Sheppard DN, Ostedgaard LS, Winter MC, Welsh MJ: Mechanism of dysfunction of two nucleotide binding domain mutations in cystic fibrosis transmembrane conductance regulator that are associated with pancreatic sufficiency. EMBO J 1995;14:876–883.

51 Sheppard DN, Rich DP, Ostedgaard LS, Gregory RJ, Smith AE, Welsh MJ: Mutations in CFTR associated with mild-disease-form Cl⁻ channels with altered pore properties. Nature 1993;362:160–164.

52 Van Oene M, Lukacs GL, Rommens JM: Cystic fibrosis mutations lead to carboxyl-terminal fragments that highlight an early biogenesis step of the cystic fibrosis transmembrane conductance regulator. J Biol Chem 2000;275:19577–19584.

53 Vankeerberghen A, Wei L, Jaspers M, Cassiman JJ, Nilius B, Cuppens H: Characterization of 19 disease-associated missense mutations in the regulatory domain of the cystic fibrosis transmembrane conductance regulator. Hum Mol Genet 1998;7: 1761–1769.

54 Vankeerberghen A, Wei L, Teng H, Jaspers M, Cassiman JJ, Nilius B, Cuppens H: Characterization of mutations located in exon 18 of the CFTR gene. FEBS Lett 1998;437: 1–4.

55 Andersson C, Dragomir A, Hjelte L, Roomans GM: Cystic fibrosis transmembrane conductance regulator (CFTR) activity in nasal epithelial cells from cystic fibrosis patients with severe genotypes. Clin Sci (Lond) 2002; 103:417–424.

56 Fajac I, Hubert D, Bienvenu T, Richaud-Thiriez B, Matran R, Kaplan JC, Dall'Ava-Santucci J, Dusser DJ: Relationships between nasal potential difference and respiratory function in adults with cystic fibrosis. Eur Respir J 1998;12:1295–1300.

57 Ho LP, Samways JM, Porteous DJ, Dorin JR, Carothers A, Greening AP, Innes JA: Correlation between nasal potential difference measurements, genotype and clinical condition in patients with cystic fibrosis. Eur Respir J 1997;10:2018–2022.

58 Wallace HL, Barker PM, Southern KW: Nasal airway ion transport and lung function in young people with cystic fibrosis. Am J Respir Crit Care Med 2003;168:594–600.

59 Fajac I, Hubert D, Guillemot D, Honore I, Bienvenu T, Volter F, Dall'Ava-Santucci J, Dusser DJ: Nasal airway ion transport is linked to the cystic fibrosis phenotype in adult patients. Thorax 2004;59:971–976.

60 Mall M, Grubb BR, Harkema JR, O'Neal WK, Boucher RC: Increased airway epithelial Na⁺ absorption produces cystic fibrosis-like lung disease in mice. Nat Med 2004;10: 487–493.

61 Veeze HJ, Halley DJ, Bijman J, de Jongste JC, de Jonge HR, Sinaasappel M: Determinants of mild clinical symptoms in cystic fibrosis patients. Residual chloride secretion measured in rectal biopsies in relation to the genotype. J Clin Invest 1994;93:461–466.

62 Hirtz S, Gonska T, Seydewitz HH, Thomas J, Greiner P, Kuehr J, Brandis M, Eichler I, Rocha H, Lopes AI, Barreto C, Ramalho A, Amaral MD, Kunzelmann K, Mall M: CFTR Cl⁻ channel function in native human colon correlates with the genotype and phenotype in cystic fibrosis. Gastroenterology 2004;127: 1085–1095.

63 Bronsveld I, Mekus F, Bijman J, Ballmann M, de Jonge HR, Laabs U, Halley DJ, Ellemunter H, Mastella G, Thomas S, Veeze HJ, Tummler B: Chloride conductance and genetic background modulate the cystic fibrosis phenotype of Delta F508 homozygous twins and siblings. J Clin Invest 2001;108:1705–1715.

64 Bronsveld I, Mekus F, Bijman J, Ballmann M, Greipel J, Hundrieser J, Halley DJ, Laabs U, Busche R, De Jonge HR, Tummler B, Veeze HJ: Residual chloride secretion in intestinal tissue of deltaF508 homozygous twins and siblings with cystic fibrosis. The European CF Twin and Sibling Study Consortium. Gastroenterology 2000;119:32–40.

65 Fulmer SB, Schwiebert EM, Morales MM, Guggino WB, Cutting GR: Two cystic fibrosis transmembrane conductance regulator mutations have different effects on both pulmonary phenotype and regulation of outwardly rectified chloride currents. Proc Natl Acad Sci USA 1995;92:6832–6836.

66 Mickle JE, Milewski MI, Macek M Jr, Cutting GR: Effects of cystic fibrosis and congenital bilateral absence of the vas deferens-associated mutations on cystic fibrosis transmembrane conductance regulator-mediated regulation of separate channels. Am J Hum Genet 2000;66: 1485–1495.

67 Masciovecchio MV, Gabbarini J, Vega M, Drittanti L: The interactivity between the CFTR gene and cystic fibrosis would be limited to the initial phase of the disease. Genet Med 2000;2:124–130.

68 Davidson DJ, Porteous DJ: Genetics and pulmonary medicine. 1. The genetics of cystic fibrosis lung disease. Thorax 1998;53: 389–397.

69 Campbell PW 3rd, Parker RA, Roberts BT, Krishnamani MR, Phillips JA 3rd: Association of poor clinical status and heavy exposure to tobacco smoke in patients with cystic fibrosis who are homozygous for the F508 deletion. J Pediatr 1992;120:261–264.

70 Gauderman WJ, Avol E, Gilliland F, Vora H, Thomas D, Berhane K, McConnell R, Kuenzli N, Lurmann F, Rappaport E, Margolis H, Bates D, Peters J: The effect of air pollution on lung development from 10 to 18 years of age. N Engl J Med 2004;351:1057–1067.

71 Gilliland FD, Gauderman WJ, Vora H, Rappaport E, Dubeau L: Effects of glutathione-S-transferase M1, T1, and P1 on childhood lung function growth. Am J Respir Crit Care Med 2002;166:710–716.

72 Gilliland FD, Li YF, Dubeau L, Berhane K, Avol E, McConnell R, Gauderman WJ, Peters JM: Effects of glutathione S-transferase M1, maternal smoking during pregnancy, and environmental tobacco smoke on asthma and wheezing in children. Am J Respir Crit Care Med 2002;166:457–463.

Julian Zielenski, PhD
Program in Genetics and Genomic Biology
The Hospital for Sick Children
555 University Avenue
Toronto, Ont. M4G 1X8 (Canada)
E-Mail jziel@genet.sickkids.on.ca

Bush A, Alton EWFW, Davies JC, Griesenbach U, Jaffe A (eds): Cystic Fibrosis in the 21st Century.
Prog Respir Res. Basel, Karger, 2006, vol 34, pp 69–76

Diagnosis of Cystic Fibrosis, *CFTR*-Related Disease and Screening

Barbara A. Karczeski[a]   Garry R. Cutting[a,b]

[a]Johns Hopkins University DNA Diagnostic Lab and [b]Johns Hopkins University School of Medicine, Baltimore, Md., USA

Abstract

Diagnosis of classic cystic fibrosis (CF) is based on a combination of specific clinical features and evidence of dysfunction of the CF transmembrane conductance regulator (CFTR). The latter is usually documented by elevated chloride concentration in the sweat although reduced nasal epithelial chloride secretion in response to beta adrenergic agents or presence of two disease-associated *CFTR* mutations are acceptable alternatives. Identification of *CFTR* mutations in a variety of disorders that manifest only one or a few features of CF presents a diagnostic challenge for the clinician. Differentiating CF from CFTR-related disorders may require extensive clinical evaluation and ancillary testing. CF is diagnosed in a growing number of patients in the newborn period due to government mandated or voluntary newborn screening programs. Diagnosis of CF in the newborn period has been associated with improved growth, but improvement in other features of CF is less clear. Antenatal carrier screening for CF has been introduced in the United States although its impact on CF incidence has not been formally evaluated. While newborn and population-based carrier screening programs are likely to identify a number of patients with CF, clinicians will have to remain vigilant for this life-limiting disorder, especially for cases with late or unusual presentations.

Introduction

The diagnosis of cystic fibrosis (CF) was based upon clinical findings alone until testing for the concentration of sweat chloride was introduced in 1953 [1]. In 2003, the median age of diagnosis was 7 months, with 50% being diagnosed by age 6 months and 90% by age 8 months [2, 3]. The typical CF patient has one or two characteristic clinical symptoms (e.g. pancreatic insufficiency, recurrent pneumonia) and an abnormal sweat test [4]. Early and accurate diagnosis is medically important [5] and allows avoidance of unnecessary testing, prompt access to specialized services, appropriate therapy, prognostic information and genetic counseling [4]. Accurate diagnosis of the atypical CF patient is becoming increasingly difficult in light of an expanding phenotype associated with the disorder [5]. The emphasis of health care systems around the world is one of early diagnosis and the provision to families of the opportunity to avoid the birth of additional affected children. Though CF alleles will always be with us, current population screening programs may have a significant effect on the incidence of CF in the future.

The Diagnosis of CF

Classic CF is defined as progressive obstructive lung disease, pancreatic insufficiency and congenital absence of the vas deferens in males. Diagnostic criteria for classic CF [4] are listed in table 1. Nonclassic CF is CF in the absence of the classic triad of symptoms [6]. Generally, these patients are pancreatic sufficient [3] with mild or absent lung disease and normal or borderline sweat chloride levels [7]. Nonclassic CF represents 2% to 10% [7] of all CF diagnoses. Most nonclassic CF patients meet the same

Table 1. Diagnostic criteria for CF [4]

'Risk factor' for CF		Lab evidence of CFTR dysfunction
≥1 phenotypic symptom *or* Positive NBS result *or* Positive history of CF in a sibling	*plus*	Positive sweat test *or* Positive NPD *or* 2 mutations[a] in CFTR

[a] CF mutations must be known to be disease causing.

diagnostic criteria as classic CF patients, however they may have a later presentation or have involvement of only one or two organ systems [7]. The majority of patients with non-classic CF carry one severe and one mild or two mild *CFTR* mutations, with mild mutations categorized as those allowing varying degrees of residual CFTR function [3, 8, 8a]. For a description of *CFTR* mutation classes, see the genotype/phenotype review by Zielenski [9] or elsewhere in this publication. Recognition of nonclassic forms of CF is important for prognostic purposes since many patients follow a significantly different clinical course [7].

Recommended Tests and Evaluations

Sweat testing by quantitative pilocarpine iontophoresis is the gold standard test for confirming the diagnosis of CF [4, 5, 10, 11]. Quantitation of chloride concentration is required, though sodium, conductivity and osmolality may also be measured [10]. Based upon studies of classic CF patients, it was established that a sweat test in which chloride concentration is >60 mmol/l is diagnostic for CF in the context of characteristic clinical findings [4, 10]. Obligate CF carriers (parents of CF patients) do not have elevated sweat chloride concentrations [12, 13]. About 98% of patients have a sweat chloride concentration in the diagnostic range; however, 2% of patients with convincing clinical evidence of CF have borderline chloride concentration (between 40 and 60 mmol/l) [10]. Sweat testing has a false-positive and false-negative rate of 15 and 12%, respectively [5]. Sweat chloride concentration increases with age [10, 13], so sweat tests must be interpreted in an age-specific fashion [4, 5, 13]. They must be performed at least twice in each patient [5], preferably several weeks apart. Sweat testing should be performed only by experienced technicians using standardized protocols (http://www.acb.org.uk/Guidelines/sweat.htm) [4, 5]. Sources of error in sweat testing include evaporation, condensation, contamination, and technical or interpretive mistakes [10].

Molecular analysis of the *CFTR* gene alone neither establishes nor excludes the diagnosis of CF. Two disease-associated *CFTR* mutations in the context of the appropriate clinical or family history are needed to establish the diagnosis [4, 5]. The finding of one or no *CFTR* mutations on molecular analysis does not exclude the diagnosis of CF [4]: Mekus et al. [14] and Groman et al. [8, 8a] reported patients with nonclassic CF and no evidence of *CFTR* mutation after exhaustive evaluation. The presence of complex genotypes, modifying factors and ameliorating mutations necessitates that most cases of CF be diagnosed on the strength of clinical findings [15]. DNA-based testing for CF is useful when sweat testing is not feasible or results are borderline or highly variable [4]. DNA testing should be performed in molecular genetic laboratories with regulatory oversight and an established sensitivity and specificity for their *CFTR* mutation panel for the major racial and ethnic populations served [16]. The American College of Medical Genetics (ACMG) published guidelines for laboratories regarding informed consent, reports and sensitivity and has recommended a basic mutation panel [16–18]. Extended mutation panels, gene scanning or gene sequencing should be for diagnostic purposes only since the disease association of many rare *CFTR* variants is unknown [16, 17].

Nasal potential difference (NPD) is a sensitive test of electrolyte transport that can be used to support or refute a diagnosis of CF [19]. NPD, like sweat testing, must be performed on infants greater than 48 h old [10] using validated equipment and a well-defined standard protocol [4, 19] and must be repeated at least once in order to be diagnostic [5]. NPD is usually only available at highly specialized CF care centers and may be difficult to perform in infants and small children who cannot cooperate with the investigator. Ancillary testing (table 2) is available to assist in the diagnosis of CF, assess disease severity, and plan appropriate patient-specific treatment [4, 5, 20].

Though the general approach to establish CFTR dysfunction remains sweat testing any patient in whom CF is a clinical suspicion, followed by NPD and/or molecular analysis in uncertain cases [2], there are special populations in which molecular analysis, NPD or other tests are of primary importance. Identification of echogenic bowel on fetal ultrasound raises the suspicion of CF that can be addressed by molecular analysis of the fetus and/or parents [21–23]. Adult patients represent a growing proportion of patients receiving a primary diagnosis of CF [24], and

Table 2. Ancillary testing available to aid in the diagnosis of CF or assess disease severity [4, 5, 20]

Ancillary evaluations	Test methods
Pancreatic function	Fecal: 72 h fecal fat, chymotrypsin activity, elastase 1 level, immunoreactive lipase Blood: IRT, pancreas-associated protein Urine: Chymotrypsin and pancreatic esterase activities using synthetic peptides and excretion ratios Direct: pH, bicarbonate levels, and enzyme activities of chymotrypsin, trypsin, amylase, lipase and carboxypeptidase measured in duodenal juice
Lung function/airway involvement	Spirometry, chest X-ray, sinus radiographs, bacterial cultures
Fertility	Genitourinary evaluation by physical exam, ultrasound and/or semen analysis

Table 3. Causes of elevated sweat chloride levels [adapted from 10]

Metabolic disorders
    CF
    Mucopolysaccharidosis, type 1
    Pseudohypoaldosteronism
    Glycogen storage disease, type 1
    Glucose-6-phosphate dehydrogenase deficiency
Dermatologic conditions
    Ectodermal dysplasia
    Atopic dermatitis
Environment
    Protein calorie malnutrition
    Psychosocial failure to thrive
Other causes
    Klinefelter syndrome
    Autonomic dysfunction
    Nephrosis

typically present with pancreatitis, chronic sinusitis or male infertility. Molecular analysis and NPD may be more valuable in this population than sweat testing, which may be normal or borderline.

Several other disorders can present with clinical findings similar to those seen in CF patients. Primary ciliary dyskinesia (PCD; OMIM#242650) is a disorder that involves recurrent respiratory infections, bronchiectasis and chronic sinusitis, though about half of the patients also have visceral mirror image arrangement. Sweat chloride concentration is normal in PCD patients, thus distinguishing this disorder from CF in almost all cases. Ultrastructural and functional studies of cilia can establish PCD, but these tests are not commonly available. Shwachman-Diamond syndrome (SDS; OMIM#260400) is an autosomal recessive disorder involving pancreatic insufficiency as a prominent feature, but patients have normal sweat chloride levels and most have hematologic abnormalities such as neutropenia. As the gene for this condition has now been identified [25], both sweat chloride testing and DNA analysis can help distinguish SDS and CF. Patients with Young syndrome (OMIM#279000) have obstructive azospermia, chronic sinopulmonary infections and bronchiectasis, but have normal sweat chloride levels, NPDs and pancreatic function. A number of causes of elevated sweat chloride concentration have been documented and should be considered in patients lacking characteristic clinical features of CF (table 3) [10].

CFTR-Related Disease

CF-related disease can be broken into two categories: monosymptomatic (single organ) manifestations of CF and conditions that are not CF but in which patients demonstrate a higher frequency of *CFTR* mutations than expected by chance. Male subfertility due to congenital bilateral absence of the vas deferens (CBAVD) is a nearly invariant finding in males with classic CF and can be a presenting feature [26]. A separate autosomal recessive condition with precisely the same findings in the male reproductive tract in the absence of other organ disease seen in CF patients has been recognized since 1980 (CBAVD; OMIM #277180). Intriguingly, most but not all cases of isolated CBAVD are caused by mutation in the *CFTR* gene. The CF mutations identified in men with CBAVD differ from those seen in the CF population, and include an increased frequency of the 5T variant of intron 8 and mild *CFTR* mutations [27]. Several studies [26, 27] have reported an increased rate of mutations in CBAVD patients, including both 'CF mutation/5T' and '5T only' genotypes. Isolated CBAVD and CF can be difficult to distinguish since some men with CBAVD have borderline sweat chloride concentration and mild sinopulmonary involvement [7]. The latter cases probably represent the mildest end of the CF spectrum. It is not known whether these individuals will develop life-limiting lung disease. Thus, prudence dictates that these men be followed routinely for signs of CF lung disease.

Pancreatitis can be a clinical presentation in CF patients who are pancreatic sufficient [28, 29]. These patients usually have 2 *CFTR* mutations and other clinical findings suggestive of CF. Isolated chronic pancreatitis is also seen in patients without CF, though elevated sweat chloride levels have been associated with the disease [28]. Pancreatitis can be associated with certain metabolic conditions [30], but idiopathic chronic pancreatitis (ICP) and alcohol-induced pancreatitis have an increased frequency of *CFTR* mutations. There is dissent over whether the 5T variant occurs at a higher frequency in this population [28, 30]. Audrezet et al. [31] studied *CFTR*, *PRSS1* (cationic trypsinogen) and *PSTI* (pancreatic secretory trypsin inhibitor) and suggested possible synergistic involvement or multiple susceptibility loci contributing to ICP. Finally, Cohn et al. [29] in 2002 identified 9/39 ICP patients with two *CFTR* mutations, at least one of the two mutations being classified as mild in each case. Thus, *CFTR* mutation analysis may be warranted in cases of chronic pancreatitis of unknown etiology.

Disseminated bronchiectasis is a feature of disorders such as CF, α_1-antitrypsin deficiency, PCD and immunodeficiencies [32]. Girodon et al. [32] identified *CFTR* mutations in 11/32 disseminated bronchiectasis patients studied, including four compound heterozygotes, and determined that the frequency of *CFTR* mutations was higher than expected in the general population, however many of the sequence variations identified in this study were novel changes and not bona fide CF disease-causing mutations. They concluded that environmental factors and other genetic factors also played a role. Allergic bronchopulmonary aspergillosis (ABPA) is a hypersensitivity to *Aspergillus fumigatus* and is most commonly identified in atopic asthmatics [33]. It is also a complication in 2–11% of CF patients [34]. A US study [35] identified a higher than expected rate of CF carriers after *CFTR* gene sequencing in 11 ABPA/asthma patients and demonstrated a significant difference in carrier rate between this group and both the general population and patients with chronic bronchitis. A study from New Zealand [33] found a higher than expected number of carriers of common CF mutations among 31 ABPA patients while a Belgian study [34] found a significantly higher frequency of *CFTR* mutations in ABPA patients as compared to control patients and allergic asthmatics. Thus there is a higher rate of *CFTR* alleles than expected by chance in this patient population, but there likely exists a significant environmental contribution [3]. There is some evidence that dysfunctional CFTR may play a role in asthma and chronic obstructive pulmonary disease [36]. A study of 55 adult onset nasal polyposis patients did not find a significantly higher rate of *CFTR* mutations [37].

Sinusitis is both an invariant feature of CF [38] and a common finding in the general population. Though there are a several proposed causes, genetic factors may play a significant role [3, 38]. Several studies [3, 38–40] have identified an excess of *CFTR* mutations and the polymorphic *CFTR* variant M470V in the chronic sinusitis population. Some of these patients have borderline sweat chloride levels and may have a second undetected mutation [39]. Knowledge of CF status in sinusitis patients allows closer vigilance for CF pathogens such as *Pseudomonas aeruginosa* [39].

Newborn Screening for CF

Newborn screening (NBS) programs for CF were originally proposed in the 1970s [41]. Advances such as immunoreactive trypsinogen (IRT) measurements (first employed in New Zealand in 1979) and DNA testing (available after 1989) [41] facilitated implementation of NBS for CF, making it easier and more cost-effective, and leading to an increase in existing programs. Most programs currently employ a protocol that combines IRT measurement followed by DNA testing for a limited number of mutations and achieve a sensitivity of near 90% [5, 41–43]. NBS for CF has been rationalized based upon lagging growth and head circumference of CF patients diagnosed after the newborn period compared to those diagnosed as newborns [42, 44].

Victoria, Australia introduced CF screening in 1989, and currently employs a first level screen of IRT and ΔF508 typing [45]. Infants with positive IRT and no evidence of ΔF508 receive expanded DNA testing using a 16 mutation panel. Victoria has screened more than 635,000 infants and is reporting a sensitivity of 94% [45]. Comparing cohorts of CF patients born pre- and post-screening, Waters et al. [46] report that the screened cohort was longer and heavier at diagnosis, heavier at age 1, taller at age 5, had significantly better pulmonary function tests, higher Shwachman clinical scores and fewer inpatient hospital stays than the clinically diagnosed cohort. The screened group also demonstrated better nutritional status. More patients in the screened group were pancreatic sufficient at diagnosis, but by age 10, the number of PS patients in each group was similar. There was no difference in chest X-rays at age 10 between the two groups.

The Wisconsin, USA, program has published the only randomized controlled clinical trial of NBS for CF [47]. More than 650,000 newborns were screened by IRT and IRT/DNA protocols. Though the screened group had a higher frequency of ΔF508 and pancreatic insufficient patients, patients still demonstrated increased height, weight and head circumference, with differences persisting after treatment was initiated in the control (nonscreened) group. No differences in lung function were identified. Cognitive function in this study population was also examined in mid-childhood: of children exhibiting vitamin E insufficiency at the time of diagnosis, those identified through NBS had significantly higher cognitive skills index scores than their traditionally diagnosed counterparts, most likely due to the prevention of a prolonged malnourished state [48].

Worldwide, NBS for CF is represented by a patchwork of local and regional programs with a few universal programs. National programs exist in New Zealand, Scotland, Wales, France, and Denmark [49]. Regional or local programs exist through portions of the US, Australia, Austria, Belgium, Italy, Poland, and most of the UK (which is organizing a national program) [49]. Young et al. [50] provide a list of infrastructure elements required to properly support a NBS program.

Diagnostic costs per child with CF have been reported as USD 7,500 per case for infants diagnosed by NBS and USD 11,400 per case diagnosed clinically [1]. In France (at a cost of USD 2 per test) the cost per affected infant identified was reported as USD 6,800, with other countries spending USD 3,000–32,000 per affected child [50]. Of course, the ultimate cost-effectiveness of any program is highly dependent upon the screening protocols employed [50].

Affected children are referred to specialized care more rapidly than with standard clinical diagnosis [42]. Families are alerted to their risks as early as possible allowing for more informed reproductive decision-making [1]. NBS may also cause less stress in families than conventional diagnosis [1]. Parsons and Bradley [51] concluded that delayed or incorrect clinical diagnosis presented a greater source of stress to families than presymptomatic diagnosis. NBS also provides one of the only opportunities to study the natural history of CF and establishes the population needed to study the efficacy of presymptomatic therapies [44].

Other issues regarding NBS have been identified and need to be addressed by NBS programs.

(1) Because DNA testing is involved, informed consent is required [1].
(2) The lack of genotype-phenotype relationships, especially regarding the severity of lung disease, makes prognosis difficult in screen-positive infants [1, 44].
(3) Clinically, presymptomatic patients detected through NBS may be at risk for overaggressive treatment with antibiotics and pancreatic enzymes [42] as well as early

and frequent exposure to pathogens at CF centers [42, 52].
(4) A child with a false-negative NBS result is likely to be part of an atypical population and represent a more atypical case, and may experience a severe delay in diagnosis [53].
(5) An unwanted byproduct of NBS is the detection of healthy carriers [42, 44], which removes the child's right to choose or not choose to learn about carrier status later in life [51].
(6) As more data is gathered on monosymptomatic presentation in CF carriers, these issues must be addressed with parents so families are fully informed [52].
(7) Even with extensive educational efforts, complete parental understanding of false-positive NBS results in a carrier infant is in doubt [51].
(8) On a programmatic level, most health care systems are not focused on preventative interventions in presymptomatic children [42], and nonuniform programs create disparities in detection [53].

Population-Based/Antenatal Carrier Screening for CF

Population screening for carriers of single gene disorders has been in place for over 20 years [54–56] with spectacular successes and ignominious failures. Carrier testing for CF has historically been reserved for those with a family history of CF (siblings and reproductive partners of CF patients). Population-based CF carrier screening has been carried out in Scotland, UK, and the US. After 5 years of population screening in Scotland, Cunningham and Marshall [57] reported 76% uptake. A decrease in the number of CF cases in their pediatric population from 4.6 cases per year to 1.6 cases was noted over the 5-year period, although this decrease could not be attributed to the termination of affected pregnancies alone. The report concluded that antenatal screening avoids the anxiety of NBS and false-positive results and offers the opportunity to terminate any affected pregnancy and should be offered in the absence of pressure to proceed to prenatal diagnosis and pregnancy termination.

In the US, the National Institutes of Health (NIH) [58] issued a recommendation to expand carrier testing to include couples seeking prenatal care or planning a pregnancy in which one or both members are Caucasian, and to make couples of other ethnicities aware of the existence of population screening for CF carrier status. Since few studies of the benefits or risks of carrier screening [59] or the impact these recommendations would have on health care services [60] have been published, the American College of Obstetricians and Gynecologists (ACOG) and ACMG issued guidelines for implementation of CF carrier screening and laboratory standards for testing facilities [17]. These documents discuss the target population, screening models, criteria for mutation selection, potential value and correct use of extended testing panels, a policy on testing for 'mild' CF mutations, recommendations for reporting, interpretation and genetic counseling as well as laboratory quality assurance and quality control. The ACMG proposed a *CFTR* mutation panel for population-based carrier screening in the US [18].

Models for CF Carrier Screening

Preconception versus Prenatal Screening
Preconception screening is preferred [12, 17] because it offers couples time to process their risk and the option of avoiding pregnancy if desired. However, it is recognized that the majority of couples will be screened once a pregnancy is under way. These couples have a more immediate interest in the results [12]. Couples who are already pregnant are also more available to screening programs since they have become part of a health care delivery system for prenatal care. The high rate of unplanned pregnancy [61] means preconception screening is not available to a significant proportion of the target population [62].

Couple versus Sequential Screening
Couple-based programs will only test if both members of the couple are available, and only report the results and risks in terms of the couple as a screening unit. Sequential screening begins with the female member of the couple and tests the male only if the female is a carrier. Sequential screening is the model preferred by many because it allows full disclosure of individual results. In theory, sequential screening allows a carrier to notify the extended family for additional carrier screening [63], though this was not the case in one study by Brock [63] where only 10% of relatives sought testing. Full disclosure will identify couples with a minimally elevated risk (one individual tests positive and one tests negative) but provide no additional options for risk assessment or diagnosis [1, 44]. The couple model is favored by some [60] because testing would only proceed if both samples were available, however screening is not an option for pregnancies in which the male is unavailable, and rescreening is required with each new partner [62]. Miedzybrodzka et al.

[64] conducted a randomized trial of these two models and concluded that sequential screening was the better overall model. Over 5 years' experience in Scotland, Brock [63] evaluated both models as well, concluding that couple screening proved to be the better theoretical model, as both members of the couple were needed to obtain information regarding the risk of CF in any pregnancy.

Issues Regarding Population Screening for CF Carriers

Carrier screening for CF differs from other population screening programs in several ways.
(1) CF does not involve CNS or musculoskeletal abnormalities, and there is an increase in survival rates and treatment options [59].
(2) Screening relies solely on molecular testing – there is no biochemical marker useful for detection of carrier status [1, 12]. Reliance on mutation identification makes these programs difficult to implement because of the heterogeneous nature of *CFTR* mutations, the considerable ethnic variation in mutation groups and frequencies, and the wide range of disease severity with very little information available from genotype alone [12].
(3) The sensitivity of CF carrier screening falls short of the preferred 95% cutoff for other clinical lab tests [12].
(4) Ethnicity of the patient is vital to result interpretation and risk assessment. It can be difficult to confirm ethnicity, especially in heterogeneous populations like the United States.
(5) CF carrier screening of the prenatal population offers only the intervention of prenatal diagnosis followed by pregnancy termination if affected [12]. Brock [63] reported pregnancy outcomes over a 5-year period of antenatal carrier screening in Scotland.

Pretest education and informed consent are of paramount importance. The information provided by a carrier test needs to be interpreted by the laboratory and health care provider and understood by the couple in light of the ethnicities and a priori risk involved [65]. These issues are compounded by a lack of trained genetics professionals to handle this often complex counseling and educational process [12]. There are psychological costs associated with the process of prenatal diagnosis and pregnancy termination [62] and this option is simply not acceptable to many couples. There is also a risk that CF carriers could experience anxiety, stigmatization, and actual or feared discrimination [62].

In an editorial review of the impact of CF carrier screening on clinical practice in the US, Farrell and Fost [59] argued that truly informed consent was difficult to obtain in routine practice, and that patients demonstrate poor retention of information and confusion over their carrier status and residual risk after a negative test. For consent to be fully informed, issues of nonpaternity and possible genetic discrimination should be part of the counseling process. ACOG's recommendation of a written decline of testing (not the norm for 'routine' tests in pregnancy) risks implying to the patient an artificially inflated idea of the importance of testing. The costs, both monetary and psychological, of population-based CF carrier screening remain a matter of debate [1, 42, 59, 60, 62].

recognized triad of organ disease, and this expansion presents a challenge to today's clinicians. Accurate diagnosis is even more important in light of the wide range of outcomes associated with this newly defined CF spectrum: from life-limiting classic CF to the chronic, though usually milder, CFTR-related disorders. Screening for CF in both the newborn and antenatal setting will identify an increasing number of patients who might benefit from early diagnosis. However these screening programs will create a new diagnostic challenge by contributing to the cohort of patients who are at risk of underdiagnosis: the adult CF patient, the nonclassic CF patient, and the patient with CFTR-related disease, each of whom may escape detection by screening programs designed to detect classic forms of CF.

Conclusions

The breadth of disorders associated with CFTR dysfunction has expanded significantly beyond the historically

References

1 Bonham JR, Downing M, Dalton A: Screening for cystic fibrosis: The practice and the debate. Eur J Pediatr 2003;162(suppl 1): S42–S45.
2 Ratjen F, Doring G: Cystic fibrosis. Lancet 2003;361:681–689.
3 Noone PG, Knowles MR: 'CFTR-opathies': Disease phenotypes associated with cystic fibrosis transmembrane regulator gene mutations. Respir Res 2001;2:328–332.
4 Rosenstein BJ, Cutting GR: The diagnosis of cystic fibrosis: A consensus statement. J Pediatr 1998;132:589–595.
5 Wang L, Freedman SD: Laboratory tests for the diagnosis of cystic fibrosis. Am J Clin Pathol 2002;117(suppl):S109–S115.
6 Knowles MR, Durie PR: What is cystic fibrosis? N Engl J Med 2002;347:439–442.
7 Boyle MP: Nonclassic cystic fibrosis and CFTR-related diseases. Curr Opin Pulm Med 2003;9:498–503.
8 Groman JD, Meyer ME, Wilmott RW, Zeitlin PL, Cutting GR: Variant cystic fibrosis phenotypes in the absence of CFTR mutations. N Engl J Med 2002;347:401–407.
8a Groman JD, Karczeski B, Sheridan M, Robinson T, Fallin D, Cutting GR: Phenotypic and genetic characterization of patients with features of "nonclassic" forms of cystic fibrosis. J Pediatr 2005;146:675–680.
9 Zielenski J: Genotype and phenotype in cystic fibrosis. Respiration 2000;67:117–133.
10 LeGrys VA: Sweat testing for the diagnosis of cystic fibrosis: Practical considerations. J Pediatr 1996;129:892–897.
11 Baumer JH: Evidence based guidelines for the performance of the sweat test for the investigation of cystic fibrosis in the UK. Arch Dis Child 2003;88:1126–1127.

12 Grody WW: Cystic fibrosis: Molecular diagnosis, population screening, and public policy. Arch Pathol Lab Med 1999;123:1041–1046.
13 Farrell PM, Koscik RE: Sweat chloride concentrations in infants homozygous or heterozygous for F_{508} cystic fibrosis. Pediatrics 1996;97:524–528.
14 Mekus F, Ballmann M, Bronsveld I, Dörk T, Bijman J, Tummler B: Cystic-fibrosis-like disease unrelated to the cystic fibrosis transmembrane conductance regulator. Hum Genet 1998;102:582–586.
15 Chmiel JF, Drumm ML, Konstan MW, Ferkol TW, Kercsmar CM: Pitfall in the use of genotype analysis as the sole diagnostic criterion for cystic fibrosis. Pediatrics 1999;103: 823–826.
16 Richards CS, Bradley LA, Amos J, Allitto B, Grody WW, Maddalena A, McGinnis MJ, Prior TW, Popovich BW, Watson MS, Palomaki GE: Standards and guidelines for CFTR mutation testing. Genet Med 2002;4: 379–391.
17 Grody WW, Cutting GR, Klinger KW, Richards CS, Watson MS, Desnick RJ: Laboratory standards and guidelines for population-based cystic fibrosis carrier screening. Genet Med 2001;3:149–154.
18 Watson MS, Cutting GR, Desnick RJ, Driscoll DA, Klinger K, Mennuti M, Palomaki GE, Popovich BW, Pratt VM, Rohlfs EM, Strom CM, Richards CS, Witt DR, Grody WW: Cystic fibrosis population carrier screening: 2004 revision of American College of Medical Genetics mutation panel. Genet Med 2004;6: 387–391.
19 Schuler D, Sermet-Gaudelus I, Wilschanski M, Ballmann M, Dechaux M, Edelman A, Hug M, Leal T, Lebacq J, Lebecque P,

Lenoir G, Stanke F, Wallemacq P, Tummler B, Knowles MR: Basic protocol for transepithelial nasal potential difference measurements. J Cyst Fibros 2004;3(suppl 2): 151–155.
20 Leus J, Van Biervliet S, Robberecht E: Detection and follow up of exocrine pancreatic insufficiency in cystic fibrosis: A review. Eur J Pediatr 2000;159:563–568.
21 Muller F, Simon-Bouy B, Girodon E, Monnier N, Malinge MC, Serre JL: Predicting the risk of cystic fibrosis with abnormal ultrasound signs of fetal bowel: Results of a French molecular collaborative study based on 641 prospective cases. Am J Med Genet 2002;110: 109–115.
22 Al Kouatly HB, Chasen ST, Streltzoff J, Chervenak FA: The clinical significance of fetal echogenic bowel. Am J Obstet Gynecol 2001;185:1035–1038.
23 Nicholls CM, Nelson PV, Poplawski NK, Chin SJ, Fong BA, Solly PB, Fietz MJ, Fletcher JM: Analysis of CFTR mutation screening in cases of isolated fetal echogenic bowel in the South Australian population. Prenat Diagn 2003;23: 1023–1025.
24 Yankaskas JR, Marshall BC, Sufian B, Simon RH, Rodman D: Cystic fibrosis adult care: Consensus conference report. Chest 2004; 125:1S–39S.
25 Boocock GR, Morrison JA, Popovic M, Richards N, Ellis L, Durie PR, Rommens JM: Mutations in SBDS are associated with Shwachman-Diamond syndrome. Nat Genet 2003;33:97–101.
26 Dork T, Dworniczak B, Aulehla-Scholz C, Wieczorek D, Bohm I, Mayerova A, Seydewitz HH, Nieschlag E, Meschede D, Horst J, Pander HJ, Sperling H, Ratjen F,

Passarge E, Schmidtke J, Stuhrmann M: Distinct spectrum of CFTR gene mutations in congenital absence of vas deferens. Hum Genet 1997;100:365–377.

27 Mak V, Zielenski J, Tsui LC, Durie P, Zini A, Martin S, Longley TB, Jarvi K: Proportion of cystic fibrosis gene mutations not detected by routine testing in men with obstructive azoospermia. JAMA 1999;281:2217–2224.

28 Cohn JA, Friedman KJ, Noone PG, Knowles MR, Silverman LM, Jowell PS: Relation between mutations of the cystic fibrosis gene and idiopathic pancreatitis. N Engl J Med 1998;339:653–658.

29 Cohn JA, Noone PG, Jowell PS: Idiopathic pancreatitis related to CFTR: Complex inheritance and identification of a modifier gene. J Investig Med 2002;50:247S–255S.

30 Sharer N, Schwarz M, Malone G, Howarth A, Painter J, Super M, Braganza J: Mutations of the cystic fibrosis gene in patients with chronic pancreatitis. N Engl J Med 1998;339: 645–652.

31 Audrezet MP, Chen JM, Le Marechal C, Ruszniewski P, Robaszkiewicz M, Raguenes O, Quere I, Scotet V, Ferec C: Determination of the relative contribution of three genes – the cystic fibrosis transmembrane conductance regulator gene, the cationic trypsinogen gene, and the pancreatic secretory trypsin inhibitor gene – to the etiology of idiopathic chronic pancreatitis. Eur J Hum Genet 2002;10: 100–106.

32 Girodon E, Cazaneuve C, Lebargy F, Chinet TC, Costes B, Ghanem N, Martin J, Lemay S, Scheid P, Housset B, Bignon J, Goossens M: CFTR gene mutations in adults with disseminated bronchiectasis. Eur J Hum Genet 1997;5:149–155.

33 Eaton TE, Weiner MP, Garrett JE, Cutting GR: Cystic fibrosis transmembrane conductance regulator gene mutations: Do they play a role in the aetiology of allergic bronchopulmonary aspergillosis? Clin Exp Allergy 2002;32: 756–761.

34 Marchand E, Verellen-Dumoulin C, Mairesse M, Delaunois L, Brancaleone P, Rahier JF, Vandenplas O: Frequency of cystic fibrosis transmembrane conductance regulator gene mutations and 5T allele in patients with allergic bronchopulmonary aspergillosis. Chest 2001;119:762–767.

35 Miller PW, Hamosh A, Macek M Jr, Greenberger PA, MacLean J, Walden SM, Slavin RG, Cutting GR: Cystic fibrosis transmembrane conductance regulator (CFTR) gene mutations in allergic bronchopulmonary aspergillosis. Am J Hum Genet 1996;59: 45–51.

36 Tzetis M, Efthymiadou A, Strofalis S, Psychou P, Dimakou A, Pouliou E, Doudounakis S, Kanavakis E: CFTR gene mutations – including three novel nucleotide substitutions – and haplotype background in patients with asthma, disseminated bronchiectasis and chronic obstructive pulmonary disease. Hum Genet 2001;108:216–221.

37 Irving RM, McMahon R, Clark R, Jones NS: Cystic fibrosis transmembrane conductance regulator gene mutations in severe nasal polyposis. Clin Otolaryngol 1997;22:519–521.

38 Wang X, Moylan B, Leopold DA, Kim J, Rubenstein RC, Togias A, Proud D, Zeitlin PL, Cutting GR: Mutation in the gene responsible for cystic fibrosis and predisposition to chronic rhinosinusitis in the general population. JAMA 2000;284:1814–1819.

39 Raman V, Clary R, Siegrist KL, Zehnbauer B, Chatila TA: Increased prevalence of mutations in the cystic fibrosis transmembrane conductance regulator in children with chronic rhinosinusitis. Pediatrics 2002;109:E13.

40 Coste A, Girodon E, Louis S, Pruliere-Escabasse V, Goossens M, Peynegre R, Escudier E: Atypical sinusitis in adults must lead to looking for cystic fibrosis and primary ciliary dyskinesia. Laryngoscope 2004;114: 839–843.

41 Gregg RG, Simantel A, Farrell PM, Koscik R, Kosorok MR, Laxova A, Laessig R, Hoffman G, Hassemer D, Mischler EH, Splaingard M: Newborn screening for cystic fibrosis in Wisconsin: Comparison of biochemical and molecular methods. Pediatrics 1997;99: 819–824.

42 Farrell MH, Farrell PM: Newborn screening for cystic fibrosis: Ensuring more good than harm. J Pediatr 2003;143:707–712.

43 Wilcken B, Wiley V: Newborn screening methods for cystic fibrosis. Paediatr Respir Rev 2003;4:272–277.

44 Castellani C: Evidence for newborn screening for cystic fibrosis. Paediatr Respir Rev 2003;4:278–284.

45 Massie RJ, Olsen M, Glazner J, Robertson CF, Francis I: Newborn screening for cystic fibrosis in Victoria: 10 years' experience (1989–1998). Med J Aust 2000;172:584–587.

46 Waters DL, Wilcken B, Irwing L, Van Asperen P, Mellis C, Simpson JM, Brown J, Gaskin KJ: Clinical outcomes of newborn screening for cystic fibrosis. Arch Dis Child Fetal Neonatal Ed 1999;80:F1–F7.

47 Farrell PM, Kosorok MR, Rock MJ, Laxova A, Zeng L, Lai HC, Hoffman G, Laessig RH, Splaingard ML: Early diagnosis of cystic fibrosis through neonatal screening prevents severe malnutrition and improves long-term growth. Pediatrics 2001;107:1–13.

48 Koscik RL, Farrell PM, Kosorok MR, Zaremba KM, Laxova A, Lai HC, Douglas JA, Rock MJ, Splaingard ML: Cognitive function of children with cystic fibrosis: Deleterious effect of early malnutrition. Pediatrics 2004; 113:1549–1558.

49 Southern KW, Littlewood JM: Newborn screening programmes for cystic fibrosis. Paediatr Respir Rev 2003;4:299–305.

50 Young SS, Kharrazi M, Pearl M, Cunningham G: Cystic fibrosis screening in newborns: Results from existing programs. Curr Opin Pulm Med 2001;7:427–433.

51 Parsons EP, Bradley DM: Psychosocial issues in newborn screening for cystic fibrosis. Paediatr Respir Rev 2003;4:285–292.

52 Wagener JS, Sontag MK, Accurso FJ: Newborn screening for cystic fibrosis. Curr Opin Pediatr 2003;15:309–315.

53 Bobadilla JL, Macek M, Fine JP, Farrell PM: Cystic fibrosis: A worldwide analysis of CFTR mutations – correlation with incidence data and application to screening. Hum Mutat 2002;19:575–606.

54 Modell B, Kuliev A: The history of community genetics: The contribution of the haemoglobin disorders. Community Genet 1998;1: 3–11.

55 D'Souza G, McCann CL, Hedrick J, Fairley C, Nagel HL, Kushner JD, Kessel R: Tay-Sachs disease carrier screening: A 21-year experience. Genet Test 2000;4:257–263.

56 Vallance H, Ford J: Carrier testing for autosomal-recessive disorders. Crit Rev Clin Lab Sci 2003;40:473–497.

57 Cunningham S, Marshall T: Influence of five years of antenatal screening on the paediatric cystic fibrosis population in one region. Arch Dis Child 1998;78:345–348.

58 Genetic testing for cystic fibrosis: National Institutes of Health Consensus Development Conference Statement on genetic testing for cystic fibrosis. Arch Intern Med 1999;159: 1529–1539.

59 Farrell PM, Fost N: Prenatal screening for cystic fibrosis: Where are we now? J Pediatr 2002;141:758–763.

60 Haddow JE: Prenatal screening for cystic fibrosis in the United States – time to re-evaluate implementation policies. J Med Screen 2003;10:105–106.

61 Besculides M, Laraque F: Unintended pregnancy among the urban poor. J Urban Health 2004;81:340–348.

62 Miedzybrodzka Z, Haites N, Hall M, Templeton A, Marteau T, Dean J, Kelly K, Russell I: Antenatal cystic fibrosis carrier screening – whether, when and how? Paediatr Perinat Epidemiol 1993;7:368–375.

63 Brock DJ: Prenatal screening for cystic fibrosis: 5 years' experience reviewed. Lancet 1996;347:148–150.

64 Miedzybrodzka ZH, Hall MH, Mollison J, Templeton A, Russell IT, Dean JC, Kelly KF, Marteau TM, Haites NE: Antenatal screening for carriers of cystic fibrosis: Randomised trial of stepwise v couple screening. BMJ 1995;310:353–357.

65 Lyon E, Miller C: Current challenges in cystic fibrosis screening. Arch Pathol Lab Med 2003;127:1133–1139.

Garry Cutting, MD
Johns Hopkins University
733 N. Broadway
BRB Suite 551
Baltimore, MD 21287 (USA)
Tel. +1 410 955 1773
Fax +1 410 614 0211
E-Mail gcutting@jhmi.edu

     Karczeski/Cutting

Bush A, Alton EWFW, Davies JC, Griesenbach U, Jaffe A (eds): Cystic Fibrosis in the 21st Century.
Prog Respir Res. Basel, Karger, 2006, vol 34, pp 77–83

CF Modifier Genes

Uta Griesenbach   Eric W.F.W. Alton   Jane C. Davies

Department of Gene Therapy, NHLI, Imperial College and Departments of Paediatric Respiratory
Medicine, The UK Cystic Fibrosis Gene Therapy Consortium and Thoracic Medicine,
Royal Brompton Hospital, London, UK

Abstract

The correlation of genotype and phenotype in cystic fibrosis (CF) is organ-specific and is high in the pancreas. However, in the lung this correlation is very poor, likely due to a significant contribution of other genes and/or the environment. A large number of studies has been undertaken to identify putative modifier genes with the aim of increasing understanding of disease pathogenesis and prognosis and ultimately to lead to the development of novel treatments. However, the search for conclusive CF modifiers has so far proven difficult. This is most likely due to heterogenous patient populations, difficult to characterize phenotypes, changes in clinical treatment over time and underpowered studies. In addition, it is unlikely that a single modifier gene will be found, but more probable that several modifier genes in combination may contribute, which in itself presents a major challenge. The relative contribution of modifier genes may also be very small and variability in CF lung disease may be mainly due to environmental factors. Most recently, the need for larger scale studies has been recognized, which may help to overcome some of the problems encountered so far.

What Is Genetic Variation?

The DNA sequence of any two unrelated individuals is approximately 99.9% identical. The remaining 0.1% contains variations (polymorphisms), which have the ability to influence the likelihood or severity of disease and perhaps

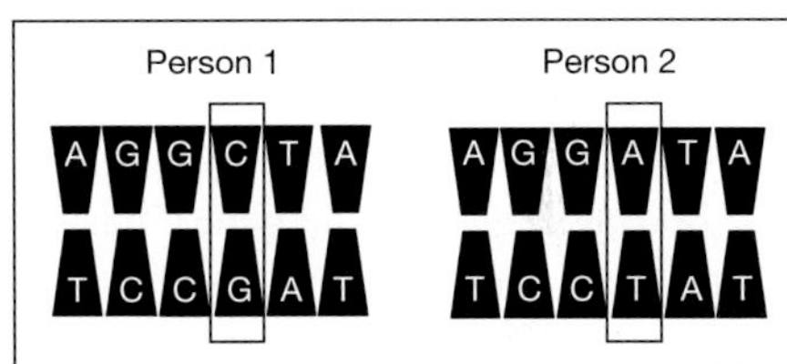

Fig. 1. Single nucleotide polymorphism. The DNA sequence of two individuals differs in the one nucleotide (see boxed area). A = Adenosine; T = thymidine; G = guanosine; C = cytidine.

alter responses to drugs. The term polymorphism includes single-base nucleotide substitutions (also known as single nucleotide polymorphisms or SNPs, fig. 1), small-scale, multi-base deletions or insertions (also called deletion insertion polymorphisms or DIPs), and repeat variations (also called short tandem repeats or STRs). In contrast to mutations, which occur at low frequencies (less than 1%) in the general population, polymorphisms are common genetic variants, occurring more frequently. Polymorphisms can lead to changes in either or both of gene expression and protein function. Recently, SNPs have received most attention, mainly due to the development of high throughput analysis [1]. In 2002 the international HapMap project (www.HapMap.org) was founded, which aims to generate a SNP haplotype map of the human genome within the next 3 years. A similar project is being carried out in different strains of mice (www.jax.org).

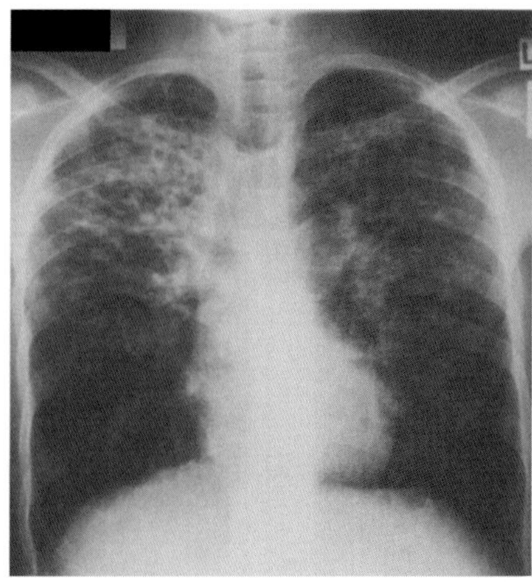

Fig. 2. Chest X-rays from two adolescents both homozygous for the ΔF508 mutation. The patient on the left was clear of chronic infection, had normal lung function, and chest X-ray shows only mild hyperinflation. In contrast the X-ray on the right, showing widespread bronchiectasis and mucus plugging, was from a patient with chronic *P. aeruginosa* infection and significantly reduced lung function.

It is estimated that approximately 10 million SNPs exist in the human genome, but the number of SNPs per gene varies widely, ranging from one or two, to several hundred. However, all SNPs may not be equally important, and analysis is often restricted to the following: (1) those within exons (the coding region of the gene), particularly if leading to an amino acid substitution, (2) those in regulatory regions, which may affect binding of transcription factors, (3) those at exon/intron boundaries, which may influence splicing and (4) those in regions that are conserved among different species, as this usually implies a significant function.

Many different technologies are used to study SNPs. The main differences are related to throughput and costs. A detailed review of the technologies is outside the remit of this chapter, but they have been extensively reviewed elsewhere [2].

Putative Cystic Fibrosis Modifier Genes

As described in chapter 8 in this book the overall correlation between genotype and phenotype in cystic fibrosis (CF) is organ-specific and is highest in the pancreas. However, in the lung this correlation is very poor, likely due to a significant contribution of other genetic factors and/or the environment (fig. 2). The hypothesis that genes outside the *CFTR* locus can affect CF disease is supported by several studies. The first evidence was generated in CF knockout mice and pointed to a gene locus on mouse chromosome 7, affecting severity of intestinal disease [3]. To date the gene responsible has not been identified; however, a similar gene locus has been located on human chromosome 19, a region syntenic to the mouse locus, and correlates with occurrence of meconium ileus in CF babies [4].

Studies in monozygotic (share 100% of genes and significant amount of pre- and postnatal environment) and dizygotic twins (share 50% of genes and a significant amount of environment, both pre- and postnatal) and other siblings (share 50% of genes and less pre- and postnatal environment) are very powerful in gaining an understanding of the relative contribution of environmental and genetic factors to disease (table 1). The European Twin and Sibling Study was initiated several years ago to make use of this powerful resource. In this study, monozygous twins showed a higher degree of concordance for nutritional and pulmonary status than dizygous twins or siblings, indicating that CF disease might be modulated by modifier genes [5, 6]. More recently, a twin study was also initiated in the USA. In contrast to the European study, a comparison of pulmonary function in monozygotic and dizygotic twins does not currently support a significant contribution of genetic modifiers to the progression of lung disease [7]. However, the numbers in this study need to be increased before this finding is accepted.

In addition to polymorphisms in modifier genes outside the *CFTR* locus, an increasing number of polymorphisms have been identified within the *CFTR* gene. These will be discussed in chapter 9 of this book. Overall, the interaction between mutations and polymorphisms within *CFTR*, polymorphisms in other genes, and the environment leads not only to significant symptom variability in patients with classical CF, but possibly also to 'non-classical' CF (see chapter 9) and a growing number of monosymptomatic

Table 1. Twin studies

	Genes shared %	Degree of environment shared	
		prenatal	postnatal
Monozygotic twins	100	High	High
Dizygotic twins	50	High	Lower
Siblings	50	Lower	Lower

Table 2. Putative CF modifier genes

Category	Examples
Inflammatory/anti-inflammatory genes	Cytokines (TNFα, IL-8, IL-10)
Ion/water transport	Alternative chloride channels, ENaC
Immunity/airway defence	MBL, HLA locus, NO synthase
Proteases/antiproteases	Neutrophil elastase, α_1-AT
Posttranslational processing/chaperons	Heat shock proteins (hsp70)
Growth factors	TGF-β
Oxidants/anti-oxidants	Glutathione S-transferase
Airway reactivity	β_2-AR
Mucus/mucins	Muc5A

diseases such as obstructive azospermia, idiopathic pancreatitis, disseminated bronchiectasis and chronic rhinosinusitis. The relative contributions of individual genetic factors and the environment to disease phenotype are currently unclear, but it is most likely that multiple CF modifier genes are involved.

Most studies in humans to date have focussed on identifying modifier genes of lung disease, although a few studies related to intestinal and liver disease have also been carried out. Based on our current knowledge of pathogenesis, putative genetic modifiers of CF may fall into different categories (table 2).

Inflammatory Genes, Antiproteases and Antioxidants as Modifiers of Respiratory Disease

Polymorphisms in pro-inflammatory cytokines may play in an important role in regulating the chronic lung inflammation. TNF-α, a cytokine found in high concentrations within the CF airway, is thought to be pivotal in the promotion of the neutrophil-dominated inflammatory response and has been inversely correlated with lung function. In 1998, Hull et al. published results of a study on 53 CF children. Twenty possessed a polymorphism in the promoter region of the TNF-α gene known to be associated with higher (constitutive and induced) levels of the cytokine. These children had a significantly lower forced expiratory volume in 1 s (FEV_1) and poorer nutritional status than the children without the polymorphism. In a larger study including both adults and children, Arkwright et al. [8] found no such association when studying the same promoter polymorphism, and in 269 adult patients, we have also found no link with any marker of severity [9]. Potential reasons for the discrepancies in these studies will be discussed later in this chapter.

In the CF airway, chronic inflammation results in an excess of destructive proteases such as neutrophil elastase. These overwhelm their inhibitors, the antiproteases, of which α_1-antitrypsin (α_1-AT) is one of the most abundant. Patients with inherited forms of α_1-AT deficiency are at risk of emphysema, and thus, it was postulated that CF patients with co-existing α_1-AT deficiency would demonstrate a more severe pulmonary phenotype. Studies examining effects on severity have reached conflicting conclusions: increased *Pseudomonas aeruginosa* infection without an adverse effect on pulmonary function, a beneficial effect on lung function (which, interestingly was also observed for the related gene, α_1-antichymotrypsin [10]), and most recently, in the largest group, no effect on lung function, age at acquisition of *P. aeruginosa*, requirement for transplantation or death [11]. The glutathione S-transferase M1 allele, an enzyme involved in oxidative stress, has been reported to be linked to more severe chest radiograph and Shwachman score in homozygous children.

Antigen Presentation and Host Defense Mechanisms as Modifiers of Respiratory Disease

The human leucocyte antigen (HLA) region is the most polymorphic in the human genome, encoding hundreds of genes including the major histocompatibility complexes. Major histocompatibility complexes class II molecules are critical in antigen presentation and the ensuing inflammatory response. Polymorphisms have been reported in association with a number of diseases including autoimmune conditions, asthma and allergy. Two reports have linked certain haplotypes with complications in CF. Aron et al. [12] found an association between DR7, high IgE levels and risk of *P. aeruginosa* infection and allergic bronchopulmonary aspergillosis and the DR2 allele [13].

Mannose-binding lectin (MBL), a liver-derived serum protein involved in innate defence, exerts its effects both by direct opsonization of pathogens and by activation of the complement system [14]. Polymorphisms have been identified within both the structural and promoter regions of human *MBL-2*, which result in mutant forms of the protein incapable of forming functional, high order oligomers and low circulating protein levels. Low levels of MBL have been shown to relate to a variety of infective processes including recurrent respiratory infections, and so the *MBL-2* gene was considered a likely candidate as a modifier in the CF lung. In the first study to examine such a link, Garred et al. [14] reported that both FEV_1 and forced vital capacity (FVC) were significantly lower in subjects with either one or two structural *MBL-2* mutations, but only following chronic *P. aeruginosa* infection. Although numbers were small, the authors also reported an increased risk of infection with *Burkholderia cepacia*. In contrast, a second study found a significant reduction in lung function only in patients possessing two variant alleles [15]. In support of this, Buranawuti et al. [16] have reported a significant survival disadvantage in American CF patients with 2 mutations. As part of a large study in almost 600 patients, we have recently found that adult patients possessing two structural mutations, but not heterozygotes, have significantly impaired lung function, oxygen saturations and raised inflammatory markers [17]. In contrast to the data from Garred et al. [14], this was not seen in our paediatric age group, in whom the majority had well-preserved lung function. This difference highlights an important point. Treatment regimes have evolved greatly over the last few years and, for example, a polymorphism that was significant in the era before the widespread use of antipseudomonal antibiotics may be less relevant now. Thus, in some cases historical data comparison might be difficult.

The antimicrobial effects of nitric oxide (NO) are being increasingly recognized. High numbers of the AAT trinucleotide repeat sequence in the *NOS-1* gene are associated with low levels of exhaled NO, and were found in CF patients to confer an increased risk of infection with both *P. aeruginosa* and *Aspergillus fumigatus*, although this did not lead to a more rapid decline in lung function [18]. This group has also recently related a polymorphism in the *NOS-3* gene with risk of infection [19]. This gene, expressed in vascular endothelium, respiratory epithelium and neutrophils, contains a functionally important polymorphism (894G/T), which affects the resistance of NOS-3 to proteolysis. Interestingly, previous work has highlighted gender differences, by demonstrating that circulating oestrogen increases the levels of NOS-3 in the vascular endothelium. Grasemann et al. [19] reported higher exhaled NO levels and decreased frequency of *P. aeruginosa* infection in association with the 894T polymorphism in females only. What is not clear from these two related studies is whether *NOS-1* and *NOS-3* are independent modifiers, or whether there is a confounding effect of one upon the other [20]. Further, these genes may merely be 'hitchhiking' with another gene of relevance, rather than be directly involved in disease modification themselves [21].

Growth Factors as Modifiers for Respiratory Disease

Polymorphisms leading to high levels of the pleotropic cytokine, tissue growth factor (TGF)-β, have previously been associated with increased pulmonary fibrosis after chemo- and radiotherapy and organ transplantation. Arkwright et al. [29] studied TGF-β in a cohort of 171 CF patients (children and adults) from Manchester, UK. Patients with known 'mild' *CFTR* mutations were excluded from analysis. Subjects with at least one high-expressing haplotype had a significantly faster rate of decline in both FEV_1 and FVC than those with low-expressing variants. Interestingly, this would appear to contrast with another study demonstrating that TGF-β expression was highest in the CF patients with the mildest lung disease.

Airway Hyperresponsiveness as Modifiers for Respiratory Disease

Polymorphisms in the β-adrenergic receptor (β-AR) have previously been related to severity of asthma and response to treatment with $β_2$-agonist drugs [22]. β-AR are important regulators of cAMP in the airway, recent in vitro data demonstrating that ion transport via protein kinase-regulated CFTR can be activated by $β_2$-agonists [23]. Buscher et al. [24] studied the effects of three polymorphisms in the *β-AR* in 87 young adults and children with CF. Subjects with either one or two copies of Gly^{16}, an amino acid change leading to downregulation of the receptor, had significantly reduced lung function and faster rates of decline than those patients homozygous for Arg^{16}. These differences were more marked when only ΔF508 homozygotes were studied. In addition, Gly^{16} was significantly less common in the CF population than in several groups of healthy controls, possibly implying a survival disadvantage. Unfortunately, the group was not large enough for subgroup analysis on the basis of age, which might have helped support this hypothesis. There were no differences in bronchodilator responsiveness, but in an in vitro assay, lymphocytes from these subjects showed a blunted cAMP

 Griesenbach/Alton/Davies

response to isoproteronol stimulation suggesting that the clinical findings may relate to differences in the level of CFTR function between the two groups.

Gastrointestinal and Liver Disease

Between 10 and 15% of CF patients are born with meconium ileus. As mentioned above a putative gene locus has been identified on human chromosome 19, although the protein encoded at this site remains unidentified. Recently, this locus has also been linked to liver disease [25]. Most recently, polymorphisms in the calcium-activated chloride channel (CaCC) gene locus have been linked to chloride transport in the intestinal mucosa. CaCC-mediated chloride secretion may in part substitute for the altered CFTR-mediated chloride transport [26]. This is the first published example of an alternative chloride channel being implicated in CF disease.

Familial clustering of portal hypertension suggests that liver disease in CF may be under genetic influence, although few associations have been found with *CFTR* genotype. One of the major problems with studies of liver disease is that of phenotypic definition, which differs widely in clinical practice. Duthie et al. performed a multicentre study on 274 unrelated children and adults examining the effect of HLA status. Almost 30% of patients had evidence of chronic liver disease, a higher proportion than reported from most centres. DQ6 was found in 66% of patients with, but only in 33% of those without, liver disease. Two other antigens in strong linkage disequilibrium with this locus, DR15 and B7, were also significant risk factors. When portal hypertension, as defined by either splenomegaly, oesophageal varices or portosystemic collateral vessels, was used as a marker of chronicity, these markers were found to be significant for males only, but there was no association with age of onset. MBL deficiency was shown by one group to be a risk factor for the development of CF liver disease [27], although we could not replicate this [28]. Arkwright et al. [8, 29] have reported associations between liver disease and both high expressing TGF-β haplotypes and angiotensin-converting enzyme (an enzyme involved in TGF-β activation) polymorphisms. Finally, Mekus et al. [30] identified loci in a partially imprinted region 3′ of *CFTR* as modifiers of both nutritional and pulmonary phenotype in 34 highly concordant or discordant sib pairs. This region includes both the leptin gene and a candidate for Russell-Silver dwarf syndrome, and is thus likely involved in growth, food intake and energy expenditure.

Ongoing Studies

Preliminary results of several unpublished studies have been described above and will be expanded upon here. Large numbers with clearly defined phenotype are crucial for polymorphism studies. To address this, a multi-centre study has been initiated in the US, with the aim of collecting a large number of homozygous ΔF508 CF patients to study modifiers of lung disease. Currently 864 subjects have been enrolled and initial results have been presented. A TGF-β$_1$ gene polymorphism, known to be involved in airway remodelling, and affecting airway disease in asthma and chronic obstructive pulmonary disease likely due to its anti-inflammatory and profibrotic effects, has been strongly associated with severely impaired lung function [31]. More recently, a Canadian CF modifier study has been started and aims to collect 3,000 families from 36 centres over the next 2 years [32].

Increased sodium absorption through epithelial cell-specific sodium channels (ENaC) may be an important component of CF pathophysiology. Interestingly, sequence alterations in the β-ENaC gene have recently been reported in patients with non-classical CF [33], stimulating speculations that ENaC may also be a modifier gene.

Issues Related to Study Design

As noted above, studies of the same gene have often reached different conclusions. To some extent this may reflect true differences in populations, but it seems equally likely, if not more so, that much of this discrepancy relates to study design. A detailed discussion of this is outside the scope of this chapter, but has recently been reviewed [34]. Here, we briefly summarize the major points: (1) Power calculations for CF modifier studies are difficult, because generally the effect size for a given modifier gene is not known. However, in general, the effects of rare alleles or modifiers that are only relevant in a particular environment will be more difficult to pick up as group sizes will be small. Thus, the larger the study population the better, particularly when subgroup analysis (males vs. female, different age bands, different genotypes) will be carried out. However, inconsistency of treatments and data collection can be a significant problem for multi-centre studies. Although multi-centre studies are more likely to generate the large numbers that may overcome some of the problems listed in table 3, data analysis may be significantly hampered. Hypothesis-driven studies, where the downstream effect of individual polymorphisms are known, will also be

Table 3. Potential confounding factors for modifier studies

Problem	Solution
Effect size of individual modifiers is unknown	High numbers required
Modifier may only be important in a particular environment	High numbers required to allow well-powered subgroup analysis and detailed descriptive data on subjects required
Modifier may only be important in a particular age range	High numbers required
Action of a modifier may depend on a particular treatment	Detailed descriptive data on subjects required

more powerful. Unlike the 'fishing expeditions' assessing multiple genes with multiple variant alleles, grouping of polymorphisms on the basis of function (for example into high- and low-expressing mutants) reduces the need for comparisons within multiple, small groups and the statistical correction factors which further reduce power. (2) A robust and accurate definition of phenotype for discrete outcomes (e.g. pancreatic status or infection with specific pathogens) and continuous outcomes (e.g. lung function) is absolutely crucial for these studies and this may have affected interpretation of some of the studies. (3) In cases where several genes and several polymorphisms per gene are analyzed, the risk of generating false-positive results due to multiple comparison is high and statistical advice should be sought before undertaking the study. (4) All studies have to be confirmed in an independent study population ('second setting').

In conclusion, this is a rapidly growing field. There are significant pitfalls, which are becoming apparent in the discrepant results being published, but identification of these will hopefully be of benefit for future studies. Ultimately, such studies should broaden our understanding of disease pathogenesis in CF, and possibly lead to the rational design of novel therapeutic agents.

References

1 Chen X, Sullivan PF: Single nucleotide polymorphism genotyping: Biochemistry, protocol, cost and throughput. Pharmacogenomics J 2003;3:77–96.

2 Weiner MP, Hudson TJ: Introduction to SNPs: Discovery of markers for disease. Biotechniques 2002;10(suppl 4–7):12–13.

3 Rozmahel R, Wilschanski M, Matin A, Plyte S, Oliver M, Auerbach W, Moore A, Forstner J, Durie P, Nadeau J, Bear C, Tsui LC: Modulation of disease severity in cystic fibrosis transmembrane conductance regulator deficient mice by a secondary genetic factor. Nat Genet 1996;12:280–287.

4 Zielenski J, Corey M, Rozmahel R, Markiewicz D, Aznarez I, Casals T, Larriba S, Mercier B, Cutting GR, Krebsova A, Macek M Jr, Langfelder-Schwind E, Marshall BC, DeCelie-Germana J, Claustres M, Palacio A, Bal J, Nowakowska A, Ferec C, Estivill X, Durie P, Tsui LC: Detection of a cystic fibrosis modifier locus for meconium ileus on human chromosome 19q13. Nat Genet 1999;22:128–129.

5 Bronsveld I, Mekus F, Bijman J, Ballmann M, Greipel J, Hundrieser J, Halley DJ, Laabs U, Busche R, de Jonge HR, Tummler B, Veeze HJ: Residual chloride secretion in intestinal tissue of deltaF508 homozygous twins and siblings with cystic fibrosis. The European CF Twin and Sibling Study Consortium. Gastroenterology 2000;119:32–40.

6 Mekus F, Ballmann M, Bronsveld I, Bijman J, Veeze HJ, Tummler B: Categories of deltaF508 homozygous cystic fibrosis twin and sibling pairs with distinct phenotypic characteristics. Twin Res 2000;3:277–293.

7 Cutting GR: CF modifiers: Comparing variations between siblings. Pediatr Pulmonol Suppl 2004;27:140.

8 Arkwright PD, Pravica V, Geraghty PJ, Super M, Webb AK, Schwarz M, Hutchinson IV: End-organ dysfunction in cystic fibrosis: Association with angiotensin I converting enzyme and cytokine gene polymorphisms. Am J Respir Crit Care Med 2003;167:384–389.

9 Low T, Lymnay PA, Davies JC, Griesenbach U, Shen N, Escudero-Garcia SI, Welsh KI, Geddes DM, du Bois RM, Alton EWFW: The effect of polymorphism in inflammatory mediators on clinical phenotype in CF. Pediatr Pulmonol Suppl 2003;25:219.

10 Mahadeva R, Sharples L, Ross-Russell RI, Webb AK, Bilton D, Lomas DA: Association of alpha(1)-antichymotrypsin deficiency with milder lung disease in patients with cystic fibrosis. Thorax 2001;56:53–58.

11 Frangolias DD, Ruan J, Wilcox PJ, Davidson AG, Wong LT, Berthiaume Y, Hennessey R, Freitag A, Pedder L, Corey M, Sweezey N, Zielenski J, Tullis E, Sandford AJ: Alpha 1-antitrypsin deficiency alleles in cystic fibrosis lung disease. Am J Respir Cell Mol Biol 2003;29:390–396.

12 Aron Y, Polla BS, Bienvenu T, Dall'Ava J, Dusser D, Hubert D: HLA class II polymorphism in cystic fibrosis. A possible modifier of pulmonary phenotype. Am J Respir Crit Care Med 1999;159:1464–1468.

13 Aron Y, Bienvenu T, Hubert D, Dusser D, Dall'Ava J, Polla BS: HLA-DR polymorphism in allergic bronchopulmonary aspergillosis. J Allergy Clin Immunol 1999;104:891–892.

14 Garred P, Pressler T, Madsen HO, Frederiksen B, Svejgaard A, Hoiby N, Schwartz M, Koch C: Association of mannose-binding lectin gene heterogeneity with severity of lung disease and survival in cystic fibrosis. J Clin Invest 1999;104:431–437.

15 Gabolde M, Guilloud-Bataille M, Feingold J, Besmond C: Association of variant alleles of mannose binding lectin with severity of pulmonary disease in cystic fibrosis: Cohort study. BMJ 1999;319:1166–1167.

16 Buranawuti K, Cheng S, Merlo C, Zeitlin P, Rosenstein B, Mogayzel P, Wang X, Boyle M, Cutting GR: Variants in mannose binding lectin gene modify survival of cystic fibrosis patients. Pediatr Pulmonol Suppl 2003;25:215.

17 Davies JC, Potter M, Bush A, Rosenthal M, Geddes DM, Alton EWFW: Bone marrow stem cells do not repopulate the healthy upper respiratory tract. Pediatr Pulmonol 2002;34:251–256.

18 Grasemann H, Knauer N, Buscher R, Hubner K, Drazen JM, Ratjen F: Airway nitric oxide levels in cystic fibrosis patients are related to a polymorphism in the neuronal nitric oxide synthase gene. Am J Respir Crit Care Med 2000;162:2172–2176.

19 Grasemann H, van's Gravesande KS, Buscher R, Knauer N, Silverman ES, Palmer LJ, Drazen JM, Ratjen F: Endothelial nitric oxide

synthase variants in cystic fibrosis lung disease. Am J Respir Crit Care Med 2003;167: 390–394.
20 Accurso FJ, Sontag MK: Seeking modifier genes in cystic fibrosis. Am J Respir Crit Care Med 2003;167:289–290.
21 Mekus F, Tummler B: Cystic fibrosis and NOS3. Am J Respir Crit Care Med 2004;169: 319–320.
22 Ramsay CE, Hayden CM, Tiller KJ, Burton PR, Goldblatt J, Lesouef PN: Polymorphisms in the beta2-adrenoreceptor gene are associated with decreased airway responsiveness. Clin Exp Allergy 1999;29:1195–1203.
23 Taouil K, Hinnrasky J, Hologne C, Corlieu P, Klossek JM, Puchelle E: Stimulation of beta 2-adrenergic receptor increases cystic fibrosis transmembrane conductance regulator expression in human airway epithelial cells through a cAMP/protein kinase A-independent pathway. J Biol Chem 2003;278:17320–17327.
24 Buscher R, Eilmes KJ, Grasemann H, Torres B, Knauer N, Sroka K, Insel PA, Ratjen F: Beta2 adrenoceptor gene polymorphisms in cystic fibrosis lung disease. Pharmacogenetics 2002;12:347–353.
25 Deering R, Algire M, McWilliams R, Lai T, Naughton K, Beck S, Hoover-Fong J, Hamosh A, Chakravarti A, Cutting GR: Meconium ileus and liver disease: An analysis of the CF twin and sibling study. Pediatr Pulmonol Suppl 2004;27:224.
26 Ritzka M, Stanke F, Jansen S, Gruber AD, Pusch L, Woelf S, Veeze HJ, Halley DJ, Tummler B: The CLC4 gene locus as a modulator of the gastrointestinal basic defect in cystic fibrosis. Hum Genet 2004;115:483–491.
27 Gabolde M, Hubert D, Guilloud-Bataille M, Lenaerts C, Feingold J, Besmond C: The mannose binding lectin gene influences the severity of chronic liver disease in cystic fibrosis. J Med Genet 2001;38:310–311.
28 Davies J, Johnson MM, Booth C, Fidler K, Bush A, Geddes M, Alton EWFW, Turner M, Klein N: Age-specific effect of the cystic fibrosis modifier gene, MBL-2. Pediatr Pulmonol Suppl 2002;23:223.
29 Arkwright PD, Laurie S, Super M, Pravica V, Schwarz MJ, Webb AK, Hutchinson IV: TGF-beta(1) genotype and accelerated decline in lung function of patients with cystic fibrosis. Thorax 2000;55:459–462.
30 Mekus F, Laabs U, Veeze H, Tummler B: Genes in the vicinity of CFTR modulate the cystic fibrosis phenotype in highly concordant or discordant F508del homozygous sib pairs. Hum Genet 2003;112:1–11.
31 Knowles M, Konstan M, Schluchter M, Handler A, Bucur C, Pace R, Yankaskas J, Zou F, Wright F, Goddard K, Drumm M: CF gene modifiers: Comparing variation between unrelated individuals with different pulmonary phenotypes. Pediatr Pulmonol Suppl 2004;27: 139.
32 Zielenski J, Sandford A, Corey M, Dorfman R, Deng G, Patel M, Markiewicz D, Evtoushenko I, Yuan X, Master A, Tan M, Li F, Aznarez I, Cassidy J, Rousseau R, Christofi M, Frangolias D, van Spall M, Berthiaume I, Pare P, Tsui LC, Duri P: Canadian Consortium for the Study of Genetic Modifiers in Cystic Fibrosis. Pediatr Pulmonol Suppl 2004;27:225.
33 Sheridan MB, Groman JD, Fong P, Conrad C, Flume P, Diaz R, Harris C, Knowles M, Cutting GR: Mutations in the epithelial sodium channel cause a novel phenotype resembling non-classic cystic fibrosis. Pediatr Pulmonol Suppl 2004;27:222.
34 Davies J, Griesenbach U, Alton EWFW: Modifier genes in cystic fibrosis. Pediatr Pulmonol 2005;39:383–391.

Jane C. Davies
Department of Gene Therapy
Imperial College London
Manresa Rd.
London SW3 6LR (UK)
E-Mail j.c.davies@imperial.ac.uk

Bush A, Alton EWFW, Davies JC, Griesenbach U, Jaffe A (eds): Cystic Fibrosis in the 21st Century.
Prog Respir Res. Basel, Karger, 2006, vol 34, pp 84–92

Animal Models

Julia R. Dorin

MRC Human Genetics Unit, Western General Hospital, Crewe Road South, Edinburgh, UK

Abstract

An animal model for cystic fibrosis (CF) that mirrors the electrophysiological defect of the disease and replicates the lung susceptibility to infection would be valuable for disease research. The ability to create predetermined genetic alterations in the mouse has allowed the creation of many strains with a variety of disruptions in the murine *Cftr* gene. The phenotype of the CF mice has been predominantly apparent in the gut rather than the airway and somewhat inappropriate for the lung-based symptoms present in humans. The phenotypic variation of CF mouse models is affected by both mutation and genetic background. Strategies to improve the models both genetically and physiologically are discussed.

Introduction

An understanding of the way the cystic fibrosis transmembrane conductance regulator (CFTR) functions can be achieved to some degree, using cell-based assays. These can provide detailed dissection of cell autonomous systems and give clues as to whether a pharmacological or gene therapy strategy may work. Polarized epithelial cultures at air/liquid interfaces (ALI) are an elegant way to mimic the in vivo situation. However, cystic fibrosis (CF) is a disease of whole organs and as such animal models must be regarded as valuable tools for studying disease pathogenesis and evaluating novel therapeutic strategies.

Creation of Mutant Mouse Models by Gene Targeting

Three years after the human *CFTR* gene was cloned, several groups reported mouse models for CF [1–3]. These models were achieved using homologous recombination or gene targeting in mouse embryonic stem (ES) cells. This technique was appropriate for CF as it is a recessive disease and the endogenous CFTR needs to be mutated to see a phenotype. Briefly, this strategy creates gene disruptions in a target gene using vectors containing fragments of the gene to direct the recombination. Clones of cells containing the targeting vector are screened to determine which have been targeted correctly and have disrupted the gene of interest. The level of homologous recombination (correct targeting) will be relatively infrequent, but around 1% (or higher) of clones that have a stably integrated vector will have undergone a targeted replacement or insertional event. Clones of embryonic stem cells that carry the mutation can then be injected into developing 3.5-day blastocysts removed from pregnant mice. These blastocysts are then introduced back into the uterus of pseudo-pregnant mice and allowed to develop as normal. Providing the ES cells have been kept under ideal cell culture conditions they will be able to contribute to all organs of the developing embryo. When the mouse pups are born, after a few days it is possible to see whether the injected cells have contributed to the mouse on the basis of the colour of the fur. The blastocysts and ES cells will come from different coat

Fig. 1. ES cell germline transmission indicated by coat colour. Litter of mice with a chimaeric father. The chimaera was created by injection of ES cells carrying the chinchilla mutation (c^{ch}) at the tyrosinase locus into blastocysts from C57Bl/6 mice (wild type at the tyrosinase locus). The chimaera was mated with an albino outbred female – coat colour due to being mutant at the tyrosinase locus (c/c) – which resulted in offspring derived from sperm from the host C57Bl/6 blastocyst being black (c/+) and offspring derived from sperm from the ES cells being grey (c^{ch}/c).

colour mutants to enable the visualization of the ES contribution (fig. 1). In order to pass the created mutation on, and to create homozygous mutants, the chimaeric mice will be crossed to a normal mouse and germline transmission of the mutation will be determined again on the basis of coat colour. Mice that display a coat colour compatible with being derived from the ES cell genome will either carry the chromosome with the mutation or the wild-type chromosome that was not targeted. A genetic screen for the mutation will elucidate which ES-cell-derived mice carry the mutation, and intercrossing will generate homozygous mutants, heterozygotes and wild types at a ratio of 1:2:1 respectively.

Different Types of *Cftr* Gene Disruption

Gene Knockouts

Table 1 lists all the models to date and divides them into the type of mutation they contain. As can be seen from the table, the importance of the disease has resulted in many different models being created. The first CF mouse models disrupted the normal gene expression by either replacement or insertion.

Replacement Mutants

The 'null' mutant mice completely disrupt gene expression from the *Cftr* allele by replacing a portion of the gene with the targeting vector construct and disrupting gene expression [2, 4]. These mice all show severe reduction in survival, with the vast majority dying either at birth or at weaning, with only 5% of mice surviving to maturity [5].

Insertional Mutants

In these mice, the mutation occurs by insertion of the targeting vector. As no genetic material is lost, insertional mutants can be a bit leaky and are sometimes called 'residual function'. Indeed the *Cftr* insertional mice have a mutation that results in the majority of the message being absent but exon skipping in the insertion means that a low level of wild-type CFTR is still produced [6, 7]. In *Cftr^tm1Bay*, the residual level of CFTR is estimated to be only 2% of normal, while in the insertional mutant mouse *Cftr^tm1Hgu*, the level is around 10% of normal.

The residual level of CFTR in the insertional mice means that the phenotypes are significantly different from the 'null' models. The low level of CFTR is sufficient to allow the intestinal phenotype to be so mild that survival is hardly affected in the *Cftr^tm1Hgu* mutant, although there are still some histological abnormalities and mucus accumulation in the gut. The *Cftr^tm1Bay* insertional mice with less residual CFTR have a 70% reduction in body weight and only 40% survival.

Introducing Human CFTR Mutations into the Mouse Gene

Around 70% of CF chromosomes carry the common ΔF508 mutation in exon 10. This results in a mutant form of the CFTR protein being produced which lacks a phenylalanine residue at position 508 in the first nucleotide-binding fold. The consequence of this is that the majority of the protein is not correctly transported to the membrane and is degraded at a high rate. A mouse model with this specific mutation was important for testing pharmaceuticals that may be able to promote correct processing of the defective gene product. Several groups have used gene targeting to introduce precise mutations into the mouse *Cftr* locus [8–10]. The region of the *Cftr* gene that encodes the first nucleotide-binding fold that includes residue 508 is very well conserved between man and mouse. The gene targeting was again carried out using different strategies.

Replacement Mutants

The first groups to report mice carrying the ΔF508 deletion used a gene-targeting vector which replaced exon 10

Table 1. Survival of different CF mouse models

Mouse	Exon mutated	Survival	Reference
Gene knockouts			
Replacement (nulls)			
Cftr[tm1Unc]	10	<5% to maturity	2
Cftr[tm1Cam]	10	<5% to maturity	4
Cftr[tm3Bay]	2	40% at 1 month	48
Cftr[tm1Hsc]	1	25% to maturity	17
Insertional (residual function)			
Cftr[tm1Bay]	3	40% by day 7	7
Cftr[tm1Hgu]	10	no significant reduction	1
Cftr[tm1Hgu]/*Cftr*[tm1Unc]	10	no significant reduction	19
Introducing human mutations			
Replacement			
Cftr[tm2Cam]	ΔF508 in exon 10	<5% to maturity	8
Cftr[tm1Kth]	ΔF508 in exon 10	40% to maturity	9
Cftr[tm1G551D]	G551D in exon 11	27–67% by day 35[1]	13
Cftr[tm2Uth]	R117H in exon 4	no significant reduction	14
Cftr[tm3Uth]	Y122X in exon 4	24% to maturity	
'Hit and run'			
Cftr[tm2Hgu]	ΔF508 in exon 10	no significant reduction	10
Cftr[tm2Hgu]	G480C in exon 10	no significant reduction	24

[1]Survival 27% in standard animal facility, 67% in specific-pathogen-free facility.

with a mutant exon containing the three base pair deletion. The dominant marker required to select for cells that had taken up the vector was included in the adjacent intron [8, 11]. Both these models (*Cftr*[tm2Cam] and *Cftr*[tm1Kth]) produce the murine mutant ΔF508 protein. However, the transcription level of the mutant gene is severely reduced, to only 15% of wild-type levels in the *Cftr*[tm2Cam] mutant. This is presumably due to the presence of the selectable marker within the gene. The phenotype of these mice was very similar to the 'null' animals.

Some *CFTR* mutations are considered mild in terms of pancreatic disease e.g. R117H, and patients with the G551D allele appear to have less incidence of meconium ileus [12]. These genotype/phenotype correlations have been investigated in the mouse. G551D is a mutation that is correctly localized but does not function correctly. It is a mutation predominant in Celtic populations and will require a different approach to that in ΔF508 to increase its activity. Mice with the G551D mutation have been created, again using a gene replacement strategy [13]. The mutant transcript levels were around 53% those of a wild-type allele and the homozygous mutant animals replicated the reduction in intestinal disease.

Other clinical mutations that have been expressed in the mouse by replacement gene targeting into the *Cftr* locus are the 'mild' mutation R117H and the stop mutation Y122X [14]. The R117H mutation reported by van Heeckeren et al. [14], when replicated in the mouse, did demonstrate a reduction in body weight but survival was not as severely affected as in animals homozygous for the Y122X 'null' mutation. These mutations were both created using the replacement technique leaving the selectable marker in the intron and reducing the transcript level considerably. Thus the R117H phenotype may be more severe than G480C (see below) due to a combination of message reduction and the mutation.

'Hit and Run' Mutants

A more elegant approach was used by a third group in the *Cftr*[tm1Eur] mouse, where the mutation was introduced by homologous recombination using an insertional vector (hit) and then all vector sequences (and the wild-type exon) were removed by intrachromosomal recombination and identi-

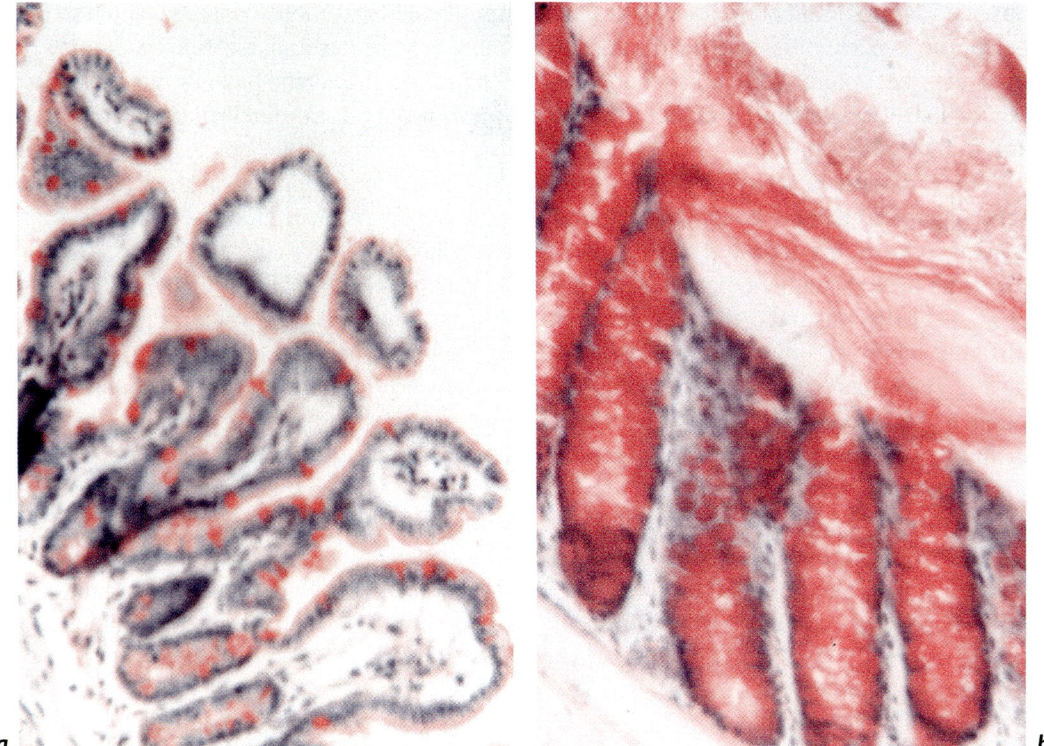

Fig. 2. Histological abnormalities in the intestine of CF mice. Sections from the colon of normal (**a**) and *Cftr*^{tm1Unc} mice (**b**) stained with periodic acid-Schiff reagent to stain mucus pink. The mutant mouse has an increased number of mucus-containing goblet cells and there is obvious accumulation of mucus in the intestinal lumen.

fied by negative selection against the vector sequence (run) [10]. This 'hit and run' strategy results in the only change being the predetermined, introduced mutation. Thus, vector sequences are not present to interfere with gene transcription. The *Cftr*^{tm1Eur} Δ508 mutant allele had normal levels of transcription (unlike the *Cftr*^{tm2Cam} ΔF508 mutant) and a much less severe phenotype, implying that in the mouse, the mutant ΔF508 protein had sufficient activity to ameliorate the intestinal phenotype. This is not true in individuals with CF, as people homozygous for the ΔF508 mutation have a severe disease, indistinguishable from that evident in patients homozygous for 'null' mutations.

The severe G480C mutation was also created by 'hit and run' targeting in ES cells. Again transcript levels for the mutant allele were equivalent to those of the wild-type allele. G480C is another severe mutation in CF patients and the protein has been demonstrated to be defective in its intracellular processing. However, when expressed in *Xenopus* oocytes, the G480C CFTR showed similar characteristics to that of wild-type CFTR [15]. In the mouse the majority of the G480C CFTR was abnormally processed and no mutant protein was detected in the apical

membranes of the intestinal cell from homozygous animals by immunohistochemistry. The phenotype of these animals was very mild with no reduction in weight or survival.

Classical Transgenics

Crossing a transgenic mouse created by pronuclear injection, with the *Cftr*^{tm1Unc} 'null' created a further model with a specific CFTR mutation [16]. The transgene was the human *CFTR* with the G542X mutation driven by the fatty-acid-binding protein promoter. Thus expression was limited to the intestine and was specifically created to examine the use of aminoglycoside suppression of this premature stop codon.

Phenotype of the CF mice

Intestinal Disease

As mentioned above the most apparent phenotype in the CF mice is intestinal blockage and death. The pups die either at birth or at weaning. The histological abnormalities include mucus accumulation, goblet cell hyperplasia and crypt dilation [5] (fig. 2). The resulting intestinal obstruction, predominantly located at the caecum, leads to perforation, peritonitis and death. Although more severe than the meconium ileus abnormality in CF patients, it obviously reflects a similar defect in chloride ion transport and fluid regulation.

The variation in survival of the different CF mice is somewhat confusing. Genetic background obviously plays a major part in this and until each model is maintained on the same background comparisons are difficult. Rozmahel et al. [17] demonstrated that backcrossing the mice onto different inbred laboratory mouse strains altered the rate of death. They identified a major locus on chromosome 7 likely to be responsible for this effect and linkage analysis in human CF patients identified a genetic modifier locus for meconium ileus on the conserved syntenic region of chromosome 19 [18]. In addition, the survival rate of the G551D mice varies whether they are housed in a specific pathogen free environment or the regular animal house [13]. Food and husbandry conditions can also affect CF mouse survival. The 'null' CF mice are sometimes maintained on a liquid diet to avoid death at weaning.

The level of CFTR, however, does seem to be the major factor affecting survival. The 'nulls' or specific-mutant alleles with low level expression mostly die before, or at weaning. However, the residual function ($Cftr^{tm1Hgu}$) mouse and mice with normal levels of a specific-mutant allele ($Cftr^{tm1Eur}$, $Cftr^{tm2Hgu}$) have much improved survival. This most likely reflects that for a recessive disease, a very low level of normal function is required to achieve a normal phenotype and the 5% of wild-type message expressed in $Cftr^{tm1Hgu}/Cftr^{tm1Unc}$ compound heterozygotes is sufficient to achieve this [19]. High levels of mutant alleles presumably have enough residual function to also reduce the level of intestinal disease (R117H, G480C and ΔF508).

The bioelectric phenotype in the gut of the CF mice seems to correlate with the level of mutant or wild-type protein. All the mice demonstrate a defect in the baseline (reduced) and cAMP-mediated chloride ion response (reduced). Interestingly the calcium-related chloride ion response appears unaffected in the $Cftr^{tm1Eur}$ mice, but it is reduced, as in humans with CF, in all the other mice where it has been tested [5]. The G480C mutant mice have a clear reduction of cAMP and calcium reponses in the jejunum but in the caecum the response to forskolin (cAMP agonist) is normal. This may explain why they do not die from caecal obstruction.

Pancreatic Disease

Human patients with severe CF mutations have pancreatic insufficiency. Interestingly, this is not predominant in the CF mice and is attributed to less *Cftr* expression in the mouse pancreas and the presence of an alternative calcium-regulated fluid secretory pathway [20]. Defects in the pancreas have only been reported in mice fed on a liquid diet. Luminal dilatation and accumulation of zymogen granules at the apical pole of the ductal epithelial cells have been reported in the $Cftr^{tm1Unc}$ mice and corrected by treatment with docosahexaenoic acid [21].

Fertility

CF mice are fertile, but both males and females have reduced fertility, particularly the females. This is in direct contrast with CF patients where males normally are sterile due to absence of the vas deferens but females can bear children. The $Cftr^{tm1Hgu}$ residual function and $Cftr^{tm2Hgu}$ G480C mutants have only slightly impaired fertility.

Tooth Abnormalities

The teeth of the 'null' CF mice have significantly reduced enamel mineral content and an elevated level of magnesium ions compared with wild-type mice. Sui et al. [22] suggest that that CFTR plays a role in pH regulation during enamel development and that a reduced pH results in a lack of calcium influx during enamel maturation and hypomineralization of the CF incisor enamel. The $Cftr^{tm1Hgu}$ and $Cftr^{tm2Hgu}$ mice do not show this abnormality and neither do CF patients.

Bone Disease

Osteoporosis is becoming recognized as a common complication in CF patients. In the CF 'knock-out' mice, both cortical and trabecular bone formation is significantly reduced [23]. Because the CF mice are not suffering from lung infections in contrast to human patients, CF mice do not require pharmacological treatment (e.g. antibiotics and glucocorticoids), and so the study of bone biology can be separated from the confusion of these compounds.

Lung Disease

Lung disease is the primary concern of CF. Initial characterization of the mouse models revealed little evidence of gross pulmonary abnormalities. However, the mice were young and children with CF are apparently born with normal histology in their airways. The bioelectric phenotype in the airway was encouraging, however, with all the CF mice showing a similar, characteristic electrophysiological profile to CF patients, particularly in the nose [5]. The trachea showed some significant differences between the mice and individuals with CF.

In the nose, there was an increased baseline, presumably reflecting increased sodium absorption and this was supported by an increased response to the sodium channel blocker amiloride. The cAMP or low chloride response was reduced or absent in most mice. The 'null' mice completely lose their forskolin response and the residual function $Cftr^{tm1Hgu}$ mice display a 70% reduction in the response to cAMP agonists. The G480C mutant animals display a 50% reduction in response to low chloride [24]. In fact, the only mice not showing a nasal chloride ion transport defect were the $Cftr^{tm1Eur}$ ΔF508 animals, which appear to have a CFTR-like, response to a chloride gradient in the nose [10]. These animals do however display a raised baseline and an increased amiloride response. This difference could be due to the genetic background on which the mice are maintained or may reflect a human/mouse difference in ΔF508 function.

The calcium-related chloride ion response was either unchanged as in CF patients or increased. The $Cftr^{tm1Unc}$ and $Cftr^{tm1Hsc}$ 'null' animals display an increase in a calcium-activated chloride channel in the nose. The lack of severe lung pathology in CF mouse models has been explained by dominant expression of this channel [25]. This channel can be stimulated by the cAMP agonist forskolin, providing potentially misleading results and producing cAMP stimulated chloride currents in the absence of CFTR [26].

Bioelectric measurements in the trachea of individuals with CF reveal an increase in amiloride-sensitive sodium transport. The same is not observed in the trachea of the CF mice where the amiloride sensitive component of the baseline PD is reported either to be unchanged compared to normals or reduced. The cAMP induced chloride transport is reduced in the tracheal region of the respiratory tract of CF patients, but is reported to be unchanged in the $Cftr^{tm1Unc}$ mice, but partially reduced in other CF mice (table 2) [5].

The lack of similarity of the bioelectric profile in the trachea of CF mice to that observed in CF sufferers has been suggested to account for the lack of lung disease in the mice. CF mice do not appear to be more susceptible to chronic infection than wild-type animals. Some abnormalities have been reported in the absence of infection and these include excessive inflammation in $Cftr^{tm1Hgu}$ mice [27], an increase in goblet cells and decreased volume of airway surface liquid in the nasal epithelium of $Cftr^{tm1Unc}$ mice [28]. In addition both $Cftr^{tm1Hgu}$ and $Cftr^{tm1G551D}$ animals displayed submucosal glands that descended further down the trachea [29], bone marrow macrophages derived from the bone marrow of $Cftr^{G551D}$ mice were hypersensitive to bacterial lipopolysaccharide [30], and abnormalities in iNOS expression were described in 'null' CF mutants [31].

Some abnormalities consistent with a defect in mucociliary clearance and infection can be observed in CF mice. The first demonstration came from mice chronically exposed to bacteria by aerosolization. The $Cftr^{tm1Hgu}$ mice

Table 2. Comparison of the CF respiratory bioelectrics in man and mouse

	CF in humans	CF in mice
Nose		
Baseline PD	↑	↑
Amiloride response	↑	↑
cAMP-mediated or low chloride response	↓	↓ or ↔
Ca^{2+}-related chloride response	↔	↑ or ↔
Trachea		
Baseline PD	↔ or ↑	↔ or ↓
Amiloride response	↑	↔ or ↓
cAMP stimulation	↓	↔ or ↓
Ca^{2+}-related chloride response	↔	↔ or ↑

Arrows indicate bioelectric response is raised (↑), unchanged (↔) or reduced (↓) in CF individuals compared to wildtype.

displayed defective clearance of *Staphylococcus aureus* and *Burkholderia cepacia* and developed severe, pathogen-dependent lung pathology in response to extended exposure [32]. Reducing the level of *Cftr* by creating compound heterozygotes with $Cftr^{tm1Hgu}$ and $Cftr^{tm1Unc}$ alleles did not increase the severity of the lung pathology following *S. aureus* exposure [33]. This supports the fact that unlike pancreatic disease, the severity of lung disease in CF patients is as severe in 'null' patients as in those with some residual function.

The defect in mucociliary clearance was replicated in the $Cftr^{tm1Unc}$ mice on exposure to *B. cepacia* [34]. However, initially, the most characteristic pathogen for CF *P. aeruginosa*, did not seem to precipitate abnormalities in clearance or pathology [26, 35]. Recently, however, Coleman et al. [36] found that wild-type, heterozygous, and homozygous CF mice housed in the same cage became chronically colonized in the oropharynx with environmental *P. aeruginosa* when the bacterium was present in drinking water. Elimination of *P. aeruginosa* from drinking water resulted in clearance in most wild-type and CF heterozygotes, but not homozygous mice. A pathologic picture indicative of chronic lung infection could be seen in the CF mice. In some instances, mucoid isolates of *P. aeruginosa* were recovered from lungs, indicating conditions were present for conversion to mucoidy. In addition, CF mice demonstrated decreased bacterial clearance of *P. aeruginosa* after 4.5 h [37]. Cannon et al. [38] demonstrated that wild-type mice had apoptotic cells in their lungs after *P. aeruginosa* infections, whereas mice homozygous for the ΔF508 or G551D *Cftr* alleles showed little apoptosis in response to acute infection. Pseudomonal infection induced expression of CD95 and CD95 ligand, a response that was also delayed in cells homozygous for the mutant *Cftr* alleles.

Thus, manipulation and careful use of the cystic fibrosis animal models may result in useful *P. aeruginosa* colonization for investigation of pathophysiology of the disease and development of mucoidy and drug or gene therapy protection.

The agar bead model of *P. aeruginosa* chronic colonization was used in mice bearing different mutations in the murine *Cftr* gene [14]. In this model, bacteria are incorporated into agar beads and these are implanted into the animals airway. In this way, a stable infection can be produced and thus provide information of relevance to the host response to established infection. The mutations, R117H, *Cftr*^{tm1Unc}, Y122X, and ΔF508 (*Cftr*^{tm1Kth}), were all backcrossed to the C57BL/6J background to congenicity and were compared with respect to growth and their ability to respond to lung infection. One caveat is that the precise mutations were made by replacement gene targeting so the mutant alleles were not expressed at wild-type levels although the R117H does show increased survival compared to the other more severe mutations. However, interestingly the inflammatory responses to *P. aeruginosa*-laden agarose beads were comparable in mice of all four *Cftr* mutant genotypes with respect to absolute and relative cell counts in bronchoalveolar lavage fluid, and cytokine levels (TNF-α, IL-1β, IL-6, macrophage inflammatory protein-2, and keratinocyte chemoattractant) and eicosanoid levels (PGE$_2$ and LTB$_4$) in epithelial lining fluid.

The lung phenotype of an animal model for CF is crucial to accurately mimic the human disease. The models described to date provide some aspects of pulmonary disease but clearly do not offer the susceptibility to pathogens experienced by individuals with CF. Species differences such as basic lung architecture and physiology, airway cell composition, less widespread submucosal gland distribution, alternative chloride channels or difference in channel dominance and differences in innate airway defence might make further manipulation necessary to improve the model.

Elegant Refinements

The early death of the *Cftr* mutants that do not express *CFTR* has made studying lung disease in these mice particularly difficult. Death around weaning appears to be as a result of the intake of solid food and caecal blockage. One solution has been to feed the mice with a liquid diet and this

certainly prolongs the lifespan [39]. A problem with this is the possibility that mice fed in this way may suffer malnutrition and confuse the subsequent phenotype [40]. An elegant solution has been to express human *CFTR* as a transgene in the *Cftr*^{tm1Unc} 'null' animals (*Cftr*^{tm1Unc}-*TgN*$^{(FABPCFTR)}$) [41]. The human gene is under the control of the gut-specific fatty-acid-binding protein (FABP) gene promoter and so the phenotype is rescued in the intestine but not in the airway. Manson et al. also created transgenic mice, this time carrying a 320 kb YAC with the intact human *CFTR* gene [42, 43]. These mice were crossed onto the *Cftr*^{tm1Cam} 'null' mice. Mice carrying this transgene and expressing no mouse *Cftr* appeared normal and survived well, in marked contrast to the control 'null' mice, where 50% died by approximately 5 days after birth. The chloride secretory responses in the double transgenics were as large or larger than in wild-type tissues. Interestingly the expression of the transgene-at sites where there are differences between mouse and human, the YAC transgene followed the mouse pattern. In a final complementation experiment, Rozmahel et al. [43] disrupted expression of the endogenous *Cftr* gene and replaced with a human *CFTR* cDNA by a gene targeted 'knock-in' event. Animals homozygous for the gene replacement failed to show improved intestinal pathology or survival when compared to mice completely lacking CFTR despite the human *CFTR* sequence being transcribed from the targeted allele in the respiratory and intestinal epithelial cells. These results emphasize the difference between the tissue-specific expression and regulation of CFTR function in human and mouse.

Epithelial Sodium Channel Transgenics

In the normal lung, the airway surface liquid (ASL) together with beating cilia, helps to remove foreign particles including microbes. Matsui et al. [28] have convincingly demonstrated that the depth of the ASL (periciliary fluid and a mucin–rich top layer) reflects sodium absorption and anion secretion. CFTR normally suppresses the activity of the epithelial sodium channel (ENaC). Therefore one consequence of disruption of CFTR is an increase in the amiloride-sensitive sodium absorption. Indeed, individuals with CF have a characteristically raised baseline PD in the airway and an increased amiloride response, which are attributable to an increased sodium channel activity. CF mice also display an increased baseline and change in response to amiloride in the nose, but not in the trachea, where the baseline is either unchanged or reduced compared to wild-type animals and the amiloride response is either unchanged or reduced. In an effort to create a superior model for CF lung disease, Mall et al. [44] over expressed the three subunits of ENaC in the airways of transgenic mice. Surprisingly, mice expressing only the β-subunit showed a 3- to 4-fold increase in sodium currents across the trachea in Ussing chamber studies. At birth, the airways were histologically normal, but after several weeks the airways had a reduced periciliary liquid depth in the lower airways. Despite being reared in sterile conditions, the mice had goblet cell metaplasia, delayed mucus transport and adhesion to airway surfaces, neutrophil inflammation and increased proinflammatory cytokines in the ASL. This latter abnormality was not present in lung homogenates from newborn pups, or in media from primary cultures of airway cells, possibly indicating that this was downstream of the genetic manipulation. When exposed to bacteria relevant to CF (*Haemophilus influenzae* and *P. aeruginosa*) the ENaC β-subunit over-expressing mice demonstrated reduced ability to clear these pathogens. Whether these mice can develop chronic *P. aeruginosa* infections such as is found in individuals with CF remains to be seen, but the development is very encouraging.

Other Animal Models

The mouse is an attractive model system due to the ease with which the genome can be altered and the relative cheapness of maintaining a colony of animals. The physiology of the mouse is however not always the most comparable to human disease. The airways of some larger mammals, e.g. the pig and sheep, may make them a better model. With the discovery of cloning, pig, sheep, rabbits and ferrets are all being considered as more likely to reproduce the CF disease phenotype. The ovine CFTR gene is very similar to the human gene and has regulatory mechanisms in common [45].

The domestic ferret has certain attributes that potentially will make it a good model for CF. It has marked similarities to human in terms of lung physiology, airway morphology and cell types. In addition, expression of CFTR in the ferret submucosal gland is identical to that in humans [46]. The level of amino acid identity of the human CFTR protein to that of the ferret is 97%, which is as high as that found in non-human primates and significantly higher than the identity to mice (80%). Finally, a major consideration is that the ferret also has both husbandry and gestational advantages over large animals. Progress in both gene targeting into ferret somatic cells and nuclear transfer and cloning has been made, but the efficiency of both is poor, and a CF ferret has not been achieved to date [47].

In conclusion, CF mice have provided a useful tool for CF research; they are a convenient model for testing novel therapies such as gene therapy and new pharmaceuticals. However, their lack of overt lung disease means the model can and no doubt will be improved.

References

1   Dorin JR, Dickinson P, Alton EW, Smith SN, Geddes DM, Stevenson BJ, Kimber WL, Fleming S, Clarke AR, Hooper ML: Cystic fibrosis in the mouse by targeted insertional mutagenesis. Nature 1992;359:211–215.
2   Snouwaert JN, Brigman KK, Latour AM, Malouf NN, Boucher RC, Smithies O, Koller BH: An animal model for cystic fibrosis made by gene targeting. Science 1992;257: 1083–1088.
3   Colledge WH, Ratcliff R, Foster D, Williamson R, Evans MJ: Cystic fibrosis mouse with intestinal obstruction. Lancet 1992;340:680.
4   Ratcliff R, Evans MJ, Cuthbert AW, MacVinish LJ, Foster D, Anderson JR, Colledge WH: Production of a severe cystic fibrosis mutation in mice by gene targeting. Nat Genet 1993;4:35–41.
5   Davidson DJ, Dorin JR: The CF mouse: An important tool for studying cystic fibrosis. Expert Rev Mol Med 2001;2001:1–27.
6   Dorin JR, Stevenson BJ, Fleming S, Alton EW, Dickinson P, Porteous DJ: Long-term survival of the exon 10 insertional cystic fibrosis mutant mouse is a consequence of low level residual wild-type *Cftr* gene expression. Mamm Genome 1994;5:465–472.
7   O'Neal WK, Hasty P, McCray PB Jr, Casey B, Rivera-Perez J, Welsh MJ, Beaudet AL, Bradley A: A severe phenotype in mice with a duplication of exon 3 in the cystic fibrosis locus. Hum Mol Genet 1993;2:1561–1569.
8   Colledge WH, Abella BS, Southern KW, Ratcliff R, Jiang C, Cheng SH, MacVinish LJ, Anderson JR, Cuthbert AW, Evans MJ: Generation and characterization of a delta F508 cystic fibrosis mouse model. Nat Genet 1995;10:445–452.
9   Zeiher BG, Eichwald E, Zabner J, Smith JJ, Puga AP, McCray PB Jr, Capecchi MR, Welsh MJ, Thomas KR: A mouse model for the delta F508 allele of cystic fibrosis. J Clin Invest 1995;96:2051–2064.
10 van Doorninck JH, French PJ, Verbeek E, Peters RH, Morreau H, Bijman J, Scholte BJ: A mouse model for the cystic fibrosis delta F508 mutation. EMBO J 1995;14:4403–4411.
11 Bijman J, Veeze H, Kansen M, Tilly B, Scholte B, Hoogeveen A, Halley D, Sinaasappel M, de Jonge H: Chloride transport in the cystic fibrosis enterocyte. Adv Exp Med Biol 1991; 290:287–294.
12 Hamosh A, King TM, Rosenstein BJ, Corey M, Levison H, Durie P, Tsui LC, McIntosh I, Keston M, Brock DJ: Cystic fibrosis patients bearing both the common missense mutation Gly→Asp at codon 551 and the delta F508 mutation are clinically indistinguishable from delta F508 homozygotes, except for decreased risk of meconium ileus. Am J Hum Genet 1992;51:245–250.
13 Delaney SJ, Alton EW, Smith SN, Lunn DP, Farley R, Lovelock PK, Thomson SA, Hume DA, Lamb D, Porteous DJ, Dorin JR, Wainwright BJ: Cystic fibrosis mice carrying the missense mutation G551D replicate human genotype-phenotype correlations. EMBO J 1996;15:955–963.
14 Van Heeckeren AM, Schluchter MD, Drumm ML, Davis PB: Role of Cftr genotype in the response to chronic *Pseudomonas aeruginosa* lung infection in mice. Am J Physiol Lung Cell Mol Physiol 2004;287: L944–L952.

15 Smit LS, Strong TV, Wilkinson DJ, Macek M Jr, Mansoura MK, Wood DL, Cole JL, Cutting GR, Cohn JA, Dawson DC: Missense mutation (G480C) in the CFTR gene associated with protein mislocalization but normal chloride channel activity. Hum Mol Genet 1995;4:269–273.

16 Du M, Jones JR, Lanier J, Keeling KM, Lindsey JR, Tousson A, Bebok Z, Whitsett JA, Dey CR, Colledge WH, Evans MJ, Sorscher EJ, Bedwell DM: Aminoglycoside suppression of a premature stop mutation in a Cftr−/− mouse carrying a human CFTR-G542X transgene. J Mol Med 2002;80:595–604.

17 Rozmahel R, Wilschanski M, Matin A, Plyte S, Oliver M, Auerbach W, Moore A, Forstner J, Durie P, Nadeau J, Bear C, Tsui LC: Modulation of disease severity in cystic fibrosis transmembrane conductance regulator deficient mice by a secondary genetic factor. Nat Genet 1996;12:280–287.

18 Zielenski J, Corey M, Rozmahel R, Markiewicz D, Aznarez I, Casals T, Larriba S, Mercier B, Cutting GR, Krebsova A, Macek M Jr, Langfelder-Schwind E, Marshall BC, DeCelie-Germana J, Claustres M, Palacio A, Bal J, Nowakowska A, Ferec C, Estivill X, Durie P, Tsui LC: Detection of a cystic fibrosis modifier locus for meconium ileus on human chromosome 19q13. Nat Genet 1999;22:128–129.

19 Dorin JR, Farley R, Webb S, Smith SN, Farini E, Delaney SJ, Wainwright BJ, Alton EW, Porteous DJ: A demonstration using mouse models that successful gene therapy for cystic fibrosis requires only partial gene correction. Gene Ther 1996;3:797–801.

20 Gray MA, Winpenny JP, Porteous DJ, Dorin JR, Argent BE: CFTR and calcium-activated chloride currents in pancreatic duct cells of a transgenic CF mouse. Am J Physiol 1994;266:C213–C221.

21 Freedman SD, Blanco PG, Shea JC, Alvarez JG: Analysis of lipid abnormalities in CF mice. Methods Mol Med 2002;70:517–524.

22 Sui W, Boyd C, Wright JT: Altered pH regulation during enamel development in the cystic fibrosis mouse incisor. J Dent Res 2003;82:388–392.

23 Dif F, Marty C, Baudoin C, de Vernejoul MC, Levi G: Severe osteopenia in CFTR-null mice. Bone 2004;35:595–603.

24 Dickinson P, Smith SN, Webb S, Kilanowski FM, Campbell IJ, Taylor MS, Porteous DJ, Willemsen R, de Jonge HR, Farley R, Alton EW, Dorin JR: The severe G480C cystic fibrosis mutation, when replicated in the mouse, demonstrates mistrafficking, normal survival and organ-specific bioelectrics. Hum Mol Genet 2002;11:243–251.

25 Clarke LL, Grubb BR, Yankaskas JR, Cotton CU, McKenzie A, Boucher RC: Relationship of a non-cystic fibrosis transmembrane conductance regulator-mediated chloride conductance to organ-level disease in Cftr(−/−) mice. Proc Natl Acad Sci USA 1994;91: 479–483.

26 Scholte BJ, Davidson DJ, Wilke M, de Jonge HR: Animal models of cystic fibrosis. J Cyst Fibros 2004;3(suppl 2):183–190.

27 Zahm JM, Gaillard D, Dupuit F, Hinnrasky J, Porteous D, Dorin JR, Puchelle E: Early alterations in airway mucociliary clearance and inflammation of the lamina propria in CF mice. Am J Physiol 1997;272:C853–C859.

28 Matsui H, Grubb BR, Tarran R, Randell SH, Gatzy JT, Davis CW, Boucher RC: Evidence for periciliary liquid layer depletion, not abnormal ion composition, in the pathogenesis of cystic fibrosis airways disease. Cell 1998;95:1005–1015.

29 Borthwick DW, West JD, Keighren MA, Flockhart JH, Innes BA, Dorin JR: Murine submucosal glands are clonally derived and show a cystic fibrosis gene-dependent distribution pattern. Am J Respir Cell Mol Biol 1999;20:1181–1189.

30 McMorran BJ, Palmer JS, Lunn DP, Oceandy D, Costelloe EO, Thomas GR, Hume DA, Wainwright BJ: G551D CF mice display an abnormal host response and have impaired clearance of Pseudomonas lung disease. Am J Physiol Lung Cell Mol Physiol 2001;281: L740–L747.

31 Kelley TJ and Drumm ML: Inducible nitric oxide synthase expression is reduced in cystic fibrosis murine and human airway epithelial cells. J Clin Invest 1998;102:1200–1207.

32 Davidson DJ, Dorin JR, McLachlan G, Ranaldi V, Lamb D, Doherty C, Govan J, Porteous DJ: Lung disease in the cystic fibrosis mouse exposed to bacterial pathogens. Nat Genet 1995;9:351–357.

33 Davidson DJ, Webb S, Teague P, Govan JR, Dorin JR: Lung pathology in response to repeated exposure to Staphylococcus aureus in congenic residual function cystic fibrosis mice does not increase in response to decreased CFTR levels or increased bacterial load. Pathobiology 2004;71:152–158.

34 Sajjan U, Thanassoulis G, Cherapanov V, Lu A, Sjolin C, Steer B, Wu YJ, Rotstein OD, Kent G, McKerlie C, Forstner J, Downey GP: Enhanced susceptibility to pulmonary infection with Burkholderia cepacia in Cftr(−/−) mice. Infect Immun 2001;69:5138–5150.

35 McCray PB Jr, Zabner J, Jia HP, Welsh MJ, Thorne PS: Efficient killing of inhaled bacteria in deltaF508 mice: Role of airway surface liquid composition. Am J Physiol 1999;277: L183–L190.

36 Coleman FT, Mueschenborn S, Meluleni G, Ray C, Carey VJ, Vargas SO, Cannon CL, Ausubel FM, Pier GB: Hypersusceptibility of cystic fibrosis mice to chronic Pseudomonas aeruginosa oropharyngeal colonization and lung infection. Proc Natl Acad Sci USA 2003; 100:1949–1954.

37 Schroeder TH, Reiniger N, Meluleni G, Grout M, Coleman FT, Pier GB: Transgenic cystic fibrosis mice exhibit reduced early clearance of Pseudomonas aeruginosa from the respiratory tract. J Immunol 2001;166: 7410–7418.

38 Cannon CL, Kowalski MP, Stopak KS, Pier GB: Pseudomonas aeruginosa-induced apoptosis is defective in respiratory epithelial cells expressing mutant cystic fibrosis transmembrane conductance regulator. Am J Respir Cell Mol Biol 2003;29:188–197.

39 Kent G, Oliver M, Foskett JK, Frndova H, Durie P, Forstner J, Forstner GG, Riordan JR, Percy D, Buchwald M: Phenotypic abnormalities in long-term surviving cystic fibrosis mice. Pediatr Res 1996;40:233–241.

40 Yu H, Nasr SZ, Deretic V: Innate lung defenses and compromised Pseudomonas aeruginosa clearance in the malnourished mouse model of respiratory infections in cystic fibrosis. Infect Immun 2000;68: 2142–2147.

41 Zhou L, Dey CR, Wert SE, DuVall MD, Frizzell RA, Whitsett JA: Correction of lethal intestinal defect in a mouse model of cystic fibrosis by human CFTR. Science 1994;266: 1705–1708.

42 Manson AL, Trezise AE, MacVinish LJ, Kasschau KD, Birchall N, Episkopou V, Vassaux G, Evans MJ, Colledge WH, Cuthbert AW, Huxley C: Complementation of null CF mice with a human CFTR YAC transgene. EMBO J 1997;16:4238–4249.

43 Rozmahel R, Gyomorey K, Plyte S, Nguyen V, Wilschanski M, Durie P, Bear CE, Tsui LC: Incomplete rescue of cystic fibrosis transmembrane conductance regulator deficient mice by the human CFTR cDNA. Hum Mol Genet 1997;6:1153–1162.

44 Mall M, Grubb BR, Harkema JR, O'Neal WK, Boucher RC: Increased airway epithelial Na$^+$ absorption produces cystic fibrosis-like lung disease in mice. Nat Med 2004;10: 487–493.

45 Broackes-Carter FC, Mouchel N, Gill D, Hyde S, Bassett J, Harris A: Temporal regulation of CFTR expression during ovine lung development: Implications for CF gene therapy. Hum Mol Genet 2002;11:125–131.

46 Sehgal A, Presente A, Engelhardt JF: Developmental expression patterns of CFTR in ferret tracheal surface airway and submucosal gland epithelia. Am J Respir Cell Mol Biol 1996;15:122–131.

47 Li Z, Engelhardt JF: Progress toward generating a ferret model of cystic fibrosis by somatic cell nuclear transfer. Reprod Biol Endocrinol 2003;1:83.

48 Hasty P, O'Neal WK, Liu KQ, Morris AP, Bebok Z, Shumyatsky GB, Jilling T, Sorscher EJ, Bradley A, Beaudet AL: Severe phenotype in mice with termination mutation in exon 2 of cystic fibrosis gene. Somat Cell Mol Genet 1995;21:177–187.

Julia R. Dorin
MRC Human Genetics Unit
Western General Hospital
Crewe Road South
Edinburgh EH4 2XU (UK)
Tel. +44 131 467 8411
Fax +44 131 467 8456
E-Mail Julia@hgu.mrc.ac.uk

Bush A, Alton EWFW, Davies JC, Griesenbach U, Jaffe A (eds): Cystic Fibrosis in the 21st Century.
Prog Respir Res. Basel, Karger, 2006, vol 34, pp 93–101

In Vitro/Ex Vivo Fluorescence Assays of CFTR Chloride Channel Function

A.S. Verkman[a]   Luis J.V. Galietta[b]

[a]Departments of Medicine and Physiology, Cardiovascular Research Institute, University of California,
San Francisco, Calif., USA; [b]Laboratorio di Genetica Molecolare, Istituto Giannina Gaslini, Genova, Italy

Abstract

Assays utilizing chloride-sensitive fluorescent indicators have been applied to study chloride transport by CFTR in native and transfected cell culture models, in ex vivo tissues in human gene therapy trials, and in high-throughput screening applications for discovery of CFTR inhibitors and activators. Our laboratory has developed a series of small molecule and genetically encoded chloride/halide-sensing fluorescent indicators to measure cellular chloride transport. Small molecule indicators, such as SPQ, MQAE and LZQ, require loading of cell cytoplasm and washing of excess extracellular dye. The genetically encoded indicators are expressed transiently or stably in cell cultures or in vivo by transfection, infection or gene targeting procedures. The general experimental strategy to assay CFTR-mediated chloride transport is to follow cell/tissue fluorescence continuously in response to solution exchanges that create chloride ionic gradients, and/or add CFTR activators or inhibitors. Fluorescence is followed by cuvette fluorimetry, microscopy/imaging or plate reader methods. The optimal choice of indicator, solution exchange protocol, and measurement method depends on the system, level of CFTR function, and questions to be addressed. Although fluorescent chloride sensors cannot provide information about single channel electrical properties, they have been very useful in quantifying CFTR function in analysis of CFTR mutations, protein-protein interactions, gene transfer, and high-throughput screening.

Introduction

There is an ongoing need for assays of CFTR-mediated chloride transport in elucidating CFTR regulatory mecha-nisms and in the evaluation and development of therapies to treat the underlying chloride transport defect in cystic fibrosis. Fluorescence-based assays of CFTR chloride transport are very attractive for many applications. With the availability of relatively inexpensive cameras for low light level detection, it is possible by microscopy to assess CFTR chloride transport function at the single cell level in many cells at the same time in cell culture and ex vivo cells/tis-sues. Measurements on homogeneous cell populations can be made using commercial automated multiwell plate read-ers for high-throughput screening applications. Unlike measurements of transepithelial electrical properties such as short-circuit current, fluorescence measurements can be made in nonelectrically tight populations of heterogeneous cells. Fluorescence measurements are technically simple compared to single-channel electrophysiology, and provide statistically meaningful data rapidly in large cell popula-tions. A recent application of fluorescence-based readout of CFTR function is in drug discovery [1].

Small Molecule Chloride-Sensitive Fluorescent Indicators

Quinolinium, Acridinium, and Aminopyrido[2,1-h]-Pteridin Compounds

Several classes of small molecule chloride-sensitive flu-orescent indicators have been identified, in which their sen-sitivity to chloride and other halides results from a collisional quenching mechanism without static indicator-halide binding. Collisional quenching predicts a linear Stern-Volmer relation, $F_o/F = 1 + K_{hal}[hal]$, where F_o is the fluorescence in the absence of halide (hal), F is the

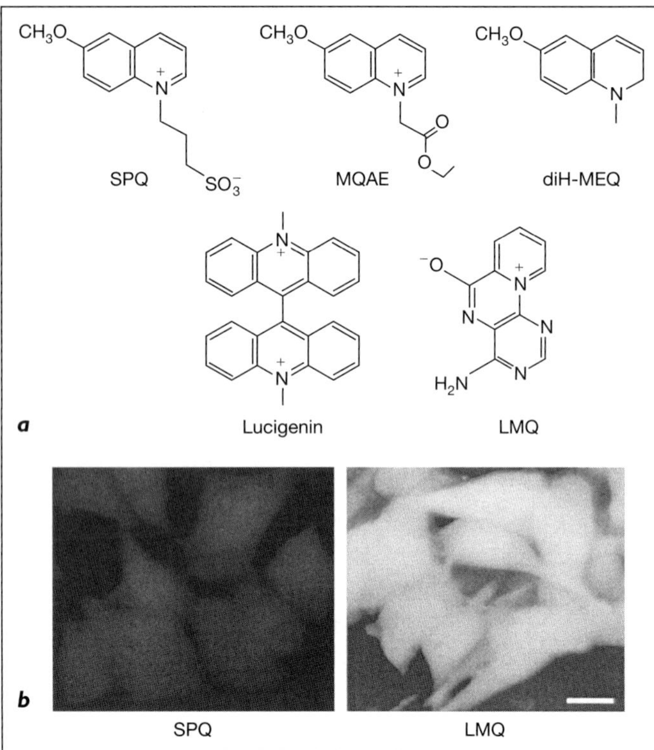

Fig. 1. Small molecule halide indicators. *a* Chemical structures of fluorescent halide indicators. *b* Fluorescence micrographs of Swiss 3T3 fibroblasts after labeling with SPQ and LMQ. Scale bar: 10 μM.

fluorescence in the presence of halide, and K_{hal} is the Stern-Volmer quenching constant (in M^{-1}). Fluorescence quenching by chloride probably results from transient formation of a charge-transfer complex [2, 3]. Because the kinetics of collisional quenching are diffusion-limited, the indicators provide submillisecond time resolution. SPQ is the first and most widely used fluorescent chloride indicator [4] (fig. 1a). SPQ fluorescence is excited at ultraviolet wavelengths (340–360 nm) and produces blue fluorescence (420–470 nm) as shown in figure 1b for an SPQ-stained cell. The Stern-Volmer constant for quenching of SPQ by chloride in aqueous buffers is $118 M^{-1}$, giving a 50% decrease in fluorescence at 8 mM chloride in aqueous solutions, though as mentioned below, its chloride sensitivity in cytoplasm is much lower. SPQ fluorescence is not altered by cations, phosphate, nitrate and sulfate, but is quenched weakly by other monovalent anions including citrate, acetate, gluconate and bicarbonate.

One of the more widely used indicators is MQAE (fig. 1a), which has very similar properties to SPQ, but somewhat better cell loading/retention because it is N-substituted with an ester moiety [5]. However, a general limitation of the quinolinium-based indicators is their high polarity and hence the necessity of invasive or slow diffusive cell loading. Although slow loading is not restrictive in most cell culture measurements, rapid loading is preferable to study freshly isolated cells ex vivo as in gene therapy trials. For this purpose a cell-permeable quinolinium indicator was engineered by masking the positive charge on the nitrogen by reduction of the quinolinium moiety to the uncharged 1,2-dihydroquinoline [6]. Reduction of the chloride-sensitive fluorescent compound 6-methoxy-N-ethylquinolinium (MEQ) with NaBH₄ gives 6-methoxy-N-ethyl-1,2-dihydroquinoline (diH-MEQ) (fig. 1a), which is nonpolar and enters cells rapidly where it is oxidized to the membrane-impermeable compound MEQ.

A limitation of SPQ, MQAE and MEQ is their single wavelength chloride detection that precludes determination of absolute chloride concentration by ratio imaging. To overcome this limitation fluorescent dual-wavelength chloride indicators were synthesized in which chloride sensitive and insensitive chromophores were linked by suitable spacers and the conjugate was designed to be cell permeable and remain entrapped in cells [7]. One such compound (bis-DMXPQ) consists of the blue fluorescing chloride-sensing chromophore 6-methoxyquinolinium (as in SPQ) and the green fluorescing chloride-insensitive chromophore 6-aminoquinolinium linked by a rigid *trans*-1,2-bis(4-pyridyl)ethylene spacer. This compound was used in ratiometric measurement of chloride in cultured cells.

Other limitations of quinolinium-type chloride indicators include the need for ultraviolet excitation and their relatively dim fluorescence emission. Ultraviolet excitation can be associated with significant autofluorescence background, photobleaching and photodynamic cell injury, which makes measurements using plate readers challenging. A series of long-wavelength polar fluorophores was screened to identify compounds with high chloride or iodide sensitivity in cells, bright fluorescence, low toxicity, uniform loading of cell cytoplasm, chemical stability, and minimal leakage out of cells. A class of 9-substituted acridinium compounds was identified, one of which (lucigenin) is shown in figure 1 [8, 9]. These compounds have bright green fluorescence and very high chloride sensitivity with Stern-Volmer quenching constants up to $390 M^{-1}$. However, although these compounds were found to work well for measurement of chloride transport in artificial systems such as liposomes, they are unstable in cytoplasm because of a 9-hydroxylation reaction, which renders them chloride insensitive. However, they are stable in cellular organelles, which has permitted measurements of chloride concentration in endosomes and Golgi in living cells [10, 11]. Of the screened compounds, the best indicator identified for use in cytoplasm was 4-aminopyrido[2,1-h]-pteridin-11-ium-6-olate (luminarine, LMQ) [12]. LMQ has bright yellow fluorescence (fig. 1a, b) that is quenched strongly by iodide (Stern-Volmer constant $70\,M^{-1}$), and its fluorescence is not pH-sensitive or quenched by other halides, nitrate or cations. LMQ and the related compound LZQ are probably the best available small molecule fluorescent halide indicators for functional CFTR measurements in cells.

CFTR Function in Cells Measured using Small Molecule Chloride Indicators

Fluorescent indicators for measurement of CFTR function in living cells should be nontoxic, chemically stable, distributed uniformly in the cytosolic compartment, sensitive and specific for cytoplasmic chloride or iodide, and able to be loaded and trapped in cytoplasm. In general, the available indicators are nontoxic, uniformly distributed, and stable in cells, with the exception of 9-substituted acridiniums as mentioned above. Quinolinium-type indicators are significantly less sensitive to chloride in cytoplasm versus solution, and factors other than chloride concentration can affect indicator fluorescence (see below). LMQ and LZQ are selective for iodide and not affected by other cytoplasmic components.

SPQ and related indicators with low membrane permeability like LMQ can be loaded into cells by slow passive diffusion (e.g. $5\,\mathrm{m}M$ SPQ for 4–12 h in culture medium), hypotonic shock (e.g. 50% hypotonic medium containing $5\,\mathrm{m}M$ SPQ for 3 min), or direct microinjection. Addition of $5\,\mathrm{m}M$ SPQ to a cell culture for a few hours generally allows enough intracellular uptake of SPQ for fluorescence measurements, yet SPQ leakage out of cells over the time course of experiments is minimal because of its intrinsic low membrane permeability. SPQ and LMQ can also be loaded into many cell types rapidly by hypotonic shock, which probably creates transient pores in the cell plasma membrane. However, there are potential concerns that hypotonic shock may activate cellular signaling cascades and volume-activated ion channels. Alternatively, loading of diH-MEQ, which undergoes intracellular oxidation to the cell-impermeant indicator MEQ, can be accomplished by brief (2–10 min) incubation at low concentrations (25–100 μM) in a physiological buffer. Several published studies have utilized diH-MEQ [13–15].

CFTR chloride transport measurements using fluorescent indicators are generally made on cell layers or individual cells grown or immobilized on a solid transparent support. Glass is a preferred support because of its transparency and minimal autofluorescence, although many plastics used in multi-well plates are suitable. The cell layer on the transparent support is positioned on the stage of an epifluorescence microscope in a chamber designed for rapid solution exchange [16] or in a plate reader. Also, polarized epithelial cells can be grown on transparent, low-autofluorescence permeable supports and mounted in a double perfusion chamber in which the apical and basolateral cell surfaces are perfused independently [17]. Cell fluorescence is detected by a low light level camera such as a cooled charged coupled device camera, or when single cell data are not required, by a photomultiplier or avalanche photodiode. Illumination with very low light intensities is important to minimize photobleaching and cell injury, particularly when ultraviolet light is used.

A number of experimental strategies have been used to study CFTR-mediated transport using fluorescent indicators. Figure 2 shows a commonly used strategy in which cells are bathed initially in a saline solution containing physiological concentrations of chloride. The fluorescence

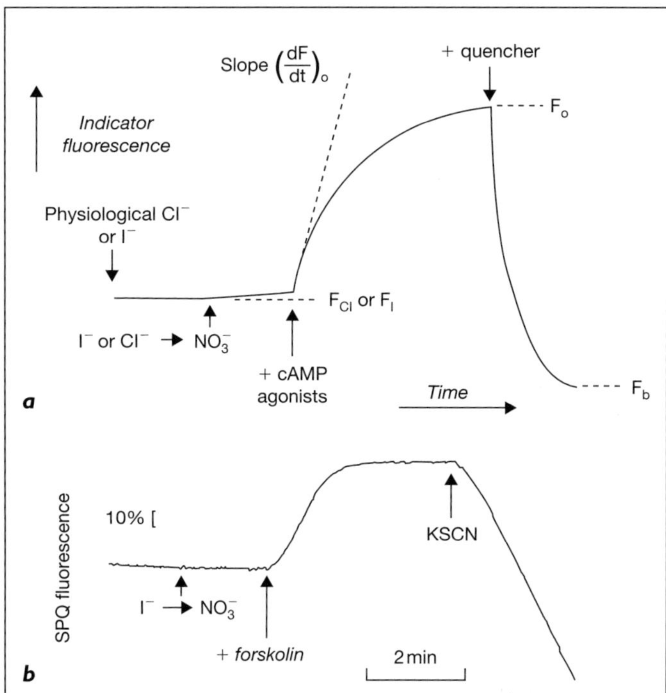

Fig. 2. Measurement of CFTR-mediated halide transport using a small molecule fluorescent indicator. *a* Schematic of measurement strategy. Time course of intracellular indicator fluorescence in response to chloride or iodide/nitrate exchange, addition of cAMP agonists, and quenching of indicator fluorescence. F_o = Fluorescence at zero chloride or iodide; F_{Cl} = fluorescence at initial intracellular chloride or iodide; F_b = background fluorescence; $(dF/dt)_o$ = initial rate of fluorescence increase after cAMP agonist addition. *b* Example of experimental measurement.

of the cytoplasmic chloride indicator is initially F_{Cl}. The saline solution is switched to an isosmolar solution in which chloride is replaced by nitrate, an anion that is transported by CFTR but does not quench SPQ fluorescence. If CFTR is the principal membrane chloride transporter but is not active, minimal chloride efflux will occur and fluorescence will increase little. After addition of cAMP agonist(s) such as forskolin and isobutylmethylxanthine (IBMX), activation of CFTR results in chloride efflux and nitrate influx producing an increase in fluorescence. Fluorescence will increase to F_o, corresponding to zero intracellular chloride. To compute absolute chloride flux, a second reference fluorescence value can be used. For example, the nitrate-containing solution can be replaced by a solution containing KSCN, which is highly permeable and strongly quenches SPQ fluorescence to give background fluorescence (F_b). Intracellular chloride concentration is $[(F_o - F_b)/(F_{Cl} - F_b) - 1]/K_{Cl}$, and the initial rate of chloride efflux in response to solution exchange of chloride for nitrate (J_{Cl}, in mM/s) is given by: $J_{Cl} = -(dF/dt)_o(F_o - F_b)/[K_{Cl}(F_{Cl} - F_b)^2]$, where $(dF/dt)_o$ is the slope of the fluorescence versus time curve, and K_{Cl} (in M^{-1}) is the Stern-Volmer constant for quenching of intracellular indicator chloride [16]. Various approaches have been used to measure intracellular K_{Cl}, such as intracellular calibration of fluorescence versus chloride using the ionophores tributyltin and nigericin in high potassium solutions to equalize intracellular and solution chloride concentration.

It is assumed in this formulation that cell volume remains constant and that only changes in chloride affect indicator fluorescence. As discussed in Chao et al. [16], the Stern-Volmer constant for SPQ in cells (generally $12–20\,M^{-1}$) is considerably less than that in aqueous solutions ($118\,M^{-1}$) because of quenching of indicator fluorescence by nonchloride intracellular anions including proteins and organic solutes. Thus it should be noted that for SPQ, even in the absence of chloride the fluorescence signal is mildly sensitive to changes in cell volume (which alters concentration of nonchloride anions) and pH (which alters intracellular protein charge).

Alternatively, iodide/nitrate rather than chloride/nitrate exchange has been used to assess CFTR function. Iodide has potential advantages over chloride in that it is transported by CFTR, but not by many non-CFTR chloride transporters. In addition, iodide quenches chloride indicators more strongly than does chloride, producing larger fluorescence signal changes. The sensitivity of LMQ to iodide, but not to chloride, permits the use of an iodide/chloride exchange protocol, generally giving large signal changes with excellent signal-to-noise ratios [12].

Chloride indicators have been used in numerous published studies of chloride transport in cells and tissues. Measurements have been made on cells expressing endogenous CFTR [17–19], and transfected with wild-type CFTR [20, 21] and various CFTR mutants [22, 23]. Functional measurements of various disease-causing and engineered CFTR deletion mutants have established the significance of specific subdomains of the CFTR molecule [24, 25]. Small molecule chloride indicators have also been useful in the screening and characterization of pharmacological agents and conditions for correction of the cystic fibrosis phenotype [26–32]. Cell culture models were also used to demonstrate the possibility of correction of the cystic fibrosis defect by CFTR protein transfer [33].

SPQ has also been used in human clinical trials to measure chloride transport in freshly isolated cells obtained by nasal and bronchial brushings. Two studies were designed to test the feasibility of distinguishing chloride transport in cells derived from normal and cystic fibrosis individuals [34, 35]. Small differences in the SPQ response were demonstrated between non-CF and CF cells, though these were relatively small compared to cell culture studies. Similar methodology was used in several gene therapy trials [36–38], though only small changes of unclear significance were reported. Additional optimization of methodology is needed for application of small molecule chloride indicators to detect gain-of-function in ex vivo tissues after *CFTR* gene transfer or pharmacological activation.

Green Fluorescent Protein-Based Halide Indicators

The green fluorescent protein (GFP) is an intrinsically fluorescent protein of ~30 kDa originally isolated from the jellyfish *Aequoria victoria* [for review, see 39]. Its chromophore is generated by autocatalytical cyclization and oxidation of three amino acid residues contained within the GFP primary sequence. Accordingly, expression of GFP in the target cell gives the fluorescence signal, so that addition of cofactors (like coelenterazine for aequorin, or luciferin for luciferase) is not needed. Amino acids throughout wild-type GFP, including its chromophore, have been mutated to generate an array of fluorescent proteins with altered spectral and biophysical properties [39]. For example, the yellow fluorescent proteins (YFP), the GFP variant from which halide-sensitive probes are derived, was generated by replacing the threonine at position 203 with tyrosine. This modification extends the π system of the chromophore, lowering its excited state energy and consequently increasing its emission wavelength. YFPs also contain mutations to increase folding efficiency (S65G, V68L, S72A) and hence brightness in transfected cells. The GFP sequence has also been modified to introduce motifs that allow targeting to specific cellular compartments, and that confer sensitivity to pH, ion concentrations, enzyme activities, and second messenger levels [40]. Compared to classical chemical probes, genetically encoded fluorescent proteins derived from GFP allow stable, bright, noninvasive staining with little or no cellular toxicity.

An intrinsic characteristic of YFPs is their fluorescence quenching by halides [41]. The tertiary structure of the YFP-H148Q mutant contains a pocket for the binding of anions including iodide and chloride [42]. Anion binding shifts the pK_a of the chromophore so that protonation is favored. Since the protonated chromophore is not fluorescent, an increase in halide concentration quenches YFP-H148Q fluorescence.

Fluorescence titration of purified recombinant YFP-H148Q indicated a pK_a of ~7 in the absence of chloride that increased to ~8 at 150 mM chloride [43]. At pH 7.5, YFP-H148Q fluorescence decreased with increasing chloride and iodide concentration such that the protein is 50% quenched by 100 mM chloride or 21 mM iodide. The anion selectivity sequence for YFP-H148Q quenching was: fluoride > iodide > nitrate > chloride > bromide. The difference in affinity for iodide versus chloride suggested the use of YFP-H148Q for cell-based assays for the measurement of anion transport mediated by ion channels and transporters. For example, CFTR halide transport function could be measured from the time course of cytoplasmic YFP-H148Q fluorescence in response to exchanging the extracellular halide (e.g. iodide for chloride).

YFP-H148Q was expressed in cultured cells to test its utility in cell-based assays of chloride transport [43, 44]. As reported for other GFP mutants, YFP-H148Q fluorescence was observed throughout the cell cytoplasm and nucleus. In vivo calibration experiments using ionophore-treated cell cultures indicated that the pH and chloride sensitivities of YFP-H148Q were similar in cells and aqueous solutions. In cells expressing CFTR, partial replacement of extracellular chloride with iodide (100 out of 137 mM) caused a decrease of cell fluorescence that was markedly accelerated by increasing CFTR activity with the cAMP-elevating agent forskolin. This effect was reversible since restoration of extracellular chloride caused near complete recovery of original fluorescence levels.

Using such chloride/iodide exchange protocols, large changes in fluorescence intensity of 40–50% were documented in various cell types. This substantial signal change is comparable to that observed for LZQ, the best chemical halide indicator developed to date [12]. After generation of stable cell lines coexpressing CFTR and YFP-H148Q, it was possible to perform assays in high-throughput screening format [44]. Cells were plated at high density in 96-well microplates and assayed after confluency in a microplate reader equipped with optical filters for YFP excitation and emission, and syringe pumps for fluid addition. CFTR activity was modulated by including in each well different concentrations of cAMP-elevating agents and/or direct CFTR activators. The functional assay consisted of a continuous reading of cell fluorescence in each well before and after addition of an iodide-rich saline solution (100 mM final iodide concentration). The maximum slope of the corresponding fluorescence decrease was determined to calculate the rate of iodide influx. As explained below, assays based on YFP-H148Q and mutants thereof have been used to perform at high speed large screens of chemical libraries to identify novel high affinity CFTR activators and inhibitors [45–47].

In developing cell-based assays, we found that the relatively low affinity of YFP-H148Q for iodide (K_d ~20 mM at physiological pH) may be limiting in the detection of small increases in halide transport such as those associated with human disease-causing CFTR mutants. Therefore, a more sensitive YFP was required. To generate derivatives of YFP-H148Q with improved halide sensitivity, we modified the polarity and/or size of the halide-binding cavity [48]. Degenerate oligonucleotide primers were used to generate YFP-H148Q plasmid libraries containing mutations in the residue pairs val150/ile152, val163/phe165, and leu201/tyr203. The library was expressed in bacteria and an efficient screening protocol was developed to measure halide sensitivities of the corresponding YFP mutants (fig. 3a). Screening of >1,000 colonies from the libraries yielded YFP-H148Q mutants with significantly different chloride and iodide affinities compared to YFP-H148Q [48]. The mutant I152L showed a significant improvement in iodide sensitivity with a K_d of only 2 mM (fig. 3a), an almost 10-fold increase in affinity compared to YFP-H148Q. In addition, the K_d for nitrate in I152L was significantly different from that for chloride, permitting cell-based assays involving nitrate/chloride exchange.

YFP-H148Q/I152L was introduced into a mammalian expression vector to test its applicability in cellular anion exchange measurements (fig. 3b). The higher sensitivity of this mutant was demonstrated by replacement of only 20 mM extracellular chloride by iodide. There was a maximum fluorescence decrease of ~50%, much greater than that of <10% in identical experiment performed using

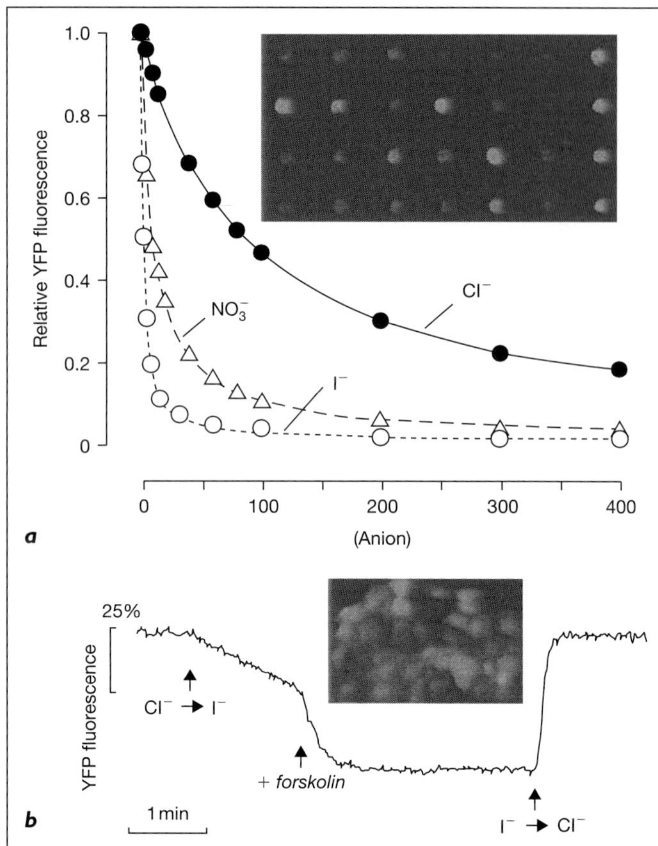

Fig. 3. Halide-sensing YFP. *a* YFP-H148Q was mutagenized at specific amino acid residues to novel sensors with improved anion sensitivity. Bacterial colonies expressing the YFP mutants (inset) were arrayed and grown individually to generate crude lysates that were titrated with different concentrations of halides and nitrate. Anion sensitivity of YFP-H148Q/I152L is shown, with ~10-fold greater iodide sensitivity than YFP-H148Q. *b* YFP-H148Q/I152L was expressed in epithelial cells together with wild-type CFTR (inset). Time course of fluorescence in response to exchange of 20 mM extracellular chloride for iodide and addition of forskolin to stimulate CFTR [adapted from 48].

YFP-H148Q. The feasibility of chloride/nitrate exchange protocols to detect CFTR activity was also demonstrated.

High-Throughput Screening Using Halide-Sensitive YFPs

Halide-sensitive fluorescent proteins are useful for rapid, automated screening of chemical libraries to identify novel pharmacological modulators of anion channels or transporters. Generation of a stable cell line coexpressing the target protein and the fluorescent sensor allows efficient scale-up without the limitations associated with small molecule fluorescent indicators, such as probe cost, cell loading and washing, and dye leakage. Recent studies have demonstrated the potential of halide-sensitive YFPs for the identification of novel activators and inhibitors of CFTR activity.

To discover new classes of CFTR modulators with improved potency and selectivity, an initial screening was carried out on 60,000 drug-like compounds using a cell line expressing wild-type CFTR and the YFP-H148Q sensor [45]. The assay consisted of short-term stimulation of cells with 10 μM test compound and 0.5 μM forskolin followed by iodide challenge. The primary screening, followed by secondary analysis using electrophysiological techniques and biochemical assays, led to the identification of 14 compounds which activated CFTR without involving cAMP elevation or phosphatase inhibition, suggesting direct CFTR interaction. The chemical structures of such molecules were different from previously known CFTR openers. These findings showed the utility of high-throughput screening in identifying novel classes of potent CFTR activators.

After this proof of principle study, attention was focused on the identification of small molecules able to correct the primary defect associated with CF mutations. The most frequent mutation among CF patients, ΔF508, causes two types of functional defects: mistrafficking of CFTR protein resulting in its retention at the endoplasmic reticulum, and a gating defect resulting in abnormally short time spent by the CFTR channel in the open state. A pharmacological approach to target the ΔF508 mutation therefore requires the identification of two classes of compounds, 'correctors' to overcome the trafficking defect, and 'potentiators' to stimulate CFTR channel opening. High-throughput screening was carried out using FRT cells expressing the YFP-H148Q/I152L probe to identify potentiators [47]. Cells were incubated at 27°C for 24 h to rescue the ΔF508-CFTR mistrafficking and then exposed to compounds contained in a collection of 100,000 diverse drug-like small molecules. Forskolin at maximal concentration was also included to maximally stimulate CFTR phosphorylation. Rapid iodide influx was found for some of the 100,000 test compounds. We identified >30 compounds that potentiated ΔF508-CFTR chloride channel activity with submicromolar affinity, with most of them belonging to six distinct chemical classes that are unrelated structurally to known CFTR activators or inhibitors.

Sublibraries containing more than 1,000 compounds with structural similarity to each class of ΔF508-CFTR potentiators were screened to establish structure-activity relationships and to identify analogues with improved

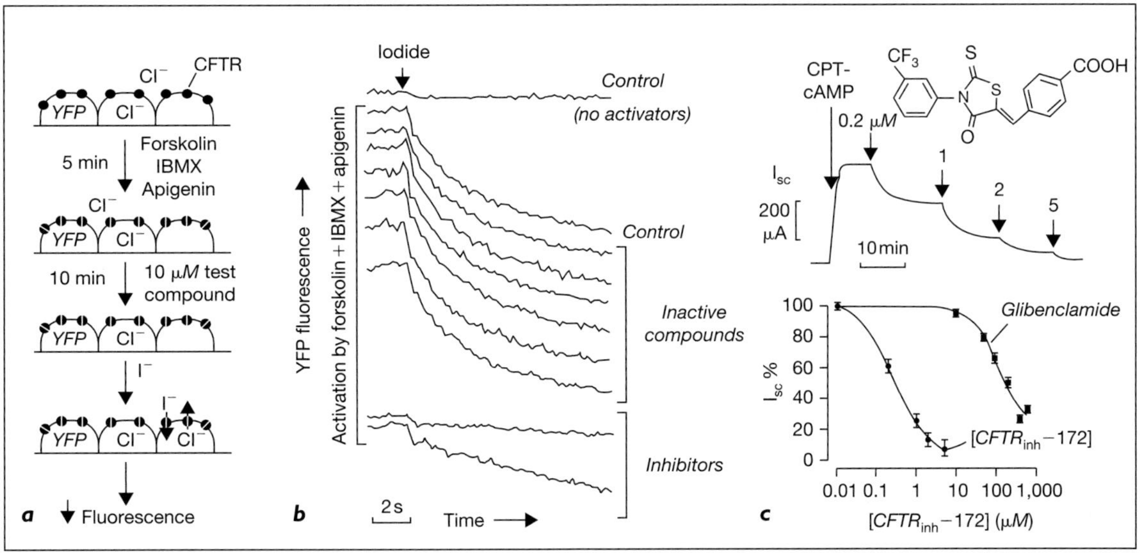

Fig. 4. Identification of CFTR inhibitors by high-throughput screening. *a* Cells coexpressing CFTR and a halide-sensitive YFP were seeded in 96-well microplates. Cells were treated with an agonist cocktail (forskolin, apigenin, IBMX) to fully activate CFTR in the presence and absence of test compounds. CFTR activity was measured in a plate reader by adding an iodide-containing solution and measuring the rate of fluorescence decrease. *b* Representative data from individual wells containing test compounds. *c* Structure of the $CFTR_{inh}$-172 (top) shown with dose-response data from short-circuit current analysis [adapted from 46].

characteristics. Secondary analysis confirmed that the novel potentiators stimulated chloride current in transfected FRT cells and in native human airway epithelial cells expressing ΔF508-CFTR [47]. Currently, we have identified potentiators with strong activity at <50 nM concentration. Experiments to identify correctors of defective ΔF508-CFTR intracellular processing are in progress.

Screenings have also been carried out to identify CFTR inhibitors (fig. 4). Selective inhibitors are useful to probe CFTR function and to produce the CF phenotype in cells, organs, and whole animals. They also have a potential therapeutic application for the treatment of secretory diarrhea. The available CFTR blockers, such as glibenclamide and DPC, have poor specificity and potency [49], requiring high micromolar concentrations at which they block other anion and cation channels. To identify new CFTR inhibitors, a collection of 50,000 drug-like molecules was initially screened [42]. FRT cells coexpressing human CFTR and YFP-H148Q were stimulated by a CFTR-activating cocktail and then subjected to an iodide gradient (fig. 4a). Inhibitors ('active compounds') were identified from the reduction in the CFTR-dependent iodide transport (fig. 4b). The best inhibitor identified by screening was the 2-thioxo-4-thiazolinone compound $CFTR_{inh}$-172 (fig. 4c, top). $CFTR_{inh}$-172 inhibited CFTR function at submicro-

molar concentrations, with a 500-fold improvement in potency with respect to glibenclamide when studied under similar conditions (fig. 4c, middle and bottom). $CFTR_{inh}$-172 did not affect several other chloride channels (calcium and volume-activated) or ABC transporters (MDR-1, SUR). Patch-clamp analysis showed voltage-independent inhibition of CFTR chloride conductance with prolonged mean channel closed time and without change in unitary conductance [46, 50]. The antidiarrheal efficacy of the thiazolidinone $CFTR_{inh}$-172 was tested in a mouse closed ileal loop model [51]. A single intraperitoneal injection of $CFTR_{inh}$-172 inhibited the cholera toxin-induced fluid accumulation. Orally administered $CFTR_{inh}$-172 was also effective in blocking intestinal fluid secretion after oral cholera toxin in an open-loop model.

More recently, screening of an additional 100,000 small drug-like compounds yielded a novel class of glycine hydrazide CFTR inhibitors [52]. The best compound, GlyH-101, has much greater water solubility than $CFTR_{inh}$-172. Interestingly, GlyH-101 appears to block CFTR currents by interacting with the channel pore, possibly from the extracellular side. Its inhibition mechanism is thus different from that of $CFTR_{inh}$-172, which probably acts by allosteric effects on channel gating. The two CFTR inhibitor classes identified by high-throughput screening may be useful in the generation of CF animal models and to produce the CF phenotype in ex vivo human tissues, as was done in a study of airway submucosal gland function [53].

Directions

Fluorescence assays have become the preferred approach for drug discovery by high-throughput screening because of their sensitivity, robustness, technical simplicity, and relatively low cost. In the examples described here, halide-sensitive GFP mutants were optimized for cell-based kinetic assays, and applied to the discovery of drug-like inhibitors of wild-type CFTR and activators of wild-type and ΔF508 CFTR. There are several other chloride channel targets suitable for drug discovery by high-throughput screening using the methodology described here, such as inhibitors of ClC-7 for therapy of osteoporosis [54, 55] or of calcium-activated chloride channels for reduction of airway mucus production [56–58]. Small molecule chloride-sensitive fluorescence indicators have been widely used in cell culture studies of chloride transporter regulation and mutational analysis, though further refinements are needed to realize their potential in ex vivo tissue and in vivo measurements of chloride channel function.

Acknowledgements

This study was supported by a CF Drug Discovery Grant from the Cystic Fibrosis Foundation, and grants DK72517, HL73856, EB00415, HL59198, EY13574 and DK35124 from the National Institutes of Health (A.S.V.) and grant GP0296Y01 from Telethon-Italy (L.J.G.).

References

1 Verkman AS: Drug discovery in academia. Am J Physiol 2004;286:C465–C474.

2 Mac M, Wirz J: Deriving intrinsic electron-transfer rates from nonlinear Stern-Volmer dependencies for fluorescence quenching of aromatic molecules by inorganic anions in acetonitrile. Chem Phys Lett 1993;211:20–26.

3 Jayaraman S, Verkman AS: Quenching mechanism of quinolinium-type chloride-sensitive fluorescent indicators. Biophys Chem 2000; 85:49–57.

4 Illsley NP, Verkman AS: Membrane chloride transport measured using a chloride-sensitive fluorescent probe. Biochemistry 1987;26: 1215–1219.

5 Verkman AS, Sellers MC, Chao AC, Leung T, Ketcham R: Synthesis and characterization of improved chloride-sensitive fluorescent indicators for biological applications. Anal Biochem 1989;178:355–361.

6 Biwersi J, Verkman AS: Cell-permeable fluorescent indicator for cytosolic chloride. Biochemistry 1991;30:7879–7883.

7 Jayaraman S, Biwersi J, Verkman AS: Synthesis and characterization of dual-wavelength chloride sensitive fluorescent indicators for ratio imaging. Am J Physiol 1999;276:C747–C757.

8 Legg KD, Hercules DM: Quenching of lucigenin fluorescence. J Phys Chem 1970;74: 2114–2121.

9 Biwersi J, Tulk B, Verkman AS: Long-wavelength chloride-sensitive fluorescent indicators. Anal Biochem 1994;219:139–143.

10 Sonawane ND, Thiagarajah JR, Verkman AS: Chloride concentration in endosomes measured using a ratioable fluorescent Cl⁻ indicator: Evidence for chloride accumulation during acidification. J Biol Chem 2002;277: 5506–5513.

11 Sonawane ND, Verkman AS: Determinants of [Cl⁻] in recycling and late endosomes and Golgi complex measured using fluorescent ligands. J Cell Biol 2003;160:1129–1138.

12 Jayaraman S, Teitler L, Skalski B, Verkman AS: Long-wavelength iodide-sensitive fluorescent indicators for measurement of functional CFTR expression in cells. Am J Physiol 1999;277:C1008–C1018.

13 Inglefield JR, Schwartz-Bloom RD: Confocal imaging of intracellular chloride in living brain slices: Measurement of GABA receptor activity. J Neurosci Methods 1997;75:127–135.

14 Wallace DP, Grantham JJ, Sullivan LP: Chloride and fluid secretion by cultured human polycystic kidney. Kidney Int 1996;50: 1327–1336.

15 Woll E, Gschwentner M, Furst J, Hofer S, Buemberger G, Jungwirth A, Frick J, Deetjen P, Paulmichl M: Fluorescence-optical measurements of chloride movements in cells using the membrane potential dye diH-MEQ. Pflügers Arch 1996;432:486–493.

16 Chao AC, Dix JA, Sellers MC, Verkman AS: Fluorescence measurement of chloride transport in monolayer cultured cells. Mechanisms of chloride transport in fibroblasts. Biophys J 1989;56:1071–1081.

17 Verkman AS, Chao AC, Hartmann T: Hormonal regulation of Cl⁻ transport in polar airway epithelia measured by a fluorescent indicator. Am J Physiol 1992;262:C23–C31.

18 Ram SJ, Kirk KL: Cl⁻ permeability of human sweat duct cells monitored with fluorescence-digital imaging microscopy: Evidence for reduced plasma membrane Cl⁻ permeability in cystic fibrosis. Proc Natl Acad Sci USA 1989;86:10166–10170.

19 Sermet-Gaudelus I, Vallee B, Urbin I, Torossi T, Marianovski R, Fajac A, Feuillet MN, Bresson JL, Lenoir G, Bernaudin JF, Edelman A: Normal function of the cystic fibrosis conductance regulator protein can be associated with homozygous ΔF508 mutation. Pediatr Res 2002;52:628–635.

20 Rich DP, Anderson MP, Gregory RJ, Cheng SH, Paul S, Jefferson DM, McCann JD, Klinger KW, Smith AE, Welsh MJ: Expression of cystic fibrosis transmembrane conductance regulator corrects defective chloride channel regulation in cystic fibrosis airway epithelial cells. Nature 1990;347:358–363.

21 Rommens JM, Dho S, Bear CE, Kartner N, Kennedy D, Riordan JR, Tsui LC, Foskett JK: cAMP-inducible chloride conductance in mouse fibroblast lines stably expressing the human cystic fibrosis transmembrane conductance regulator. Proc Natl Acad Sci USA 1991;88:7500–7504.

22 Cheng SH, Rich DP, Marshall J, Gregory RJ, Welsh MJ, Smith AE: Phosphorylation of the R domain by cAMP-dependent protein kinase regulates the CFTR chloride channel. Cell 1991;66:1027–1036.

23 Rich DP, Berger HA, Cheng SH, Travis SM, Saxena M, Smith AE, Welsh MJ: Regulation of the cystic fibrosis transmembrane conductance regulator Cl⁻ channel by negative charge in the R domain. J Biol Chem 1993; 268:20259–20267.

24 Delaney SJ, Rich DP, Thomson SA, Hargrave MR, Lovelock PK, Welsh MJ, Wainwright BJ: Cystic fibrosis transmembrane conductance regulator splice variants are not conserved and fail to produce chloride channels. Nat Genet 1993;4:426–431.

25 Xie J, Drumm ML, Ma J, Davis PB: Intracellular loop between transmembrane segments IV and V of cystic fibrosis transmembrane conductance regulator is involved in regulation of chloride channel conductance state. J Biol Chem 1995;270:28084–28091.

26 Andersson C, Servetnyk Z, Roomans GM: Activation of CFTR by genistein in human

   Verkman/Galietta

airway epithelial cell lines. Biochem Biophys Res Commun 2003;308:518–522.

27 Brown CR, Hong-Brown LQ, Biwersi J, Verkman AS, Welch WJ: Chemical chaperones correct the mutant phenotype of the delta F508 cystic fibrosis transmembrane conductance regulator protein. Cell Stress Chaperones 1996;1:117–125.

28 Cheng SH, Fang SL, Zabner J, Marshall J, Piraino S, Schiavi SC, Jefferson DM, Welsh MJ, Smith AE: Functional activation of the cystic fibrosis trafficking mutant delta F508-CFTR by overexpression. Am J Physiol 1995; 268:L615–L624.

29 Jiang C, Fang SL, Xiao YF, O'Connor SP, Nadler SG, Lee DW, Jefferson DM, Kaplan JM, Smith AE, Cheng SH: Partial restoration of cAMP-stimulated CFTR chloride channel activity in DeltaF508 cells by deoxyspergualin. Am J Physiol 1998;275:C171–C178.

30 Yang Y, Devor DC, Engelhardt JF, Ernst SA, Strong TV, Collins FS, Cohn JA, Frizzell RA, Wilson JM: Molecular basis of defective anion transport in L cells expressing recombinant forms of CFTR. Hum Mol Genet 1993;2: 1253–1261.

31 Howard M, Frizzell RA, Bedwell DM: Aminoglycoside antibiotics restore CFTR function by overcoming premature stop mutations. Nat Med 1996;2:467–469.

32 Clancy JP, Bebok Z, Ruiz F, King C, Jones J, Walker L, Greer H, Hong J, Wing L, Macaluso M, Lyrene R, Sorscher EJ, Bedwell DM: Evidence that systemic gentamicin suppresses premature stop mutations in patients with cystic fibrosis. Am J Respir Crit Care Med 2001; 163:1683–1692.

33 Marshall J, Fang S, Ostedgaard LS, O'Riordan CR, Ferrara D, Amara JF, Hoppe HT, Scheule RK, Welsh MJ, Smith AE, Cheng SH: Stoichiometry of recombinant cystic fibrosis transmembrane conductance regulator in epithelial cells and its functional reconstitution into cells in vitro. J Biol Chem 1994;269: 2987–2995.

34 Stern M, Munkonge FM, Caplen NJ, Sorgi F, Huang L, Geddes DM, Alton EW: Quantitative fluorescence measurements of chloride secretion in native airway epithelium from CF and non-CF subjects. Gene Ther 1995;2:766–774.

35 McLachlan G, Ho LP, Davidson-Smith H, Samways J, Davidson H, Stevenson BJ, Carothers AD, Alton EW, Middleton PG, Smith SN, Kallmeyer G, Michaelis U, Seeber S, Naujoks K, Greening AP, Innes JA, Dorin JR, Porteous DJ: Laboratory and clinical studies in support of cystic fibrosis gene therapy using pCMV-CFTR-DOTAP. Gene Ther 1996; 3:1113–1123.

36 Gill DR, Southern KW, Mofford KA, Seddon T, Huang L, Sorgi F, Thomson A, MacVinish LJ, Ratcliff R, Bilton D, Lane DJ, Littlewood JM, Webb AK, Middleton PG, Colledge WH, Cuthbert AW, Evans MJ, Higgins CF, Hyde SC: A placebo-controlled study of liposome-mediated gene transfer to the nasal epithelium of patients with cystic fibrosis. Gene Ther 1997;4:199–209.

37 Porteous DJ, Dorin JR, McLachlan G, Davidson-Smith H, Davidson H, Stevenson BJ, Carothers AD, Wallace WA, Moralee S, Hoenes C, Kallmeyer G, Michaelis U, Naujoks K, Ho LP, Samways JM, Imrie M, Greening AP, Innes JA: Evidence for safety and efficacy of DOTAP cationic liposome mediated CFTR gene transfer to the nasal epithelium of patients with cystic fibrosis. Gene Ther 1997;4:210–218.

38 Alton EW, Stern M, Farley R, Jaffe A, Chadwick SL, Phillips J, Davies J, Smith SN, Browning J, Davies MG, Hodson ME, Durham SR, Li D, Jeffery PK, Scallan M, Balfour R, Eastman SJ, Cheng SH, Smith AE, Meeker D, Geddes DM: Cationic lipid-mediated CFTR gene transfer to the lungs and nose of patients with cystic fibrosis: A double-blind placebo-controlled trial. Lancet 1999;353:947–954.

39 Tsien RY: The green fluorescent protein. Annu Rev Biochem 1998;67:509–544.

40 Haggie P, Verkman AS: GFP sensors. Top Fluorescence 2005; in press.

41 Wachter RM, Remington SJ: Sensitivity of the yellow variant of green fluorescent protein to halides and nitrate. Curr Biol 1999;9: R628–R629.

42 Wachter RM, Yarbough D, Kallio K, Remington SJ: Crystallographic and energetic analysis of binding of selected anions to the yellow variants of green fluorescent protein. J Mol Biol 2000;301:157–171.

43 Jayaraman S, Haggie P, Wachter RM, Remington SJ, Verkman AS: Mechanism and cellular application of a green fluorescent protein-based halide sensor. J Biol Chem 2000; 275:6047–6050.

44 Galietta LJV, Jayaraman S, Verkman AS: Cell-based assay for high-throughput quantitative screening of CFTR chloride transport agonists. Am J Physiol 2001;281:C1734–C1742.

45 Ma T, Vetrivel L, Yang H, Pedemonte N, Zegarra-Moran O, Galietta LJ, Verkman AS: High-affinity activator of cystic fibrosis transmembrane conductance regulator (CFTR) chloride conductance identified by high-throughput screening. J Biol Chem 2002;277: 37235–37241.

46 Ma T, Thiagarajah JR, Yang H, Sonawane ND, Folli C, Galietta LJ, Verkman AS: Thiazolidinone CFTR inhibitor identified by high-throughput screening blocks cholera toxin-induced intestinal fluid secretion. J Clin Invest 2002;110:1651–1658.

47 Yang H, Shelat AA, Guy RK, Gopinath VS, Ma T, Du K, Lukacs GL, Taddei A, Folli C, Pedemonte N, Galietta LJ, Verkman AS: Nanomolar-affinity small-molecular activators of ΔF508-CFTR chloride channel gating. J Biol Chem 2003;278:35079–35085.

48 Galietta LJ, Haggie P, Verkman AS: Green fluorescent protein-based halide indicators with improved chloride and iodide affinities. FEBS Lett 2001;499:220–224.

49 Schultz BD, Singh AK, Devor DC, Bridges RJ: Pharmacology of CFTR chloride channel activity. Physiol Rev 1999;79:S109–S144.

50 Taddei A, Folli C, Zegarra-Moran O, Fanen P, Verkman AS, Galietta LJV: Altered channel gating mechanism for CFTR inhibition by a high-affinity thiazolidinone blocker. FEBS Lett 2004;558:52–56.

51 Thiagarajah JR, Broadbent T, Hsieh E, Verkman AS: Prevention of toxin-induced intestinal ion and fluid secretion by a small-molecule CFTR inhibitor. Gastroenterology 2004;126:511–519.

52 Muanprasat C, Sonawane ND, Salinas D, Taddei A, Galietta LJV, Verkman AS: Discovery of glycine hydrazide pore-occluding CFTR inhibitors: Mechanism, structure-activity analysis and in vivo efficacy. J Gen Physiol 2004;124:125–137.

53 Thiagarajah JR, Song Y, Haggie P, Verkman AS: A small molecule CFTR inhibitor produces cystic fibrosis-like submucosal gland fluid secretions in normal airways. FASEB J 2004;18:875–877.

54 Kornak U, Kasper D, Bosl MR, Kaiser E, Schweizer M, Schulz A, Friedrich W, Delling G, Jentsch TJ: Loss of the ClC-7 chloride channel leads to osteopetrosis in mice and man. Cell 2001;104:205–215.

55 Schaller S, Henriksen K, Sveigaard C, Heegaard AM, Helix N, Stahlhut M, Ovejero MC, Johansen JV, Solberg H, Andersen TL, Hougaard D, Berryman M, Shiodt CB, Sorensen BH, Lichtenberg J, Christophersen P, Foged NT, Delaisse JM, Engsig MT, Karsdal MA: The chloride channel inhibitor n53736 prevents bone resorption in ovariectomized rats without changing bone formation. J Bone Miner Res 2004;19:1144–1153.

56 Nakanishi A, Morita S, Iwashita H, Sagiya Y, Ashida Y, Shirafuji H, Fujisawa Y, Nishimura O, Fujino M: Role of gob-5 in mucus overproduction and airway hyperresponsiveness in asthma. Proc Natl Acad Sci USA 2001;98: 5175–5180.

57 Zhou Y, Dong Q, Louahed J, Dragwa C, Savio D, Huang M, Weiss C, Tomer Y, McLane MP, Nicolaides NC, Levitt RC: Characterization of a calcium-activated chloride channel as a shared target of Th2 cytokine pathways and its potential involvement in asthma. Am J Respir Cell Mol Biol 2001;25:486–491.

58 Hoshino M, Morita S, Iwashita H, Sagiya Y, Nagi T, Nakanishi A, Ashida Y, Nishimura O, Fujisawa Y, Fujino M: Increased expression of the human Ca^{2+}-activated Cl^- channel 1 (CaCC1) gene in the asthmatic airway. Am J Respir Crit Care Med 2002;165:1132–1136.

Alan S. Verkman, MD, PhD
1246 Health Sciences East Tower,
Cardiovascular Research Institute
University of California
San Francisco, CA 94143–0521 USA
Tel. +1 415 476 8530, Fax +1 415 665 3847
E-Mail verkman@itsa.ucsf.edu
http://www.ucsf.edu/verklab

Bush A, Alton EWFW, Davies JC, Griesenbach U, Jaffe A (eds): Cystic Fibrosis in the 21st Century.
Prog Respir Res. Basel, Karger, 2006, vol 34, pp 102–108

In Vivo Measurement of Airway Potential Difference to Assess CFTR Function in Man

Peter Middleton[a]   Eric W.F.W. Alton[b]

[a]Cystic Fibrosis Unit, Department of Respiratory Medicine, Westmead Hospital, Westmead, Australia;
[b]Department of Gene Therapy, National Heart and Lung Institute, Imperial College, London, UK

Abstract

In vivo measurement of airway potential difference (PD) has been shown to reflect the cystic fibrosis electrophysiological changes which can be dissected in more detail in vitro. It can also provide useful diagnostic information for the individual patient. Techniques for measuring PD in the human airway have been established and standardized. The challenges for this technology are now to determine (1) what further information regarding the control of epithelial ion transport can be discerned using this system, and (2) how changes in this surrogate end point predict more clinically relevant assays.

Overview of Respiratory Epithelial Ion Transport

The epithelial cells of the respiratory tract, like all mammalian cells, consist of a lipid bilayer surrounding an aqueous (hydrophilic) cell interior. As the lipid bilayer is relatively impermeable to the passage of charged ions, most ion transport occurs via specialized membrane structures such as ion channels, cotransporters, exchangers and pumps, which are summarized in figure 1a.

Ion channels form hydrophilic pores in the cell membrane, allowing charged ions to move across the hydrophobic cell membrane. The importance of ion channels in cellular homeostasis is suggested by the relatively large numbers of ions moved (10^6–10^7/s), many orders of magnitude larger than the pumps or cotransporters. Each cell has many hundreds or thousands of these ion channels with

particular types localized to the apical and basolateral membranes.

Cotransporters are membrane proteins which can link the movement of one ion down its electrochemical gradient with the movement of others against an unfavourable gradient. One example is the basolateral Na^+-K^+-$2Cl^-$ cotransporter, which uses the large inwardly directed force for the movement of Na^+, to move K^+ and Cl^- ions into the cell against unfavourable electrochemical gradients. Exchange mechanisms also exist which link the movement of one ion down its electrochemical gradient with that of another ion in the opposite direction, against its electrochemical gradient.

Pumps are distinguished by energy consumption, usually in the form of adenosine triphosphate (ATP), to move ions against their electrochemical gradient. Thus, for example, the Na^+-K^+-ATPase pump located on the basolateral membrane of most epithelial cells creates a large gradient for Na^+ ions across both the basolateral and apical membranes. In turn, these large electrochemical gradients for Na^+ drive other ion movements in the cell, such as the basolateral cotransporter discussed above.

Airway Epithelial Electrophysiology

Following the first demonstration of epithelial ion transport across the frog skin [1], the cellular models for salt absorption and secretion were proposed by Koefoed-Johnsen and Ussing [2] and are applicable to the respiratory epithelium. The model for salt absorption describes a

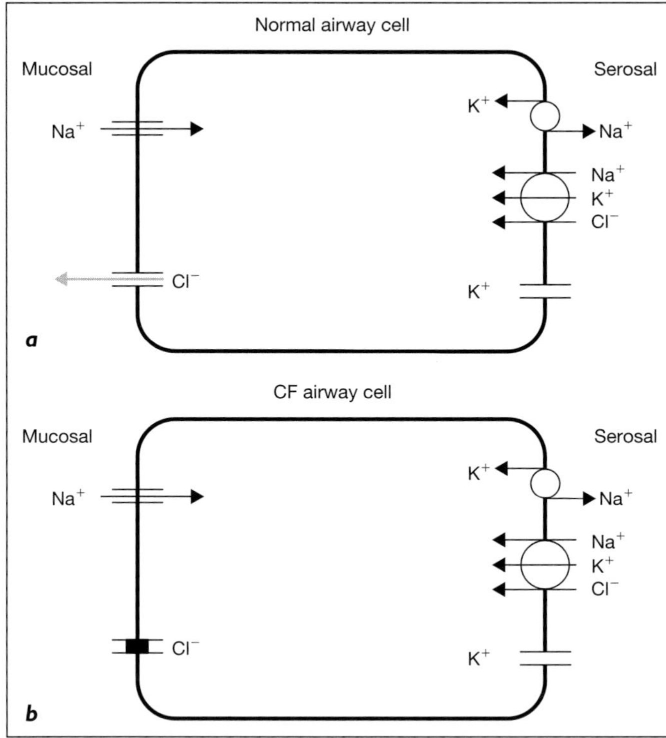

Fig. 1. The principal ion transport mechanisms in a **a** typical human airway epithelial cell **b** typical CF human airway epithelial cell.

two-step process as shown in figure 1. Sodium is actively transported out of a cell across the basolateral membrane by the Na^+-K^+-ATPase pump, providing a favourable electrochemical gradient for apical Na^+ absorption. In absorptive cells, such as the respiratory epithelium, the Na^+ that enters the cell across the apical membrane is accompanied by paracellular movement of Cl^-, resulting in net absorption of NaCl from the airway lumen. Salt secretion is dependent on Cl^- movement from the airway epithelium into the airway lumen. This requires the accumulation of Cl^- within the cell through the action of the Na^+-K^+-$2Cl^-$ cotransporter located in the basolateral membrane. Chloride then exits the cell across the apical Cl^- channels, coupled with Na^+ moving paracellularly to maintain electroneutrality. With salt absorption and secretion, water follows passively – the magnitude of net salt (and fluid) movement across an epithelium is determined, in part, by the balance between these opposing forces and in part by the relative leakiness of the tight junctions. As a consequence of these ion movements, the epithelium develops a potential difference (PD) or voltage that can be measured across the epithelium with simple electrodes. This chapter will focus on in vivo measurements of human airway ion

transport, and the abnormalities in airway epithelial ion transport characteristically found in cystic fibrosis (CF), as summarized in figure 1b. As the PD across the human airway is lumen negative, for simplicity the terms 'increased and decreased PD' will use the convention where this relates to changes in the absolute PD, which nearly always remains lumen negative.

Airway PD: CF versus Non-CF

Sodium Transport

Knowles et al. [3] first showed that both the nasal and tracheobronchial epithelium of CF subjects generated an increased baseline PD. This abnormality was also seen in CF neonates [4] suggesting that the elevated PD was related to the genetic defect, and not a result of secondary infection. Perfusion of the nasal epithelium with the sodium channel blocker amiloride induced a greater reduction in the nasal PD of CF than non-CF subjects [3], suggesting that the elevated baseline PD related to increased Na^+ transport. As similar ion transport defects were found in excised CF nasal polyps in vitro, with an elevated baseline PD and a greater amiloride response [5], this confirmed that the abnormal ion transport was inherent to the epithelium itself, and not due to hormonal or other factors. Subsequent studies localized the increased PD to an increased apical membrane Na^+ permeability, which relates to CFTR-ENaC interactions described in chapter 5. It is now generally agreed that the CF nose and lower airways demonstrate similar abnormalities of ion transport. The simplicity of measuring PD in the nose retains its place as an important model for both the investigation of CF pathophysiology and the development of new treatments.

Chloride Transport

The characteristic Cl^- impermeability found in the apical membrane of the CF respiratory epithelium was also first demonstrated using measurements of nasal PD [6]. Following pretreatment with amiloride to block Na^+ absorption, subsequent perfusion of the nasal mucosa with a solution containing reduced Cl^- concentration (to induce a driving force for Cl^- exit) produced a larger response in the non-CF subjects compared with CF subjects. Nasal polyps from CF subjects also showed a smaller response in vitro to perfusion with low Cl^- solution [5]. Subsequently the abnormality in Cl^- conductance was also localized to the apical membrane of the CF airway cells using microelectrodes [7]. Further delineation of the defect in apical Cl^- conductance was provided by the comparison of responses to different secretagogues: the response to iso-prenaline (isoproterenol), which stimulates Cl^- secretion through cAMP pathways, was also markedly diminished [8, 9], whilst pathways of Cl^- secretion induced by elevation of Ca^{2+} functioned normally in the CF cells. Thus, in vivo measurements of PD reflect the more detailed in vitro electrophysiological analyses described in chapter 5. Both systems can demonstrate the Na^+ and Cl^- abnormalities found in the apical membrane of CF respiratory epithelial cells.

Methods to Measure PD

Two techniques using similar, but slightly different sets of equipment are available. Importantly, both use similar protocols and achieve very comparable results, the choice of equipment and protocol simply relating to operator preference.

(1) Human airway epithelial PD has been measured in vivo for more than 20 years, using a system developed by Knowles et al. [10] at the University of North Carolina (UNC), which was modified from techniques to measure epithelial PD across the gut lumen. This comprises two electrodes filled with a physiological saline solution, connected via agar bridges to calomel electrodes which allow connection of a fluid-filled system to a measuring voltmeter. The reference electrode is inserted subcutaneously, and the exploring electrode, a fine double-lumen plastic tube, is placed in contact with the airway lumen allowing measurement of PD and drug perfusion. In the nose, the exploring electrode is moved along the undersurface of the inferior turbinate to the point of maximum PD. Initial attempts by other groups to replicate this work showed the requirement for considerable technical expertise, attention to detail and motivation by both the investigator and the subject, each highlighted during measurements in uncooperative patients [11]. Finally, the use of a subcutaneous reference electrode is somewhat uncomfortable.

(2) To address some of these technical difficulties, we developed a system for measuring the baseline PD along the floor of the nose using a modified Foley urinary catheter [12]. Positioning of the catheter along the nasal floor gives qualitatively similar results, but is more comfortable and technically easier than against the inferior turbinate. The system includes silver-silver chloride wires replacing the calomel electrodes, with the reference electrode placed over an area of abraded skin, removing the need for a subcutaneous electrode [13]. The two original lumina of the Foley catheter are used to measure PD and perfuse the epithelium with drug solutions [14, 15].

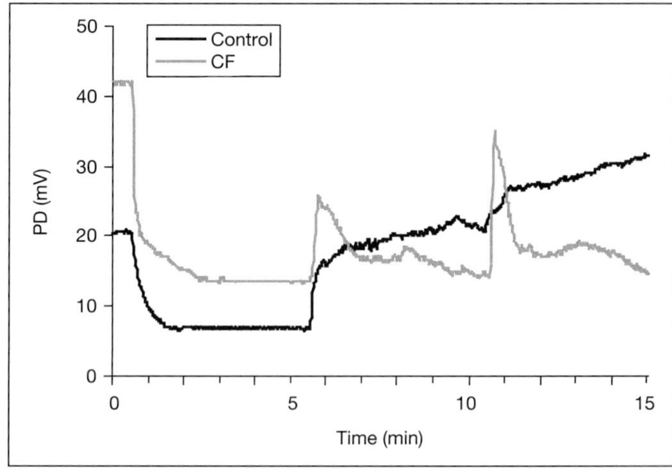

Fig. 2. Typical non-CF and CF nasal PD responses. Following the initial stabilization in normal physiological saline buffer, amiloride is added at time 0, low Cl^- solution at time = 5 min, and finally low Cl^-/isoproterenol at time = 10 min.

Both techniques use a standardized set of solutions to determine the Na^+ and Cl^- transport of the airway [8, 9]. Typical responses for non-CF and CF individuals are shown in figure 2. The protocol starts with perfusion of a physiological saline solution to allow measurement of the unstimulated (also called resting or baseline) PD, which predominantly reflects Na^+ absorption. Following stabilization of PD, the epithelium is then perfused with amiloride (0.1 mM) to block Na^+ absorption. The combination of the baseline PD and the response to amiloride represent the two measures of airway Na^+ absorption in vivo. Stimulation of Cl^- secretion requires the development of an electrochemical gradient for Cl^- ions to move from the epithelium to the airway surface. This is usually provided by perfusion of the airway surface with a low or zero Cl^- solution, achieved by replacing the permeable anion, Cl^-, with the impermeable anion gluconate. This produces a large chemical gradient ($>$100 mM) to induce Cl^- movement across the apical membrane into the airway lumen, reflected by an increase in (more negative) PD across the airway. To measure the effect of (nominally) zero Cl^- versus low (6 mM) Cl^- we compared the nasal PD responses in 6 normal subjects. Following amiloride pretreatment, there was an average of 2 mV greater response to the zero Cl^- compared with the 6 mM Cl^-.

The PD changes seen following low Cl^- perfusion likely represent Cl^- movement across channels that are constitutively open at the time of measurement. Clearly, these may include both CFTR as well as calcium-regulated and other chloride channels. To measure cAMP-regulated Cl^- secretion, the characteristic defect in CF, the airway is subsequently perfused with 0.01 mM isoproterenol. This serves to open CFTR and other protein kinase A-regulated chloride channels and increases the distinction between CF and non-CF subjects seen following the low chloride perfusion. There is agreement that the combined low chloride/isoproterenol response provides the optimal discrimination between CF and non-CF subjects. Finally, if the investigator wishes to exclude epithelial damage or equipment failure as a cause of absent chloride secretion, the perfusion of ATP (0.1 mM), allows confirmation that the airway can respond to a stimulus of Ca^{2+}-regulated Cl^- secretion [9].

Variables Influencing the Technique

Perfusion Rate

The drug and electrolyte solutions can be perfused rapidly to maximize solution change and to minimize the total time taken for the different responses. However, excessive fluid flow becomes increasingly uncomfortable for the subject, causing difficulty co-ordinating breathing and swallowing [8, 14]. Both groups have concluded that 4–5 ml/min provides the optimal balance for PD recordings in the adult nose.

Duration of Perfusion

There is general agreement that the key issue is for the investigator to 'interact' with the measurement being undertaken, so that a new solution is only perfused when a stable tracing has been achieved with the prior intervention. Broadly, this will occur after about 3–4 min for amiloride, and 4–5 min for low chloride and isoproterenol.

Solution Temperature

Boyle et al. [16] compared the use of room temperature (ambient = 22°C) solutions with those warmed to body temperature (37°C). They demonstrated that 37°C solutions had similar baseline PD and amiloride responses to those at ambient temperature. Whilst the zero Cl^- response showed a non-significant trend to smaller responses with warmed solutions (-8.0 ± 7.1 vs. -10.0 ± 8.0 mV), the isoproterenol response was significantly increased when tested with 37°C solutions (13.3 ± 8.8 vs. 6.9 ± 6.4 mV). The combined Cl^- response (of both low Cl^- and isoprenaline) was significantly greater with 37°C solutions (21.3 ± 11.9 mV) compared with 22°C solutions (16.9 ± 9.5 mV). The authors, therefore, suggested that the nasal PD solutions should be perfused at 37°C to maximize Cl^- responses. However, this paper also demonstrates the variability of the responses, with the difference between 22 and 37°C solutions for individual subjects ranging from $+25$ to -40 mV. This suggests that the inherent noise in the technique is much greater than the 5 mV difference in average response, likely in part to relate to the increased complexity of warming solutions. At present, the US Cystic Fibrosis Foundation Therapeutic Development Network (TDN) recommends warming, whilst the UK groups use ambient temperature solutions.

Multicentre Studies

Recently, the TDN co-ordinated the development of a standardized operating procedure for the measurement of nasal PD. The aim of this standardized operating procedure was to facilitate the multicentre use of the nasal PD as an end point in clinical trials. In a preliminary study, Ahrens et al. [17] confirmed that different sites across the US could generate similar results using a standardized procedure. Whilst overall, the responses generated using different individuals at the different centres were similar, the responses did show site-to-site variability. For example, the baseline PD and amiloride response showed complete separation of CF and non-CF subjects in some centres, but overlap was demonstrated in others. A follow-up study has suggested further refinements both to the technique and investigator training [18]. Unsurprisingly, it is clear that there is no substitute for experienced investigators using identical equipment.

Problems with the Technique

Site of Measurement

The nasal PD only provides a reliable measurement from an area of healthy respiratory epithelium (pseudostratified columnar). The squamous epithelium at the anterior vestibule generates a low baseline PD, and a markedly reduced or absent response to amiloride and low Cl^- solutions. There are well-documented ways to deduce the position of the catheter from the profile of the PD tracing using either of the above techniques. Both nasal polyps and previous surgery produce squamous metaplasia, and are considered relative exclusion criteria for PD measurements.

In the standard protocols, the exploring electrode is placed at the site of maximal baseline PD to optimize the difference between CF and non-CF subjects for diagnostic purposes. However, this may work against the nasal PD being able to detect changes produced by novel therapies. In the therapeutic situation, those cells demonstrating some degree of electrophysiological sodium correction will have their baseline PD reduced towards the normal range. In turn, this site may not be chosen as the site for further PD measurements; rather the catheter may be moved to an area of higher (uncorrected) PD. Thus, a therapeutic effect may be missed. One way to help ease this problem is to make measurements at a given distance into the nasal cavity, irrespective of the baseline PD profile encountered.

Epithelial Damage

The sensitivity of the nasal PD to minor trauma was demonstrated by Knowles et al. [10] in the original studies, where rubbing the nasal surface with a cotton bud was found to almost completely abolish the PD. This has led to generally accepted exclusion criteria, such as deferring recordings within a few weeks of an infection, or after nasal brushings to obtain epithelial samples. The presence of sinusitis is not necessarily associated with nasal conducting airway alterations, and is not, therefore, regarded as an exclusion criterion.

Uncooperative Subjects

The technique depends on the subject sitting relatively still for 20 min. Whilst this is generally not an issue in adults, children particularly between the ages of 2 and 5 may not be amenable to these measurements. Up to the age of approximately 2 years, parental or nursing restraint generally allows for PD recordings. Recently, Southern et al. [19] have described a system with extremely low flow rates that is sufficiently comfortable to be used in newborns whilst asleep.

Use of PD for Diagnostic Purposes

Sweat tests and genotyping remain the cornerstones of CF diagnosis. However, nasal PD measurements are now part of the accepted criteria for CF diagnosis. In general, in the difficult-to-diagnose patients, where a sweat test may be indeterminate, nasal PDs are also less clear cut, and require careful separate assessment of sodium and chloride components. Thus, for example in a group of infertile men with congenital bilateral absence of the vas deferens baseline and amiloride responses were within normal limits, whilst the low Cl^-/isoprenaline responses were reduced compared with non-CF, though not to the same level as in CF patients [20]. A relationship between PD and disease severity can be proposed, where a small reduction in CFTR function may alter the chloride component of the PD, leaving the sodium response unaltered. As CFTR function

reduces further, sodium also becomes abnormal, and lung disease becomes evident.

Use of PD for Prognostic Purposes

In the presence of an established CF diagnosis, studies have assessed the prognostic value of different components of PD measurements. We demonstrated that disease severity, as measured by FEV_1, correlated inversely with the baseline nasal PD [13] in a group of patients with a spectrum of CFTR mutations. Fajac et al. [21] divided a group of 95 adult CF patients dependent on their baseline nasal PD. Those subjects with the smallest (least negative) baseline PDs showed better lung function than those with PDs more in the typical CF range. To address the potential confounding factor of differing mutations in this response, Fajac et al. [21] also analyzed their data for those subjects with a defined genotype, showing that those with a more normal nasal PD response still demonstrated better lung function. This group has expanded these findings to PDs produced during the perfusion protocols outlined above [22]. Again, patients with poor lung function were more likely to have an enhanced amiloride response. Interestingly, pancreatic insufficiency was, however, predicted by a more abnormal response to low chloride perfusion. However, Wallace et al. [23] compared clinical phenotype with a variety of different nasal PD measures, including both the Na^+ and Cl^- components. They found no correlation between these measures of airway ion transport and either clinical or radiological scores. More recently, we assessed the degree of residual Cl^- secretion in the airways of $\Delta F508$ homozygous subjects showing a significant correlation, in males only, between FEV_1 and the isoproterenol response [24].

Use of PD to Assess Physiological Processes Relevant to CF

We have recently demonstrated that differences in the electrolyte content of the fluid bathing the airway surface can have profound effects on the airway ion transport of the underlying epithelium [25, 26]. Thus, the normal human airway epithelium responds to the addition of extra sodium chloride to the airway surface with a large reversible decrease in nasal PD [25]. Interestingly, there is little net change in nasal PD following hypertonic mannitol, suggesting that it is the ionic content of the saline that induces the response, and not the hypertonicity. In a separate series

     Middleton/Alton

of experiments, we have also shown that removal of divalent ions allows the induction of Cl^- secretory responses in CF subjects [26]. Nasal PD measurements thus provide an important in vivo assay for assessing ion transport mechanisms in the human airway.

Use of PD to Assess Effects of Novel Therapies

Nasal PD has been used as a surrogate end point in many studies of both small molecule and gene therapy. Both types of agents have produced changes in the Cl^- response in CF subjects, typically restoring 20–25% of the normal Cl^- response. The key question is, of course, how this relates to any clinical benefit to the patient. To date, no study of novel therapy has measured both PD and clinical benefit. This relates largely to the need for the therapy to be administered repeatedly to have any realistic expectation of effecting clinical improvement. In turn this means large-scale clinical studies with the attendant costs to patients, investigators and bank balances. The UK Cystic Fibrosis Gene Therapy Consortium aims to move forward into such a large scale study using the current best buy gene therapy package. This should provide data on if, and how, changes in PD correlate with measurements such as inflammatory markers, imaging, or lung physiology.

Experiments of nature provide a further potential source of data. A number of CF mutations such as A455E and R117H are generally associated with mild lung disease. Although there are relatively few measurements of nasal PD in these patients, the trend is for values to be only slighter better than in patients with severe lung disease [27]. Thus, it can be argued that the changes of 20–25% noted above for small molecule and gene therapy may have already exceeded that associated with mild lung disease and that we have a number of exciting new treatments already present in our armoury. One practical caveat is that these patients with these milder mutations clearly have these CFTR levels in every cell. Whilst small molecule therapy should also achieve this, gene therapy may not, and hence the relationship may differ for the two treatment classes. Also of relevance to gene therapy is the relationship between PD measured across an epithelium and the number of cells which contribute to this PD. Because cells are linked in parallel, the total PD measured is equivalent to the PD of any one of the cells. This analysis becomes complicated when some cells have differing PDs to others, as might be the case following gene therapy. However, there is a theoretical possibility that a change in measured PD may only represent changes in one or a few cells, thus overestimating the significance of the measurement. Clearly, from the above it is possible to be optimistic, pessimistic or neutral about the data from studies so far carried out. This, in turn, should perhaps encourage the field to take the bold step noted above, namely to find out whether changes in PD, and other surrogate markers, relate to alterations in more clinically relevant parameters.

Measurement of Lower Airway PD

Similar techniques to those described for the nose have been used to measure PD in the lower airways [3, 28] via a bronchoscope. One important technical caveat is that the use of lignocaine as a topical local anaesthetic during bronchoscopy markedly alters PD recordings. Thus, wherever possible measurements should be made whilst subjects are under general anaesthesia. However, the same sodium and chloride-related abnormalities can be detected in the large airways. Interestingly, values for the baseline PD become smaller as the exploring electrode is passed more distally down the airways. This may either reflect a lower rate of sodium absorption or more 'leaky' airways. Clearly, it would be of interest to make such measurements at sites closer to the small airways, likely the key region in CF. To this end adaptations are being made to the exploring electrodes to reduce external diameter. However, compromises need to be made in the perfusion protocol, since the perfused solutions readily pool in the smaller airways. This makes it difficult to assess separately an amiloride response (in normal chloride solution), followed by a low chloride perfusion. One solution is, following measurement of the baseline PD, to next perfuse with low chloride and subsequently to assess the effect of amiloride, albeit now dissolved in a low chloride solution. We used this protocol to assess non-viral mediated gene transfer to the lower airways [29]. Interestingly, similar magnitude of changes in the low chloride response were seen in both the nose and lung in this study. Finally, it is children who are most likely to benefit from novel therapies. With this in mind, we have recently shown the feasibility of making these measurements in the lower airways of children as young as a year old [30].

Conclusions

The technical aspects of PD measurements have largely been established in the nose, and are increasingly being assessed in the lower airways. Over the last 20 years PD has become an accepted investigation for the diagnosis of CF in appropriate individuals. It also provides important information concerning the physiology of the normal human airway, and the abnormalities found in CF. Finally, the technique is routinely used in the assessment of new therapies aimed at the basic CF defect. The major challenge for this assay is now to understand the relationship between changes in PD, and more clinically relevant markers.

References

1 Ussing HH, Zerahn K: Active transport of sodium as the source of electric current in the short-circuited isolated frog skin. Acta Physiol Scand 1951;23:110–127.
2 Koefoed-Johnsen V, Ussing HH: The nature of the frog skin potential. Acta Physiol Scand 1958;42:298–308.
3 Knowles M, Gatzy J, Boucher R: Increased bioelectric potential difference across respiratory epithelia in cystic fibrosis. N Engl J Med 1981;305:1489–1495.
4 Gowen CW, Lawson EE, Gingras-Leatherman J, et al: Increased nasal potential difference and amiloride sensitivity in neonates with cystic fibrosis. J Pediatr 1986;108:517–521.
5 Knowles MR, Stutts MJ, Spock A, et al: Abnormal ion permeation through cystic fibrosis respiratory epithelium. Science 1983; 221:1067–1070.
6 Knowles M, Gatzy J, Boucher R: Relative ion permeability of normal and cystic fibrosis nasal epithelium. J Clin Invest 1983;71: 1410–1417.
7 Willumsen NJ, Davis CW, Boucher RC: Cellular CI-transport in cultured cystic fibrosis airway epithelium. Am J Physiol 1989; 256:C1045–C1053.
8 Middleton PG, Geddes DM, Alton EWFW: Protocols for in vivo measurement of the ion transport defects in cystic fibrosis nasal epithelium. Eur Respir J 1994;7:2050–2056.
9 Knowles MR, Paradiso AM, Boucher RC: In vivo nasal potential difference: Techniques and protocols for assessing efficacy of gene transfer in cystic fibrosis. Hum Gene Ther 1995;6:445–455.
10 Knowles MR, Carson JL, Collier AM, et al: Measurements of nasal transepithelial electric potential differences in normal human subjects in vivo. Am Rev Respir Dis 1981;124: 484–490.
11 Hay JG, Geddes DM: Transepithelial potential difference in cystic fibrosis. Thorax 1985;40: 493–496.
12 Alton EWFW, Hay JG, Munro C, et al: Measurement of nasal potential difference in adult cystic fibrosis, Young's syndrome, and bronchiectasis. Thorax 1987;42:815–817.
13 Alton EWFW, Currie D, Logan-Sinclair R, et al: Nasal potential difference: A clinical diagnostic test for cystic fibrosis. Eur Respir J 1990;3:922–926.
14 Middleton PG: Cystic Fibrosis Ion Transport and the Effect of CFTR Gene Transfer. PhD thesis, University of London, 1995.
15 Middleton PG, Geddes DM, Alton EWFW: Effect of amiloride and saline on nasal mucociliary clearance and potential difference in cystic fibrosis and normal subjects. Thorax 1993;48:812–816.
16 Boyle MP, Diener-West M, Milgram L, et al: A multicenter study of the effect of solution temperature on nasal potential difference measurements. Chest 2003;124:482–489.
17 Ahrens RC, Standaert TA, Launspach J, et al: Use of nasal potential difference and sweat chloride as outcome measures in multicenter clinical trials in subjects with cystic fibrosis. Pediatr Pulmonol 2002;33:142–150.
18 Standaert TA, Boitano L, Emerson J, et al: Standardized procedure for measurement of nasal potential difference: An outcome measure in multicenter cystic fibrosis clinical trials. Pediatr Pulmonol 2004;37:385–392.
19 Southern KW, Noone PG, Bosworth DG, et al: A modified technique for measurement of nasal transepithelial potential difference in infants. J Pediatr 2001;139:353–358.
20 Osborne LR, Lynch M, Middleton PG, et al: Nasal epithelial ion transport and genetic analysis of infertile men with congenital bilateral absence of the vas deferens. Hum Mol Genet 1993;2:1605–1609.
21 Fajac I, Hubert D, Bienvenu T, et al: Relationships between nasal potential difference and respiratory function in adults with cystic fibrosis. Eur Respir J 1998;12: 1295–1300.
22 Fajac I, Hubert D, Guillemot D, et al: Nasal airway ion transport is linked to the cystic fibrosis phenotype in adult patients. Thorax 2004;59:916–917.
23 Wallace HL, Barker PM, Southern KW: Nasal airway ion transport and lung function in young people with cystic fibrosis. Am J Respir Crit Care Med 2003;168:594–600.
24 Thomas SR, Jaffe A, Geddes DM, Alton EWFW: Pulmonary disease severity in men with deltaF508 cystic fibrosis and residual chloride secretion. Lancet 1999;353:984–985.
25 Middleton PG, Pollard KA, Wheatley JR: Hypertonic saline alters ion transport across the normal airway epithelium. Eur Respir J 2001;17:195–199.
26 Middleton PG, Pollard KA, Donohoo E, et al: Airway surface liquid calcium modulates chloride permeability in the cystic fibrosis airway. Am J Respir Crit Care Med 2003;168: 1223–1226.
27 Walker LC, Venglarik CJ, Aubin G, et al: Relationship between airway ion transport and a mild pulmonary disease mutation in CFTR. Am J Respir Crit Care Med 1997;155: 1684–1689.
28 Alton EWFW, Khagani A, Taylor RFH, et al: Effect of heart-lung transplantation on airway potential difference in patients with and without cystic fibrosis. Eur Respir J 1991;4:5–9.
29 Alton EWFW, Stern M, Farley R, et al: Cationic lipid-mediated CFTR gene transfer to the lungs and nose of CF patients: A double-blind placebo-controlled trial. Lancet 1999;353: 947–954.
30 Davies JC, Davies M, McShane D, et al: Potential difference measurements in the lower airways of children with and without cystic fibrosis. Am J Respir Crit Care Med 2005;171:1015–1019.

Peter Middleton, Assoc. Prof.
Cystic Fibrosis Unit
Department of Respiratory Medicine
Westmead Hospital
Sydney (Australia)
Tel. +61 2 9845 6797
Fax +61 2 9845 7286
Email: peterm@westgate.wh.usyd.edu.au

Bush A, Alton EWFW, Davies JC, Griesenbach U, Jaffe A (eds): Cystic Fibrosis in the 21st Century.
Prog Respir Res. Basel, Karger, 2006, vol 34, pp 109–114

Arrays and Proteomics

Varrie C. Ogilvie   Gordon MacGregor

Medical Genetics Section, Department of Medical Sciences, Edinburgh University, Edinburgh;
The UK Cystic Fibrosis Gene Therapy Consortium, Edinburgh/London/Oxford, UK

Abstract

Microarrays (or arrays) and proteomics are powerful new technologies that have emerged in the post-genomic era and their applications in biomedical research fields are rapidly expanding. We aim in this chapter to encapsulate advances made in cystic fibrosis (CF) research through use of these technologies and highlight their respective contributions to our understanding of disease pathogenesis. The chapter is divided into two parts to give each technology justice; however, some overlap cannot be avoided. Each part begins with a brief introduction in which fundamental aspects of microarray and proteomic techniques will be described. We then describe how these new technologies are furthering our current understanding of CF, focusing on aspects of pulmonary disease. Key publications in this area will be discussed in more detail. Further, a description of exciting future applications of microarrays and proteomics in CF concludes this chapter.

Microarrays: New Tools for Analysis of the Genome

Microarrays are genome-wide analysis tools used predominantly for differential mRNA expression profiling [1] but also increasingly for other applications including genotyping [2] (fig. 1). Completion of the human genome mapping project was the driving force behind the development of microarrays, providing accurate sequence information on all possible coding and non-coding regions within the

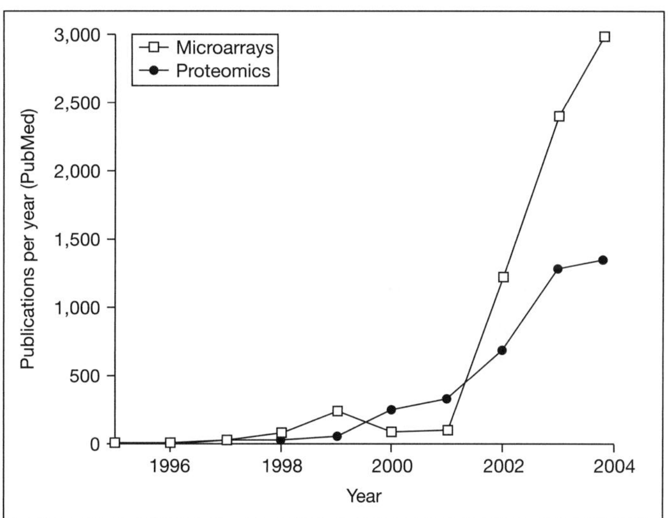

Fig. 1. Graph illustrating the rate at which published articles in the microarray and proteomic fields has increased in recent years. The number of articles retrieved from PubMed searches for each field is plotted per year.

genome. The fabrication of microarrays quickly followed with the timely development of new technologies for automated synthesis and deposition of nucleic acids at very high densities [3].

The advent of microarrays has enabled the study of the structure and function of the genome on a scale that was previously inconceivable. Microarrays comprise tens of thousands of different probes (oligonucleotides or PCR-generated

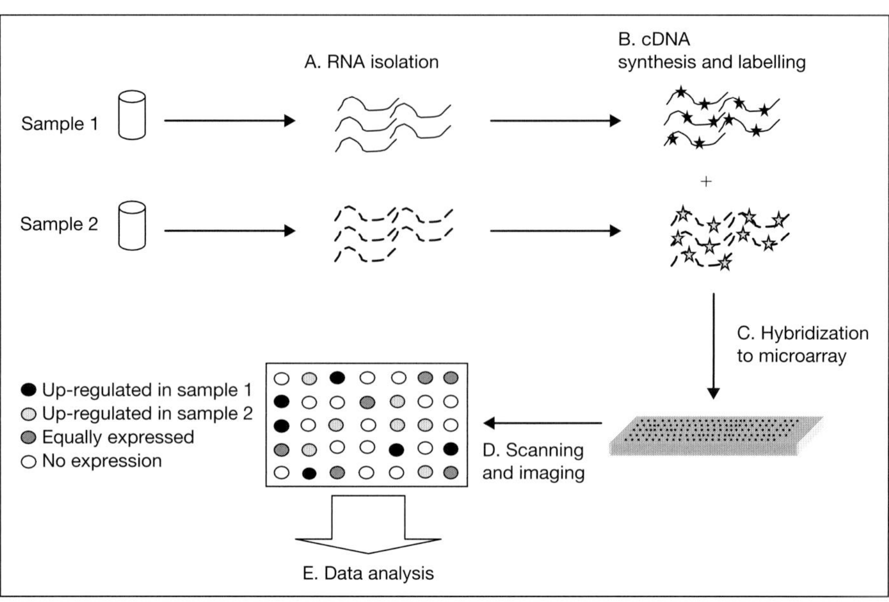

Fig. 2. Schematic representation of microarray analysis of gene expression. The process generally involves the following steps: A, isolation of RNA (total or messenger RNA) from samples; B, synthesis of fluorescently labelled complementary (c)DNA in a reverse transcription reaction; C, hybridization of labelled cDNA to a microarray, followed by washing to remove excess; D, scanning using a laser scanner to generate an image of the hybridized array for signal quantification; E, data acquisition, normalization and statistical analysis are then performed using specialised software to identify differentially expressed genes.

cDNAs) arranged in pre-defined positions on a solid support, such as a glass slide, and their miniaturized format allows the nucleic acid (namely RNA) content of small biological samples to be interrogated on a global scale in a single experiment. So far, the most common usage of microarrays is in the parallel analysis of gene expression in diseased and healthy tissue. Cancer profiling has been one of the great successes of this technology [4].

The primary application of microarrays in CF is also in the study of gene expression, e.g. to identify differences between CF and non-CF samples. For experiments such as these, information on relative transcript abundance is usually obtained by hybridizing fluorescently labelled cDNA populations, synthesized from template mRNA extracted from two samples of interest, to a single microarray in a competitive hybridization reaction [5]. The microarray is then scanned using a laser to quantify the fluorescent signals for each probe on the array, thereby generating a large data matrix consisting of gene expression values that reflect the relative abundance of transcripts in each sample. Data generated are then normalized or scaled across replicate experiments and analysed using appropriate statistics to identify up- or down-regulated genes. Figure 2 outlines the microarray analysis process.

The size of data sets generated by microarrays (and proteomics) poses many challenges, setting a steep learning curve for scientists endeavouring to use these high-throughput technologies. The main difficulties have arisen in the management, analysis and interpretation of expression profiles. Although these issues are being addressed [6] they are beyond the scope of this chapter and will not be discussed further.

Microarrays and Cystic Fibrosis

Transcriptional Consequences of the CFTR Defect

Despite significant advances in our understanding of CF over the last 15 years, there remains a deficit in our understanding of the mechanisms connecting mutations in the *CFTR* gene to the plethora of symptoms characteristic of the disease (see chapter 8). Microarrays provide an opportunity to bridge this gap by elucidating the gene expression pathways that lead to disease. A recent microarray study by Xu et al. [7] took the first few steps towards addressing this gap by examining the transcriptional consequences of CFTR defects in vivo. The authors compared mRNA profiles of lung tissue from normal mice and knockout mice lacking CFTR expression and identified 54 differentially regulated genes. Interestingly, these genes could be classified into a wide variety of functional categories; including, transcriptional regulation, inflammatory response, ion transport, signal transduction, intracellular protein trafficking and protein degradation. These findings support the concept that CFTR has multiple roles in the cell and suggest that complex adaptive changes in gene expression occur to maintain pulmonary tissue homeostasis in the absence of CFTR. The significance of these data to CF patient airways remains to be demonstrated. However, it is clear from this study that altered transcriptional activity correlates with defects in CFTR and likely plays a role in disease pathogenesis.

Pharmacogenomics in Cystic Fibrosis

Pharmacogenomics is the study of the effects of drugs or therapies on transcription. This is a growing field in CF research as its utility in drug discovery and biological end-point identification has been recognized [8]. A recent example of this kind of application of microarrays is reported by Srivastava et al. [9]. In this study, microarrays were used to compare gene expression profiles from a CF lung epithelial cell line treated with either digitoxin (a potent inhibitor of the IL-8 inflammatory pathway) or CFTR gene therapy. They observed that digitoxin treatment replicates the effects of CFTR gene therapy and modulates the transcriptional activity of similar sets of genes. Confirming Xu et al. [7], these genes are involved in diverse cellular pathways and functions. The study is the first of its kind to report specifically on the transcriptional effects of restored cellular expression of CFTR, analogous to gene therapy, albeit in vitro. From a drug discovery point of view, the authors concluded from their investigation that digitoxin should be considered as a candidate drug for ameliorating IL-8-dependent CF lung inflammation.

Disease Biomarkers

As inflammation is a key component of the CF lung phenotype, identification of genes involved in the pro- and anti-inflammatory responses to infection will be important for further understanding disease pathogenesis and the development of therapeutic (diagnostic and prognostic) tools. Historically, IL-8 has been considered the gold standard marker of inflammation in CF lung disease [10]. However, the number of CF-related genes steadily increases and to date well over 150 genes are considered potential biomarkers of CF lung disease. Many of these genes have known roles in inflammation, protein processing or trafficking, ion channel function and tissue remodelling. Comparisons of gene expression in healthy individuals and CF patients and in CF mouse models using microarrays are either currently under way or published [11]. These studies will further aid CF biomarker discovery. Analysis of gene expression in other inflammatory lung diseases [12, 13] will also help to define genes specifically associated with CF lung disease (before, during and after inflammation) and genes specifically expressed in the lung.

Microarray Analysis of Pseudomonas aeruginosa Pathogenesis

P. aeruginosa is the most common opportunistic pathogen found to infect the airways of CF patients. Understanding the genes and cellular pathways that are activated or repressed in both the host and bacteria during intermittent and chronic infection will shed light on the processes involved in disease pathogenesis. Microarrays are considerably adding to our knowledge of these processes. For example, a key microarray study investigating global changes in gene expression between non-mucoid and mucoid forms of P. aeruginosa (see chapter 18) identified a new set of virulence factors associated with the mucoid form [14]. These genes included the major protease elastase protein and the toxic factor cyanide synthase whose expression likely contributes to the declining respiratory function and poor prognosis of CF patients chronically infected with P. aeruginosa. Microarrays are also being used to understand how P. aeruginosa adapts to the CF lung environment [15] and how lung epithelial cells respond to P. aeruginosa infection in vitro [16]. One mechanism used by P. aeruginosa to evade detection by the host immune system and persist in the CF lung is activation of the quorum-sensing gene network (associated with biofilm formation in chronically infected CF patient lungs) and repression of the microbial motility protein flagellin [15]. Clues to the transcriptional changes induced in bacteria by interspecies communication are also being revealed by microarrays [17]. Again, some of the genes were found to encode known virulence factors, such as exotoxin A and elastase, but also multidrug transporter proteins.

Proteomics and Novel Protein Discovery

Proteomics is the genome-wide study of proteins. The completion of the human genome project was a key point in the evolution of the field of proteomics [18]. Since then, in the post-genomic era, work on the predicted human proteome (identification of the full human complement of proteins) has rapidly progressed.

The human proteome is much more complex than its corresponding genome as over 500,000 proteins are generated from some 42,000 genes. This complexity in part reflects different protein isoforms due to differential splicing of genes and post-translational modifications such as glycosylation, phosphorylation and cleavage [19]. These modifications may lead to a poor correlation between mRNA and protein levels and function. Additional sources of complexity, and a particular focus of this field of study, are the many protein-protein interactions and changes in protein expression that occur during disease. The human proteome should be thought of as a highly complex and dynamic protein network, generated from a static genomic blueprint.

Characterization of the human proteome in its different compartments such as serum and plasma has begun. There are large-scale collaborative plans to complete study of the proteome of all human biological fluids. This field of study has evolved by relying on two types of technology. Polyacrylamide gel electrophoresis (PAGE)-based techniques and, more recently, highly sensitive mass spectrometry (MS) techniques have been applied. These techniques are most powerful when used in combination, offering protein identification through highly sensitive detection by MS, gel electrophoresis techniques for protein isolation and digestion. By searching of the human proteomic databases, the majority of the proteins can be identified [20].

Proteomics in Cystic Fibrosis

CFTR Protein

Initial proteomic research in CF started at the time of discovery of the CFTR gene and the ΔF508 mutation. Its tertiary structure was predicted and compared to other proteins in its class of ABC transporters. Understanding of CFTR function progressed with the identification of the phosphorylation-dependent regulatory domain [21]. The technique matrix-assisted laser desorption/ionization time-of-flight (MALDI-TOF) MS was applied to identify the relevant phosphorylation sites [22]. Understanding of the structure and function relationship of CFTR should be further improved as newer magnetic resonance imaging techniques become available.

CFTR Interactome

CFTR functions, not related to chloride channel activity, have been studied. Proteins that interact with a given protein such as CFTR can be referred to as its interactome. Several proteins are known to interact with the C terminus of CFTR, such as EBP50/NHERF1 and ezrin. Whilst their function is poorly understood, it is known that these molecules also interact with other intracellular proteins such as actin [23]. It is conceivable that proteomic studies will reveal CFTR functions unrelated to chloride efflux, which may be vital for the progression of CF lung disease.

Proteomics and P. aeruginosa Infection

Proteomic analysis of *P. aeruginosa* has revealed multiple proteins which may be suitable targets for novel therapeutics. Comparison of the proteome expressed by *P. aeruginosa* before and after biofilm formation have identified many pathways which are differentially regulated and therefore may be potential therapeutic targets [24]. Membrane proteins responsible for the antibiotic resistance and adaptivity of *P. aeruginosa* have been identified using liquid chromatography and MS [25]. Proteomic studies have also informed on the host defences of human airway surface fluid, showing relative bactericidal activity of the cationic peptides lysozyme, lactoferrin and secretory leukoprotease inhibitor in nasal fluid [26].

Protein Biomarkers of CF Lung Disease

Another exciting application of proteomics in CF is the potential to identify novel biomarkers of disease [27]. Such markers could ease diagnosis, predict severity and progression of lung disease, as well as the response to drug/non-drug therapies. Differential expression analysis of serum, the airway surface fluid and respiratory epithelial cells in CF and non-CF subjects is ongoing in our laboratory. Difficulties using proteomic technology, which still need to be addressed, include identification and quantification of candidate biomarkers. High throughput MS techniques are likely to advance this field significantly.

Methods and Current Applications

Proteomics research is based on two main technologies: PAGE and mass spectrometry [28]. MS-based technologies offer a far greater sensitivity than gel-based techniques, reportedly to attomolar concentrations. The MS techniques most commonly used are electrospray ionization (ESI) and MALDI-based technologies. These techniques vaporize and ionize the sample before presenting it for mass analysis. A detailed description of these technologies is outside the remit of this chapter, but has been reviewed by Aebersold and Mann [20].

As biological fluids are generally complex with mixtures of high abundance and low abundance proteins, steps to enrich for possibly scarce compounds of interest are required. Methods available include liquid chromatography-column-based approaches or enrichment directly onto the mass spectrometer chip surface – such as surface-enhanced laser desorption/ionization time-of-flight (SELDI-TOF) MS (see fig. 3 for an outline of the procedure for SELDI-TOF protein analysis). Both techniques can enrich

     Ogilvie/MacGregor

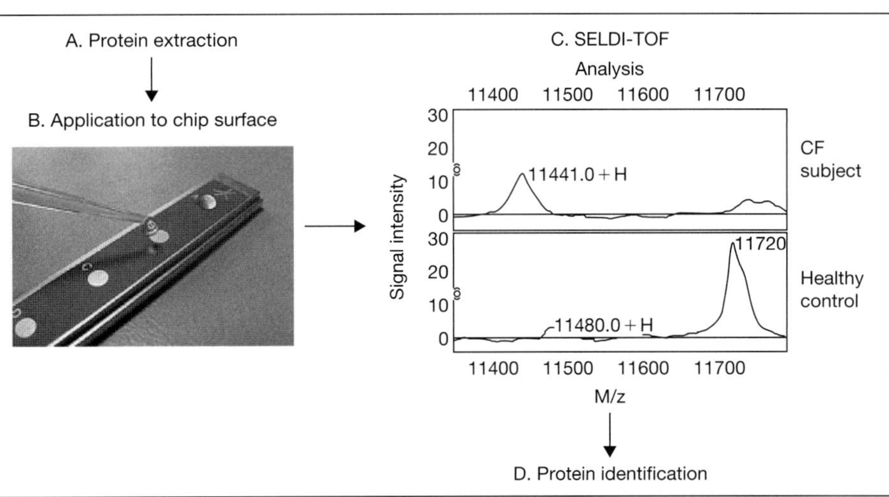

Fig. 3. Illustration of the steps involved in protein profiling using SELDI-TOF. A. Proteins are solubilized from samples of interest. B. The protein lysate is applied to SELDI chip surface, and bound to different chemically active surfaces. The unbound solution is then washed off, before applying an ionising matrix. C. SELDI-TOF analysis generates a protein mass spectrum. Profiles pictured are from BAL fluid collected from a CF patient and a healthy control. Peaks indicate different proteins. Protein masses are expressed as a mass-to-charge ratio (M/z).

for proteins based on their electrophoresis point or properties such as their hydrophobicity or affinity to bind to various metals or antibodies.

Both gel based and MS-based techniques have their strengths e.g. PAGE is particularly sensitive at detecting larger proteins above 20 kDa, whereas most MS techniques are more sensitive for smaller proteins. A combinatorial approach is generally required for complete proteomic coverage.

The Future for Microarrays and Proteomics in Cystic Fibrosis

One of the most exciting and ambitious goals for microarrays, and proteomics, is direct clinical application. Continuing progress in improving gene sequence information [29] and array manufacture will likely lead to the development of more advanced gene expression profiling tools suitable for application in the clinic, such as in the molecular diagnosis and prognosis of disease in the individual patient. There are huge benefits in utilizing arrays in this way as clinicians will be able to outline risk factors for susceptible patients, predict how well a patient will respond to a particular treatment and monitor improvements made during the course of treatment [30, 31]. It is easy to imagine that protein-profiling techniques could be used in a similar fashion in the future. In terms of genetic testing, microarrays are already being developed in the CF field for this purpose.

In conclusion, the future of microarray and proteomic research in CF should reveal further the basic mechanisms of the disease, facilitate surrogate endpoint identification, will give us a greater understanding of CF lung disease progression and may lead to the discovery of therapeutic targets for the individual patient. High throughput differential analysis of mRNA and proteins within diseased subjects with different phenotypes and genotypes, as well as comparison to healthy control subjects, could also lead to improved predictors of well-being and report on patients' need for additional therapies and interventions. The development of array and proteomic profiles and analysis of multiple biomarkers in combination will undoubtedly change the way that CF patients are managed in the future.

References

1 Lockhart DJ and Winzeler EA. Genomics, gene expression and DNA arrays. Nature 2000;405:827–836.

2 Wang DG, Fan JB, Siao CJ, Berno A, Young P, Sapolsky R, Ghandour G, Perkins N, Winchester E, Spencer J, Kruglyak L, Stein L, Hsie L, Topaloglou T, Hubbell E, Robinson E, Mittmann M, Morris MS, Shen N, Kilburn D, Rioux J, Nusbaum C, Rozen S, Hudson TJ, Lipshutz R, Chee M, Lander ES: Large-scale identification, mapping, and genotyping of single-nucleotide polymorphisms in the human genome. Science 1998;280: 1077–1082.

3 Fodor SP, Rava RP, Huang XC, Pease AC, Holmes CP, Adams CL: Multiplexed biochemical assays with biological chips. Nature 1993; 364:555–556.

4 Golub TR, Slonim DK, Tamayo P, Huard C, Gaasenbeek M, Mesirov JP, Coller H, Loh ML, Downing JR, Caligiuri MA, Bloomfield CD, Lander ES: Molecular classification of cancer: Class discovery and class prediction by gene expression monitoring. Science 1999; 286:531–537.

5 Freeman WM, Robertson DJ, Vrana KE: Fundamentals of DNA hybridization arrays for

gene expression analysis. Biotechniques 2000;29:1042–1055.

6 Quackenbush J: Computational analysis of microarray data. Nat Rev Genet 2001;2:418–427.

7 Xu Y, Clark JC, Aronow BJ, Dey CR, Liu C, Wooldridge JL, Whitsett JA: Transcriptional adaptation to cystic fibrosis transmembrane conductance regulator deficiency. J Biol Chem 2003;278:7674–7682.

8 Pollard HB, Eidelman O, Jacobson KA, Srivastava M: Pharmacogenomics of cystic fibrosis. Mol Interv 2001;1:54–63.

9 Srivastava M, Eidelman O, Zhang J, Paweletz C, Caohuy H, Yang Q, Jacobson KA, Heldman E, Huang W, Jozwik C, Pollard BS, Pollard HB: Digitoxin mimics gene therapy with CFTR and suppresses hypersecretion of IL-8 from cystic fibrosis lung epithelial cells. Proc Natl Acad Sci USA 2004;101:7693–7698.

10 McElvaney NG, Nakamura H, Birrer P, Hebert CA, Wong WL, Alphonso M, Baker JB, Catalano MA, Crystal RG: Modulation of airway inflammation in cystic fibrosis. In vivo suppression of interleukin-8 levels on the respiratory epithelial surface by aerosolization of recombinant secretory leukoprotease inhibitor. J Clin Invest 1992;90:1296–1301.

11 Norkina O, Kaur S, Ziemer D, De Lisle RC: Inflammation of the cystic fibrosis mouse small intestine. Am J Physiol Gastrointest Liver Physiol 2004;286:G1032–G1041.

12 Zuo F, Kaminski N, Eugui E, Allard J, Yakhini Z, Ben-Dor A, Lollini L, Morris D, Kim Y, DeLustro B, Sheppard D, Pardo A, Selman M, Heller RA: Gene expression analysis reveals matrilysin as a key regulator of pulmonary fibrosis in mice and humans. Proc Natl Acad Sci USA 2002;99:6292–6297.

13 Sheppard D, Roger S: Mitchell lecture. Uses of expression microarrays in studies of pulmonary fibrosis, asthma, acute lung injury, and emphysema. Chest 2002;121:21S–25S.

14 Firoved AM, Deretic V: Microarray analysis of global gene expression in mucoid *Pseudomonas aeruginosa*. J Bacteriol 2003;185:1071–1081.

15 Wolfgang MC, Jyot J, Goodman AL, Ramphal R, Lory S: *Pseudomonas aeruginosa* regulates flagellin expression as part of a global response to airway fluid from cystic fibrosis patients. Proc Natl Acad Sci USA 2004;101:6664–6668.

16 Ichikawa JK, Norris A, Bangera MG, Geiss GK, van 't Wout AB, Bumgarner RE, Lory S: Interaction of *Pseudomonas aeruginosa* with epithelial cells: Identification of differentially regulated genes by expression microarray analysis of human cDNAs. Proc Natl Acad Sci USA 2000;97:9659–9664.

17 Duan K, Dammel C, Stein J, Rabin H, Surette MG: Modulation of *Pseudomonas aeruginosa* gene expression by host microflora through interspecies communication. Mol Microbiol 2003;50:1477–1491.

18 Lander ES, Linton LM, et al: Initial sequencing and analysis of the human genome. Nature 2001;409:860–921.

19 Mann M, Jensen ON: Proteomic analysis of post-translational modifications. Nat Biotechnol 2003;21:255–261.

20 Aebersold R, Mann M: Mass spectrometry-based proteomics. Nature 2003;422:198–207.

21 Cheng SH, Rich DP, Marshall J, Gregory RJ, Welsh MJ, Smith AE: Phosphorylation of the R domain by cAMP-dependent protein kinase regulates the CFTR chloride channel. Cell 2003;66:1027–1036.

22 Neville DC, Rozanas CR, Price EM, Gruis DB, Verkman AS, Townsend RR: Evidence for phosphorylation of serine 753 in CFTR using a novel metal-ion affinity resin and matrix-assisted laser desorption mass spectrometry. Protein Sci 1997;6:2436–2445.

23 Haggie PM, Stanton BA, Verkman AS: Increased diffusional mobility of CFTR at the plasma membrane after deletion of its C-terminal PDZ binding motif. J Biol Chem 2004;279:5494–5500.

24 Sauer K, Camper AK, Ehrlich GD, Costerton JW, Davies DG: *Pseudomonas aeruginosa* displays multiple phenotypes during development as a biofilm. J Bacteriol 2002;184:1140–1154.

25 Guina T, Purvine SO, Yi EC, Eng J, Goodlett DR, Aebersold R, Miller SI: Quantitative proteomic analysis indicates increased synthesis of a quinolone by *Pseudomonas aeruginosa* isolates from cystic fibrosis airways. Proc Natl Acad Sci USA 2003;100:2771–2776.

26 Cole AM, Liao HI, Stuchlik O, Tilan J, Pohl J, Ganz T: Cationic polypeptides are required for antibacterial activity of human airway fluid. J Immunol 2002;169:6985–6991.

27 Hirsch J, Hansen KC, Burlingame AL, Matthay MA: Proteomics: Current techniques and potential applications to lung disease. Am J Physiol Lung Cell Mol Physiol 2004;287:L1–L23.

28 Mann M, Hendrickson RC, Pandey A: Analysis of proteins and proteomes by mass spectrometry. Annu Rev Biochem 2001;70:437–473.

29 Finishing the euchromatic sequence of the human genome. Nature 2004;431:931–945.

30 Petricoin EF 3rd, Hackett JL, Lesko LJ, Puri RK, Gutman SI, Chumakov K, Woodcock J, Feigal DW Jr, Zoon KC, Sistare FD: Medical applications of microarray technologies: A regulatory science perspective. Nat Genet 2002;32 (suppl):474–479.

31 Kiechle FL, Zhang X: The postgenomic era: Implications for the clinical laboratory. Arch Pathol Lab Med 2002;126:255–262.

Varrie C. Ogilvie
Medical Genetics Section, Department of Medical Sciences, Edinburgh University
Molecular Medicine Centre, Western General Hospital, Crewe Road
Edinburgh EH4 2XU (UK)
Tel. +44 131 651 1049
Fax +44 131 651 1059
E-Mail v.c.ogilvie@ed.ac.uk

The Airway

Bush A, Alton EWFW, Davies JC, Griesenbach U, Jaffe A (eds): Cystic Fibrosis in the 21st Century.
Prog Respir Res. Basel, Karger, 2006, vol 34, pp 116–121

Pathogenesis of Pulmonary Disease in Cystic Fibrosis

Marcus Mall[a]   Richard C. Boucher[b]

[a]Department of Pediatrics III, Pediatric Pulmonology and Cystic Fibrosis Center,
University of Heidelberg, Heidelberg, Germany;
[b]Cystic Fibrosis/Pulmonary Research and Treatment Center, School of Medicine,
The University of North Carolina at Chapel Hill, Chapel Hill, N.C., USA

Abstract

Chronic lung disease is the major cause of mortality and morbidity in cystic fibrosis (CF) patients. Mutations in the cystic fibrosis transmembrane conductance regulator (*CFTR*) gene result in defective epithelial cAMP-dependent chloride secretion and increased sodium absorption in airway epithelia. The link between these ion transport defects and the pathogenesis of CF lung disease has been the subject of intensive investigation. Data emerging from in vitro and recent in vivo studies of a novel animal model with CF-like lung disease demonstrate that these altered ion transport processes cause airway surface liquid (ASL) volume depletion, delayed mucus clearance, and mucus adhesion to airway surfaces. This pathogenetic sequence results in airway mucus obstruction and sets the stage for chronic airways inflammation and bacterial infection. These studies demonstrate that mucus transport is a key component of innate lung defense and suggest that novel treatment strategies that restore proper ASL volume will likely be successful in the treatment of CF lung disease.

Introduction

While the mechanisms linking mutations in the cystic fibrosis transmembrane conductance regulator (*CFTR*) gene to cystic fibrosis (CF) lung disease have been the subject of intensive research and debate [1–8], there is general agreement that innate host defense mechanisms are fundamentally altered in the CF lung. Important components of innate defense on airway surfaces include: (1) a chemical shield of antimicrobial peptides/proteins, such as lysozyme, lactofer-

rin, and other antimicrobial peptides, and (2) mechanical clearance of mucus and inhaled particles and/or pathogens by the mucociliary apparatus [9, 10]. Data from in vitro studies of primary CF airway cultures suggested that altered ion transport properties in CF airway epithelia may lead to defective antimicrobial activity of airway surface liquid (ASL; chemical shield or high-salt hypothesis) or dysregulation of ASL volume and deficient mucus clearance (low-volume hypothesis) [1–7]. Adding to the complexity are hypotheses that predict that CF pathogenesis reflects deficient epithelial binding and phagocytosis of bacteria [11, 12], abnormal submucosal gland function [13, 14], or an intrinsically enhanced inflammatory response in CF airways [15] (chapter 16). In this chapter, we will focus on recent in vivo data, which strongly support the concept that proper ASL volume regulation is critical for normal mucus clearance, and that increased electrolyte absorption and ASL volume depletion are likely the initiating event in the pathogenesis of CF lung disease. We will then follow the steps that link dysregulation of ASL volume, deficient mucus clearance, mucus plugging and adherence to airway surfaces, chronic neutrophilic inflammation, goblet cell metaplasia and mucus hypersecretion, and increased susceptibility to bacterial infection to produce chronic CF lung disease.

Clinicopathological Description of CF Lung Disease: Hints to the Pathogenesis

While the lungs of CF patients appear normal at birth, postmortem studies of lungs from infants who died of meconium ileus identified mucus plugging in terminal bronchioles with emphysema secondary to obstruction of mucus-filled bronchioles, but without signs of infection or inflammation [16]. These observations suggest that mucus plugging of small airways is an early event in CF lung disease. In addition to this evidence, several other early observations indicated that abnormal ASL volume regulation and failure of mucus transport may play a crucial role in this complex pathogenesis of CF lung disease. First, CF mucus is more concentrated (i.e., has less water content) than mucus from healthy individuals, indicating that mucus hydration on airway surfaces is abnormal in CF [17]. Second, in vivo lower airway mucus clearance, as determined from airway clearance of radiolabeled particles as measured by external gamma camera scanning, is significantly delayed in CF patients [9, 18, 19].

The Link between Altered Ion Transport and Deficient Mucus Transport in CF Airways

Following the cloning of the *CFTR* and *ENaC* genes, it was demonstrated by co-expression studies in heterologous cells that activation of wild-type CFTR led to inhibition of ENaC activity [20–22]. This observation led to the hypothesis that CFTR acts not only as a cAMP-dependent Cl^- channel, but also as a regulator of ENaC, and that loss of this regulatory function results in increased ENaC activity in CF airway epithelia (see also chapter 6). Because airway epithelia are highly water permeable [23], the volume of the thin film of liquid covering airway surfaces (ASL) is determined by the mass of salt on airway surfaces, which is tightly regulated by normal airway epithelia via two opposing active ion transport systems, i.e. ENaC-mediated Na^+ absorption, and Cl^- secretion mediated by CFTR and Ca^{2+}-activated Cl^- channels (CaCC) [7]. Accordingly, increased ENaC activity and defective CFTR-mediated Cl^- secretion in CF airway epithelia were predicted to produce ASL volume depletion in CF airway epithelia.

A major breakthrough in the understanding of how abnormal ion transport and related changes in ASL volume contribute to CF airway pathogenesis came from the development of highly differentiated primary cultures of normal and CF airway epithelia that preserve the active ion transport properties and mucus transport properties of native tissues [3, 4, 24].

The ASL is partitioned into two compartments. The mucus layer entraps inhaled particles and pathogens. The periciliary liquid layer (PCL) provides a mucin-free low viscosity solution to facilitate ciliary beating and a lubricant layer to separate the mucus layer from the mucins tethered to the cell surface to facilitate cough clearance [7, 9]. It was demonstrated in this primary airway culture model that normal airway epithelia have the capability to regulate the volume of ASL by setting the height of the periciliary liquid (PCL) to approximately the height of the extended cilium ($\sim 7\,\mu M$). In contrast, this PCL autoregulation fails in CF cultures. PCL height is significantly reduced and cilia are flattened/collapsed onto cell surfaces [3, 4, 7, 24] (fig. 1a). Importantly, PCL volume depletion in CF cultures is associated with significantly reduced mucus transport rates in CF compared to normal airway cultures (fig. 1b). Taken together with ion transport studies in native CF airways, these ex vivo studies suggested a direct link between abnormal ion transport (i.e., Na^+ hyperabsorption and reduced Cl^- secretion) and deficient mucus clearance in the airways of CF patients.

Increased Airway Na⁺ Absorption Produces ASL Volume Depletion and CF-Like Lung Disease In Vivo

To further elucidate the role of accelerated electrolyte absorption in the in vivo pathogenesis of CF airway disease, we have recently generated a transgenic mouse model with airway-specific overexpression of ENaC [25, 26]. Interestingly, although ENaC is composed of three individual subunits (α, β and γ), overexpression of the β subunit of ENaC (Scnn1b) alone was sufficient to produce a ~ 3-fold increase in airway Na^+ absorption, suggesting that expression levels of *Scnn1b* are rate limiting for Na^+ absorption in the airways. Similar to primary human CF airway cultures, accelerated Na^+ absorption in the airways of *Scnn1b*-transgenic mice caused ASL volume depletion with a reduced height of the PCL (fig. 1a), increased mucus concentration (i.e., decreased water content), and significantly slowed mucus clearance in vivo (fig. 1b). Interestingly, deficient mucus transport in the *Scnn1b*-overexpressing mouse produced a disease phenotype strikingly similar to the clinico-pathological description of early CF lung disease, including mucus plugging and mucus adhesion to airway surfaces (fig. 2a, b), neutrophilic inflammation, goblet cell metaplasia, and mucus hypersecretion [8, 16, 27].

Pathogenetic Sequence of the Development of CF Airway Disease Deduced from a Mouse Model with ASL Volume Depletion

The data from the *Scnn1b*-overexpressing mouse strongly suggest that increased electrolyte absorption and

Fig. 1. Increased airway epithelial Na$^+$ absorption causes ASL volume depletion and reduced mucus clearance. **a**, **b** The height of the PCL in primary human airway cultures and bronchi from mice was visualized by electron microscopy. Scale bar: 2.5 μm. **a** PCL height is substantially reduced from ~7 μm in normal airway epithelia to ~3 μm in CF. **b** A similar PCL volume reduction was observed in native bronchi from *Scnn1b*-overexpressing mice compared to wild-type littermates. Na$^+$ hyperabsorption and PCL volume depletion resulted in significantly slowed mucus transport in human CF airway cultures **c**, and in vivo mucus clearance in the *Scnn1b*-overexpressing mouse **d** [adapted from 3, 25].

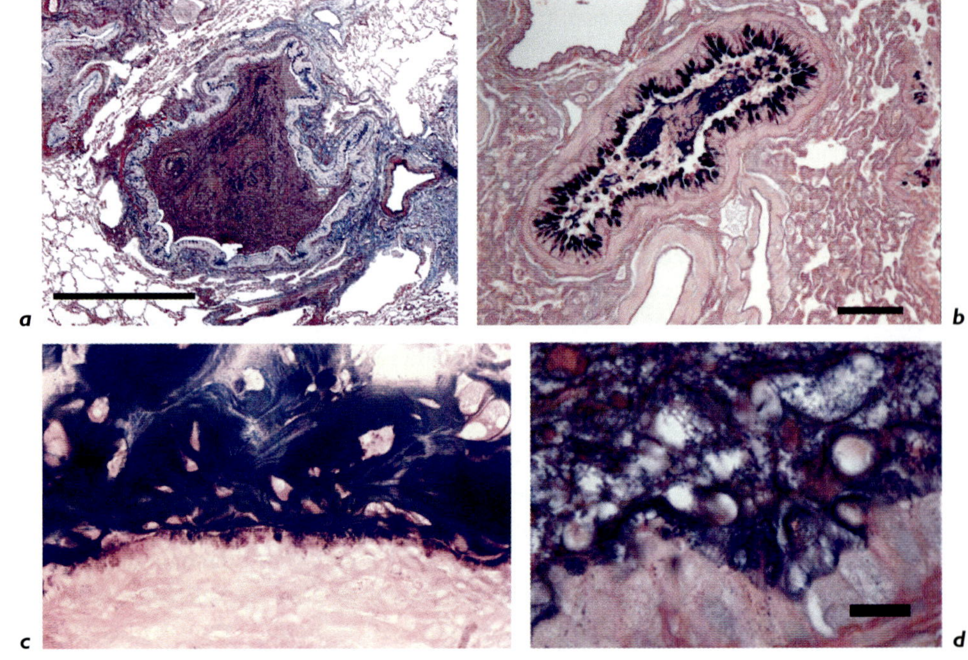

Fig. 2. Excessive ASL volume absorption and slowed mucus transport cause mucus plugging and adhesion to airway surfaces. Low-power light micrographs of lung sections from a CF patient obtained at transplant **a** and a *Scnn1b*-overexpressing mouse killed at 4 weeks of age **b**. Staining for mucins identifies plaque formation with airway narrowing and mucus plugging of airways. AB/PAS. Scale bar: 100 μm. Note the similarity in mucus obstruction between the human CF lung and the *Scnn1b*-overexpressing mouse. High-resolution light micrographs demonstrate that the thickened mucus covering the airways of a CF patient **c** and a *Scnn1b*-overexpressing mouse **d** is adherent to cell surfaces. Scale bar: 10 μm [adapted from 7, 25].

ASL volume depletion are the initiating events in the pathogenesis of CF-like lung disease. The hypothetical pathogenetic sequence triggered by ASL hyperabsorption is summarized in figure 3. In normal airways, ASL volume is adjusted by a balance of ENaC-mediated Na$^+$ absorption and CFTR- and CaCC-mediated Cl$^-$ secretion, and proper ASL volume provides for normal function of the mucociliary apparatus and efficient mucus clearance (fig. 3a). In the *Scnn1b*-overexpressing mouse, airway mucus is more concentrated and PCL height on airway surfaces is reduced, with flattening of cilia on airway surfaces. This volume depletion of airway surfaces slows mucus transport and clearance of inhaled particulates and pathogens (fig. 3b), and it produces mucus adhesion to airway surfaces (fig. 2d) and mucus plugging of airways (fig. 2b).

Na$^+$ absorption
(ENaC)

Anion
secretion
(CFTR + CaCC)

- Mucus
- PCL
- Epithelial cell
} ASL

Volume
autoregulation

a

- Mucus
- PCL
} ASL
volume
depletion

Volume
depletion

b

Entrapment of particles/irritants ⟶ Inflammation + GCM

Entrapped particles,
pathogens

Macrophage

Proinflammatory
cytokines
(Mip-2, KC)
IL-8

PMN

Elastase

Mucus
hypersecretion

Goblet cell
metaplasia

Concentrated autocrine or
paracrine factors

c

Fig. 3. Schematic model of the pathogenetic sequence hypothesized to link altered ASL volume homeostasis to chronic airway disease in CF patients. *a* Normal airway surfaces are covered by a thin mucus layer that entraps inhaled particles and pathogens. Proper ASL volume regulation provides a low viscosity PCL that facilitates efficient vectorial mucus clearance and removal of inhaled particulates and pathogens from the lung. *b* In CF airways, hyperabsorption of Na$^+$ and deficient CFTR-mediated Cl$^-$ secretion depletes ASL volume, cilia collapse in the shallow PCL, and concentrated mucus adheres to airway surfaces. *c* Mucostasis-induced plaque formation. Once mucostasis is established, multiple mechanisms perpetuate chronic inflammation, goblet cell metaplasia, and mucus hypersecretion.

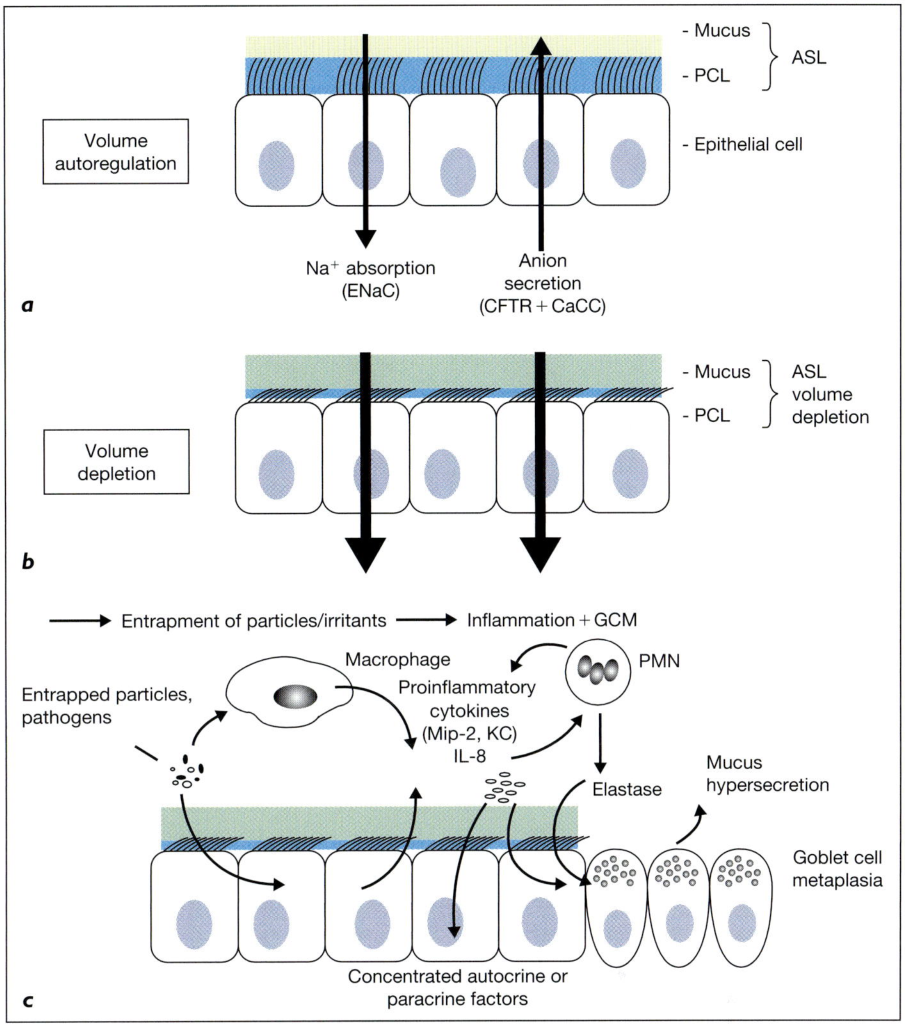

As the next step, we speculate that failure to clear inhaled environmental particles and irritants triggers release of proinflammatory chemokines, like IL-8, from airway epithelia and/or macrophages patrolling airway surfaces [28] (fig. 3c). These cytokines become concentrated in thickened mucus, recruit neutrophils into airway lumens, and trigger airway inflammation in the absence of bacterial infection (fig. 3c).

Several concomitant mechanisms may cause the goblet cell metaplasia observed in the airways of *Scnn1b*-overexpressing mice and contribute to airway mucus obstruction. First, poor mucus clearance may concentrate neutrophil products in the airway lumen, including neutrophil elastase, which trigger goblet cell metaplasia and mucin hypersecretion [29–31]. Elastase can also induce chemokine release by airway epithelia [32], producing a self-perpetuating inflammatory cycle. Goblet cell metaplasia may occur when autocrine and/or paracrine factors that are normally secreted in the airway lumen, e.g. TGF-α, become concentrated as a result of poor clearance [33] (fig. 3c).

As a result of all these processes, a microenvironment is created that is characterized by mucostasis and accumulation of thickened mucus that adheres to airway walls. It is this microenvironment of thickened, stationary mucus that is the nidus for chronic bacterial infection.

Lessons from a Mouse Model with ASL Volume Depletion for the Pathogenesis of CF Airway Disease

The data from the *Scnn1b*-overexpressing mouse have important implications for several controversial issues related to the pathogenesis of CF airway disease. In recent years, it has been debated whether the link between abnormal CF ion transport by the superficial airway epithelium and abnormalities in the protective properties of ASL reflect a change in ASL ion composition (high-salt or chemical shield hypothesis) or ASL volume (low-volume hypothesis) [1–3]. Further, because of the known Cl⁻ channel function of CFTR [27, 34], the role of enhanced airway Na⁺ transport [3, 35, 36] in ASL volume homeostasis and CF pathogenesis remained controversial [1–3, 35, 36]. The phenotype of the *Scnn1b*-overexpressing mouse strongly supports the hypothesis that airway Na⁺ hyperabsorption produces ASL volume depletion and contributes to the primary pathogenesis of CF. Indeed, the parallels between the phenotypes of human CF lung and that of the βENaC transgenic mouse are strong, with both exhibiting ~3 fold increase in airway Na⁺ absorption [35, 36], increased mucus concentration [17], and severe obstructive lung disease [27]. In contrast, mice deficient in components of the antimicrobial shield, i.e. defensins, do not exhibit a pulmonary disease phenotype [10].

Second, it has been hypothesized that CF lung disease is initiated by a dysfunction of submucosal glands rather than abnormal ion transport in the superficial epithelium [13, 14], and the absence of glands in the intraparenchymal airways has been offered as an explanation for the failure of the gene-targeted CF mouse models (i.e., *Cftr*–/– mice) to develop lung disease [37, 38]. Although the pulmonary disease phenotype in the *Scnn1b*-transgenic mouse does not rule out a role for glands in human CF, the data suggest that accelerated electrolyte absorption can produce mucus plugging in the small, nonglandular airways that are thought to be the site of disease initiation in CF infants [16].

Finally, the finding that ASL volume depletion caused neutrophilic inflammation in airways of *Scnn1b*-overexpressing mice without evidence of bacterial infection is highly relevant to another issue pertinent to CF pathogenesis. Neutrophils and IL-8 are harvested in BAL fluid from a significant number of CF infants without culture-positive bacterial infection [39]. Accordingly, it has been debated whether this constellation of findings is caused by a failure to culture bacteria that are shielded in mucus plugs, or a CFTR-mediated state of intrinsic airway epithelial hyperinflammation (see chapter 16). However, the findings of a sterile neutrophilic inflammation in the airways of the *Scnn1b*-transgenic mouse strongly suggest a third possibility, i.e., that the CF disease-dependent slowing of mucus clearance can initiate and maintain particle- or irritant-mediated chronic inflammation.

Conclusion

Emerging evidence from recent in vitro studies on primary human CF airway cultures and in vivo studies on the *Scnn1b*-overexpressing mouse strongly suggests a mechanistic link amongst alterations in airway ion transport, dysregulation of ASL volume, and a dysfunction in the mucus clearance component of innate lung defense. These data demonstrate that increased Na⁺ absorption and a reduced capacity to secrete Cl⁻ in CF airways leads to inadequate airway surface volume (hydration), which in turn allows mucus to adhere to airway surfaces and form a nidus for bacterial infection. Once mucus infection and host inflammatory responses become established, a complex spectrum of other factors may contribute to and modulate the severity of lung disease. The *Scnn1b*-overexpressing mouse will allow an in vivo evaluation of such factors and of novel therapeutic interventions in CF airway disease. It is predicted that novel therapeutic strategies that target this pathogenetic sequence to restore proper ASL volume to airway surfaces or reduce mucus accumulation may be successful in preventing and/or ameliorating lung disease in CF patients.

References

1 Smith JJ, Travis SM, Greenberg EP, Welsh MJ: Cystic fibrosis airway epithelia fail to kill bacteria because of abnormal airway surface fluid. Cell 1996;85:229–236.

2 Zabner J, Smith JJ, Karp PH, Widdicombe JH, Welsh MJ: Loss of CFTR chloride channels alters salt absorption by cystic fibrosis airway epithelia in vitro. Mol Cell 1998;2:397–403.

3 Matsui H, Grubb BR, Tarran R, Randell SH, Gatzy JT, Davis CW, Boucher RC: Evidence for periciliary liquid layer depletion, not abnormal ion composition, in the pathogenesis of cystic fibrosis airways disease. Cell 1998;95:1005–1015.

4 Tarran R, Grubb BR, Parsons D, Picher M, Hirsh AJ, Davis CW, Boucher RC: The CF salt controversy: In vivo observations and therapeutic approaches. Mol Cell 2001;8:149–158.

5 Wine JJ: The genesis of cystic fibrosis lung disease. J Clin Invest 1999;103:309–312.

6 Guggino WB: Cystic fibrosis and the salt controversy. Cell 1999;96:607–610.

7 Knowles MR, Boucher RC: Mucus clearance as a primary innate defense mechanism for mammalian airways. J Clin Invest 2002;109:571–577.

8 Gibson RL, Burns JL, Ramsey BW: Pathophysiology and management of pulmonary infections in cystic fibrosis. Am J Respir Crit Care Med 2003;168:918–951.

9 Wanner A, Salathe M, O'Riordan TG: Mucociliary clearance in the airways. Am J Respir Crit Care Med 1996;154:1868–1902.

10 Bals R, Weiner DJ, Wilson JM: The innate immune system in cystic fibrosis lung disease. J Clin Invest 1999;103:303–307.

11 Imundo L, Barasch J, Prince A, Al-Awqati Q: Cystic fibrosis epithelial cells have a receptor for pathogenic bacteria on their apical surface. Proc Natl Acad Sci USA 1995;92:3019–3023.
12 Pier GB, Grout M, Zaidi TS: Cystic fibrosis transmembrane conductance regulator is an epithelial cell receptor for clearance of *Pseudomonas aeruginosa* from the lung. Proc Natl Acad Sci USA 1997;94:12088–12093.
13 Jayaraman S, Joo NS, Reitz B, Wine JJ, Verkman AS: Submucosal gland secretions in airways from cystic fibrosis patients have normal [Na$^+$] and pH but elevated viscosity. Proc Natl Acad Sci USA 2001;98:8119–8123.
14 Verkman AS, Song Y, Thiagarajah JR: Role of airway surface liquid and submucosal glands in cystic fibrosis lung disease. Am J Physiol 2003;284:C2–C15.
15 Heeckeren A, Walenga R, Konstan MW, Bonfield T, Davis PB, Ferkol T: Excessive inflammatory response of cystic fibrosis mice to bronchopulmonary infection with *Pseudomonas aeruginosa*. J Clin Invest 1997;100:2810–2815.
16 Zuelzer WW, Newton WA Jr: The pathogenesis of fibrocystic disease of the pancreas. A study of 36 cases with special reference to the pulmonary lesions. Pediatrics 1949;4:53–69.
17 Matthews LW, Spector S, Lemm J, Potter JL: Studies on pulmonary secretions. I. The overall chemical composition of pulmonary secretions from patients with cystic fibrosis, bronchiectasis, and laryngectomy. Am Rev Respir Dis 1963;88:199–204.
18 Regnis JA, Robinson M, Bailey DL, Cook P, Hooper P, Chan HK, Gonda I, Bautovich G, Bye PT: Mucociliary clearance in patients with cystic fibrosis and in normal subjects. Am J Respir Crit Care Med 1994;150:66–71.
19 Yeates DB, Sturgess JM, Kahn SR, Levison H, Aspin N: Mucociliary transport in trachea of patients with cystic fibrosis. Arch Dis Child 1976;51:28–33.
20 Stutts MJ, Canessa CM, Olsen JC, Hamrick M, Cohn JA, Rossier BC, Boucher RC: CFTR as a cAMP-dependent regulator of sodium channels. Science 1995;269:847–850.
21 Mall M, Hipper A, Greger R, Kunzelmann K: Wild type but not delta F508 CFTR inhibits Na$^+$ conductance when coexpressed in *Xenopus* oocytes. FEBS Lett 1996;381:47–52.
22 Letz B, Korbmacher C: cAMP stimulates CFTR-like Cl$^-$ channels and inhibits amiloride-sensitive Na$^+$ channels in mouse CCD cells. Am J Physiol 1997;272:C657–C666.
23 Matsui H, Davis CW, Tarran R, Boucher RC: Osmotic water permeabilities of cultured, well-differentiated normal and cystic fibrosis airway epithelia. J Clin Invest 2000;105:1419–1427.
24 Tarran R, Grubb BR, Gatzy JT, Davis CW, Boucher RC: The relative roles of passive surface forces and active ion transport in the modulation of airway surface liquid volume and composition. J Gen Physiol 2001;118:223–236.
25 Mall M, Grubb BR, Harkema JR, O'Neal WK, Boucher RC: Increased airway epithelial Na$^+$ absorption produces cystic fibrosis-like lung disease in mice. Nat Med 2004;10:487–493.
26 Frizzell RA, Pilewski JM: Finally, mice with CF lung disease. Nat Med 2004;10:452–454.
27 Welsh MJ, Ramsey BW, Accurso FJ, Cutting GR: Cystic fibrosis; in Scriver CR, Beaudet AL, Sly WS, Valle D (eds): The Metabolic and Molecular Bases of Inherited Disease, ed 8. New York, McGraw-Hill, 2001, pp 5121–5188.
28 Fujii T, Hayashi S, Hogg JC, Vincent R, van Eeden SF: Particulate matter induces cytokine expression in human bronchial epithelial cells. Am J Respir Cell Mol Biol 2001;25:265–271.
29 Breuer R, Christensen TG, Lucey EC, Stone PJ, Snider GL: An ultrastructural morphometric analysis of elastase-treated hamster bronchi shows discharge followed by progressive accumulation of secretory granules. Am Rev Respir Dis 1987;136:698–703.
30 Takeyama K, Agusti C, Ueki I, Lausier J, Cardell LO, Nadel JA: Neutrophil-dependent goblet cell degranulation: Role of membrane-bound elastase and adhesion molecules. Am J Physiol 1998;275:L294–L302.
31 Voynow JA, Young LR, Wang Y, Horger T, Rose MC, Fischer BM: Neutrophil elastase increases MUC5AC mRNA and protein expression in respiratory epithelial cells. Am J Physiol 1999;276:L835–L843.
32 Nakamura H, Yoshimura K, McElvaney NG, Crystal RG: Neutrophil elastase in respiratory epithelial lining fluid of individuals with cystic fibrosis induces interleukin-8 gene expression in a human bronchial epithelial cell line. J Clin Invest 1992;89:1478–1484.
33 Takeyama K, Dabbagh K, Lee HM, Agusti C, Lausier JA, Ueki IF, Grattan KM, Nadel JA: Epidermal growth factor system regulates mucin production in airways. Proc Natl Acad Sci USA 1999;96:3081–3086.
34 Anderson MP, Gregory RJ, Thompson S, Souza DW, Paul S, Mulligan RC, Smith AE, Welsh MJ: Demonstration that CFTR is a chloride channel by alteration of its anion selectivity. Science 1991;253:202–205.
35 Knowles MR, Stutts MJ, Spock A, Fischer N, Gatzy JT, Boucher RC: Abnormal ion permeation through cystic fibrosis respiratory epithelium. Science 1983;221:1067–1070.
36 Boucher RC, Stutts MJ, Knowles MR, Cantley L, Gatzy JT: Na$^+$ transport in cystic fibrosis respiratory epithelia. Abnormal basal rate and response to adenylate cyclase activation. J Clin Invest 1986;78:1245–1252.
37 Grubb BR, Boucher RC: Pathophysiology of gene-targeted mouse models for cystic fibrosis. Physiol Rev 1999;79:S193–S214.
38 Snouwaert JN, Brigman KK, Latour AM, Malouf NN, Boucher RC, Smithies O, Koller BH: An animal model for cystic fibrosis made by gene targeting. Science 1992;257:1083–1088.
39 Berger M: Lung inflammation early in cystic fibrosis. Bugs are indicted, but the defense is guilty. Am J Respir Crit Care Med 2002;165:857–858.

Marcus Mall
Department of Pediatrics III
Pediatric Pulmonology and
Cystic Fibrosis Center
University of Heidelberg
Im Neuenheimer Feld 153
DE–69120 Heidelberg (Germany)
Tel. +49 6221 56 8840
Fax +49 6221 56 8806
E-Mail Marcus.Mall@med.uni-heidelberg.de

Bush A, Alton EWFW, Davies JC, Griesenbach U, Jaffe A (eds): Cystic Fibrosis in the 21st Century.
Prog Respir Res. Basel, Karger, 2006, vol 34, pp 122–130

Infection versus Inflammation

Assem G. Ziady   Pamela B. Davis

Department of Pediatrics, Case Western Reserve University, Cleveland, Ohio, USA

Abstract

Infection and inflammation are critical in the progression of cystic fibrosis (CF) lung disease, and are the cause of death of most patients. Therefore, understanding their origins and mechanisms is critical to designing effective interventions. Bacterial infection in CF occurs in the wake of ion transport abnormalities that thwart the usual mucociliary clearance mechanisms, but may also be promoted by excess adhesion of particular bacteria to the CF airway and by protracted inflammatory responses to prior viral infections, leaving the airway vulnerable. Once initiated by infection (or perhaps, even before infection sets in), the inflammatory process in CF airways is excessive relative to the stimulus applied, protracted, and fails to resolve appropriately, even if the infectious stimulus is removed. Whether this is driven by epithelial responses, the responses of other immune cells, or, most likely, some interaction of the vulnerable cell types in the lung, is not yet clear. Moreover, the precise intracellular abnormalities that produce this phenotype are still being elucidated, and their relationship to CFTR dysfunction is similarly unclear. Nevertheless, infection and inflammation remain the two most attractive independent therapeutic targets in CF, short of the basic defect itself. In this chapter, we examine the relationship between infection and inflammation in CF lung disease.

Infection

Although viral infections occur in all infants, in those with CF *Haemophilus influenzae* and *Staphylococcus aureus* are frequently recovered from deep throat cultures.

Later in life, most CF patients become infected with *Pseudomonas aeruginosa*, which, once established, persists, usually for life. More than 80% of adult CF patients are chronically infected [1]. Thus, investigators have studied the relationship between infection with *P. aeruginosa* and the inflammatory response (fig. 1).

Is There Inflammation Prior to Infection in the Cystic Fibrosis Airway?

In the few published studies of CF fetuses and newborns [2, 3], the lungs appear histologically largely normal, with the possible exception of distended mucus glands in the upper airways. Other investigators have observed inflammatory cells and mediators in bronchoalveolar lavage (BAL) fluid in infants with CF in the absence of apparent infection, as assessed by the insensitive measure of culture of the diluted BAL fluid, and later by the more sensitive PCR methodology [4, 5]. However, it cannot be excluded that these infants had been infected earlier, but the infection cleared at the time of testing. Clinical studies, therefore, do not distinguish inflammatory responses in the absence of infection from prolonged inflammatory responses to prior infection, which has since resolved.

Since CF mouse models do not spontaneously acquire lung infection or display CF-like lung disease, they also do not define the relationship between infection and inflammation. Therefore, investigators have studied xenografts of CF human tracheal and bronchial epithelial cells or fetal airways, implanted in immunocompromised (SCID) mice [6–8]. Tirouvanziam et al. [6] and Escotte et al. [7] used CF

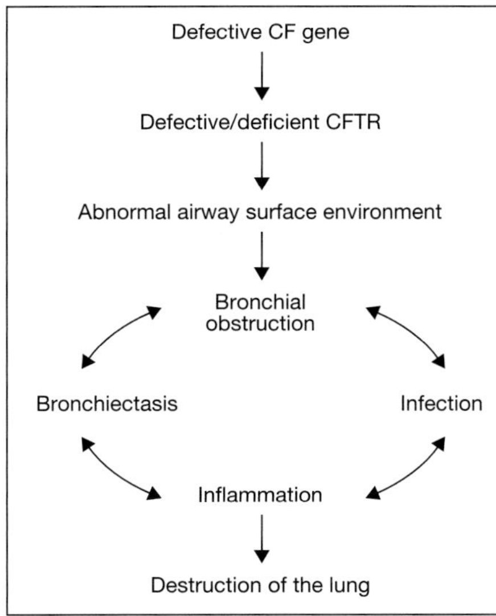

Fig. 1. Pathogenesis and disease progression in the CF lung. The temporal and causal relationships of events involved in CF lung disease have not been fully delineated, but it is clear that a vicious cycle of inflammation and infection plays a central role in lung deterioration.

human fetal grafts in SCID mice to examine tissues not yet exposed to infection. Although these grafts were histologically normal, they produced airway surface liquid that contained higher levels of proinflammatory cytokines and chemokines compared to normal xenografts, suggesting some proinflammatory predisposition prior to infection [6]. However, as the migratory cells are of murine origin, and the immune system in these mice is impaired, these studies may not accurately model infection in CF patients. Moreover, the airways are separated from other cells in the lung and are not ventilated or vascularized in normal fashion. Cohn et al. [8] found that despite a higher susceptibility to infection, CF xenografts did not differ from controls in infection severity following exposure to *P. aeruginosa*. They concluded that while xenografts may be useful for the study of initial infection, this model system may not replicate conditions in the CF lung sufficiently to permit conclusions on whether inflammation is present prior to infection in CF. Thus, the current model systems are insufficient to prove whether inflammation can occur prior to infection in the CF airway.

Is the Inflammatory Response Increased, Dysregulated, Prolonged, or All of These?

Studies in Humans

Shortly after birth, the lungs of CF patients become infected by a variety of bacterial pathogens. This is followed by massive neutrophilic infiltration recruited by elevated cytokine and chemokine production by preexisting neutrophils and epithelial cells in the small airways [9]. Muhlebach et al. [10] and Kazachkov et al. [11] compared CF patients age 1–2 years to age-matched children with other chronic respiratory problems [10, 11]. The ratio of either neutrophil number or interleukin 8 (IL-8, a proinflammatory chemokine) to bacteria or concentration of lipopolysaccharide (LPS) was significantly elevated in the CF patients [10]. Other investigators found that culture-negative CF infants had increased IL-8 and neutrophils even compared to culture-positive non-CF children [5, 10], but it is not possible to distinguish spontaneous inflammation from inflammatory responses that reverberate after infection is contained in these studies. Older patients with CF display marked increases in proinflammatory mediators, including elastase and interleukin (IL)-8, IL-1 [12], IL-2 [13], IL-9 [14], and tumor necrosis factor alpha (TNF-α) [15] in BAL fluid, and IL-6 and TNF-α in the blood [16] compared to healthy controls, but appropriate disease controls with similar bacterial burden are difficult to obtain, so these data are difficult to interpret. Compared to many control groups, CF subjects show elevations of arachidonic acid in many tissues, and high levels are seen in the CF airway [17]. Its metabolites, leukotrienes B_4 and E_4 (LTB$_4$ and LTE$_4$, respectively) have been found in urine [18], sputum [18] and even breath condensate [19] in excess in CF patients. These mediators cause bronchoconstriction, increased microvascular permeability, and are potent mediators of inflammation in CF.

Despite the high levels of other cytokines, IL-10, an anti-inflammatory cytokine, was found to be significantly decreased in CF compared to normal controls [15], supporting the hypothesis that regulation of the inflammatory response is aberrant in CF. IL-10 modulates the inflammatory functions of monocytes/macrophages, lymphocytes, and neutrophils in the airway [20]. Furthermore, it promotes resolution of nuclear factor-kappa B (NF-κB)-mediated inflammation [21]. In IL-10 knockout mice, Pseudomonas infection resulted in a prolonged inflammatory response similar to that observed in CF mice and humans [22]. These knockout mice were unable to regenerate degraded IκBα (an inhibitor of NF-κB, a proinflammatory transcription factor) in a timely fashion. Failure to produce sufficient IL-10 in the CF lung could result in an inability to control or shorten the excessive inflammatory response.

Lipoxin A_4 (LXA$_4$) can inhibit neutrophil chemotaxis, adherence, transmigration, and activation [23]. An analog of this lipoxin suppresses neutrophilic inflammation and reduces the severity of disease in mice with chronic airway infection with *P. aeruginosa*. Recently, Karp et al. [24] showed that LXA$_4$ is present in significantly lower quantities in BAL fluid from CF patients compared to controls with inflammatory lung diseases. Thus, studies in humans indicate that CF patients have excessive proinflammatory mediators in lung, but insufficient quantities of mediators that contribute to the resolution of inflammation.

Studies in Cystic Fibrosis Knockout Mice

Investigators have also examined the inflammatory response in CF mouse models [25–32]. Although studies in CF mice should be interpreted with caution, since these mice do not exhibit spontaneous infection or inflammation, they have proven very useful in the study of subacute inflammation [25–29] and bacterial clearance [30–32]. Embedding *P. aeruginosa* in agar beads prevents its rapid clearance, and the resulting infection displays many characteristics of CF lung infection [25]. All the groups that have compared CF mice with wild-type controls in this model have observed increased inflammatory responses to Pseudomonas agar beads, although some of the details differ. Gosselin et al. [26] found that CF knockout mice exhibited higher mortality and increased bacterial burden, but no significant differences in lung histology or neutrophil count in BAL fluid. Van Heeckeren et al. [27] also observed significantly increased mortality in CF knockout mice, significant increases in proinflammatory cytokines and chemokines such as TNF-α, and macrophage inflammatory protein (MIP)-2 and KC/N51 (both analogs of human IL-8) in BAL fluid 3 days post infection, as well as marked increase in inflammatory foci and mucus plugging, but no difference in bacterial burden compared to wild-type littermates. Moreover, all the different mutant strains of CF mice tested (R117H, S489X, Y122X, or ΔF508 on a C57BL/6 background) had similar increases in cytokines, neutrophils, and eicosanoid levels in BAL fluid, compared to wild-type mice [28], implicating loss of CFTR function, rather than specific properties of individual mutations, in the excess inflammation. McMorran et al. [29] studied chronic infection in a CF mouse model of the G551D mutant, and reported both an increase in cytokine production and bacterial burden, as well as an impaired clearance of Pseudomonas in the CF mice compared with normal littermates.

Although attempts to infect CF mice with naked *P. aeruginosa*, even dosed repeatedly, have failed to yield chronic infection, CF mice challenged with repeated administration of *S. aureus* [30, 31] or *Burkholderia cepacia* [31, 32] exhibited mucus retention, lung disease and progressive impairment in bacterial clearance compared with normal littermates (wild-type or CF heterozygote). *B. cepacia* isolates from CF patients, administered repeatedly [32] produced significantly higher levels of viable bacteria, and severe bronchopneumonia in CF mice compared with normal littermates.

While both the agar bead and repeated administration of naked bacteria models mimic some aspects of CF pathology, no one model shows complete concordance with the human condition. Nevertheless, all the studies indicate that once infection is induced, inflammation is greater in CF mice, regardless of *cftr* genotype or mouse genetic background [25–32]. Interestingly, in some but not all studies this excessive inflammation persists in the face of comparable or even reduced bacterial burden in CF mice.

Studies in Cell Culture

Cultured airway epithelial cells are an appealing model for the study of inflammatory responses, since these cells are the first encountered by inhaled pathogens and are a primary site of the CF defect. In addition, cultures provide a level of control not afforded in intact animals or patients. However, this is not a perfect model system, either. In culture, epithelial cells are removed from the influences of the many other cell types in the lung, and they are forced to adapt to grow in culture. Immortalization of cell lines may cause responses that are not relevant to the native state. Culture conditions (e.g., serum content of the media) profoundly influence the production of proinflammatory cytokines [33]. Careful matching of CF and non-CF controls is critical. Similar genetic background except for CFTR is essential. If CFTR is restored by transfection to CF cells, care must be taken to avoid its overexpression or alteration of cell properties by the gene transfer agent itself. If CFTR is inhibited to create the CF member of the pair, similar considerations apply to the mechanism of inhibition. The comparison of cells from different individuals, CF and non-CF, is confounded by the effects of variation in inflammation-related genes apart from CFTR. The level of differentiation of the cells may be important. Highly differentiated model systems derived from primary cultures are appealing because they appear similar to the native epithelium, but matching of cultures for genes other than CFTR is difficult. The inflammatory stimulus applied (live bacteria, bacterial products, cytokines, viruses) may be crucial. The variable and sometimes conflicting results reported for these studies may derive from these differences and concerns. Nevertheless, commonalities emerge.

In well-matched airway epithelial cell lines, Kube et al. [34] reported increases in the secreted proinflammatory cytokines and chemokines, IL-8, IL-6, and GM-CSF in CF

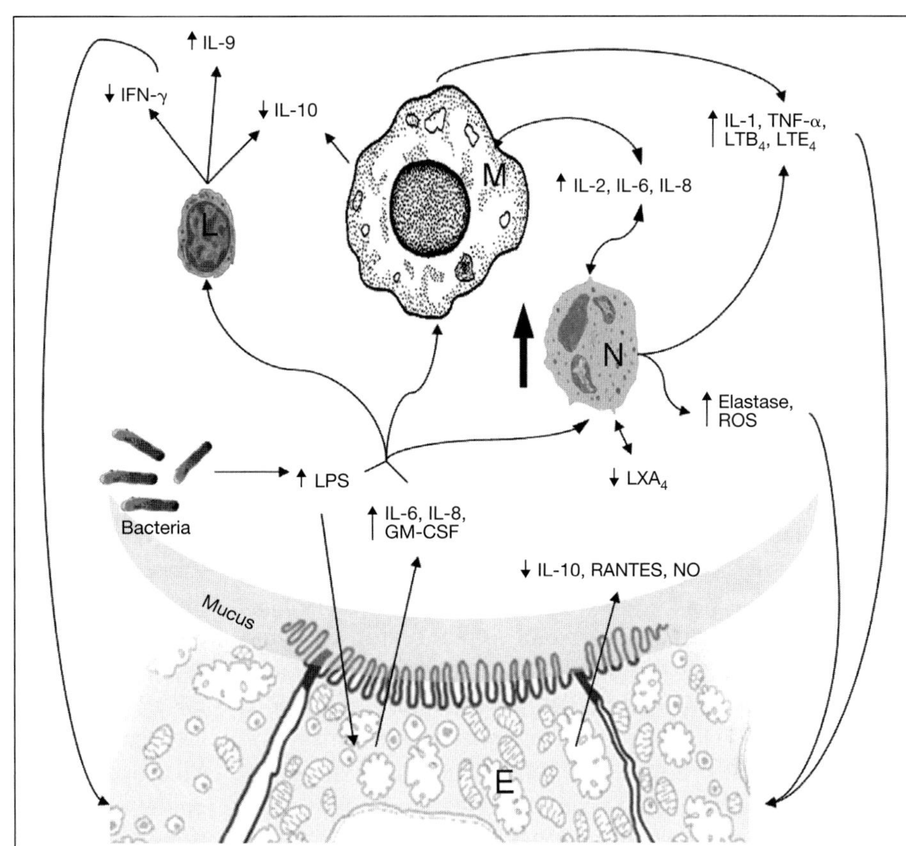

Fig. 2. Cellular cross-talk involved in the CF inflammatory response. Modulators of inflammation are produced by a number of cells in the CF lung. Levels of pro-inflammatory cytokines, chemokines and arachidonic acid leukotriene derivatives are markedly increased (depicted by up arrows) in CF compared to normal. Conversely, a significant decrease (depicted by down arrows) in anti-inflammatory mediators is observed. M = Macrophage; N = neutrophil; L = lymphocyte; E = airway epithelial cell; NO = nitric oxide; ROS = reactive oxygen species.

phenotype cell lines, in response to Pseudomonas or cytokines. Immortalized CF airway epithelial cells are reported to have increased IL-8, both at baseline and in response to stimulation, compared with their CFTR-corrected counterparts [35]. In these studies, the increases were minimal at early time points, but the CF-non-CF differences increased with time. Perez et al. [pers. commun.] have shown that similar increases from normal controls are achieved in well-differentiated human primary tracheal epithelial cells grown at an air-liquid interface (ALI) when CFTR activity is inhibited with the I172 inhibitor of channel activity.

However, other studies in well-differentiated cells in primary culture at the ALI give mixed results. CF-non-CF differences in response to bacterial products are modest, but submerged cultures may show greater differences related to genotype than those maintained at ALI. Becker et al. [36] showed no significantly increased production of IL-8 or IL-6 by CF cells grown at ALI when stimulated by IL-1β or *S. aureus*, but when *P. aeruginosa* products were used, increased responses were only observed in the presence of serum. Other investigators point out the differences in applying Pseudomonas to apical vs. basolateral surfaces of polarized epithelial cells [37].

Some investigators assert that the increased inflammatory response in CF airway epithelial cells arises from increased adhesion of *P. aeruginosa* [38]. However, others have demonstrated that even at comparable or reduced levels of *P. aeruginosa* binding, the cytokine responses of CF phenotype cells are enhanced [59]. Moreover, differences in bacterial adherence cannot explain the increased response of CF airway epithelia to mediators such as TNF-α or IL-1β.

Non-Epithelial Cell Type Participation in the Inflammatory Response

Although many investigators have studied the link between inflammation and CFTR dysfunction in epithelial cells, other cell types may prove to be important contributors to the inflammatory response in the airways. Figure 2 demonstrates cross-talk between cells that contribute to the inflammatory response. Neutrophils are clearly important effector cells in the lung, and their products, both destructive enzymes and oxidants, contribute to the ongoing damage in the airway and have been suggested as therapeutic targets. Oxidants may be particularly noxious in the CF airway because glutathione levels are reduced [39]. Most studies demonstrate no differences between CF and non-CF neutrophil function when conditions are properly controlled. However, alterations in other cells of the immune system such as macrophages and lymphocytes may also play a role in CF inflammation [13, 40, 41]. These cells' contribution to the initiation of the excessive inflammatory response has been largely ignored, as the primary defect in CF has generally been expected to be most pronounced in cells primarily involved in ion transport. However, CFTR is indeed expressed at low levels in lung fibroblasts, cultured cells of lymphocytic lineage U-937, HL-60, and K-562, and freshly isolated blood lymphocytes, monocytes, neutrophils, and alveolar macrophages [42]. Functional studies in lymphocytes confirmed CFTR-mediated chloride conductance during the G_1 phase of the cell cycle and changes in inflammatory responses have been reported in CF lymphocytes, including decreased IL-10 secretion [40], and reduced interferon-gamma (IFN-γ) secretion by CD4+ T lymphocytes [41]. Alveolar macrophages also may not function normally. Soltys et al. [43] found that expression of B7, a proinflammatory stimulant, was reduced on CF macrophages, possibly due to the reduced IL-10. Taken together, the data suggest that CF lymphocytes and macrophages may not regulate inflammatory responses appropriately. If CF lymphocytes and macrophages had an intrinsic defect one would expect infection/alterations in other organs, but this does not seem to be the case.

All in all, it seems likely that the increased inflammatory response observed consistently in CF patients and CF mice, regardless of genotype or genetic background, in response to bacterial challenge, may result from exaggerated responses to bacterial stimulation, but that the failure to resolve the inflammatory process in a timely fashion may also contribute. To the extent that the seeds of resolution are contained in the initial response to infection, one can speculate that the failure of counter-regulatory mechanisms may be primarily responsible for the excess inflammation. However, the precise mechanisms by which this occurs, in which cell types, and how they are tied to CFTR dysfunction have not been elucidated.

Signaling Pathways Activated in the Cystic Fibrosis Inflammatory Process

Excess inflammation in CF may be derived from aberrant activation of cell-signaling pathways. Although some have argued that the misprocessing of ΔF508 leads to an endoplasmic reticulum stress response and excess inflammation, and others suggest that dysfunctional signaling in CF stems from the lack of interaction between CFTR and other molecules at the membrane [44, 45], the bulk of the data indicate that markedly reduced CFTR function is associated with the signaling aberrations. Although the link between the loss of CFTR function and the abnormal signaling is not understood, a number of pathways central to inflammation and responses to infection have been characterized [46–57]. Alterations in epithelial cell inflammatory signaling are presented in figure 3 and table 1 and will be briefly described below.

An important common transcription factor shared by many of the mediators overexpressed in the CF lung, is NF-κB [reviewed in ref. 46], which is required for maximal transcription of a number of proinflammatory molecules, including many cytokines (IL-1β, TNF-α, IL-6) and chemokines (IL-8). In CF, NF-κB appears to be persistently activated rather than overexpressed [47–49]. Activators of NF-κB, such as reactive oxygen species (ROS) and LPS, are persistently present in elevated quantities in CF [50]. Moreover, bacterial interaction with surface receptors on epithelial cells is a potent activator of NF-κB. The activation of macrophages, with release of TNF-α and IL-1β allows these cytokines to interact with their receptors at the surface of epithelial cells and activate proinflammatory cascades. Thus, there is convergence of many potent stimuli to activate NF-κB in airway epithelial cells (and likely other cell types as well). Conversely, inhibitors of NF-κB, such as IL-10, are downregulated in CF [21]. Indeed, IL-10 knockout mice, exhibit similar inflammation and pathology to CF mice, when chronically infected with Pseudomonas [22]. An anti-inflammatory transcription factor, the peroxisome proliferation activator receptor (PPAR), which has reciprocal action with NF-κB, appears to be reduced in CF mice [51]. Thus, stimuli align to increase activation of NF-κB, and the usual 'brakes' on the system may not be fully functional.

Other pathways that modulate NF-κB transcriptional activation of proinflammatory cytokine synthesis are also dysregulated in CF. Although signal transducer and activator of transcription 1 (STAT-1) is over-expressed in CF phenotype cells [52] and in CF mice, increased expression of protein inhibitor of activated STAT-1 (PIAS-1) interferes with its ability to translocate to the nucleus and impairs STAT-1 signaling. Since STAT-1 is a major signal induced by IFN-γ, this reduced activity will impair STAT-1-dependent IFN responses, and quite likely accounts for the increased symptomatic response of CF patients to viral

    Ziady/Davis

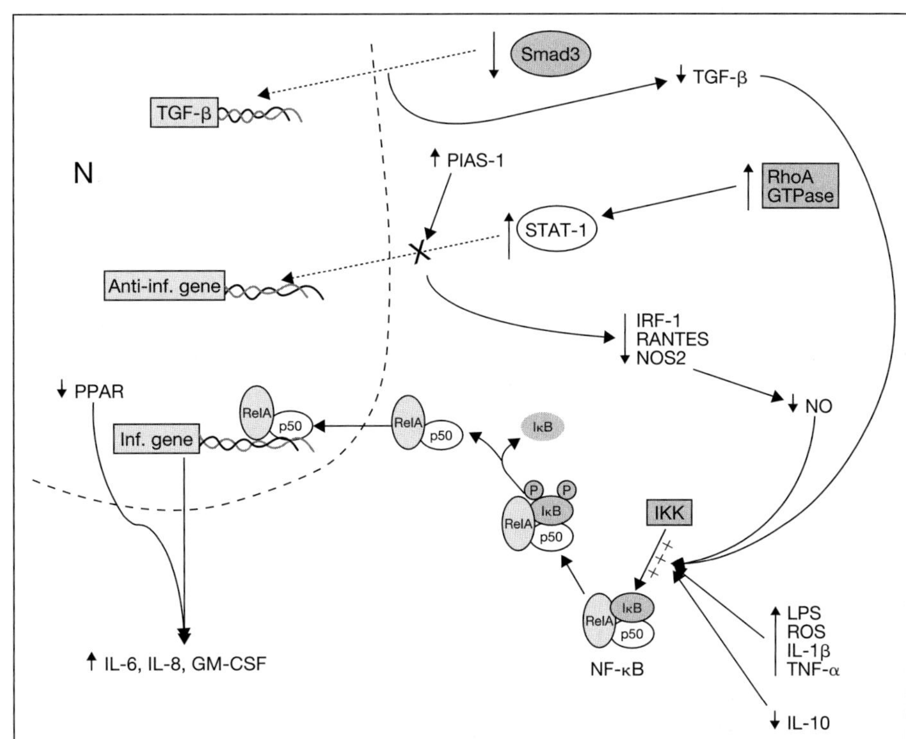

Fig. 3. Inflammatory signaling in CF airway epithelia. Alterations in cell signaling and protein expression in CF epithelia results in the excessive production of pro-inflammatory mediators and the decreased production of anti-inflammatory mediators. N = Nucleus; NOS2 = nitric oxide synthase; NO = nitric oxide; PIAS = protein inhibitor of activated stat.

Table 1. Altered cell signaling and its consequences in CF

Alteration	Effect	Consequence	Ref.
↑Prenylation of Smad3	↑Degradation of Smad3	↓Production of TGF-β, which leads to increased activation of NF-κB	56, 57
↑Prenylation of RhoA GTPase	↑Activation of RhoA GTPase	↑Expression and activation of STAT-1	52, 53
↑PIAS-1	Inhibition of STAT-1 activity, which overcomes activation by RhoA GTPase	↓Transcription of anti-inflammatory proteins such as IRF-1, RANTES, and NOS2	52
↑LPS, ROS, IL-1β, TNF-α, ↓IL-10, NO	↑Activation of IκB kinase, or direct destabilization of IκB	↑Dissociation of IκB from NF-κB complex and increased activation of NF-κB, leading to increased transcription and expression of pro-inflammatory genes	21, 22, 47–50, 54
↑COX1 and altered modification of 5-LO and FLAP	↑Production of LTB₄ & LTE₄	↑Pro-inflammatory and diminished anti-inflammatory activity	17–19, 23, 24, 58
↓15-LO, COX2	↓Production of LXA₄		
↓PPAR	↓Control of the production of anti-inflammatory genes	↑Production of pro-inflammatory genes	51

infections. Impaired STAT-1 signaling results in the down-regulation of a variety of proteins, including nitric oxide synthase 2 (NOS2), interferon regulatory factor 1 (IRF-1), and regulated on activation normal T cell expressed and secreted (RANTES) factor [53]. Decreased NOS2 decreases the production of NO, a bactericidal agent, impairing early host defense against infection. NO also promotes IκB stability, probably by decreasing IκB kinase activity [54]. Recent reports suggest that increased RhoA GTPase expression in CF is responsible for aberrations in STAT-1 expression and signaling [55].

At the same time, the quantity of the signaling and tran-scription factor Smad3 is reduced in CF [56, 57]. Smad3 is an important intermediary in the response to transforming growth factor beta (TGF-β), which participates in resolu-tion of inflammation, so reduction could contribute to the proinflammatory CF diathesis. Both the activation of RhoA and the degradation of Smad3 are controlled by prenyla-tion, the addition of lipid moieties to the protein to alter their function. The lipids for prenylation are produced in the cholesterol synthetic pathway, and this branch of the pathway may be upregulated in CF [57]. Indeed, applica-tion of statins to block this pathway reverses many of the excess inflammatory responses in epithelial cell models. However, once again, the precise mechanism or connection to CFTR dysfunction is not clear.

Eicosanoid-related pathways also exhibit dysregulation in CF [17–19, 23, 24, 58]. In a study measuring the mRNA levels of enzymes associated with these pathways in patients with CF or asthma, Kuitert et al. [58] demon-strated increases in cyclooxygenase 1 (COX1) in both patients groups compared to normals. No increases in 5-lipoxygenase (5-LO) or 5-LO activating protein (FLAP) were observed in CF, and no mRNA was detected for 15-LO or COX2. As 5-LO and FLAP play a role in leuko-triene production, these results failed to explain a direct source of their elevated production in CF. However, although COX1 upregulation and altered posttranslational modification of 5-LO or FLAP that alters their activity in CF may account for increased LTB$_4$ production, this has not yet been evaluated. Decreases in LXA$_4$ also need to be reconciled with increases in leukotriene production. Decreases in 15-LO and COX2, both involved in the pro-duction of LXA$_4$ by oxygenating the C15 position may explain decreases in the production of this anti-inflamma-tory lipoxin [24, 58].

References

1 Cystic Fibrosis Foundation Patient Registry: 2001 Annual Data Report to the Center Directors. Bethesda, Cystic Fibrosis Founda-tion, 2002.
2 Ornoy A, Arnon J, Katznelson D, Granat M, Caspi B, Chemke J: Pathological confirmation of cystic fibrosis in the fetus following prena-tal diagnosis. Am J Med Genet 1987;28: 935–947.
3 Chow CW, Landau LI, Taussig LM: Bronchial mucous glands in the newborn with cystic fibrosis. Eur J Pediatr 1982;139:240–243.
4 Khan TZ, Wagener JS, Bost T, Martinez J, Accurso FJ, Riches DW: Early pulmonary inflammation in infants with cystic fibrosis. Am J Respir Crit Care Med 1995;151: 1075–1082.
5 Rosenfeld M, Gibson RL, McNamara S, Emerson J, Burns JL, Castile R, Hiatt P, McCoy K, Wilson CB, Inglis A, Smith A, Martin TR, Ramsey BW: Early pulmonary infection, inflammation, and clinical out-comes in infants with cystic fibrosis. Pediatr Pulmonol 2001;32:356–366.
6 Tirouvanziam R, de Bentzmann S, Hubeau C, Hinnrasky J, Jacquot J, Peault B, Puchelle E: Inflammation and infection in naive human cystic fibrosis airway grafts. Am J Respir Cell Mol Biol 2000;23:121–127.
7 Escotte S, Danel C, Gaillard D, Benoit S, Jacquot J, Dusser D, Triglia JM, Majer-Teboul

C, Puchelle E: Fluticasone propionate inhibits lipopolysaccharide-induced proinflammatory response in human cystic fibrosis airway grafts. J Pharmacol Exp Ther 2002;302:1151–1157.
8 Cohn LA, Weber A, Phillips T, Lory S, Kaplan M, Smith A: *Pseudomonas aeruginosa* infec-tion of respiratory epithelium in a cystic fibro-sis xenograft model. J Infect Dis 2001;183: 919–927.
9 Konstan MW, Hilliard KA, Norvell TM, Berger M: Bronchoalveolar lavage findings in cystic fibrosis patients with stable, clinically mild lung disease suggest ongoing infection and inflammation. Am J Respir Crit Care Med 1994;150:448–454.
10 Muhlebach MS, Stewart PW, Leigh MW, Noah TL: Quantitation of inflammatory responses to bacteria in young cystic fibrosis and control patients. Am J Respir Crit Care Med 1999;160:186–191.
11 Kazachkov MY, Muhlebach MS, Livasy CA, Noah TL: Lipid-laden macrophage index and inflammation in bronchoalveolar lavage fluids in children. Eur Respir J 2001;18:790–795.
12 Wilmott RW, Kassab JT, Kilian PL, Benjamin WR, Douglas SD, Wood RE: Increased levels of interleukin-1 in bronchoalveolar washings from children with bacterial pulmonary infec-tions. Am Rev Respir Dis 1990;142:365–368.
13 Hubeau C, Le Naour R, Abely M, Hinnrasky J, Guenounou M, Gaillard D, Puchelle E:

Dysregulation of IL-2 and IL-8 production in circulating T lymphocytes from young cystic fibrosis patients. Clin Exp Immunol 2004; 135:528–534.
14 Hauber HP, Manoukian JJ, Nguyen LH, Sobol SE, Levitt RC, Holroyd KJ, McElvaney NG, Griffin S, Hamid Q: Increased expres-sion of interleukin-9, interleukin-9 receptor, and the calcium-activated chloride channel hCLCA1 in the upper airways of patients with cystic fibrosis. Laryngoscope 2003;113: 1037–1042.
15 Bonfield TL, Panuska JR, Konstan MW, Hilliard KA, Hilliard JB, Ghnaim H, Berger M: Inflammatory cytokines in cystic fibrosis lungs. Am J Respir Crit Care Med 1995;152: 2111–2118.
16 Ionescu AA, Nixon LS, Luzio S, Lewis-Jenkins V, Evans WD, Stone MD, Owens DR, Routledge PA, Shale DJ: Pulmonary function, body composition, and protein catabolism in adults with cystic fibrosis. Am J Respir Crit Care Med 2002;165:495–500.
17 Meyer KC, Sharma A, Brown R, Weatherly M, Moya FR, Lewandoski J, Zimmerman JJ: Function and composition of pulmonary sur-factant and surfactant-derived fatty acid pro-files are altered in young adults with cystic fibrosis. Chest 2000;118:164–174.
18 Sampson AP, Spencer DA, Green CP, Piper PJ, Price JF: Leukotrienes in the sputum and urine

of cystic fibrosis children. Br J Clin Pharmacol 1990;30:861–869.

19 Carpagnano GE, Barnes PJ, Geddes DM, Hodson ME, Kharitonov SA: Increased leukotriene B_4 and interleukin-6 in exhaled breath condensate in cystic fibrosis. Am J Respir Crit Care Med 2003;167:1109–1112.

20 Moore KW, de Waal Malefyt R, Coffman RL, O'Garra A: Interleukin-10 and the interleukin-10 receptor. Annu Rev Immunol 2001;19:683–765.

21 Schottelius AJ, Mayo MW, Sartor RB, Baldwin AS Jr: Interleukin-10 signaling blocks inhibitor of kappaB kinase activity and nuclear factor kappaB DNA binding. J Biol Chem 1999;274:31868–31874.

22 Chmiel JF, Konstan MW, Saadane A, Krenicky JE, Lester KH, Berger M: Prolonged inflammatory response to acute Pseudomonas challenge in interleukin-10 knockout mice. Am J Respir Crit Care Med 2002;165:1176–1181.

23 Serhan CN: Lipoxins and aspirin-triggered 15-epi-lipoxin biosynthesis: An update and role in anti-inflammation and pro-resolution. Prostaglandins Other Lipid Mediat 2002;68–69:433–455.

24 Karp CL, Flick LM, Park KW, Softic S, Greer TM, Keledjian R, Yang R, Uddin J, Guggino WB, Atabani SF, Belkaid Y, Xu Y, Whitsett JA, Accurso FJ, Wills-Karp M, Petasis NA: Defective lipoxin-mediated anti-inflammatory activity in the cystic fibrosis airway. Nat Immunol 2004;5:388–392. Epub 2004 Mar 21.

25 Stotland PK, Radzioch D, Stevenson MM: Mouse models of chronic lung infection with *Pseudomonas aeruginosa*: Models for the study of cystic fibrosis. Pediatr Pulmonol 2000;30:413–424.

26 Gosselin D, Stevenson MM, Cowley EA, Griesenbach U, Eidelman DH, Boule M, Tam MF, Kent G, Skamene E, Tsui LC, Radzioch D: Impaired ability of *Cftr* knockout mice to control lung infection with Pseudomonas aeruginosa. Am J Respir Crit Care Med 1998;157(4 pt 1):1253–1262.

27 Heeckeren A, Walenga R, Konstan MW, Bonfield T, Davis PB, Ferkol T: Excessive inflammatory response of cystic fibrosis mice to bronchopulmonary infection with *Pseudomonas aeruginosa*. J Clin Invest 1997;100:2810–2815.

28 van Heeckeren AM, Schluchter MD, Drumm ML, Davis PB: Role of *Cftr* genotype in the response to chronic *Pseudomonas aeruginosa* lung infection in mice. Am J Physiol Lung Cell Mol Physiol 2004;287:L944–L952.

29 McMorran BJ, Palmer JS, Lunn DP, Oceandy D, Costelloe EO, Thomas GR, Hume DA, Wainwright BJ: G551D CF mice display an abnormal host response and have impaired clearance of Pseudomonas lung disease. Am J Physiol Lung Cell Mol Physiol 2001;281:L740–L747.

30 Davidson DJ, Webb S, Teague P, Govan JR, Dorin JR: Lung pathology in response to repeated exposure to *Staphylococcus aureus* in congenic residual function cystic fibrosis mice does not increase in response to

31 Davidson DJ, Dorin JR, McLachlan G, Ranaldi V, Lamb D, Doherty C, Govan J, Porteous DJ: Lung disease in the cystic fibrosis mouse exposed to bacterial pathogens. Nat Genet 1995;9:351–357.

32 Sajjan U, Wu Y, Kent G, Forstner J: Preferential adherence of cable-piliated *Burkholderia cepacia* to respiratory epithelia of CF knockout mice and human cystic fibrosis lung explants. J Med Microbiol 2000;49:875–885.

33 Jayme DW, Blackman KE: Culture media for propagation of mammalian cells, viruses, and other biologicals. Adv Biotechnol Processes 1985;5:1–30.

34 Kube D, Sontich U, Fletcher D, Davis PB: Proinflammatory cytokine responses to *P. aeruginosa* infection in human airway epithelial cell lines. Am J Physiol Lung Cell Mol Physiol 2001;280:L493–L502.

35 Eidelman O, Srivastava M, Zhang J, Leighton X, Murtie J, Jozwik C, Jacobson K, Weinstein DL, Metcalf EL, Pollard HB: Control of the proinflammatory state in cystic fibrosis lung epithelial cells by genes from the TNF-alphaR/NFkappaB pathway. Mol Med 2001;7:523–534.

36 Becker MN, Sauer MS, Muhlebach MS, Hirsh AJ, Wu Q, Verghese MW, Randell SH: Cytokine secretion by cystic fibrosis airway epithelial cells. Am J Respir Crit Care Med 2004;169:645–653.

37 Fleiszig SM, Evans DJ, Do N, Vallas V, Shin S, Mostov KE: Epithelial cell polarity affects susceptibility to *Pseudomonas aeruginosa* invasion and cytotoxicity. Infect Immun 1997;65:2861–2867.

38 Scheid P, Kempster L, Griesenbach U, Davies JC, Dewar A, Weber PP, Colledge WH, Evans MJ, Geddes DM, Alton EW: Inflammation in cystic fibrosis airways: Relationship to increased bacterial adherence. Eur Respir J 2001;17:27–35.

39 Roum JH, Buhl R, McElvaney NG, Borok Z, Crystal RG: Systemic deficiency of glutathione in cystic fibrosis. J Appl Physiol 1993;75:2419–2424.

40 Moss RB, Bocian RC, Hsu YP, Dong YJ, Kemna M, Wei T, Gardner P: Reduced IL-10 secretion by CD4+ T lymphocytes expressing mutant cystic fibrosis transmembrane conductance regulator (CFTR). Clin Exp Immunol 1996;106:374–388.

41 Moss RB, Hsu YP, Olds L: Cytokine dysregulation in activated cystic fibrosis (CF) peripheral lymphocytes. Clin Exp Immunol 2000;120:518–525.

42 Yoshimura K, Nakamura H, Trapnell BC, Chu CS, Dalemans W, Pavirani A, Lecocq JP, Crystal RG: Expression of the cystic fibrosis transmembrane conductance regulator gene in cells of non-epithelial origin. Nucleic Acids Res 1991;19:5417–5423.

43 Soltys J, Bonfield T, Chmiel J, Berger M: Functional IL-10 deficiency in the lung of cystic fibrosis (cftr−/−) and IL-10 knockout mice causes increased expression and function of B7 costimulatory molecules on alveolar

macrophages. J Immunol 2002;1684:1903–1910.

44 Vankeerberghen A, Cuppens H, Cassiman JJ: The cystic fibrosis transmembrane conductance regulator: An intriguing protein with pleiotropic functions. J Cyst Fibros 2002;1:13–29.

45 Guggino WB, Banks-Schlegel SP: Macromolecular interactions and ion transport in cystic fibrosis. Am J Respir Crit Care Med 2004;170:815–820.

46 Barnes PJ, Karin M: Nuclear factor-B: A pivotal transcription factor in chronic inflammatory diseases. N Engl J Med 1997;336:1066–1071.

47 Weber A, Soong G, Bryan B, Saba S, Prince A: Activation of NFkappaB in airway epithelial cells is dependent on CFTR trafficking and Cl-channel function. Am J Physiol Lung Cell Mol Physiol 2001;281:L71–L78.

48 DiMango E, Ratner AJ, Bryan R, Tabibi S, Prince A: Activation of NFkappaB by adherent *Pseudomonas aeruginosa* in normal and cystic fibrosis respiratory epithelial cells. J Clin Invest 1998;101:2598–2605.

49 Venkatakrishnan A, Stecenko A, King G, Blackwell TR, Brigham KL, Christman JW, Blackwell TS: Exaggerated activation of nuclear factorkB and altered IkB-B process in cystic fibrosis bronchial epithelial cells. Am J Respir Cell Mol Biol 2000;23:396–403.

50 van der Vliet A, Cross CE: Phagocyte oxidants and nitric oxide in cystic fibrosis: New therapeutic targets? Curr Opin Pulm Med 2000;6:533–539.

51 Ollero M, Junaidi O, Zaman MM, Tzameli I, Ferrando AA, Andersson C, Blanco PG, Bialecki E, Freedman SD: Decreased expression of peroxisome proliferators activated receptor gamma in cftr−/− mice. J Cell Physiol 2004;200:235–244.

52 Kreiselmeier NE, Kraynack NC, Corey DA, Kelley TJ: Statin-mediated correction of STAT1 signaling and inducible nitric oxide synthase expression in cystic fibrosis epithelial cells. Am J Physiol Lung Cell Mol Physiol 2003;285:L1286–L1295.

53 Koller DY, Nething I, Otto J, Urbanek R, Eichler I: Cytokine concentrations in sputum from patients with cystic fibrosis and their relation to eosinophil activity. Am J Respir Crit Care Med 1997;155:1050–1054.

54 Kroncke KD: Nitrosative stress and transcription. Biol Chem 2003;384:1365–1377.

55 Kraynack NC, Corey DA, Elmer HL, Kelley TJ: Mechanisms of NOS2 regulation by Rho GTPase signaling in airway epithelial cells. Am J Physiol Lung Cell Mol Physiol 2002;283:L604–L611.

56 Kelley TJ, Elmer HL, Corey DA: Reduced Smad3 protein expression and altered transforming growth factor-beta1-mediated signaling in cystic fibrosis epithelial cells. Am J Respir Cell Mol Biol 2001;25:732–738.

57 Lee JY, Elmer HL, Ross KR, Kelley TJ: Isoprenoid-mediated control of SMAD3 expression in a cultured model of cystic fibrosis epithelial cells. Am J Respir Cell Mol Biol 2004;31:234–240.

58 Kuitert LM, Newton R, Barnes NC, Adcock IM, Barnes PJ: Eicosanoid mediator expression in mononuclear and polymorphonuclear cells in normal subjects and patients with atopic asthma and cystic fibrosis. Thorax 1996;51:1223–1228.

59 Kube DM, Fletcher D, Davis PB: Relation of exaggerated cytokine responses of CF airway epithelial cells to PAO1 adherence. Resp Res 2005;6:69 (epub ahead of print).

Pamela B. Davis, MD, PhD
Department of Pediatrics
Case Western Reserve University
2109 Adelbert rd. BRB #826
Cleveland, Ohio, 44106–4948 (USA)
Tel. +1 216 368 4370, Fax +1 216 368 4223
E-Mail pamela.davis@case.edu

Ziady/Davis

Bush A, Alton EWFW, Davies JC, Griesenbach U, Jaffe A (eds): Cystic Fibrosis in the 21st Century.
Prog Respir Res. Basel, Karger, 2006, vol 34, pp 131–137

Pseudomonas aeruginosa: Clinical Research

David Armstrong

Department of Paediatrics, Monash University and Department of Respiratory and Sleep Medicine, Monash Medical Center, Clayton, Australia

Abstract

Clinical studies have demonstrated the importance of *Pseudomonas aeruginosa* infection as a major cause of morbidity and mortality in cystic fibrosis (CF) patients. Chronic *P. aeruginosa* lower airway infection is associated with the development of antibiotic-resistant mucoid strains, which cannot be cleared. It is, therefore, important to accurately determine the timing of initial *P. aeruginosa* infection, to maximize the chances of successfully eliminating the organism. The use of oropharyngeal cultures, serum anti-pseudomonas antibodies, and flexible bronchoscopy with bronchoalveolar lavage in nonexpectorating CF infants has shown that initial infection may occur in the preschool years. *P. aeruginosa* strains causing early infection usually have a nonmucoid phenotype and are relatively antibiotic-sensitive. Aggressive treatment of these early infecting strains in infancy often successfully eradicates infection. Although most infants and young children presumably acquire *P. aeruginosa* from their environment, cross-infection has been demonstrated in several CF centers. This has led to an increased focus on infection control guidelines, and the best methods for detecting the presence of cross-infection and preventing further spread of these transmissible strains.

Introduction

Chronic lower airway infection by the opportunistic pathogen *Pseudomonas aeruginosa* is one of the hallmarks of lung disease in cystic fibrosis (CF), and acquisition and persistence of *P. aeruginosa* within the lower airway is associated with increased morbidity and mortality [1, 2].

Initial isolates have a nonmucoid colonial appearance and are antibiotic sensitive, suggesting environmental acquisition [3]. These early infecting strains can be eradicated by aggressive antibiotic treatment [4–6]. However, over time, a mucoid antibiotic-resistant phenotype develops, which is associated with an accelerated decline in pulmonary function [7, 8] and increased risk of death [9]. Chronic mucoid *P. aeruginosa* infection is usually impossible to eradicate, and the goal of antibiotic treatment is then to suppress, rather than eliminate the pathogen.

Identifying Early *Pseudomonas* Pulmonary Infection in Infants and Young Children with CF

The best opportunity to successfully eradicate lower airway *P. aeruginosa* infection in infants or young children with CF is at the time of initial infection by nonmucoid, antibiotic-sensitive strains [3, 5, 6]. However, infants and young children with CF are usually unable to produce sputum, and therefore the only ways to obtain respiratory tract specimens for culture in this age group are either upper airway cultures [using oropharyngeal (OP) or 'cough' swabs] or bronchoalveolar lavage (BAL) samples. Although recent studies have confirmed the usefulness of induced sputum in nonexpectorating patients [10–12], the technique is seldom useful in children under 5 years of age.

There are advantages and disadvantages in the use of both OP and BAL cultures for detection of early *P. aeruginosa* infection in nonexpectorating CF infants, and there is some controversy as to which of these two methods is of most practical use.

OP Cultures for Detection of P. aeruginosa *Respiratory Infection in Young Children with CF*

A retrospective study reported that 33% of infants diagnosed with CF had a positive OP culture for *P. aeruginosa* by 1 year, and this rose to 49% at 2 years [13]. Children in this study attended a specialist CF center and were diagnosed with CF after presentation with typical clinical features (rather than by a newborn screening program); therefore, the rate of positive *Pseudomonas* culture is likely to be higher than if the whole population of CF infants (i.e. including those without respiratory symptoms) had been studied. The use of neonatal screening for CF has allowed prospective studies of asymptomatic CF infants, to determine the natural history of early lung disease. In a prospective study of 42 screened CF infants, *Staphylococcus aureus* was the organism most frequently cultured from the oropharynx in the first 12 months [14]. However, at a mean follow-up age of 27 months, 19% of infants already had serial OP cultures positive for *P. aeruginosa*, with the first positive culture detected at a mean age of 21 months. Infants with *P. aeruginosa* had more frequent respiratory symptoms, lower chest radiographic scores and elevated levels of circulating immune complexes. A prospective study of 76 infants with CF diagnosed by newborn screening in Italy found that 11% of infants had a positive throat culture for *P. aeruginosa* at diagnosis, increasing to 21% at 1 year and 42% at 2 years [15]. However, there is some controversy as to whether OP cultures accurately predict lower airway bacteriology in these patients. Although OP cultures do not accurately identify the pathogens causing bacterial pneumonia [16], they may be of more use in CF patients, who are more likely to have *P. aeruginosa* present in the upper airway [17]. If OP cultures have poor specificity (i.e. an increased risk of a false-positive result), and are used to determine therapy, then antibiotic treatment may be prescribed in the absence of lower airway *P. aeruginosa* infection. This increases the risk of the organism developing antibiotic resistance [18], and in the longer term may be associated with the emergence of new pathogenic organisms [19].

Combination of OP Cultures with Measurement of Serum Antibodies for Detection of P. aeruginosa *in Infants and Young Children with CF*

Two recent studies have investigated the utility of serum antibody responses to *P. aeruginosa* antigens in conjunction with OP cultures. West et al. [20] found that antibody responses were present 6–12 months before a positive OP culture for *Pseudomonas*. Burns et al. [3] performed a longitudinal study of OP and BAL cultures, together with serum antibody responses in a cohort of 40 CF infants in the first 3 years of life. When culture and serology results were combined, 97.5% of children had evidence of *P. aeruginosa* infection by 3 years of age. However, serum antibodies lacked specificity and thus, using OP cultures and serum antibody responses may overestimate the prevalence of *P. aeruginosa* lower respiratory infection, potentially leading to unnecessary antibiotic treatment.

Flexible Bronchoscopy and BAL for Detection of P. aeruginosa *Infection in Young Children with CF*

A cross-sectional study of BAL in Melbourne, Australia examined 45 newly diagnosed CF infants at a mean age of 2.6 months [21]. BAL culture showed that 38% of these infants already had lower airway infection, with *S. aureus* as the major pathogen. None were positive for *P. aeruginosa*. In a later study, the same researchers showed that the prevalence of a positive BAL culture for *P. aeruginosa* increased from 2% of infants <12 months old to 23% of those over 2 years of age [22]. Burns et al. [3] have also used BAL to demonstrate an increasing prevalence of *P. aeruginosa* lower airway infection with age, increasing from 18% (7/40) at 1 year of age to 33% (11/33) at 3 years of age.

Diagnostic Accuracy of OP versus BAL Cultures for Nonexpectorating Infants and Young Children with CF

Four studies have attempted to evaluate the diagnostic accuracy of OP cultures in nonexpectorating CF patients, by comparing culture results for simultaneously collected OP and BAL fluid samples. In each of these studies, the BAL culture has been regarded as the 'gold standard' for lower respiratory infection. When considering the diagnostic accuracy of any test, the sensitivity, specificity and positive and negative predictive value of the test must be assessed (table 1). Unlike sensitivity and specificity, predictive values are influenced by prevalence. This means that the predictive value of OP cultures is likely to be less useful in the youngest CF children, who are likely to have the lowest prevalence of *P. aeruginosa* infection. Ramsey et al. [23] studied 42 CF patients. Throat swab culture for *S. aureus* and *P. aeruginosa* was specific but not sensitive for the presence of these organisms in the lower airway, and the positive predictive accuracy was 91 and 83%, respectively. Wood [24] performed a cross-sectional study (with longitudinal measures in some cases) in 44 infants with CF in the first 3 years of life and found *S. aureus* in 65%, *Haemophilus influenzae* in 25% and *P. aeruginosa* in 25% of BAL specimens. Cultures were considered 'significant' if >10% of lavaged cells were neutrophils. A good correlation

Table 1. Definitions of sensitivity, specificity and predictive values for OP cultures

	BAL culture	
	positive	negative
OP culture		
Positive	a	c
Negative	b	d

Sensitivity = a/a + c; specificity = d/b + d; positive predictive value = a/a + b; negative predictive value = d/c + d.

between OP cultures and BAL was found in less than 50% of infected and 75% of uninfected patients. Armstrong et al. [22] obtained 150 simultaneous OP and BAL samples over a 31-month period from 75 CF infants diagnosed by newborn screening. The mean age at the time of study was 17 months (range 1–52 months). The most frequently isolated pathogens from BAL fluid were *S. aureus* and *P. aeruginosa*, present in 19 and 11% of cultures, respectively. The most common upper airway pathogens were *S. aureus* (47%), *Escherichia coli* (23%), *H. influenzae* (15%) and *P. aeruginosa* (11%). The sensitivity, specificity, and positive and negative predictive values of OP cultures for *P. aeruginosa* lower airway infection (defined as $\geq 10^5$ CFU/ml of BAL fluid) were 71, 93, 57 and 96%, respectively. Finally, Rosenfeld et al. [25] reported the results of 286 simultaneous OP and BAL cultures in 141 subjects over a 7-year period. This study combined the Melbourne data [22] with that of 4 CF centers in the USA. The sensitivity, specificity, and positive and negative predictive values of OP cultures for the presence of any growth of *P. aeruginosa* in subjects ≤ 18 months of age were 44, 95, 44 and 95%, respectively.

Therefore, in preschool-aged children with CF, the specificity and negative predictive value of OP cultures for *P. aeruginosa* are high, whilst the sensitivity and positive predictive value are low. The measurement of serum antibodies to various *P. aeruginosa* antigens may not increase the specificity of OP cultures. Thus, a negative throat culture may be helpful in excluding lower respiratory *P. aeruginosa* infection. In contrast, however, a positive OP culture does not reliably indicate the presence of *P. aeruginosa* in the lower airway, and BAL will be required to identify infection. Finally, it should be noted that the question of whether the use of BAL to direct therapy in young nonexpectorating CF infants leads to improved outcomes is unknown. However a multicenter, prospective, randomized,

controlled study of the utility of BAL in CF infants currently underway in Australia should help to answer this question [26].

How and When Is *P. aeruginosa* Acquired in CF Patients?

Recent studies of CF infants using flexible bronchoscopy and BAL [3, 22, 27] or throat cultures with serology [3, 20] have shown that *P. aeruginosa* infection, although uncommon at the time of diagnosis by newborn screening [21], occurs earlier in life than once thought. Although initial isolates from CF infants and young children usually display an environmental phenotype [3], there are few studies of early CF lung disease in the literature, and no studies conclusively showing that the environment is the major source for early infection. Another potential source of initial *Pseudomonas* infection in infants and young children with CF is via person-to-person transmission from older CF patients with established chronic infection. However, although siblings with CF usually share the same strain of *P. aeruginosa* [28, 29], most unrelated patients harbor unique isolates over many years [30–32]. Therefore, in the past, person-to-person transmission of *P. aeruginosa* has been thought to be rare. However, a number of clinical and epidemiological studies suggest that acquisition of *P. aeruginosa* by cross-infection may be more common than previously thought. Various risk factors have been identified for *P. aeruginosa* lower respiratory infection in CF [13, 20, 33–36]. These include presentation with meconium ileus [13], number of days in hospital [13, 33, 35], and care of CF infants in centers with older, chronically infected CF patients [20, 36]. Each of these factors would lead to increased exposure of CF infants and young children to older CF patients with chronic *Pseudomonas* infection. Thus, these studies provide circumstantial, rather than direct evidence of *P. aeruginosa* cross-infection in CF.

Previous studies asserting the existence of *P. aeruginosa* cross-infection in CF patients [1, 36] have been criticized for using phenotypic characteristics such as colonial morphology and antibiotic susceptibility profiles to determine differences between bacterial strains from individual patients [37]. More recently, studies using analysis of the bacterial genome by techniques such as pulsed-field gel electrophoresis have shown that genotypically identical strains can express different phenotypic characteristics [38–41]. For analysis by pulsed-field gel electrophoresis, *P. aeruginosa* strains from different patients undergo two-dimensional electrophoresis after digestion by a rare cutting

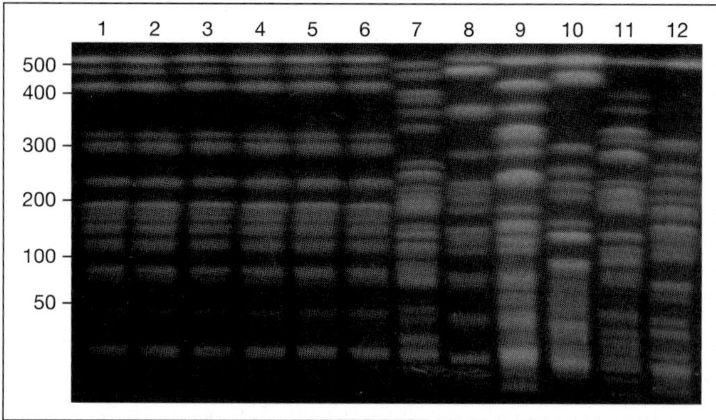

Fig. 1. Pulsed-field gel electrophoresis (*Spe*I macrorestriction profile) assay of epidemic (lanes 1–6) and sporadic (lanes 7–12) *P. aeruginosa* isolates from unrelated CF patients.

enzyme such as *Spe*I. The digest usually produces 10–15 large fragments of DNA, which can then be resolved on a single gel. Figure 1 shows the results of a *Spe*I digest of *P. aeruginosa* isolates from 12 unrelated CF patients [41]. The banding patterns for patients 1–6 are identical, indicating that these patients are infected with a genotypically identical or 'epidemic' strain. In contrast, the banding patterns in patients 7–12 are all different from one another, indicating that these patients are each infected with unique or 'sporadic' strains. Several recent studies have used molecular analysis to more accurately determine the extent and consequences of *P. aeruginosa* cross-infection, with contradictory findings. A large longitudinal epidemiological study from Vancouver, Canada analyzed *P. aeruginosa* isolates obtained over a 20-year period and found a very low risk of *P. aeruginosa* cross-infection, apart from that occurring between siblings [42]. In contrast, several cross-sectional studies have found unequivocal evidence of *P. aeruginosa* cross-infection. The first demonstrated widespread cross-infection in a pediatric CF clinic in Liverpool, UK [38]. The epidemic strain showed increased antibiotic resistance, but otherwise did not appear to have increased virulence compared to nonepidemic strains. Different epidemic strains were subsequently identified in an adult CF center in Manchester [39], and a pediatric clinic in Leeds [40], UK. Once again, these epidemic strains showed increased antibiotic resistance. More recently, widespread cross-infection by an epidemic, antibiotic-resistant *P. aeruginosa* strain has also been reported in a pediatric CF clinic in Melbourne, Australia [41]. This strain was first identified after the unexpected deaths of five preschool CF children, all of whom were found to have infection with the epidemic strain. A subsequent cross-sectional survey of 326 patients attending the clinic found that 55% of patients with *P. aeruginosa* infection harbored the epidemic strain. Subsequently, the same strain has been identified in patients cared for in CF clinics in Sydney and Brisbane [43]. Worryingly, cross-infection with the Melbourne epidemic *P. aeruginosa* strain to a non-CF patient has also been reported [44].

Following the identification of the Melbourne epidemic strain of *P. aeruginosa*, and because of concerns that this strain was also more virulent than nonepidemic strains, cohort segregation according to *P. aeruginosa* culture status and genotype during hospitalization and clinic attendance was introduced. To determine if these strategies had interrupted cross-infection within the clinic, patients from the initial study were followed prospectively [45]. Three years after the introduction of cohort segregation, the epidemic strain prevalence had decreased from 21% of the entire CF clinic to 14% (p = 0.03), while the proportion of patients with nonepidemic *P. aeruginosa* strains was unchanged. Increased mortality or transfer to another clinic did not explain this difference. These findings suggest that the Melbourne epidemic strain is spread by patient-to-patient contact, probably via droplet transmission [46]. The overall clinical status of children from Melbourne with either epidemic or sporadic isolates was similar. Nevertheless, epidemic isolates had increased antibiotic resistance and this may have contributed to longer hospitalization for respiratory exacerbations and less frequent use of inhaled antibiotics compared to patients with other strains [47]. Longitudinal data from a cohort of children recruited at the time of diagnosis by newborn screening [5, 22, 48] are currently being analyzed to further examine over a much longer period whether the Melbourne epidemic strain is associated with increased mortality and morbidity.

All clinics that have reported the presence of transmissible *P. aeruginosa* strains now recommend segregation of these patients from others, and ongoing molecular surveillance of all *P. aeruginosa* isolates [38–41, 46, 47]. It is not acceptable to use phenotypic characteristics (such as multiple antibiotic resistance) as the basis of segregation, as these phenotypic features are unstable over time [49] and do not discriminate transmissible epidemic strains from sporadic strains [38–41]. Currently available infection control guidelines for CF will need to be modified to deal with the existence of epidemic transmissible *P. aeruginosa* strains. The UK CF Research Trust recommends segregation of all patients with chronic *P. aeruginosa* infection from remaining patients [50]. However, this policy is unnecessary if patients are infected with unique sporadic strains, which by definition are not transmissible. On the

Table 2. Sources and implications of genetically identical *Pseudomonas* strains

Source	Implications
Patient-to-patient transmission (within or outside center, excluding CF siblings)	Cohort segregation based upon bacterial genotyping and sputum culture to prevent further spread Ongoing molecular surveillance of all *P. aeruginosa* isolates Emphasize importance of universal precautions (particularly hand washing) Discourage social contact inside and outside hospital
Patient acquisition from a common environmental source (within or outside center)	Eliminate environmental source Emphasize importance of universal precautions (particularly hand washing)
Acquisition from staff/equipment	Reinforce universal precautions, particularly hand washing Review disinfection/cleaning protocols for medical equipment Consider ongoing molecular surveillance
Acquisition from independent sources (within or outside center)	Segregate source from all CF patients, emphasize universal precautions
Not really identical	Discourage use of phenotypic characteristics for strain identification Molecular typing of all *P. aeruginosa* isolates

other hand, if an epidemic strain is present, the grouping of all *P. aeruginosa*-infected patients together will encourage further spread of the organism as well as increasing the risk of superinfection [51]. Similarly, recent infection control guidelines from the United States Cystic Fibrosis Foundation (CFF) [52], which emphasize segregation of patients with multiresistant *P. aeruginosa* from other CF patients, are not likely to prevent spread of transmissible strains. There have been no recent molecular epidemiological studies of *P. aeruginosa* infection in US CF centers, which may explain why the CFF Guidelines currently recommend molecular surveillance only for *Burkholderia cepacia* complex infection.

The implications of detection of a transmissible strain of *P. aeruginosa* in the CF clinic differ depending upon the source of the strain and the way in which it is identified (table 2). These recommendations again highlight the importance of universal precautions (and in some situations cohort segregation) to prevent transmission, and the use of molecular rather than phenotypic typing to detect and monitor the presence of transmissible strains. Therefore, if transmissible strains are to be detected, molecular surveillance must be adopted more widely amongst CF centers as an essential component of infection control. Using these methods, a transmissible strain originally identified in Melbourne and Sydney was also found in Brisbane [43], and the Liverpool epidemic strain has been shown to be widespread in CF clinics throughout the UK [53]. However, molecular surveillance will add considerably to the costs of patient care, and cohort segregation presents numerous challenges, particularly in large clinics. Although some transmissible strains also appear to show increased virulence [41, 47], more research needs to be undertaken to better understand the relationship between transmissibility, virulence and the effects on morbidity and mortality [54, 55]. There is a risk of 'stigmatization by sputum culture', although one study has shown broad acceptance of cohort segregation by CF children and parents [56]. Therefore, universal precautions, particularly hand washing, will remain the foundation of infection control policy in all CF centers. However, if molecular studies of *P. aeruginosa* indicate the presence of a transmissible strain, cohort segregation and continued surveillance will be required to prevent further spread of clonal strains.

References

1 Frederiksen B, Koch C, Hoiby N: Changing epidemiology of *Pseudomonas aeruginosa* infection in Danish Cystic Fibrosis patients (1974–1995). Pediatr Pulmonol 1999;28: 159–166.

2 Emerson J, Rosenfeld M, McNamara S, Ramsey B, Gibson RL: *Pseudomonas aeruginosa* and other predictors of mortality and morbidity in young children with cystic fibrosis. Pediatr Pulmonol 2002;34:91–100.

3 Burns JL, Gibson RL, McNamara S, Yim D, Emerson J, Rosenfeld M, Hiatt P, McCoy K, Castile R, Smith AL, Ramsey B: Longitudinal assessment of *Pseudomonas aeruginosa* in young children with cystic fibrosis. J Infect Dis 2001;183:444–452.

4 Frederiksen B, Koch C, Hoiby N: Antibiotic treatment of initial colonization with *Pseudomonas aeruginosa* postpones chronic infection and prevents deterioration of pulmonary function in cystic fibrosis. Pediatr Pulmonol 1997;23:330–335.

5 Nixon GM, Armstrong DS, Carzino R, Carlin JB, Olinsky A, Robertson CF, Grimwood K: Clinical outcome after early *Pseudomonas aeruginosa* infection in cystic fibrosis. J Pediatr 2001;138:699–704.

6 Gibson RL, Emerson J, McNamara S, Burns JL, Rosenfeld M, Yunker A, Hamblett N, Accurso F, Dovey M, Hiatt P, Konstan MW, Moss R, Retsch-Bogart G, Wagener J, Waltz D, Wilmott R, Zeitlin PL, Ramsey B: Significant microbiological effect of inhaled tobramycin in young children with cystic fibrosis. Am J Respir Crit Care Med 2003; 167:841–849.

7 Pedersen SS, Hoiby N, Espersen F, Koch C: Role of alginate in infection with *Pseudomonas aeruginosa* in cystic fibrosis. Thorax 1992;47:6–13.

8 Navarro J, Rainisio M, Harms HK, Hodson ME, Koch C, Mastella G, Strandvik B, McKenzie SG: Factors associated with poor pulmonary function: Cross-sectional analysis of data from the ECRF. Eur Respir J 2001;18:298–305.

9 Henry RL, Mellis CM, Petrovic L: Mucoid *Pseudomonas aeruginosa* is a marker of poor survival in cystic fibrosis. Pediatr Pulmonol 1992;12:158–161.

10 De Boeck K, Alifier M, Vandeputte S: Sputum induction in young cystic fibrosis patients. Eur Respir J 2000;16:91–94.

11 Sagel SD, Sontag MK, Wagener JS, Kapsner RK, Osberg I, Accurso FJ: Induced sputum inflammatory measures correlate with lung function in children with cystic fibrosis. J Pediatr 2002;141:811–817.

12 Suri R, Marshall LJ, Wallis C, Metcalfe C, Shute JK, Bush A: Safety and use of sputum induction in children with cystic fibrosis. Pediatr Pulmonol 2003;35:309–313.

13 Kerem E, Corey M, Stein R, Gold R, Levison H: Risk factors for *Pseudomonas aeruginosa* colonization in cystic fibrosis patients. Pediatr Infect Dis J 1990;9:494–498.

14 Abman SH, Ogle JW, Harbeck RJ, Butler-Simon N, Hammond KB, Accurso FJ: Early bacteriologic, immunologic and clinical courses of young infants with cystic fibrosis identified by neonatal screening. J Pediatr 1991;119:211–217.

15 Faraguna D, Giglio L, D'Orazio C, Pederzini F, de Vonderweid U, Mastella G: Is clinical status at diagnosis a prognostic factor in CF infants identified by neonatal screening? Pediatr Pulmonol Suppl 1991;7:46–51.

16 Silverman M, Stratton D, Diallo A, Egler LJ: Diagnosis of acute bacterial pneumonia in Nigerian children. Arch Dis Child 1977;52:925–931.

17 Woods DE, Bass JA, Johanson WG, Straus DC: Role of adherence in the pathogenesis of *Pseudomonas aeruginosa* lung infection in cystic fibrosis patients. Infect Immun 1980;30:694–699.

18 Ryan G, Mukhopadhyay S, Singh M: Nebulised anti-pseudomonal antibiotics for cystic fibrosis. Cochrane Database Syst Rev 2003;3:CD001021.

19 Stutman HR, Lieberman JM, Nussbaum E, Marks MI: Antibiotic prophylaxis in infants and young children with cystic fibrosis: A randomized controlled trial. J Pediatr 2002;140:299–305.

20 West SHE, Zeng L, Lee BL, Kosorok MR, Laxova BS, Rock MJ, Spalingard MJ, Farrell PM: Respiratory infections with *Pseudomonas aeruginosa* in children with cystic fibrosis: Early detection by serology and assessment of risk factors. JAMA 2002;287:2958–2967.

21 Armstrong DS, Grimwood K, Carzino R, Carlin JB, Olinsky A, Phelan PD: Lower respiratory infection and inflammation in newly diagnosed cystic fibrosis infants. BMJ 1995;310:1571–1572.

22 Armstrong DS, Grimwood K, Carlin JB, Carzino R, Olinsky A, Phelan PD: Broncho-alveolar lavage or oropharyngeal cultures to identify lower respiratory pathogens in infants with cystic fibrosis. Pediatr Pulmonol 1996;21:267–275.

23 Ramsey BW, Wentz KR, Smith AL, Richardson M, Williams-Warren J, Hedges DL, Gibson R, Redding GJ, Lent K, Harris K: Predictive value of oropharyngeal cultures for identifying lower airway bacteria in cystic fibrosis patients. Am Rev Respir Dis 1991;144:331–337.

24 Wood RE: Treatment of cystic fibrosis lung disease in the first two years. Pediatr Pulmonol Suppl 1989;4:68–70.

25 Rosenfeld M, Emerson J, Accurso F, Armstrong D, Castile R, Grimwood K, Hiatt P, McCoy K, McNamara S, Ramsey B, Wagener J: Diagnostic accuracy of oropharyngeal cultures in infants and young children with cystic fibrosis. Pediatr Pulmonol 1999;28:321–328.

26 Wainwright C, Carlin JB, Cooper C, Byrnes C, Whitehead B, Martin J, Cooper D, Armstrong D, Robertson C: Early infection with *Pseudomonas aeruginosa* can be cleared in young children with cystic fibrosis. Pediatr Pulmonol Suppl 2002;24:300.

27 Khan TZ, Wagener JS, Bost T, Martinez J, Accurso FJ, Riches DWH: Early pulmonary inflammation in infants with cystic fibrosis. Am J Respir Crit Care Med 1995;151:1075–1082.

28 Grouthes D, Kopman U, von der Hardt H, Tummler B: Genome fingerprinting of *Pseudomonas aeruginosa* indicates colonization of cystic fibrosis siblings with closely related strains. J Clin Microbiol 1988;26:1973–1977.

29 Wolz C, Kiosz G, Ogle JW, Vasil ML, Schaad U, Botzenhart K, Doring G: *Pseudomonas aeruginosa* cross-colonization and persistence in patients with cystic fibrosis: Use of a DNA probe. Epidemiol Infect 1989;102:205–214.

30 Speert DP, Campbell ME, Henry DA, Milner R, Taha F, Gravelle A, Davidson AGF, Wong LTK, Mahenthiralingam E: Epidemiology of *Pseudomonas aeruginosa* in cystic fibrosis in British Columbia, Canada. Am J Respir Crit Care Med 2002;166:988–999.

31 Ogle JW, Janda JM, Woods DE, Vasil ML: Characterization and use of a DNA probe as an epidemiological marker for *Pseudomonas aeruginosa*. J Infect Dis 1987;155:119–126.

32 Speert DP, Campbell ME, Farmer SW, Volpel K, Joffe AM, Paranchych W: Use of a pilin gene probe to study molecular epidemiology of *Pseudomonas aeruginosa*. J Clin Microbiol 1989;27:2589–2593.

33 Maselli JH, Sontag MK, Norris JM, MacKenzie T, Wagener JS, Accurso FJ: Risk factors for initial acquisition of *Pseudomonas aeruginosa* in children with cystic fibrosis identified by newborn screening. Pediatr Pulmonol 2003;35:257–262.

34 Mahadeva R, Webb K, Westerbeek RC, Carroll NR, Dodd ME, Bilton D, Lomas DA: Clinical outcome in relation to care in centers specializing in cystic fibrosis: Cross-sectional study. BMJ 1998;316:1771–1775.

35 Armstrong D, Grimwood K, Carlin JB, Carzino R, Hull J, Olinsky A, Phelan PD: Severe viral respiratory infections in infants with cystic fibrosis. Pediatr Pulmonol 1998;26:371–379.

36 Farrell PM, Shen G, Splaingard M, Colby CE, Laxova A, Kosorok MR, Rock MJ, Mischler EH: Acquisition of *Pseudomonas aeruginosa* in children with cystic fibrosis. Pediatrics 1997;100:E2.

37 Saiman L, Corey M: Further insights into early acquisition of *Pseudomonas aeruginosa*. Pediatr Pulmonol 1998;26:79–80.

38 Chen K, Smyth RL, Govan JR, Doherty C, Winstanley C, Denning N, Heaf DP, van Saene H, Hart CA: Spread of a β-lactam-resistant *Pseudomonas aeruginosa* in a cystic fibrosis clinic. Lancet 1996;348:639–642.

39 Jones AM, Govan JR, Doherty CJ, Dodd ME, Isalaka BJ, Stanbridge TN, Webb AK: Spread of a multiresistant strain of *Pseudomonas aeruginosa* in an adult cystic fibrosis clinic. Lancet 2001;358:557–558.

40 Denton M, Kerr K, Mooney L, Keer V, Rajgopal A, Brownlee K, Arundel P, Conway S: Transmission of a colistin-resistant *Pseudomonas aeruginosa* between patients attending a pediatric cystic fibrosis center. Pediatr Pulmonol 2002;34:257–261.

41 Armstrong DS, Nixon GM, Carzino R, Bigham A, Carlin JB, Robins-Browne R, Grimwood K: Detection of a widespread clone of *Pseudomonas aeruginosa* in a pediatric cystic fibrosis clinic. Am J Respir Crit Care Med 2002;166:983–987.

42 Speert DP, Campbell ME, Henry DA, Milner R, Taha F, Gravelle A, Davidson AGF, Kwong LTK, Mahenthiralingham E: Epidemiology of *Pseudomonas aeruginosa* in cystic fibrosis in British Columbia, Canada. Am J Respir Crit Care Med 2002;166:988–993.

43 Armstrong D, Bell S, Robinson M, Bye P, Rose B, Harbour C, Lee C, Sevice H, Nissen M, Syrmis M, Wainwright C: Evidence for spread of a clonal strain of *Pseudomonas aeruginosa* among cystic fibrosis clinics. J Clin Microbiol 2003;41:2266–2267.

44 Robinson P, Carzino R, Armstrong D, Olinsky A: *Pseudomonas* cross-infection from cystic fibrosis patients to non-cystic fibrosis patients: Implications for inpatient care of respiratory patients. J Clin Microbiol 2003;41:5741.

45 Griffiths AL, Jamsen K, Carlin JB, Grimwood K, Carzino R, Robinson PJ, Massie J, Armstrong DS: Effects of segregation upon an epidemic *Pseudomonas aeruginosa* strain in a cystic fibrosis clinic. Am J Respir Crit Care Med 2005;171:1020–1025.

46 Jones AM, Govan JRW, Doherty CJ, Dodd ME, Isalska BJ, Stanbridge TN, Webb AK: Identification of airborne transmission of epidemic strains of *Pseudomonas aeruginosa* at a CF center during a cross infection outbreak. Thorax 2003;58:525–527.

47 Jones AM, Dodd ME, Doherty CJ, Govan JRW, Webb AK: Increased treatment requirements of patients with cystic fibrosis who harbour a highly transmissible strain of *Pseudomonas aeruginosa*. Thorax 2002;57:924–925.

48 Armstrong DS, Grimwood K, Carlin JB, Carzino R, Gutierrez JP, Hull J, Olinsky A, Phelan EM, Robertson CF, Phelan PD: Lower airway inflammation in infants and young children with cystic fibrosis. Am J Respir Crit Care Med 1997;156:1197–1204.

49 Davies G, McShane D, Davies JC, Bush A: Multiresistant *Pseudomonas aeruginosa* in a pediatric cystic fibrosis center: Natural history and implications for segregation. Pediatr Pulmonol 2003;35:253–256.

50 Cystic Fibrosis Trust Control of Infection Group: *Pseudomonas aeruginosa*: Infection in People with Cystic Fibrosis: Suggestions for Prevention and Infection Control. Kent, Cystic Fibrosis Trust, 2001.

51 McCallum SJ, Corkill J, Gallagher M, Ledson MJ, Hart CA, Walshaw MJ: Superinfection with a transmissible strain of *Pseudomonas aeruginosa* in adults with cystic fibrosis chronically colonized by *P. aeruginosa*. Lancet 2001;358:558–560.

52 Saiman L, Siegel J: Infection control recommendations for patients with cystic fibrosis: Microbiology, important pathogens, and infection control practices to prevent patient-to-patient transmission. Am J Infect Control 2003;31:S9–S62.

53 Scott F, Pitt L: Identification and characterization of transmissible *Pseudomonas aeruginosa* strains in UK cystic fibrosis centers. J Cystic Fibrosis 2003;2(suppl 1):S39.

54 Geddes DM: Of isolates and isolation: *Pseudomonas aeruginosa* in adults with cystic fibrosis. Lancet 2001;358:522–523.

55 Ramsey BW: To cohort or not to cohort: How transmissible is *Pseudomonas aeruginosa*? Am J Respir Crit Care Med 2002;166: 906–907.

56 Griffiths AL, Armstrong D, Carzino R, Robinson P: Cystic fibrosis patients and families support cross-infection measures. Eur Respir J 2004;24:449–452.

David Armstrong
Department of Paediatrics
Monash Medical Center
Locked Bag 29
Clayton, Vic. 3168 (Australia)
Tel. +613 9594 2900, Fax +613 9594 6311
E-Mail
david.armstrong@southernhealth.org.au

Bush A, Alton EWFW, Davies JC, Griesenbach U, Jaffe A (eds): Cystic Fibrosis in the 21st Century.
Prog Respir Res. Basel, Karger, 2006, vol 34, pp 138–144

Pseudomonas aeruginosa: Basic Research

F.W. Scott   T.L. Pitt

Laboratory of HealthCare Associated Infection, Centre for Infections, Health Protection Agency,
London, UK

Abstract

Recent research has shed much light on the mechanism of attachment of *Pseudomonas aeruginosa* in the cystic fibrosis (CF) lung and identified the pivotal role of surface adhesion structures such as type IV pili and flagella and their involvement in the establishment of the biofilm mode of growth. Comparative genomic studies utilizing microarray technology have informed the basis of genetic variation of isolates in infections and the environment and identified genes differentially expressed in planktonic and biofilm-grown cells. Four distinct genomic islands and two pathogenicity islands of varying size contain numerous open reading frames, many of unknown function, and further work is necessary to determine their significance to the infectious process. The recent finding of hypermutating *P. aeruginosa* strains in CF and their relative absence from isolates from other infections suggest that this mechanism may select populations with increased genetic fitness and promote their survival and proliferation in the lung.

Introduction

Pseudomonas aeruginosa is the most prevalent infection of the lungs of cystic fibrosis (CF) patients. Its presence is linked with a poorer prognosis and thus much basic research in recent years has focussed on the mechanisms by which the organism is acquired, how it avoids clearance by the host, and how it progresses to establish permanent populations in the chronically infected state.

It is widely held that CF patients are first colonized by *P. aeruginosa* organisms from the natural environment. As non-CF and CF patients are similarly exposed to this environment it follows that CF subjects are in some way predisposed to colonization. Several suggestions have been made to explain this association include increased binding of *P. aeruginosa* to airway epithelia [1], failure of epithelial cells to internalize and kill the bacteria [2], inactivation of natural antibacterial cationic peptides (defensins) by the high salt content of airway secretions [3], disruption of the ciliary escalator by pyocyanin, the phenazine pigment characteristic of *P. aeruginosa* [4], and reduction of the volume of airway surface fluid leading to reduced mucus clearance and stagnation of secretions on the airway surface with the proliferation of microbes in general [5]. Many of these features have been extensively explored, but it is worthwhile highlighting some of the more recent observations to impact in this area.

Attachment of *P. aeruginosa*

Type IV Pili and Other Fimbriae

Pili are hair-like proteins extending from the surface of the bacterium into its environment that mediate attachment of the bacterium to a host cell. The bacterium is able to retract the pili into the cell thus drawing the bacterium into close contact with the host cell; they also promote the phenomenon of 'twitching motility'. The pili of *P. aeruginosa* are members of a class of polar type-4 pili composed of subunits of pilin protein [6]. The N-terminal amino acids of the pilin are conserved but the C terminus displays amino acid sequence diversity [7]. The pili bind to the eukaryotic glycolipid receptor, asialoGM$_1$[1]. A single structural gene *pilA* is responsible for pilin synthesis in *P. aeruginosa* and so diversity in pilin sequences probably occurs through random mutation within a strain or by horizontal recombination with pilin genes from other species. Two pilin groups (I and II) can be distinguished by DNA sequence analysis [8] and a third pilin sequence with partial amino acid homology to the N-terminal sequence of group I pilins has recently been identified in the highly virulent strain PA14 [9]. Post-translational modification of a group I pilin by *O*-glycosylation has been reported and this modification is encoded by *pilO* [renamed *tfpO*, see ref.10] which is part of an operon containing *pilA*. The glycan was shown to have antigenic homology with lipopolysaccharide (LPS) of serotype O7 *P. aeruginosa* as antibodies to the modified pilin cross-reacted with purified LPS from this strain [11]. The glycan (composed of 3 sugars and serine) is located close to the pilin disulphide loop which is a primary attachment site for the pilus and functions as a major B-cell epitope [11].

The potential of a vaccine based on type IV pili would depend greatly on the diversity of pilin antigens. This prompted Kus et al. [10] to investigate the distribution of pilin alleles in a large series of environmental and clinical (including CF) isolates. They found 5 distinct phylogenetic groups of alleles and each allele was stringently associated with characteristic distinct accessory genes thus facilitating the specific identification of the allele by PCR. Group I pilin genes were identified in almost half of all isolates and were significantly more frequent among CF isolates than non-CF (81% versus 61%). This perhaps indicates that group I pilins, which can be modified by glycosylation, may confer an acquisition or persistence advantage for *P. aeruginosa* in the CF host. Interestingly this allele was detected in over two-thirds of paediatric CF isolates suggesting that this feature may be important in early infection.

It is clear that fimbrial adhesins distinct from type IV pili are also involved in adherence of *P. aeruginosa* to cell surfaces and the establishment of biofilm. Many gram-negative bacteria form fimbriae on their surfaces using a system called the 'chaperone/usher pathway (cup). Chaperone proteins similar to immunoglobulins in structure escort intracellular subunits of macromolecules to an usher protein complex which facilitates their translocation and assembly of these subunits across the outer membrane [12]. At least three cup systems (CupA, CupB and CupC) have been identified in the genome of the *P. aeruginosa* reference strain PAO1 and a cluster of genes in the *cupA* cluster has been implicated in the development of biofilm formation in an experimental strain deficient in structural pili [13]. The *cupA* cluster, but not the *cupB* or *cupC* cluster, was required for biofilm formation on abiotic surfaces and *cupA* was negatively regulated by the gene *mvaT* (originally identified as a regulator of mevalonate catabolism in another species of pseudomonad) which is likely part of a complex regulatory network involved in biofilm formation [14].

Flagella and Adhesion

P. aeruginosa binds not only to eukaryotic cell surfaces but also to respiratory mucin. An association between mucin binding and the expression of some flagellar genes was clearly demonstrated by Simpson et al. [15], who differentiated between the genes controlling attachment of *P. aeruginosa* to epithelial cells and mucin. Mutants defective in two flagellar genes, *fliF* and *fliO*, proved to be non-motile and non-adhesive while a *fliC* mutant which is non-motile and does not make flagellin continued to be adherent to mucin. A specific protein FliD on the tip of the flagellar filament was subsequently identified to be responsible for the adhesion to mucin and a mutation in the *fliD* gene abolished both motility and adhesion; this was termed the 'flagellar cap protein' [16]. Sequencing of the *fliD* region of eight laboratory and clinical strains of *P. aeruginosa* revealed two distinct types of FliD protein, named A and B. The A type was characteristic of strain PAK a widely used reference strain producing pili and type B was found in strain PAO1. The two types were antigenically distinct and Western blotting analysis of a large number of isolates confirmed that all had either A or B FliD flagellin proteins [17].

Biofilm Formation

Although a strict aerobe, *P. aeruginosa* is able to grow in anaerobic conditions owing to its ability to cleave oxygen from nitrate due to the production of a nitrate reductase. In the CF lung the organisms are often located within hypoxic mucopurulent masses in airway lumens rather than the epithelial cell surface and excess alginate polysaccharide is formed as a response to this hypoxic stress leading to the establishment of biofilm containing coalesced microcolonies [18]. However, reduced oxygen tension is not the only factor leading to hyperproduction of alginate as Mathee et al. [19] showed that polymorphonuclear leukocytes and their oxygen radicals (H_2O_2) can stimulate the conversion of non-mucoid *P. aeruginosa* to the mucoid form. The switch to mucoidy is thought to be a pivotal event in the maturation of the biofilm. Motile *P. aeruginosa* attach and spread across a surface by type IV pili-driven twitching motility. Following aggregation and microcolony formation, the biofilm develops into mushroom-shaped or flat structures, and this is dependent on the energy and carbon source [20]. The cells in the stalk of the mushroom are apparently stationary but some cells progress up the stalk by twitching motility to form a cap on the mushroom. Quorum sensing (QS) mutants do not form stalks readily and grow as flat layers on the surface. The synthesis of rhamnolipids which have surfactant properties and are thought to be important in the development of the mushroom structures is controlled by QS mechanisms. Recent work suggests that rhamnolipid synthesis occurs after the stalks are formed but prior to the formation of the cap of the mushroom structure [21].

P. aeruginosa cells communicate with each other by at least two QS systems *(las* and *rhl)* involving the production of small diffusible signal molecules, acyl-homoserine lactones (HSLs). These HSLs are produced during lung infection and mutations in either of these systems resulted in reduced virulence in an experimental strain in a mouse lung infection model [22, 23]. These systems are clearly functional in CF patients who are chronically infected with *P. aeruginosa* as gene transcripts accumulate in the sputum of these patients [24]. A survey of 200 clinical isolates of *P. aeruginosa* including a selection from CF patients showed that only five were deficient in lacking LasA and LasB activities and produced either none or very low levels of HSLs [25].

Genomics and Proteomics

Genetic analysis of *P. aeruginosa* has progressed significantly in the last few decades, developing from determination of the sequence and function of single genes to the elucidation and publication of the entire genome sequence of two strains PAO1 [26] and PA14 (GenBank Accession No. AABQ00000000). At 6.3 Mb, the PAO1 genome is one of the largest bacterial genomes to be sequenced to date; 5,570 ORFs were predicted from the sequence the majority of which have a high G + C content. Some regions, however, have a significantly lower G + C content, which may be due to the acquisition of these regions by recent horizontal transfer from other organisms. The *Pseudomonas* genome project (www.pseudomonas.com) is a multinational effort to complete the annotation of the genome sequence. When the sequence was first published in 2001, 54% of the genes were assigned to a functional class, nearly 7% of which were associated with LPS biosynthesis, virulence factors, motility and adherence. The PA14 sequence

is still being edited, but the initial results show that this is slightly larger than PAO1 at 6.5 Mb.

Genome Sequence Comparison

A number of molecular methods have been successfully applied to characterize *P. aeruginosa* and extend our understanding of its pathogenicity and virulence factors. Choi et al. [9] used representational difference analysis, involving subtractive hybridization and kinetic enrichment to identify genomic differences between PAO1 and PA14 and to identify which differences corresponded to increased virulence. They identified twenty PA14 genes that were either missing from PAO1 or substantially different in sequence, 14 of those were homologous to known genes while five had no significant homology to known genes. This approach was complemented by Spencer et al. [27] who used whole genome sequence sampling to compare the genomes of sequential late stage clinical *P. aeruginosa* isolates from CF patients, an environmental isolate and PAO1. They found that, despite a conserved genetic backbone, two segments from PAO1, one a 17-kb region, the other corresponding to the O-antigen biosynthetic locus region, were missing from the clinical isolates. They also demonstrated that most of the variation observed between sequences was due to hypervariable regions at specific loci, including pyoverdine recognition and synthesis, as well as genes associated with flagellin biosynthesis and the presence of strain-specific genomic islands.

The elucidation of whole genome sequence data allows for the development of powerful tools to further characterize isolates and their cellular mechanisms. For example the PAO1 sequence was used to develop a genome wide, random insertion transposon mutant library. This tool will allow researchers to identify which genes are necessary to genetic pathways by knocking out individual genes in turn [28].

Microarrays

A number of *P. aeruginosa* microarrays are now available, some, based on the sequence of PAO1, can be used for genomic and comparative expression profiling, others incorporate known virulence determinants and can be used for molecular typing and genomic comparisons [for a review of microarrays, see ref. 29].

Ernst et al. [30] used microarray analysis to look for genomic variation in *P. aeruginosa* isolates from paediatric CF patients and the environment. They showed that 38 gene islands including those encoding O-antigen biosynthetic loci, DNA polymerase II and type IV pili were absent or divergent between the genomes. Some of these islands had a relatively low G + C content, suggesting that they were sites of recent horizontal transfer in PAO1. These islands are nonetheless not CF specific and are also found in the environmental isolates. Moreover, *P. aeruginosa* isolates from CF patients undergo frequent large chromosomal inversions that may aid the bacterium in evading the host immune system [31].

These studies highlighted above demonstrate that whilst *P. aeruginosa* retains a common genetic backbone the genome evolves quickly, is highly plastic and readily acquires new genetic elements by horizontal transfer. Such plasticity appears to be highly beneficial to the organism in establishing itself in the CF lung and then persisting long term despite the hostile nature of the environment. For example, it has been repeatedly shown that genes associated with O-antigen biosynthesis are commonly subject to mutation, an event which may help the organism evade the host immune system.

Microarray analysis has been used to compare the expression of genes between *P. aeruginosa* growing in planktonic form and in a biofilm; only 1% of genes are differentially regulated between the two growth modes [32]. Firoved and Deretic [33] applied microarray analysis to look for genes that were co-induced, upon conversion to the mucoid phenotype, with the AlgU sigmulon (AlgU-dependent promoters). Co-induction of a specific subset of known virulence determinants was demonstrated. Changing levels of expression of genes after incubation of *P. aeruginosa* PAO1 with primary normal human epithelial cells has been demonstrated using microarrays [34]. After 4 h of incubation, 41 genes were differentially expressed, up-regulated genes included putative proteins involved in membrane transport, and probable transcriptional regulators. Down-regulated genes were associated with iron acquisition by pyoverdine and a pyochelin biosynthesis protein. After 12 h of incubation, differential expression was observed in 121 genes, including significant activation of *plc*N, a non-haemolytic phospholipase C which degrades phosphatidyl-choline a constituent of lung surfactant. When *P. aeruginosa* converts to a mucoid phenotype there is also a substantial and preferential induction of genes encoding bacterial lipoproteins that can contribute to the induction of inflammation in the CF lung [35].

A microarray has been used to determine levels of transcription of genes associated with QS and shown that its regulation is more complex than originally thought and involves a hierarchical cascade of regulation [36].

Genomic Islands

The *P. aeruginosa* genome is organized into areas of core sequences and genomic islands which may differ significantly in their G + C content. Four genomic islands (PAGI-1–4) have been described in *P. aeruginosa*. PAGI-1 is a 48.9-kb island found in 85% of isolates from sepsis and urinary tract infections [37]. PAGI-2(C) and PAGI-3 (SG) were identified in two epidemiologically unrelated clone C strains [38]. These islands consist of approximately 7-kb that are identical in both strains and unique DNA of approximately 105 kb and 103 kb in size, respectively. PAGI-2(C) contains 111 ORFs, whereas PAGI-3(SG) has 106 ORFs of which more than 60% of the genes are either conserved hypothetical genes of unknown function or have no apparent homology to any reported sequences. In both strains these hypothetical ORFs are clustered in gene islands, which are partitioned into two blocks: the first block consists of genes which are specific to each strain, the function of which could be predicted. The second block consists mainly of hypothetical ORFs of which 47 have homologues in both gene islands. The cargo genes may confer the strains with additional metabolic features and transport resistance capabilities, e.g. the expression of genes for cytochrome c biogenesis encoded by PAGI-2(C) could facilitate iron uptake and inactivation of peroxides and thus may confer an advantage for the bacteria to persist in the CF lung where they are exposed to oxidative stress and iron limitation. PAGI-4 is an approximately 16 kb DNA segment in PAK, which carries genes for the glycosylation of α-flagellin among others. Two other islands not integrated into the tRNA gene have been identified [39].

Two *P. aeruginosa* pathogenicity islands (PAPI-1 and PAPI-2) are carried by PA14, a highly virulent mouse strain. PAPI-1 is a 108-kb island, not found in PAO1, which is inserted in a hypervariable region of the PA14 genome near the *pil*A gene. PAPI-1 is associated with tRNA genes, has seven genes linked with mobility encoding integrases and tranposases and carries 115 predicted ORFs of which 75 are unrelated to any previously identified proteins or functional domains and at least 19 encode virulence factors. Paradoxically, over 80% of the sequence of PAPI-1 is unique; however, the translated sequences of 40 ORFs showed homology to proteins from a number of bacterial species. PAPI-2 is smaller than PAPI-1, 11 kb; however, a small part overlaps with PAO1. PAPI-2 is also located in a hypervariable region of the genome near the genes encoding pyocyanin and it too contains several predicted mobility genes including an integrase, transposases and a nearly complete IS222 element. PAPI-2 consists of 15 predicted ORFs, seven of which are predicted to encode hypothetical proteins of unknown function; half of PAPI-2 is homologous to PA0977–0987 a PAO1 genomic island [40].

Proteomics

Proteomics involves separating; characterising and identifying proteins associated with a cell, there are many different methods which are reviewed by Washburn and Yates [41]. Proteomic analysis has been applied to the characterization of *P. aeruginosa* isolates to extend understanding of the cellular processes associated with virulence and pathogenicity. Analysis of genetically identical mucoid and non-mucoid CF isolates showed differences in the outer-membrane proteins expressed by these organisms [42]. Wehmhöner et al. [43] used proteome analysis to determine the level of heterogeneity amongst clinical isolates of *P. aeruginosa* compared to PAO1. Two-dimensional gel analysis showed similar patterns between the cellular extracts, suggesting they share a core genome; however, the patterns of the secretome (extracellular subproteome) showed obvious differences between the isolates. They also found that the profile of the cellular extract did not vary significantly in relation to the incubation time, but the number of proteins in the secretome increased significantly as the incubation period increased. Further comparisons were made using three isogenic morphotypes of *P. aeruginosa* with identical genomic fingerprints. This again showed similar patterns between the cellular extracts and marked differences in the secretome, suggesting that this variation is due to differential regulation of protein expression. One morphotype investigated was hyperpiliated, this isolate differentially regulated 45 proteins including proteins associated with the type III secretion system and proteins known to be modulated by QS.

Hypermutation

The basic Darwinian tenet holds that in evolution genetically fit variants are selected by mutation and or recombination and that these survive better in a changing and hostile environment than their static counterparts. Bacteria that are able to adapt to changes in their environment may survive and persist in the human host better than those unable to do so. By evolving mechanisms to produce high mutation rates in specific regions of their genomes that result in rapid generation of variants, they maximize the diversity in their populations, thus increasing the likelihood of survival and persistence. These 'hypermutators' may have fitness benefits for colonizing specific environmental or anatomic niches [44].

Hypermutation may arise through two basic mechanisms, (1) mutation localized to contingency loci or (2) generalized hypermutation involving those strains that are intrinsically hypermutable.

The best characterized mechanism of hypermutation involves mutation in the mechanisms for DNA replication and repair. Disruption of DNA replication increases the mutation rate by elevating the replication error rate. However, disruption of DNA repair is more complex as bacteria possess a number of intrinsic mechanisms to detect and repair errors in DNA including, nucleotide excision, base excision and methyl-directed mismatch repair (MMR) [45, 46]. MMR is a bidirectional system that repairs cellular DNA, protects against replication errors and acts as a barrier to recombination between divergent chromosomes (homologous recombination). In *Escherichia coli,* MMR consists of three main components: MutS, MutL and MutH [for details, see ref. 46, 47].

Mutations in the *mut*S locus, in particular, are common in isolates with a hypermutator phenotype [48, 49]. Mutations in the MMR loci can result in MMR-negative variants, a hypermutator phenotype in which homologous recombination occurs 1,000-fold more frequently than in wild-type bacteria [50]. Although hypermutators may have an advantage under certain conditions, as they can quickly obtain beneficial adaptive mutations, the ongoing accumulation of mutations will eventually lead to a loss of fitness and thus a survival disadvantage. One hypothesis that explains this paradox is that MMR-negative bacteria are able to restore MMR function by recombination with wild-type organisms whilst retaining mutations acquired during the hypermutator period. This also helps to explain why it is mutations in *mut*S and other loci related to MMR function rather than other genes involved in DNA repair that are associated with the hypermutator phenotype [51, 52].

Hypermutation in Cystic Fibrosis P. Aeruginosa *Isolates*

Hypermutation has been demonstrated in isolates from CF patients. A recent study in Spain showed that approximately 36% of CF patients were colonized by hypermutable *P. aeruginosa* [49]. DNA fingerprinting of these strains showed that patients were colonized by one or a few dominant but distinct lineages and levels of antibiotic resistance were higher in the hypermutator strains. PCR and sequencing analysis showed a number of mutations in the *mut*S gene including an approximately 1.5 kb deletion, *mut*L and *uvr*D loci that resulted in frameshifts, insertions, deletions and substitutions in the amino acid sequence of the translated proteins. Complementation of these strains with the respective wild-type genes restored the mutation frequencies to the level of PAO1. Four isolates failed to amplify the *mut*Y gene (involved in oxidative damage repair). Further complementation studies with *mut*S demonstrated that a 54-bp deletion between 8-bp repeat regions was responsible for the increased mutation frequency observed. However, there may also be an interaction between the deleted region and a 3.3-kb inserted sequence [53]. Disruption of *mut*S has also been shown to drive the emergence of diverse colony morphotypes, compared to the PAO1 wild type, these *mut*S morphotypes showed enhanced motility on swarming plates, increased pyocyanin and pyoverdine production but impaired elastolytic activity [54].

Screening for hypermutable *P. aeruginosa* strains is performed by plating out successive serial dilutions of a cell suspension on media containing either 500 μg/ml streptomycin or 300 μg/ml rifampicin. These antibiotics are commonly chosen as resistance may arise through a single step mutation. The number of colonies growing in the presence of high levels of antibiotics can then be compared with the number of colonies grown on non-selective agar and the number of hypermutable colonies calculated. The minimum inhibitory concentration of these antibiotics is also determined to confirm that these strains are truly hypermutable and not simply highly resistant to that agent. Recently, simpler screening methods have been developed for a number of antibiotics using disk diffusion assays and Etests [55, 56], this should allow for more rapid screening for hypermutability.

The nature and treatment of CF lung disease may promote the development of hypermutable strains. Patients colonized by *P. aeruginosa* receive frequent courses of inhaled antibiotics; however, it is widely accepted that it is not possible to achieve uniform distribution of the drug in the lungs. Therefore regions exposed to sub-therapeutic/inhibitory concentrations may become hotspots for the development of hypermutable strains. It has been suggested, given the high numbers of bacteria present, the likely presence of hypermutable strains and the inherent propensity for *P. aeruginosa* to adapt to its environment, that mutant isolates which are resistant to most antipseudomonal agents should be anticipated to be present in high numbers in CF patients with chronic infections prior to treatment [49, 57].

References

1 Saiman L, Prince A: *Pseudomonas aeruginosa* pili bind to asialoGM1 which is increased on the surface of cystic fibrosis epithelial cells. J Clin Invest 1993;92: 1875–1880.

2 Pier GB, Grout M, Zaidi TS, Goldberg JB: How mutant CFTR may contribute to *Pseudomonas aeruginosa* infection in cystic fibrosis. Am J Respir Crit Care Med 1996;154: S175–S182.

3 Goldman MJ, Anderson GM, Stolzenberg ED, Kari UP, Zasloff M, Wilson JM: Human beta-defensin-1 is a salt-sensitive antibiotic in lung that is inactivated in cystic fibrosis. Cell 1997; 88:553–560.

4 Wilson R, Pitt T, Taylor G, Watson D, MacDermot J, Sykes D, Roberts D, Cole P: Pyocyanin and 1-hydroxyphenazine produced by *Pseudomonas aeruginosa* inhibit the beating of human respiratory cilia in vitro. J Clin Invest 1987;79:221–229.

5 Matsui H, Grubb BR, Tarran R, Randell SH, Gatzy JT, Davis CW, Boucher RC: Evidence for periciliary liquid layer depletion, not abnormal ion composition, in the pathogenesis of cystic fibrosis airways disease. Cell 1998;95:1005–1015.

6 Paranchych W: Molecular studies on N-methylphenylalanine pili; in Iglewski BH, Clark VL (eds): The Bacteria. San Diego, Academic Press, 1990, vol 11, pp 61–78.

7 Spangenberg C, Fislage R, Sierralta W, Tummler B, Romling U: Comparison of type IV-pilin genes of *Pseudomonas aeruginosa* of various habitats has uncovered a novel unusual sequence. FEMS Microbiol Lett 1995;125:265–273.

8 Castric PA, Deal CD: Differentiation of *Pseudomonas aeruginosa* pili based on sequence and B-cell epitope analyses. Infect Immun 1994;62:371–376.

9 Choi JY, Sifri CD, Goumnerov BC, Rahme LG, Ausubel FM, Calderwood SB: Identification of virulence genes in a pathogenic strain of *Pseudomonas aeruginosa* by representational difference analysis. J Bacteriol 2002;184:952–961.

10 Kus JV, Tullis E, Cvitkovitch DG, Burrows LL: Significant differences in type IV pilin allele distribution among *Pseudomonas aeruginosa* isolates from cystic fibrosis (CF) versus non-CF patients. Microbiology 2004;150: 1315–1326.

11 Comer JE, Marshall MA, Blanch VJ, Deal CD, Castric P: Identification of the *Pseudomonas aeruginosa* 1244 pilin glycosylation site. Infect Immun 2002;70:2837–2845.

12 Sauer FG, Futterer K, Pinkner JS, Dodson KW, Hultgren SJ, Waksman G: Structural basis of chaperone function and pilus biogenesis. Science 1999;285:1058–1061.

13 Vallet I, Olson JW, Lory S, Lazdunski A, Filloux A: The chaperone/usher pathways of *Pseudomonas aeruginosa*: Identification of fimbrial gene clusters (cup) and their involvement in biofilm formation. Proc Natl Acad Sci USA 2001;98:6911–6916.

14 Vallet I, Diggle SP, Stacey RE, Camara M, Ventre I, Lory S, Lazdunski A, Williams P, Filloux A: Biofilm formation in *Pseudomonas aeruginosa*: Fimbrial *cup* gene clusters are controlled by the transcriptional regulator MvaT. J Bacteriol 2004;186:2880–2890.

15 Simpson DA, Ramphal R, Lory S: Genetic analysis of *Pseudomonas aeruginosa* adherence: Distinct genetic loci control attachment to epithelial cells and mucins. Infect Immun 1992;60:3771–3779.

16 Arora SK, Ritchings BW, Almira EC, Lory S, Ramphal R: The *Pseudomonas aeruginosa* flagellar cap protein, FliD, is responsible for mucin adhesion. Infect Immun 1998;66: 1000–1007.

17 Arora SK, Dasgupta N, Lory S, Ramphal R: Identification of two distinct types of flagellar cap proteins, FliD, in *Pseudomonas aeruginosa*. Infect Immun 2000;68:1474–1479.

18 Worlitzsch D, Tarran R, Ulrich M, Schwab U, Cekici A, Meyer KC, Birrer P, Bellon G, Berger J, Weiss T, Botzenhart K, Yankaskas JR, Randell S, Boucher RC, Doring G: Effects of reduced mucus oxygen concentration in airway *Pseudomonas* infections of cystic fibrosis patients. J Clin Invest 2002;109:317–325.

19 Mathee K, Ciofu O, Sternberg C, Lindum PW, Campbell JI, Jensen P, Johnsen AH, Givskov M, Ohman DE, Molin S, Hoiby N, Kharazmi A: Mucoid conversion of *Pseudomonas aeruginosa* by hydrogen peroxide: A mechanism for virulence activation in the cystic fibrosis lung. Microbiology 1999;145:1349–1357.

20 Klausen M, Heydorn A, Ragas P, Lambertsen L, Aaes-Jorgensen A, Molin S, Tolker-Nielsen T: Biofilm formation by *Pseudomonas*

aeruginosa wild type, flagella and type IV pili mutants. Mol Microbiol 2003;48:1511–1524.

21 Lequette Y, Greenberg EP: Timing and localization of rhamnolipid synthesis gene expression in *Pseudomonas aeruginosa* biofilms. J Bacteriol 2005;187:37–44.

22 Wu H, Song Z, Hentzer M, Andersen JB, Heydorn A, Mathee K, Moser C, Eberl L, Molin S, Hoiby N, Givskov M: Detection of N-acylhomoserine lactones in lung tissues of mice infected with *Pseudomonas aeruginosa*. Microbiology 2000;146:2481–2493.

23 Wu H, Song Z, Givskov M, Doring G, Worlitzsch D, Mathee K, Rygaard J, Hoiby N: *Pseudomonas aeruginosa* mutations in *las*I and *rhl*I quorum sensing systems result in milder chronic lung infection. Microbiology 2001;147:1105–1113.

24 Erickson DL, Endersby R, Kirkham A, Stuber K, Vollman DD, Rabin HR, Mitchell I, Storey DG: *Pseudomonas aeruginosa* quorum-sensing systems may control virulence factor expression in the lungs of patients with cystic fibrosis. Infect Immun 2002;70:1783–1790.

25 Schaber JA, Carty NL, McDonald NA, Graham ED, Cheluvappa R, Griswold JA, Hamood AN: Analysis of quorum sensing-deficient clinical isolates of *Pseudomonas aeruginosa*. J Med Microbiol 2004;53: 841–853.

26 Stover CK, Pham XQ, Erwin AL, Mizoguchi SD, Warrener P, Hickey MJ, Brinkman FS, Hufnagle WO, Kowalik DJ, Lagrou M, Garber RL, Goltry L, Tolentino E, Westbrock-Wadman S, Yuan Y, Brody LL, Coulter SN, Folger KR, Kas A, Larbig K, Lim R, Smith K, Spencer D, Wong GK, Wu Z, Paulsen IT: Complete genome sequence of *Pseudomonas aeruginosa* PA01, an opportunistic pathogen. Nature 2000;406:959–964.

27 Spencer DH, Kas A, Smith EE, Raymond CK, Sims EH, Hastings M, Burns JL, Kaul R, Olson MV: Whole-genome sequence variation among multiple isolates of *Pseudomonas aeruginosa*. J Bacteriol 2003;185:1316–1325.

28 Jacobs MA, Alwood A, Thaipisuttikul I, Spencer D, Haugen E, Ernst S, Will O, Kaul R, Raymond C, Levy R, Chun-Rong L, Guenthner D, Bovee D, Olson MV, Manoil C: Comprehensive transposon mutant library of *Pseudomonas aeruginosa*. Proc Natl Acad Sci USA 2003;100:14339–14344.

29 Harrington CA, Rosenow C, Retief J: Monitoring gene expression using DNA microarrays. Curr Opin Microbiol 2000;3: 285–291.

30 Ernst RK, D'Argenio DA, Ichikawa JK, Bangera MG, Selgrade S, Burns JL, Hiatt P, McCoy K, Brittnacher M, Kas A, Spencer DH, Olson MV, Ramsey BW, Lory S, Miller SI: Genome mosaicism is conserved but not unique in *Pseudomonas aeruginosa* isolates from the airways of young children with cystic fibrosis. Environ Microbiol 2003;5: 1341–1349.

31 Kresse AU, Dinesh SD, Larbig K, Romling U: Impact of large chromosomal inversions on the adaptation and evolution of *Pseudomonas aeruginosa* chronically colonizing cystic fibrosis lungs. Mol Microbiol 2003;47: 145–158.

32 Whiteley M, Bangera MG, Bumgarner RE, Parsek MR, Teitzel GM, Lory S, Greenberg EP: Gene expression in *Pseudomonas aeruginosa* biofilms. Nature 2001;413:860–864.

33 Firoved AM, Deretic V: Microarray analysis of global gene expression in mucoid *Pseudomonas aeruginosa*. J Bacteriol 2003;185: 1071–1081.

34 Frisk A, Schurr JR, Wang G, Bertucci DC, Marrero L, Hwang SH, Hassett DJ, Schurr MJ: Transcriptome analysis of *Pseudomonas aeruginosa* after interaction with human airway epithelial cells. Infect Immun 2004;72: 5433–5438.

35 Firoved AM, Ornatowski W, Deretic V: Microarray analysis reveals induction of lipoprotein genes in mucoid *Pseudomonas aeruginosa*: Implications for inflammation in cystic fibrosis. Infect Immun 2004;72: 5012–5018.

36 Wagner VE, Gillis RJ, Iglewski BH: Transcriptome analysis of quorum-sensing regulation and virulence factor expression in *Pseudomonas aeruginosa*. Vaccine 2004; 22(suppl 1):S15–S20.

37 Liang X, Pham XQ, Olson MV, Lory S: Identification of a genomic island present in the majority of pathogenic isolates of *Pseudomonas aeruginosa*. J Bacteriol 2001;183: 843–853.

38 Larbig KD, Christmann A, Johann A, Klockgether J, Hartsch T, Merkl R, Wiehlmann L, Fritz HJ, Tummler B: Gene islands integrated into tRNA^{Gly} genes confer genome diversity on a *Pseudomonas aeruginosa* clone. J Bacteriol 2002;184:6665–6680.

39 Klockgether J, Reva O, Larbig K, Tummler B: Sequence analysis of the mobile genome island pKLC102 of *Pseudomonas aeruginosa* C. J Bacteriol 2004;186:518–534.

40 He J, Baldini RL, Deziel E, Saucier M, Zhang Q, Liberati NT, Lee D, Urbach J, Goodman HM, Rahme LG: The broad host range pathogen *Pseudomonas aeruginosa* strain PA14 carries two pathogenicity islands harboring plant and animal virulence genes. Proc Natl Acad Sci USA 2004;101:2530–2535.

41 Washburn MP, Yates JR 3rd: Analysis of the microbial proteome. Curr Opin Microbiol 2000;3:292–297.

42 Hanna SL, Sherman NE, Kinter MT, Goldberg JB: Comparison of proteins expressed by *Pseudomonas aeruginosa* strains representing initial and chronic isolates from a cystic fibrosis patient: An analysis by 2-D gel electrophoresis and capillary column liquid chromatography-tandem mass spectrometry. Microbiology 2000;146:2495–2508.

43 Wehmhöner D, Häussler S, Tümmler B, Jänsch L, Bredenbruch F, Wehland J, Steinmetz I: Inter- and intraclonal diversity of the *Pseudomonas aeruginosa* proteome manifests within the secretome. J Bacteriol 2003; 185:5807–5814.

44 Moxon ER, Rainey PB, Nowak MA, Lenski RE: Adaptive evolution of highly mutable loci in pathogenic bacteria. Curr Biol 1994;4: 24–33.

45 Bayliss CD, Moxon ER: Hypermutation and bacterial adaption. ASM News 2002;68: 549–555.

46 Fleck O, Nielsen O: DNA repair. J Cell Sci 2004;117:515–517.

47 Schofield MJ, Hsieh P: DNA mismatch repair: Molecular mechanisms and biological function. Annu Rev Microbiol 2003;57:579–608.

48 LeClerc JE, Li B, Payne WL, Cebula TA: High mutation frequencies among *Escherichia coli* and *Salmonella* pathogens. Science 1996;274: 1208–1211.

49 Oliver A, Canton R, Campo P, Baquero F, Blazquez J: High frequency of hypermutable *Pseudomonas aeruginosa* in cystic fibrosis lung infection. Science 2000;288:1251–1254.

50 Giraud A, Radman M, Matic I, Taddei F: The rise and fall of mutator bacteria. Curr Opin Microbiol 2001;4:582–585.

51 Denamur E, Lecointre G, Darlu P, Tenaillon O, Acquaviva C, Sayada C, Sunjevaric I, Rothstein R, Elion J, Taddei F, Radman M, Matic I: Evolutionary implications of the frequent horizontal transfer of mismatch repair genes. Cell 2000;103:711–721.

52 Brown EW, LeClerc JE, Li B, Payne WL, Cebula TA: Phylogenetic evidence for horizontal transfer of mutS alleles among naturally occurring *Escherichia coli* strains. J Bacteriol 2001;183:1631–1644.

53 Oliver A, Baquero F, Blazquez J: The mismatch repair system (*mut*S, *mut*L and *uvr*D genes) in *Pseudomonas aeruginosa*: Molecular characterization of naturally occurring mutants. Mol Microbiol 2002;43: 1641–1650.

54 Smania AM, Segura I, Pezza RJ, Becerra C, Albesa I, Argarana CE: Emergence of phenotypic variants upon mismatch repair disruption in *Pseudomonas aeruginosa*. Microbiology 2004;150:1327–1338.

55 Galan JC, Tato M, Baquero MR, Turrientes C, Baquero F, Martinez JL: Fosfomycin and rifampin disk diffusion tests for detection of *Escherichia coli* mutator strains. J Clin Microbiol 2004;42:4310–4312.

56 Maciá MD, Borrell N, Pérez JL, Oliver A: Detection and susceptibility testing of hypermutable *Pseudomonas aeruginosa* strains with the Etest and disk diffusion. Antimicrob Agents Chemother 2004;48:2665–2672.

57 Oliver A, Levin BR, Juan C, Baquero F, Blazquez J: Hypermutation and the preexistence of antibiotic-resistant *Pseudomonas aeruginosa* mutants: Implications for susceptibility testing and treatment of chronic infections. Antimicrob Agents Chemother 2004;48: 4226–4233.

T.L. Pitt
Laboratory of HealthCare Associated
Centre for Infections
Health Protection Agency
London NW9 5HT (UK)
Tel. +44 020 8200 4400
Fax +44 020 8200 7449
E-Mail tyrone.pitt@hpa.org.uk

Bush A, Alton EWFW, Davies JC, Griesenbach U, Jaffe A (eds): Cystic Fibrosis in the 21st Century.
Prog Respir Res. Basel, Karger, 2006, vol 34, pp 145–152

Other Gram-Negative Organisms

Burkholderia cepacia complex and *Stenotrophomonas maltophilia*

John R.W. Govan

Medical Microbiology, University of Edinburgh Medical School, Edinburgh, UK

Abstract

Since the discovery of cystic fibrosis (CF) in the 1940s, the most striking aspect of the associated lung disease, and the primary cause of this unusual inflammatory process, has been chronic pulmonary infection caused by a surprisingly limited spectrum of microbial pathogens. Historically, the dominant microbial pathogens recovered from CF airways have been bacteria, in particular *Staphylococcus aureus*, non-capsulate *Haemophilus influenzae* and *Pseudomonas aeruginosa*. The 1990s saw the emergence of the *Burkholderia cepacia* complex. Today, the CF community is faced with a Pandora's box of potential new pathogens whose common feature is their environmental origin and innate resistance to most major groups of antibiotics. This chapter will focus on the *B. cepacia* complex and *Stenotrophomonas maltophilia*: species for future consideration may include *Pandoraea apista, Ralstonia pickettii, Alcaligenes xylosoxidans* and *Chryseobacterium* species.

The *Burkholderia cepacia* Complex

B. cepacia was first identified by Francis Burkholder in 1950 [1] as the causative agent of soft rot of onions (cepia, Latin = onion). Subsequent reports identified its role as an opportunistic human pathogen in infections, which range from 'foot rot' in US marines during training in the Florida everglades to fatal pneumonia in children with chronic granulomatous disease [2]. The first reports of culture of *B. cepacia* from cystic fibrosis (CF) airways appeared in the late 1970s. In 1984, a seminal paper by Isles et al. [3]

reported increased *B. cepacia* isolations in the Toronto clinic, and the first description of 'cepacia syndrome' – a fatal necrotizing pneumonia, associated with bacteraemia in approximately 20% of colonized patients. Over the next decades, *B. cepacia* emerged as a major CF pathogen [2, 4, 5], and the cause of much anxiety amongst the CF population and CF carers due to the risk of cepacia syndrome, the organism's innate multiresistance to antibiotics and transmissibility between patients by social contact [2]. *B. cepacia* is probably the most nutritionally adaptable of all pseudomonads and provides a striking example of a soil saprophyte and phytopathogen that has emerged as an important threat to susceptible human hosts. In ironic contrast to its role as a life-threatening human pathogen, the organism is also a potent biopesticide in the protection of crops against fungal diseases, and as bioremediation agent in the breakdown of recalcitrant herbicides and pesticides in soils [6]. In 1997, polyphasic and molecular analyzes revealed that isolates previously identified as '*B. cepacia*' belong to at least ten genomic species which are referred to collectively as the *B. cepacia* complex [7–10] (table 1). For the purpose of this review, reference to publications prior to 1997 or to the *B. cepacia* complex as a whole will use the term Bcc. In other instances, and wherever possible, members of the complex will be referred to by their new species names.

Laboratory Identification

Accurate identification of *B. cepacia* complex plays a vital role in the management of CF lung disease, in particular in the implementation of infection control measures to

Table 1. Genomovars and species within the *B. cepacia* complex [see 7–10]

Burkholderia cepacia (genomovar I)
Burkholderia multivorans (genomovar II)
Burkholderia cenocepacia (genomovar III)
Burkholderia stabilis (genomovar IV)
Burkholderia vietnamiensis (genomovar V)
Burkholderia dolosa (genomovar VI)
Burkholderia ambifaria (genomovar VII)
Burkholderia anthina (genomovar VIII)
Burkholderia pyrrocinia (genomovar IX)
Burkholderia ubonensis (genomovar X)

reduce patient-to-patient spread [11–14]. Species identification within the *B. cepacia* complex is also important in the surgical and clinical management of CF patients selected for lung transplantation [15, 16] and to determine the group's habitats and diversity.

Culture and identification of *B. cepacia* complex from clinical specimens and from environmental niches is challenging [11]. Use of selective media is essential [11, 12, 17–21], followed by identification of Bcc species or genomovars by phenotypic or DNA-based methods [19]. Growth on selective media acts only as a presumptive screen for Bcc and further identification is essential [20–22].

Commercial multitest kits used in diagnostic laboratories are unreliable for identification of Bcc [11, 17, 19, 22]. API 20NE is probably the most useful as a presumptive screen. In a quality control study (unpubl. data), BBL Crystal, VITEK NFC, and VITEK GNI systems failed to identify isolates belonging to *Burkholderia doloso*, and more importantly, failed to identify virulent and highly transmissible strains of *Burkholderia cenocepacia* and *Burkholderia multivorans* (author's unpubl. data). In 2000, data from a US referral laboratory showed that of 770 isolates referred as *B. cepacia* complex, 11% were misidentified; equally alarming, of 282 isolates received as unidentified or 'not *B. cepacia* complex', 36% were confirmed as Bcc [17]. During 2003, of 282 isolates referred to the Edinburgh CF Microbiology Laboratory and Repository as '*B. cepacia*', 12% were identified as false positives (author's unpubl. data). False negatives were rare with only one isolate, received as *P. aeruginosa* subsequently identified as Bcc. Bacterial species that have been misidentified as Bcc include: *Pseudomonas aeruginosa, Burkholderia gladioli, Achromobacter xylosoxidans, Pseudomonas fluorescens, Pseudomonas putida, Stenotrophomonas maltophilia, Chryseobacterium* species, *Acinetobacter baumannii, Serratia* species, *Herbaspirillum* species.

B. cepacia-selective media [17, 18, 20] including the commercially available Mast *B. cepacia* medium [21] are useful for the isolation of Bcc from environmental and clinical specimens; however, lack of selectivity allows growth of other species resulting in the need for further time-consuming and expensive screening tests including use of API 20NE. When large numbers of potential Bcc isolates require preliminary identification, Stewart's arginine dihydrolase/glucose medium provides a useful single tube screen for all known Bcc species [22, 23].

Accurate identification of Bcc, and species within the complex, requires molecular analyzes, which may include whole-cell protein, fatty acid analyzes and DNA-DNA homology [7, 8]. In the clinical laboratory, PCR analyzes based on 16S rRNA, 16S rDNA or *Rec*A enable identification of isolates as Bcc or Bcc species [24–26]. When these techniques are not available at a local level, presumptive Bcc isolates may be sent to referral laboratories established by national CF organizations. Details of these facilities and updates on Bcc research and publications can be obtained from the national CF organizations and through the website of the International *B. cepacia* Working Group (www.goto/cepacia).

Epidemiology of Bcc Infection in CF

The prevalence of Bcc in the CF population varies from centre to centre. In the absence of epidemic spread, prevalence due to sporadic acquisition is generally <10%. All Bcc species described in table 1 have been cultured from individuals with CF. However, *B. cenocepacia* and *B. multivorans* account for approximately 90% of Bcc isolates from CF sputum [5, 27]. Epidemic spread and virulence are most closely associated with *B. cenocepacia*; however, outbreaks involving other Bcc, in particular *B. multivorans*, have been described [25, 28]. In addition, 'cepacia syndrome' is not restricted to *B. cenocepacia* [29] and can occur many years after initial colonization [30]. Prevalence of Bcc species appears to be influenced by geography, and by infection control management. For example, *B. cenocepacia* appears to predominate in North America and Italy whilst *B. multivorans* predominates in other European clinics [5, 27]. In the past, and particularly in the absence of patient segregation, prevalence has been strongly influenced by the spread of highly transmissible strains, such as the intercontinental *B. cenocepacia* lineage, known as ET12 [2]. In 1996, almost 50% of UK Bcc isolates belonged to ET12 [31]. In contrast, current UK surveillance shows that the majority of new Bcc infections are caused by genomically discrete strains of *B. multivorans* [32] (author's unpubl. results). *B. cenocepacia* ET12 possesses *cbla*A-encoded cable pili

Govan

and the *B. cepacia* epidemic strain marker, BCESM. However, these 'markers' are not found in all epidemic strains of *B. cenocepacia* and appear to be absent in *B. multivorans*. Thus, in contrast to previous advice [33], CF centres should not require the presence of these markers before introducing patient segregation [34]. Absence of human commensal carriage of Bcc suggests that most new cases of Bcc infection result from sporadic acquisition from environmental sources. Unfortunately, the diversity and widespread habitats of the Bcc make reliable identification of these bacterial sources difficult [23].

Antibiotic Therapy

Bcc species are commonly multiresistant and in some cases panresistant to the major groups of antimicrobial agents. This high level resistance derives from reduced permeability, efflux pumps, cell target alteration and enzymatic inactivation. Perhaps the most striking example of the latter is the ability of some Bcc to produce a highly inducible β-lactamase and use penicillin as a sole carbon source [35]. The combination of these mechanisms deprives the CF patient of the benefits of antimicrobial therapy. Even when susceptibility is noted in the laboratory, Bcc is seldom eradicated in vivo. Synergy testing with double and triple antibiotic combinations, multiple combination bactericidal testing, suggests that triple antibiotic combinations are most likely to be bactericidal against Bcc isolates in vitro [36]. Randomized, controlled studies in CF patients are required to indicate whether multiple combination bactericidal testing will result in more efficacious use of antibiotics in patients infected with Bcc. A further complication is that clonal isolates of Bcc from the same clinical specimen show considerable heterogeneity in antibiotic susceptibility [31]. At present, the most active antimicrobial agent against Bcc in vitro is meropenem [37] and most treatment regimens include high-dosage intravenous use of this agent in combination with one or more of the following: cotrimoxazole, piperacillin-tazobactam, doxycycline, ceftazidime, trimethoprim-sulfamethoxazole and aerosolized tobramycin.

Bcc are almost unique amongst bacteria in their resistance to defensins, the cysteine-rich antimicrobial peptides (AMPs) that constitute a key component of innate immune systems against bacterial, fungal and viral infections [38]. AMPs have a canonical 6-cysteine motif and exhibit broad, salt-sensitive antimicrobial activity against many resistant bacterial pathogens including *P. aeruginosa* and MRSA. The exception is the murine defensin Defr1 which is a product of a variant allele of the murine Defb8 defensin gene [39]. Defr1 has an unusual chemical structure for a defensin with 5 instead of 6 cysteines, and retains antimicrobial activity at salt concentrations, which totally inhibit other salt-sensitive AMPs. Defr1 is also unique in showing antimicrobial activity against highly resistant strains of *B. cenocepacia*. Although the killing mechanism of defensins is unclear, current evidence suggests that these positively charged molecules are able to 'punch' holes in microbial membranes resulting in bacterial lysis [38]. Recent evidence from chemical analyzes indicates that the antimicrobial activity of Defr1 is strongly influenced by tertiary structure including dimerization mediated by intermolecular disulfide bonds [40].

Acquisition of Bcc

The anxiety that Bcc infection arouses in the CF community results from the risk of cepacia syndrome, the problems of multiresistance and the risks of acquiring Bcc from natural environments or by cross-infection. Acquisition of Bcc from natural environments is probably unavoidable (see later section: Friend or Foe: Are Environmental and Clinical Isolates of Bcc Related?). In contrast, prevention of cross-infection either by person-to-person spread or from contaminated equipment is achievable. In 2004, the UK CF Trust Infection Control Group, the US CFF and the European CF Society produced updated guidelines on the control of Bcc infection in the light of current information, including major developments in taxonomy and epidemiology [12–14]. These documents recommend microbiological surveillance of individuals with CF to identify whether they have Bcc, and to identify cross-infection as early as possible.

Pathogenesis: An Inflammatory Tale

'The host-bacterial interaction in CF results in a highly unusual chronic respiratory inflammatory process, the nature of which has been an area of intense investigative scrutiny' [41].

A striking feature of Bcc infection in individuals with CF is the pronounced inflammatory response and rapid loss of lung function associated with cepacia syndrome. Bcc produce a range of putative virulence determinants including lipopolysaccharide (LPS, endotoxin), protease, haemolysin, catalases, siderophores, quorum sensing, type III and type IV secretion, cytotoxins, exopolysaccharide, and two transmissibility markers cable pili and BCESM, the latter recently found to encode a genomic pathogenicity island [42–44]. The role of these factors in CF lung disease or transmission is unclear although the highly transmissible and virulent *B. cenocepacia* ET12 is unique in possessing both cable pili and BCESM. In various in vivo models of infection, loss of any single factor can be shown to reduce but not eliminate virulence suggesting that virulence is multifactorial [43–46]. One potentially important virulence factor of the Bcc, which is shared with its close relative *Burkholderia pseudomallei*, the causative agent of melioidosis, is an ability to behave as a facultative intracellular pathogen. In a recent review of the cellular aspects of Bcc infection, including survival within amoebae, Mohr et al. [43] focus on the ability of *B. cenocepacia* ET12 strain J2315, to invade and survive in cultured human macrophages suggesting that Bcc may be cytotoxic to these key defensive cells of the human lung.

In view of the striking inflammatory response associated with cepacia syndrome, the role of Bcc LPS is likely to be significant. Bcc LPS is a potent endotoxin and biologically unusual [47–53]. It is more inflammatory than *P. aeruginosa* LPS [47, 48, 53], primes human neutrophils to respond to other bacteria and inflammatory agents [49], and unlike other gram-negative LPS, the inflammatory properties of Bcc LPS are not inhibited by polymyxin B [50].

B. cenocepacia appears to be the most virulent Bcc species encountered in CF patients and is associated with most mortality following transplantation [15, 16]. However, the variable clinical response in individual patients to infection by the same highly virulent *B. cenocepacia* strain [2] and the recent case of 'cepacia syndrome' caused by *B. multivorans* which occurred 9 years after initial colonization [30] provide circumstantial evidence that chronic Bcc colonization may prime a patient's immune system to respond adversely to other CF pathogens or immunostimulants and to invasive procedures such as bronchoscopy. There is little doubt that clinical outcome of Bcc infection is influenced by host-pathogen interactions within individual patients.

Friend or Foe: Are Environmental and Clinical Isolates of Bcc Related?

There is unequivocal evidence that cross-infection measures, including social and nosocomial segregation of Bcc-positive individuals, have reduced transmission of Bcc and the prevalence of the Bcc amongst the CF population. However, even the most effective measures cannot eliminate the risk of Bcc acquisition from the natural environment. This raises the issue of which environments act as reservoirs for Bcc infections and what evidence is there to show that clinical and environmental isolates (including biopesticide strains) are related. The diversity of the Bcc [23, 54] and ethical issues concerning deliberate patient exposure prevent an easy answer to the former question. However,

Fig. 1. Soft rot of onion segments caused by the *B. cepacia*-type strain ATCC 25416T (top left) isolated from rotting onions in the 1940s, and by a clinical isolate (top right) isolated from sputum from a CF patient in the UK. Clonality of the isolates was demonstrated by molecular fingerprinting techniques [55]. The bottom segment is an uninoculated control.

molecular fingerprinting and other circumstantial evidence indicate that environmental Bcc isolates and those from CF respiratory secretions can be clonal and that both exhibit phytopathogenicity (fig. 1) [55, 56]. Such diversity and pathogenic potential for both human and plant hosts is perhaps to be expected given that the *B. cenocepacia* type strain J2315 (LMG16656), belonging to the ET12 lineage, has a genome of 8.056Mb comprising three large replicons and a plasmid of 92.7kb [44] (http://www.sanger.ac.uk/ projects/B_cepacia). Interestingly, J2315 harbours a unique temperate phage DK4/BcepMu and a prophage-encoded LPS acyltransferase which appears specific to the ET12 lineage and could contribute to the inflammatory potential and resistance of this particular lineage [57, 58].

S. maltophilia

Stenotrophomonas (previously *Pseudomonas* and *Xanthomonas*) *maltophilia* is an aerobic innately resistant gram-negative bacillus found in soils, aquatic environments and vegetative matter. Historically, *S. maltophilia* has been of interest for its use in biological control of plant pathogens and bioremediation [59, 60] and for studies of the bacterial mechanisms responsible for multidrug resistance [61]. These organisms are readily cultured in hospitals and domestic homes from sinks, taps, drains, nebulizer equipment, washing machines and other sites associated with water [62, 63].

Laboratory Identification

Culture and identification of *S. maltophilia* present few difficulties [22, 62]. Selective media, biochemical tests and commercial kits used for culture and identification of these organisms from environmental sites and clinical specimens have been authoritatively reviewed elsewhere [62]. If large numbers of isolates are to be identified, e.g. during microbiological surveillance of hospital environments, Stewarts single tube arginine glucose medium (see The *Burkholderia cepacia* Complex section) provides a simple preliminary screen to discriminate *S. maltophilia* from other pseudomonads and Enterobacteriaceae [22]. Molecular methods for identification of *S. maltophilia* are increasingly available and include rRNA-directed PCR [64], PCR [65, 66] and real-time LightCycler PCR [67]. In individuals with CF, accurate identification of *S. maltophilia* in CF is vital since misidentification as Bcc will have important clinical and psychological consequences. False-positive identification would lead to unnecessary segregation and false-negative identification to the risk of 'cepacia syndrome' and spread to other patients.

S. maltophilia as a Human Pathogen

Historically, *S. maltophilia* has been considered an opportunistic pathogen of low virulence; however, these organisms are important in nosocomial infections in immunocompromised individuals and may cause life-threatening infections with limited options for antibiotic therapy [68, 69]. Use of carbapenems and other antibiotics in recent decades has been associated with increasing isolation of these innately resistant bacteria [70, 71]. An interesting example is increased isolation of *S. maltophilia* in tropical Australia during the wet season, reflecting perhaps an increased usage of ceftazidime and imipenem against *B. pseudomallei* in the treatment of melioidosis, which peaks at this season [70].

The pathogenic role of *S. maltophilia* in CF lung disease is unclear and the subject of continued debate [72–75]. Isolation of *S. maltophilia* from CF patients was first reported in Denmark in the 1970s by Frederiksen et al. [76] and in the US in 1985 by Klinger and Thomassen [77]. In the next decades, prevalence increased but varies between CF centres from 0% to as much as 30%. The reasons for this variance are also unclear but may reflect differences in antibiotic usage [78, 79], the use of selective medium and the attention given to identification of pseudomonas-like isolates. The use of genomic typing by pulsed-field gel electrophoresis, and PCR fingerprinting with BOX primers, emphasizes the diversity of both clinical and environmental isolates [80–82]. However, genomic typing of CF isolates shows little evidence of strain clustering and cross-infection [80, 81, 83]. Hence segregation of *S. maltophilia*-positive individuals would provide little benefit in infection control. Furthermore, although the organism can be cultured from the homes of CF individuals and from nebulizers, there is little evidence that these strains are responsible for pulmonary colonization [84]. The risk factors associated with *S. maltophilia* colonization in CF patients are also unclear. Denton et al. [79] reported a significant association with *S. maltophilia* colonization and prior aerosolized aminoglycoside therapy against *P. aeruginosa*. In another study, use of stepwise logistic regression showed colonization was linked to long-term treatment with antibiotics (p = 0.0016) and the number of days of intravenous antibiotic therapy (p = 0.035) [85]. Interestingly, in a multicentred trial of inhaled tobramycin in the management of chronic *P. aeruginosa* infection, the incidence of superinfection by *S. maltophilia* was found to be less in the tobramycin group (16%) compared to those inhaling an antibiotic placebo (22%) suggesting that the use of nebulizers per se may be a risk factor for colonization [86]. Recently, a case-control study showed that *S. maltophilia* isolates had increased from 3.3 to 15% during the period 1991–1999; risk factors included more than two courses of intravenous antibiotics, *Aspergillus fumigatus* isolation and oral steroid therapy [75]. Other risk factors reported include concomitant *S. aureus* colonization [73, 85].

Isolates of *S. maltophilia* from CF patients produce few obvious virulence factors and the endotoxicity of *S. maltophilia* LPS is considerably less than that of Bcc [48, 87]. It has been suggested, however, that *S. maltophilia* β-lactamases may play an indirect role by inhibiting antibiotic therapy and encouraging the growth of more virulent CF pathogens [88]. Some studies have indicated adverse clinical effects associated with long-term *S. maltophilia* colonization and bacterial densities in excess of 10^5 CFU/ml sputum [73, 77, 89–91]. However, despite its increasing prevalence and potential for long-term colonization, other studies [75, 78, 92] indicate a negligible role for *S. maltophilia* as a CF pathogen. Acute clinical deterioration, similar to cepacia syndrome, has not been reported for *S. maltophilia*.

Antibiotic Therapy

Reflecting the uncertainty on the pathogenic significance of *S. maltophilia*, the optimal antimicrobial treatment for true *S. maltophilia* infections in CF patients is unknown. This uncertainty is exacerbated by the characteristic innate resistance of these organisms to the major groups of antibiotics [93]. When the organism is considered to be colonizer, antimicrobial treatment is unnecessary and may promote superinfection with more virulent organisms. Where treatment is warranted, recent studies show most isolates (90%) to be susceptible to trimethoprim-sulfamethoxazole [69] or doxycline (80%) and that most isolates are susceptible in vitro to synergistic combinations of trimethoprim-sulfamethoxazole and ticarcillin-clavulanate [93] with the latter recommended in high doses for in vivo use [94].

A Taxonomic Caveat

Recent taxonomic analyzes based on *gyr*B RFLP [95] indicate that the majority of CF isolates of *S. maltophilia* grouped into two clusters suggesting that, as with the Bcc, specific genomic subpopulations may have an increased potential to colonize and infect the CF lung. Until further

evidence is available, considering that *S. maltophilia* can cause serious opportunistic infections in non-CF individuals, it would seem prudent to consider these organisms as potential pathogens in individual patients when the role of other CF pathogens is discounted, and pulmonary deterioration is evident.

Pandora's Box and Further Emergent CF Pathogens

At the time of writing, some isolates cultured from CF respiratory secretions remain unidentified even with state-of-the art polyphasic identification schemes. When identification is successful, the evidence suggests that genera which may merit particular scrutiny as potential emergent CF pathogens include *Ralstonia* and *Achromobacter* [96, 97]. At present, there is insufficient clinical information to judge whether these microbes are commensal or pathogenic in the CF lung. There is no doubt, however, that these emergent bacteria require close scrutiny based on their innate antibiotic resistance and the inflammatory potential of their LPS which exceeds that of *S. maltophilia* [87].

References

1 Burkholder WH: Sour skin, a bacterial rot of onion bulbs. Phytopathology 1950;40: 115–117.

2 Govan J, Hughes J, Vandamme P: *Burkholderia cepacia*: Medical, taxonomic and ecological issues. J Med Microbiol 1996;45: 395–407.

3 Isles A, MacLuskey I, Corey M, Gold R, Prober C, Fleming P, Levison H: *Pseudomonas cepacia* infection in cystic fibrosis: An emerging problem. J Pediatr 1984;104:206–210.

4 LiPuma JJ: *Burkholderia cepacia*: Management issues and new insights. Clin Chest Med 1998;19:473–486.

5 Mahenthiralingam E, Baldwin A, Vandamme P: *Burkholderia cepacia* complex infection in patients with cystic fibrosis. J Med Microbiol 2002;51:533–538.

6 Parke J, Gurian-Sherman D: Diversity of the *Burkholderia cepacia* complex and implication for the risk assessment of biological control strains. Annu Rev Phytopathol 2001;39: 225–258.

7 Vandamme P, Holmes B, Vancanneyt CT, Hoste B, Coopman R, Revets H, Lauwers S, Gillis M, Kersters K, Govan J: Occurrence of multiple genomovars of *Burkholderia cepacia* in cystic fibrosis patients and proposal of *Burkholderia multivorans* sp. nov. Int J Syst Bacteriol 1997;47:1188–1200.

8 Coenye T, Vandamme P, Govan JRW, LiPuma JJ: Taxonomy and identification of the *Burkholderia cepacia* complex. Clin Microbiol Rev 2001;39:3427–3436.

9 Vandamme P, Holmes B, Coenye T, Goris J, Mahenthiralingam E, LiPuma JJ, Govan JRW: *Burkholderia cenocepacia* sp. nov. – A new twist to an old story. Res Microbiol 2003;154: 91–96.

10 Vermis K, Coenye T, LiPuma JJ, Mahenthiralingam E, Nelis HJ, Vandamme P: Proposal to accommodate *Burkholderia dolosa* sp. nov. Int J Syst Evol Microbiol 2004; 54:689–691.

11 Miller MB, Gilligan P: Laboratory aspects of management of chronic pulmonary infections in patients with cystic fibrosis. J Clin Microbiol 2003;41:4009–4015.

12 UK Cystic Fibrosis Trust Infection Control Group: The *Burkholderia cepacia* complex. Suggestions for prevention and infection control. Bromley, Cystic Fibrosis Trust, 2004.

13 Saiman L, Siegel J: Infection control in cystic fibrosis. Clin Microbiol Rev 2004;17:57–71.

14 Doring G, Hoiby N, Consensus Study Group: Early intervention and prevention of lung disease in cystic fibrosis: A European Consensus. J Cyst Fibros 2004;3:67–91.

15 Aris RM, Routh JC, LipUma JJ, et al: Lung transplantation for cystic fibrosis patients with *Burkholderia cepacia* complex: Survival linked to genomovar type. Am J Respir Crit Care Med 2001;164:2102–2106.

16 De Soyza A, Morris K, McDowell A, Doherty C, Archer L, Perry J, Govan JRW, Corris PA, Gould K: Prevalence and clonality of *Burkholderia cepacia* complex genomovars in UK patients with cystic fibrosis referred for transplantation. Thorax 2004;59:526–528.

17 McMenamin JD, Zaccone TM, Coenye T, Vandamme P, LiPuma JJ: Misidentification of *Burkholderia cepacia* in US cystic fibrosis treatment centres: An analysis of 1051 recent sputum isolates. Chest 2000;117:1661–1665.

18 Henry DA, Campbell M, McGimpsey C, Clarke A, Louden L, Burns JL, Roe MH, Vandamme P, Speert DP: Comparison of isolation media for recovery of *Burkholderia cepacia* complex from respiratory secretions of patients with cystic fibrosis. J Clin Microbiol 1999;37:1004–1007.

19 Henry DA, Mahenthiralingam E, Vandamme P, Coenye T, Speert DP: Biochemical and molecular approaches for determining genomovar status of the *Burkholderia cepacia* complex. J Clin Microbiol 2001;39:1073–1078.

20 Vermis K, Vandamme P, Nelis HJ: *Burkholderia cepacia* complex genomovars: Utilization of carbon sources, susceptibility to antimicrobial agents and growth on selective media. J Appl Microbiol 2003;95:1191–1196.

21 Pitt TL, Govan JRW: *Pseudomonas cepacia* in cystic fibrosis patients. PHLS Microbiol Digest 1993;10:69–72.

22 Govan JRW: *Pseudomonas, Stenotrophomonas, Burkholderia*; in Collee JG, Fraser AG, Marmion BP, Simmons A (eds): Practical Medical Microbiology. Edinburgh, Churchill Livingstone, 1996, pp 413–424.

23 Vermis K, Brachkova M, Vandamme P, Nelis H: Isolation of *Burkholderia cepacia* complex from waters. Syst Appl Microbiol 2003;26: 595–600.

24 Vermis K, Coenye T, Mahenthiralingam E, Nellis HJ, Vandamme P: Evaluation of species-specific RecA-based PCR tests for genomovar identification within the *Burkholderia cepacia* complex. J Med Microbiol 2002; 51:937–940.

25 Segonds C, Heulin T, Marty N, Chabonon G: Differentiation of *Burkholderia* species by

PCR-restriction fragment length polymorphism analysis of the 16S rRNA gene and application to cystic fibrosis isolates. J Clin Microbiol 1999;37:2201–2208.

26 Seo ST, Tsuchiya K: PCR-based identification and characterization of *Burkholderia cepacia* bacteria from clinical and environmental sources. Lett Appl Microbiol 2004;39: 413–419.

27 LiPuma JJ, Spilker T, Gill LH, Campbell PW, Liu L, Mahenthiralingam E: Disproportionate distribution of *Burkholderia cepacia* complex species and transmissibility markers in cystic fibrosis. Am J Crit Care Med 2001;164: 92–96.

28 Whiteford ML, Wilkinson JD, McColl JH, Conlon FM, Michie JR, Evans TJ, Paton JY: Outcome of *Burkholderia (Pseudomonas) cepacia* colonisation in children with cystic fibrosis following a hospital outbreak. Thorax 1995;50:1194–1198.

29 Jones AM, Dodd ME, Govan JRW, Barcus V, Doherty CJ, Morris J, Webb AK: *Burkholderia cenocepacia* and *Burkholderia multivorans*: Influence on survival in cystic fibrosis. Thorax, in press.

30 Blackburn L, Brownlee K, Conway S, Denton M: 'Cepacia syndrome' with *Burkholderia multivorans*, 9 years after initial colonization. J Cyst Fibros 2004;3:133–134.

31 Pitt TL, Kaufmann ME, Patel PS, Benge LS, Gaskin S, Livermore DM: Type characterisation and antibiotic susceptibility of *Burkholderia (Pseudomonas) cepacia* isolates from patients with cystic fibrosis in the United Kingdom and Republic of Ireland. J Med Microbiol 1996;44:203–210.

32 Turton JF, Kaufmann ME, Mustafa N, Kawa S, Clode FE, Pitt TL: Molecular comparison of isolates of *Burkholderia multivorans* from patients with cystic fibrosis in the United Kingdom. J Clin Microbiol 2003;41: 5750–5754.

33 Clode FE, Kaufmann ME, Malnick H, Pitt TL: Distribution of genes encoding putative transmissibility factors among epidemic and nonepidemic strains of *Burkholderia cepacia* from cystic fibrosis patients. J Clin Microbiol 2000;38:1763–1766.

34 McDowell A, Mahenthiralingam E, Dunbar KEA, Moore JE, Crowe M, Elborn JS: Epidemiology of *Burkholderia cepacia* complex species recovered from cystic fibrosis patients: Issue related to patient segregation. J Med Microbiol 2004;53:663–668.

35 Beckman W, Lessie TG: Response of *Pseudomonas cepacia* to β-lactam antibiotics: Utilization of penicillin G as the carbon source. J Bacteriol 1979;140:1126–1128.

36 Aaron SD, Ferris W, Henry DA, Speert DP, MacDonald NE: Multiple combination bactericidal antibiotic testing for patients with cystic fibrosis infected with *Burkholderia cepacia*. Am J Respir Crit Care Med 2000; 161:1206–1212.

37 Nzula S, Vandamme P, Govan JRW: Influence of taxonomic status on the in vitro antimicrobial susceptibility of the *Burkholderia cepacia* complex. J Antimicrob Chemother 2002;50: 265–269.

38 Lehrer R: Primate defensins. Nat Rev 2004;2:727–738.

39 Morrison GM, Rolfe M, Kilanowski FM, Cross SH, Dorin JR: Identification and characterization of a novel murine beta-defensin-related gene. Mamm Genome 2002;13: 445–451.

40 Campopiano DJ, Clarke DJ, Polfera NC, Barran PE, Langley RJ, Govan JRW, Maxwell A, Dorin J: Structure-activity relationships in defensin dimers: A novel link between β-defensin tertiary structure and antimicrobial activity. J Biol Chem 2004;279:48671–48679.

41 Speert DP, Goldberg JB: *Burkholderia cepacia* complex and cystic fibrosis: In search of the smoking gun. Am J Respir Crit Care Med 2004;170:6–7.

42 Hutchison M, Govan J: Pathogenicity of microbes associated with cystic fibrosis. Microbes Infect 1999;12:1005–1014.

43 Mohr CD, Tomich M, Herfst CA: Cellular aspects of *Burkholderia cepacia* infection. Microbes Infect 2001;3:425–435.

44 Mahenthiralingam E, Urban TA, Goldberg JB: The multifarious, multireplicon *Burkholderia cepacia* complex. Nat Rev Microbiol 2005;3: 1–13.

45 Tomich M, Griffith A, Herfst CA, Burns JL, Mohr CD: Attenuated virulence of *Burkholderia cepacia* type III secretion mutant in a murine model of infection. Infect Immun 2003;71:1405–1415.

46 Conway B-A, Chu K, Bylaund J, Altman E, Speert DP: Production of exopolysaccharide by *Burkholderia cenocepacia* results in altered cell-surface interactions and altered bacterial clearance. J Infect Dis 2004;190:957–966.

47 Shaw D, Poxton I, Govan J: Biological activity of *Burkholderia (Pseudomonas) cepacia* lipopolysaccharide. FEMS Immunol Med Microbiol 1995;11:99–106.

48 Zughaier SM, Ryley HC, Jackson SK: Lipopolysaccharide (LPS) from *Burkholderia cepacia* is more active than LPS from *Pseudomonas aeruginosa* and *Stenotrophomonas maltophilia* in stimulating tumour necrosis factor alpha from human monocytes. Infect Immun 1999;67:1505–1507.

49 Hughes JE, Stewart J, Barclay GR, Poxton IR, Govan JRW: Priming of neutrophil respiratory burst activity by lipopolysaccharide from *Burkholderia cepacia*. Infect Immun 1997;65: 4281–4287.

50 Shimomura H, Matsuura M, Saito S, Hirai Y, Isshiki Y, Kawahara K: Unusual interaction of a lipopolysaccharide isolated from *Burkholderia cepacia* with polymyxin B. Infect Immun 2003;71:5225–5230.

51 Vinion-Dubiel A, Goldberg J: Lipopolysaccharide of *Burkholderia cepacia* complex. J Endotoxin Res 2003;9:201–213.

52 Govan JRW: The *Burkholderia cepacia* complex and cytokine induction: An inflammatory tale. Pediatr Res 2003;54:1–3.

53 De Soyza A, Que-Gewirth N, Ellis CD, Kalb SR, Cotter RI, Taetz RH, Khan A, Corris PA, De Hormaeche RD: *Burkholderia cenocepacia* lipopolysaccharide, lipid A and pro-inflammatory activity. Am J Respir Crit Care Med 2004;170:70–77.

54 Coenye T, Vandamme P: Diversity and significance of *Burkholderia* species occupying diverse ecological niches. Exp Microbiol 2003;5:719–729.

55 Govan JRW, Balandreau J, Vandamme P: *Burkholderia cepacia* – Friend and foe. ASM News 2000;66:124–125.

56 LiPuma JJ, Spilker T, Coenye T, Gonzalez CF: An epidemic *Burkholderia cepacia* complex strain identified in soil. Lancet 2002;359: 2002–2003.

57 Kenna DT, Barcus VA, Langley RJ, Vandamme P, Govan JRW: Lack of correlation between O-serotype, bacteriophage susceptibility and genomovar status in the *Burkholderia cepacia* complex. FEMS Immunol Med Microbiol 2003;35:87–92.

58 Summer E, Gonzalez C, Carlisle T, Mebane L, Cass A, Savva C, LiPuma J, Young R: *Burkholderia cenocepacia* phage BcepMu and a family of Mu-like phages encoding potential pathogenesis factors. J Mol Biol 2004;340: 49–65.

59 Wang X, Li B, Herman PL, Weeks DP: A three-component enzyme system catalyses the O demethylation of the herbicide dicamba in *Pseudomonas maltophilia* DI-16. Appl Environ Microbiol 1997;63:1623–1626.

60 Minkwitz A, Berg G: Comparison of antifungal activities and 16S ribosomal DNA sequences of clinical and environmental isolates of *Stenotrophomonas maltophilia*. J Clin Microbiol 2001;39:139–145.

61 Chang L-L, Chen H-F, Chang C-Y, Lee T-M, Wu W-J: Contribution of integrons and SmeABC and SmeDEF efflux pumps to multidrug resistance in clinical isolates of *Stenotrophomonas maltophilia*. J Antimicrob Chemother 2004;53:518–521.

62 Denton M, Kerr KG: Microbiological and clinical aspects of infection associated with *Stenotrophomonas maltophilia*. Clin Microbiol Rev 1998;11:57–80.

63 Denton M, Kerr KG: Molecular epidemiology of *Stenotrophomonas maltophilia* isolated from cystic fibrosis patients. J Clin Microbiol 2002;40:1884.

64 Whitby PW, Carter KB, Burns JL, Royal JA, LiPuma JJ, Stull TL: Identification and detection of *Stenotrophomonas maltophilia* by rRNA-directed PCR. J Clin Microbiol 2000;38:4305–4309.

65 Krzewinski JW, Nguyen CD, Foster JM, Burns JL: Use of random amplified polymorphic DNA PCR to examine epidemiology of *Stenotrophomonas maltophilia* and *Achromobacter (Alcaligenes) xylosoxidans* from patients with cystic fibrosis. J Clin Microbiol 2001;39: 3597–3602.

66 Da Silva Filho LV, Tateno AF, Velloso Lde F, Levi JE, Fernandes S, Bento CN, Rodrigies JC, Ramos SR: Identification of *Pseudomonas aeruginosa, Burkholderia cepacia* complex and *Stenotrophomonas maltophilia* in respiratory samples from cystic fibrosis patients using multiplex PCR. Pediatr Pulmonol 2004; 37:537–547.

67 Wellinhausen N, Wirths B, Franz AR, Karolyi L, Marre R, Reischl U: Algorithm for the identification of bacterial pathogens in

positive blood cultures by real time LightCycler polymerase chain reaction (PCR) with sequence-specific probes. Diagn Microbiol Infect Dis 2004;48:229–241.

68 Calza L, Manfredi R, Chiodo F: *Stenotrophomonas (Xanthomonas) maltophilia* as an emerging opportunistic pathogen in association with HIV infection: A 10-year study. Infection 2003;31:155–161.

69 Senol E: *Stenotrophomonas maltophilia*: The significance and role as a nosocomial pathogen. J Hosp Infect 2004;57:1–7.

70 Heath T, Currie B: Nosocomial and community acquired *Xanthomonas maltophilia* in tropical Australia. J Hosp Infect 1995;30: 309–313.

71 Sanyal SC, Mokaddas EM: The increase in carbapenem use and emergence of *Stenotrophomonas maltophilia* as an important nosocomial pathogen. J Chemother 1999;11: 28–33.

72 Denton M: *Stenotrophomonas maltophilia*: An emerging problem in cystic fibrosis patients. Rev Med Microbiol 1997;8:15–19.

73 Dobbin CJ, Bye PT: Management of an older patient with *Stenotrophomonas maltophilia*: Is it an emerging pathogen in cystic fibrosis? Intern Med J 2001;31:499–500.

74 Goss CH, Otto K, Aitken ML, Rubenfeld GD: Detecting *Stenotrophomonas maltophilia* does not reduce survival of patients with cystic fibrosis. Am J Crit Care Respir Med 2002; 166:356–361.

75 Marchac V, Equi A, Le Bihan-Benjamin C, Hodson M, Bush A: Case-control study of *Stenotrophomonas maltophilia* acquisition in cystic fibrosis patients. Eur Respir J 2004; 23:96–102.

76 Frederiksen B, Hoiby N, Koch C: *Stenotrophomonas (Xanthomonas) maltophilia* at a Danish Cystic Fibrosis Centre 1974–1993 (abstract). Pediatr Pulmonol Suppl 1975;12: 287.

77 Klinger JD, Thomassen MJ: Occurrence and antimicrobial susceptibility of gram-negative non-fermenting bacilli in cystic fibrosis patients. Diagn Microbiol Infect Dis 1985;3: 149–158.

78 Gladman G, Connor PG, Williams RF, David TJ: Controlled study of *Pseudomonas cepacia* and *Pseudomonas maltophilia* in cystic fibrosis. Arch Dis Child 1992;67:192–195.

79 Denton M, Todd NJ, Littlewood JM: Role of anti-pseudomonal antibiotics in the emergence of *Stenotrophomonas maltophilia* in cystic fibrosis patients. Eur J Clin Microbiol Infect Dis 1996;15:402–405.

80 Denton M, Todd NJ, Littlewood JM, Hawkey PM: Molecular typing of *Stenotrophomonas maltophilia* isolated from cystic fibrosis patients. J Infect 1996;32:88.

81 Denton M, Todd NJ, Kerr KG: Molecular epidemiology of *Stenotrophomonas maltophilia* isolated from clinical specimens from patients with cystic fibrosis and associated with environmental samples. J Clin Microbiol 1998;36: 1953–1958.

82 Berg G, Roskot N, Smalla K: Genotypic and phenotypic relationships between clinical and environmental isolates of *Stenotrophomonas maltophilia*. J Clin Microbiol 1999;37: 3594–3600.

83 Vu-Thein H, Moissenet D, Valcin M, Dulot C, Tournier G, Garberg-Chenon A: Molecular epidemiology of *Burkholderia cepacia, Stenotrophomonas maltophilia* and *Alcaligenes xylosoxidans* in a cystic fibrosis center. Eur J Clin Microbiol Infect Dis 1996;15:876–878.

84 Denton M, Rajgopal A, Mooney L, Qureshi A, Kerr KG, Keer V, Pollard K, Peckham DG, Conway SP: *Stenotrophomonas maltophilia* contamination of nebulizers used to deliver aerosolized therapy to inpatients with cystic fibrosis. J Hosp Infect 2003;55:180–183.

85 Talmaciu I, Varlotta L, Mortensen J, Schidlow DV: Risk factors for emergence of *Stenotrophomonas maltophilia* in cystic fibrosis. Pediar Pulmonol 2000;30:10–15.

86 Ramsey BW, Pepe MS, Quan JM, Otto KL, Montgomery AB, Williams-Warren J, Vasiljev-K M, Borowitz D, Bowman CM, Marshall BC, Marshal S, Smith A: Intermittent administration of inhaled tobramycin in patients with cystic fibrosis. N Engl J Med 1999;340:23–30.

87 Hutchison ML, Bonell EC, Poxton IR, Govan JRW: Endotoxic activity of lipopolysaccharides isolated from emergent potential cystic fibrosis pathogens. FEMS Immunol Med Microbiol 2000;27:73–77.

88 Kataoke D, Fujiwara H, Kawakami T, Tanaka Y, Tanimoto A, Ikawa S, Tanaka Y: The indirect pathogenicity of *Stenotrophomonas maltophilia*. Int J Antimicrob Agents 2003;22: 601–666.

89 Govan JRW, Nelson JW: Microbiology of lung infection in cystic fibrosis. Br Med Bull 1992; 48:912–930.

90 Karpati F, Malmborg AS, Alfredson H, Hjelte L: Bacterial colonisation with *Xanthomonas maltophilia* – A retrospective study in a cystic fibrosis patient population. Infection 1994;22: 258–263.

91 Ballestero S, Virseda I, Escobar H, Suarez L, Baquero F: *Stenotrophomonas maltophilia* in cystic fibrosis patients. Eur J Clin Microbiol Infect Dis 1995;14:728–729.

92 Demko C, Doershuk C, Stern R: Thirteen year experience with *Xanthomonas maltophilia* in patients with cystic fibrosis (abstract). Pediatr Pulmon Suppl 1995;12:244.

93 San Gabriel P, Zhou J, Tabibi S, Chen Y, Trauzzi M, Saimon L: Antimicrobial susceptibility and synergy studies of *Stenotrophomonas maltophilia* isolates from patients with cystic fibrosis. Antimicrob Agents Chemother 2004;48:168–171.

94 Gilbert DN, Moellering RC, Sande MA: The Sanford Guide to Antimicrobial Therapy, ed 33. Hyde Park, Antimicrobial Therapy, 2002, pp 50, 52–54.

95 Coenye T, Vanlaere E, LiPuma JJ, Vandamme P: Identification of genomic groups in the genus *Stenotrophomonas* using gyrB RFLP analysis. FEMS Immunol Med Microbiol 2004;40:181–185.

96 Coenye T, Vandamme P, LiPuma JJ: *Ralstonia respiraculi* sp nov. isolated from the respiratory tract of cystic fibrosis patients. Int J Syst Evol Microbiol 2003;53:1339–1342.

97 Kanellopoulou M, Pournaras S, Igleezos H, Skarmoutsou N, Papafrangas E, Maniatis AN: Persistent colonization of nine cystic fibrosis patients with an *Achromobacter (Alcaligenes) xylosoxidans* clone. Eur J Clin Microbiol Infect Dis 2004;23:336–339.

John R.W. Govan
Medical Microbiology, University of
Edinburgh Medical School
Chancellor's Building
49 Little France Crescent
Edinburgh EH16 4SB (UK)
Tel. +44 131 650 3164
Fax +44 131 650 6653
E-Mail john.r.w.govan@ed.ac.uk

Bush A, Alton EWFW, Davies JC, Griesenbach U, Jaffe A (eds): Cystic Fibrosis in the 21st Century.
Prog Respir Res. Basel, Karger, 2006, vol 34, pp 153–159

Staphylococcus aureus and MRSA

Steven Conway   Miles Denton

Consultant Paediatrician, Clinical Director Cystic Fibrosis Services, Leeds Teaching Hospitals Trust,
Leeds, UK

Abstract

Staphylococcus aureus is a common pathogen in cystic fibrosis (CF), being detected in the sputum of about 50% of children over the age of 10 years. Although it may not itself usually cause airway damage, its presence may predispose to *Pseudomonas aeruginosa* infection and so eradication is advised. Methicillin-resistant *S. aureus* (MRSA) is an emerging pathogen in CF, and current prevalence rates are 0–23%, and increasing. *S. aureus* isolation rates may be improved by using selective media. The recently described small colony variants of *S. aureus* may be more likely to persist in the airway. Nosocomial transmission and antibiotic use are associated with MRSA isolation. The choice of antibiotics for *S. aureus*, and whether prophylaxis is given as a routine, is largely on an empirical basis in the absence of large trials of therapy. Prevention of acquisition of MRSA by good hygiene and cohorting is essential. Eradication regimes in CF are largely empirical. Linezolid, a recently available oxazolidinone antibiotic, shows promise for eradication both of *S. aureus* and MRSA. However, virtually all the important questions in the therapy of all forms of *S. aureus* infection in CF require further research as a matter of urgency.

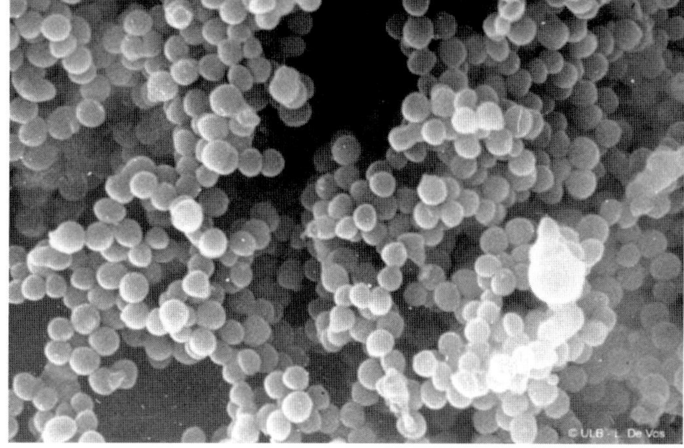

Fig. 1. Electron microscopic image of *S. aureus*. Reproduced from www.ulb.ac.be/sciences/biodic/ImBacterie) – Université Libre de Bruxelles.

Despite the development of potent antimicrobial agents, improved public health conditions, and hospital infection-control measures *Staphylococcus aureus,* a gram-positive coccus (fig. 1), is an important human pathogen. Resistance to penicillin, mediated by a β-lactamase, and to penicillinase-resistant compounds such as methicillin and flucloxacillin, mediated by an altered penicillin-binding protein (PBP2a), is common. These strains are known as 'methicillin-resistant *Staphylococcus aureus*' (MRSA).

S. aureus is still frequently isolated from sputum samples of adult cystic fibrosis (CF) patients, many of whom are chronically infected (fig. 2), and has a prevalence of approximately 50% by 10 years of age in CF children [1]. It is no longer a common cause of significant morbidity or mortality in CF [2]. However, it is suggested that *S. aureus*-induced lung damage can predispose to *P. aeruginosa*

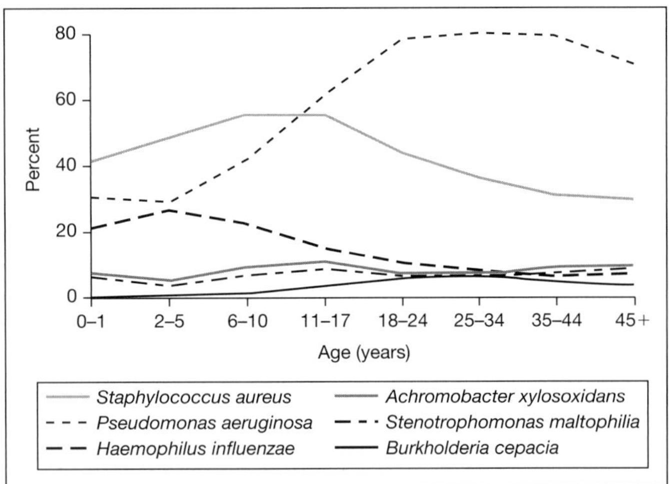

Fig. 2. Age-specific prevalence of *S. aureus* and other CF pathogens, Cystic Fibrosis Foundation registry, 2001. Published by Cystic Fibrosis Foundation Patient Registry, 2001 Annual Data Report to the Center Directors. Bethesda, Cystic Fibrosis Foundation, 2002.

infection [3]. Hence early treatment and eradication (if possible) of *S. aureus* infection are advocated.

The emergence of MRSA as a pathogen in the CF community appears to have occurred at the same rate and over the same time as the epidemic in non-CF patients [4]. Its prevalence in CF patients is likely to reflect that in the local community and hospital. MRSA has been increasingly reported from CF units [5]. Chronic infection is frequently seen as a relative contraindication to lung transplantation [4, 6].

Microbiology

Isolation of *S. aureus* from sputa of CF patients is improved by the use of selective media, such as Mannitol Salt Agar (MSA). In one study of CF sputum samples, all 64 isolates of *S. aureus* were isolated using MSA compared to 62/64 (97%) using non-selective media [7]. Some but not all have recommended using blood agar supplemented with colistin and nalidixic acid in addition to MSA [8]. A preliminary study suggested that the use of CHROMagar (a selective and differential agar utilizing colour change to distinguish *S. aureus* from other cultured bacteria) for culturing CF sputa may improve sensitivity compared to MSA [9]. Molecular techniques have been utilized to improve the detection of *S. aureus* in CF sputa, including fluorescent in situ hybridization [10] and the polymerase chain reaction [11]. They are not yet part of the routine diagnostic repertoire.

Epidemiology

Nasal carriage of *S. aureus* is significantly higher in CF than non-CF patients. Goerke et al. [12] found nasal carriage rates of 66% in patients with CF compared to 32% in non-CF patients. Interestingly, CF patients receiving anti-staphylococcal treatment at the time of sampling had carriage rates of only 29%. In 55% of families with CF patients, family members also shared the same strain of *S. aureus*. Infection can also be spread from patient to patient if there is prolonged and close contact. *S. aureus* infection can persist for several years. Kahl et al. [13] prospectively studied 72 patients with CF and found that the median length of persistence was 37 months (range 6–70 months). The development of 'small colony variant' (SCV) (fig. 3) is associated with persistence of *S. aureus* in the airways [13, 14]. The persistence of SCVs may relate to their greater resistance to host defence mechanisms and greater adherence to respiratory epithelial cells [14].

Prevalence of MRSA infection in CF centres ranges from 0% to about 23% [4, 15–19] and the infection rate is increasing [15–18, 20]. Seven percent of US CF patients have MRSA infection [19]. Routine testing and standardization of microbiological procedures are likely to identify a higher prevalence rate [16]. Givney et al. [21] reported that strains with the same phage type and genotype were distributed between CF and non-CF patients. Using pulsed field gel electrophoresis Garske et al. [22] also showed that MRSA strains in patients with CF were of the same clonal origin as the endemic hospital strain. This suggests that nosocomial transmission in hospital, rather than outside social contact, is the most significant route of transmission for MRSA [4, 21]. Nadesalingam et al. [23] ascertained that MRSA-positive patients spent significantly more days in hospital (mean 19.8 days versus 5.5 days, p = 0.0003) than MRSA-negative patients. Admission to general medical wards may increase the risk [21]. In addition, acquisition of community rather than nosocomial strains of MRSA is increasingly recognized [24] and may be a potential source of MRSA for CF patients [18].

Antibiotic use has also been associated with acquisition of MRSA by CF patients. Nadesalingam et al. [23] found that MRSA-positive patients received more treatment days of oral ciprofloxacin (43.5 days versus 13.9 days, p = 0.03) and more treatment days of oral/intravenous cephalosporins (42.7 days versus 15.4 days, p = 0.04) compared to MRSA-negative controls. Surgery and indwelling intravenous access devices have been identified as independent risk factors for MRSA infection [4, 23]. Worse pulmonary status is also a risk factor for acquisition [15, 17].

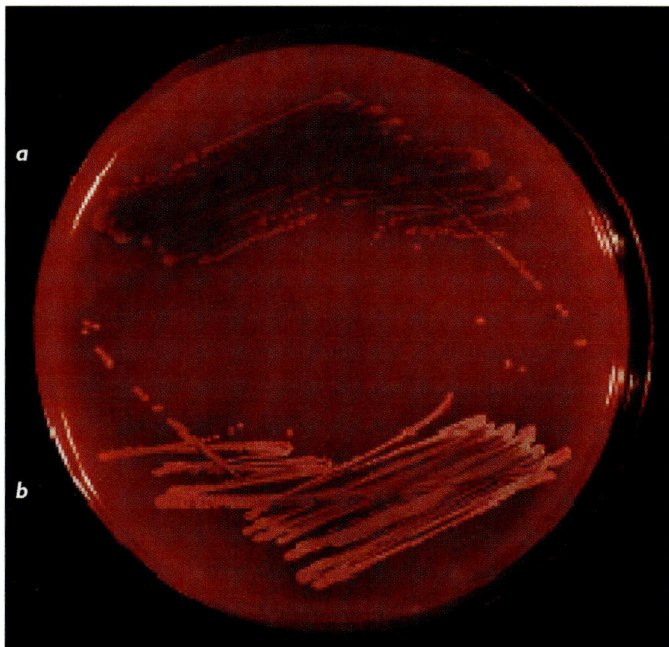

Fig. 3. *S. aureus* SCVs (*a*) and normal phenotype (*b*) cultured on sheep blood agar after 24 h of incubation at 35°C. From Seifert et al.: Emerg Infect Dis 1999;5:450–453.

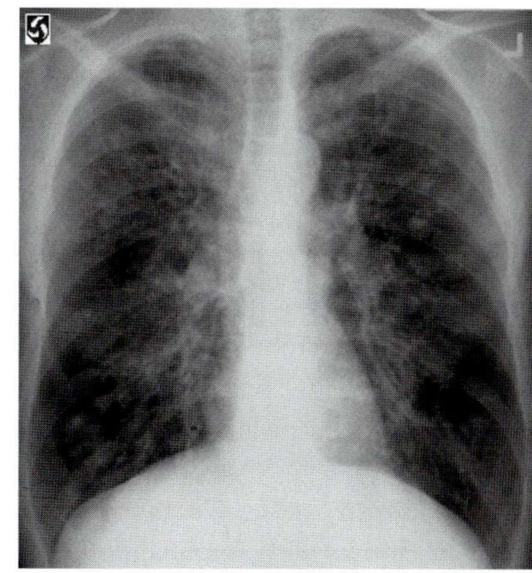

Fig. 4. Chest x-ray of CF patient with chronic MRSA infection.

Clinical Impact of S. aureus Infection in Cystic Fibrosis

Evidence for S. aureus-*Exacerbating Lung Disease in Cystic Fibrosis*

S. aureus colonization in the placebo group in the recent study by Stutman et al. [3] of continuous prophylactic antibiotic treatment was not associated with an increase in pulmonary symptoms. However, *S. aureus* infection and inflammation have been found in infants with CF as young as 3 months of age, suggesting the possibility of early lung damage [25]. Infection with both *S. aureus* and *P. aeruginosa* has been identified as a significant risk factor for a rapid decline in lung function and consideration for referral for lung transplantation [26]. Chronic infection, up to 11 years, has been documented in several other studies [4, 15, 17, 20].

Evidence that MRSA Infection Exacerbates Cystic Fibrosis Lung Disease

Although chronic MRSA infection in CF is associated with severe bronchiectasis, no definite link between MRSA infection and deterioration in lung function has been identified. However in some patients marked changes are found on chest x-ray (fig. 4). Healthier patients appear to tolerate the infection well [4]. Although some patients may show increased dyspnoea, wheeze and sputum production concurrent with MRSA isolation, Solis et al. [18] found no change in respiratory function, nutritional status, or chest x-ray appearances in children with MRSA infection, irrespective of whether or not the infection was successfully eradicated. Other studies [4, 15, 21] also suggest that MRSA infection has only a minimal effect in CF despite its undisputed pathogenic role in other patients [27, 28]. These studies were, however, short, uncontrolled and in small numbers.

Treatment of S. aureus Infection

Choice of Antibiotic

For methicillin-susceptible *S. aureus* (MSSA) the mainstays of therapy are the isoxazolyl penicillins, such as flucloxacillin. The addition of second agents, such as fusidic acid, rifampicin, or gentamicin, has been advocated for the treatment of serious infections [29]. The treatment of MSSA in patients with genuine penicillin allergy is more problematic. Erythromycin and clindamycin may be less active than penicillins. Glycopeptides such as vancomycin or teicoplanin may be useful but they are only available as intravenous agents and have significant toxicity. In serious infections with *S. aureus* consideration should therefore be given to desensitization of penicillin-allergic patients. The mainstays of therapy for MRSA are the glycopeptides. In non-CF patients there is some evidence that clinical outcome may be improved if rifampicin is added to vancomycin [30]. It is not known if combination therapy improves outcome in CF patients with MRSA infection.

Continuous or Intermittent Antibiotics

Approaches to the treatment of *S. aureus* infection include short-term antibiotic courses as a routine response to a positive sputum culture, antibiotic therapy only if the sputum culture is associated with clinical symptoms or signs of lower respiratory tract disease, or continuous prophylactic antibiotic treatment from diagnosis [31, 32]. The aim of the latter is to reduce the prevalence of *S. aureus* infection and thereby any associated inflammatory changes and lung damage, but there are concerns about increased risks of bacterial resistance and *P. aeruginosa* infection [3].

Reviewing a 15-year period of intermittent treatment for each *S. aureus* isolate whether or not the patient had clinical symptoms, the Copenhagen centre reported a chronic infection rate (>6 months continuously) of <10%. Staphylococcal infection was eradicated in approximately 74% of cases with each 14 days of antibiotic therapy. Repeated or extended treatment for 1–3 months was successful in most of the remaining cases. This success rate was maintained when treating subsequent relapses without the development of antibiotic resistance. Patients received treatment for *S. aureus* infection on average twice a year and remained infection free on average for 10 months each year [2].

A systematic review of the various clinical approaches to the treatment of *S. aureus* infection identified 13 trials (11 in children 5 or less years of age), the use of 19 different antibiotics, and a variety of laboratory and clinical outcome measures [33]. Treatment frequently cleared *S. aureus* from the sputum and thus prophylactic antibiotic administration was thought likely to be of benefit in young children. However, eradication of *S. aureus*, though desirable, is not the goal of treatment and improvement in lung function was rarely measured or observed in these studies. There was no convincing evidence of reduced symptoms, and effect on survival was not addressed. The review concluded that a large, randomized, approximately 2-year study was needed to answer the question whether prophylactic or intermittent treatment improved clinical status and/or led to earlier *P. aeruginosa* acquisition.

The Cochrane Review [34] of continuous versus no prophylactic antibiotic treatment of *S. aureus* infection in CF found only 4 studies meeting its inclusion criteria. Fewer children receiving prophylaxis had ≥1 *S. aureus* isolate when treatment was begun early in infancy and continued for up to 6 years of age. The authors, on the evidence available, were not able to reach conclusions on the clinical impact of this lower infection rate, or which approach to recommend. They also found no significant difference in the number of *Haemophilus influenzae* or *P. aeruginosa*

isolates between the two groups, nor any report of MRSA or *Burkholderia cepacia* complex infection. There was a trend for a lower cumulative *P. aeruginosa* isolation rate in the prophylaxis group after 2–3 years of treatment and towards a higher isolation rate after 4–6 years. The reviewers had insufficient data to answer the question whether this trend towards more *P. aeruginosa* infection after 4–6 years of continuous anti-staphyloccocal antibiotics was a chance finding or not, or whether this result reflected the choice of antibiotic or the duration of treatment. Their conclusions were necessarily based on the work of Stutman et al. [3], the only group to have followed children, treated with either cephalexin or placebo, for up to 6 years. Patient withdrawal (90 of 209) may have resulted in reporting bias [34]. Nonetheless, the increased frequency of *P. aeruginosa* isolates in the treatment group was noted in the first year of the study, and persisted throughout.

Ratjen et al. [35] (not included in the Cochrane Review) used the German database to assess the effect of anti-staphylococcal prophylaxis on the rate of *P. aeruginosa* acquisition. Patients receiving continuous treatment had significantly fewer *S. aureus* sputum isolates but a significantly higher rate of *P. aeruginosa* acquisition. This finding was predominantly seen in children under 6 years of age. It is, however, difficult to draw any firm conclusions from this study which is based on retrospective data. 'Continuous treatment' was in fact antibiotics for ≥200 days per year only. Only 3% of patients received flucloxacillin, the rest receiving a variety of broad-spectrum antibiotics. Nearly half had a cephalosporin, which, taken in conjunction with the findings of the study by Stutman et al., might lead one to conclude that it is cephalosporins which are particularly selective for *Pseudomonas* isolation. The authors accept that the administration of 'continuous treatment' may have been a response to increased symptoms. It is possible that these may themselves have been a result of already established early *P. aeruginosa* infection.

Prevention of MRSA

Patients with MRSA should be separated from others, taking care to minimize stigma and sense of isolation [4]. The importance of adherence to basic good hygiene, especially with invasive procedures, must be regularly and frequently emphasized [20]. A policy for the handling and cleaning of equipment should be agreed with the hospital infection control team.

The role of screening swabs for MRSA in patients with CF remains unclear. The CF Foundation Infection Control

Table 1. Published data on eradication strategies used against MRSA in patients with CF

Reference		Regimen	Duration	Outcome
Maiz et al. [44]		Aerosolized vancomycin 250 mg in 4 ml sterile H$_2$O nebulized twice daily (preceded by nebulized terbutaline 500 µg) for 10 min	17 months	reduced sputum counts of MRSA but no eradication
Solis et al. [18]		Aerosolized vancomycin 4 mg/kg/dose diluted in normal saline four times daily (preceded by nebulized salbutamol)	5 days	Successful eradication in 7 of 12 patients for mean of 12 months
	Tracheostomy	2% vancomycin cream twice daily; change foreign body		
	Nasal carriage	2% mupirocin cream four times daily or 2% vancomycin cream four times daily		
	Oropharyngeal carriage	2% vancomycin paste or 2% vancomycin gel or 5 mg vancomycin lozenges four times daily		
	Gastro-intestinal carriage	40 mg/kg/day vancomycin oral suspension in four divided doses		
	Skin carriage	4% chlorhexidine bath/ shower on alternate days		
Garske et al. [22]		rifampicin 600 mg four times per day orally plus sodium fusidate 250–500 mg twice daily orally	6 months	Successful eradication in 5 of 7 patients for a mean of 6 months

guidelines make no reference to the role of MRSA screening [20]. United Kingdom guidelines on control of MRSA make no specific mention of CF [36]. However, they do suggest that any regional referral centre should consider screening for MRSA. Screening can identify carriers who may pose a continued source of cross-infection. Throat carriage is more difficult to eradicate than nasal carriage. Sending nose swabs alone has a sensitivity of 78% whereas sending swabs taken from nose, throat and perineum has a sensitivity of 98%. There are also no universally agreed criteria for which sites should be sampled, by which method, how often, or over what time frame. It is also not certain how many negative screening swabs are needed to formally declare patients free of MRSA. The role of screening other family members requires further study. Another potential source of MRSA that needs further study is family pets, particularly dogs [37].

Eradication of MRSA

This may be more difficult in patients with CF since infection may involve the lower respiratory and gastro-intestinal tracts as well as the nose, pharynx, and skin surface. The eradication trials in CF are uncontrolled and only involved small numbers of patients (table 1). These studies also included many CF patients with established MRSA infection rather than patients presenting with their first positive culture for MRSA. Further studies are needed to determine the optimum eradication regimens.

Table 2. Recommended regimens for treating MRSA colonization/infection of non-respiratory sites – Working Party Report [36]

Nasal carriage	2% nasal mupirocin – each nostril three times daily for 5 days If 2 treatment failures (or isolate is mupirocin resistant): naseptin cream (0.5% neomycin plus 0.1% chlorhexidine)
Skin carriage	Bathe for 5 days with an antiseptic detergent. Options include: 4% chlorhexidine or 2% triclosan or 7.5% povidone-iodine Wash hair twice weekly with one of the above apply hexachlorophane powder (e.g. 0.33% SterZac) to axillae/groins

There are many published eradication regimens for management of non-lower-respiratory-tract MRSA colonization and infection. The evidence for the efficacy of these has been reviewed [36] and recommendations are shown in table 2.

Treatment of MRSA

Lower respiratory tract infection should be treated according to antibiotic sensitivity patterns. MRSA is often susceptible to a number of antibiotics. Teicoplanin is often preferred to vancomycin for intravenous treatment because of its better safety profile. Fusidic acid and rifampicin should not be used alone as resistance may develop rapidly [38]. Linezolid, a recently available oxazolidinone antibiotic, acts by inhibiting the initiation complex formation in bacterial protein synthesis. Resistance is slow to develop and because of its unique mechanism of action it has shown no cross-resistance with other antibacterial agents. It has both in vitro activity against MRSA and clinical efficacy (in non-CF patients) equivalent to, or better than vancomycin [39, 40]. Good linezolid sputum penetration has been shown in adult patients with CF following oral administration. Mean levels exceeded the required MIC for the treatment of MRSA for >80% of the dosing period for serum and for the majority of the dosing period for sputum [41]. Successful eradication of MRSA infection in adult patients with CF has been reported [42, 43].

Conclusion

S. aureus remains one of the main pathogens in CF. The optimum approach to management, particularly in relation to prophylaxis, is still an area of intense debate. The recent emergence of MRSA has created new therapeutic and infection control problems for CF units. Whilst there is still uncertainty regarding the overall clinical impact of MRSA in patients with CF, the future is likely to yield more evidence of significant long-term morbidity. It is therefore imperative that further research into optimizing preventative strategies, particularly with respect to eradication regimens, is conducted as a matter of urgency.

References

1 Beringer PM, Appleman MD: Unusual bacterial respiratory flora in cystic fibrosis: Microbiologic and clinical features. Curr Opin Pulm Med 2000;6:545–560.
2 Hoiby N, Frederiksen B: Microbiology of cystic fibrosis; in Hodson ME, Geddes DM (eds): Cystic Fibrosis, ed 2. London, Arnold, 2000.
3 Stutman HR, Lieberman JM, Nussbaum E, Marks MI: Antibiotic prophylaxis in infants and young children with cystic fibrosis: A randomized controlled trial. J Pediatr 2002;140: 299–305.
4 Rao G, Gaya H, Hodson M: MRSA in cystic fibrosis. J Hosp Infect 1998;49:179–191.
5 Cystic Fibrosis Foundation, Patient Registry 1996: Annual Data Review, Bethesda, August 1997.
6 Maurer JR, Frost AE, Estenne M, Higenbottam T, Glanville AR: International guidelines for selection of lung transplant candidates. Transplantation 1998;66:951–956.
7 Doern GV, Brogden-Torres B: Optimum use of selective plated media in primary processing of respiratory tract samples from patients with cystic fibrosis. J Clin Microbiol 1992;30: 2740–2742.
8 Gibson RL, Burns JL, Ramsey BW: Pathophysiology and management of pulmonary infections in cystic fibrosis. Am J Respir Crit Care Med 2003;68:918–951.
9 Flayhart D, Lema C, Borek A, Carroll KC: Comparison of the BBL CHROMagar Staph aureus agar medium to conventional media for detection of *Staphylococcus aureus* in respiratory samples. J Clin Microbiol 2004;42: 3566–3569.
10 Hogardt M, Trebesius K, Geiger AM, Hornef M, Rosenecker J, Heesemann J: Specific and rapid detection by fluorescent in situ hybridization of bacteria in clinical samples obtained from cystic fibrosis patients. J Clin Microbiol 2000;38:818–825.
11 Van Belkum A, Renders NH, Smith S, Overbeek SE, Verbrugh HA: Comparison of conventional and molecular methods for the detection of bacterial pathogens in sputum samples from cystic fibrosis patients. FEMS Immunol Med Microbiol 2000;27:51–57.
12 Goerke C, Kraning K, Stern M, Doring G, Botzenhart K, Wolz C: Molecular epidemiology of community-acquired *Staphylococcus aureus* in families with and without cystic fibrosis patients. J Infect Dis 2000;181:984–989.
13 Kahl BC, Duebbers A, Lubritz G, Haeberle J, Koch HG, Ritzerfeld B, Reilly M, Harms E, Proctor RA, Herrmann M, Peters G: Population dynamics of persistent *Staphylococcus aureus* isolated from the airways of cystic fibrosis patients during a 6-year prospective study. J Clin Microbiol 2003;41: 4424–4427.
14 Sadowska B, Bonar A, von Eiff C, Proctor RA, Chmiela M, Rudnicka W, Rozalska B:

Characterization of *Staphylococcus aureus*, isolated from airways of cystic fibrosis patients, and their small colony variants. FEMS Immunol Med Microbiol 2002;32: 191–197.

15 Thomas SR, Gyi KM, Gaya H, Hodson ME: Methicillin-resistant *Staphylococcus aureus* – impact at a national cystic fibrosis centre. J Hosp Infect 1998;40:203–209.

16 Burns JL, Emerson J, Stapp JR, Yim DL, Krzewinski J, Louden L, Ramsey BW, Clausen CR: Microbiology of sputum from patients at cystic fibrosis centers in the United States. Clin Infect Dis 1998;27:158–163.

17 Miall LS, McGinley NT, Brownlee KG, Conway SP: Methicillin resistant *Staphylococcus aureus* (MRSA) infection in cystic fibrosis. Arch Dis Child 2001;84:160–162.

18 Solis A, Brown D, Hughes J, Van Saene HK, Heaf DP: Methicillin-resistant *Staphylococcus aureus* in children with cystic fibrosis: An eradication protocol. Pediatr Pulmonol 2003;36:189–195.

19 Cystic Fibrosis Foundation, Patient Registry 2002: Annual data review, Bethesda 2001.

20 Saiman L, Siegel J: Cystic Fibrosis Foundation. Infection control recommendations for patients with cystic fibrosis: Microbiology, important pathogens and infection control practices to prevent patient-to-patient transmission. Infect Control Hosp Epidemiol 2003; 24(suppl 5):S6–S52.

21 Givney R, Vickery A, Holliday A, Pegler M, Benn R: Methicillin-resistant *Staphylococcus aureus* in a cystic fibrosis unit. J Hosp Infect 1997;35:27–36.

22 Garske LA, Kidd TJ, Gan R, Bunting JP, Franks CA, Coulter C, Masel PJ, Bell SC: Rifampicin and sodium fusidate reduces the frequency of methicillin-resistant *Staphylococcus aureus* (MRSA) isolations in adults with cystic fibrosis and chronic MRSA infection. J Hosp Infect 2004;56:208–214.

23 Nadesalingam K, Conway SP, Denton M: Risk factor for acquisition of methicillin-resistant *Staphylococcus aureus* (MRSA) by patients with cystic fibrosis. J Cystic Fibrosis, in press.

24 Charlebois ED, Perdreau-Remington F, Kreiswirth B, Bangsberg DR, Ciccarone D, Diep BA, Ng VL, Chansky K, Edlin B, Chambers HF: Origins of community strains of methicillin-resistant *Staphylococcus aureus*. Clin infect Dis 2004;39:47–54.

25 Armstrong DS, Grimwood K, Carzino R, Carlin JB, Olinsky A, Phelan PD: Lower respiratory infection and inflammation in infants with newly diagnosed cystic fibrosis. BMJ 1995;310:1571–1572.

26 Rosenbluth DB, Wilson K, Ferkol T, Schuster DP: Lung function decline in cystic fibrosis patients and timing for lung transplantation referral. Chest 2004;126:412–419.

27 Romero-Vivas J, Rubio M, Fernandez C, Picazo JJ: Mortality associated with nosocomial bacteraemia due to methicillin-resistant *Staphylococcus aureus*. Clin Infect Dis 1995; 21:1417–1423.

28 Rello J, Torres A, Ricart M, Valles J, Gonzalez J, Artigas A, Rodriguez-Roisin R: Ventilator-associated pneumonia by *Staphylococcus aureus*: Comparision of methicillin-resistant and methicillin-sensitive episodes. Am J Respir Crit Care Med 1994;150:1545–1549.

29 Cystic Fibrosis Trust: Antibiotic treatment for cystic fibrosis, April 2000.

30 Burnie J, Matthews R, Jiman-Fatami A, Gottardello P, Hodgetts S, D'Arcy S: Analysis of 42 cases of septicemia caused by an epidemic strain of methicillin-resistant *Staphylococcus aureus*: Evidence of resistance to vancomycin. Clin Infect Dis 2000;31: 684–689.

31 Taylor RFH, Hodson ME: Cystic fibrosis: Antibiotic prescribing practices in the UK and Eire. Respir Med 1993;87:535–539.

32 Weaver LT, Green MR, Nicholson K, Mills J, Heeley ME, Kuzemko JA, Austin S, Gregory GA, Dux AE, Davis JA: Prognosis in cystic fibrosis treated with continuous flucloxacillin from the neonatal period. Arch Dis Child 1994;70:84–89.

33 McCaffery K, Olver RE, Franklin M, Mukhopadhyay S: Systematic review of anti-staphylococcal antibiotic therapy in cystic fibrosis. Thorax 1999;54:380–383.

34 Smyth A, Walters S: Prophylactic antibiotics for cystic fibrosis. Cochrane Database Syst Rev 2003:CD001912.

35 Ratjen F, Comes G, Paul K, Posselt HG, Wagner TO, Harms K: Effect of continuous antistaphylococcal therapy on the rate of *Pseudomonas aeruginosa* acquisition in patients with cystic fibrosis. Pediatr Pulmonol 2001;31:13–16.

36 Working Party Report. Revised guidelines for the control of methcillin-resistant *Staphylo-coccus aureus* infection in hospitals. J Hosp Infect 1998;39:253–290.

37 Manian FA: Asymptomatic nasal carriage of mupirocin-resistant methicillin-resistant *Staphylococcus aureus* in a pet dog associated with infection in household contacts. Clin Infect Dis 2003;36:26–28.

38 Scheel O, Lyon DJ, Rosdahl VT, Adeyemi-Doro FA, Ling TK, Cheng AF: In-vitro susceptibility of isolates of methicillin-resistant *Staphylococcus aureus* 1988–1993. J Antimicrob Chemother 1996;37:243–251.

39 Kaplan SL, Afghani B, Lopez P, Wu E, Fleishaker D, Edge-Padbury B, Naberhuis-Stehouwer S, Bruss JB: Linezolid for the treatment of methicillin-resistant *Staphylococcus aureus* infections in children. Pediatr Infect Dis J 2003;22 (suppl):S178–S185.

40 Wunderink RG, Rello J, Cammarata SK, Croos-Dabrera RV, Kollef MH: Linezolid versus vancomycin: Analysis of two double-blind studies of patients with methicillin-resistant *Staphylococcus aureus* nosocomial pneumonia. Chest 2003;124:1789–1797.

41 Saralaya D, Peckham D, Hulme B, Tobin C, Denton M, Conway S, Etherington C: Serum and sputum levels following oral administration of linezolid in adults with cystic fibrosis. J Antimicrob Chemother 2004;53:325–328.

42 Ferrin M, Zuckerman JB, Meagher A, Blumberg EA: Successful treatment of methicillin-resistant *Staphylococcus aureus* pulmonary infection with linezolid in a patient with cystic fibrosis. Pediatr Pulmonol 2002; 33:221–223.

43 Serisier DJ, Jones G, Carroll M: Eradication of pulmonary methicillin-resistant *Staphylococcus aureus* (MRSA) in cystic fibrosis with linezolid. J Cystic Fibrosis 2004;3:61.

44 Maiz L, Canton R, Mir N, Baquero F, Escobar H: Aerosolized vancomycin for the treatment of methicillin-resistant *Staphylococcus aureus* infection in cystic fibrosis. Pediatr Pulmonol 1998;26:287–289.

Dr. Steven Conway
St. James' Hospital
Beckett Street
Leeds, LS9 7TF (UK)
Tel. +44 113 2206 4966
Fax +44 113 2206 4966
E-Mail steven.conway@leedsth.nhs.uk

Bush A, Alton EWFW, Davies JC, Griesenbach U, Jaffe A (eds): Cystic Fibrosis in the 21st Century.
Prog Respir Res. Basel, Karger, 2006, vol 34, pp 160–165

Nontuberculous Mycobacterial Lung Disease in Patients with Cystic Fibrosis

David E. Griffith

University of Texas Health Center, Tyler, Tex., USA

Abstract

Nontuberculous mycobacteria (NTM) are important emerging pathogens in patients with cystic fibrosis (CF). The two most important potential NTM pathogens are *Mycobacterium abscessus* and *Mycobacterium avium* complex (MAC). The diagnosis of NTM lung disease in CF patients in complicated by the overlap in symptoms and clinical findings between disease caused by NTM, CF and other respiratory pathogens. The impact of NTM infection on the course of CF in general is not yet determined, but for some patients, NTM infection can be a rapidly progressive process. *M. abscessus* also appears to be a more virulent respiratory pathogen than MAC. It is also unclear if treatment is necessary for all CF patients who meet diagnostic criteria for NTM lung disease, particularly MAC lung disease. Treatment of *M. abscessus* lung disease is especially difficult, and with current agents, is not likely to result in microbiologic cure. Much more study is necessary to answer important fundamental questions about the diagnosis, natural history and treatment of NTM infections in CF patients.

The impact of nontuberculous mycobacterial (NTM) infection on cystic fibrosis (CF) patients remains incompletely understood. Defining NTM disease in CF patients is challenging because of the variable yet inexorable progression of CF, the generally indolent progression of NTM disease and the considerable overlap of clinical and radiographic abnormalities between disease caused by NTM and CF. Some case reports have described an association between NTM infection and deterioration in clinical and radiographic features, whereas others note a prolonged time span between initial recovery of NTM and subsequent adverse effects [1–3]. Others suggest that NTM may be recovered from the lower airways of some CF subjects without definite or discernable adverse effect [4, 5]. Although many uncertainties remain, it is apparent that NTM are emerging pathogens in CF patients and that investigation in this area is only just beginning. A list of NTM respiratory pathogens is provided in table 1.

Epidemiology

Prior to 1990, NTM had been associated with CF patients in only a small number of reports [6–10]. Since that time, multiple centers in the US and Europe have reported the prevalence of NTM recovered from respiratory specimens to be 4–20% of the CF population screened [11]. In the largest study published to date, Olivier et al. [4, 5] performed a cross-sectional assessment of NTM prevalence over the course of a year from approximately 1,000 subjects age 10 and above from 21 US CF centers. The overall prevalence of NTM isolated from sputum was 13% – *Mycobacterium avium* complex (MAC) in 72%; *Mycobacterium abscessus* in 16%. The majority of patients, 90/128 (70%), had only one of three AFB positive cultures and AFB smears were positive in only 33 (26%) of the culture-positive

Table 1. Nontuberculous mycobacterial respiratory pathogens

Common	Uncommon
M. avium complex (MAC)[1]	*M. asiaticum*
M. avium	*M. celatum*
M. intracellulare	*M. chelonae*
M. abscessus[1]	*M. fortuitum*
M. kansasii	*M. haemophilum*
M. malmoense	*M. scrofulaceum*
M. xenopi	*M. Shimodii*
	M. simiae
	M. smegmatis
	M. szulgai

[1]Commonly isolated from patients with CF.

patients. The prevalence was highly correlated with age, approaching 40% in patients over age 40, although some recent reports suggest a high prevalence in younger subjects as well [3, 12]. Compared to NTM culture-negative individuals, CF subjects with NTM seemed to have milder lung disease as reflected by a higher FEV_1, a higher frequency of *Staphylococcus aureus*, and lower frequency of *Pseudomonas aeruginosa* isolated from sputum [4].

Pathophysiology

CF patients are at risk for NTM infection, presumably because of underlying structural airway disease and altered mucociliary clearance. A common form of MAC lung disease in older, female non-CF patients is characterized radiographically by the presence of nodules and bronchiectasis and termed nodular/bronchiectatic MAC lung disease [13]. It is unknown whether bronchiectasis in these patients is a result of or a predisposition for the MAC infection. An increased prevalence of mutations in CFTR has been reported in these patients, suggesting a possible role of CFTR for predisposing this older population to NTM infection [14], although these findings were not reproducible in another study [11].

The presence of bronchiectasis per se, may be an important factor that predisposes CF patients to NTM, as infection has also been described in primary ciliary dyskinesia [15]. Once bronchiectasis develops in older female patients with nodular/bronchiectatic MAC lung disease, they are at risk for repeated MAC infection (reinfection) by new MAC strains [16].

Similarly to other infected patient populations, a definite source for acquisition of NTM by CF patients has not yet been identified. No evidence has emerged of person-to-person transmission of NTM in non-CF patient populations. Two single-center studies and a large multicenter study using molecular epidemiologic techniques have also failed to show any evidence of person-to-person transfer of NTM in CF patients [3, 4, 17]. Nosocomial acquisition of NTM from institutional water supplies is also a concern [18]. Clinical observations suggest, however, no greater exposure to medical water reservoirs among NTM culture positive CF patients compared to NTM culture negative patients [3, 4]. Institutional water reservoirs remain potential sources of concern for selected patients; however, as illustrated in a recent study of an *Mycobacterium simiae* pseudo-outbreak which included a CF patient with multiple AFB-smear-positive isolates [19].

Clinical Impact of Nontuberculous Mycobacterial Infection in Cystic Fibrosis Patients

Olivier et al. [5] described the clinical impact of NTM isolates from 60 CF patients. The majority (57%) had only 1 positive sputum culture, 26% had 2 or 3 positive cultures and 17% had more than three. The more positive sputum cultures for AFB, the more likely a patient would have specimens that were AFB smear positive. Thirty-seven percent met current American Thoracic Society (ATS) dianositic criteria for NTM lung disease. MAC was recovered from the majority (57%) with *M. abscessus* next most common (12%), although the latter group was more likely to meet ATS diagnostic criteria for NTM lung disease. Comparing 99 CF patients without positive NTM cultures (controls) and NTM culture positive patients who did and did not meet ATS disease criteria, there were no clinical differences, including number of days with hemoptysis and FEV_1 decline over the period of the study. Two or more HRCT findings suggesting NTM disease were more common in NTM-positive patients but also occurred in 19% of controls. Similarly, progressive abnormalities were more commonly seen in NTM culture positive patients, but they were also seen in controls.

The two studies from Olivier et al. [4, 5] suggest that NTM are: (a) relatively frequently isolated from CF patients, (b) not generally associated with severe disease, and (c) are not associated with a detectable deleterious clinical impact of NTM on the CF patients studied [4, 5]. These studies also have significant limitations including the short duration (15 months), which may be inadequate to evaluate a process that can take years or decades to progress to a clinically significant level in non-CF patients. Additionally, the study analysis grouped together patients infected with different NTM species, MAC and *M. abscessus*, that probably impact CF patients in significantly different ways.

There are other studies that expand on and help clarify the findings by Olivier et al. [4, 5]. Tomashefski et al. [1] retrospectively studied at autopsy the lungs from 18 CF patients who had antemortem respiratory cultures positive for NTM [1]. Twelve patients had only 1 positive NTM culture, and only 3 patients had multiple positive cultures for the same NTM. These same 3 patients were the only ones with multiple AFB smear positive specimens. 'M. chelonae group' (*M. abscessus*) was isolated from 10 patients, *M. fortuitum* from 6 patients, MAC from 4 patients and 'rapidly growing Mycobacteria' from 3 patients. Clinically important mycobacterial disease was suspected before death in only 3 patients, the 3 with multiple positive cultures. At autopsy, necrotizing granulomatous inflammation with AFB smear positive specimens was found in 2 of the 3 patients with multiple positive cultures for *M. chelonae* group. No patients with a single positive NTM culture had pathologic evidence of NTM infection. The authors concluded that the patients who had only 1 positive sputum culture and no significant granulomatous inflammation were not clinically infected. A second study by Cullen et al. [2] described a case of *M. abscessus* lung disease in a patient with CF and an indolent course. In a more recent report, Sermet-Gaudelus et al. [3] recovered *M. abscessus* from 15 of 296 patients screened with serial sputum specimens. Ten of these had ≥3 positive cultures and 6 had positive smears. Only 4 of these patients had disease documented by temporal clinical decline responding to specific antimycobacterial treatment or histopathologic findings and death.

These studies demonstrate some very important aspects of NTM disease in CF patients [1–3]. First, some CF patients unquestionably have significant NTM infection, especially due to *M. abscessus*, based on the presence of granulomatous inflammation in tissue associated with clinical decline not explained by other infecting organisms. Clearly also, not all patients with a single positive NTM culture isolated from sputum have clinical deterioration related to NTM or pathological evidence of mycobacterial infection. Alternatively, patients with AFB smear positive specimens and multiple positive AFB cultures for NTM are more likely to have clinically significant NTM infection. These studies also suggest that *M. abscessus* is more virulent for CF patients than other NTM, specifically MAC.

For some patients with progressive *M. abscessus* disease, there is little doubt that clinical decline is a consequence of *M. abscessus* infection and that some patients benefit from therapy against *M. abscessus* [1–3]. Even with a relatively virulent organism such as *M. abscessus*, the course of NTM infection can be unpredictable and sometimes indolent. It may be that MAC lung disease in CF is so indolent that the relatively limited life expectancy of CF patients almost precludes the appearance of significant MAC disease for all but a few patients with more aggressive MAC disease [4, 5].

Diagnosis

A critically important short-term goal must be establishing rigorous, uniform and easily applied diagnostic criteria for NTM lung disease in patients with CF. As noted above, sputum AFB analysis is an essential element of the diagnostic evaluation, but the results must still be interpreted in the context of less well-characterized diagnostic tests such as spirometry, chest radiographs and HRCT scans [4, 5]. Another potentially useful test is bronchoscopy with transbronchial biopsy. The major advantage of this approach is that it allows the identification of pathological changes associated with significant mycobacterial infection within the lung, i.e. granulomatous inflammation. There are, however, significant risks of this approach so that the role of bronchoscopy as a routine diagnostic tool in this setting is not determined.

Although they are not optimal, the diagnostic criteria suggested by the ATS for NTM lung disease are generally applicable to CF patients [13]. The most recently published ATS NTM guidelines suggest that in the appropriate clinical setting and with other potential causes of lung disease excluded, recovery of the same species of NTM with multiple cultures is the core of the diagnostic criteria [13]. It is noteworthy that these diagnositic criteria are currently undergoing revision.

Unfortunately, even following the ATS guidelines, the diagnosis of NTM lung disease in CF patients can be quite difficult due to overlapping symptoms and radiographic changes attributable to the underlying CF. It can be difficult to exclude other causes of clinical deterioration given the frequent presence of other organisms such as *Pseudomonas aeruginosa* that can be associated with similar clinical disease. Aggressive treatment of bacteria usually associated

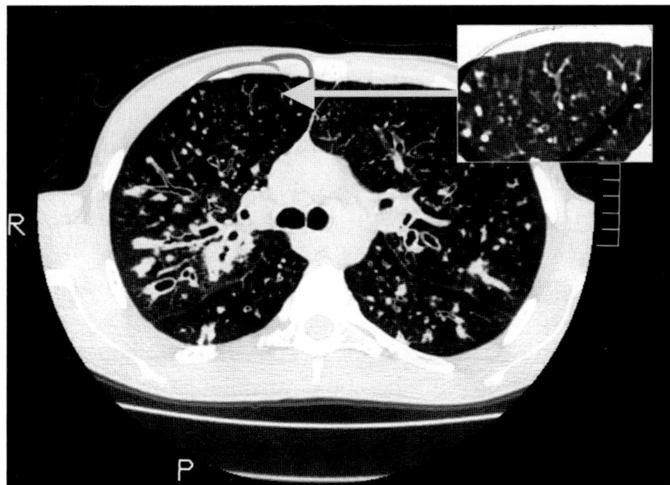

Fig. 1. High-resolution CT image from a 20-year-old man with cystic fibrosis and multiple sputum cultures strongly positive for *M. abscessus*. The CT scan demonstrates severe bronchiectasis associated with nodular densities and peripheral fine branching structures and nodular tips (magnified inset) which is the 'tree-in-bud' finding that is characteristic of NTM infection.

with CF pulmonary exacerbations should generally be pursued prior to consideration of specific antimycobacterial treatment. Persistence of signs and symptoms such as increased cough, sputum production, fever, weight loss or decline in FEV_1 from baseline despite treatment of known CF pathogens or an increase in the frequency of exacerbations despite maximal maintenance treatment with airways clearance measures, may be consistent with significant mycobacterial disease.

Plain chest radiographs are not adequate for diagnosing NTM lung disease in CF patients [5]. Rather, high-resolution CT (HRCT) scanning is recommended. HRCT findings characteristic of NTM lung disease include: (1) cystic and or cavitary parenchymal lung disease; (2) subsegmental (or larger) parenchymal consolidation; (3) single or multiple pulmonary nodules, and (4) tree-in-bud opacities (fig. 1) [20–22]. In the study by Olivier et al. [5], the findings compatible with NTM infection were more prevalent among CF subjects who met the ATS diagnostic criteria for disease than among culture-positive subjects who did not meet these criteria or culture-negative subjects. Furthermore, progression of these findings was more common over the 15-month period in subjects who met the ATS diagnostic criteria [5]. HRCT findings still overlapped with those from CF patients without microbiologic evidence of NTM lung disease.

While the majority of CF patients from whom NTM are recovered will likely not meet ATS microbiologic diagnostic criteria for disease at the time of initial culture positivity, close surveillance of these patients is warranted, including follow-up sputum mycobacterial cultures, especially if *M. abscessus* is recovered. [1–3, 6–10, 23–25]. Once treatment for NTM is initiated, it is important to obtain sputum cultures monthly to assess the effect on the burden of organisms and to document the time of sputum conversion, as this will impact the length of treatment. It is important that common CF pathogens such as *Pseudomonas aeruginosa* be maximally treated prior to initiating specific antimycobacterial treatment in order to assess clinical response. HRCT and serial measures of lung function will also be helpful in this respect. The primary microbiologic goal of therapy for NTM pulmonary disease in non-CF patients is 12 months of sputum culture negativity while on therapy [13, 26]. This goal may be very difficult to achieve in CF patients for several reasons including poor response of an organism to therapy (*M. abscessus*) or reinfection during therapy (MAC). For these complicated patients, other clinical factors such as symptoms and radiographic and spirometric results will also likely need to be included in the evaluation of treatment response.

Some organisms, such as MAC, tend to respond well to treatment regimens used in non-CF patients and therefore, following the standard recommendations by the ATS, seem appropriate (clarithromycin 1000 mg/day, ethambutol 15 mg/kg/day and rifampin 10 mg/kg/day – maximum dose 600 mg/day). This is generally well tolerated [13], although, if required, more aggressive therapy would include substitution of rifabutin 150–300 mg/day for rifampin and the addition of an injectable agent such as amikacin or streptomycin for several months early in the treatment regimen [13].

Assessment of the effect of NTM treatment on clinical parameters like FEV_1 may be confounded by the beneficial effects of macrolides in CF [27]. However, macrolide monotherapy in patients infected by MAC would result in macrolide-resistant MAC lung disease, which is difficult to treat [26], and this should therefore be avoided. Some experts also recommend screening CF patients for NTM respiratory isolates prior to initiating macrolide monotherapy in CF patients.

M. abscessus lung disease is very difficult to treat in both CF and non-CF patients [28]. No antibiotic regimens based on in vitro susceptibilities have been shown to lead predictably to long-term sputum conversion. The goal of 12 months of sputum culture negativity while on therapy may be unachievable for these patients. Alternative therapeutic goals such as symptomatic improvement, radiographic improvement or decrease in mycobacterial load are probably more realistic at this point for these patients. Monotherapy with clarithromycin or azithromycin may be associated with symptomatic improvement, but is not sufficient to produce microbiologic cure for *M. abscessus* lung disease and may lead to macrolide resistance. Combination therapy with a macrolide, amikacin and cefoxitin may produce clinical and microbiologic improvement, but cost and morbidity are significant impediments to a curative course of therapy with the parenteral agents. Medical therapy may, however, alleviate or delay patient symptoms and disease progression. For some patients, symptoms can be controlled with intermittent (6–8 weeks) periods of therapy with clarithromycin or azithromycin in combination with one or more parenteral drugs. This strategy is similar to the approach to managing *Pseudomonas* in CF. Suppressive therapy with periodic antibiotic administration to control the symptoms and progression of *M. abscessus* lung disease may be all that can be realistically be accomplished with present therapy. In general, antimycobacterial medications are well tolerated in younger patients. The effect of rifamycins, especially rifampin, on the metabolism of other drugs in a patient's regimen must be evaluated for each patient. Erratic absorption of oral drugs in pancreatic-insufficient CF patients and possible drug-drug interactions may affect the levels of these drugs in respiratory secretions [29]. Obtaining serum levels of drugs in a multidrug regimen after a few weeks of treatment and adjusting dosing based on these levels may have a beneficial response on treatment efficacy, although this approach is not proven for NTM disease [11, 29].

Surgical resection of limited NTM disease in non-CF patients, especially with *M. abscessus* disease, can contribute to cure. However, surgery in the setting of CF may be associated with an increased risk of mortality [28, 30]. Surgical resection, lobectomy or pneumonectomy, should be reserved for those who have an FEV_1 greater than 30% predicted and severe, symptomatic localized disease that fails to respond to aggressive medical therapy [30–32]. Poor control of the mycobacterial infection with medical management should be a relative contraindication for lung transplantation as the risk of overwhelming infection from these organisms in the posttransplantation period is increased [33, 34]. However, if the disease is manageable with medical therapy pretransplantation, the risk and impact of NTM recovery posttransplantation may be less of a concern [9, 35–38].

Future Studies

NTM are undoubtedly important pathogens in CF patients, but many formidable challenges remain in the evaluation and treatment of these patients. The management of NTM disease in CF is more difficult than in non-CF patients and more research is needed into every aspect of this problem. An easily applied set of diagnostic criteria is perhaps most important. Criteria to predict clinical disease activity and outcome, to identify those patients who would most benefit from therapy, would be extremely helpful. Studies are needed to determine if intervention in NTM lung disease improves the prognosis and clinical outcome of CF patients and if early intervention against NTM pathogens is meaningful or effective. Clearly, more and larger multicenter studies and interventional trials are necessary.

References

1 Tomashefski JF Jr, Stern RC, Demko CA, Doershuk CF: Nontuberculous mycobacteria in cystic fibrosis: An autopsy study. Am J Respir Crit Care Med 1996;154:523–528.

2 Cullen AR, Cannon CL, Mark EJ, Colin AA: *Mycobacterium abscessus* infection in cystic fibrosis: Colonization or infection? Am J Respir Crit Care Med 2000;161:641–645.

3 Sermet-Gaudelus I, Le Bourgeois M, Pierre-Audiger C, et al: *Mycobacterium abscessus* and children with cystic fibrosis. Emerg Infect Dis 2003;9:1587–1591.

4 Olivier KN, Weber DJ, Lee JH, Handler A, Tudor G, Molina PL, Tomashefski J, Knowles MR: Nontuberculous Mycobacteria in Cystic Fibrosis Study Group. Nontuberculous myco-

bacteria. II: Nested-cohort study of impact on cystic fibrosis lung disease. Am J Respir Crit Care Med 2003;167:835–840.

5 Olivier KN, Weber DJ, Wallace RJ Jr, Faiz AR, Lee JH, Zhang Y, Brown-Elliot BA, Handler A, Wilson RW, Schechter MS, Edwards LJ, Chakraborti S, Knowles MR: Nontuberculous Mycobacteria in Cystic Fibrosis Study Group. Nontuberculous mycobacteria. I: Multicenter prevalence study in cystic fibrosis. Am J Respir Crit Care Med 2003; 167:828–834.

6 Boxerbaum B: Isolation of rapidly growing mycobacteria in patients with cystic fibrosis. J Pediatr 1980;96:689–691.

7 Efthimiou J, Smith MJ, Hodson ME, et al: Fatal pulmonary infection with *Mycobac-

terium fortuitum* in cystic fibrosis. Br J Dis Chest 1984;78:299–302.

8 Smith MJ, Efthimiou J, Hodson ME, et al: Mycobacterial isolations in young adults with cystic fibrosis. Thorax 1984;39:369–375.

9 Kinney JS, Little BJ, Yolken RH, et al: *Mycobacterium avium* complex in a patient with cystic fibrosis: Disease vs colonization. Pediatr Infect Dis J 1989;8:393–396.

10 Kilby JM, Gilligan PH, Yankaskas JR, et al: Nontuberculous mycobacteria in adult patients with cystic fibrosis. Chest 1992;102: 70–75.

11 Olivier KN, Yankaskas JR, Knowles MR: Nontuberculous mycobacterial pulmonary disease in cystic fibrosis. Semin Repir Infect 1996;11:272–284.

12 Torrens JK, Dawkins P, Conway SP, et al: Non-tuberculous mycobacteria in cystic fibrosis. Thorax 1998;53:182–185.

13 Wallace RJ Jr, Glassroth J, Griffith DE, Olivier KN, Cook JL, Gordin F: American Thoracic Society: Diagnosis and treatment of disease caused by nontuberculous mycobacteria. Am J Respir Crit Care Med 1997; 156(suppl):S1–S25.

14 Ehrmantraut ME, Hilligoss DM, Steagall WK, et al: Pulmonary nontuberculous mycobacterial infections are highly associated with mutations in CFTR (abstract). Am J Respir Crit Care Med 2003;167: A708.

15 Noone PG, Leigh MW, Sannuti A, et al: Primary ciliary dyskinesia: Diagnostic and phenotypic features. Am J Respir Crit Care Med 2004;169:459–467.

16 Wallace RJ Jr, Zhang Y, Brown-Elliott BA, Yakrus MA, Wilson RW, Mann L, Couch L, Girard WM, Griffith DE: Repeat positive cultures in *Mycobacterium intracellulare* lung disease after macrolide therapy represent new infections in patients with nodular bronchiectasis. J Infect Dis 2002;186: 266–273.

17 Bange F-C, Brown BA, Smaczny C, et al: Lack of transmission of *Mycobacterium abscessus* among patients with cystic fibrosis attending a single clinic. Clin Infect Dis 2001; 32:1648–1650.

18 Wallace RJ Jr, Brown BA, Griffith DE: Nosocomial outbreaks/pseudo-outbreaks caused by nontuberculous mycobacteria. Annu Rev Microbiol 1998;52:453–490.

19 Conger NG, O'Connell RJ, Laurel VL, et al: *Mycobacterium simiae* pseudo-outbreak associated with a hospital water supply. Infect Control Hosp Epidem, in press.

20 Moore EH: Atypical mycobacterial infection in the lung: CT appearance. Radiology 1993; 187:777–782.

21 Tanaka E, Amitani R, Niimi A, et al: Yield of computed tomography and bronchoscopy for the diagnosis of *Mycobacterium avium* complex pulmonary disease. Am J Respir Crit Care Med 1997;155:2041–2046.

22 Fujita J, Ohtsuki Y, Suemitsu I, et al: Pathological and radiological changes in resected lung specimens in *Mycobacterium avium intracellulare* complex disease. Eur Respir J 1999;13:535–540.

23 Hjelte L, Petrini B, Kallenius G, et al: Prospective study of mycobacterial infections in patients with cystic fibrosis. Thorax 1990; 45:397–400.

24 Oliver A, Maiz L, Canton R, et al: Nontuberculous mycobacteria in patients with cystic fibrosis. Clin Infect Dis 2001;32: 1298–1303.

25 Fauroux B, Delaisi B, Clement A, et al: Mycobacterial lung disease in cystic fibrosis: A prospective study. Pediatr Infect Dis J 1997;16:354–358.

26 Wallace RJ Jr, Brown BA, Griffith DE, Girard WM, Murphy DT: Clarithromycin regimens for pulmonary *Mycobacterium avium* complex:. The first 50 patients. Am J Respir Crit Care Med 1996;153:1766–1772.

27 Saiman L, Marshall BC, Mayer-Hamblett N, et al: Azithromycin in patients with cystic fibrosis chronically infected with *Pseudomonas aeruginosa:* A randomized controlled trial. JAMA 2003;290:1749–1756.

28 Griffith DE, Girard WM, Wallace RJ Jr: Clinical features of pulmonary disease caused by rapidly growing mycobacteria: An analysis of 154 patients. Am Rev Respir Dis 1993;147: 1271–1278.

29 Gilljam M, Berning SE, Peloquin CA, et al: Therapeutic drung monitoring in patients with cystic fibrosis and mycobacterial disease. Eur Respir J 1999;14:347–351.

30 Nelson KG, Griffith DE, Brown BA, et al: Results of operation in *Mycobacterium avium-intracellulare* lung disease. Ann Thorac Surg 1998;66:325–330.

31 Hausler M, Frank E, Wendt G, et al: Pneumonectomy in CF. Pediatr Pulmonol 1999;28:376–379.

32 Smith MB, Hardin WD, Dressel DA, et al: Predicting outcomes following pulmonary resection in CF patients. J Pediatr Surg 1991; 26:655–659.

33 Trulock EP, Bolman RM, Genton R: Pulmonary disease caused by *Mycobacterium chelonae* in a heart-lung transplant recipient with obliterative bronchiolitis. Am Rev Respir Dis 1989;140:802–805.

34 Sanguinetti M, Ardito F, Fiscarelli E, et al: Fatal pulmonary infection due to multidrug-resistant *Mycobacterium abscessus* in a patient with cystic fibrosis. J Clin Microbiol 2001;39:816–819.

35 Ebert DL, Olivier KN: Nontuberculous mycobacteria in the setting of cystic fibrosis. Clin Chest Med 2002;23:655–663.

36 Flume PA, Egan TM, Paradowski LJ, et al: Infectious complications of lung transplantation. Impact of cystic fibrosis. Am J Respir Crit Care Med 1994;149:1601–1607.

37 Kesten S, Chaparro C: Mycobacterial infections in lung transplant recipients. Chest 1999;115:741–745.

38 Malouf MA, Glanville AR: The spectrum of mycobacterial infection after lung transplantation. Am J Respir Crit Care Med 1999;160: 1611–1616.

Prof. David E. Griffith, MD
University of Texas Health Center
11937 US Highway 271
Tyler, TX 75708 (USA)
Tel. +1 903 877 7267, Fax +1 903 877 5566
E-Mail david.Griffith@uthct.edu

Bush A, Alton EWFW, Davies JC, Griesenbach U, Jaffe A (eds): Cystic Fibrosis in the 21st Century.
Prog Respir Res. Basel, Karger, 2006, vol 34, pp 166–172

Respiratory Fungal Infections and Allergic Bronchopulmonary Aspergillosis

Chengli Que   Duncan Geddes

Department of Respiratory Medicine, Royal Brompton Hospital, London, UK

Abstract

Fungal colonization of the lungs in CF is common and leads to clinical deterioration when allergy develops. Diagnostic and treatment guidelines for allergic bronchopulmonary aspergillosis are useful in management. Fungal infection with tissue invasion is probably rare but may be underdiagnosed; antifungal treatment is conventional and usually successful.

Introduction

Fungal isolation in sputum is common in cystic fibrosis (CF). *Aspergillus fumigatus* can be isolated in 40–60% of patients [1], and *Candida albicans* in 75–90% [2, 3]. Recently, a prospective 5-year study showed that *Scedosporium apiospermum* was recovered from the sputum in 8.6% of patients with CF, becoming the second most frequent filamentous fungus after *A. fumigatus* [4]. In another retrospective review, *Scedosporium prolificans* isolates were recovered from airways in 53% of CF patients, although the clinical status did not worsen.

The presence of necrotic tissue and lung cysts in CF provides an ideal environment for the colonization of *A. fumigatus* and long-term antibiotic therapy may also be a predisposing factor. *Aspergillus* spp. usually only colonize but can cause asthma, allergic bronchopulmonary aspergillosis (ABPA), aspergilloma and rarely infection with true tissue invasion [5]. The most common species include *A. fumigatus* and *Aspergillus clavatus*. The most common species causing invasive disease is *A. fumigatus* (90% in some series) with rare isolates of *Aspergillus flavus,*

Aspergillus niger, Aspergillus terreus, and *Aspergillus nidulans.*

In CF the diagnosis is often difficult due to the overlap of clinical features of aspergillosis with CF lung disease. Consensus guidelines are now available on the diagnosis and treatment of ABPA in CF (table 1).

Epidemiology and Natural History

ABPA is the most common *Aspergillus*-related disease in CF, while invasive aspergillosis and aspergilloma are relatively rare. ABPA occurs in up to 10% of CF patients and accounts for approximately 10% of pulmonary exacerbations [6–8]. CF patients do not usually develop invasive aspergillosis despite high doses of corticosteroids, unless they undergo lung transplantation. Only isolated cases of invasive aspergillosis in CF patients have been reported [9, 10], and even fewer for semi-invasive aspergillosis [11]. There have been no good epidemiological surveys giving the prevalence of aspergilloma, although clinically it is not an uncommon finding. With the improvement in survival of CF patients and increased use of CT scanning, aspergilloma and semi-invasive aspergillosis are expected to be detected more frequently.

ABPA is a hypersensitivity disease of the lung related to *A. fumigatus*, which is seen both in patients with CF and in non-CF patients with poorly controlled asthma. It is difficult to estimate the true prevalence of ABPA because of poorly standardized diagnostic criteria and problems in distinguishing between ABPA and CF. Recently, the North America Epidemiologic Study of Cystic Fibrosis (ESCF)

Table 1. Consensus conference-proposed diagnostic criteria for ABPA in CF [from 14]

Classic case
Acute or subacute clinical deterioration not attributable to another etiology
Serum total IgE >1,000 IU/ml
IgE antibody to *A. fumigatus* or positive skin prick test
Precipitating antibodies to *A. fumigatus* or serum IgG antibody to *A. fumigatus*
New shadowing on chest radiography/CT not clearing with antibiotics and standard physiotherapy

Minimal diagnostic criteria
Acute or subacute deterioration not attributable to another etiology
Total IgE >500 IU/ml; if total IgE level is 200–500 IU/ml, repeat test in 1–3 months
IgE antibody to *A. fumigatus* or positive skin prick test
One of the following: (1) precipitins to *A. fumigatus* or IgG antibody to *A. fumigatus* or (2) new or recent abnormalities on chest radiography (infiltrates or mucus plugging) or chest CT that have not cleared with antibiotics and standard physiotherapy

Screening for ABPA has been suggested by determining the total serum IgE concentration annually in CF patients older than 6 years of age, with further diagnostic tests if >500 IU/ml and repeat measurement if 200–500 IU/ml. However, IgE cannot be used in isolation to make a decision because of its nonspecificity.

Table 2. Prevalence of ABPA in CF by age [from 14]

Age years	Number of patients with ABPA	Reported prevalence %
5–10	62 (22)	1.3
11–15	78 (28)	2.4
16–20	65 (23)	2.9
21–25	37 (13)	2.3
>25	39 (14)	1.5
Total	281 (100)	

Figures in parentheses represent percentage.

showed a prevalence of ABPA of 2% [7] in 14,210 patients with CF. By using slightly less strict diagnostic criteria, the European Epidemiologic Registry of Cystic Fibrosis (ERCF), consisting of 12,447 patients, showed an overall prevalence of 7.8%, ranging from 2% in Sweden to 14% in Belgium [8]. An Italian epidemiological study revealed a prevalence of 6%, while other small single-center studies have reported prevalence ranging from 1 to 15%.

The data from ESCF also suggested that there was increased prevalence of ABPA in males, adolescents, and subjects with worse pulmonary function, wheeze, asthma and positive culture of *Pseudomonas*. Data from USA and Europe show a low prevalence among those younger than 6 years of age increasing with age (table 2).

There is evidence that FEV_1 and FEF_{25-75} are lower in CF patients sensitized to *A. fumigatus* and that a raised total IgE level is associated with a more rapid decline in lung function [12], suggesting that sensitization to *A. fumigatus* per se may contribute to progressive respiratory disease. However, at present it is still unclear whether the deterioration in lung function in CF is related to the immune response to *A. fumigatus* or requires the full-blown syndrome of ABPA.

Predisposing Factors for ABPA in CF

The lungs of CF patients are a favorable environment for the development of ABPA as the spores of *A. fumigatus* are 3 μm, small enough to penetrate the airways, and the fungus grows best at 37°C. Frequent use of antibiotics has been suggested to pave the way for fungal colonization and there is an increase in isolation of fungi during prolonged intermittent use of inhaled tobramycin. Atopy is an important risk factor for the development of ABPA in patients with CF. Forty-three percent of CF patients fulfilled the criteria for fungal atopy (total serum IgE >100 kU/l, fungus-specific radioimmunoassay ≥grade 1, and a positive skin prick test ≥3 mm to the same fungus). Over half of this group of patients were allergic to more than one fungus [13]. ABPA occurred in 22% of atopic patients with CF but in only 2% of nonatopic patients [14]. The high frequency of fungal allergy in CF is unexplained but is more likely to be due to antigen access through the damaged airways than to the CFTR mutation itself. Nevertheless, the frequency of CFTR mutations is higher than expected in non-CF patients with ABPA [15]. It has been suggested that HLA-DR molecules DR2, DR5, and possibly DR4 or DR7 contribute to susceptibility, whereas HLA-DQ2 contributes to resistance, and a combination of these may determine the outcome of ABPA in CF and asthma [14].

Immunopathogenesis of ABPA

The allergic inflammatory response in CF patients with ABPA appears to be quantitatively greater than that in *Aspergillus*-sensitive atopic asthmatics and patients with CF but without ABPA [14] (fig. 1). In addition to type I and type III immune responses initiated by *Aspergillus*

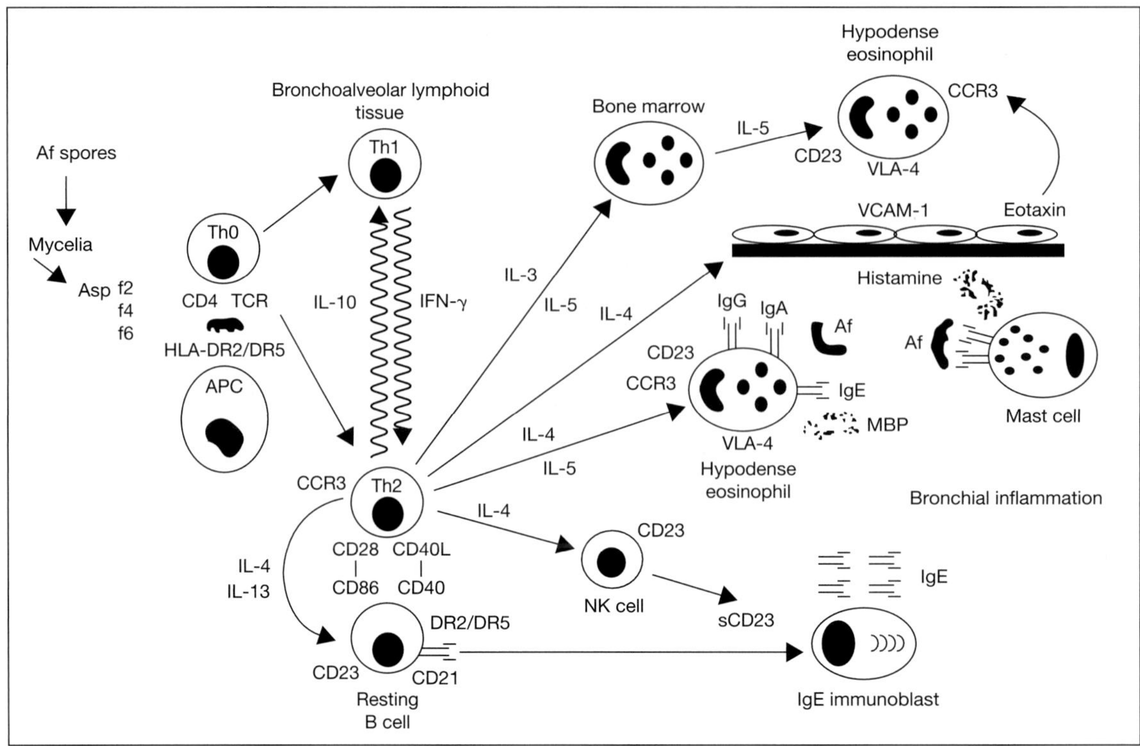

Fig. 1. Model of pathogenesis of ABPA [from 7]. Af = *A. fumigatus*; APC = antigen-presenting cell; MBP = major basic protein; NK = natural killer; s = soluble; TCR = T cell receptor; VCAM = vascular cell adhesion molecule; VLA = very late antigen [from 14].

antigens, ABPA is also associated with an abnormal T lymphocyte cellular response, skewed towards a T helper type 2 with the production of Th2 cytokines, causing an allergic inflammatory pattern. Quantitative increases in the Th2 CD4+ cell responses to *Aspergillus* in both bronchoalveolar lymphoid tissue and systemic immune systems characterize ABPA. There is an immunogenetic susceptibility to develop ABPA that resides within the HLA-DR antigen-T-cell receptor signaling of the T cells toward a Th2 CD4+ cell response [14].

Clinical Features and Diagnostic Criteria for ABPA in CF

ABPA is an episodic and recurrent disorder, sharing many clinical and laboratory findings with CF (table 1). Airway obstruction, fleeting pulmonary infiltrates (fig. 2) and central cystic or varicose bronchiectasis are the three classical findings of ABPA in both CF and non-CF patients. A flare-up of bronchiectasis sometimes with hemoptysis may herald either ABPA or a bacterial infection in CF.

However, CF patients with ABPA appear to have more severe pulmonary disease and lower FEV_1 than those without ABPA. ABPA in CF is also associated with higher rates of microbial colonization, pneumothorax, massive hemoptysis and with higher IgG serum levels and poorer nutritional status [8]. Although infiltrates are common in both ABPA and CF lung disease, they are more likely to respond to steroids in CF with ABPA than those without ABPA [14]. About 30% of asthmatic patients with ABPA have radiographic evidence of high-attenuation mucus plugs but this has not been specifically reported in patients with CF.

ABPA should be suspected in any CF patient with deteriorating pulmonary disease. The diagnosis relies principally on immunological evidence, in particular specific IgE and IgG to *A. fumigatus* and skin reactivity. However, *A. fumigatus* sensitization alone, in the absence of clinical features of ABPA, is not enough for diagnosis [8]. In general, high ($>1,000/mm^3$) eosinophil count or elevated IgE titers greater than 500 IU/ml suggest ABPA. In patients with baseline high IgE level, a 2-fold rise in IgE may antedate an ABPA flare. Reduction in IgE by 50% after

Que/Geddes

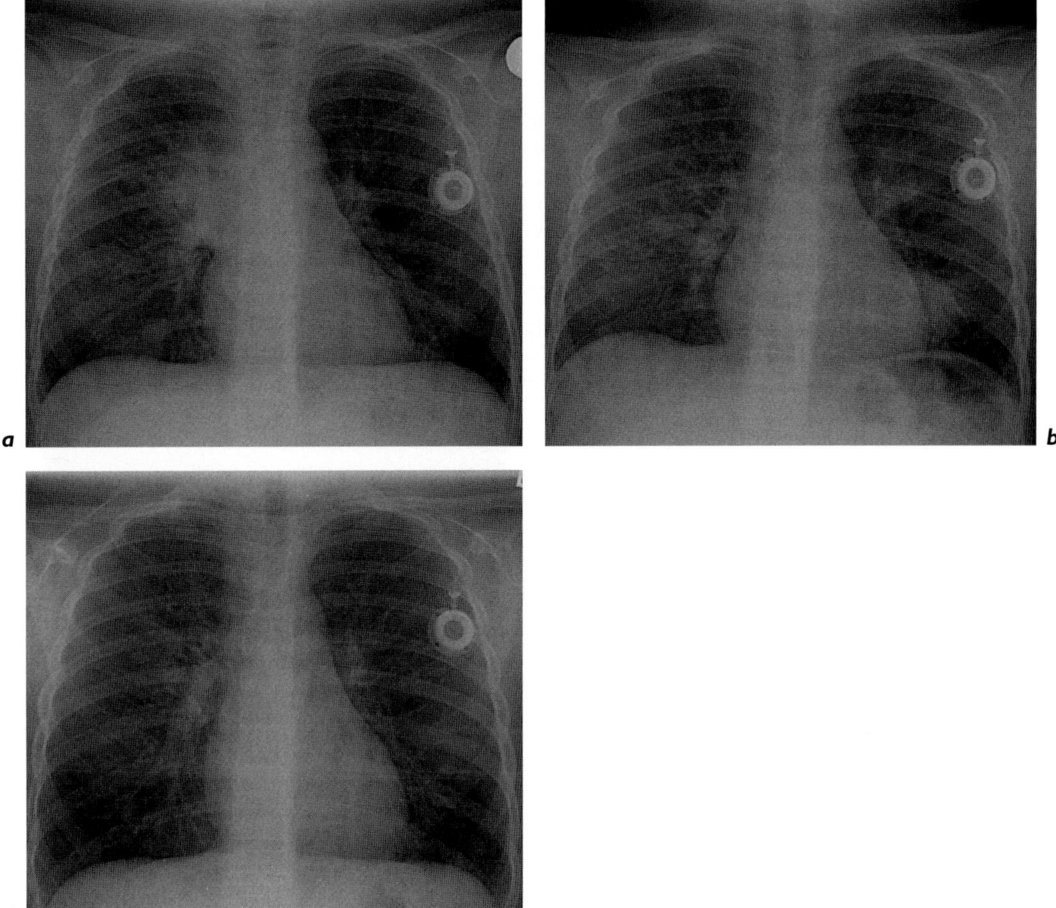

Fig. 2. Chest x-ray showing fleeting infiltrates in ABPA. Note the patient has a Portacath. *a* July 2003. There is a large right upper zone infiltrate which is based at the hilum, with smaller infiltrates in the right lower zone. *b* September 2003. The previous right-sided changes have largely cleared, although there are small right midzone infiltrates. However, there are now predominantly left-sided changes in the mid- and lower-zones. *c* April 2004. The infiltrates have largely cleared leaving a minimal right midzone infiltrate close to the hilum. There is still evidence of widespread CF airway disease.

high-dose systemic corticosteroid therapy also supports the diagnosis [5].

Recent work suggests that a limited panel of recombinant allergens, used either for skin tests or as the antigen in RAST testing, improve specificity for the detection of sensitization to *A. fumigatus*, as well as for the detection of ABPA [16, 17].

Rarely, allergic bronchopulmonary mycosis may be caused by an *Aspergillus* sp. other than *A. fumigatus*, such as *A. niger*, or by *Candida* spp., *Curvularia* spp., *Dreschlera* spp. and *Pseudallescheria boydii*. In these instances, the serologic tests for *A. fumigatus* are negative [18]. Specific tests for the other fungi are not routinely used.

Treatment of ABPA in CF Patients

Treatment guidelines are based on uncontrolled case series and expert opinion and clinical trials are few (tables 3, 4). Corticosteroids attenuate inflammation and immunological activity, while antifungal therapy with oral itraconazole [14] reduces the antigenic burden. Itraconazole is usually reserved for relapsed and steroid-dependent CF patients.

It is important to differentiate a bacterial exacerbation of CF from an ABPA flare before embarking on corticosteroid treatment. Because treatment of ABPA requires relatively high-dose corticosteroids, it may be prudent to initiate

Table 3. Treatment recommendations for ABPA in CF in different clinical situations

Total serum IgE, IU/ml	Pulmonary symptoms and/or worsening PFT results	New infiltrates on CXR/CT	Positive serology	Diagnosis of ABPA and treatment recommendations
>1,000 or >2-fold rise from baseline	Yes	Yes	Yes	Definite: treat for ABPA
>1,000 or >2-fold rise from baseline	No	No	Yes	At risk: no treatment; monitor IgE, CXR, PFT
>1,000 or >2-fold rise from baseline	No	Yes	Yes	Possible: treat for CF-related infection; consider treatment for ABPA if no response
>1,000 or >2-fold rise from baseline	Yes	No	Yes	Possible: consider treatment for ABPA, CF-related infection, and/or asthma
>500 in the past; no change from baseline	Yes	Yes	Yes	Possible: treat for CF-related infection; consider treatment for ABPA or asthma if no response
500–1,000	Yes (failure of therapy for CF)	Yes	Yes	Most probable: treat for ABPA

PFT = Pulmonary function tests; CXR = chest x-ray.

therapy for ABPA while treating concomitantly for CF-related infection. Ideally corticosteroid therapy should be withdrawn in 2–3 months. If there is no improvement, the diagnosis of ABPA [5] should be questioned.

If the patient relapses during the corticosteroid taper, corticosteroid dosages should be increased and/or itraconazole added. When clinical parameters improve, corticosteroid should be retapered. Only a small percentage of patients requires long-term therapy. There is insufficient evidence to allow any conclusion about the value of inhaled corticosteroids in CF ABPA but they are usually given.

In ABPA without CF oral itraconazole improves pulmonary function, chest radiography and serum IgE levels [19, 20]. In CF patients with ABPA, itraconazole alone or in combination with systemic corticosteroids seemed effective [21] and several small studies found a reduction in prednisone dose, number of exacerbations, and spirometry [22]. Itraconazole should be added if there is a slow or poor response to corticosteroids, for relapse, and in cases of corticosteroid toxicity. The therapy is usually continued for 3–6 months unless it is still clinically required [14]. Because itraconazole is an inhibitor of the drug-metabolizing enzyme CYP3A4, doses of concomitant medications such as cyclosporine, tacrolimus, oral hypoglycemics, and methylprednisolone should be adjusted. There have been two case reports of adrenal suppression and iatrogenic Cushing's syndrome due to the inhaled budesonide or fluticasone combined with itraconazole. There is insufficient evidence to recommend other oral or inhaled antifungal agents.

Que/Geddes

 Treatment for ABPA [from 14]

Treatment, factor	Explanation
Corticosteroids	
Indications	All patients except those with steroid toxicity (grade II-3)
Initial	0.5–2.0 mg/kg/day p.o. prednisone equivalent, maximum 60 mg/day, for 1–2 weeks
Begin taper	0.5–2.0 mg/kg/day every other day for 1–2 weeks
Taper off	Attempt to taper off within 2–3 months
Relapse	Increase corticosteroids, add itraconazole, taper corticosteroids when clinical parameters improve
Itraconazole	
Indications	Slow or poor response to corticosteroids, relapse, corticosteroid-dependent, or corticosteroid toxicity (grade III)
Dosing	5 mg/kg/day, maximum dose 400 mg/day p.o. unless itraconazole levels determined; b.i.d. dosing required when daily dose exceeds 200 mg
Duration	3–6 months
Monitor	Liver function tests for all cases; itraconazole serum concentrations if concern of adequate absorption, lack of response, possible drug-drug interaction; serum concentrations of concomitant drugs with potential for drug-drug interaction
Adjunctive therapy	
Inhaled corticosteroids, bronchodilators, other antiasthma drugs	No evidence for use in ABPA; may be used for the asthma component of ABPA (grade III)
Environmental manipulation	Attempt to search for and modify mold spore exposure in refractory cases (grade III)

In refractory cases, attempt should be made to search for and modify mould spore exposure (e.g. stables, rotting vegetation), and regular replacement of nebulizer tubing and filters is crucial to the prevention of mould growth [5].

Other Forms of Fungal Infection in CF

Aspergilloma seldom causes symptoms and can usually be left alone (fig. 3). Rarely surgery may be considered for associated life-threatening hemoptysis. It is not clear

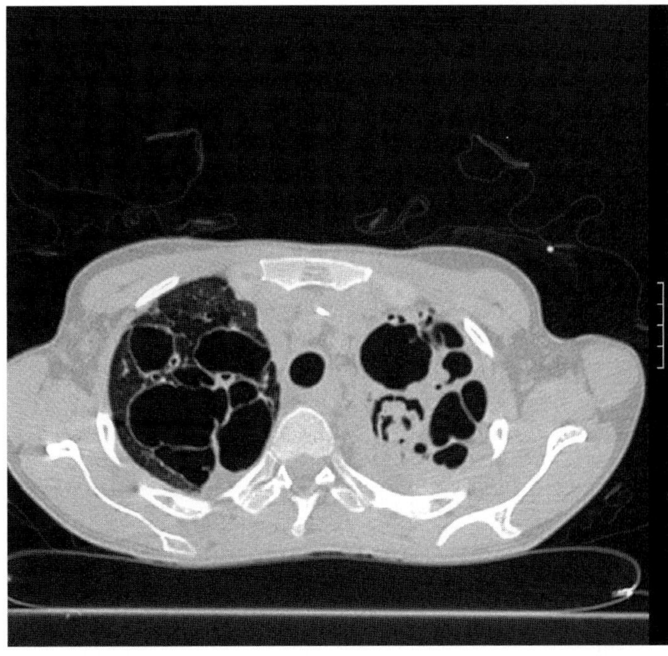

Fig. 3. CT showing an aspergilloma in a CF patient, situated posteriorly on the left. The characteristic 'air crescent' sign is well shown, differentiating the mycetoma from the large right-sided bronchiectatic cavities which are relatively empty and featureless.

whether antifungal therapy is helpful. The elimination of an aspergilloma in CF prior to transplantation by percutaneous instillation of amphotericin B has also been described [23].

Although infrequent, invasive infection with *Aspergillus* or *Candida* does occur. Voriconazole is probably the best treatment [24] although there are no published trial data in CF. In non-CF immunocompromised patients with invasive aspergillosis survival with voriconazole (70.8%) was better than with amphotericin B (57.9%) [25]. Voriconazole is given intravenously for over 1 week and then orally for 2–3 months and is also licensed for fluconazole-resistant serious invasive *Candida* infection or serious fungal infections caused by *Scedosporium* spp. or *Fusarium* spp. in patients aged more than 2 years. For patients refractory or intolerant to other therapies, caspofungin is the alternative, which is at least as effective as and better tolerated than amphotericin B for primary treatment of invasive candidiasis [24].

References

1 Chen KY, Ko SC, Hsueh PR, et al: Pulmonary fungal infection: Emphasis on microbiological spectra, patient outcome and prognostic factors. Chest 2001;120:177–184.
2 Bakare N, Rickerts V, Bargon J, Just-Nubling G: Prevalence of *Aspergillus fumigatus* and other fungal species in the sputum of adult patients with cystic fibrosis. Mycoses 2003; 46:19–23.
3 Maiz L, Cuevas M, Quirce S, Canon JF, Pacheco A, Sousa A, Escobar H: Serologic IgE immune responses against *Aspergillus fumigatus* and *Candida albicans* in patients with cystic fibrosis. Chest 2002;121:782–788.
4 Cimon B, Carrere J, Vinatier JF, Chazalette JP, Chabasse D, Bouchara JP: Clinical significance of *Scedosporium apiospermum* in patients with cystic fibrosis. Eur J Clin Microbiol Infect Dis 2000;19:53–56.
5 Hanley-Lopez J, Clement LT: Allergic bronchopulmonary aspergilliosis in cystic fibrosis. Curr Opin Pulm Med 2000;6:540–544.
6 Moss RB: Allergic bronchopulmonary aspergillosis. Clin Rev Allergy Immunol 2002;23: 87–104.
7 Geller DE, Kaplowitz H, Light MJ, Colin AA: Allergic bronchopulmonary aspergillosis in cystic fibrosis: Reported prevalence, regional distribution, and patient characteristics. Chest 1999;116:639–646.
8 Mastella G, Rainisio M, Harms HK, Hodson ME, Koch C, Navarro J, Strandvik B, McKenzie SG: Allergic bronchopulmonary aspergillosis in cystic fibrosis: A European epidemiological study. Eur Respir J 2000;16: 464–471.
9 Mastella G, Rainisio M, Harms HK, et al: Allergic bronchopulmonary aspergillosis in cystic fibrosis. Eur Respir J 2001;17: 1052–1053.
10 Brown K, Rosenthal M, Bush A: Fatal invasive aspergillosis in an adolescent with cystic fibrosis. Pediatr Pulmonol 1999;27:130–133.
11 Grahame-Clarke CN, Roberts CM, Empey DW: Chronic necrotizing pulmonary aspergillosis and pulmonary phycomycosis in cystic fibrosis. Respir Med 1994;88:465–468.
12 Wojnarowski C, Eilchler I, Gartner C, Gotz M, Renner S, Koller DY, Frischer T: Sensitization to *Aspergillus fumigatus* and lung function in children with cystic fibrosis. Am J Respir Crit Care Med 1997;155:1902–1907.
13 Henry M, Bennett DM, Keily J, Kelleher N, Bredin CP: Fungal atopy in adult cystic fibrosis. Respir Med 2000;94:1092–1096.
14 Stevens DA, Moss RB, Kurup VP, Knutsen AP, Greenberger P, Judson MA, Denning DW, Crameri R, Brody AS, Light M, Skov M, Maish W, Mastella G: Allergic bronchopulmonary aspergillosis in cystic fibrosis – state of the art: Cystic Fibrosis Foundation Consensus Conference. Clin Infect Dis 2003; 37(suppl 3):S225–S264.
15 Marchand E, Verellen-Dumoulin C, Mairesse M, Delaunois L, Brancaleone P, Rahier JF, Vandenplas O: Frequency of cystic fibrosis transmembrane conductance regulator gene mutations and 5T allele in patients with allergic bronchopulmonary aspergillosis. Chest 2001;119:762–767.
16 Nikolaizik WH, Weichel M, Blaser K, Crameri R: Intracutaneous tests with recombinant allergens in cystic fibrosis patients with allergic bronchopulmonary aspergillosis and Aspergillus allergy. Am J Respir Crit Care Med 2002;165:916–921.
17 Knutsen AP, Hutcheson PS, Slavin RG, Kurup VP: IgE antibody to *Aspergillus fumigatus* recombinant allergens in cystic fibrosis patients with allergic bronchopulmonary aspergillosis. Allergy 2004;59:198–203.
18 Miller MA, Greenberger PA, Amerian R, Toogood JH, Noskin GA, Roberts M, Patterson R: Allergic bronchopulmonary mycosis caused by *Pseudallescheria boydii*. Am Rev Respir Dis 1993;148:810–812.
19 Stevens DA, Schwartz HJ, Lee JY, Moskovitz BL, Jerome DC, Catanzaro A, Bamberger DM, Weinmann AJ, Tuazon CU, Judon MA, Platts-Mills TA, Degraff AC Jr: A randomized trial of itraconazole in allergic bronchopulmonary aspergillosis. N Engl J Med 2000; 342:756–762.
20 Wark PA, Hensley MJ, Saltons N, Boyle MJ, Toneguzzi RC, Epid GD, Simpson JL, McElduff P, Gibson PG: Anti-inflammatory effect of itraconazole in stable allergic bronchopulmonary aspergillosis: A randomized controlled trial. J Allergy Clin Immunol 2003; 111:952–957.
21 Skov M, Hoiby N, Koch C: Itraconazole treatment of allergic bronchopulmonary aspergillosis in patients with cystic fibrosis. Allergy 2002;57:723–728.
22 Nepomuceno IB, Esrig S, Moss RB: Allergic bronchopulmonary aspergillosis in cystic fibrosis: Role of atopy and response in itraconazole. Chest 1999;115:364–370.
23 Ryan PJ, Stableforth DE, Reynolds J, Muhdi KM: Treatment of pulmonary aspergilloma in cystic fibrosis by percutaneous instillation of amphotericin B via indwelling catheter. Thorax 1995;50:809–810.
24 Maschmeyer G, Ruhnke M: Update on antifungal treatment of invasive Candida and Aspergillus infections. Mycoses 2004;47: 263–276.
25 Herbrecht R, Denning DW, Patterson TF: Voriconazole versus amphotericin B for invasive aspergillosis. N Engl J Med 2002;347: 408–416.

Dr. Duncan Geddes
Department of Respiratory Medicine
Royal Brompton Hospital
Sydney Street
London SW3 6NP (UK)
Tel. +44 20 7351 8182
Fax +44 20 7351 8999
E-Mail d.geddes@rbh.nthames.nhs.uk

Que/Geddes

Bush A, Alton EWFW, Davies JC, Griesenbach U, Jaffe A (eds): Cystic Fibrosis in the 21st Century.
Prog Respir Res. Basel, Karger, 2006, vol 34, pp 173–179

Advanced Disease Management and Advances in Transplant Medicine

Martin R. Carby[a] Margaret E. Hodson[b]

[a]Consultant Respiratory and Transplant Physician, Royal Brompton & Harefield NHS Trust,
Harefield Hospital, Harefield, Uxbridge, and [b]Medicine and Honorary Consultant Physician,
Royal Brompton & Harefield, Royal Brompton Hospital, London, UK

Abstract

Respiratory disease is a major cause of morbidity and mortality in cystic fibrosis (CF). Recent advances in disease management include improved management of respiratory failure, non-invasive ventilation and terminal care. Lung transplantation offers a chance of improved survival and quality of life. Recent advances in the management of lung transplant recipients include refinements in recipient selection and prioritization, decreasing post-transplantation morbidity associated with chronic rejection and drug induced renal disease, improved anti-infective and immunosuppressant drug regimens and attention to post-operative rehabilitation.

Lung transplantation has demonstrated both a survival [1] (fig. 1) and quality of life benefit [2] for selected patients. However, due to a shortage of organ donors as many as 40% of potential lung transplant recipients die from their underlying disease while awaiting surgery. This mandates careful transplant recipient selection and timing of treatment. Advances in medical care are being made for those individuals who undergo lung transplantation.

One hundred and fifty lung transplantations were carried out in the UK in 2003. Without lung transplantation it is highly likely that many of these individuals would have perished. One- and five-year survival rates following lung transplantation now exceed 70% and 50%, respectively. The major complications affecting early mortality are graft failure and infection while the major determinants of late

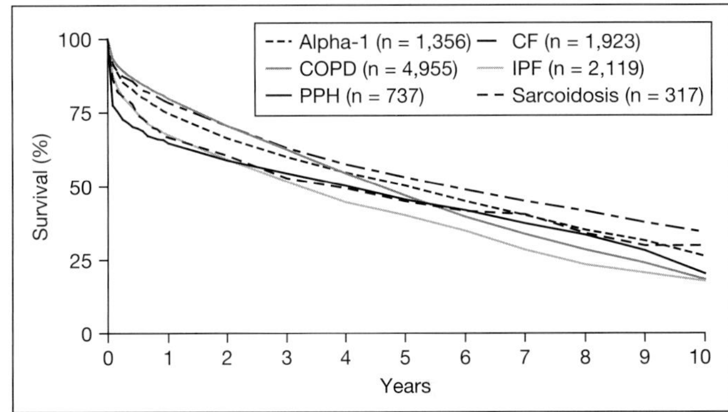

Fig. 1. Kaplan-Meier survival curves for adult lung transplantation (transplantations: January 1990 to June 2002). Note that despite the residual burden of infection in the sinuses, the CF figures are at least as good as those for other indications.

mortality are lung infections complicating chronic allograft rejection manifest as bronchiolitis obliterans.

Advanced Disease Management

Management of Respiratory Failure

Hypoxic patients with cystic fibrosis (CF) generally function well for many years, but hypoxia with or without hypercapnia is a cause for concern (fig. 2). Thought should be given to a long-term plan for these patients including discussion of lung transplantation as a potential treatment. A small

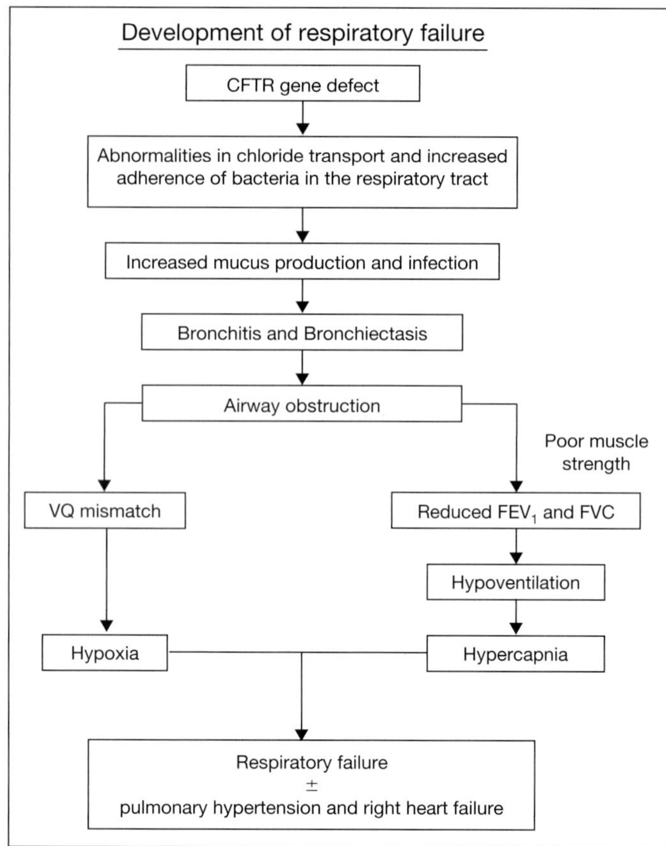

Fig. 2. This demonstrates the progression from a single gene defect to respiratory failure.

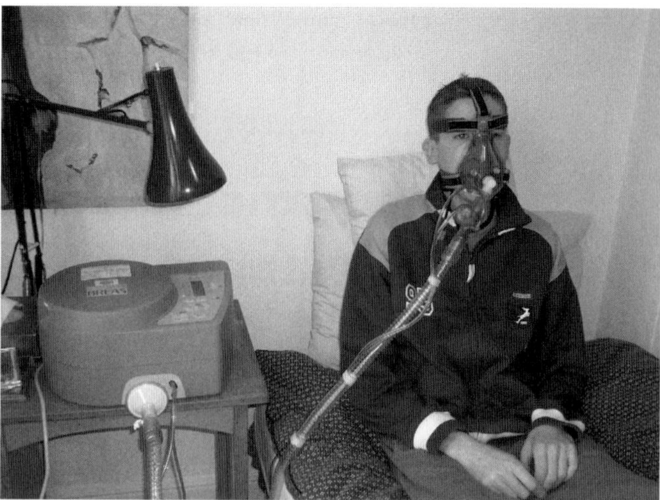

Fig. 3. A young man with CF and end stage lung disease using non-invasive positive pressure ventilation at home.

Table 1. The advantages of NIV compared to intubation in respiratory failure

It is easy to commence
Patients can talk, perform chest physiotherapy, move limbs
Patients can eat and drink
There is no risk of tracheal stenosis
No sedation is needed
Results following transplantation may be better than after intubation
It is cost effective

proportion of patients may develop cardiac problems, but this is unusual. Right-ventricular failure is easily managed with oxygen and diuretics. Many patients will be stable for many years and others will deteriorate rapidly. When the patient develops chronic respiratory failure, the clinician must make sure that all conventional treatment is being given.

Hypoxia should be treated with controlled oxygen therapy. Initially, this may only be needed at night. Overnight oximetry is helpful to determine the amount of oxygen required to raise the PaO_2 without causing an undesirable rise in $PaCO_2$. As the disease progresses, patients may need continuous oxygen therapy.

Ventilation

Some patients may benefit from continuous positive airway pressure (CPAP) or establishment of non-invasive ventilation (NIV) [3]. During CPAP, patients continue to breathe normally, whereas during NIV, the spontaneous inspiratory pressure is augmented by the ventilator. Ventilators can be used during an acute episode or as a bridge to transplantation (fig. 3). There is no justification for using ventilation to prolong the process of dying if a patient does not want or is not suitable for transplantation. NIV can be used very successfully as a bridge to transplantation [4]. NIV has many advantages over intubation (table 1). In many patients where it is said that NIV has failed it is because the distending pressure of the CF lung is very high and in some patients ventilators may not produce the required pressure. Either a ventilator generating high pressure should be used or a volume-cycled machine.

Results with intubation and ventilation in patients with acute on chronic and with end-stage disease are variable. Patients may develop chronic sepsis and renal failure probably due to pseudomonas toxaemia. This is rarely seen in patients treated with NIV. Results of transplantation following NIV are also encouraging. Some centres have had success with transplantation following intubation [8]. Controlled clinical trials in this area are few partly because it is difficult when dealing with such sick patients [5].

Symptomatic Treatment – Terminal Care

In spite of maximal medical treatment, patients who are in respiratory failure may continue to deteriorate and may not want or be suitable for transplantation. Other patients may be listed for transplantation but deteriorate before an organ becomes available. It is essential that good symptom control is provided. The most distressing symptoms are cough, sputum retention, breathlessness and fear. It is the clinician's responsibility to co-ordinate the activities of the multi-disciplinary team. Full psychological and social support should be given as well as appropriate medication. Small doses of morphine linctus, initially 2.5 mg four-hourly, together with prochlorperazine can control dyspnoea and to some extent anxiety. If more sedation is needed a low-dose midazolam infusion can be helpful. Gentle chest physiotherapy should be continued to clear secretions, in order to add to the patient's comfort. The amount given should be appropriate for the patient's clinical status and will naturally be reduced as the patient's condition deteriorates. Relatives and friends should be encouraged to participate in caring for the patient. These end-of-life issues are more fully discussed in standard texts.

Advances in Transplant Medicine

Recipient Selection

Recipient selection criteria are well documented [7]. Growing experience allows optimal allocation of scarce donor organs maximizing the survival benefit to individuals. The relevance to patient selection of factors such *Burkholderia cenocepacia*, 6-min walk test (6 MWT), compliance and CF-related diabetes (CFRDM) have all recently been highlighted. Experience is growing in living lobe donor lung transplantation.

B. cenocepacia has been considered by many transplant programmes to be an absolute contraindication to lung transplantation. This view (as well as the nomenclature of the organism!) has evolved over time. Previously, *B. cepacia* was demonstrated to be responsible for excess morbidity and mortality after lung transplantation [8]. More recently, the excessive mortality due to *B. cepacia* was attributed solely to genomovar III (*B. cenocepacia*) [9]. Moreover, data from some transplant units suggest that modification of antimicrobial testing and treatment can lead to equivalent post-transplantation outcome even in those infected pre-operatively with *B. cenocepacia* [10].

The 6 MWT is a self-paced test of the distance that can be covered in 6 min. This simple test has predictive value in terms of post-operative outcome in patients with CF [11, 12]. It can be used in both quantifying risk and prioritization in candidate selection. Patients achieving greater distances during the 6 MWT had shorter ITU and hospital stays and greater functional mobility at all post-operative intervals. Lower 6 MWT distances have been correlated with shorter survival on the waiting list.

The importance of strict adherence to drug protocols cannot be overemphasized in the maintenance of the lung allograft and the prevention of complications. Non-adherence to the prescribed drug regimen is a common cause of death for children and adolescents with CF after transplantation [13]. Psychosocial morbidity before transplantation is a risk factor for non-compliant behaviours post-transplantation and interventions need to be undertaken prospectively with these individuals in order to improve compliance with the treatment protocol after transplantation.

CFRDM before transplantation can result in more severe nutritional deficiency, lung disease and earlier death than in patients with CF alone. Insulin therapy appears to palliate the decline in lung function and nutritional status. In the context of lung transplantation, CFRDM has most often been thought not to affect outcome. However, recent analysis has concluded that there is a higher rate of hospitalization for infection or allograft rejection than in non-diabetic recipients and also an earlier onset of bronchiolitis obliterans syndrome (BOS) [14]. Given that pre-transplantation, earlier identification and treatment of CFRDM may decrease morbidity and mortality, it seems warranted to maximize treatment of diabetes after transplantation, not only in the peri-operative period but also in the context of the long-term follow-up. This requires close involvement of the patient and input from a multi-disciplinary team, including a dietician and physician to work in parallel and monitor for complications and in some cases a clinical psychologist to help with issues of compliance.

Living lobe donor lung transplantation has developed because of the shortage of available organs. Two donors have a lobectomy to remove one or other of their lower lobes. The patient receives a right lower lobe from one donor and a left lower lobe from the other. Recent data concerning 84 adults following living lobar lung transplantation showed recipient survival rates at 1, 3 and 5 years of 70, 54 and 45% respectively. Living lobe transplantation provided comparable pulmonary function and exercise capacity to cadaveric lung transplantation [15].

Post-Transplantation Morbidity

Post-transplantation complications vary according to the length of time that has passed since transplant surgery.

Mortality during the immediate post-operative period is largely due to acute infection. Later, although infection remains prevalent, malignancy and chronic rejection play a more major role [16]. Primary graft failure has recently been readdressed as a cause for early mortality. Advances are being made in the treatment of late complications such as renal impairment, anaemia and hypertension.

A working group under the auspices of the International Society of Heart and Lung Transplantation has recently examined the definition, risk factors, outcome and treatment of primary graft failure. It is hoped that this variable phenomenon can be approached in a standardized manner in order to be able to better understand and treat it in the future to avoid some of the early post-operative deaths ascribed to it.

Renal impairment and indeed end-stage renal failure due to calcineurin-inhibitor-based immunosuppressive regimen is a well-recognized complication following lung transplantation. However, the rate of decline of renal function is not well known and independent risk factors that can be addressed are uncertain. Some studies have variously highlighted increasing age, hypertension, diabetes, levels of calcineurin inhibitors and peri-transplant incidence of acute renal failure as factors predicting progression to renal failure. A recent study has refuted all but hypertension as a risk factor [17]. What seems clear is that there is a diminished long-term survival among individuals progressing to renal failure following lung transplantation [18]. Therefore, strategies to protect renal function and limit the toxicity of immunosuppressive drugs are extremely important.

In order to slow the progression of calcineurin-inhibitor-related renal toxicity, not only risk factors such as diabetes and hypertension must be addressed, but also immunosuppressive drugs themselves. The minimization of exposure to calcineurin inhibitors seems important (see later). In this way it has been possible to lower the total drug exposure while maintaining adequate protection against episodes of acute rejection [19]. Immunosuppression regimens that do not include a calcineurin inhibitor are being tested. The 'target of rapamycin' (TOR) inhibitor sirolimus has been shown to provide adequate protection against lung allograft rejection in the short term [20], although more long-term data are needed. Withdrawal of a calcineurin inhibitor has led to renal recovery and therefore, at least in the early stages of drug-induced renal impairment, there seems to be a reversible element [20].

Anaemia related to immunosuppression is prevalent in transplant recipients. A haemoglobin level of less than 11 mg/dl has been shown to have clinically relevant effects on health-related quality of life as demonstrated by the

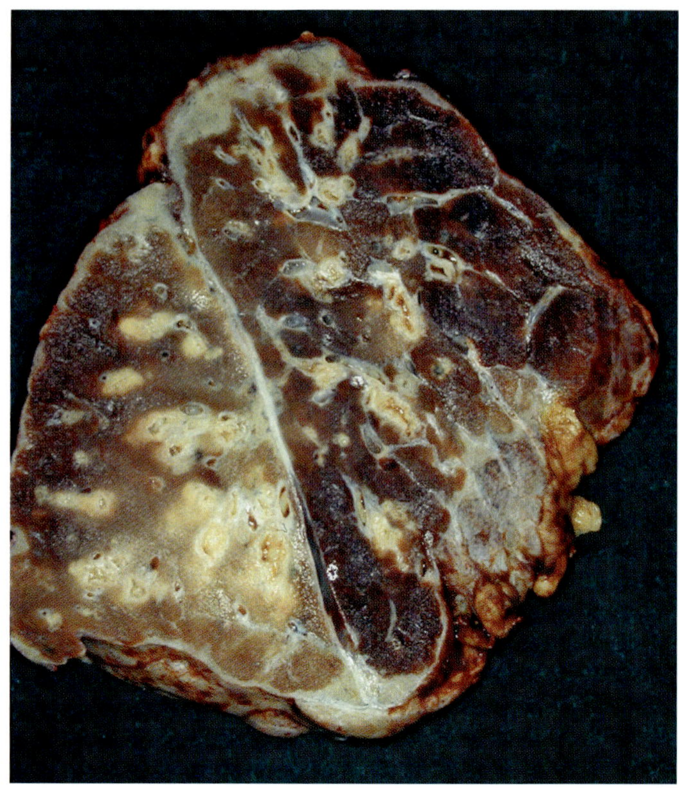

Fig. 4. Macroscopic appearances of BOS: widespread and obvious airway thickening revealed on cutting into the lung.

SF-36 questionnaire [21]. This supports greater efforts towards maintenance of a haemoglobin level greater than 11 mg/dl to achieve these quality-of-life benefits.

Chronic Rejection

BOS is considered to be synonymous with chronic rejection since an alloimmune reaction is considered to play a major role in pathogenesis. It is the major late cause of morbidity and mortality following lung transplantation (fig. 4, 5). Risk factors for developing chronic rejection include recurrent episodes of acute rejection and infections by cytomegalovirus (CMV). However, other possible risk factors such as gastro-oesophageal reflux (GER) disease, respiratory syncytial virus and the involvement of antibodies in mediating allograft rejection are being researched. Azithromycin seems to have a therapeutic effect, but the mechanism for this is not yet established.

GER is prevalent following lung transplantation and may be asymptomatic. Recent evidence demonstrates that GER may contribute to development of BOS via direct injury and by precipitating alloimmune injury [22]. Surgical treatment of GER has been shown to prevent further

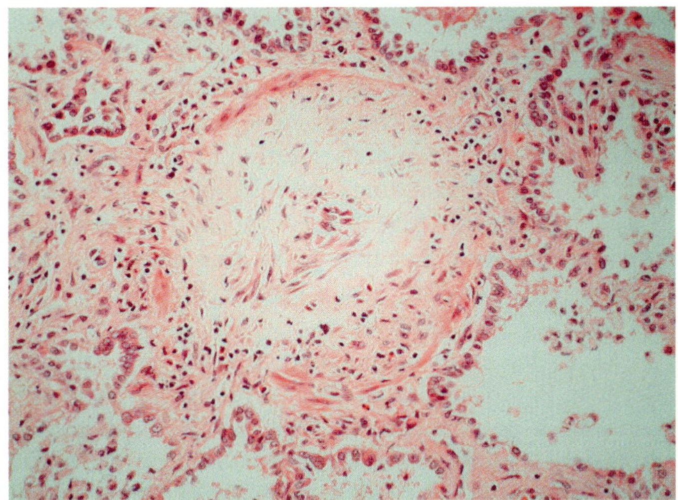

Fig. 5. Microscopic appearances of BOS. The lumens of the small airways are virtually completely obliterated.

decline in lung function [23]. Studies are under way to determine if treatment can stabilize or reverse changes in lung function both before and after transplantation. Other groups have determined that the presence of bile acids in broncho-alveolar lavage (BAL) fluid can be used as a marker for pulmonary aspiration [24].

Respiratory viral infections are common in immuno-compromised patients and following lung transplantation are associated with secondary bacterial infection and with acute and chronic allograft rejection through immune up-regulation [25]. A recent study has shown a high sensitivity of indirect fluorescent antibody testing on nasopharyngeal and throat swabs for the presence of community-acquired viral infections. These results were available within 24 h [25]. The availability of a rapid method of assessment of a viral-type illness in the context of the known deleterious short- and long-term effects of viral infection with agents such as RSV, adenovirus, influenza and parainfluenza viruses has prompted further evaluation of therapies such as ribavirin and immunoglobulins [26].

CMV is a known risk factor for development of BOS, and infection remains a serious problem in lung transplant recipients. However, current prophylactic, treatment and diagnostic strategies may have a significant impact on the outcome of CMV in these patients. Indeed, a recent study following markers of BOS over 8 years has shown that donor/recipient CMV mismatching is not a risk factor for onset or severity of BOS using modern diagnostic and treatment strategies [27]. It therefore seems hopeful that improving strategies for diagnosis and treatment of CMV

and perhaps other viral infections may already be improving the outlook for lung transplant recipients.

Lung transplantation in patients with preformed HLA antibodies is contraindicated since it leads to rapid deterioration in graft function. This can be avoided by the cross-matching of organs in those individuals found to have a positive panel-reactive antibody test. It is becoming increasingly recognized that de novo production of antibodies (HLA or non-HLA) after transplantation can be disadvantageous to the allograft. Techniques of detecting circulating antibodies and furthermore relating these antibodies to graft dysfunction rather than merely reflecting immunoactivation are improving [28]. Therapeutic intervention in terms of reducing antibody production or removal of circulation antibody requires further evaluation.

Some patients with BOS show a sustained improvement in airway dysfunction following treatment with azithromycin [29]. The mechanism of this effect is uncertain but may be via antimicrobial activity, faster gastric transit diminishing GER, or via anti-inflammatory effects either directly on neutrophils or through quorum-sensing signals.

Drugs

Pharmacological immunosuppression is essential after transplantation to prevent allograft rejection. Infectious complications and unwanted side effects of these drugs are common. Advances are being made in limiting drug toxicity and in treating infectious complications.

Therapeutic drug monitoring is an essential part of post-transplantation care in order to gain efficacious suppression of alloimmune responses while minimizing drug toxicity. Cyclosporin, the mainstay of many immunosuppressive regimens for lung transplant recipients is traditionally monitored by the trough (12 h post dose) level. Some transplantation centres have started to use the 2-hour post-dose level (C2) to judge the total oral dosage of cyclosporin to prescribe. The C2 levels have been shown to provide a better representation of the total exposure to cyclosporin (the area under the curve when blood levels are plotted against time) than does the trough level. Some investigators have found it possible to reduce the total oral dose and the total cyclosporin exposure but retain control of acute rejection episodes by prescribing according to target C2 drug levels rather than trough levels [19]. It is hoped this will lead to a lower prevalence of renal and infective complications in the long term. However, in a different randomized clinical trial unpredictable decreases in C2 levels were seen in almost half of patients particularly in the early post-transplantation period [30]. Another study also found significant differences in the time to peak drug blood concentration after dosing and found it difficult to achieve the proposed target ranges for C2 drug levels [31]. Furthermore there are concerns about the differences in absorption in the CF population and in the effect of drug interactions on the profile of cyclosporin blood levels. Therefore although C2 monitoring shows potential benefit concerns remain about complete conversion to this method of drug monitoring. Another potential new advance which may reduce cyclosporin toxicity is the use of an aerosolized preparation [32].

The TOR inhibitors, everolimus (Certican; Novartis) and sirolimus (Rapamune; Wyeth) are newer immunosuppressants that are under investigation with regard to their utility in lung transplantation. The main utility of sirolimus thus far in lung transplantation has been as a renal-sparing agent when renal dysfunction has become apparent in the early post-operative period [20]. Concerns regarding undesirable effects on surgical wounds and airway anastomoses have limited its utility de novo following lung transplantation but these effects have not been found when sirolimus has been instituted more than 3 months following transplantation [33]. In short-term studies it has appeared efficacious in preventing acute rejection but so far no direct comparison has been made with other immunosuppressive regimens in the prevention of BOS.

Trials of everolimus have shown a trend towards decreased incidence of BOS at 12 months but these were not continued at 24 months [34]. There was a significant decrease in patients treated for acute rejection while treated with everolimus compared to azathioprine in combination with cyclosporin and steroids. However, there was a significant discontinuation in the everolimus group as a result of adverse events in terms of infection and reduced renal function. A further study is planned to investigate if therapeutic drug monitoring of everolimus can reduce these complications while maintaining efficacy compared to Myfortic (mycophenolate; Novartis).

Opportunistic fungal infections, usually due to *Aspergillus* and *Candida* species are common in lung transplant recipients and carry a high mortality. Clinical presentation is highly variable and the diagnosis is challenging and often treatment is started on empirical grounds in a patient with fever and no response to antibacterial agents. The diagnosis may be aided by advances in the ability of the laboratory to detect circulating fungal antigens and DNA. Secondary prophylaxis in those individuals colonized with aspergillus has been shown to prevent progression to invasive disease [35]. Newer antifungal drugs are emerging (Voriconazole, Pfizer; Caspofungin, MSD) which expand the therapeutic armamentarium and may have fewer drug interactions and lower toxicity. However, there is limited clinical experience and the drugs are expensive.

CMV is the most common opportunistic viral pathogen after lung transplantation. Regular screening for viral antigenaemia or positive plasma CMV PCR is standard. Preemptive therapy when circulating virus is detected is used to prevent progression to CMV disease and its complications. Patients are traditionally treated with intravenous ganciclovir, with or without CMV immunoglobulin. Recently, valganciclovir, the pro-drug of ganciclovir has been shown to be efficacious and safe in this context as an alternative treatment strategy [36]. However, regular blood counts need to be performed to monitor for development of leucopenia. This strategy has the advantage that valganciclovir is in oral formulation and thus avoids the complication of central venous catheter insertion and the financial and quality of life costs of hospitalization.

Post-Transplantation Rehabilitation
Peripheral skeletal muscle work capacity is reduced following lung transplantation and is the largest contributing factor responsible for the limitation of exercise performance seen following lung transplantation [37]. This has recently been correlated with the cumulative dose of prednisolone received after transplantation [38]. However, a physical training programme has been shown to improve exercise capacity through improved muscle reconditioning and improved pulmonary function tests [39]. Therefore despite the apparent reduced capacity of muscles following transplantation this further confirms that the muscle dysfunction after transplantation is mostly due to detraining rather than to the immunosuppressant therapy and is reversible. This information highlights the need for rehabilitation following as well as before transplantation.

References

1 Liou TG, Adler FR, Cahill BC, FitzSimmons SC, Huang D, Hibbs JR, Marshall BC: Survival effect of lung transplantation among patients with cystic fibrosis. JAMA 2001;286: 2683–2689.

2 Squier H, Ries A, Kaplan R, Prewitt LM, Smith CM, Kriett JM, Jamieson SW: Quality of well-being predicts survival in lung transplantation candidates. Am J Respir Crit Care Med 1995;152:2032–2036.

3 Simons AK: Non-Invasive Respiratory Support. A Practical handbook, ed 2. Arnold, London, 2001.

4 Madden BP, Kariyawasam H, Siddiqi AJ, Machin A, Pryor JA, Hodson ME:

Noninvasive ventilation in cystic fibrosis patients with acute or chronic respiratory failure. Eur Respir J 2002;19:310–313.

5 Flume PA, Egan TM, Westerman JH, Paradowski LJ, Yankaskas JR, Detterbeck FC, Mill MR: Lung transplantation for mechanically ventilated patients. J Heart Lung Transplant 1994;13(1 pt 1):15–21; discussion 22–23.

6 Moran F, Bradley J: Non-invasive ventilation for cystic fibrosis. Cochrane Database Syst Rev 2003;2:CD002769.

7 Yankaskas JR, Mallory GB Jr: Lung transplantation in cystic fibrosis: Consensus conference statement. Chest 1998;113:217–226.

8 Chaparro C, Maurer J, Gutierrez C, Krajden M, Chan C, Winton T, Keshavjee S, Scavuzzo M, Tullis E, Hutcheon M, Kesten S: Infection with *Burkholderia cepacia* in cystic fibrosis: Outcome following lung transplantation. Am J Respir Crit Care Med 2001;163:43–48.

9 De Soyza A, McDowell A, Archer L, Dark JH, Elborn SJ, Mahenthiralingam E, Gould K, Corris PA: *Burkholderia cepacia* complex genomovars and pulmonary transplantation outcomes in patients with cystic fibrosis. Lancet 2001;358:1780–1781.

10 Aris RM, Routh JC, LiPuma JJ, Heath DG, Gilligan PH: Lung transplantation for cystic fibrosis patients with *Burkholderia cepacia* complex. Survival linked to genomovar type. Am J Respir Crit Care Med 2001;164: 2102–2106.

11 L'Abbe J, Loadman JM, Lau S, Bentley M, Lien D: Predictive value of the 6-minute walk test in lung transplant outcomes. J Heart Lung Transplant 2004;23(suppl):S92–S93.

12 De Soyza A, Dark J, Corris P: Single centre prognostic modelling of cystic fibrosis patients referred for lung transplantation. J Heart Lung Transplant 2004:23(suppl):S93.

13 Lunnon-Wood T, Aurora P, Whitmore P, Fenton M, Radley-Smith R, Elliott M: Non-adherence to therapy is a common cause of death in adolescence following heart or lung transplantation. J Heart Lung Transplant 2004;23(suppl):S77–S78.

14 Zamora M, Edwards L, Weill D, Astor T, Nicolls M: Impact of cystic fibrosis-related diabetes (CFRD) on lung transplant outcomes. J Heart Lung Transplant 2004;23(suppl):S93.

15 Starnes V, Bowdish M, Woo M, Barbers R, Schenkel F, Horn M, Pessotto R, Sievers E, Baker C, Cohen R, Bremner R, Wells W, Barr M: A decade of living lobar lung transplantation: Recipient outcomes. J Thorac Cardiovasc Surg 2004;127:114–122.

16 Trulock-Elbert P, Edwards Leah B, Taylor-David O, Boucek Mark M, Keck Berkeley M, Hertz Marshall I: The Registry of the International Society for Heart and Lung Transplantation: Twenty-first official adult heart transplant report – 2004. J Heart Lung Transplant 2004;23:804–815.

17 Kunst H, Thompson D, Hodson M: Hypertension as a marker for later development of end-stage renal failure after lung and heart-lung transplantation: A cohort study. J Heart Lung Transplant 2004;23:1182–1188.

18 Lake K, Ojo A, Christensen L, Bustami R, Merion R: Chronic renal failure following lung transplantation. J Heart Lung Transplant 2004;23(suppl):S68.

19 Glanville A, Morton J, Aboyoun C, Plit M, Malouf M: Cyclosporin C2 monitoring improves renal dysfunction after lung transplantation. J Heart Lung Transplant 2004;23: 1170–1174.

20 Snell G, Levvey B, Chin W, Kotsimbos T, Whitford H, Waters K, Richardson M, Williams T: Sirolimus allows renal recovery in lung and heart transplant recipients with chronic renal impairment. J Heart Lung Transplant 2002;21:540–546.

21 Swarbrick D, Corris MP: Anaemia in lung transplant recipients: Effect on health related quality of life. J Heart Lung Transplant 2004;23(suppl):S93–S94.

22 Hartwig M, Cantu E, Appel J, Woreta H, Palmer S, Davis R: Non-alloimmune injury mediated by gastroesophageal reflux precipitates alloimmune injury in lung transplant patients. J Heart Lung Transplant 2004; 23(suppl):S43.

23 Davis R, Lau C, Eubanks S, Messier R, Hadjiliadis D, Steele M, Palmer S: Improved lung allograft function after fundoplication in patients with gastroesophageal reflux disease undergoing lung transplantation. J Thorac Cardiovasc Surg 2003;125:533–542.

24 D'Ovidio F, Mura M, Waddell T, Pierre A, Hutcheon M, Hadjiliadis D, Singer L, Miller L, Darling G, de Perrot M, Shargall Y, Keshavjee S: Bile acids in bronchoalveolar lavage after lung transplantation as a marker of pulmonary aspiration associated with alveolar neutrophilia. J Heart Lung Transplant 2004;23(suppl):S42.

25 Hopkins P, Plit L, Carter I, Chhajed P, Malouf M, Glanville A: Indirect fluorescent antibody testing of nasopharyngeal swabs for influenza diagnosis in lung transplant recipients. J Heart Lung Transplant 2003;22:161–168.

26 Zamora M, Hodges T, Nicolls M, Astor T, Marquesen J, Weill D: Impact of respiratory syncytial virus pneumonia following lung transplantation: A case-controlled study. J Heart Lung Transplant 2004;23(suppl): S43–S44.

27 Glanville A, Valentine V, Aboyoun C, Malouf M: CMV mismatch is not a risk factor for survival or severe bronchiolitis obliterans syndrome after lung transplantation. J Heart Lung Transplant 2004;23(suppl):S43.

28 Rose M: De novo production of antibodies after heart or lung transplantation should be regarded as an early warning system. J Heart Lung Transplant 2004;23:385–395.

29 Verleden G, Dupont L: Azithromycin therapy for patients with bronchiolitis obliterans syndrome after lung transplantation. Transplantation 2004;77:1465–1467.

30 Cornelissen J, Trull A, Parameshwar J, Bellm S, Charman S, Wallwork J: Difficulty in managing cyclosporin C_2 monitoring in de novo lung transplant recipients. J Heart Lung Transplant 2004;23(suppl):S139.

31 Thekkudan J, Fildes J, Sivaprakasam R, Khasati N, Datta S, Martyszczuck R, Maachal A, Keevil B, Leonard C, Yonan N: Caveats of cyclosporine dosing based on C2 monitoring. J Heart Lung Transplant 2004;23(suppl): S139.

32 Iacono AT, Corcoran TE, Griffith BP, et al: Aerosol cyclosporin therapy in lung transplant recipients with bronchiolitis obliterans. Eur Respir J 2004;23:384–390.

33 Bhorade S, Ahya V, Kotloff R, Baz M, Valentine V, Arcasoy S, Love R, Young R, Vigneswaran W, Garrity E: Comparison of sirolimus versus azathioprine in a tacrolimus-based immunosuppressive regimen in lung transplantation. J Heart Lung Transplant 2004;23(suppl):S113.

34 Snell G, Valentine V, Love R, Vitulo P, Glanville A, Pirron U: Two-year results of an international, randomized, double-blind study of everolimus (RAD) vs azathioprine to inhibit the decline of pulmonary function in stable lung transplant recipients. J Heart Lung Transplant 2004;23(suppl):S45.

35 Mehrad B, Paciocco G, Martinez FJ, Ojo TC, Iannettoni MD, Lynch JP 3rd: Spectrum of *Aspergillus* infection in lung transplant recipients: Case series and review of the literature. Chest 2001;119:169–175.

36 Aigner C, Jaksch P, Winkler G, Czebe C, Devyatko E, Taghavi S, Wisser W, Klepetko W: Initial experience with oral valgancyclovir for pre-emptive cytomegalovirus therapy after lung transplantation. J Heart Lung Transplant 2004;23(suppl):S95.

37 Lands LC, Smountas AA, Mesiano G, Brosseau L, Shennib H, Charbonneau M, Gauthier R: Maximal exercise capacity and peripheral skeletal muscle function following lung transplantation. J Heart Lung Transplant 1999;18:113–120.

38 Estenne M, Pinet C, Scillia P, Cassart M, Lamotte M, Knoop C, Mélot C: Function and bulk of respiratory and limb muscles after lung transplantation for cystic fibrosis. J Heart Lung Transplant 2004;23(suppl):S96.

39 Tegtbur U, Guetzlaff E, Niedermeyer J, Pethig K, Warnecke G, Kugler C, Strueber M, Busse M, Haverich A: A controlled trial of exercise rehabilitation after lung transplantation. J Heart Lung Transplant 2004:23(suppl):S95.

Dr. Martin Carby
Royal Brompton & Harefield NHS Trust
Harefield Hospital, Harefield, Uxbridge
UB9 6JH (UK)
Tel. +44 1895 828562
Fax +44 1895 828975
E-Mail m.carby@rbh.nthames.nhs.uk

Bush A, Alton EWFW, Davies JC, Griesenbach U, Jaffe A (eds): Cystic Fibrosis in the 21st Century.
Prog Respir Res. Basel, Karger, 2006, vol 34, pp 180–186

Current and Novel Antimicrobial Approaches

Jane C. Davies

Imperial College, London, Honorary Consultant in Paediatric Respiratory Medicine,
Royal Brompton Hospital, London, UK

Abstract

Antibiotics are the mainstay of treatment for pulmonary
disease in CF. Along with improved airway clearance techniques and nutrition, the availability (and aggressive use) of
effective antimicrobials has been a major contributing factor
in the improved health and prognosis of patients today. However, significant problems still exist: bacteria develop resistance, either via genetic mutation or biofilm formation; the
huge numbers of antibiotics administered to patients pose a
burden on them, both in terms of time and toxicity; and most
of the current strategies treat rather than prevent infection.
For all these reasons, new approaches would be highly desirable. This chapter will review recent developments in conventional antimicrobial approaches and go on to highlight some
of the research areas with promise.

Conventional Antimicrobial Treatment

Antibiotics are used in four distinct situations in patients
with CF: (a) long-term treatment for *Staphylococcus
aureus*, either prophylactically or after infection, (b) in an
attempt to eradicate early *Pseudomonas aeruginosa* infection, (c) as suppressive therapy in chronic pseudomonal
infection and (d) as acute treatment for respiratory exacerbations. For each of these scenarios, there remain unan-
swered questions, with centres making many clinical
management decisions in the absence of clear-cut evidence.
Some of these are highlighted in table 1. The background to
conventional treatment regimen has been well reviewed [1],

Table 1. Areas of uncertainty in conventional clinical management

---

Anti-staphylococcal drugs
Should anti-staphylococcal prophylactic antibiotics be initiated from
 diagnosis?
Is the risk of *P. aeruginosa* increased with all long-term anti-
 staphylococcal antibiotics, or only cephalosporins?
What is the optimal treatment regimen and when should treatment
 be stopped in an uninfected child?

Eradication of early *P. aeruginosa* infection
What is the optimal drug regimen and duration of administration?

Chronic suppressive anti-pseudomonal therapy
Is there a superior nebulized antibiotic?
Will the development of aminoglycoside resistance matter in the
 longer-term?

Treatment of multiresistant organisms
Would the availability of synergy testing improve outcome?

---

and this chapter will largely focus on advances over the last
5–8 years.

Anti-Staphylococcal Antibiotics

As discussed in previous chapters, *S. aureus* is often the
first bacterium to be cultured from the respiratory tract in
infants and children with CF. Several studies have
addressed the benefits and risks of the early initiation of
anti-staphylococcal prophylactic antibiotics, some of which
were included in a systematic review in 2003 [2]. A major
problem with interpretation of the data is the extreme heterogeneity in study design, indications for starting drugs (at

the time of diagnosis or only after detection of the pathogen), choice of drug and outcome measures. Conclusions on the benefit of prophylaxis could not be reached from this meta-analysis, and although treatment did reduce the rate of positive cultures, clinical benefit could not be determined with certainty in those already infected with the organism. More recently, a large retrospective review of German patients on the European database was reported by Ratjen et al. [3]. Of 693 children who were *P. aeruginosa* negative, 48.2 and 40.4% had received anti-staphylococcal treatment, either continuously or intermittently, respectively. Continuous treatment reduced the risk of *S. aureus* culture positivity, but worryingly, appeared to be associated with an increased frequency of *P. aeruginosa* infection. Continuous treatment did not confer clinical benefit over intermittent use. Importantly, unlike in the UK, where flucloxacillin is the drug used most commonly, almost half the patients had been treated with an oral cephalosporin, and only 4.6% with flucloxacillin. Another study has raised concerns over the use of cephalosporins in this context, pseudomonal infection being more common in children receiving long-term cephalexin as part of a randomized, placebo-controlled trial [4]. This may be a concern only for this group of drugs therefore, although insufficient data are available on other drugs. Several reports have been published recently of linezolid, a new, potent anti-staphylococcal antibiotic, which has been used with some success for methicillin-resistant *Staphylococcus aureus* (MRSA) [5, 6].

Eradication of Early P. aeruginosa Infection

Initial infecting strains of *P. aeruginosa* are usually non-mucoid, antibiotic susceptible and thus more amenable to treatment than their mucoid counterparts, which appear with chronic infection and form biofilms offering further protection against clearance (see chapter 18). There are accumulating data in support of the benefits of early detection and aggressive treatment, to prevent or delay the onset of chronic infection. Recent data demonstrating significant deterioration in lung function after the bacteria have become mucoid [7] lend further support to attempts to eradicate early. Pioneered by the Danish clinics following the success of a small randomized, controlled trial in the early 1990s [8], many studies since have been uncontrolled, limiting interpretation of the data. The same group has recently reported longer-term results, demonstrating that their strategy of combined oral ciprofloxacin and nebulized colomycin treatment prevented or delayed chronic infection in over 75% of patients during the 3.5-year follow-up [9]. Eradication has also been reported with a combination of intravenous and nebulized agents [10]. The duration of

eradication was between 2 months and 2 years, and interestingly, the majority of patients presented with a genetically distinct organism. Gibson et al. [11] have reported a significant rate of eradication with inhaled tobramycin, which led to early termination of their trial. After 28 days treatment, 0/8 children had positive *P. aeruginosa* cultures from broncho-alveolar lavage (BAL), compared with clearance in only 1/13 of the placebo group. A multicentre study of the use of tobramycin for inhalation (TOBI™) has started in Europe. It seems clear despite the limited amount of data from randomized, controlled trials that eradication can be successfully achieved after early infection. What remains less clear is the best type of therapy and the duration for which it should be administered.

Chronic Suppressive Therapy

Nebulized antibiotics, most commonly the polymixin, colomycin, have been used in this context for many years in European centres. Although individual studies were small, a published meta-analysis confirmed benefit with very few side effects [12] and the approach has been the subject of a recent Cochrane Review [13]. Recent trials have demonstrated clinical benefit with long-term nebulized tobramycin, although the induction of resistance means that the drug is best used only on alternate months [14]. Resistance does not seem to translate into lack of clinical efficacy, although whether this will remain so in the longer term is unknown. Systemic absorption is low, limiting the risk of toxicity. No good trial has compared different nebulized drugs. One study compared colomycin with tobramycin over 1 month (not a complete month on, month off TOBI cycle), demonstrating a significant improvement in lung function with the latter drug only [15]. However, all patients were already on colomycin, which is likely to have affected the results. A retrospective study showing that inhaled tobramycin was associated with an increased risk of death was almost certainly influenced by patient selection, the more severely affected patients receiving treatment [16]. One significant disadvantage of the nebulized route is the time taken to administer treatment, which can be up to an hour daily. Alternatives to inhaled antibiotics include the long-term use of macrolides (see chapter 25) and regular 3-monthly courses of intravenous antibiotics. These approaches can be combined. The benefit of the latter approach has not been demonstrated for the majority of patients [17]

Acute Treatment for Pulmonary Exacerbations

Intermittent treatment for exacerbations is most commonly administered as either short courses of oral or intravenous antibiotics. Decisions must be guided by microbiological sensitivities, although in vitro susceptibility testing does not always predict clinical response [18].

In general, two intravenous antibiotics with different mechanisms of action (e.g. a β-lactam and an aminoglycoside) will be used in combination. There are few recent advances or changes to conventional management, which has been recently reviewed [1]. However, one such change is to once-daily high dose (as opposed to three times daily) intravenous aminoglycosides [19–21]. This regimen achieves a higher peak (upon which bacterial killing is dependent) whilst providing a longer period for the post-antibiotic effects of the drug [22]. There does not appear to be an increased risk of toxicity, although levels must be monitored routinely. Multi-resistant organisms, such as *Burkholderia cepacia*, can pose a serious management problem. Synergy testing may reveal a combination of useful antibiotics, despite lack of in vitro sensitivity to each of the agents individually, and may be helpful in certain cases [23]. However, this is not widely available and the impact of such a service on clinical outcomes has not been determined.

Novel Treatments

It will be clear from the above that there are significant limitations to current anti-infective approaches: the development of resistance to conventional antibiotics; the reduced efficacy of most drugs on chronic, mucoid *P. aeruginosa* growing in biofilms; the burden of treatment to the patient; and the risk of toxicity with certain intravenous antibiotics, which necessitates invasive monitoring. Furthermore, there are no treatments that have conclusively been shown to prevent initial infection, in particular with *P. aeruginosa*. Clearly, new strategies are needed. The next section of this chapter will discuss the possible future approaches to preventing and treating infection in the CF lung, some of which may eventually become clinically useful. Much research is beginning to focus on genomic- and proteomic-based strategies [24–26] made possible by the recent mapping of the *Pseudomonas* genome (www. pseudomonas.com).

Immunization

Active Immunization. One of the paradoxes of CF pathophysiology is that despite the exuberant immune response mounted by CF patients upon infection with *P. aeruginosa*, it is insufficient to clear the infection. One suggestion is that this reflects low affinity or effector functions of natural antibodies to various bacterial components. Advocates of active immunization argue that vaccines (particularly conjugates) could generate more effective immune responses and prevent early infection. The subject has been the focus of a recent Cochrane Review [27]. Preclinical animal studies have demonstrated enhanced clearance of *P. aeruginosa* from the lungs of mice and increased survival following immunization against a variety of epitopes, in particular components of lipopolysaccharide (LPS) [reviewed in ref. 28]. The earliest human studies involved vaccines with unacceptable side effects; the progress from these to more recent studies has been well reviewed [29], and the safety concerns appear, at least in part, to have been resolved in human studies [30, 31]. In targeting LPS, one problem is the wide variation between strains of the O-antigen, which would lead to a narrow spectrum of protection. In an attempt to combat this, Lang et al. [32] developed a conjugate vaccine comprising an octavalent O-polysaccharide and an exotoxin A. They reported short-term safety and immunogenicity of the vaccine in CF children without prior *P. aeruginosa* infection. Over the next 10 years, the patients have received annual immunization, and have recently been compared with controls, who have received otherwise identical management, although this follow-up was not conducted in a blinded or placebo-controlled fashion. Fewer immunized patients were infected, and in fewer of these had the organism converted to a mucoid phenotype, leading to better preservation of lung function. The vaccine has since been granted Orphan Drug status and is being evaluated in a large multicentre European trial (http://www.bernabiotech. com/news/ archive/news_full/id.25/).

Passive Immunization. Kollberg et al. [33] administered IgY (derived from the yolks of *P. aeruginosa*-immunized hens) as a topical gargle in a small, open-label study. CF patients had their first isolate of *P. aeruginosa* eradicated with conventional antibiotics, after which the IgY was administered. The treatment was well tolerated, and compared to a non-treated group, a second positive culture was significantly delayed. A similar approach has been employed for other infective diseases, such as gingivitis and enteritic infections, although, clearly larger, controlled studies will be required before benefit is confirmed in CF. Monoclonal antibodies derived from transgenic mice are another route under investigation, conferring complete protection from fatal infection in neutropenic mice [34].

Preventing Adherence

Although organisms such as *P. aeruginosa* have been shown to adhere in increased numbers to CF epithelia [35], the relevance of this observation to disease pathogenesis is unclear, and it is in fact unclear whether the bacteria adhere to the mucosa or the mucus in the lumen. Strategies aimed directly at the cell surface receptor, a disaccharide moiety contained within asialylated glycolipids, are unlikely to be successful; in vitro studies have demonstrated a prompt immune response to ligation of this receptor with antibody as well as bacterial components [36]. Non-specific strategies may, however, be more applicable. Sajjan et al. [37] demonstrated that either dextran or xylitol significantly decreased the adherence of *B. cepacia* to explanted tracheas, possibly by removing or thinning surface mucus. The former agent has also been shown to improve mucus clearance [38], which could be an additional clinical benefit. Heparin, which has been administered topically to the human airway, may also possess this ability [39], in addition to its postulated anti-inflammatory or anti-allergic properties [40]. As discussed earlier in this book, the mechanism of action of the macrolide group of antibiotics in CF is incompletely understood. Baumann et al. [41] have reported decreased adherence of *P. aeruginosa* to the buccal epithelium after treatment with azithromycin, which may be relevant in the early stages of oropharyngeal infection.

Anti-Biofilm Strategies

As discussed in chapter 18, biofilm formation is a survival strategy employed by many bacteria, most importantly, in the context of the CF airway *P. aeruginosa*. In addition to the physical barrier posed by the biofilm matrix, a number of other properties render these organisms less susceptible to conventional antibiotics and many of the commonly used drugs (e.g. aminoglycosides and β-lactams) are poorly effective on slow-growing organisms [42]. The steps in the initiation and progression of biofilm formation are becoming clearer [43, 44], further understanding of which may lead to novel treatment approaches. Early infection occurs with low numbers of planktonic bacteria. These bacteria employ a mechanism of 'quorum sensing' via the production of freely diffusible compounds, acyl-homoserine lactones (AHLs), which have been detected in CF sputum [45]. The concentration of these rises with increases in bacterial numbers, and at high concentration induce changes in gene expression leading to a switch from a planktonic to a biofilm mode of growth. Bacteria, which have previously been motile, become sessile and dormant. Synthetic furanones, compounds able to inhibit AHLs, were highly protective in a murine model of *P. aeruginosa* pneumonia [46]. Rogan et al. [47] demonstrated that lactoferrin, a component of the innate immune system present in the lung,

prevents biofilm formation by enhancing 'twitching motility' and preventing the formation of the clusters required in the early stage of biofilm development. It achieves this by chelating iron, leading to the suggestion that this molecule or other iron chelators could be useful in preventing biofilm formation [48]. In the presence of mature biofilm, alternative approaches will be required. Alginate lyase [49], and the application of electrical currents [50] have both enhanced the effect of conventional antibiotics by facilitating diffusion through disrupted biofilm, although the clinical applications of the latter are unclear. Although the exact mechanism(s) by which macrolides achieve clinical benefit is unknown, they may be acting to inhibit formation of, or break down, biofilm [51].

Novel Antimicrobial Agents

A full review of all microbial agents under development is outside the scope of this chapter. However, several groups of compounds merit a mention. Peptide antimicrobial agents, including defensins and cathelicidins are naturally occurring substances, produced by inflammatory cells and epithelia [52]. Although the suggestion that defensins fail to function in the CF airway because of abnormalities in salt concentration [53] is thought by most to be incorrect [54], the development of defensin-like molecules may hold some promise. They are broad spectrum and do not induce resistance, two highly desirable properties for CF infections. Analogues have reached the clinical trial stage for other conditions [55]. Histatins, similar peptides produced by salivary glands, have potent in vitro anti-pseudomonal activity, although certain of these are inhibited by purulent CF sputum [56]. Cathelicidins were reported to have rapid in vitro bactericidal properties, and importantly were also effective on multiresistant organisms such as *B. cepacia* [57].

Dry-Powder Antibiotics

Although nebulized antibiotics are effective, disadvantages include the time required for administration, maintenance of the equipment and environmental contamination. The development of dry powder formulations would overcome all of these issues. A small, proof-of-principle study has been published using a dry powder device with colistin sulphate [58]. CF patients, unlike their healthy counterparts experienced moderately severe cough and some deterioration in lung function. Colistemethate sodium, which is thought to be less irritant to the airway has also been formulated for a dry powder device and was confirmed to be safe and well tolerated in adults and children with CF [59]. Phase III trials are planned. Other antibiotics, such as

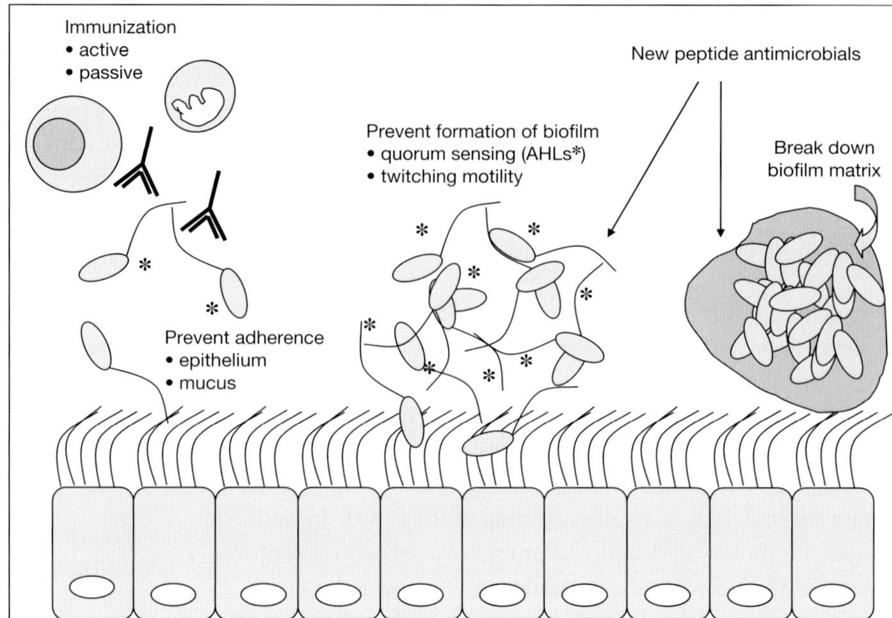

Fig. 1. Novel antimicrobial approaches. Investigators have considered strategies at all stages of the infective process as targets for new treatments, including prevention (vaccination and anti-adherence agents), prevention of chronic, biofilm infection (inhibiting signalling molecules or encouraging twitching motility), breakdown of biofilm matrix and novel synthetic peptides based on naturally occurring molecules, such as defensins.

aztreonam are also under development. Advances could provide significant quality of life benefits to CF patients with respect to the burden of current treatment.

Which, and how many, of the approaches mentioned above will make it through to clinically applicable treatments remains to be seen. In the meantime, more well-conducted clinical research to answer the outstanding questions on basic treatment regimens and to lead to a consensus on management would be extremely useful.

References

1   Gibson RL, Burns JL, Ramsey BW: Pathophysiology and management of pulmonary infections in cystic fibrosis. Am J Respir Crit Care Med 2003 15;168:918–951.
2   Smyth A, Walters S: Prophylactic antibiotics for cystic fibrosis. Cochrane Database Syst Rev 2003;3:CD001912.
3   Ratjen F, Comes G, Paul K, Posselt HG, Wagner TO, Harms K, German Board of the European Registry for Cystic Fibrosis (ERCF): Effect of continuous antistaphylococcal therapy on the rate of *P. aeruginosa* acquisition in patients with cystic fibrosis. Pediatr Pulmonol 2001;31:13–16.
4   Stutman HR, Lieberman JM, Nussbaum E, Marks MI: Antibiotic prophylaxis in infants and young children with cystic fibrosis: A randomized controlled trial. J Pediatr 2002;140:299–305.
5   Ferrin M, Zuckerman JB, Meagher A, Blumberg EA: Successful treatment of methicillin-resistant *Staphylococcus aureus* pulmonary infection with linezolid in a patient with cystic fibrosis. Pediatr Pulmonol 2002;33:221–223.
6   Serisier DJ, Jones G, Carroll M: Eradication of pulmonary methicillin-resistant *Staphylococcus aureus* (MRSA) in cystic fibrosis with linezolid. J Cyst Fibros 2004;3:61.

7   Li Z, Kosorok MR, Farrell PM, Laxova A, West SE, Green CG, Collins J, Rock MJ, Splaingard ML: Longitudinal development of mucoid *Pseudomonas aeruginosa* infection and lung disease progression in children with cystic fibrosis. JAMA 2005;293:581–588.
8   Valerius NH, Koch C, Hoiby N: Prevention of chronic *Pseudomonas aeruginosa* colonisation in cystic fibrosis by early treatment. Lancet 1991;338:725–726.
9   Frederiksen B, Koch C, Hoiby N: Antibiotic treatment of initial colonization with *Pseudomonas aeruginosa* postpones chronic infection and prevents deterioration of pulmonary function in cystic fibrosis. Pediatr Pulmonol 1997;23:330–335.
10 Munck A, Bonacorsi S, Mariani-Kurkdjian P, Lebourgeois M, Gerardin M, Brahimi N, Navarro J, Bingen E: Genotypic characterization of *Pseudomonas aeruginosa* strains recovered from patients with cystic fibrosis after initial and subsequent colonization. Pediatr Pulmonol 2001;32:288–292.
11 Gibson RL, Emerson J, McNamara S, Burns JL, Rosenfeld M, Yunker A, Hamblett N, Accurso F, Dovey M, Hiatt P, Konstan MW, Moss R, Retsch-Bogart G, Wagener J, Waltz D, Wilmott R, Zeitlin PL, Ramsey B: Cystic Fibrosis Therapeutics Development Network

Study Group: Significant microbiological effect of inhaled tobramycin in young children with cystic fibrosis. Am J Respir Crit Care Med 2003;67:841–849.
12 Mukhopadhyay S, Singh M, Cater JI, Ogston S, Franklin M, Olver RE: Nebulised antipseudomonal antibiotic therapy in cystic fibrosis: A meta-analysis of benefits and risks. Thorax 1996;51:364–368.
13 Ryan G, Mukhopadhyay S, Singh M: Nebulised anti-pseudomonal antibiotics for cystic fibrosis. Cochrane Database Syst Rev 2003;3:CD001021.
14 Ramsey BW, Pepe MS, Quan JM, Otto KL, Montgomery AB, Williams-Warren J, Vasiljev-KM, Borowitz D, Bowman CM, Marshall BC, Marshall S, Smith AL: Intermittent administration of inhaled tobramycin in patients with cystic fibrosis. Cystic Fibrosis Inhaled Tobramycin Study Group. N Engl J Med 1999;340:23–30.
15 Hodson ME, Gallagher CG, Govan JR: A randomised clinical trial of nebulised tobramycin or colistin in cystic fibrosis. Eur Respir J 2002;20:658–664.
16 Rothman KJ, Wentworth CE 3rd: Mortality of cystic fibrosis patients treated with tobramycin solution for inhalation. Epidemiology 2003;14:55–59.

17 Elborn JS, Prescott RJ, Stack BH, Goodchild MC, Bates J, Pantin C, Ali N, Shale DJ, Crane M: Elective versus symptomatic antibiotic treatment in cystic fibrosis patients with chronic *Pseudomonas* infection of the lungs. Thorax 2000;55:355–358.

18 Smith AL, Fiel SB, Mayer-Hamblett N, Ramsey B, Burns JL: Susceptibility testing of *Pseudomonas aeruginosa* isolates and clinical response to parenteral antibiotic administration: Lack of association in cystic fibrosis. Chest 2003;123:1495–1502.

19 Whitehead A, Conway SP, Etherington C, Caldwell NA, Setchfield N, Bogle S: Once-daily tobramycin in the treatment of adult patients with cystic fibrosis. Eur Respir J 2002;19:303–309.

20 Aminimanizani A, Beringer PM, Kang J, Tsang L, Jelliffe RW, Shapiro BJ: Distribution and elimination of tobramycin administered in single or multiple daily doses in adult patients with cystic fibrosis. J Antimicrob Chemother 2002;50:553–559.

21 Smyth A, Tan KH, Hyman-Taylor P, Mulheran M, Lewis S, Stableforth D, Prof Knox A, TOPIC Study Group. Once versus three-times daily regimens of tobramycin treatment for pulmonary exacerbations of cystic fibrosis – the TOPIC study: A randomised controlled trial. Lancet 2005;365:573–578.

22 Contopoulos-Ioannidis DG, Giotis ND, Baliatsa DV, Ioannidis JP: Extended-interval aminoglycoside administration for children: A meta-analysis. Pediatrics 2004;114:111–118.

23 Aaron SD, Ferris W, Henry DA, Speert DP, Macdonald NE: Multiple combination bactericidal antibiotic testing for patients with cystic fibrosis infected with *Burkholderia cepacia*. Am J Respir Crit Care Med 2000; 161:1206–1212.

24 Beckmann C, Brittnacher M, Ernst R, Mayer-Hamblett N, Miller SI, Burns JL: Use of phage display to identify potential *Pseudomonas aeruginosa* gene products relevant to early cystic fibrosis airway infections. Infect Immun 2005;73:444–452.

25 Arevalo-Ferro C, Hentzer M, Reil G, Gorg A, Kjelleberg S, Givskov M, Riedel K, Eberl L: Identification of quorum-sensing regulated proteins in the opportunistic pathogen *Pseudomonas aeruginosa* by proteomics. Environ Microbiol 2003;5:1350–1369.

26 Hassett DJ, Limbach PA, Hennigan RF, Klose KE, Hancock RE, Platt MD, Hunt DF: Bacterial biofilms of importance to medicine and bioterrorism: Proteomic techniques to identify novel vaccine components and drug targets. Expert Opin Biol Ther 2003;3:1201–1207.

27 Keogan MT, Johansen HK: Vaccines for preventing infection with *Pseudomonas aeruginosa* in people with cystic fibrosis. Cochrane Database Syst Rev 2000;2:CD001399.

28 Holder IA: *Pseudomonas* immunotherapy: A historical overview. Vaccine 2004;22:831–839.

29 Pier GB: Promises and pitfalls of *Pseudomonas aeruginosa* lipopolysaccharide as a vaccine antigen. Carbohydr Res 2003;338: 2549–2556.

30 Baumann U, Mansouri E, von Specht BU: Recombinant OprF-OprI as a vaccine against *Pseudomonas aeruginosa* infections. Vaccine 2004;22:840–847.

31 Larbig M, Mansouri E, Freihorst J, Tummler B, Kohler G, Domdey H, Knapp B, Hungerer KD, Hundt E, Gabelsberger J, von Specht BU: Safety and immunogenicity of an intranasal *Pseudomonas aeruginosa* hybrid outer membrane protein F-I vaccine in human volunteers. Vaccine 2001;19:2291–2297.

32 Lang AB, Rudeberg A, Schoni MH, Que JU, Furer E, Schaad UB: Vaccination of cystic fibrosis patients against *Pseudomonas aeruginosa* reduces the proportion of patients infected and delays time to infection. Pediatr Infect Dis J 2004;23:504–510.

33 Kollberg H, Carlander D, Olesen H, Wejaker PE, Johannesson M, Larsson A: Oral administration of specific yolk antibodies (IgY) may prevent *Pseudomonas aeruginosa* infections in patients with cystic fibrosis: A phase I feasibility study. Pediatr Pulmonol 2003;35: 433–440.

34 Hemachandra S, Kamboj K, Copfer J, Pier G, Green LL, Schreiber JR: Human monoclonal antibodies against *Pseudomonas aeruginosa* lipopolysaccharide derived from transgenic mice containing megabase human immunoglobulin loci are opsonic and protective against fatal pseudomonas sepsis. Infect Immun 2001;69:2223–2229.

35 Saiman L, Prince A: *Pseudomonas aeruginosa* pili bind to asialoGM1 which is increased on the surface of cystic fibrosis epithelial cells. J Clin Invest 1993;92: 1875–1880.

36 DiMango E, Zar HJ, Bryan R, Prince A: Diverse *Pseudomonas aeruginosa* gene products stimulate respiratory epithelial cells to produce interleukin-8. J Clin Invest 1995;96: 2204–2210.

37 Sajjan U, Moreira J, Liu M, Humar A, Chaparro C, Forstner J, Keshavjee S: A novel model to study bacterial adherence to the transplanted airway: Inhibition of *Burkholderia cepacia* adherence to human airway by dextran and xylitol. J Heart Lung Transplant 2004;23:1382–1391.

38 Feng W, Garrett H, Speert DP, King M: Improved clearability of cystic fibrosis sputum with dextran treatment in vitro. Am J Respir Crit Care Med 1998;157:710–714.

39 Thomas R, Brooks T: Common oligosaccharide moieties inhibit the adherence of typical and atypical respiratory pathogens. J Med Microbiol 2004;53:833–840.

40 Stelmach I, Jerzynska J, Stelmach W, Majak P, Brzozowska A, Gorski P, Kuna P: The effect of inhaled heparin on airway responsiveness to histamine and leukotriene D_4. Allergy Asthma Proc 2003;24:59–65.

41 Baumann U, Fischer JJ, Gudowius P, Lingner M, Herrmann S, Tummler B, von der Hardt H: Buccal adherence of *Pseudomonas aeruginosa* in patients with cystic fibrosis under long-term therapy with azithromycin. Infection 2001;29:7–11.

42 Chernish RN, Aaron SD: Approach to resistant gram-negative bacterial pulmonary infections in patients with cystic fibrosis. Curr Opin Pulm Med 2003;9:509–915.

43 Yoon SS, Hassett DJ: Chronic *Pseudomonas aeruginosa* infection in cystic fibrosis airway disease: Metabolic changes that unravel novel drug targets. Expert Rev Anti Infect Ther 2004;2:611–623.

44 Drenkard E, Ausubel FM: Pseudomonas biofilm formation and antibiotic resistance are linked to phenotypic variation. Nature 2002 18;416:740–743.

45 Singh PK, Schaefer AL, Parsek MR, Moninger TO, Welsh MJ, Greenberg EP: Quorum-sensing signals indicate that cystic fibrosis lungs are infected with bacterial biofilms. Nature 2000;407:762–764.

46 Wu H, Song Z, Hentzer M, Andersen JB, Molin S, Givskov M, Hoiby N: Synthetic furanones inhibit quorum-sensing and enhance bacterial clearance in *Pseudomonas aeruginosa* lung infection in mice. J Antimicrob Chemother 2004;53:1054–1061.

47 Rogan MP, Taggart CC, Greene CM, Murphy PG, O'Neill SJ, McElvaney NG: Loss of microbicidal activity and increased formation of biofilm due to decreased lactoferrin activity in patients with cystic fibrosis. J Infect Dis 2004;190:1245–1253.

48 Weinberg ED: Suppression of bacterial biofilm formation by iron limitation. Med Hypotheses 2004;63:863–865.

49 Hatch RA, Schiller NL: Alginate lyase promotes diffusion of aminoglycosides through the extracellular polysaccharide of mucoid *Pseudomonas aeruginosa*. Antimicrob Agents Chemother 1998;42:974–977.

50 Caubet R, Pedarros-Caubet F, Chu M, Freye E, de Belem Rodrigues M, Moreau JM, Ellison WJ: A radio frequency electric current enhances antibiotic efficacy against bacterial biofilms. Antimicrob Agents Chemother 2004;48:4662–4664.

51 Schultz MJ: Macrolide activities beyond their antimicrobial effects: Macrolides in diffuse panbronchiolitis and cystic fibrosis. J Antimicrob Chemother 2004;54:21–28.

52 Hiemstra PS, Fernie-King BA, McMichael J, Lachmann PJ, Sallenave JM: Antimicrobial peptides: Mediators of innate immunity as templates for the development of novel anti-infective and immune therapeutics. Curr Pharm Des 2004;10:2891–2905.

53 Goldman MJ, Anderson GM, Stolzenberg ED, Kari UP, Zasloff M, Wilson JM: Human beta-defensin-1 is a salt-sensitive antibiotic in lung that is inactivated in cystic fibrosis. Cell 1997; 88:553–560.

54 Donaldson SH, Boucher RC: Update on pathogenesis of cystic fibrosis lung disease. Curr Opin Pulm Med 2003;9:486–491.

55 Giles FJ, Rodriguez R, Weisdorf D, Wingard JR, Martin PJ, Fleming TR, Goldberg SL, Anaissie EJ, Bolwell BJ, Chao NJ, Shea TC, Brunvand MM, Vaughan W, Petersen F, Schubert M, Lazarus HM, Maziarz RT, Silverman M, Beveridge RA, Redman R, Pulliam JG, Devitt-Risse P, Fuchs HJ, Hurd DD: A phase III, randomized, double-blind, placebo-controlled, study of iseganan for the reduction of stomatitis in patients receiving stomatotoxic chemotherapy. Leuk Res 2004; 28:559–565.

56 Sajjan US, Tran LT, Sole N, Rovaldi C, Akiyama A, Friden PM, Forstner JF, Rothstein DM: P-113D, an antimicrobial peptide active against *Pseudomonas aeruginosa*, retains activity in the presence of sputum from cystic fibrosis patients. Antimicrob Agents Chemother 2001;45:3437–3444.

57 Saiman L, Tabibi S, Starner TD, San Gabriel P, Winokur PL, Jia HP, McCray PB Jr, Tack BF: Cathelicidin peptides inhibit multiply antibiotic-resistant pathogens from patients with cystic fibrosis. Antimicrob Agents Chemother 2001;45:2838–2844.

58 Le Brun PP, de Boer AH, Mannes GP, de Fraiture DM, Brimicombe RW, Touw DJ, Vinks AA, Frijlink HW, Heijerman HG: Dry powder inhalation of antibiotics in cystic fibrosis therapy. 2. Inhalation of a novel colistin dry powder formulation: A feasibility study in healthy volunteers and patients. Eur J Pharm Biopharm 2002;54: 25–32.

59 Davies JC, Hall P, Francis J, Scott S, Geddes DM, Conway S, Phillips G, Stramik SJ, Girdwood K, Goldman MH: Dry powder formulation of colistimethate sodium is safe and well tolerated in adults and children with CF. Pediatr Pulmonol 2004;283:S27.

Jane C. Davies
Department of Gene Therapy
National Heart and Lung Institute
Imperial College, Emmanuel Kaye Building
Manresa Rd
London SW3 6LR (UK)
Tel. +44 207 351 8398
Fax +44 207 351 8340
E-Mail j.c.davies@imperial.ac.uk

Davies

Bush A, Alton EWFW, Davies JC, Griesenbach U, Jaffe A (eds): Cystic Fibrosis in the 21st Century.
Prog Respir Res. Basel, Karger, 2006, vol 34, pp 187–194

Anti-Inflammatory Agents

A Clinical Perspective

T.N. Hilliard   I.M. Balfour-Lynn

Department of Paediatric Respiratory Medicine, Royal Brompton Hospital, London, UK

Abstract

Neutrophil-dominated airway inflammation begins early in cystic fibrosis (CF) and is integral to progressive lung damage. It is, therefore, logical to attempt to modify this inflammatory response, ideally as early as possible, and before sustained severe inflammation is established. Oral corticosteroids have been shown to slow the progression of lung disease, but have unacceptable side effects for long-term use. Inhaled corticosteroids, although widely prescribed, have not been shown to be an effective anti-inflammatory agent in CF, and they may not be as safe as previously assumed. High-dose oral ibuprofen has proven efficacy, but its potential adverse effects have limited its uptake in clinical practice. Antileukotrienes have had limited study in CF, but as yet no significant benefit has been demonstrated. There were encouraging early phase trials of aerosolized antiproteases but they have not been developed further, and the focus is now on novel synthetic anti-elastase agents. Nebulized recombinant DNase, despite concerns of a possible pro-inflammatory effect, in fact appears to stabilize airway inflammation in mild lung disease. Macrolide antibiotics have been shown to have a small but significant beneficial effect on pulmonary function, and although the mechanism is unclear, it is likely that it relates to an anti-inflammatory effect. Other anti-inflammatory therapies that have been tried in difficult asthma, e.g. cyclosporine, methotrexate and intravenous immunoglobulin, have had very limited study, and their effect in CF is unproven. Finally, there are a number of novel agents, which target individual inflammatory mediators and pathways, which are being studied in the laboratory setting. Although it is quite likely their actions are too specific to cope with the generalized inflammatory response in CF, there is hope that an agent will be found that significantly limits inflammation, but that also has an acceptable side-effect profile and a suitable route of delivery.

Introduction

It is well established that lung inflammation that starts early in life and persists, whatever the clinical state of the patient, has an important pathological role in cystic fibrosis (CF) lung disease. Although a normal inflammatory response is a beneficial host defence mechanism, the exaggerated response seen in CF contributes to the morbidity and ultimately the mortality associated with the disease. The search for a suitable anti-inflammatory strategy has gone on for over a decade, but unfortunately the ideal agent has not yet been identified. There are concerns that substantial inhibition of inflammation could impair immunity and make the patient more susceptible to infection [1]. The difficulty has been to find a drug that is effective, easy to administer at any age, and safe, since it is likely that treatment would need to start early (even at diagnosis) and continue lifelong. The recent Cystic Fibrosis Foundation Therapeutic Drugs Network Anti-Inflammatory Markers (CFF TDN AIM) program has been set up to conduct short (30-day) studies to look at reduction in inflammatory mediators measured in induced sputum [2]. This will act as a screening tool to select the best drugs to study in larger

Table 1. Categories of anti-inflammatory agents

Oral corticosteroids
Inhaled corticosteroids
Ibuprofen
Agents directed against arachidonic acid metabolites
Antiproteases
rhDNase
Macrolide antibiotics
Miscellaneous (cyclosporine, methotrexate, anti-oxidants, IVIG)

long-term trials, concentrating on decline in lung function. The validity of this approach depends on knowing which are the key modulators of inflammation to target, and the time course of any response. Although many studies have suggested interleukin-8 (IL-8) is important, it is by no means certain that this is the pivotal mediator. If the wrong mediators are assessed in these rapid trials, then potentially valuable agents may be discarded. This chapter summarizes the present position as regards anti-inflammatory therapy (table 1). We have attempted to include the latest research, and as well as using the standard electronic search methods, have hand-searched abstract books for the European and North American CF Conferences held since 2000.

Oral Corticosteroids

The Cochrane systematic review on use of oral corticosteroids was last updated in November 2003. It included all 3 randomized controlled trials (RCTs) [3]. Thus 354 patients (children and adults) were included, 2 studies were long term (4 years) whilst one was for 12 weeks only. The conclusion was that prednisolone at a dose of 1–2 mg/kg on alternate days appeared to slow down progression of lung disease. However, significant adverse events reported in the studies, which included impaired glucose metabolism, development of cataracts and impaired growth, renders this form of therapy unjustifiable for long-term use in the majority of patients. Furthermore, follow-up of the children from one study beyond 18 years of age showed the difference in height was permanent amongst the steroid-treated boys [4]. There have been no studies reported since 1995. What is now required is a trial of truly low-dose oral corticosteroids given at equivalent doses to that used for chronic severe asthma (e.g. 5–10 mg per day). Published studies of use during acute exacerbations are lacking, but a small pilot study (reported in abstract form) of prednisolone 2 mg/kg given for 5 days during pulmonary exacerbations had no effect on lung function [5].

Inhaled Corticosteroids

The use of inhaled corticosteroids (ICS), which has been increasing in children and adults with CF, may be justified as a form of symptomatic prophylaxis for those with recurrent wheezing or CF asthma [6]. However, there appears to be a perception (as yet unfounded) that ICS may be able to treat CF lung inflammation. There have been several studies of ICS, incorporated into the Cochrane systematic review, which was last updated in February 2004 [7]. The review included 10 trials, which had studied 293 adults and children for between 4 weeks and 2 years, and were of variable quality. The conclusion was unchanged from the original 1999 review, namely that there was insufficient evidence of benefit, and no significant reduction in inflammatory markers demonstrated in the trials that measured them [8]. However, a small uncontrolled study (hence not included in the systematic review) in children under 13 years showed a decrease in neutrophil concentration in bronchoalveolar lavage after 2 months of therapy with beclomethasone dipropionate, although there was no change in IL-8 concentration [9]. An abstract has reported data from the Epidemiologic Study of CF (ESCF) as showing that consistent long-term use (5 years) of ICS was associated with a small decrease in the rate of decline in lung function [10]. The Cochrane review also concluded there was no evidence of harm, but a recent small study in Belgium (not yet published) has shown a significant slowing in linear growth in children receiving 500 µg twice daily of fluticasone propionate over 12 months, compared to the placebo group [pers. commun., de Boeck]. Over the last few years several case series have been published highlighting the occasional dangers of high-dose ICS in asthmatic children, in terms of adrenal failure, so clearly these drugs should not be used if there is no evidence of benefit [11]. Finally, a large UK multicentre study (CF WISE) is evaluating the outcome of randomized withdrawal of ICS therapy in 171 children and adults already taking ICS, and this will be reported in 2005 (protocol 01PRT/31 available on www.thelancet.com).

Ibuprofen

It is over 10 years since Konstan et al. [12] showed that high-dose ibuprofen led to a significant slowing in decline in lung function in CF patients with mild disease, which was most noticeable in the younger patients. Use of this therapy has been somewhat limited however. This is mostly due to concern over potential renal adverse effects

        Hilliard/Balfour-Lynn

(especially when the patient is also receiving other nephrotoxic drugs), as well as gastrointestinal problems. Its proponents have looked at the US CFF Registry data, and reported no statistically significant increase in renal failure requiring dialysis, peptic ulcer disease, and gastrointestinal haemorrhage compared to CF patients not on ibuprofen [13]. Nevertheless, the incidence of renal failure was more than doubled (0.16 vs. 0.07%), as was the incidence of gastrointestinal haemorrhage (0.44 vs. 0.17%) [13]. There is high interindividual variability in ibuprofen pharmacokinetics, and therapeutic drug monitoring with repeated venepuncture is necessary in order to get the dosing right [14]. The Cochrane systematic review concluded that whilst there is preliminary evidence to suggest that nonsteroidal anti-inflammatory drugs may prevent pulmonary deterioration in those with mild lung disease, currently their routine use cannot be recommended and larger studies are needed to further assess safety and efficacy [15]. Since the Cochrane review, a large Trans-Canadian trial of 145 children with mild disease has been conducted, but full results have not yet been published. Initial analysis, however, showed it was well tolerated and the rate of decline in FVC was significantly lower in the ibuprofen group, but there was no difference in FEV_1 [16]. Most recently, a further retrospective review of CFF Registry data in patients under 18 years old showed a reduced rate of decline in FEV_1 in those on ibuprofen compared to those who never had it (18% reduction in slope), although amongst those taking ibuprofen the average discontinuation rate was 23% per year [17]. Alternative non-steroidal anti-inflammatory drugs offer some potential as alternatives to ibuprofen, e.g. selective cyclo-oxygenase-2 (COX-2) inhibitors. A study is planned by the CFF TDN AIM program to compare ibuprofen versus celecoxib versus placebo. Recent reports of associations between COX-2 inhibitors and cardiovascular events should be noted.

Arachidonic Acid Metabolites

Elevated levels of leukotrienes (LT), particularly the non-cysteinyl leukotriene LTB_4, have been found in the airways of patients with CF [18]. LTB_4 is an important neutrophil chemoattractant, and the 5-lipoxygenase inhibitor, zileuton, decreases its production but has not been studied in CF, possibly because of concerns over its effect on liver function. Recently, BIIL 284 BS, which is a specific LTB_4 receptor antagonist, has been studied in a large phase II trial. Unfortunately this was stopped prematurely (with 418 patients enrolled) due to an excess of significant adverse events (increased hospitalization presumed due to exacerbations) in the active group (Konstan, presentation at 2004 North American CF Conference). The metabolism of arachidonic acid can be diverted from LTB_4 to LTB_5 with omega-3 fatty acids (eicosapentaenoic acid and docosahexaenoic acid) derived from fish oil [19]. Omega-3 fatty acid supplementation has been studied in a small number of patients with CF over a short duration and there is currently insufficient evidence to draw firm conclusions [20], but it is likely that the amount that would need to be ingested for any chance of therapeutic benefit would not be tolerated [19].

Receptor antagonists of the cysteinyl leukotrienes (LTC_4, D_4 and E_4), such as montelukast and zafirlukast, may have a role in CF therapy, particularly in those with asthma-like symptoms. Montelukast has been studied in a small randomized crossover (21 day) trial in 16 CF children [21]. There was a reduction in serum eosinophilic cationic protein (ECP) and eosinophil counts, but no changes in either nasal ECP, nasal and serum IL-8, nor clinical symptom scores. In addition, a small uncontrolled study of montelukast given for 2 weeks to 11 adult CF patients showed an improvement in subjective symptom score and absolute peak flow rates as well as peak flow variability [22]. Montelukast appears to have similar pharmacokinetics in patients with CF compared to healthy controls [23]. Clearly an RCT needs to be conducted before conclusions can be drawn, although individual therapeutic trials may be justified, particularly in those with a tendency to wheeze.

Finally, the finding of low concentrations of lipoxin A_4 (an arachidonic acid metabolite which promotes resolution of inflammation), in CF bronchoalveolar lavage fluid (BALF) has suggested that there might be a defect in lipoxin-mediated anti-inflammatory activity in the CF lung [24]. Administration of a lipoxin analogue in a mouse model of chronic airway inflammation and infection (as seen in CF) led to suppression of neutrophilic inflammation, decreased pulmonary bacterial burden and attenuated disease severity [24].

Antiproteases

The antiprotease defence is overwhelmed within CF airways by the huge accumulation of harmful neutrophil-derived proteases [25]. Supplementation with exogenous antiproteases might, therefore, restore the balance and perhaps limit lung damage. Intravenous replacement with α_1-antitrypsin does not produce high enough levels within the airways to match the protease load [25]. An initial pilot study of aerosolized α_1-antitrypsin in 12 adults with CF [26] was followed by a phase I study in 22 adult patients [27]. Results were encouraging but it was not taken any further, since the source of α_1-antitrypsin was pooled human serum. Sheep-derived transgenic α_1-antitrypsin has been developed by PPL Therapeutics Ltd., and in a placebo-controlled phase II multicentre study (reported in abstract form only) was given to 131 CF patients aged over 12 years for 6 months [28]. It was well tolerated, and there was a trend to improvement in terms of exacerbations and lung function. Bayer Biological Products now has control of the product and it is not known if the work in CF will continue.

Recombinant secretory leukoprotease inhibitor (rSLPI) was used over 10 years ago in a pilot study of 17 adult patients, and the reduction in inflammatory markers after only 1 week of therapy was encouraging [29]. However, it has never been subject to a further trial and the reason is unclear. Although rSLPI is only deposited in well-ventilated areas of the lung in CF patients [30], this is probably the case with all inhaled therapies in CF, especially in more advanced lung disease. Exogenous antiproteases need to penetrate beneath the airway mucus layer, since that is where most of the excess proteases exist. In addition, a large amount of aerosolized antiprotease is needed to inhibit the huge elastase activity. A novel method to target delivery of antiproteases to the respiratory epithelium has recently been described, using a tracheal xenograft model, whereby intravenous α_1-antitrypsin was bound to single-chain Fv antibody directed against secretory component [31]. Finally, a number of synthetic anti-elastase agents have been developed (e.g. DMP777, FK706), but none have reached phase III studies.

rhDNase

Despite the beneficial effect nebulized rhDNase (Dornase alfa) has on the lung function and exacerbation rate of many patients, there has been concern that rhDNase may have a pro-inflammatory effect. This is because the breakdown of DNA that has bound and inactivated free IL-8 and neutrophil elastase may lead to their release in the airways. However, in a crossover trial of daily versus alternate day rhDNase versus hypertonic saline (in 12-week treatment periods), inflammatory markers did not increase in 28 children compared to baseline, except for a significant increase in free IL-8 in sputum with the alternate day rhDNase [32]. Recently, the effect of rhDNase as a long-term anti-inflammatory agent has been studied in an RCT of 105 patients (mean age 12 years) with mild lung disease

($FEV_1 > 80\%$) [33]. Patients had a bronchoalveolar lavage performed at baseline and after 18 and 36 months. Patients with a lower percentage of neutrophils in BALF at baseline were not randomized (to rhDNase or no rhDNase) and served as a further control group. The percentage of neutrophils in pooled BALF increased in non-treatment and non-randomized groups but was unchanged in the rhDNase-treated group. Similar patterns were found for neutrophil elastase activity and IL-8 concentration in BALF over the 3-year study period. This important study, therefore, provides evidence that long-term therapy with rhDNase can have a stabilizing effect on airway inflammation in patients with mild lung disease, although it did not lead to an actual reduction in inflammation. Within this study, in a proportion of children, concentrations of matrix metalloproteinase (MMP)-8 and MMP-9 were found to be decreased after 18 months compared to baseline in those treated with rhDNase, with an increase in those not treated [34]. MMPs may contribute to destruction of airway tissue in CF, so this suggests a further anti-inflammatory effect of rhDNase.

Macrolides

Using antibiotics to target bacterial infections that are stimulating airway inflammation may in itself have some anti-inflammatory effect. Macrolide antibiotics may have intrinsic anti-inflammatory properties as well, and it is likely that this contributes to their clinical effectiveness [35]. Several in vitro and animal studies have shown that macrolides have effects on neutrophil phagocytosis, degranulation, migration and apoptosis [35, 36]. They also suppress the production of pro-inflammatory cytokines, including IL-1β, IL-6, IL-8 and TNF-α [35, 36]. In addition, macrolides have an effect on formation of biofilms, as they inhibit the production of alginate and the expression of flagellin by *Pseudomonas aeruginosa* [35, 37]. Finally, it has been suggested that macrolides have an effect on reducing sputum viscosity [38] and *P. aeruginosa* adherence [39].

Three RCTs have studied azithromycin in CF (table 2), and a recent Cochrane systematic review concluded that there is clear evidence of a small but significant positive effect on lung function [40]. In the first, 60 Australian adults received either 250 mg azithromycin daily or placebo for 3 months [41]. There was a mean decline in FEV_1 of 3.6% predicted in the placebo group (a worryingly high figure) compared to the treatment group, with fewer courses of intravenous antibiotics and an increased quality of life score in those receiving azithromycin. In a UK study,

Hilliard/Balfour-Lynn

Table 2. Summary of published clinical trials of azithromycin

Centre (Ref. no.)	Trial design	Azithromycin dose	n	Length of study months	FEV$_1$ drug vs. placebo	Other clinical outcomes	Inflammatory markers	Adverse effects
Australia: two centres [41]	parallel	250 mg daily	60 adults	3	mean relative difference +3.6%	↓ intravenous antibiotics ↑ quality of life	↓ CRP	nil
UK: single centre [42]	crossover	250 or 500 mg daily	41 children >8 years	6	median relative difference +5.4%	↓ oral antibiotics	no difference sputum IL-8, neutrophil elastase	nil
USA: multicentre [43]	parallel	250 or 500 mg 3 times/week	185 children >6 years and adults	6	mean relative difference +6.2%	↓ non-quinolone antibiotics ↓ exacerbation ↑ weight	modest ↓ sputum elastase, no difference IL-8	nausea, diarrhoea, wheezing

41 children with CF aged 8 years and above received daily azithromycin (250 or 500 mg, depending on body weight) or placebo in a 6-month crossover trial [42]. The median relative difference in FEV$_1$ between azithromycin and placebo was 5.4%, and a greater number of children had fewer courses of oral antibiotics while on active treatment. There was no change in sputum inflammatory markers. However, this was only assessed in the minority of the study group who could spontaneously expectorate sputum. In a large multicentre US study, 185 patients aged over 6 years and chronically infected with *P. aeruginosa* received either azithromycin (250 or 500 mg depending on weight) 3 days per week or placebo for 6 months [43]. At the end of the study, the azithromycin group had a mean relative increase in FEV$_1$ of 6.2%, had less risk of an exacerbation and weighed a mean of 0.7 kg more than the placebo group. However, the improvement in FEV$_1$ was not sustained at 4 weeks after the end of therapy. A modest but statistically significant reduction in sputum elastase activity was found, but IL-8 concentrations were not different. In all 3 trials, azithromycin was well tolerated, although there was statistically more nausea, diarrhoea and wheezing in the azithromycin group in the US study. It should be noted that there was marked individual variation in benefit; in the UK paediatric study 50% children had an increase in FEV$_1$ of more than 10%, and in the US study, around 15% had an increase of more than 15%, but many in both studies showed no benefit. Individual response cannot be predicted from baseline characteristics, and the detailed mechanism of any benefit is not known.

Miscellaneous Agents

There are a number of other agents which have been tried in patients with CF for their potential anti-inflammatory effects. However, none has been subject to proper RCTs in CF although they have been used in other conditions. It is difficult to recommend them on the basis of case reports/series, but nevertheless they may be considered for individual patients on the basis of an 'n of 1' trial. Cyclosporine is an immunosuppressant, which has been used as an alternative anti-inflammatory therapy in difficult asthma. A small case series has described its use in 6 children who had been unable to be weaned from oral corticosteroid therapy [44]. The dose of cyclosporine was adjusted to maintain blood trough levels at 100–150 mg/l, and 4 patients appeared to benefit with eventual discontinuation of oral steroid therapy. In the future, nebulized cyclosporine might be an option although the molecule needs to be altered to enhance absorption and reduce airway irritability [45]. The use of low-dose methotrexate (10–20 mg/m^2 once per week) has been described in 5 patients with CF, with an increase in pulmonary function in the year after starting methotrexate compared to the year before [46]. However, both cyclosporine and methotrexate are potentially toxic with well-recognized adverse effects, and close monitoring with frequent venepuncture is required. Increased oxidant levels contribute to lung damage, and there is some evidence of reduced levels of anti-oxidants, in particular glutathione, in the airway in CF. In a recent uncontrolled study of 17 CF patients, aerosolized reduced glutathione leads to an increase in pulmonary function and glutathione levels after therapy, but no improvement in markers of oxidative injury in BALF [47]. In the past, intravenous immuno-globulin (IVIG) has been given as immunotherapy to CF patients during chest exacerbations to help combat infection. More recently, subsequent to reports of its use to reduce corticosteroid therapy in severe asthma, it has been tried as an anti-inflammatory agent in CF children with severe small airways disease. In a retrospective uncontrolled series of 16 children, 1 g/kg of IVIG at monthly intervals (on average for 7 months) was associated with an improvement in pulmonary function and reduction in oral corticosteroid dose, and was generally well-tolerated [48]. However, it is an expensive and time-consuming therapy, and as a blood product there is always the potential for significant adverse effects.

The Future

In recent years several new agents have been developed which target specific pathways of the inflammatory cascade, with a view to achieving greater therapeutic activity while limiting adverse effects. Some of these may be relevant for CF lung disease but it is unlikely that blocking a single pathway will successfully limit the extensive inflammatory response seen in CF. Etanercept, a soluble recombinant TNF receptor fusion protein, and infliximab, a monoclonal antibody against TNF-α, have proven efficacy in rheumatoid arthritis, but have not yet been tried in CF, perhaps because of concerns over reactivation of tuberculosis [49]. Other agents have been directed against cytokines, including monoclonal anti-IL-5 [50] and anti-IL-8 [51]. None of these agents has been formally studied in CF although an anti-IL-5 antibody has been used in asthma. A study of the use of interferon-γ in CF is currently in progress [19]. Administration of the anti-inflammatory cytokine IL-10 was associated with improved morbidity in CF mice with *P. aeruginosa* infection [52], but this has not as yet undergone a clinical trial.

An alternative approach is to modulate intracellular signalling by limiting the transcription of pro-inflammatory molecules. Using a CF cell line, anti-inflammatory gene

Table 3. Evidence of benefit and potential for harm of anti-inflammatory agents

Therapeutic agent	Evidence of benefit	Potential for harm
Oral corticosteroids	+	+ + + +
Inhaled corticosteroids	+	+ +
Ibuprofen	+ +	+ +
Antileukotrienes		
Cysteinyl	0	0
LTB$_4$	0	+ +
Antiproteases	+	0
rhDNase	+ +	0
Macrolides	+ +	+

Range is from minimal (0) to + + + +.

therapy to provide overexpression of a NFκB inhibitor has been shown in vitro to reduce NFκB activity and IL-8 secretion [53]. Similarly, transfection with oligonucleotide decoys reduced IL-8 secretion in vitro [53], probably due to the low efficiency of cell entry, but had no effect in an animal model of lung inflammation [54]. Early work with an inhibitor of p38 MAP kinase modified neutrophil inflammatory response without reducing bacterial killing [55].

Another avenue is the inhibition of phosphodiesterases (PDE), a family of intracellular enzymes that degrade cyclic nucleotides and have effects on a variety of cellular responses [56]. An inhibitor of PDE4, cilomilast, inhibits in vitro fibroblast-mediated collagen degradation, and may have a role as a regulator of tissue remodelling [56]. Initial work with heparin has shown that it can inhibit transendothelial migration in vitro with the same potency as cilomilast, and heparin may have a role as an anti-inflammatory agent [57]. Indeed nebulized heparin sulphate has been tried in 6 CF adults colonized with *Burkholderia cepacia* for 7 days, and led to a reduction in sputum and serum IL-6 and IL-8 [58].

Conclusions

It is well established that excessive lung inflammation is harmful to patients with CF. Anti-inflammatory therapy seems a good idea, albeit that a degree of inflammation is beneficial in combating chronic bacterial infection. The ideal therapeutic agent has not yet been found whereby benefit far outweighs harm (table 3).

References

1 Koehler DR, Downey GP, Sweezey NB, Tanswell AK, Hu J: Lung inflammation as a therapeutic target in cystic fibrosis. Am J Respir Cell Mol Biol 2004;31:377–381.

2 Berger M: Anti-inflammatory therapy for CF: Mechanism of action and adverse effects. Pediatr Pulmonol 2004;38:170–172.

3 Cheng K, Ashby D, Smyth R: Oral steroids for cystic fibrosis (Cochrane Review). The Cochrane Library, Issue 2. Chichester, Wiley, 2000.

4 Lai HC, FitzSimmons SC, Allen DB, Kosorok MR, Rosenstein BJ, Campbell PW, Farrell PM: Risk of persistent growth impairment after alternate-day prednisone treatment in children with cystic fibrosis. N Engl J Med 2000;342:851–859.

5 Dovey M, Aitken ML, Emerson J, McNamara S, Dorman D, Gibson RL: A randomized, double-blind, placebo controlled trial of oral corticosteroid therapy in cystic fibrosis patients hospitalized for pulmonary exacerbations. Pediatr Pulmonol 2004;(suppl 27):301.

6 Balfour-Lynn IM, Elborn JS: 'CF asthma': What is it and what do we do about it? Thorax 2002;57:742–748.

7 Balfour-Lynn I, Walters S, Dezateux C: Inhaled corticosteroids for cystic fibrosis (Cochrane Review). The Cochrane Library, Issue 4. Chichester, Wiley, 2004.

8 Bisgaard H, Pedersen SS, Nielsen KG, Skov M, Laursen EM, Kronborg G, Reimert CM, Hoiby N, Koch C: Controlled trial of inhaled budesonide in patients with cystic fibrosis and chronic bronchopulmonary *Pseudomonas aeruginosa* infection. Am J Respir Crit Care Med 1997;156:1190–1196.

9 Wojtczak HA, Kerby GS, Wagener JS, Copenhaver SC, Gotlin RW, Riches DW, Accurso FJ: Beclomethasone diproprionate reduced airway inflammation without adrenal suppression in young children with cystic fibrosis: A pilot study. Pediatr Pulmonol 2001;32:293–302.

10 Ren CL, Pasta DJ, Konstan MW, Wagener JS, Morgan WJ: Inhaled corticosteroid (ICS) use is associated with a slower rate of decline in CF lung disease. Pediatr Pulmonol 2003;(suppl 25):324.

11 Russell G: The use of inhaled corticosteroids during childhood: Plus ca change. Arch Dis Child 2004;89:893–895.

12 Konstan MW, Byard PJ, Hoppel CL, Davis PB: Effect of high-dose ibuprofen in patients with cystic fibrosis. N Engl J Med 1995;332:848–854.

13 Konstan MW, FitzSimmons SC: Clinical use of ibuprofen for cystic fibrosis lung disease: Data from the 1996 CFF Patient Registry. Pediatr Pulmonol 1997;(suppl 14):322.

14 Arranz I, Martin-Suarez A, Lanao JM, Mora F, Vazquez C, Escribano A, Juste M, Mercader J, Ripoll E: Population pharmacokinetics of high dose ibuprofen in cystic fibrosis. Arch Dis Child 2003;88:1128–1130.

15 Lands LC, Dezateux C, Crighton A: Oral non-steroidal anti-inflammatory drug therapy for cystic fibrosis (Cochrane Review). The Cochrane Library, Issue 4. Chichester, Wiley, 2004.

16 Lands LC, Corey M, Milner R, Kilcullen A, Cantin AM: High dose ibuprofen in CF children: The trans-Canadian Trial. Pediatr Pulmonol 2002;(suppl 24):276.

17 Schlucter MD, Konstan MW, Xue L, Davis PB: Relationship between high-dose ibuprofen use and rate of decline in FEV1 among young patients with mild lung disease in the CFF registry. Pediatr Pulmonol 2004;38:322.

18 Konstan MW, Walenga RW, Hilliard KA, Hilliard JB: Leukotriene B4 markedly elevated in the epithelial lining fluid of patients with cystic fibrosis. Am Rev Respir Dis 1993;148:896–901.

19 Konstan MW, Davis PB: Pharmacological approaches for the discovery and development of new anti-inflammatory agents for the treatment of cystic fibrosis. Adv Drug Deliv Rev 2002;54:1409–1423.

20 Beckles WN, Elliott TM, Everard ML: Omega-3 fatty acids (from fish oils) for cystic fibrosis (Cochrane Review). The Cochrane Library, Issue 3. Chichester, Wiley, 2002.

21 Schmitt-Grohe S, Eickmeier O, Schubert R, Bez C, Zielen S: Anti-inflammatory effects of montelukast in mild cystic fibrosis. Ann Allergy Asthma Immunol 2002;89:599–605.

22 Morice AH, Kastelik JA, Aziz I: Montelukast sodium in cystic fibrosis. Thorax 2001;56:244–245.

23 Graff GR, Weber A, Wessler-Starman D, Smith AL: Montelukast pharmacokinetics in cystic fibrosis. J Pediatr 2003;142:53–56.

24 Karp CL, Flick LM, Park KW, Softic S, Greer TM, Keledjian R, Yang R, Uddin J, Guggino WB, Atabani SF, Belkaid Y, Xu Y, Whitsett JA, Accurso FJ, Wills-Karp M, Petasis NA: Defective lipoxin-mediated anti-inflammatory activity in the cystic fibrosis airway. Nat Immunol 2004;5:388–392.

25 Balfour-Lynn IM: The protease-antiprotease battle in the cystic fibrosis lung. J R Soc Med 1999;92(suppl 37):23–30.

26 McElvaney NG, Hubbard RC, Birrer P, Chernick MS, Caplan DB, Frank MM, Crystal RG: Aerosol alpha 1-antitrypsin treatment for cystic fibrosis. Lancet 1991;337:392–394.

27 Berger M, Konstan MW, Hilliard J; CF Prolastin Study Group: Aerosolized Prolastin (a1-protease inhibitor) in CF. Pediatr Pulmonol 1995;(suppl 12):421.

28 Bilton D, Elborn S, Conway S, Edgar J, Redmond A: A phase II trial to assess the clinical efficacy of transgenic alpha-1-antitrypsin (tg-hAAT) as an effective treatment of cystic fibrosis. Pediatr Pulmonol 1999;(suppl 19):246.

29 McElvaney NG, Nakamura H, Birrer P, Hebert CA, Wong WL, Alphonso M, Baker JB, Catalano MA, Crystal RG: Modulation of airway inflammation in cystic fibrosis. In vivo suppression of interleukin-8 levels on the respiratory epithelial surface by aerosolization of recombinant secretory leukoprotease inhibitor. J Clin Invest 1992;90:1296–1301.

30 Stolk J, Camps J, Feitsma HI, Hermans J, Dijkman JH, Pauwels EK: Pulmonary deposition and disappearance of aerosolised secre-

tory leucocyte protease inhibitor. Thorax 1995;50:645–650.

31 Ferkol T, Cohn LA, Phillips TE, Smith A, Davis PB: Targeted delivery of antiprotease to the epithelial surface of human tracheal xenografts. Am J Respir Crit Care Med 2003;167:1374–1379.

32 Suri R, Marshall LJ, Wallis C, Metcalfe C, Bush A, Shute JK: Effects of recombinant human DNase and hypertonic saline on airway inflammation in children with cystic fibrosis. Am J Respir Crit Care Med 2002;166:352–355.

33 Paul K, Rietschel E, Ballmann M, Griese M, Worlitzsch D, Shute J, Chen C, Schink T, Doring G, van Koningsbruggen S, Wahn U, Ratjen F: Effect of treatment with dornase alpha on airway inflammation in patients with cystic fibrosis. Am J Respir Crit Care Med 2004;169:719–725.

34 Ratjen F, Hartog CM, Paul K, Wermelt J, Braun J: Matrix metalloproteases in BAL fluid of patients with cystic fibrosis and their modulation by treatment with dornase alpha. Thorax 2002;57:930–934.

35 Schultz MJ: Macrolide activities beyond their antimicrobial effects: Macrolides in diffuse panbronchiolitis and cystic fibrosis. J Antimicrob Chemother 2004;54:21–28.

36 Jaffe A, Bush A: Anti-inflammatory effects of macrolides in lung disease. Pediatr Pulmonol 2001;31:464–473.

37 Kawamura-Sato K, Iinuma Y, Hasegawa T, Yamashino T, Ohta M: Postantibiotic suppression effect of macrolides on the expression of flagellin in *Pseudomonas aeruginosa* and *Proteus mirabilis*. J Infect Chemother 2001;7:51–54.

38 Baumann U, King M, App EM, Tai S, Konig A, Fischer JJ, Zimmermann T, Sextro W, von der HH: Long term azithromycin therapy in cystic fibrosis patients: A study on drug levels and sputum properties. Can Respir J 2004;11:151–155.

39 Baumann U, Fischer JJ, Gudowius P, Lingner M, Herrmann S, Tummler B, von der HH: Buccal adherence of *Pseudomonas aeruginosa* in patients with cystic fibrosis under long-term therapy with azithromycin. Infection 2001;29:7–11.

40 Southern KW, Barker PM, Solis A: Macrolide antibiotics for cystic fibrosis (Cochrane Review). The Cochrane Library, Issue 3. Chichester, Wiley, 2004.

41 Wolter J, Seeney S, Bell S, Bowler S, Masel P, McCormack J: Effect of long term treatment with azithromycin on disease parameters in cystic fibrosis: A randomised trial. Thorax 2002;57:212–216.

42 Equi A, Balfour-Lynn IM, Bush A, Rosenthal M: Long term azithromycin in children with cystic fibrosis: A randomised, placebo-controlled crossover trial. Lancet 2002;360:978–984.

43 Saiman L, Marshall BC, Mayer-Hamblett N, Burns JL, Quittner AL, Cibene DA, Coquillette S, Fieberg AY, Accurso FJ, Campbell PW 3rd: Azithromycin in patients

with cystic fibrosis chronically infected with *Pseudomonas aeruginosa*: A randomized controlled trial. JAMA 2003;290:1749–1756.

44 Bhal GK, Maguire SA, Bowler IM: Use of cyclosporin A as a steroid sparing agent in cystic fibrosis. Arch Dis Child 2001;84:89.

45 Fukaya H, Iimura A, Hoshiko K, Fuyumuro T, Noji S, Nabeshima T: A cyclosporin A/maltosyl-alpha-cyclodextrin complex for inhalation therapy of asthma. Eur Respir J 2003;22: 213–219.

46 Ballmann M, Junge S, von der HH: Low-dose methotrexate for advanced pulmonary disease in patients with cystic fibrosis. Respir Med 2003;97:498–500.

47 Griese M, Ramakers J, Krasselt A, Starosta V, van Koningsbruggen S, Fischer R, Ratjen F, Mullinger B, Huber RM, Maier K, Rietschel E, Scheuch G: Improvement of alveolar glutathione and lung function but not oxidative state in cystic fibrosis. Am J Respir Crit Care Med 2004;169:822–828.

48 Balfour-Lynn IM, Mohan U, Bush A, Rosenthal M: Intravenous immunoglobulin for cystic fibrosis lung disease: A case series of 16 children. Arch Dis Child 2004;89: 315–319.

49 Olsen NJ, Stein CM: New drugs for rheumatoid arthritis. N Engl J Med 2004;350: 2167–2179.

50 Leckie MJ: Anti-interleukin-5 monoclonal antibodies: Preclinical and clinical evidence in asthma models. Am J Respir Med 2003;2: 245–259.

51 Yang XD, Corvalan JR, Wang P, Roy CM, Davis CG: Fully human anti-interleukin-8 monoclonal antibodies: Potential therapeutics for the treatment of inflammatory disease states. J Leukoc Biol 1999;66:401–410.

52 Chmiel JF, Konstan MW, Knesebeck JE, Hilliard JB, Bonfield TL, Dawson DV, Berger M: IL-10 attenuates excessive inflammation in chronic Pseudomonas infection in mice. Am J Respir Crit Care Med 1999;160:2040–2047.

53 Griesenbach U, Scheid P, Hillery E, de Martin R, Huang L, Geddes DM, Alton EW: Anti-inflammatory gene therapy directed at the airway epithelium. Gene Ther 2000;7:306–313.

54 Griesenbach U, Cassady RL, Cain RJ, duBois RM, Geddes DM, Alton EW: Cytoplasmic deposition of NFkappaB decoy oligonucleotides is insufficient to inhibit bleomycin-induced pulmonary inflammation. Gene Ther 2002;9:1109–1115.

55 Nick JA, Saavedra M, Poch K, Walker TS, Malcolm K, Avdi NJ, Accurso FJ, Higgins LS, Vasil ML, Worthen GS: Selective modification of the response of human neutrophils to *Pseudomonas aeruginosa* by inhibition of p38 MAP kinase. Pediatr Pulmonol 2003;36:269.

56 Kohyama T, Liu X, Zhu YK, Wen FQ, Wang HJ, Fang Q, Kobayashi T, Rennard SI: Phosphodiesterase 4 inhibitor cilomilast inhibits fibroblast-mediated collagen gel degradation induced by tumor necrosis factor-alpha and neutrophil elastase. Am J Respir Cell Mol Biol 2002;27:487–494.

57 Broughton-Head VJ, Shute JK: Heparin inhibits neutrophil transendothelial migration with the same potency as the PDE-4 inhibitor cilomilast. J Cystic Fibrosis 2004;3:S27.

58 Ledson M, Gallagher M, Hart CA, Walshaw M: Nebulized heparin in *Burkholderia cepacia* colonized adult cystic fibrosis patients. Eur Respir J 2001;17:36–38.

I.M. Balfour-Lynn
Department of Paediatric Respiratory
Medicine
Royal Brompton Hospital, Sydney Street
London SW3 6NP (UK)
Tel. +44 207 351 8509
Fax +44 207 351 8763
E-Mail i.balfourlynn@imperial.ac.uk

Bush A, Alton EWFW, Davies JC, Griesenbach U, Jaffe A (eds): Cystic Fibrosis in the 21st Century.
Prog Respir Res. Basel, Karger, 2006, vol 34, pp 195–204

Recent Advances in Infant and Pre-School Lung Function

Sarath Ranganathan

Department of Respiratory Medicine, Royal Children's Hospital and Department of Paediatrics,
University of Melbourne, Parkville, Australia

Abstract

Infant pulmonary function testing has been performed in CF patients for many years, but studies have generally been small, and with inadequate controls. Flow-volume curves are routinely used in older children, but until the advent of the raised volume, thoraco-abdominal compression technique, could not be replicated in infants. Recent studies using this technique have shown that there is evidence of airway obstruction shortly after diagnosis in CF infants, with no evidence of 'catch-up' growth despite specialist treatment, over a six-month period. This technique appears to be more sensitive than the standard thoracoabdominal compression at end-inspiration. The pre-school years have traditionally been a 'black hole', in which neither active nor passive co-operation is likely, and lung function measurements very difficult. Techniques such as forced oscillation and the interrupter method have been used with some success, and recent research has lead to the application of other methods. Spirometry can be performed even in some two years olds, with the aid of skilled and experienced paediatric lung function technicians. Indices of gas mixing such as lung clearance index show great promise, and the normal range appears to be independent of age and size, making it a very powerful tool. The development of these powerful techniques will be a stimulus to better monitoring of children in whom clinicians hitherto had to rely on history and physical examination alone. It will be important in future work to make these techniques accessible as routine clinical tools.

Measurement of lung function is a central part of the clinical assessment of older children and adults with cystic fibrosis (CF). Serial tests provide longitudinal information about the extent of abnormality, progression of disease and individual response to treatment. Although neutrophil-dominated inflammation similar to that seen in older subjects has been identified in the lungs of affected infants [1, 2], the evolution of airway pathology in early infancy remains poorly understood. As evidence accumulates that chronic pulmonary disease commences in infancy, the need to evaluate lung function early is self-evident. The limiting factors, however, are that infant lung function is technically difficult to test, expensive, time-consuming and requires sedation, which makes it difficult to recruit healthy subjects as controls. Furthermore, between infancy and school age, toddlers are too old to sedate and too young to co-operate with most testing.

Many studies are difficult to interpret due to the small numbers of subjects, lack of appropriate control data from prospectively studied healthy infants and the use of relatively insensitive techniques [3]. Consequently, it is uncertain whether airway function is impaired at or shortly after diagnosis and whether diminished airway function precedes the onset of lower respiratory symptoms. Recent studies, however, are just beginning to bridge the 'silent years' of lung function testing before children are old enough to perform spirometry.

In older subjects with CF, spirometry, for example forced expiratory volume in 1 s (FEV_1) and maximal expiratory flow at low lung volumes (e.g. MEF_{25}) is most

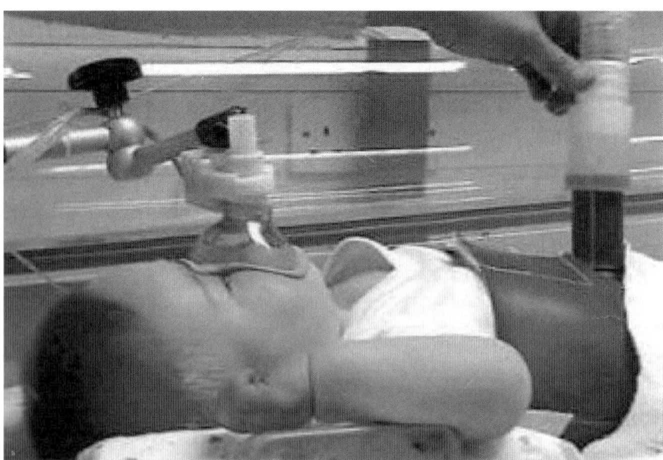

Fig. 1. A sedated infant undergoing rapid thoraco-abdominal compression during tidal breathing. At end-tidal inspiration, inflation of the jacket provides a 'squeeze' and forces expiration. Flow and volume are measured using a pneumotachometer.

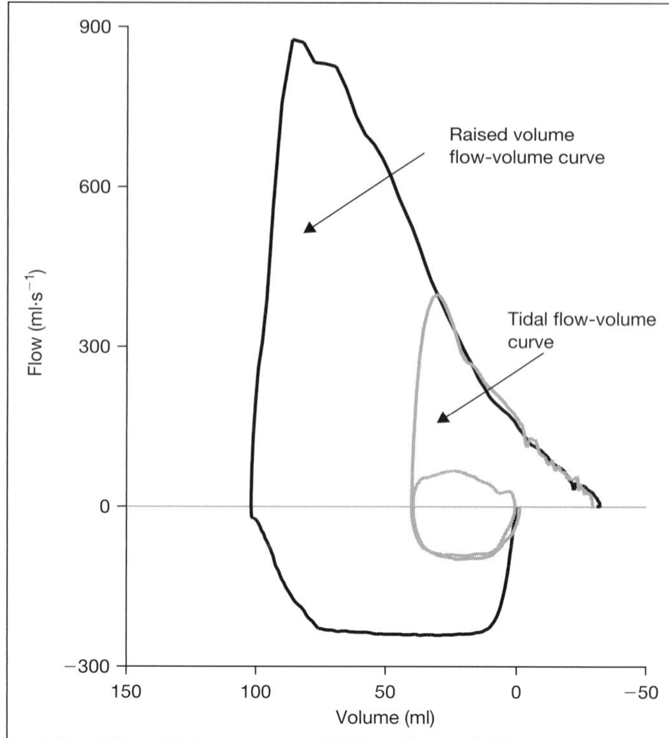

Fig. 2. Flow volume curves obtained in an infant with CF using the techniques of thoraco-abdominal compression from raised lung volume and tidal breathing, respectively. Parameters similar to those obtained using spirometry such as forced expiratory flows and volumes can be calculated from the raised volume curve.

frequently used to assess airway function. Partial forced expiratory manoeuvres have been performed in infants for over 20 years using the tidal rapid thoraco-abdominal compression technique (RTC). In this technique an inflatable jacket is used to apply a rapidly rising external pressure to the chest wall and abdomen at the end of a tidal inspiration in order to generate forced expiratory flow-volume curves (fig. 1). Methods to assess forced expiration over an extended volume range in infants have only been described recently [4, 5]. This raised volume rapid thoraco-abdominal compression technique (raised volume technique; RVRTC) involves using a pump or augmented manual inflations to increase the volume of inspiration prior to forcing expiration. From the RVRTC flow-volume curves (fig. 2) it is possible to derive parameters comparable to those obtained in older subjects such as forced expiratory volume in 0.4 or 0.5 s ($FEV_{0.4}$ or $FEV_{0.5}$) and MEF_{25} [6]. This technique has been applied recently to infants with CF in studies which have prospectively recruited control infants.

Recent Studies in Infants

Evidence for Diminished Airway Function in Infants Newly Diagnosed with Cystic Fibrosis

Two papers were published recently as part of the London Collaborative Cystic Fibrosis Study [7, 8]. Airway function of infants newly diagnosed with CF was measured in order to test the hypothesis that lung function is diminished shortly after diagnosis independent of clinically recognized lower respiratory illness. RTC and RVRTC were used to assess the

airway function of 47 infants newly diagnosed with CF and 137 healthy infants. The airway function of infants with CF was compared with that of healthy infants and expressed as standard deviation scores (z-scores). The respiratory status of the CF infants was assessed by their specialist physicians who were blinded to the results of airway function.

The association between $FEV_{0.4}$ and length according to disease status is shown in figure 3. Multiple linear regression was used to assess the influence of CF on $FEV_{0.4}$ after accounting for differences in body size, gender and exposure to maternal smoking measured by salivary cotinine level. On average the decrement in $FEV_{0.4}$ in those with CF was 40 ml (p < 0.001) when their median (range) age was 30 (6–93) weeks. This was irrespective of clinical evidence of prior lower respiratory illness (LRI) as this particular subgroup of infants (of median age 12 weeks) had a similar decrement in airway function.

Approximately a third of infants with CF had an $FEV_{0.5}$ below the 2.5th centile [8]. In 17 CF infants assessed as having normal clinical respiratory status by their specialist, the mean z-score for $FEV_{0.5}$ was −1.1 and there were 4 infants

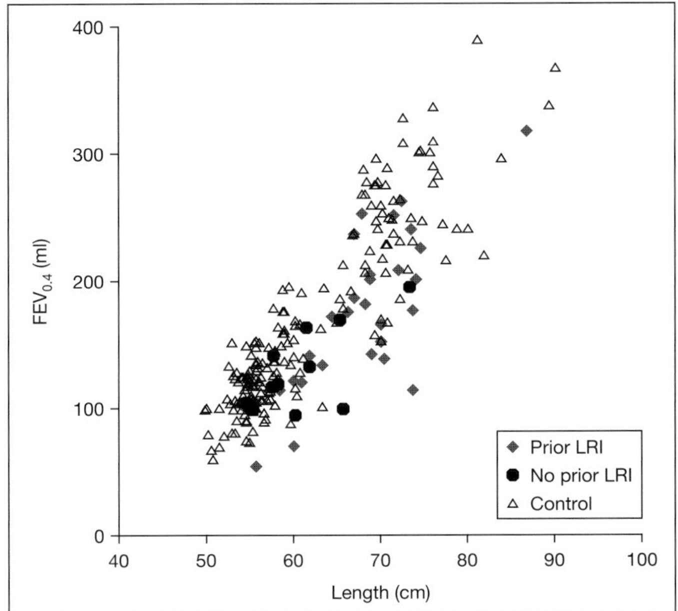

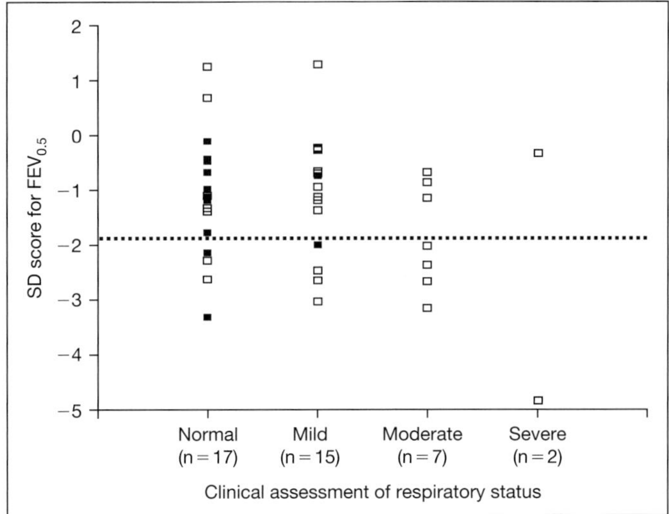

Fig. 3. $FEV_{0.4}$ plotted against length according to disease status. Crosses indicate airway function of healthy infants and open and closed squares indicate CF infants with and without prior lower respiratory illness, respectively. There is no significant difference between the two groups of CF infants, both of whom have a significantly lower $FEV_{0.4}$ than normal.

Fig. 4. SD score for $FEV_{0.5}$ in CF infants with normal and abnormal clinical assessments of respiratory status made by the CF specialist. The dotted line is plotted at -1.96 standard deviations and equates to the 2.5th centile. Closed squares indicate infants with non-respiratory presentations, no respiratory exacerbations or evidence of infection. The clinical assessment bears no relation to the lung function findings.

whose airway function was below the 2.5th centile for healthy infants (fig. 4). RVRTC identified diminished airway function in more infants with CF than RTC.

Thus, airway function was diminished early in the course of disease in those with CF, was more likely to be detected by the raised volume technique than the tidal forced expiratory technique and occurred irrespective of clinical evidence of prior LRI and even when specialist physicians identified infants with CF as having normal respiratory status. In this study, infants were diagnosed with CF following a clinical presentation of the disease and so those without clinical evidence of prior LRI usually had presented with meconium ileus in the newborn period, or had evidence of failure to thrive when airway function was assessed. Whether reductions in airway function are evident during the first months of life in those diagnosed with CF by neonatal screening and without failure to thrive remains unknown.

Are Simpler Methods Useful in Determining Airway Function of Infants with Cystic Fibrosis?

Since the Tucson study suggested that time to reach peak tidal expiratory flow in relation to total expiratory time (the tidal breathing ratio, $t_{PTEF}:t_E$) was diminished in symptom

free male infants who subsequently wheezed, several groups have examined tidal flow patterns in both healthy infants and those with airway disease [9–11]. Recording of tidal breathing is technically simple and potentially allows measurements to be undertaken at the patient's bedside during natural quiet sleep, without the need for sedation. Consequently, despite ongoing controversy regarding their interpretation [12, 13], evaluation of the 'tidal breathing ratio' and other related parameters has continued to attract interest. The relationship between tidal breathing parameters and $FEV_{0.4}$ was assessed in the London cohort and compared with a prospectively recruited population of healthy infants.

There was no difference in $t_{PTEF}:t_E$ and tidal volume between healthy infants and those with CF. Minute ventilation was significantly greater in infants with CF due to a mean (95% CI) increase in respiratory rate of 5.8 (3.2, 8.4) min^{-1}. Thirteen (28%) infants with CF had a respiratory rate elevated by more than 2 z-scores. Unlike $FEV_{0.4}$, there was clear discrimination in respiratory rate between those with and without prior LRI, suggesting that an elevated respiratory rate may be due in part to ventilation inhomogeneities [14]. As no association between respiratory rate and $FEV_{0.4}$ could be identified, respiratory rate was poorly predictive of diminished airway function measured by forced expiration. The use of more sophisticated techniques

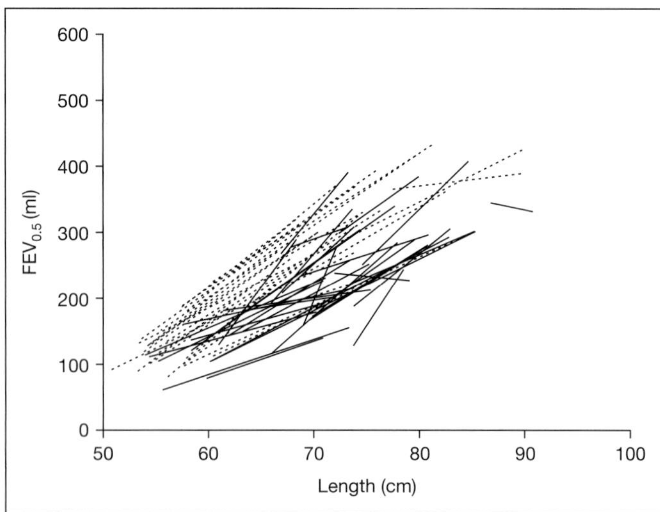

Fig. 5. Association of FEV$_{0.5}$ with length in a prospective, longitudinal study. The dashed and continuous lines indicate airway function of healthy infants and infants with CF respectively. There is no evidence of any catch-up growth in the CF infants.

such as multiple breath inert gas washout (see below), which can also be performed during tidal breathing, but which give detailed information on ventilation distribution, could be used in the future to evaluate the relative contribution of impaired gas mixing to alterations in respiratory rate in infants with CF [15].

Does Initial Impairment in Airway Function Noted in Infants following Clinical Diagnosis of Cystic Fibrosis Persist Despite Treatment in Specialist Centres?

The evolution of airway function in the London cohort was assessed by measuring FEV$_{0.5}$ using RVRTC soon after diagnosis (median age 28 weeks), 6 months later in subjects with CF, and on two occasions 6 months apart (median age 7.4 and 33.7 weeks) in healthy infants. Repeated measurements were successful in 34 CF and 32 healthy subjects. After adjustment for age, length, sex and exposure to maternal smoking, mean FEV$_{0.5}$ was significantly lower in infants with CF both shortly after diagnosis and at second test, with no significant difference in rate of increase in FEV$_{0.5}$ with growth between the two groups (fig. 5). When compared with published reference data, FEV$_{0.5}$ was reduced by an average of 2 z-scores on both test occasions in those with CF. Subjects with CF experienced a mean (95% CI) reduction in FEV$_{0.5}$ of 20% [16].

Data from the London Collaborative study, therefore, suggest that airway function is diminished soon after diagnosis in infants with CF and does not catch up (or deteriorate) over a 6-month period during infancy and early childhood despite treatment in centres specializing in the management of CF. This result needs confirmation, but does highlight the potential value of such measurements for future clinical research and the implications for early interventions in CF.

What Is the Association between Early Lung Function and Markers of Inflammation or Infection?

As the gold standard for determining airway inflammation or lower respiratory tract infection is bronchoscopy (FOB) and broncho-alveolar lavage (BAL), usually under general anaesthesia, few investigators have combined FOB and infant lung function in the same study. In one recent study, specific respiratory system compliance was measured using the single breath occlusion technique and lung volumes by nitrogen washout in 22 children with CF of median age 23 months [17]. Diminished lung function correlated with BAL markers of both inflammation [interleukin-8 (IL-8) and neutrophil percentage] and infection. However, for the majority of subjects, lung function remained within the normal range and only very little of the variability in lung function was explained by either inflammation or infection. In another study, recruitment commenced in infancy, but measurements in 40 subjects were made initially at a mean of 13 months of age [18]. No association between lung function and IL-8 concentration, neutrophil density, or pathogen load was demonstrated. However, due to the invasive nature of the assessments, not all parameters were measured simultaneously, BAL being performed annually and infant lung function every 6 months. An association of borderline significance between FEV$_{0.5}$ measured using RVRTC and infection (defined as at least 10^5 colony-forming units of bacterial respiratory pathogens per millilitre of BAL fluid) and no association between FEV$_{0.5}$ and pulmonary inflammation was identified in a third study in 36 CF children during the second year of life [19].

No studies have been able to address this question during the first year of life. Although it is highly likely that the early functional abnormalities identified are the result of inflammation and infection, as is the case in older subjects (fig. 6), further studies are required in order to demonstrate this in infants with CF.

Are Early Functional Changes Obstructive in Nature?

In older subjects, one physiological result of chronic inflammation and infection is airway obstruction. An obstructive process in infancy would suggest similar pathophysiology. This could be indicated by a reduction in the

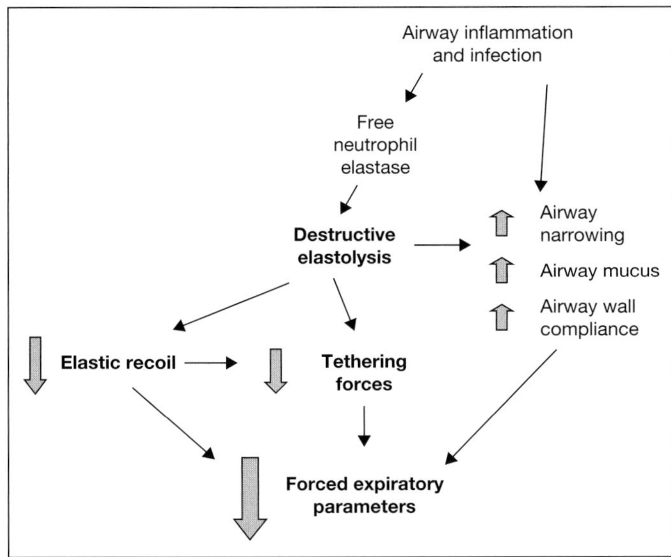

Fig. 6. Schematic of pathogenesis of early lung function abnormalities in infants with CF.

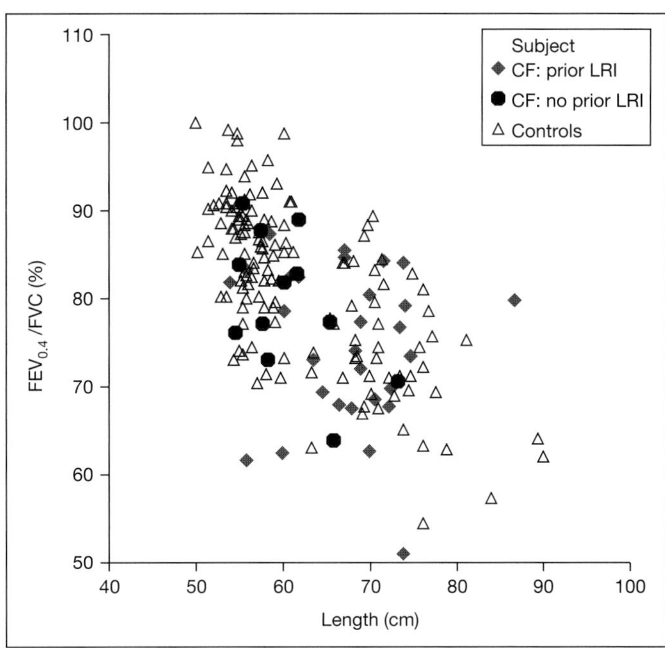

Fig. 7. The relationship of $FEV_{0.4}/FVC$ (%) with length according to disease status. $FEV_{0.4}/FVC$ is diminished in infants with CF, but for the majority of CF infants the ratio remains within the normal range.

ratio of forced expiratory volume to forced vital capacity. Although this appears to be the case (fig. 7), in the London Collaborative study the mean reduction in $FEV_{0.4}/FVC$ was 3.2% in infants with CF (p = 0.015), such a small reduction is unlikely to be clinically significant and clearly does not definitively confirm an important obstructive process. The ratio was in the normal range in the vast majority of infants with CF, even those in whom $FEV_{0.4}$ was reduced. Moreover, interpretation of this ratio in infancy is especially difficult due to the rapid increase in the forced expiratory time with age or length and the fact that $FEV_{0.4}/FVC$ is so close to unity in early life.

An elegant study recently assessed gas-trapping (as an indication of airway obstruction) by three methods in infants with CF [20]:

(1) A combination of plethysmography to determine functional residual capacity and the raised volume technique to determine forced vital capacity (FVC) and the expiratory reserve volume was used. After an RVRTC manoeuvre, lung volume is assumed to have reached residual volume and the volume expired is the FVC. After a period of apnoea, spontaneous breathing resumes and the end-expiratory level rises until a stable level is re-established. The expiratory reserve volume is assumed to be the difference between residual volume (the lung volume at the end of the forced expiratory manoeuvre) and the volume at which the end-expiratory level is re-established. Therefore, the equations determining (a) residual volume and (b) total lung capacity are:

(a) residual volume = functional residual capacity − expiratory reserve volume;

(b) total lung capacity = forced vital capacity + residual volume.

The ratio of residual volume/total lung capacity can then be expressed and used to determine whether a reduction in FVC is due to reduced lung volume or due to an increase in residual volume because of airway closure and gas trapping.

(2) The second method involved comparing values of functional residual capacity made in a plethysmograph (which measures all intrathoracic gas including trapped gas) with those made by a nitrogen washout gas mixing technique (which does not measure gas in alveoli ventilated with a long time constant), the volume of trapped gas being the difference between the two values. It should be noted that plethysmography and other techniques such as the resistance interrupter technique (see later section), which are predicated on the assumption that pressure at the mouth is close to alveolar pressure when measurements are made, inevitability lead to over-estimation of lung volume in the presence of airflow obstruction, due to the collapse of this assumption.

(3) After a nitrogen washout procedure, augmented inflations in 100% oxygen were given in order to open up

lung units and liberate any trapped nitrogen gas. The volume of trapped gas was calculated from the total additional amounts of nitrogen expired following lung inflations.

Gas-trapping was indicated by an increase in the ratio of residual volume to total lung capacity in those with CF. However, no evidence for gas-trapping was identified using the second or third method. It is possible that active elevation of end-expiratory level in response to increased peripheral airway resistance reduces gas-trapping in early CF pulmonary disease and results in a compensatory increase in functional residual capacity, making the second method less effective at demonstrating trapped gas. Recent data, however, published as an abstract, also employing the second method, suggest that gas trapping can be detected in this way and is common in infants with CF [21]. Collectively, these data suggest that an obstructive rather than a restrictive pattern of airway functional abnormality occurs in infants with CF.

Is Genotype or Phenotype Correlated with Infant Lung Function?

Infants with pulmonary hyperinflation had a different genotype than those without in one study [22]: infants who were compound heterozygotes 3905insT/ΔF508 (n = 9) had a mean z-score for functional residual capacity of 8.4 (SD 4.9) above the mean of a reference population, whereas 8 infants who were compound heterozygotes for the nonsense mutation R553X were not significantly different from normal. However, as weight gain was abnormally low in the former (58% predicted), the elevated values for functional residual capacity may at least in part reflect inappropriate comparison in underweight infants with CF to the reference population of healthy infants. In any case, the numbers studied are small. In another study, in which infants were grouped into those homo- or heterozygous for ΔF508, a washout technique and the single breath technique were used to assess lung volume and respiratory mechanics, respectively [23]. The single breath technique is based on the ability to invoke the Hering-Breuer inflation reflex in infants and young children. At end-inspiration this vagally mediated reflex results in cessation of inspiration and of respiratory muscle activity. An occlusion at the airway opening is performed at end-inspiration. Provided there is complete absence of respiratory muscle activity during the airway occlusion and rapid equilibration can be reached during periods of no flow (unlikely in those with airway disease), the relaxation pressure at the airway opening represents alveolar pressure, which in turn represents the summed elastic recoil pressure of the lung and chest wall during periods of muscle relaxation. This pressure (P) can be related to changes in volume (V) and flow (V') in order to calculate the compliance (V/P) and resistance (P/V') of the respiratory system, that is, the combined resistance of the airways, lung tissue and the chest wall. Infants homozygous for ΔF508 had elevated respiratory system resistance measured in this way and a positive bronchodilator response. The studies of the London Collaborative group, however, failed to demonstrate any association between CF genotype and results of infant lung function, probably due to insufficient power. Much larger multicentre collaborations are required before such associations can be confirmed in infants with CF.

Probably also due to lack of power, no studies in infants have confirmed impairment of lung function in association with infection with *Pseudomonas aeruginosa*. However, exposure to maternal smoking during pregnancy and/or post-natally is independently associated with diminished airway function, the effect size being approximately 50% of that associated with CF [7, 8].

Recent Studies in Pre-School Children

The long-term implications of early lung function findings in those with CF will only be known when lung function is measured longitudinally until school-age when conventional measurements using spirometry become possible. Too old to sedate routinely, considered fickle and uncooperative by many, the years between 2 and 6 provide several challenges to those wishing to measure lung function. In many centres, pre-school children with CF do not undergo lung function testing until they are able to perform reproducible forced expiratory manoeuvres. In order to bridge the 'silent years', recent studies have focused on pre-school subjects with CF by skilfully employing tests requiring minimal cooperation or by using incentives to effectively coerce the subject.

The resistance interrupter technique (Rint) requires minimal cooperation and can be performed in the ambulatory setting in pre-school children [24]. Rint is a non-invasive method of measuring airway resistance, first described in 1927. It is based on the assumption that during an imperceptibly brief interruption of airflow during tidal breathing, the pressure changes at the airway opening can be used to determine the alveolar pressure (P) at the moment of interruption, and, hence, knowing the flow immediately prior to interruption (V'), resistance can be calculated as P/V'. Resistance measured by this technique was significantly elevated (z-score of 1.31 compared with 0.19 in healthy controls) in 40 subjects with CF aged between 3 and 8 years

[25]. Those with a history of prior LRI or exposure to environmental tobacco smoke had significantly elevated expiratory resistance. Although measurements in those with CF were no more variable than those made in healthy subjects, the overall variability of the resistance measurements was high. Expiratory resistance measured by the interrupter technique depends on the proximal airways and so physiologically it would not appear to be the ideal test to measure peripheral airway function associated with early pulmonary disease. Furthermore, as with plethysmography (above), alveolar pressure cannot be assumed to track mouth pressure in the presence of airflow obstruction.

In recent years, computer animation programs have been developed in order to instruct and stimulate young children in maximal forced expiratory manoeuvres. Children are asked to blow out candles, to make aeroplanes fly or to blow up balloons on the computer screen. These programs help children to focus on the task and provide a visual motivation to accomplish an optimal manoeuvre. A wide range of such programs is now available. The type of incentive software and the age and experience of the subject are likely to determine the reproducibility of measurements obtained in this way [26]. Preliminary data from the London cohort using incentive spirometry suggest that tracking of lung function is likely to persist through the pre-school years (that is, those with the lowest airway function in infancy maintain this position with growth). When the children were retested at a mean age of 3.9 years, mean (SD) z-scores were -0.55 (0.85) for $FEV_{0.5}$, with only 1 child with CF still having an abnormal $FEV_{0.5}$. This could either suggest that airway function may improve through the pre-school years or alternatively that incentive spirometry is less able to identify diminished airway function in pre-school subjects than RVRTC in infants [27].

Contrary to popular notions, skilled paediatric pulmonary function technicians, who regularly work with pre-school children, may find greater success using conventional spirometry if the subject is appropriately coached. In a recent study, 33 of 38 children with CF (including 4 of 6 three year-olds and 9 of 11 four year-olds) were able to perform either two or three conventional spirometry manoeuvres (without incentive) and fulfil study acceptability and reproducibility criteria [28]. Children with CF had significantly lower FVC, FEV_1 and MEF_{75-25} than healthy subjects. Those homozygous for $\Delta F508$ had significantly lower FVC and FEV_1 than heterozygotes. Subjects were well at the time of lung function testing and so the feasibility and variability of performing spirometry during exacerbations are not known.

RVRTC had been performed previously in 14 subjects during infancy in this study and again tracking of lung function occurred through to the pre-school assessments [28]. In those with lung function in the normal range as infants (z-score for $MEF_{75-25} \geq -2$) there was no significant change in z-score for the same parameter measured using spirometry. However, z-scores increased significantly between infancy and childhood in 4 subjects with diminished MEF_{75-25} as infants. Similarly, these data would suggest that either airway function genuinely improves with age or that the raised volume technique is better able to detect diminished airway function in infants than spirometry in pre-school children. The reasons for improvement could include the result of treatment, normal lung development, or an increase in airway luminal diameter such that obstruction is less severe in older subjects. The process of augmenting inflations, the effects of sedation and testing in the supine opposed to upright position in infants may alter airway and lung mechanics, increasing detection of functional abnormalities by RVRTC when compared with older children tested using spirometry.

One caveat to these studies is that the reliability of measurements of forced expiration depends on strict quality control. No internationally agreed guidelines exist for pre-school children currently and adult criteria do not apply [29]. Further assessment of those with CF is limited by the need for standardization of spirometry in this age group.

Newer Lung Function Techniques

Properties of an ideal test of airway function are summarized in table 1. It should be stated that no such test exists for use in infants and that RVRTC fulfils few of these requirements. The wider acceptability of tests of airway function in infancy is limited due to the necessity for sedation of the subject. The repeatability of measurements following a short interval, such as 2 weeks, is unknown and will be difficult to determine because of reluctance to repeat sedation. This will have an impact on study design where short-term interventions, for example, a 2-week course of intravenous antibiotics, are being assessed. Two newer techniques may prove more suitable for the early detection of lung function abnormalities in CF, fulfilling more of the requirements for an ideal test and offering further insights into the pathophysiology of pulmonary disease.

Multiple Breath Inert Gas Washout
This exciting technique is performed during tidal breathing and therefore requires minimal cooperation. Subjects

Table 1. Requirements for an ideal test of infant lung function

Safe
Cheap
Easy to perform
Quick to perform
Does not require sedation
Non-invasive
Portable
Sensitive to early changes
Repeatable
Measures parameters which track longitudinally
Measures parameters which are unaffected by body size or sex

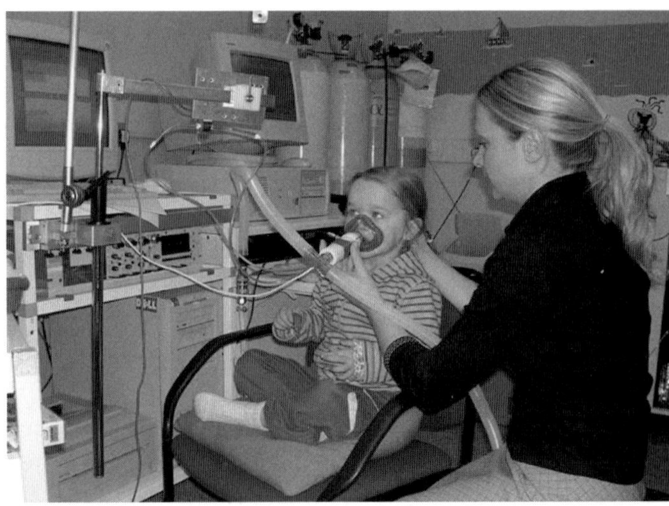

Fig. 8. Multiple breath inert gas washout. Distracted by television, a pre-school child inhales an inert gas (wash-in phase). Gas flow and concentration are monitored on the computer screen. The child is disconnected from the gas source when equilibration is reached. The lung clearance index essentially measures the volume of air needed to wash out and remove the inert gas. It is expressed as a ratio to the functional residual capacity of the subject.

breathe in inert gases, such as helium and sulphur hexafluoride, until wash-in is complete when gas concentrations within the lung reach a steady state. Following disconnection from the gas source, the concentrations of the gases are evaluated as they are washed-out during ensuing tidal breathing. Several parameters can be measured, the commonest being measures of the efficiency of expiration such as the lung clearance index (calculated as the cumulative expired volume, expressed as a ratio to functional residual capacity, required to reduce the concentration of the inert gas 40-fold, arbitrarily, from a starting concentration of 4% down to 0.1%) and the mixing index (fig. 8). The technique has the advantage that it is quick and easy to perform, that the lung clearance index appears to be stable throughout life in healthy subjects and that the use of multiple gases of different densities may provide insights into the site of inhomogeneity within the lung. During inspiration, two mechanisms contribute to gas transport: convection (bulk flow) predominates in the conducting airways as far as the bronchioles, and diffusion predominates in the alveolus. Between the conducting airways and alveoli there is a transition zone termed the diffusion front. Because of their different molecular weights, helium (light) and sulphur hexafluoride (heavy) have different diffusivities. For helium the diffusion front is thought to be located near the terminal and respiratory bronchioles (pre-acinar) and that for sulphur hexafluoride around the alveolar ducts (intra-acinar). The location of the diffusion front contributes to the phase III slope of a standard gas washout curve, another measure of inhomogeneity. Theoretically, comparing the phase III slopes for gases of different molecular weights could be used to estimate the anatomical site affected by disease.

As reductions in flow from obstructed lung regions appear to be compensated by increases in flow from unobstructed regions (a concept known as 'kinematic interdependence of regional expiratory flows'), during measurements of forced expiration upstream (peripheral) non-uniformities may be masked [30]. For this reason, the maximum forced expiratory flow-volume curve is not considered the ideal tool for the detection of early non-uniform airway disease whilst the multiple breath inert gas washout technique should be more sensitive at detecting such inhomogeneities.

Measurements using the technique have been made in CF subjects of all ages. Inert gas washout disclosed airway dysfunction in the majority of children with CF aged 3–18 years who had normal lung function assessed by spirometry, suggesting that multiple-breath inert gas washout is indeed of greater value than spirometry in detecting early CF lung disease [31]. These findings have recently been confirmed in younger subjects [15, 32, 33]. Although multiple breath inert gas washout appears to be the most sensitive test of early functional dysfunction in those with CF, it is not known if it will be more or less useful than parameters of forced expiration to monitor lung function longitudinally or the response to therapeutic interventions and so further studies are required in order to determine its role. As measurements in infants have so far been incorporated into studies with protocols including more invasive and complex tests, they have been performed following sedation of the infant. It is possible that when used alone, a period of natural sleep without sedation will be sufficient to enable successful performance of the test. This would significantly enhance the ability of investigators to obtain longitudinal data during the first months of life.

Low Frequency Forced Oscillation Technique

The low frequency forced oscillation technique LFFOT was adapted for use in infants in 1996 [34]. The airway is occluded at the end of an augmented inflation in order to invoke the Hering-Breuer inflation reflex, resulting in a short period of apnoea during which a pressure wave is applied to the airway opening. Impedance is calculated as the ratio of pressure to flow measured at the airway opening at each frequency of the applied pressure. Modelling in animals has confirmed both frequency-dependent and independent components of impedance which are considered to provide separate estimates of airway and parenchymal function. Although no advantage of the LFFOT over conventional spirometry has been identified in older subjects with CF [35, 36], LFFOT has the theoretical advantage of being able to determine early peripheral functional abnormalities in the alveolar ducts, alveoli and interstitium and its progression more centrally into those parts of the respiratory tree where bulk-flow occurs (airway opening to the terminal bronchioles) and could therefore provide further information regarding the pathophysiology and location of very early functional abnormalities in CF. LFFOT has been used to study growth and development of the respiratory system of healthy and wheezy infants [37, 38]. However, in one longitudinal study, in which LFFOT, Rint and plethysmographic R_{aw} were measured serially in preschool children with CF, only R_{aw} was abnormal in the CF group [39]. No technique showed any evidence of catch-up growth over a 3-year period, confirming the London Cystic Fibrosis Collaboration study results [27]. The utility of these techniques, and comparison with LCI, await further study.

Conclusions

Early detection of pre-symptomatic changes in lung function, together with the ability objectively to assess response to treatment, should strengthen our competence at evaluating the effectiveness of therapeutic interventions which aim to minimize or prevent lung damage in subjects with CF during the critical period of growth and development in infancy. This in turn could increase longevity and contribute to an improved quality of life for these children. Full forced expiratory manoeuvres provide useful information regarding airway status in infants with CF. Standardization and further evaluation are required to determine the appropriateness and feasibility of their use as outcome measures in multicentre studies. Forced expiratory manoeuvres also provide insight into structural/functional relationships of the lung and allow us to generate hypotheses regarding the pathogenesis of early functional abnormalities in the early stages of CF. The development of exciting newer techniques, however, herald future advances in the assessment of infants and pre-school children with CF.

References

1 Khan TZ, Wagener JS, Bost T, Martiniez J, Accurso FJ, Riches DWH: Early pulmonary inflammation in infants with cystic fibrosis. Am J Respir Crit Care Med 1995;151: 1075–1082.
2 Balough K, McCubbin M, Weinberger M, Smits W, Ahrens R, Fick R: The relationship between infection and inflammation in the early stages of lung disease from cystic fibrosis. Pediatr Pulmonol 1995;20:63–70.
3 Gappa M, Ranganathan SC, Stocks J: Lung function testing in infants with cystic fibrosis: Lessons from the past and future directions. Pediatr Pulmonol 2001;32:228–245.
4 Turner DJ, Stick SM, LeSouëf KL, Sly PD, LeSouëf PN: A new technique to generate and assess forced expiration from raised lung volume in infants. Am J Respir Crit Care Med 1995;151:1441–1450.
5 Feher A, Castile R, Kisling J, Angelicchio C, Filbrun D, Flucke R, Tepper R: Flow limitation in normal infants: A new method for forced expiratory maneuvers from raised lung volumes. J Appl Physiol 1996;80:2019–2025.
6 Ranganathan SC, Hoo AF, Lum SY, Goetz I, Castle RA, Stocks J: Exploring the relationship between forced maximal flow at functional residual capacity and parameters of forced expiration from raised lung volume in healthy infants. Pediatr Pulmonol 2002;33: 419–428.
7 Ranganathan S, Dezateux CA, Bush A, Carr SB, Castle R, Madge SL, Price JF, Stroobant J, Wade AM, Wallis CE, Stocks J: Airway function in infants newly diagnosed with cystic fibrosis. Lancet 2001;358:1964–1965.
8 Ranganathan SC, Bush A, Dezateux C, Carr SB, Hoo AF, Lum S, Madge S, Price J, Stroobant J, Wade A, Wallis C, Wyatt H, Stocks J: Relative ability of full and partial forced expiratory maneuvers to identify diminished airway function in infants with cystic fibrosis. Am J Respir Crit Care Med 2002;166:1350–1357.
9 Dezateux CA, Stocks J, Dundas I, Jackson EA, Fletcher ME: The relationship between tPTEF:tE and specific airways conductance in infancy. Pediatr Pulmonol 1994;18:299–307.
10 Banovcin P, Seidenberg J, von der Hardt H: Assessment of tidal breathing patterns for monitoring of bronchial obstruction in infants. Pediatr Res 1995;38:218–220.
11 Bates J, Schmalisch G, Filbrun D, Stocks J: Tidal breath analysis for infant pulmonary function testing. Eur Respir J 2000;16: 1180–1192.
12 Clarke J, Silverman M: Infant lung function and tidal breathing patterns. Pediatr Pulmonol 1995;20:135–136.
13 Rusconi F, Gagliardi L, Aston H, Silverman M: Changes in respiratory rate affect tidal expiratory flow indices in infants with airway obstruction. Pediatr Pulmonol 1996;21: 236–240.
14 Ranganathan SC, Goetz I, Hoo AF, Lum S, Castle R, Stocks J: Assessment of tidal breathing parameters in infants with cystic fibrosis. Eur Respir J 2003;22:761–766.
15 Van Muylem A, Baran D: Overall and peripheral inhomogeneity of ventilation in patients with stable cystic fibrosis. Pediatr Pulmonol 2000;30:3–9.

16 Ranganathan SC, Stocks J, Dezateux C, Bush A, Wade A, Carr S, Castle R, Dinwiddie R, Hoo AF, Lum S, Price J, Stroobant J, Wallis C: The evolution of airway function in early childhood following clinical diagnosis of cystic fibrosis. Am J Respir Crit Care Med 2004; 169:928–933.

17 Dakin CJ, Numa AH, Wang H, Morton JR, Vertzyas CC, Henry RL: Inflammation, infection, and pulmonary function in infants and young children with cystic fibrosis. Am J Respir Crit Care Med 2002;165:904–910.

18 Rosenfeld M, Gibson RL, McNamara S, Emerson J, Burns JL, Castile R, Hiatt P, McCoy K, Wilson CB, Inglis A, Smith A, Martin TR, Ramsey BW: Early pulmonary infection, inflammation, and clinical outcomes in infants with cystic fibrosis. Pediatr Pulmonol 2001;32:356–366.

19 Nixon GM, Armstrong DS, Carzino R, Carlin JB, Olinsky A, Robertson CF, Grimwood K, Wainwright C: Early airway infection, inflammation, and lung function in cystic fibrosis. Arch Dis Child 2002;87:306–311.

20 Castile RG, Iram D, McCoy KS: Gas trapping in normal infants and in infants with cystic fibrosis. Pediatr Pulmonol 2004;37:461–469.

21 Ljungberg H, Hulskamp G, Hoo A-F, Pillow J, Lum S, Gustafsson P, Stocks J: Estimates of plethysmographic FRC exceed those by gas dilution in infants with cystic fibrosis but ot in healthy controls. Thorax 2002;57(suppl iii):S72.

22 Kraemer R, Birrer P, Liechti-Gallati S: Genotype-phenotype association in infants with cystic fibrosis at the time of diagnosis. Pediatr Res 1998;44:920–926.

23 Mohon RT, Wagener JS, Abman SH, Seltzer WK, Accurso FJ: Relationship of genotype to early pulmonary function in infants with cystic fibrosis identified through neonatal screening. J Pediatr 1993;122:550–555.

24 Bridge PD, Ranganathan S, McKenzie SA: Measurement of airway resistance using the interrupter technique in preschool children in the ambulatory setting. Eur Respir J 1999;13: 792–796.

25 Beydon N, Amsallem F, Bellet M, Boule M, Chaussain M, Denjean A, Matran R, Pin I, Alberti C, Gaultier C: Pulmonary function tests in preschool children with cystic fibrosis. Am J Respir Crit Care Med 2002;166: 1099–1104.

26 Gracchi V, Boel M, van der Laag J, van der Ent CK: Spirometry in young children: Should computer-animation programs be used during testing? Eur Respir J 2003;21:872–875.

27 Kozlowska WJ, Aurora P, Lum S, Saunders C, Ranganathan S, Castle R, Stocks J: Longitudinal assessment of lung function in infants and pre-school children with cystic fibrosis. Arch Dis Child 2004;89(suppl I):A38.

28 Marostica PJ, Weist AD, Eigen H, Angelicchio C, Christoph K, Savage J, et al: Spirometry in 3- to 6-year-old children with cystic fibrosis. Am J Respir Crit Care Med 2002;166:67–71.

29 Aurora P, Stocks J, Oliver C, Saunders C, Castle R, Chaziparasidis G, Bush A: Quality control for spirometry in preschool children with and without lung disease. Am J Respir Crit Care Med 2004;169:1152–1159.

30 McNamara J, Castile R, Ludwig M: Interdependent regional emptying during forced expiration. J Appl Physiol 1994;76:356–360.

31 Gustafsson PM, Aurora P, Lindblad A: Evaluation of ventilation maldistribution as an early indicator of lung disease in children with cystic fibrosis. Eur Respir J 2003;22:972–979.

32 Aurora P, Gustafsson P, Bush A, Lindblad A, Oliver C, Wallis C, Stocks J: Multiple-breath inert gas washout as a measure of ventilation distribution in children with cystic fibrosis. Thorax 2004;59:1068–1073.

33 Ljungberg H, Hulskamp G, Hoo A, Pillow J, Aurora P, Gustafsson P, Stocks J:Abnormal lung clearance index is more common than reduced $FEV_{0.5}$ in infants with CF. Am J Respir Crit Care Med 2003;167:A41.

34 Sly PD, Hayden MJ, Peták F, Hantos Z: Measurement of low-frequency respiratory impedance in infants. Am J Respir Crit Care Med 1996;154:161–166.

35 Hellinckx J, De Boeck K, Demedts M: No paradoxical bronchodilator response with forced oscillation technique in children with cystic fibrosis. Chest 1998;113:55–59.

36 Lebecque P, Stanescu D: Respiratory resistance by the forced oscillation technique in asthmatic children and cystic fibrosis patients. Eur Respir J 1997;10:891–895.

37 Hall GL, Hantos Z, Petak F, Wildhaber JH, Tiller K, Burton PR, Sly PD: Airway and respiratory tissue mechanics in normal infants. Am J Respir Crit Care Med 2000;162: 1397–1402.

38 Hall GL, Hantos Z, Sly PD: Altered respiratory tissue mechanics in asymptomatic wheezy infants. Am J Respir Crit Care Med 2001;164:1387–1391.

39 Nielsen KG, Presslet T, Klug B, Koch C, Bisgaard H: Serial lung function and responsiveness in cystic fibrosis during early childhood. Am J Respir Crit Care Med 2004;169: 1209–1216.

Dr. Sarath Ranganathan MB ChB MRCP FRCPCH PhD
Department of Respiratory Medicine
Royal Children's Hospital and
Department of Paediatrics
University of Melbourne, Flemington Road
Parkville VIC 3040 (Australia)
Tel. +61 3 9345 5818, Fax +61 3 9349 1289
E-Mail sarath.ranganathan@rch.org.au

Bush A, Alton EWFW, Davies JC, Griesenbach U, Jaffe A (eds): Cystic Fibrosis in the 21st Century.
Prog Respir Res. Basel, Karger, 2006, vol 34, pp 205–211

Recent Advances in Imaging

Anastasia Oikonomou[a]   David M. Hansell[b]

[a]Department of Radiology, Alexandroupolis University Hospital, Dragana, Alexandroupolis, Greece;
[b]Department of Radiology, Royal Brompton Hospital, London, UK

Abstract

Pulmonary involvement in cystic fibrosis (CF) is heterogeneous, especially in the early stages, and so may not be accurately depicted by global measures of pulmonary function. Imaging studies, in particular computed tomography (CT), can display regional variation in CF lung disease. Chest radiography remains one of the most widely used diagnostic methods for assessing progression of CF lung disease. Thin-section or high-resolution computed tomography (HRCT) has been shown to have significantly higher sensitivity in depicting lung airway abnormalities than chest radiography. Spirometrically triggered HRCT technique can be used to ensure comparable images for serial evaluations, but is not generally used in clinical practice. Controlled ventilation technique is recommended in young infants who cannot cooperate with breath holding. Low-dose thin-section CT can achieve a 40–50% reduction of radiation dose without affecting image quality. HRCT scoring systems have been proposed that may aid the earlier detection of lung damage, objective evaluation of the progression of disease, and as a research tool in the assessment of response to newly developed treatments. Although the role of conventional proton density MRI in CF patients is limited, preliminary studies with hyperpolarized ^{3}He MRI in CF suggest that early functional abnormalities may be more readily detected, before morphologic changes on HRCT are evident.

Chest Radiography

The lungs of a newborn child with cystic fibrosis (CF) are radiographically normal but in those with progression of the disease within the first months of life substantial small airway mucus plugging, inflammation and infection develops, which is reflected radiographically as hyperinflation of the lungs [1]. Peribronchial cuffing of end-on bronchi (bronchial wall thickening) and mucus plugging of small airways are manifest as a nodular or reticulonodular pattern on chest radiography [2]. With more advanced disease bronchiectasis of the larger airways, the hallmark of established CF, develops. Severe peribronchial wall thickening is common in bronchiectasis and may precede bronchial dilatation. Abnormal wall thickening allows visualization of the airways in the periphery of the lung beyond the level of the segmental bronchi. With further progression of bronchiectasis larger cysts form, mainly representing bronchiectatic cysts [3]. Mucoid impaction of the bronchi appears as rounded or band-like opacities of increased density that follow the course of the dilated bronchi. Lobar and segmental atelectasis may result, most frequently in the upper and middle lobes. Lobar pneumonia is unusual compared to small patches representing peribronchial consolidation, except in end-stage disease.

Standard posteroanterior radiographs are necessary for evaluating and excluding complications such as atelectasis, consolidation, pneumothorax or cor pulmonale on regular follow-ups or during exacerbations. Conventional radiography by being simple, inexpensive, widely available and involving relatively little radiation exposure remains the primary imaging technique for following the progression of the disease and assessing serial change. However, it may be less sensitive at detecting early disease compared with other modalities such as computed tomography (CT) scanning.

Scoring Systems

Multiple scoring systems based solely on chest radiography or on a combination of chest radiography and clinical status [4–6] have been developed to grade the severity of CF [7–11]. There has been only limited evaluation of the reproducibility and reliability of such scores and of their correlation with other indicators of the extent of lung disease [10, 12]. The value of scoring systems in clinical practice and research has been questioned, especially given the increasing availability of newer imaging techniques such as high-resolution computed tomography (HRCT).

Computed Tomography

Technique

Although chest radiography is used in the evaluation of CF, the superiority of HRCT over chest radiography and conventional CT in providing extremely detailed evaluation of the lungs in CF has long been recognized [13–18].

Advances in CT technology, particularly the development of helical and multislice CT scanners, have resulted in faster scan acquisition times, enabling images of sufficient diagnostic quality to be obtained in children during quiet respiration obviating the need for sedation [19–21]. The new scanners are also quieter, which helps children remain calm and so reduces the need for sedation and improves image quality. However, since the greatest impediment to high-quality HRCT studies in children is respiratory and other motion, sedation may be required in uncooperative children between the ages of 6 months and 6 years. Children less than 6 months old usually fall asleep after a feed and sedation is rarely required after the age of 6 years.

The HRCT technique involves the use of the narrowest available sections. Published protocols for HRCT examinations in children for conventional CT scanners vary from 1- to 1.5-mm slice thickness, 1- to 2-second acquisition time, 120 to 140 kVp and 100 to 280 mA [22]. Section spacing may differ slightly to that used in adult examinations (i.e. 10 mm) due to the smaller thoracic volume that has to be scanned. It is, therefore, recommended to obtain inspiratory examinations with 5 mm interspacing for children less than 2 years of age, 7 mm interspacing in children from 2 to 10 years, and 10 mm interspacing in children over 10 years old (as for adults) [20]. A set of expiratory images can be obtained using twice the interval of the inspiratory images [38]; some authorities recommend only three expiratory sections, one in the upper lobes, one in the middle lobe/lingula and one in the lower lobes [23]. A high-frequency reconstruction algorithm (bone or lung) is used to increase edge enhancement and improve visualization of fine parenchymal detail. The use of the smallest field of view possible optimizes spatial resolution. Scan acquisition time is crucial, and needs to be as short as possible in order to minimize motion artifacts [20]. Electron beam CT (EBCT) scanners with a 0.1-second scan time and the newer multisection helical CT scanners with subsecond scanning are optimal in this respect and are particularly useful for imaging children [19]. It is important to bear in mind that image noise (giving a granular appearance) produced by the use of thinner slices is higher with helical CT than with conventional axial technique [23] and, therefore, some authors advocate axial (nonhelical) acquisition [24].

As children are more radiosensitive than adults and have a longer life span in which to develop radiation-induced disease, great care must be taken with the use of radiation and the HRCT protocol should be designed to provide the best image quality at the lowest possible radiation dose. It has been shown that as compared to a 180-mAs technique, a lower-dose HRCT technique results in a significant dose reduction of 72% for 50 mAs and 80% for 34 mAs; good quality images were obtained with 50 mAs in uncooperative children and with 34 mAs in cooperative children and young adult patients [25]. Low dose HRCT has been reported to have a radiation dose as low as that required for several chest radiographs [26]. The use of even lower doses in children such as the use of 25 mA and 1-second scan time [21, 23] has been reported.

Special HRCT Techniques

EBCT, although not widely available, is a valuable technique in pediatric radiology as it allows routine use of a 0.1-second scan time preventing respiratory motion artifact and allowing a relatively low radiation dose in comparison to conventional CT scanners [27]. Drawbacks include spatial resolution inferior to that of helical scanners and the lack of commercial development of EBCT [20]. Protocols for EBCT scanners differ slightly in that acquisition times of as little as 100 ms in young children may require thicker sections (3 mm) to allow sufficient photons to reach the detector system in order to produce an acceptable image. A typical protocol for an EBCT scanner is 3-mm sections every 6 mm, 150 kVp, 650 mA with an acquisition time of 100 ms, the milliamperes being fixed and dose reduction being achieved by reducing exposure time [19].

In spirometric gating the CT scan is triggered, and airflow is mechanically inhibited at a predetermined user-selected point in the respiratory cycle [28]. Spirometric triggering of scan acquisition enables HRCT to be reproducibly obtained at preselected lung volumes and is recommended to ensure anatomically standardized images for comparative and serial evaluations and for objective CT quantification in progressive lung diseases [27, 29–32]. Although the method provides highly reproducible results, drawbacks include the time-consuming scanning as well as the preparatory phase for the supine measurements of static lung volumes, the exclusion of young children (generally younger than 7 years) due to their inability to cooperate with breath holding, and the scarcity of the commercial apparatus [27, 28]. Recently spirometric-triggered HRCT has been used for an automated approach of quantification of air trapping (AT) in patients with mild CF [31, 32] and it has been reported that AT measured in an objective automated manner may be a highly sensitive means of assessing early lung disease. However, there are no studies that compare spirometer-triggered with non-spirometer-triggered images for quantification of AT so as to justify the method.

Uncooperative children younger than 5–6 years of age cannot maintain breath holding at full lung inflation and end expiration necessary for high-quality HRCT images. Although motion artifacts are improved with the use of EBCT and multidetector CT, these methods do not address the limitations of imaging during shallow tidal breathing. Full-inspiration and end-expiration controlled ventilation technique aims to overcome these difficulties. The respiration of a sedated child is taken over using positive pressure applied by a face mask, and then a transient respiratory pause can be induced for the duration of CT image acquisition. During the induced respiratory pause the lungs can be scanned at either full-lung inflation, or end expiration [33, 34]. However, it is important to standardize technique across CF centers if research comparisons are to be made.

HRCT Scoring Systems

In 1991 Bhalla et al. [16] reported the first HRCT scoring system designed to quantify structural lung abnormalities in patients with CF by using HRCT. Since then a number of modified semiquantitative HRCT scoring systems have been proposed and have been evaluated as indicators of the presence and severity of disease at different ages [14, 17, 18, 35–40]. Numerous studies have shown moderate to strong correlations between HRCT scoring systems and pulmonary function tests (PFTs) [14, 17, 18, 35, 36, 39, 41, 42]. Other studies, however, have reported weak correlations, in part because HRCT can identify lung disease in patients with normal PFTs [18, 37, 40, 43–45]. The inference that can be drawn from this limited evidence is that PFTs are global measures reflecting the overall status of both lungs, while CT, which provides excellent anatomic localization, identifies abnormalities that affect only a small portion of the lung (and may therefore be functionally silent).

Many series have reported good interobserver and intraobserver agreement of HRCT scores, indicating that such quantitation is robust with relatively little noise from observer variation [16, 17, 36, 37, 41]. Limitations of HRCT scoring systems are the lower reproducibility with longer interval (1–2 months) between repeat measurements, the increased interobserver variability when the HRCT scores are low and the lack of clear definitions of some of the variables [41].

HRCT and Early Detection

CF lung disease begins very early in life, even within the first 10 weeks of life [46]. The challenge of CF care is to preserve the normal airway morphology that is present at birth. It is, therefore, crucial to detect airway changes as early as possible so as to initiate more aggressive treatment at a stage when abnormalities are still reversible. Although the later phase of CF is easily observed clinically and monitored with PFTs, HRCT has been suggested to be more sensitive and to have a role in the early detection of CF lung disease [41, 44].

AT has been found to be one of the earliest HRCT manifestations of CF lung disease [37, 39, 46]. It is related to small airways disease that cannot be affirmed on HRCT otherwise. It represents air in the lung that cannot be expired and is characterized by low attenuation areas on expiratory CT. In younger patients with CF with mild or no functional lung abnormalities and with little or no bronchiectasis, it has been reported to be the most common HRCT finding [32, 37, 39]. It has thus been suggested that detection of AT should initiate more aggressive treatment before irreversible damage has been established [37]. Expiratory CT technique is a necessary adjunct for the evaluation of AT since it is accentuated or sometimes can only be detected on expiratory scans [29, 31, 47]. More recent HRCT studies based on spirometric triggering have shown that quantification of AT with an objective automated approach could be a powerful discriminator of early CF lung disease in contrast to PFTs that failed to detect any early abnormalities. It has also been reported that the volume, at which expiratory scans are obtained if spirometric triggering is to be used, significantly affects the results of quantitative AT measurements. Lung volumes close to residual volume are recommended to avoid overdiagnosis of the extent of AT [32].

Bronchiectasis is the cardinal HRCT feature of established CF [14] and invariably develops with age [43].

However, HRCT studies with controlled ventilation technique have shown that large airways of minimally symptomatic infants and young children with CF have thicker walls and are more dilated than those of healthy control infants [48]. Furthermore, it was suggested that their airways increase in size at a greater rate than in normal children. These findings suggest that the structural airway changes that lead to widespread bronchiectasis in patients with CF begin very early in life [37, 46, 48].

Airway wall thickening may be seen independent of frank bronchiectasis and is generally more evident than bronchial dilatation in patients who have early disease [14, 35, 39, 45, 48]. Airway wall thickening and AT at end expiration were the most frequently identified HRCT abnormalities (75 and 69% accordingly) in infants with CF with an age range from 10 to 285 weeks [46].

Longitudinal HRCT Studies and Serial Changes

There have been a few longitudinal studies investigating the role of serial CT in response to treatment [29, 30, 42, 49–51], as well as in understanding the natural history of CF [44, 52]. HRCT can accurately depict radiographic improvement following treatment during an acute exacerbation with statistical correlation being reported between HRCT and PFT changes. Imaging features in CF that may show reversibility after an acute exacerbation are mucous plugging, centrilobular nodules, peribronchial thickening and presence of air-fluid levels in bronchiectatic cavities [42]. Mucus plugging of small airways is thought to precede that of large airways in the progress of CF [15]. Both forms of plugging are clearly reversible as judged by HRCT before and after an acute exacerbation [29, 42, 43, 49]. Airway wall thickening has been reported to be a major determinant of the loss of FEV_1 [38] and a reversible HRCT feature after an acute infective exacerbation [42, 49]. AT may be a reversible phenomenon in the younger population with CF – perhaps as a result of bronchial hyperreactivity from inflammation or increased airway compliance in this age group [50]. However, AT was not a reversible finding when studied at an older age group [42].

HRCT has the potential to assess the progression of structural changes in CF lung disease over time. Morphologic components that were most significantly shown to progress with age were severity and extent of bronchiectasis, mucus plugging and mosaic perfusion representing AT [44, 52]. Interestingly, CT seemed to have an advantage over spirometry (FEV_1, FVC, FEV_1/FVC, $FEF_{25–75\%}$, RV, RV/TLC, TLC) in the depiction of sequential morphologic changes. In contrast to the significant progression of morphologic changes on serial CT, spirometric parameters remained unchanged or showed only minimal improvement [44, 52].

Monitoring CF with HRCT

The most active area of current investigation in regard to the use of CT in CF is its use as an outcome surrogate for research studies [37, 39, 41, 44, 49, 51, 52]. Many clinical studies have shown the potential of HRCT to detect short-term treatment effect and have justified its use as a tool to test the efficacy of new therapeutic regimens. The need for improved and more sensitive outcome measures in early CF lung disease has led to the combination of CT and PFTs in a composite score to investigate whether there would be even better results to this effect [30]. A recent study has shown that a composite CT/PFT score was a more sensitive outcome measure in discriminating a treatment effect in young patients with CF, who have normal or mildly reduced pulmonary function than PFT or CT alone. The composite scores may offer a promising new assessment tool that could advance the field of outcomes research in patients with CF, but clearly needs further study for future evaluation [30, 37, 41, 44, 49, 50].

Monitoring CF patients using HRCT carries potential risk due to the associated radiation exposure and the higher probability of cancer induction in the pediatric age group. A recent clinical gene therapy study showed no difference at HRCT obtained 90 days after the initial dose [51]. This finding probably suggests that follow-up scanning should be done earlier than 3 months if a treatment effect is to be determined. Because of concern about radiation since life expectancy is increasing in patients with CF, CF clinical centers are performing biennial rather than annual HRCT examinations to assess the progression of disease [41]. However, Helbich et al. [52] reported that morphologic changes in CF progress more slowly within the first 18 months of the follow-up periods and, therefore, they recommended a follow-up period of less than 18 months so that effective therapies could be induced before irreversible lung damage would be established. The benefits from employing serial HRCT scanning must be balanced against the radiation risk. New ways to further reduce the radiation dose at HRCT scanning have to be found and further research to determine the frequency of scanning for monitoring CF is required [41].

Magnetic Resonance Imaging

Initial reports of magnetic resonance imaging (MRI) in patients with CF suggested that MRI would be a useful

Oikonomou/Hansell

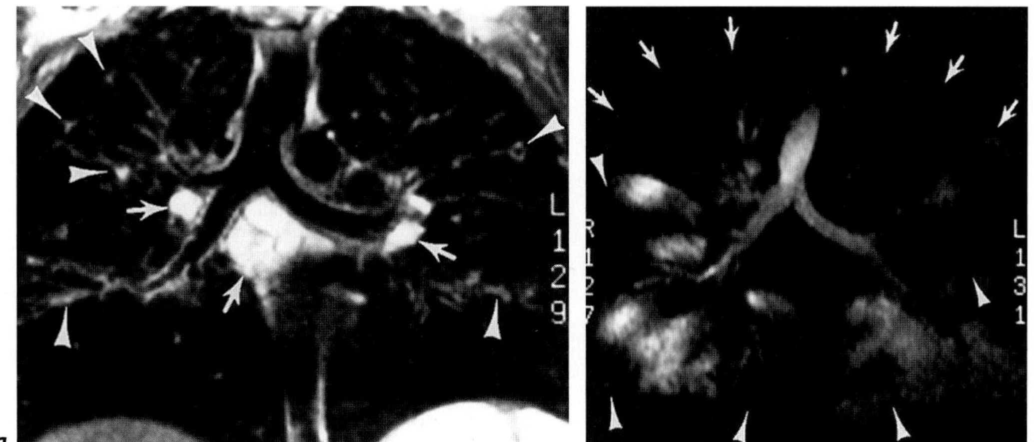

Fig. 1. *a* Coronal conventional fast SE MR image at the level of the trachea shows areas of mildly increased linear and punctate nodular signal intensity (arrowheads) within the upper and middle lung zones bilaterally, as well as bilateral hilar and subcarinal lymphadenopathy (arrows). *b* Coronal ^{3}He gradient echo MR image shows complete (left) absence of signal intensity within the upper lungs (arrows). There are also large areas of absent signal intensity within the middle and lower lungs (arrowheads) [reprint with permission from 58].

technique in the assessment of CF due to its ability to detect early mucus plugs and to differentiate mucoid impaction from atelectasis [53] or hilar adenopathy from enlarged hilar vessels [54]. Subsequently it was shown that chest radiography was superior to MRI in its ability to assess air-containing structures and hyperinflation [55]. More recently a single report suggested that MRI might be suitable solely for the monitoring of atelectasis and pulmonary infiltrates in CF [56].

Hyperpolarized ^{3}He MRI is a relatively new imaging method that allows the depiction of both lung function and morphology. Because this tracer gas is administered by inhalation and is not absorbed by the lung, it allows precise visualization of the ventilated airspaces. Imaging with a hyperpolarized ventilation contrast agent generates a high signal within the pulmonary airspaces, offering a potential increase in signal-to-noise ratio of 100,000 times that

obtained by conventional proton imaging [57–59]. There is limited experience with hyperpolarized ^{3}He MRI in the assessment of patients with CF. Nevertheless, two studies have shown multiple ventilation defects in patients with CF that increased in number with increasing disease severity [58, 60]. Abnormalities of both ventilation and morphology were readily detected in several lung zones, and severe ventilation defects were seen despite minimal or no morphologic abnormalities (fig. 1). The additional advantage of a lack of irradiation and the combination of both functional and morphologic information concerning the pulmonary status of patients with CF in a safe, fast and reproducible manner makes ^{3}He MRI an attractive and promising, alternative tool for serial evaluation of patients with CF [58]. However, further investigation is necessary to establish whether there are genuine clinical applications for this technique.

References

1 Wood BP: Cystic fibrosis: 1997. Radiology 1997;204:1–10.
2 Ruzal-Shapiro C: Cystic fibrosis: An overview. Radiol Clin North Am 1998;36:143–161.
3 Tomashefski JF Jr, Bruce M, Stern RC, Dearborn DG, Dahms B: Pulmonary air cysts in cystic fibrosis: Relation of pathologic features to radiologic findings and history of pneumothorax. Hum Pathol 1985;16:253–261.
4 Shwachman H, Kulczycki LL: Long-term study of one hundred five patients with cystic fibrosis; studies made over a five- to fourteen-year period. AMA J Dis Child 1958;96:6–15.
5 Matouk E, Ghezzo RH, Gruber J, Hidvegi R, Gray-Donald K: Internal consistency reliability and predictive validity of a modified N. Huang clinical scoring system in adult cystic fibrosis patients. Eur Respir J 1997;10:2004–2013.
6 Taussig LM, Kattwinkel J, Friedewald WT, Di Sant'Agnese PA: A new prognostic score and clinical evaluation system for cystic fibrosis. J Pediatr 1973;82:380–390.

7 Brasfield D, Hicks G, Soong S, Tiller RE: The chest roentgenogram in cystic fibrosis: A new scoring system. Pediatrics 1979;63:24–29.

8 Chrispin AR, Norman AP: The systematic evaluation of the chest radiograph in cystic fibrosis. Pediatr Radiol 1974;2:101–105.

9 O'Laoide RM, Fahy J, Coffey M, Ward K, Malone D, Fitzgerald MX, Masterson J: A chest radiograph scoring system in adult cystic fibrosis: Correlation with pulmonary function. Clin Radiol 1991;43:308–310.

10 Sawyer SM, Carlin JB, DeCampo M, Bowes G: Critical evaluation of three chest radiograph scores in cystic fibrosis. Thorax 1994; 49:863–866.

11 Weatherly MR, Palmer CG, Peters ME, Green CG, Fryback D, Langhough R, Farrell PM: Wisconsin cystic fibrosis chest radiograph scoring system. Pediatrics 1993;91:488–495.

12 Conway SP, Pond MN, Bowler I, Smith DL, Simmonds EJ, Joanes DN, Hambleton G, Hiller EJ, Stableforth DE, Weller P: The chest radiograph in cystic fibrosis: A new scoring system compared with the Chrispin-Norman and Brasfield scores. Thorax 1994;49: 860–862.

13 Jacobsen LE, Houston CS, Habbick BF, Genereux GP, Howie JL: Cystic fibrosis: A comparison of computed tomography and plain chest radiographs. Can Assoc Radiol J 1986;37:17–21.

14 Hansell DM, Strickland B: High-resolution computed tomography in pulmonary cystic fibrosis. Br J Radiol 1989;62:1–5.

15 Lynch DA, Brasch RC, Hardy KA, Webb WR: Pediatric pulmonary disease: Assessment with high-resolution ultrafast CT. Radiology 1990; 176:243–248.

16 Bhalla M, Turcios N, Aponte V, Jenkins M, Leitman BS, McCauley DI, Naidich DP: Cystic fibrosis: Scoring system with thin-section CT. Radiology 1991;179:783–788.

17 Nathanson I, Conboy K, Murphy S, Afshani E, Kuhn JP: Ultrafast computerized tomography of the chest in cystic fibrosis: A new scoring system. Pediatr Pulmonol 1991;11:81–86.

18 Maffessanti M, Candusso M, Brizzi F, Piovesana F: Cystic fibrosis in children: HRCT findings and distribution of disease. J Thorac Imaging 1996;11:27–38.

19 Copley SJ, Padley SP: High-resolution CT of paediatric lung disease. Eur Radiol 2001;11: 2564–2575.

20 Kuhn JP, Brody AS: High-resolution CT of pediatric lung disease. Radiol Clin North Am 2002;40:89–110.

21 Garcia-Pena P, Lucaya J: HRCT in children: Technique and indications. Eur Radiol 2004; 14:L13–L30.

22 Seely JM, Effmann EL, Muller NL: High-resolution CT of pediatric lung disease: Imaging findings. Am J Roentgenol 1997;168: 1269–1275.

23 Brody AS: Thoracic CT technique in children. J Thorac Imaging 2001;16:259–268.

24 Donnelly LF, Frush DP: Pediatric multidetector body CT. Radiol Clin North Am 2003;41: 637–655.

25 Lucaya J, Piqueras J, Garcia-Pena P, Enriquez G, Garcia-Macias M, Sotil J: Low-dose high-resolution CT of the chest in children and young adults: Dose, cooperation, artifact incidence, and image quality. Am J Roentgenol 2000;175:985–992.

26 Mayo JR, Jackson SA, Muller NL: High-resolution CT of the chest: Radiation dose. Am J Roentgenol 1993;160:479–481.

27 Robinson TE, Leung AN, Moss RB, Blankenberg FG, Al-Dabbagh H, Northway WH: Standardized high-resolution CT of the lung using a spirometer-triggered electron beam CT scanner. Am J Roentgenol 1999; 172:1636–1638.

28 Goldin GG: Quantitative CT of the lung. Radiol Clin North Am 2002;40:145–162.

29 Robinson TE, Leung AN, Northway WH, Blankenberg FG, Bloch DA, Oehlert JW, Al-Dabbagh H, Hubli S, Moss RB: Spirometer-triggered high-resolution computed tomography and pulmonary function measurements during an acute exacerbation in patients with cystic fibrosis. J Pediatr 2001; 138:553–559.

30 Robinson TE, Leung AN, Northway WH, Blankenberg FG, Chan FP, Bloch DA, Holmes TH, Moss RB: Composite spirometric-computed tomography outcome measure in early cystic fibrosis lung disease. Am J Respir Crit Care Med 2003;168:588–593.

31 Goris ML, Zhu HJ, Blankenberg F, Chan F, Robinson TE: An automated approach to quantitative air trapping measurements in mild cystic fibrosis. Chest 2003;123: 1655–1663.

32 Bonnel AS, Song SM, Kesavarju K, Newaskar M, Paxton CJ, Bloch DA, Moss RB, Robinson TE: Quantitative air-trapping analysis in children with mild cystic fibrosis lung disease. Pediatr Pulmonol 2004;38:396–405.

33 Long FR, Castile RG, Brody AS, Hogan MJ, Flucke RL, Filbrun DA, McCoy KS: Lungs in infants and young children: Improved thin-section CT with a noninvasive controlled-ventilation technique – initial experience. Radiology 1999;212:588–593.

34 Long FR, Castile RG: Technique and clinical applications of full-inflation and end-exhalation controlled ventilation chest CT in infants and young children. Pediatr Radiol 2001;31: 413–422.

35 Santis G, Hodson ME, Strickland B: High resolution computed tomography in adult cystic fibrosis patients with mild lung disease. Clin Radiol 1991;44:20–22.

36 Santamaria F, Grillo G, Guidi G, Rotondo A, Raia V, de Ritis G, Sarnelli P, Caterino M, Greco L: Cystic fibrosis: When should high-resolution computed tomography of the chest be obtained? Pediatrics 1998;101:908–913.

37 Brody AS, Klein JS, Molina PL, Quan J, Bean JA, Wilmott RW: High-resolution computed tomography in young patients with cystic fibrosis: Distribution of abnormalities and correlation with pulmonary function tests. J Pediatr 2004;145:32–38.

38 Oikonomou A, Manavis J, Karagianni P, Tsanakas J, Wells AU, Hansell DM, Papadopoulou F, Efremidis SC: Loss of FEV1 in cystic fibrosis: Correlation with HRCT features. Eur Radiol 2002;12:2229–2235.

39 Helbich TH, Heinz-Peer G, Eichler I, Wunderbaldinger P, Gotz M, Wojnarowski C, Brasch RC, Herold CJ: Cystic fibrosis: CT assessment of lung involvement in children and adults. Radiology 1999;213:537–544.

40 Stiglbauer R, Schurawitzki H, Eichler I, Vergesslich KA, Gotz M: High resolution CT in children with cystic fibrosis. Acta Radiol 1992;33:548–553.

41 de Jong PA, Ottink MD, Robben SG, Lequin MH, Hop WC, Hendriks JJ, Pare PD, Tiddens HA: Pulmonary disease assessment in cystic fibrosis: Comparison of CT scoring systems and value of bronchial and arterial dimension measurements. Radiology 2004;231: 434–439.

42 Shah RM, Sexauer W, Ostrum BJ, Fiel SB, Friedman AC: High-resolution CT in the acute exacerbation of cystic fibrosis: Evaluation of acute findings, reversibility of those findings, and clinical correlation. Am J Roentgenol 1997;169: 375–380.

43 Helbich TH, Heinz-Peer G, Fleischmann D, Wojnarowski C, Wunderbaldinger P, Huber S, Eichler I, Herold CJ: Evolution of CT findings in patients with cystic fibrosis. Am J Roentgenol 1999;173:81–88.

44 de Jong PA, Nakano Y, Lequin MH, Mayo JR, Woods R, Pare PD, Tiddens HA: Progressive damage on high resolution computed tomography despite stable lung function in cystic fibrosis. Eur Respir J 2004;23:93–97.

45 Tiddens HA: Detecting early structural lung damage in cystic fibrosis. Pediatr Pulmonol 2002;34:228–231.

46 Castile R, Long F, Flucke R, Goldstein A, Filbrun D, Brody A, McCoy K: High resolution computed tomography of the chest in infants with cystic fibrosis. Pediatr Pulmonol 1999;19(suppl):277–278.

47 Dorlochter L, Nes H, Fluge G, Rosendahl K: High resolution CT in cystic fibrosis – the contribution of expiratory scans. Eur J Radiol 2003;47:193–198.

48 Long FR, Williams RS, Castile RG: Structural airway abnormalities in infants and young children with cystic fibrosis. J Pediatr 2004; 144:154–161.

49 Brody AS, Molina PL, Klein JS, Rothman BS, Ramagopal M, Swartz DR: High-resolution computed tomography of the chest in children with cystic fibrosis: Support for use as an outcome surrogate. Pediatr Radiol 1999;29: 731–735.

50 Nasr SZ, Kuhns LR, Brown RW, Hurwitz ME, Sanders GM, Strouse PJ: Use of computerized tomography and chest X-rays in evaluating efficacy of aerosolized recombinant human DNase in cystic fibrosis patients younger than age 5 years: A preliminary study. Pediatr Pulmonol 2001;31:377–382.

51 Moss RB, Rodman D, Spencer LT, Aitken ML, Zeitlin PL, Waltz D, Milla C, Brody AS, Clancy JP, Ramsey B, Hamblett N, Heald AE: Repeated adeno-associated virus serotype 2 aerosol-mediated cystic fibrosis transmembrane regulator gene transfer to the lungs of patients with cystic fibrosis: A multicenter, double-blind, placebo-controlled trial. Chest 2004;125:509–521.

52 Helbich TH, Heinz-Peer G, Fleischmann D, Wojnarowski C, Wunderbaldinger P, Huber S, Eichler I, Herold CJ: Evolution of CT findings in patients with cystic fibrosis. Am J Roentgenol 1999;173:81–88.
53 Gooding CA, Lallemand DP, Brasch RC, Wesbey GE, Davis B: Magnetic resonance imaging in cystic fibrosis. J Pediatr 1984;105:384–388.
54 Kinsella D, Hamilton A, Goddard P, Duncan A, Carswell F: The role of magnetic resonance imaging in cystic fibrosis. Clin Radiol 1991;44:23–26.
55 Fiel SB, Friedman AC, Caroline DF, Radecki PD, Faerber E, Grumbach K: Magnetic resonance imaging in young adults with cystic fibrosis. Chest 1987;91:181–184.
56 Hebestreit A, Schultz G, Trusen A, Hebestreit H: Follow-up of acute pulmonary complications in cystic fibrosis by magnetic resonance imaging: A pilot study. Acta Paediatr 2004;93:414–416.
57 Kauczor HU: Hyperpolarized helium-3 gas magnetic resonance imaging of the lung. Top Magn Reson Imaging 2003;14:223–230.
58 Donnelly LF, MacFall JR, McAdams HP, Majure JM, Smith J, Frush DP, Bogonad P, Charles HC, Ravin CE: Cystic fibrosis: Combined hyperpolarized ^{3}He-enhanced and conventional proton MR imaging in the lung-preliminary observations. Radiology 1999;212:885–889.
59 Altes TA, de Lange EE: Applications of hyperpolarized helium-3 gas magnetic resonance imaging in pediatric lung disease. Top Magn Reson Imaging 2003;14:231–236.
60 Altes TA, Froh DK, Salerno M: Hyperpolarized helium-3 MR imaging of lung ventilation changes following airway mucus clearance treatment in cystic fibrosis. Proceedings of the International Society for Magnetic Resonance Medicine, 9th Meeting, Glasgow, 2001.

David M. Hansell, MD, FRCP, FRCR
Department of Radiology
Royal Brompton Hospital
Sydney Street
London SW3 6NP (UK)
Tel. +44 207 352 8121
Fax +44 207 351 8098
E-Mail d.hansell@rbh.nthames.nhs.uk

Bush A, Alton EWFW, Davies JC, Griesenbach U, Jaffe A (eds): Cystic Fibrosis in the 21st Century.
Prog Respir Res. Basel, Karger, 2006, vol 34, pp 212–220

New Pharmacological Approaches for Treatment of Cystic Fibrosis

Ronald C. Rubenstein

University of Pennsylvania School of Medicine, Division of Pulmonary Medicine,
Children's Hospital of Philadelphia, Philadelphia, Pa., USA

Abstract

Since the identification of the CFTR gene in 1989, tremendous understanding of the cellular and molecular functions of CFTR has been achieved, as well as insight into the mechanism of dysfunction of disease-causing CFTR mutants. These data have suggested the hypothesis that pharmacologic agents can improve mutant CFTR function, a strategy that has been coined 'protein repair therapy'. With the understanding of CFTR's physiologic function as a chloride transporter, other chloride transporters that can potentially replace some or all of the missing function of CFTR have also become potential pharmacological targets. Finally, with the recognition that CFTR also influences the function of a number of other cellular proteins and processes, such proteins and processes have also recently become potential targets for pharmaceutical intervention in cystic fibrosis.

Strategies to Repair or Restore CFTR Function

There are currently over 1,000 disease-causing mutations described in the CFTR gene (http://www.genet.sickkids.on.ca/cftr/). Unlike strategies to replace a defective CFTR in patients with cystic fibrosis (CF) through gene transfer, a novel therapeutic approach, coined 'protein repair therapy', has emerged as a potential means for therapeutically improving the function of mutant CFTRs in a mutation-specific fashion [1, 2]. Mutant CFTRs can be assigned to one of five classes based on the molecular characteristic of the protein's defect [adapted from 3]. Thus, mutant CFTRs are considered class I if protein production is defective, class II if intracellular trafficking is aberrant, class III if regulation of chloride transport is defective, class IV if chloride transport function is reduced, but not absent, and class V if the protein functions normally but has reduced expression. Mutations belonging to classes I, II or III tend to be associated with a more severe CF phenotype with pancreatic insufficiency, as CFTR function is absent. In contrast, mutations belonging to either class IV or class V tend to lead to a milder CF phenotype with pancreatic sufficiency due to reduced rather than absent CFTR function. In general, protein repair strategies are directed at restoring protein production in class I mutants, correcting aberrant intracellular trafficking for class II mutants and improving function and/or regulation of chloride transport for class III mutants. Class IV or V mutants may also be amenable to improvement of function by overexpression or pharmacological activation [1].

Repair of Class I CFTR Mutations

Any mutation of the CFTR gene that leads to absence of CFTR protein production from that allele is considered a class I mutation. Such mutations that result in a nonsense or premature termination codon in the gene sequence (for example G542X, R553X, W1282X, and W1316X) may be amenable to reparative pharmacological therapy with aminoglycoside antibiotics such as gentamicin or G418 (Geneticin®). This approach is discussed in more detail in chapter 1 by Nissim-Rafinia et al., and will be only briefly summarized here. In vitro, these agents cause expression of a full-length, functional CFTR protein from G542X [4], R553X [4], R1162X [5] and W1282X [5] alleles. Interestingly, tobramycin, an aminoglycoside with wider clinical use in CF than gentamicin, was less effective at suppressing these stop mutations and did not allow expression of a repaired, full length CFTR from these 'X' alleles [4]. Similar effects were noted in a murine model. Gentamicin improved intestinal CFTR functional expression and animal survival in *cftr*−/− knockout mice where a human CFTR-G542X allele was expressed under control of an intestine-specific promoter. Again, tobramycin was less effective than gentamicin [6].

These in vitro and animal model observations have been extended to ex vivo and in vivo pilot human trials. Ex vivo exposure of airway epithelia harboring a stop mutation allele to gentamicin improved CFTR expression by both immunocytochemical and functional measures of chloride transport [7]. A pilot clinical study of parenterally administered (intravenous) gentamicin to subjects with CF and a single stop mutation also suggested improved chloride transport by nasal potential difference (NPD) measurements, an in vivo technique for assessment of CFTR and ENaC function in the nasal epithelia. However, these improvements in NPD did not approach the NPD pattern observed in non-CF subjects. The chloride transport response remained less than in the non-CF subjects, and there was no change in the ENaC-mediated NPD [7]. Wilschanski et al. [8] also observed improvements in NPD after 2 weeks of topical delivery of gentamicin to the nasal epithelia of CF subjects with 1 or 2 stop mutations in both an unblinded experiment, and a more extensive randomized, placebo-controlled, double-blind crossover study. Interestingly, in the crossover study, there were also small improvements in NPD measures of ENaC-mediated sodium transport, although again these improvements did not recapitulate a non-CF pattern. These data support the potential utility of this strategy for repair of mutant CFTR due to nonsense mutations, although these data also suggest that additional repair to further activate these repaired mutants may be necessary to have clinical utility.

Repair of Class II CFTR Mutations

Missense mutations in CFTR that allow synthesis of a full-length CFTR protein that does not reach its appropriate intracellular location at the apical plasma membrane of epithelia are referred to as class II mutations. The most common mutation of the CFTR, which occurs on ∼70% of mutant alleles, is the deletion of phenylalanine 508 (ΔF508). ΔF508 is a temperature-sensitive trafficking

Table 1. Physical and pharmacological manipulations that improve ΔF508 trafficking

Model	Treatment	Ref. no.
In vitro	Reduced temperature	23, 92
	Chemical chaperones	
	Glycerol	93, 94
	Trimethyl amine oxide (TMAO)	93
	Dimethyl sulfoxide (DMSO)	95
	Osmolytes	
	NaCl	93
	Betaine	96
	Taurine	96
	Myoinositol	96
	Pharmaceuticals	
	Butyrate	12, 97
	4PBA	16
	Deoxyspergualin	98
	Milrinone	99
	Doxorubicin	100
	1,3-Dipropyl-8-cyclopentylxanthine (CPX)	101
	S-Nitrosoglutathione (GSNO)	102, 103
	Benzo(c)quinolizinium compounds	104
	Thaspigargin	29
	Curcumin	28
ΔF508 mice	Milrinone	105
	TMAO	106
	4PBA	28
	Thaspigargin	29
	Curcumin	28

mutant [9] that is retained in the endoplasmic reticulum (ER) [10] where it interestingly maintains function as a cAMP-regulated chloride channel [11].

ΔF508's ability to transport chloride in the ER [11] has led to a number of groups to hypothesize that 'rescue' of ΔF508 from the ER will improve CFTR function in CF epithelia. In fact, the intracellular trafficking of ΔF508 can be improved by a number of physical or pharmacological manipulations in model systems (table 1), although the mechanism by which these improvements occur remains unclear. It is unclear whether these agents improve ΔF508 trafficking by the same mechanisms, or whether these agents will act similarly on other less common class II mutations such as ΔI507 and N1303K.

Butyrate improves the function of ΔF508 in a heterologous expression system in vitro [12]. While butyrate has been used as an investigational agent for increasing fetal hemoglobin expression in patients with β-hemoglobinopathies [13], it has a number of shortcomings as a therapeutic agent such as poor oral bioavailability and an extremely short half-life necessitating dosage by continuous intravenous infusion. It also smells of rancid milk.

4-Phenylbutyrate (4PBA) is an orally bioavailable butyrate analog with a much more tolerable odor, and only minimal side effects at standard doses (20 g/day) that yield millimolar serum levels [14, 15]. Like butyrate, 4PBA at millimolar concentrations led to improved ΔF508 intracellular trafficking and CFTR function in CF epithelial cells [16]. The mechanism by which 4PBA causes improvement in ΔF508 trafficking is the subject of active investigation, with recent evidence suggesting it may regulate the expression of molecular chaperones such as Hsc70 [17, 18] and Hsp70 [19], which are hypothesized to be important in targeting proteins for intracellular degradation and promoting protein folding, respectively.

We demonstrated that 4PBA corrects ΔF508's trafficking defect and restores CFTR function at the plasma membrane of cultured CF epithelial cells [16]. These initial in vitro observations and the benign experience with 4PBA in patients with urea cycle deficiencies enabled us to perform a pilot clinical trial with 4PBA in ΔF508-homozygous CF subjects [20] where 18 ΔF508-homozygous subjects with CF were given 19 g/day 4PBA or placebo (9/group) in a randomized, placebo-controlled, double-blind fashion for 1 week. Minimal side effects were reported and were similar between the groups. NPD and sweat chloride concentration were assessed as physiologic outcome measures. There were small, statistically significant improvements in chloride transport of the subjects treated with 4PBA compared to the placebo group, without significant changes in epithelial sodium transport (as reflected by the basal NPD and amiloride-sensitive change in NPD) or sweat chloride concentration. These data served as a proof-of-principle that a pharmacological agent could improve ΔF508 function in vivo, but the small improvement in NPD, which was similar to that observed in the pilot clinical trials of gentamicin in subjects with class I nonsense mutations [7, 8, 21], may not be clinically significant.

That repair of ΔF508's trafficking defect alone only minimally improved nasal epithelial chloride transport and did not alter nasal epithelial sodium transport was not unexpected. ΔF508 protein also has lower chloride channel activity due to a decreased open probability relative to that of the wild-type CFTR [22, 23]. 4PBA also may not increase the trafficking of ΔF508 to the level of wild-type CFTR, or the 'repaired' ΔF508 protein may be more rapidly removed from the cell surface than the wild type [24]. Finally, ΔF508 seems deficient in other functions attributable to CFTR, such as regulation of the epithelial sodium channel, ENaC [25, 26] and bicarbonate transport [27].

To attempt to effect further improvement in chloride transport in response to 4PBA, a 4PBA dose escalation trial was recently performed [14]. Unfortunately, there was no significant further improvement in NPD at 30 g 4PBA/day (compared with 20 g/day), and there were significant side effects (headaches, nausea) at 40 g/day. These data also suggest that alternate strategies to activate further ΔF508 are required if a trafficking 'corrector' such as 4PBA is used.

Enteral administration of 4PBA to ΔF508-CF mice leads to changes in the epithelial ion transport of the mice (as assessed by NPD) similar to that of the human trials, namely improved chloride transport without change in the aberrant sodium transport [28]. In this regard, recent reports that Ca-ATPase pump (SERCA) inhibitors such as thapsigaring [29] and curcumin [28] can improve ΔF508 trafficking are intriguing. Administration of either thaspigargin or curcumin to ΔF508-CF mice resulted in normalization of both nasal epithelial chloride and sodium transport as assessed by NPD [28, 29], although conflicting data regarding curcumin has recently been published [30]. While the actual mechanism by which these agents cause this improvement is not yet proven (thaspigargin is also a potent stimulus of the unfolded protein response), these data suggest that their mechanism of action to repair ΔF508 may differ from other correctors such as 4PBA.

High concentrations of 4PBA and curcumin are required to achieve repair of ΔF508 trafficking. While such concentrations are achievable in vivo [14, 20, 28], achieving them in vivo will require patients to take large quantities of the respective drug. Such considerations have led to recent high-throughput screening protocols to identify more potent correctors of ΔF508 trafficking [31, 32]. These efforts are primarily being conducted at Vertex Pharmaceuticals and in a separate collaborative effort centered at the University of California, San Francisco (UCSF) and are based on cell lines engineered to express ΔF508. Using sophisticated robotics, these cells are exposed to compound libraries at micromolar concentrations in multiwell plates, and chloride transport is assessed by changes in fluorescence that result from changes in membrane polarization (Vertex) or intracellular halide concentration (UCSF). Lead compounds, which can correct ΔF508 trafficking or potentiate its chloride transport function, are then optimized for potency and predicted minimization of potential toxicity by combinatorial chemistry. A number of candidate compounds that can activate ΔF508 after reduced temperature correction have recently been reported [31].

Repair of Class III and Class IV CFTR Mutations

Mutations in CFTR that lead to synthesis of a full-length protein with appropriate intracellular localization at the apical plasma membrane of epithelia, but markedly reduced chloride transport function are referred to as class III mutations. Mutations leading to class III defects are typically missense changes found in regulatory regions of CFTR, including the nucleotide binding domains (NBDs) [33]. The most common class III mutation is G551D, which is within CFTR's first NBD, is present on ~2.2% of mutant alleles and is associated with a severe phenotype and pancreatic insufficiency because of its severely limited function [3].

Initial attempts at pharmacological activation of G551D met with limited success. More recently, genistein, an isoflavone that is present in quantities of milligrams per kilogram in tofu and other soy products, has garnered significant interest. Genistein, in micromolar concentrations, enhances the chloride channel activity of both wild-type and mutant CFTRs [34], including G551D [35]. Topical perfusion of genistein onto the nasal epithelia increased chloride transport, as assessed by NPD, in non-CF subjects, as well as improved CFTR chloride transport activity (also as assessed by NPD) in vivo in 5 CF subjects with the G551D mutation (three ΔF508/G551D compound heterozygotes, one G551D homozygote, and one G551D compound heterozygote with an unidentified second allele) [35]. Thus, genistein can be thought of as a prototype of a mutant CFTR 'potentiator'.

One of genistein's disadvantages is that its effect requires relatively high concentrations (~30–50 μM). While such concentrations may be deliverable topically, these concentrations exceed the serum levels found in people eating soy-rich diets (average 150–180 nM) by 2–3 orders of magnitude [36]. These considerations have fueled a search for more potent potentiators by automated high-throughput screening in the efforts discussed above [37]. A number of promising, higher potency compounds have recently been published [38–40], and their entry into clinical trials awaits appropriate preclinical and animal testing.

There are a number of less common missense mutations of CFTR, such as R117H, that have normal intracellular localization but reduced chloride transport function [41]. Such mutations are typically referred to as class IV mutations. Compared to class III mutations, where CFTR function is absent or severely diminished, class IV mutations are usually associated with a milder clinical phenotype and pancreatic sufficiency [41]. Conceptually, class IV mutations may also be amenable to repair or improved function with CFTR potentiators such as genistein, although such data are not yet available.

Combinations of Protein Repair Agents

In addition to their trafficking defects, the class II mutations ΔF508 and N1303K may also have defects in their ability to regulate chloride transport. ΔF508 may have a reduced channel open probability [22, 23], although others have not concurred in this finding [42], while N1303K interferes with an intrinsic adenylate kinase activity of NBD-2 that regulates CFTR gating [43]. These considerations suggest a hypothesis that, in order to restore CFTR-mediated chloride transport to normal levels in CF epithelia harboring a class II allele with pharmacological agents, therapy with both a trafficking 'corrector', such as 4PBA, and chloride transport 'potentiator', such as genistein, may be necessary, as has recently been suggested [44, 45]. Similar combination strategies may also be required to enhance the repair of class I 'X' mutations by aminoglycoside antibiotics, where gentamicin can be viewed as the 'corrector'. Pilot studies to test this hypothesis are in progress in the author's group, and are examining the ability of genistein to augment the improvement in nasal epithelia chloride transport caused by 4PBA in ΔF508 homozygous and ΔF508 compound heterozygous subjects.

CFTR has many functions and it is not clear that repair of CFTR-mediated chloride transport alone will be sufficient to completely correct the dysfunction of CF epithelia; restoration of these other regulatory interactions may be required. In particular, recent evidence suggests that repair of CFTR-regulated HCO_3^- and Na^+ may be especially critical considerations.

In general, mutant CFTRs that are defective in regulation of Cl^- transport, like ΔF508, are also defective in regulation of HCO_3^- transport. Mutations associated with milder clinical phenotypes and pancreatic sufficiency, like R117H, maintained some ability to regulate Cl^- and HCO_3^- transport [27]. However, this study also presented data regarding a rare mutation, I148T, which is associated with a severe CF phenotype and pancreatic insufficiency. Interestingly, I148T was defective in regulation of HCO_3^- transport, but regulated Cl^- transport in a manner similar to wild-type CFTR [27]. These data argue that repair of Cl^- transport dysfunction alone may not be sufficient to ameliorate the CF phenotype.

Neither ΔF508 (even when its trafficking has been corrected) nor G551D regulate ENaC appropriately [25, 26, 46]. We hypothesized that augmentation of ΔF508 and G551D function with a potentiator would improve the regulatory interactions between ENaC and these mutant CFTRs. Our data were consistent with genistein, the prototype potentiator, significantly improving the regulatory interactions of ENaC with ΔF508 [25] and G551D [47] in *Xenopus* oocytes, which are maintained at reduced temperature and allow trafficking of ΔF508 to the oocyte membrane. These data also support the notion that combination therapy with a corrector and a potentiator may be required to effect full functional repair of mutant CFTR.

Strategies to Bypass Absent CFTR-Mediated Chloride Transport

Another novel strategy for treatment of CF is to target and activate other apical epithelial chloride channels, with the hypothesis that activation of these alternate channels will bypass the CFTR defect. Thus far, three distinct pathways or targets have generated considerable interest, namely purinergic receptor agonists, which can stimulate chloride transport in airway epithelia, a calcium-activated chloride channel (CaCC or ClCA), and the ClC-2 chloride channel.

Purinergic Receptor Agonists

Increases in chloride transport in response to P_2Y_2 agonists are presumably mediated by CaCC/ClCA channels [48, 49], but non-Ca^{2+}-regulated chloride transport pathways may exist as well [50]. This response to agonist is limited to the apical epithelia; basolateral exposure to agonist does not increase chloride transport [49]. Unfortunately, the metabolism of nucleotides such as ATP at the apical epithelial surface by kinases, phosphatases, and nucleotidases is very active [51–55], and likely leads to ATP having a very short $t_{1/2}$ when applied to the apical surface [56]. Longer-acting analogs, such as nonhydrolyzable homologues of ATP or UTP, may be necessary. One such analog, UTPγS, is resistant to hydrolysis and stimulates chloride transport in both CF and non-CF epithelia [57].

Initial human studies have also suggested potential efficacy of this strategy. Inhalation of UTP (INS316) caused small improvements in mucociliary clearance in adult subjects with chronic bronchitis [58, 59] and in cough clearance in subjects with primary ciliary dyskinesia [59]. In subjects with CF, a phase I dose escalation trial of a chemically stable UTP analog, INS365, suggested safety of a single dose, with the tendency of a few adult subjects to produce more sputum. Side effects included cough, wheeze, chest tightness, and a transient decrease in FEV_1 at higher doses [60]. INS37217 [P(1)-(uridine 5′)-P(4)- (2′-deoxycytidine 5′)tetraphosphate, tetrasodium salt] is a newer, metabolically stable, potent dinucleotide P_2Y_2 agonist that increases mucus transport in animal models for up to 8 h [61], and clinical trials using this compound in CF are in progress [62].

CaCC/ClCA Activators

As discussed above, CaCC/ClCA is likely the chloride channel that is ultimately stimulated in the airway epithelia in response to purinergic stimulation. These channels are also expressed at high levels in the respiratory epithelia of *cftr*−/− knockout mice, where they are hypothesized to compensate for the absence of CFTR and prevent them from developing severe pulmonary disease [63]. Thus, direct activation of CaCC/ClCA is an attractive potential target of novel therapies to bypass the CFTR defect in CF airway epithelia.

Duramycin (Moli1901) is a peptide antibiotic that increases intracellular calcium concentration in airway epithelial cells [64], as well as chloride transport [65]. When applied to the nasal epithelia in vivo in a pilot trial, duramycin acutely altered NPD consistent with increased chloride transport, although this effect was more transient and variable in subjects with CF than in non-CF subjects [66]. These data are consistent with CaCC/ClCA being a viable target of novel drugs to bypass CFTR.

ClC-2 Activators

ClC-2 is a member of the ClC family of chloride channels. It is localized in the apical membrane of the respiratory epithelia where its peak expression occurs during fetal development [67]. Overexpression of ClC-2 can improve chloride transport in CF epithelial cells in vitro [68], and ClC-2-mediated chloride transport can be pharmacologically activated in a number of airway epithelial cells by arachadonic acid and acid-activated omeprazole [69]. Lubiprostone (SPI-0211) is a novel bicyclic fatty acid that is being developed for the treatment of bowel dysfunction. Recent data suggest that this compound is also a highly potent activator of ClC-2 [70].

If ClC-2 activation either can or does compensate for the absence of CFTR, then it is reasonable to hypothesize that absence of ClC-2 should increase mortality from CF. Initial data addressing this hypothesis were recently published, and were somewhat surprising. In *cftr*−/− knockout mice expressing a ΔF508 allele (ΔF508 mice), additional knockout of ClC-2 actually improved the survival of the ΔF508 mice but had no effect on the survival of *cftr*−/− knockout mice [71]. The mechanism by which this occurs in not readily apparent, but further development of ClC-2 as a target for novel pharmaceuticals will need to bear these observations in mind.

Strategies to Inhibit ENaC

As discussed above, and in chapter 15, ENaC hyperactivity in the apical plasma membrane of airway epithelia is hypothesized to be a critical determinant of the pathophysiology of the CF airway [72–75]. Thus, inhibition of ENaC-mediated sodium transport should improve airway clearance in the CF airway. This hypothesis is supported by data suggesting that benzamil, a specific blocker of ENaC, can increase mucociliary transport in porcine airways ex vivo where liquid secretion from submucosal glands has been pharmacologically inhibited [76].

The availability of amiloride, a specific blocker of ENaC and approved agent for use as a diuretic, allowed initial early human trials aimed at inhibiting ENaC in the CF airway. The early results were disappointing [77, 78] and, in retrospect, were likely due to disfavorable pharmacokinetics of amiloride delivery to the airway by nebulization, where a $t_{1/2}$ for clearance of amiloride was ~23 min [77]. The rapid clearance of amiloride from the airway may have also prevented a further increase in the rate of mucociliary clearance when aerosolized amiloride was added to aerosolized UTP [79]. These data have spurred interest in the development of amiloride analogs such as benzamil, which has a ~2-fold longer $t_{1/2}$ for inhibition of sodium transport in the nasal epithelia than does amiloride [80], although this prolongation may still be insufficient for clinical use.

An alternate strategy for inhibiting ENaC activity has emerged from novel, recent observations that ENaC-mediated sodium transport is activated by proteolysis of ENaC's extracellular domains by ENaC channel-activating proteases (CAPs) [81–84], and the human homologue of CAP, prostasin [85]. Such proteolysis and activation of ENaC may also occur during trafficking of ENaC to the apical plasma membrane by the intracellular protease furin [86, 87], or by extracellular proteases such as trypsin and elastase [88], the latter of which is especially abundant in the CF airway as a result of the chronic neutrophilic inflammation.

Prostasin is endogenously expressed in airway epithelia [89, 90], and thus is a logical potential target for pharmaceutical intervention. The Kunitz-type serine protease inhibitor, aprotinin, inhibits prostasin, and decreases ENaC-mediated current in both CF and non-CF airway epithelia [85]. A novel, recombinant Kunitz-type serine protease inhibitor, bikunin (BAY 39-9437) has recently been developed, and, like aprotinin, also inhibits ENaC-mediated sodium transport in CF and non-CF airway epithelia [91].

Conclusions

Investigation of the molecular, cellular, and airway pathophysiology of CF has yielded a significantly greater understanding of the function of CFTR. Such an understanding has allowed rational identification of potential targets for pharmaceutical intervention to either repair the broken protein or bypass the consequences of its absence. Such novel strategies are thus directed at repairing the more fundamental causes of CF pathophysiology, rather than the more symptomatic therapies in common use today, and have tremendous promise of potential future benefit.

Acknowledgements

The author's research is supported by grants from the National Institutes of Health/National Institute of Diabetes and Digestive and Kidney Diseases (R01 DK58046), the Cystic Fibrosis Foundation, and an Established Investigator Award from the American Heart Association.

References

1 Rubenstein RC, Zeitlin PL: Use of protein repair therapy in the treatment of cystic fibrosis. Curr Opin Pediatr 1998;10:250–255.

2 Zeitlin PL: Novel pharmacologic therapies for cystic fibrosis. J Clin Invest 1999;103:447–452.

3 Welsh MJ, Smith AE: Molecular mechanisms of CFTR chloride channel dysfunction in cystic fibrosis. Cell 1993;73:1251–1254.

4 Howard M, Frizzell RA, Bedwell DM: Aminoglycoside antibiotics restore CFTR function by overcoming premature stop mutations. Nat Med 1996;2:467–469.

5 Bedwell DM, Kaenjak A, Benos DJ, Bebok Z, Bubien JK, Hong J, Tousson A, Clancy JP, Sorscher EJ: Suppression of a CFTR premature stop mutation in a bronchial epithelial cell line. Nat Med 1997;3:1280–1284.

6 Du M, Jones JR, Lanier J, Keeling KM, Lindsey JR, Tousson A, Bebok Z, Whitsett JA, Dey CR, Colledge WH, Evans MJ, Sorscher EJ, Bedwell DM: Aminoglycoside suppression of a premature stop mutation in a Cftr−/− mouse carrying a human CFTR-G542X transgene. J Mol Med 2002;80:595–604.

7 Clancy JP, Bebok Z, Ruiz F, King C, Jones J, Walker L, Greer H, Hong J, Wing L, Macaluso M, Lyrene R, Sorscher EJ, Bedwell DM: Evidence that systemic gentamicin suppresses premature stop mutations in patients with cystic fibrosis. Am J Respir Crit Care Med 2001;163:1683–1692.

8 Wilschanski M, Yahav Y, Yaacov Y, Blau H, Bentur L, Rivlin J, Aviram M, Bdolah-Abram T, Bebok Z, Shushi L, Kerem B, Kerem E: Gentamicin-induced correction of CFTR function in patients with cystic fibrosis and

CFTR stop mutations. N Engl J Med 2003; 349:1433–1441.

9   Denning GM, Anderson MA, Amara JF, Marshall J, Smith AE, Welsh MJ: Processing of mutant cystic fibrosis transmembrane regulator is temperature-sensitive. Nature 1992; 358:761–764.

10 Cheng SH, Gregory RJ, Marshall J, Paul S, Souza DW, White GA, O'Riordan CR, Smith AE: Defective intracellular transport and processing of CFTR is the molecular basis of most cystic fibrosis. Cell 1990;63:827–834.

11 Pasyk EA, Foskett JK: Mutant (delta F508) cystic fibrosis transmembrane conductance regulator Cl⁻ channel is functional when retained in endoplasmic reticulum of mammalian cells. J Biol Chem 1995;270: 12347–12350.

12 Cheng SH, Fang SL, Zabner J, Marshall J, Piraino S, Schiavi SC, Jefferson DM, Welsh MJ, Smith AE: Functional activation of the cystic fibrosis trafficking mutant delta F508-CFTR by overexpression. Am J Physiol 1995; 268:L615–L624.

13 Perrine SP, Ginder GD, Faller DV, Dover GJ, Ikuta T, Witkowski HE, Cai S-P, Vichinsky EP, Oliveri NF: A short-term trial of butyrate to stimulate fetal-globin-gene expression in the β-globin disorders. N Engl J Med 1993;328: 81–86.

14 Zeitlin PL, Diener-West M, Rubenstein RC, Boyle MP, Lee CK, Brass-Ernst L: Evidence of CFTR function in cystic fibrosis after systemic administration of 4-phenylbutyrate. Mol Ther 2002;6:119–126.

15 Brusilow SW: Phenylacetylglutamine may replace urea as a vehicle for waste nitrogen excretion. Pediatr Res 1991;29:147–150.

16 Rubenstein RC, Egan ME, Zeitlin PL: In vitro pharmacologic restoration of CFTR-mediated chloride transport with sodium 4-phenylbutyrate in cystic fibrosis epithelial cells containing delta F508-CFTR. J Clin Invest 1997; 100:2457–2465.

17 Rubenstein RC, Zeitlin PL: Sodium 4-phenylbutyrate downregulates Hsc70: Implications for intracellular trafficking of DeltaF508-CFTR. Am J Physiol Cell Physiol 2000;278: C259–C267.

18 Rubenstein RC, Lyons BM: Sodium 4-phenylbutyrate downregulates HSC70 expression by facilitating mRNA degradation. Am J Physiol Lung Cell Mol Physiol 2001;281:L43–L51.

19 Choo-Kang LR, Zeitlin PL: Induction of HSP70 promotes DeltaF508 CFTR trafficking. Am J Physiol Lung Cell Mol Physiol 2001;281:L58–L68.

20 Rubenstein RC, Zeitlin PL: A pilot clinical trial of oral sodium 4-phenylbutyrate (Buphenyl) in deltaF508-homozygous cystic fibrosis patients: Partial restoration of nasal epithelial CFTR function. Am J Respir Crit Care Med 1998;157:484–490.

21 Wilschanski M, Famini C, Blau H, Rivlin J, Augarten A, Avital A, Kerem B, Kerem E: A pilot study of the effect of gentamicin on nasal potential difference measurements in cystic fibrosis patients carrying stop mutations. Am J Respir Crit Care Med 2000;161: 860–865.

22 Dalemans W, Barbry P, Champigny G, Jallat S, Dott K, Dreyer D, Crystal RG, Pavirani A, Lecocq JP, Lazdunski M: Altered chloride ion channel kinetics associated with the delta F508 cystic fibrosis mutation. Nature 1991; 354:526–528.

23 Egan ME, Schwiebert EM, Guggino WB: Differential expression of ORCC and CFTR induced by low temperature in CF airway epithelial cells. Am J Physiol 1995;268(1 Pt 1): C243–C251.

24 Sharma M, Benharouga M, Hu W, Lukacs GL: Conformational and temperature-sensitive stability defects of the delta F508 cystic fibrosis transmembrane conductance regulator in post-endoplasmic reticulum compartments. J Biol Chem 2001;276:8942–8950.

25 Suaud L, Li J, Jiang Q, Rubenstein RC, Kleyman TR: Genistein restores functional interactions between Delta F508-CFTR and ENaC in Xenopus oocytes. J Biol Chem 2002; 277:8928–8933.

26 Mall M, Hipper A, Greger R, Kunzelmann K: Wild type but not deltaF508 CFTR inhibits Na⁺ conductance when coexpressed in Xenopus oocytes. FEBS Lett 1996;381/1–2: 47–52.

27 Choi JY, Muallem D, Kiselyov K, Lee MG, Thomas PJ, Muallem S: Aberrant CFTR-dependent HCO₃⁻ transport in mutations associated with cystic fibrosis. Nature 2001; 410:94–97.

28 Egan ME, Pearson M, Weiner SA, Rajendran V, Rubin D, Glockner-Pagel J, Canny S, Du K, Lukacs GL, Caplan MJ: Curcumin, a major constituent of turmeric, corrects cystic fibrosis defects. Science 2004;304:600–602.

29 Egan ME, Glockner-Pagel J, Ambrose C, Cahill PA, Pappoe L, Balamuth N, Cho E, Canny S, Wagner CA, Geibel J, Caplan MJ: Calcium-pump inhibitors induce functional surface expression of Delta F508-CFTR protein in cystic fibrosis epithelial cells. Nat Med 2002;8:485–492.

30 Song Y, Sonawane ND, Salinas D, Qian L, Pedemonte N, Galietta LJ, Verkman AS: Evidence against rescue of defective DeltaF508-CFTR cellular processing by curcumin in cell culture and mouse models. J Biol Chem 2004;279:40629–40633.

31 Yang H, Shelat AA, Guy RK, Gopinath VS, Ma T, Du K, Lukacs GL, Taddei A, Folli C, Pedemonte N, Galietta LJ, Verkman AS: Nanomolar affinity small molecule correctors of defective Delta F508-CFTR chloride channel gating. J Biol Chem 2003;278: 35079–35085.

32 Gonzalez JE, Oades K, Leychkis Y, Harootunian A, Negulescu PA: Cell-based assays and instrumentation for screening ion-channel targets. Drug Discov Today 1999;4: 431–439.

33 Logan J, Hiestand D, Daram P, Huang Z, Muccio DD, Hartman J, Haley B, Cook WJ, Sorscher EJ: Cystic fibrosis transmembrane conductance regulator mutations that disrupt nucleotide binding. J Clin Invest 1994;94: 228–236.

34 Hwang TC, Wang F, Yang IC, Reenstra WW: Genistein potentiates wild-type and delta F508-CFTR channel activity. Am J Physiol 1997;273:C988–C998.

35 Illek B, Zhang L, Lewis NC, Moss RB, Dong JY, Fischer H: Defective function of the cystic fibrosis-causing missense mutation G551D is recovered by genistein. Am J Physiol 1999; 277:C833–C839.

36 Urban D, Irwin W, Kirk M, Markiewicz MA, Myers R, Smith M, Weiss H, Grizzle WE, Barnes S: The effect of isolated soy protein on plasma biomarkers in elderly men with elevated serum prostate specific antigen. J Urol 2001;165:294–300.

37 Galietta LV, Jayaraman S, Verkman AS: Cell-based assay for high-throughput quantitative screening of CFTR chloride transport agonists. Am J Physiol Cell Physiol 2001;281: C1734–C1742.

38 Caci E, Folli C, Zegarra-Moran O, Ma T, Springsteel MF, Sammelson RE, Nantz MH, Kurth MJ, Verkman AS, Galietta LJ: CFTR activation in human bronchial epithelial cells by novel benzoflavone and benzimidazolone compounds. Am J Physiol Lung Cell Mol Physiol 2003;285:L180–L188.

39 Ma T, Vetrivel L, Yang H, Pedemonte N, Zegarra-Moran O, Galietta LJ, Verkman AS: High-affinity activators of cystic fibrosis transmembrane conductance regulator (CFTR) chloride conductance identified by high-throughput screening. J Biol Chem 2002; 277:37235–37241.

40 Galietta LJ, Springsteel MF, Eda M, Niedzinski EJ, By K, Haddadin MJ, Kurth MJ, Nantz MH, Verkman AS: Novel CFTR chloride channel activators identified by screening of combinatorial libraries based on flavone and benzoquinolizinium lead compounds. J Biol Chem 2001;276:19723–19728.

41 Sheppard DN, Rich DP, Ostedgaard LS, Gregory RJ, Smith AE, Welsh MJ: Mutations in CFTR associated with mild-disease-form Cl⁻ channels with altered pore properties. Nature 1993;362:160–164.

42 Li C, Ramjeesingh M, Reyes E, Jensen T, Chang X, Rommens JM, Bear CE: The cystic fibrosis mutation (delta F508) does not influence the chloride channel activity of CFTR. Nat Genet 1993;3:311–316.

43 Randak C, Welsh MJ: An intrinsic adenylate kinase activity regulates gating of the ABC transporter CFTR. Cell 2003;115:837–850.

44 Lim M, McKenzie K, Floyd AD, Kwon E, Zeitlin PL: Modulation of DeltaF508 CFTR trafficking and function with 4-PBA and flavonoids. Am J Respir Cell Mol Biol 2004; 31:351–357.

45 Illek B, Zhang L, Lewis NC, Moss RB, Dong JY, Fischer H: Defective function of the cystic fibrosis-causing missense mutation G551D is recovered by genistein. Am J Physiol 1999; 277:C833–C839.

46 Yan W, Samaha FF, Ramkumar M, Kleyman TR, Rubenstein RC: Cystic fibrosis transmembrane conductance regulator differentially regulates human and mouse epithelial sodium channels in Xenopus oocytes. J Biol Chem 2004;279:23183–23192.

47 Suaud L, Carattino M, Kleyman TR, Rubenstein RC: Genistein improves regula-

tory interactions between G551D-cystic fibrosis transmembrane conductance regulator and the epithelial sodium channel in Xenopus oocytes. J Biol Chem 2002;277:50341–50347.

48 Tarran R, Loewen ME, Paradiso AM, Olsen JC, Gray MA, Argent BE, Boucher RC, Gabriel SE: Regulation of murine airway surface liquid volume by CFTR and Ca^{2+}-activated Cl^- conductances. J Gen Physiol 2002; 120:407–418.

49 Paradiso AM, Ribeiro CM, Boucher RC: Polarized signaling via purinoceptors in normal and cystic fibrosis airway epithelia. J Gen Physiol 2001;117:53–67.

50 Stutts MJ, Fitz JG, Paradiso AM, Boucher RC: Multiple modes of regulation of airway epithelial chloride secretion by extracellular ATP. Am J Physiol 1994;267:C1442–C1451.

51 Donaldson SH, Picher M, Boucher RC: Secreted and cell-associated adenylate kinase and nucleoside diphosphokinase contribute to extracellular nucleotide metabolism on human airway surfaces. Am J Respir Cell Mol Biol 2002;26:209–215.

52 Picher M, Burch LH, Hirsh AJ, Spychala J, Boucher RC: Ecto 5′-nucleotidase and nonspecific alkaline phosphatase. Two AMP-hydrolyzing ectoenzymes with distinct roles in human airways. J Biol Chem 2003;278: 13468–13479.

53 Lazarowski ER, Boucher RC, Harden TK: Constitutive release of ATP and evidence for major contribution of ecto-nucleotide pyrophosphatase and nucleoside diphosphokinase to extracellular nucleotide concentrations. J Biol Chem 2000;275:31061–31068.

54 Lazarowski ER, Homolya L, Boucher RC, Harden TK: Identification of an ecto-nucleoside diphosphokinase and its contribution to interconversion of P2 receptor agonists. J Biol Chem 1997;272:20402–20407.

55 Picher M, Boucher RC: Human airway ecto-adenylate kinase. A mechanism to propagate ATP signaling on airway surfaces. J Biol Chem 2003;278:11256–11264.

56 Picher M, Burch LH, Boucher RC: Metabolism of P2 receptor agonists in human airways: Implications for mucociliary clearance and cystic fibrosis. J Biol Chem 2004; 279:20234–20241.

57 Lazarowski ER, Watt WC, Stutts MJ, Brown HA, Boucher RC, Harden TK: Enzymatic synthesis of UTP gamma S, a potent hydrolysis resistant agonist of P2U-purinoceptors. Br J Pharmacol 1996;117:203–209.

58 Bennett WD, Zeman KL, Foy C, Shaffer CL, Johnson FL, Regnis JA, Sannuti A, Johnson J: Effect of aerosolized uridine 5′-triphosphate on mucociliary clearance in mild chronic bronchitis. Am J Respir Crit Care Med 2001; 164:302–306.

59 Noone PG, Bennett WD, Regnis JA, Zeman KL, Carson JL, King M, Boucher RC, Knowles MR: Effect of aerosolized uridine-5′-triphosphate on airway clearance with cough in patients with primary ciliary dyskinesia. Am J Respir Crit Care Med 1999;160: 144–149.

60 Noone PG, Hamblett N, Accurso F, Aitken ML, Boyle M, Dovey M, Gibson R, Johnson C, Kellerman D, Konstan MW, Milgram L, Mundahl J, Retsch-Bogort G, Rodman D, Williams-Warren J, Wilmott RW, Zeitlin P, Ramsey B: Safety of aerosolized INS 365 in patients with mild to moderate cystic fibrosis: Results of a phase I multi-center study. Pediatr Pulmonol 2001;32:122–128.

61 Yerxa BR, Sabater JR, Davis CW, Stutts MJ, Lang-Furr M, Picher M, Jones AC, Cowlen M, Dougherty R, Boyer J, Abraham WM, Boucher RC: Pharmacology of INS37217 [P(1)-(uridine 5′)-P(4)- (2′-deoxycytidine 5′)tetraphosphate, tetrasodium salt], a next-generation P2Y(2) receptor agonist for the treatment of cystic fibrosis. J Pharmacol Exp Ther 2002;302:871–880.

62 Kellerman D, Evans R, Mathews D, Shaffer C: Inhaled P2Y2 receptor agonists as a treatment for patients with cystic fibrosis lung disease. Adv Drug Deliv Rev 2002;54:1463–1474.

63 Grubb BR, Vick RN, Boucher RC: Hyperabsorption of Na^+ and raised Ca^{2+}-mediated Cl^- secretion in nasal epithelia of CF mice. Am J Physiol 1994;266:C1478–C1483.

64 Cloutier MM, Guernsey L, Sha'afi RI: Duramycin increases intracellular calcium in airway epithelium. Membr Biochem 1993;10: 107–118.

65 Cloutier MM, Guernsey L, Mattes P, Koeppen B: Duramycin enhances chloride secretion in airway epithelium. Am J Physiol 1990;259: C450–C454.

66 Zeitlin PL, Boyle MP, Guggino WB, Molina L: A phase I trial of intranasal Moli1901 for cystic fibrosis. Chest 2004;125:143–149.

67 Murray CB, Morales MM, Flotte TR, McGrath-Morrow SA, Guggino WB, Zeitlin PL: ClC-2: A developmentally dependent chloride channel expressed in the fetal lung and downregulated after birth. Am J Respir Cell Mol Biol 1995;12:597–604.

68 Schwiebert EM, Cid-Soto LP, Stafford D, Carter M, Blaisdell CJ, Zeitlin PL, Guggino WB, Cutting GR: Analysis of ClC-2 channels as an alternative pathway for chloride conduction in cystic fibrosis airway cells. Proc Natl Acad Sci USA 1998;95:3879–3884.

69 Cuppoletti J, Tewari KP, Sherry AM, Kupert EY, Malinowska DH: ClC-2 Cl^- channels in human lung epithelia: Activation by arachidonic acid, amidation, and acid-activated omeprazole. Am J Physiol Cell Physiol 2001; 281:C46–C54.

70 Cuppoletti J, Malinowska DH, Tewari KP, Li QJ, Sherry AM, Patchen ML, Ueno R: SPI-0211 activates T84 cell Cl^- transport and recombinant human ClC-2 Cl^- currents. Am J Physiol Cell Physiol 2004;287: C1173–C1183.

71 Zdebik AA, Cuffe JE, Bertog M, Korbmacher C, Jentsch TJ: Additional disruption of the ClC-2 Cl^- channel does not exacerbate the cystic fibrosis phenotype of cystic fibrosis transmembrane conductance regulator mouse models. J Biol Chem 2004;279: 22276–22283.

72 Matsui H, Grubb BR, Tarran R, Randell SH, Gatzy JT, Davis CW, Boucher RC: Evidence for periciliary liquid layer depletion, not abnormal ion composition, in the pathogenesis of cystic fibrosis airways disease. Cell 1998;95:1005–1015.

73 Worlitzsch D, Tarran R, Ulrich M, Schwab U, Cekici A, Meyer KC, Birrer P, Bellon G, Berger J, Weiss T, Botzenhart K, Yankaskas JR, Randell S, Boucher RC, Doring G: Effects of reduced mucus oxygen concentration in airway Pseudomonas infections of cystic fibrosis patients. J Clin Invest 2002;109: 317–325.

74 Tarran R, Grubb BR, Parsons D, Picher M, Hirsh AJ, Davis CW, Boucher RC: The CF salt controversy: In vivo observations and therapeutic approaches. Mol Cell 2001;8: 149–158.

75 Mall M, Grubb BR, Harkema JR, O'Neal WK, Boucher RC: Increased airway epithelial Na^+ absorption produces cystic fibrosis-like lung disease in mice. Nat Med 2004;10: 487–493.

76 Ballard ST, Trout L, Mehta A, Inglis SK: Liquid secretion inhibitors reduce mucociliary transport in glandular airways. Am J Physiol Lung Cell Mol Physiol 2002;283: L329–L335.

77 Noone PG, Regnis JA, Liu X, Brouwer KL, Robinson M, Edwards L, Knowles MR: Airway deposition and clearance and systemic pharmacokinetics of amiloride following aerosolization with an ultrasonic nebulizer to normal airways. Chest 1997;112: 1283–1290.

78 Graham A, Hasani A, Alton EW, Martin GP, Marriott C, Hodson ME, Clarke SW, Geddes DM: No added benefit from nebulized amiloride in patients with cystic fibrosis. Eur Respir J 1993;6:1243–1248.

79 Olivier KN, Bennett WD, Hohneker KW, Zeman KL, Edwards LJ, Boucher RC, Knowles MR: Acute safety and effects on mucociliary clearance of aerosolized uridine 5′-triphosphate $+/-$ amiloride in normal human adults. Am J Respir Crit Care Med 1996;154:217–223.

80 Hofmann T, Stutts MJ, Ziersch A, Ruckes C, Weber WM, Knowles MR, Lindemann H, Boucher RC: Effects of topically delivered benzamil and amiloride on nasal potential difference in cystic fibrosis. Am J Respir Crit Care Med 1998;157:1844–1849.

81 Vallet V, Pfister C, Loffing J, Rossier BC: Cell-surface expression of the channel activating protease xCAP-1 is required for activation of ENaC in the Xenopus oocyte. J Am Soc Nephrol 2002;13:588–594.

82 Vallet V, Chraibi A, Gaeggeler HP, Horisberger JD, Rossier BC: An epithelial serine protease activates the amiloride-sensitive sodium channel. Nature 1997;389: 607–610.

83 Vuagniaux G, Vallet V, Jaeger NF, Hummler E, Rossier BC: Synergistic activation of ENaC by three membrane-bound channel-activating serine proteases (mCAP1, mCAP2, and mCAP3) and serum- and glucocorticoid-regulated kinase (Sgk1) in Xenopus oocytes. J Gen Physiol 2002;120:191–201.

84 Vuagniaux G, Vallet V, Jaeger NF, Pfister C, Bens M, Farman N, Courtois-Coutry N,

Vandewalle A, Rossier BC, Hummler E: Activation of the amiloride-sensitive epithelial sodium channel by the serine protease mCAP1 expressed in a mouse cortical collecting duct cell line. J Am Soc Nephrol 2000;11: 828–834.

85 Adachi M, Kitamura K, Miyoshi T, Narikiyo T, Iwashita K, Shiraishi N, Nonoguchi H, Tomita K: Activation of epithelial sodium channels by prostasin in Xenopus oocytes. J Am Soc Nephrol 2001;12:1114–1121.

86 Hughey RP, Mueller GM, Bruns JB, Kinlough CL, Poland PA, Harkleroad KL, Carattino MD, Kleyman TR: Maturation of the epithelial Na$^+$ channel involves proteolytic processing of the alpha- and gamma-subunits. J Biol Chem 2003;278:37073–37082.

87 Hughey RP, Bruns JB, Kinlough CL, Harkleroad KL, Tong Q, Carattino MD, Johnson JP, Stockand JD, Kleyman TR: Epithelial sodium channels are activated by furin-dependent proteolysis. J Biol Chem 2004;279:18111–18114.

88 Caldwell RA, Boucher RC, Stutts MJ: Serine protease activation of near-silent epithelial Na$^+$ channels. Am J Physiol Cell Physiol 2004;286:C190–C194.

89 Donaldson SH, Hirsh A, Li DC, Holloway G, Chao J, Boucher RC, Gabriel SE: Regulation of the epithelial sodium channel by serine proteases in human airways. J Biol Chem 2002; 277:8338–8345.

90 Tong Z, Illek B, Bhagwandin VJ, Verghese GM, Caughey GH: Prostasin, a membrane-anchored serine peptidase, regulates sodium currents in JME/CF15 cells, a cystic fibrosis airway epithelial cell line. Am J Physiol Lung Cell Mol Physiol 2004;287: L928–L935.

91 Bridges RJ, Newton BB, Pilewski JM, Devor DC, Poll CT, Hall RL: Na$^+$ transport in normal and CF human bronchial epithelial cells is inhibited by BAY 39-9437. Am J Physiol Lung Cell Mol Physiol 2001;281: L16–L23.

92 Denning GM, Anderson MP, Amara JF, Marshall J, Smith AE, Welsh MJ: Processing of mutant cystic fibrosis transmembrane conductance regulator is temperature-sensitive. Nature 1992;358:761–764.

93 Brown CR, Hong-Brown LQ, Biwersi J, Verkman AS, Welch WJ: Chemical chaperones correct the mutant phenotype of the delta F508 cystic fibrosis transmembrane conductance regulator protein. Cell Stress Chaperones 1996;1:117–125.

94 Sato S, Ward CL, Krouse ME, Wine JJ, Kopito RR: Glycerol reverses the misfolding phenotype of the most common cystic fibrosis mutation. J Biol Chem 1996;271: 635–638.

95 Bebok Z, Venglarik CJ, Panczel Z, Jilling T, Kirk KL, Sorscher EJ: Activation of DeltaF508 CFTR in an epithelial monolayer. Am J Physiol 1998;275:C599–C607.

96 Zhang XM, Wang XT, Yue H, Leung SW, Thibodeau PH, Thomas PJ, Guggino SE: Organic solutes rescue the functional defect in delta F508 cystic fibrosis transmembrane conductance regulator. J Biol Chem 2003;278: 51232–51242.

97 Moyer BD, Loffing-Cueni D, Loffing J, Reynolds D, Stanton BA: Butyrate increases apical membrane CFTR but reduces chloride secretion in MDCK cells. Am J Physiol 1999; 277(2 Pt 2):F271–F276.

98 Jiang C, Fang SL, Xiao YF, O'Connor SP, Nadler SG, Lee DW, Jefferson DM, Kaplan JM, Smith AE, Cheng SH: Partial restoration of cAMP-stimulated CFTR chloride channel activity in DeltaF508 cells by deoxyspergualin. Am J Physiol 1998;275(1 Pt 1): C171–C178.

99 Kelley TJ, Al Nakkash L, Cotton CU, Drumm ML: Activation of endogenous deltaF508 cystic fibrosis transmembrane conductance regulator by phosphodiesterase inhibition. J Clin Invest 1996;98:513–520.

100 Maitra R, Shaw CM, Stanton BA, Hamilton JW: Increased functional cell surface expression of CFTR and DeltaF508-CFTR by the anthracycline doxorubicin. Am J Physiol Cell Physiol 2001;280:C1031–C1037.

101 Jacobson KA, Guay-Broder C, van Galen PJ, Gallo-Rodriguez C, Melman N, Jacobson MA, Eidelman O, Pollard HB: Stimulation by alkylxanthines of chloride efflux in CFPAC-1 cells does not involve A1 adenosine receptors. Biochemistry 1995;34: 9088–9094.

102 Zaman K, McPherson M, Vaughan J, Hunt J, Mendes F, Gaston B, Palmer LA: S-Nitrosoglutathione increases cystic fibrosis transmembrane regulator maturation. Biochem Biophys Res Commun 2001;284:65–70.

103 Andersson C, Gaston B, Roomans GM: S-Nitrosoglutathione induces functional Delta F508-CFTR in airway epithelial cells. Biochem Biophys Res Commun 2002;297: 552–557.

104 Dormer RL, Derand R, McNeilly CM, Mettey Y, Bulteau-Pignoux L, Metaye T, Vierfond JM, Gray MA, Galietta LJ, Morris MR, Pereira MM, Doull IJ, Becq F, McPherson MA: Correction of delF508-CFTR activity with benzo(c)quinolizinium compounds through facilitation of its processing in cystic fibrosis airway cells. J Cell Sci 2001;114: 4073–4081.

105 Kelley TJ, Thomas K, Milgram LJ, Drumm ML: In vivo activation of the cystic fibrosis transmembrane conductance regulator mutant deltaF508 in murine nasal epithelium. Proc Natl Acad Sci USA 1997;94: 2604–2608.

106 Fischer H, Fukuda N, Barbry P, Illek B, Sartori C, Matthay MA: Partial restoration of defective chloride conductance in DeltaF508 CF mice by trimethylamine oxide. Am J Physiol Lung Cell Mol Physiol 2001;281: L52–L57.

Prof. Ronald C. Rubenstein, MD, PhD
Pulmonary Medicine-Abramson 410C
Children's Hospital of Philadelphia
34th St. and Civic Center Blvd.
Philadelphia, PA 19104 (USA)
Tel. +1 215 590 1281, Fax +1 215 590 1283
E-Mail rrubenst@mail.med.upenn.edu

Bush A, Alton EWFW, Davies JC, Griesenbach U, Jaffe A (eds): Cystic Fibrosis in the 21st Century.
Prog Respir Res. Basel, Karger, 2006, vol 34, pp 221–229

Gene and Stem Cell Therapy

A. Christopher Boyd[a,b]

[a]Medical Sciences (Medical Genetics), University of Edinburgh, Molecular Medicine Centre,
Western General Hospital, Edinburgh; [b]UK Cystic Fibrosis Gene Therapy Consortium

Abstract

Gene and stem cell therapy are being developed as novel
treatments for cystic fibrosis (CF). In gene therapy, the therapeutic nucleic acid is delivered to terminally differentiated
epithelial cells in the airways. While technically less demanding, this approach has the drawback that therapy must be continually re-administered because of target cell turnover.
Direct airway administration is also faced with powerful local
and systemic defences and the lung's own extensive armamentarium. Viral delivery strategies are attractive because of
the relative efficiency of transduction, but there is no easy
solution to the problem of avoiding immune responses to
re-administration. Such responses are less significant for non-
viral delivery agents, but these vectors are not yet as efficient,
despite improvements in their targeting properties. Augmenting delivery may be achievable through physical mobilizing
strategies and the use of adjuncts such as mucolytics. The low
expression levels of CFTR, and the episodic nature of CF clinical disease conspire to make measurement of the efficacy of
gene therapy difficult, requiring specialized direct and indirect
assays to be devised. For stem cell therapy, the aim is to correct the genetic defect in perpetually regenerating cells that
can replenish the epithelium, with the promise of lifelong
remediation in the treated organ. Current knowledge of lung
stem cell biology is scant; progress depends on an improvement in our understanding of how the lung regenerates,
and the development of relevant empirical models. Stem cell
correction by direct in vivo targeting of cells in the airway is
conceivable, but might be more efficiently achieved by ex vivo
correction of stem cells harvested from sources such as the
haematopoietic system for subsequent repopulation of the
airways.

Gene Therapy

The principle of gene therapy – that effective treatment
of a genetic disorder could be achieved by delivery of
appropriate DNA or RNA to compensate for the defective
gene – remains seductive. The fact that cystic fibrosis gene
therapy (CFGT) is yet to emerge as a clinical reality,
despite the conduct of many human trials [1] since the
demonstration that mutant cystic fibrosis transmembrane
conductance regulator (CFTR) could be complemented by
gene delivery [2], shows that converting principle to practice is difficult. Significant advances have, nonetheless,
been made in our understanding of the barriers to airway
gene delivery, with concomitant improvements to gene
transfer agents (GTAs): progress that warrants a new wave
of clinical trials.

The Target Cells

As pulmonary disease is the cause of most CF morbidity, it is clear that the priority target for somatic CFGT is
the lung. All CFGT lung trials have used lumenal administration, since access to the epithelium is most straightforward via this route, with aerosolization being favoured over
instillation both as a less invasive and less inflammatory
means of dispersal [3]. Systemic delivery has also been
evaluated pre-clinically, but this otherwise attractive route
has several problems which probably rule it out for CFGT

in the immediate future. Firstly, many non-viral GTAs are sensitive to serum; secondly, there is a risk that GTAs would reach other organs including the gonads; thirdly, access to airway epithelial cells (AECs) is extremely inefficient [4, 5].

CFTR expression is low and concentrated at the lumenal surface of polarized AECs, but its pattern of expression is highly variegated even at the cellular level [6]. The serous cells of submucosal glands, which occur predominantly in the conducting airways, display most CFTR expression [7]. Expression is also high in certain serous cells of the surface epithelium [8]. Because gland hyperplasia and blockage is an early sign of the disease, it was originally thought that CF was a consequence of gland dysfunction. However, the lack of CFTR expression in CF surface epithelium clearly also contributes, since CF lung pathology starts in the gland-free bronchioles [9]. CFTR expression is high in the maturing lung in utero, suggesting a developmental role (see chapter 7); this has raised the possibility that CFGT in adults may be unable to fully reverse the defect [10]. Gene transfer in utero to fetal sheep airways is relatively efficient [11], but profound ethical and practical considerations argue against this approach being adopted for clinical CFGT.

It is increasingly accepted that maintenance of appropriate airway surface liquid (ASL) height to promote efficient mucociliary clearance (MCC) is the principal process undermined in the CF lung [12] (see chapter 15). Planar air liquid interface cultures of well-differentiated primary AECs derived from CF patients display the ASL height diminution seen in vivo: significantly, the ASL height can be restored by transduction with viral-borne CFTR [13]. This implies that in vivo restoration of CFTR function to surface epithelium alone may be sufficient to reverse the ASL height defect, and supports the notion that topically delivered CFGT would be beneficial even if unable to penetrate deeply into diseased glands.

Most current CFGT vectors are likely to express CFTR in all cells targeted. Results from experiments in model systems predict that the relatively widely distributed expression this implies – although at variance with the observed pattern – should ameliorate the CF phenotype. Vectors providing more targeted expression may give superior correction [14–16].

Barriers

GTAs, in contrast to conventional small molecule drugs that readily diffuse through cells and their environment, are large macromolecular assemblages that are impeded by physical and biological barriers at every level. In addition, because airways have evolved specifically to attack or clear inhaled particles, CFGT must contend extracellularly with active barriers (MCC, immune surveillance, inflammatory response, opsonization) as well as passive barriers (sputum, mucus, glycocalyx) [17]. There are many complex immunological responses to GTAs in the lung. On first administration, acute non-specific inflammation, combined with the innate immune response, acts to rapidly clear GTAs and suppress promoters of viral origin. In the longer term, cellular and humoral immune responses conspire to dampen the effectiveness of subsequent gene delivery [18]. Viruses vary in the extent to which they have evolved to penetrate passive extracellular barriers: for example, Sendai virus (SeV) is far less impeded by sputum than adenovirus (Ad) [19]. In general, however, non-viral GTAs are much more prone to become entrapped in the extracellular milieu [20].

Having traversed the extracellular milieu, GTAs must enter the cell through the plasma membrane. Viruses gain access via cell surface receptors, but in AECs of the lower airways the receptors for many viruses (including Ad [21]) are sequestered on the basolateral surface beneath tight junctions and are hence lumenally inaccessible. Non-viral GTAs typically enter cells via a non-receptor-mediated uptake mechanism, but this process is particularly inefficient (at least for cationic liposomes) in polarized AECs, where both binding and uptake are reduced, possibly because of unfavourable surface charge and restricted trafficking [22, 23].

Once inside the cell, GTAs must be released from the endosome, traverse the cytoplasm and enter the nucleus. Viruses have evolved an impressive array of strategies to achieve this [24], and are thus far less susceptible to intracellular barriers. Significantly, no virus relies on diffusion alone, so part of the inefficiency of most non-viral GTAs can be attributed to their lack of cytoplasmic transport apparatus, a deficit more problematical in columnar AECs where the distance between apical surface and nucleus is relatively large. In addition, for entities the size of GTAs, nuclear entry is a complex and active process, suggesting that addition of nuclear localization signals may be required to permit efficient ingress [25, 26].

How increased knowledge of these daunting barriers is positively informing development of more effective GTAs for CFGT is outlined below.

Viral Vectors

Evolution has moulded viruses into efficient natural GTAs and appropriately modified viruses have been exploited for CFGT. They are, however, immunogenic, difficult to produce on a large scale, and pose safety concerns.

Wild-type Ad is a mildly pathogenic virus with marked trophism for AECs. Consequently, much early work and several clinical trials have involved recombinant Ad vectors for CF and other conditions [27]. Ad vectorology is highly advanced, with so-called gutless vectors allowing maximal cloning capacity and minimal cell-mediated immunogenicity [28]. However, Ad has specific drawbacks for CFGT. Firstly, the CAR receptor for entry of the commonly used serotypes Ad2 and Ad5 in target AECs is not topically accessible in undamaged tissue [21, 29, 30]. Targeting Ad [31] (or relocating CAR [32, 33]) to the apical surface are among strategies being developed to address this problem. Secondly and more importantly, repeated delivery of Ad results in the accumulation of immune effects that prevent transgene expression [34]. Gutted vectors are not a complete solution, since capsid proteins remain to provoke an immune response. Means of circumventing this problem, either by the stealth approach of coating virions with, e.g., polyethylene glycol [35], or by recruiting anti-immune molecules such as CTLA4Ig [36], are being developed. Serotype switching, while theoretically attractive, has met with limited success [37], possibly a consequence of the cross-reactivity of cytotoxic T lymphocytes generated by Ad infection [38].

Adeno-associated virus (AAV), despite its small packaging capacity, has been widely adopted as an alternative viral vector for CFGT. In its native form, AAV can latently infect cells by genomic integration at a safe site on human chromosome 19, an appealing property suggesting the potential for long-term expression. Recombinant AAV does not integrate specifically, but can nevertheless persist through random integration and episomal stability [39]. Trials with an AAV2-CFTR vector have demonstrated its safety [40–42], but in the largest scale lung trial, correction of clinically relevant measures has been limited to a slight improvement in FEV_1, and transgene *CFTR* RNA was not detected [43]. New AAV vectors are being developed using capsid serotypes that, unlike AAV2 [44], can infect via the lumenal surface: AAV5 for example targets apical 2,3-linked sialate [45, 46]. There is evidence that AAV can, unlike Ad, be delivered repeatedly without loss of transgene expression, at least in mice [45]. However, a recent comprehensive attempt to reproduce this effect was unsuccessful [Sumner-Jones, pers. commun.].

Other viral vectors are emerging for CFGT with an ability to infect apically, which is clearly an advantage. SeV is pre-eminent in its transduction efficiency in vivo [19, 47]. Related viruses such as respiratory syncytial virus [48] and human parainfluenza virus [13] also show promise but have not yet been tested in vivo. The development of lentiviruses (together with pseudotyping) may provide vectors capable of integration and hence long-term expression [49–51].

Non-Viral Vectors

Non-viral GTAs invoke less marked inflammatory and immunogenic responses than do viruses, and are also easier to modify chemically and produce in large quantities. However, they are less efficient.

The simplest non-viral GTA is naked DNA, usually in the form of plasmids. Surprisingly, naked DNA can transfect adult sheep lungs in vivo, with detectable expression of transgene-specific RNA [52], despite the reported immobility of plasmid DNA in the cytoplasm [20]. A major obstacle to the development of naked DNA for CFGT is its inability to be aerosolized intact. Most non-viral GTAs consist of DNA complexed with carrier chemicals such as lipids (liposomes) or polycations (e.g. polyethylenimine) [53–55]. Trials of CFGT using liposome delivery have largely paralleled those using viral vectors except, importantly, that repeat delivery has been shown to be possible [56]. The results – variable toxicity with modest, though patchy, efficacy [1] – emphasize the considerable scope for enhancement of non-viral GTAs [57, 58]. One promising enhancement strategy, the ultracompaction of DNA using modified polylysine to create GTA complexes potentially small enough to transit the nuclear pore complex [59] has been shown to be effective in mice [60]. Another relies on providing enhanced cell-targeting capability to the GTA, in the form of peptides that bind receptors (e.g. integrin [61] or SecR [62]) or improve membrane translocation (e.g. Tat [63]).

The use of small DNA fragments to correct mutations directly by recombination or mismatch repair is being developed for CFGT [64, 65]. Unfortunately, this attractive approach is beset by methodological pitfalls [66, 67], and many in the field have become sceptical about its ability to be efficient enough in vivo.

The DNA moiety of non-viral GTAs is another component capable of improvement. Firstly, the known immunostimulatory property of DNA (caused by non-methylated CpG sequences [68, 69]) has been successfully addressed by in vitro methylation of CpGs [70] (which however runs the risk of silencing the promoter) or by depleting the CpG content [71]. Secondly, many virally derived promoters, including the commonly used PCMV promoter/enhancer, are rapidly switched off by cytokine activity, leading to short-term transgene expression [72]. Prolonged expression has been obtained simply and effectively by the use of alternative promoters [73]: genomic context plasmids, which include natural CFTR control elements, also offer the prospect of longer-term expression, but efficient delivery of such large DNAs is challenging [14, 74, 75].

Correction of Resident Stem Cells

Terminally differentiated AECs have half-lives of approximately 90 days [76]. Even if all such cells were permanently corrected by in vivo CFGT, few would be alive a year later, an attrition underlining the present need for repeat delivery in CFGT. However, not all cells are turned over: like most organs, the lung has the capacity to replace dead cells. It is assumed that this function is initiated by currently elusive lung somatic stem cells, defined as self-renewing, extremely slowly cycling cells capable of spawning all differentiated AEC types [77].

Pulmonary stem cell biology is relatively poorly developed. Its progress is hindered by the complexity of the lung epithelium, which in humans can be divided into at least four basic compartments proximally to distally: pseudostratified with glands, pseudostratified, columnar, and alveolar [77]. Classically, stem cell properties have been ascribed to mucous-secreting and basal cells in the first two multilayered compartments, to Clara cells in the third monolayered and to type II pneumocytes in the fourth monolayered compartments [78]. This classification is an oversimplification, and current work suggests that certain subsets of some (if not all) of these candidates are the actual stem cells [77, 79, 80]. Also, the undoubtedly significant role of pulmonary neuroepithelial cells remains unclear [77, 78], and other cell types as yet undescribed may also participate [81].

Much of what we know about lung regeneration has been ascertained using models subjected to lung damage together with e.g. bromodeoxyuridine labelling to mark slowly cycling cells and trace lineages. The geography of stem cell niches in the airways is emerging from such studies [76, 80, 82], giving rise to a tentative model wherein innervated depots of stem cells are systematically arrayed throughout the conducting airways [83]. The co-ordination and signalling of regenerative responses to injury and cell death in the lung have yet to be worked out, though recently a most elegant receptor-ligand relationship that may provide part of the answer has been described [84].

If a sufficient proportion of stem cells could be corrected, repeat delivery CFGT would be unnecessary. This represents a compelling incentive to pursue the stem cell approach as an alternative to long-term repeat delivery with its regular burden on patients, and cost implications. Delivering CFGT directly to lumenally accessible resident stem cells in the distal airways in vivo could bring about unlimited expression without the need for ex vivo manipu-

lations. However, nothing is known about the relative transfectability of these cell types.

Enhancing Delivery

Gene transfer may be enhanced by modifying the physical barriers noted above. One desired effect is to increase the contact time between GTA and epithelium, which is expected to increase transfer efficiency. Various reagents in clinical use including glycopyrrolate (an anticholinergic that reduces mucus secretion), nacystelyn and recombinant human DNAse (both mucolytics). All have been shown to enhance transfection across a CF sputum layer [85–87]. There is also evidence that perfluorochemicals (used clinically for liquid-assisted ventilation) increase gene transfer, possibly through tight junction interactions [88]. Other agents such as EGTA and caprate that disrupt tight junctions, allowing access to basolateral receptors, have also proved effective [89], but their safety in the context of the CF lung has not been established. It is likely that future trials of CFGT will use one or more such adjuncts to boost the efficiency of gene delivery.

The relative immobility of non-viral GTAs is addressable by two strategies. Firstly, energy can be added to the system to physically enhance GTA kinetics [90]. This has been achieved in small-scale studies using electroporation [91], magnetofection [92] and ultrasound [93]. Secondly, GTAs can be modified to take advantage of the cell's own cytoskeletal transport system [94, 95], mimicking some viruses. All GTA-mobilizing approaches need considerable development before clinical use in the human lung can be contemplated: for example, the presence of air means that only low-frequency ultrasound has any possibility of success [Xenariou, pers. commun.]. Progressing these strategies towards the clinic requires the use of human-scale model systems such as the sheep [52].

Assays: Why and How

As discussed above, CFTR is normally expressed at low levels in AECs. It is estimated from genetics and mouse model studies that restoration to approximately 5% of normal levels in each cell would be therapeutic in patients [96, 97]. For these reasons, monitoring the clinical efficacy of CFGT is difficult, especially since it is not known what acute effects such correction would have. Given the chronic course of the disease, it is unlikely that any therapy would rapidly restore the lung to normal levels of function: bronchiectatic regions for example are probably irreversibly damaged. Traditional measures such as FEV_1 (which in any case are subject to large intra-subject fluctuation over time) may therefore not indicate efficacy for some time after treatment. There is clearly a need for other measures to assess CFGT in the acute post-administration phase [98].

Assays for CFGT can be classed as direct (measuring transgene and clinical parameters) or indirect (measuring surrogate markers). The fundamental measure of CFGT is DNA transfer to cells. This is problematical because of the difficulty of determining whether the transgene DNA being assayed in a sample is inside or outside the cell. Far better is to measure transgene RNA, which can only arise if DNA has successfully reached the nucleus. (This does not apply to viruses such as SeV which have RNA genomes and function in the cytoplasm: here it is difficult to verify transduction at the nucleic acid level [47].) Reliable quantitative methods of measuring *CFTR* RNA have been developed using real-time RT-PCR [99]. Other important direct measures are potential difference, ASL height, bacterial adherence, and immunohistochemistry. Much effort is also being expended on discovering new indirect or surrogate markers that can report on CFTR status, particularly using microarray- and proteomics-based methods (see chapter 14) [98]. The same assays and others are also invaluable for pre-clinical evaluation of GTAs in model systems such as the mouse and sheep [98].

The hope is that even modest changes in a number of independent assays will provide statistically robust evidence of productive gene transfer. This is particularly important in the context of projected multiple-dosing lung trials, where the aim is to correlate the kinetics of CFTR expression with clinical measures over an extended period of time.

Conclusion: There have been enormous advances in developing and testing GTAs, and in understanding the barriers to CFGT. Given the significant immunogenic and repeat delivery problems of viral GTAs, non-viral GTAs appear to represent the best current hope for CFGT. In truth, however, the clinical experience with CFGT is severely limited: for example, it could be that multiple dosing with an existing non-viral GTA would, of itself, provide benefit to patients. Through GTA and plasmid enhancement, pre-clinical testing in appropriate models, development and use of sensitive assays, and the adoption of adjuncts and mobilization techniques, there is good reason to be optimistic that a clinically effective CFGT regimen will emerge.

Stem Cell Therapy

Stem cell therapy is attracting much attention, particularly in the context of degenerative conditions such as Parkinson's disease [100]. CF is also potentially treatable by stem cell therapy. The difference between in vivo CFGT and stem cell therapy is that, in the latter, gene transfer would be carried out ex vivo, and treated cells returned to the lung – in essence as living vectors for the transgene – to repopulate the epithelium with progeny expressing normal CFTR.

Should it be determined unequivocally that transdifferentiation from haematopoietic lineage stem cells into human AECs occurs, what form might a stem-cell-based therapy for CF take? It can currently be set out only in general terms [81, 101], though proof of principle through repopulation of AECs in irradiated mice using retrovirally modified bone marrow stem cells has been claimed [102].

Firstly, appropriate stem cells must be identified and harvested. Embryonic stem cells may ultimately be adapted for this purpose, but their allogeneic nature would require patients to be immunosuppressed to allow subsequent re-administration. Autologous bone marrow or haematopoietic stem cells would be less problematical, with harvesting of the latter from peripheral blood being the most convenient source. Secondly, the relevant subset of cells must be purified. Flow cytometry sorting using stem cell marker antigens to classify cells is a highly developed means of achieving this in general [81], but which markers are relevant for a pulmonary epithelial target remains to be determined. Thirdly, purified cells must be genetically corrected using CFTR vectors, as in in vivo gene therapy. For correction to be permanent, transgene expression must be constitutive and uninterrupted, demanding careful promoter selection (see above). The transgene must also physically persist through subsequent cell divisions. Direct mutation reversal is one option to achieve this [65]; maintenance through genomic integration [103, 104], or as episomal artificial chromosomes [105] are others. The integration route is favoured following the demonstration that recombinant retroviral integration ex vivo is curative for X-SCID [106]. Finally, corrected cells selected and expanded to provide a homogeneous population must be introduced into patients. Since most studies of transdifferentiation in humans report only sporadic cells of donor origin, raising the efficiency of the process represents a major challenge. Carefully controlled damage or other adjunct regimes, coupled with tailored delivery (via instillation, aerosol or systemically), could increase recolonization levels. In addition, phased re-administration might lead to the accumulation of sufficient numbers of corrected stem cells to promote long-term clinical benefit.

A factor complicating this exegesis is the emergence of controversial data that suggest circulating cells can sometimes reach the epithelium and transdifferentiate into resident cell types. This is supported by studies of AECs from cell-mismatched murine and human bone marrow transplant recipients which appeared to contain donor chromosomes [101, 107, 108]: however, contrary results have also been reported [109]. The appeal of this phenomenon, if confirmed, is that bone marrow and peripheral blood are abundant, well-characterized and clinically-proven sources of stem cells and could therefore provide ideal material for ex vivo stem cell therapy for the lung and other organs. Great care is needed in interpreting data from experiments of this kind. For example, it may be that the apparently transdifferentiated cells observed result from cell fusion [110]. The crucial question is whether this putative process is sufficiently malleable that it can be engineered into an effective cell therapy for CF.

Conclusion: While research into gene-based cell therapies is entering a highly productive era, the experience of X-SCID [106], in which two treated children developed leukaemia attributable to retroviral insertion [111], emphasizes the fact that new therapies bring new risks. There is an increased danger of oncogenesis with any stem-cell-based approach, and safety measures must be built into treatment protocols: long-term follow-up is also essential. Nevertheless, despite the huge biotechnological hurdles ahead, cell-based therapy offers the beguiling prospect of a more patient-friendly and efficient route to CF treatment that may in time supplant in vivo CFGT.

Acknowledgements

Many thanks to Julia Dorin and colleagues in the UK CFGT Consortium for help and advice.

References

1 Griesenbach U, Geddes DM, Alton EW: Update on gene therapy for cystic fibrosis. Curr Opin Mol Ther 2003;5:489–494.

2 Drumm ML, Pope HA, Cliff WH, Rommens JM, Marvin SA, Tsui LC, Collins FS, Frizzell RA, Wilson JM: Correction of the cystic fibrosis defect in vitro by retrovirus-mediated gene transfer. Cell 1990;62:1227–1233.

3 Gill DR, Davies LA, Pringle IA, Hyde SC: The development of gene therapy for diseases of the lung. Cell Mol Life Sci 2004;61: 355–368.

4 Goula D, Becker N, Lemkine GF, Normandie P, Rodrigues J, Mantero S, Levi G, Demeneix BA: Rapid crossing of the pulmonary endothelial barrier by polyethylenimine/DNA complexes. Gene Ther 2000;7:499–504.

5 Fox E, Griesenbach U, Rogers DF, Huang L, Li S, Booy B, Geddes DM, Alton EWFW: Towards nucleic acid transfer to the airway epithelium via the systemic route. Mol Ther 2001;3:S197.

6 Jiang QS, Engelhardt JF: Cellular heterogeneity of CFTR expression and function in the lung: Implications for gene therapy of cystic fibrosis. Eur J Hum Genet 1998;6: 12–31.

7 Engelhardt JF, Yankaskas JR, Wilson JM: In vivo retroviral gene transfer into human bronchial epithelia of xenografts. J Clin Invest 1992;90:2598–2607.

8 Engelhardt JF, Zepeda M, Cohn JA, Yankaskas JR, Wilson JM: Expression of the cystic fibrosis gene in adult human lung. J Clin Invest 1994;93:737–749.

9 Widdicombe JH: Regulation of the depth and composition of airway surface liquid. J Anat 2002;201:313–318.

10 Larson JE, Cohen JC: Cystic fibrosis revisited. Mol Genet Metab 2000;71:470–477.

11 Peebles D, Gregory LG, David A, Themis M, Waddington SN, Knapton HJ, Miah M, Cook T, Lawrence L, Nivsarkar M, Rodeck C, Coutelle C: Widespread and efficient marker gene expression in the airway epithelia of fetal sheep after minimally invasive tracheal application of recombinant adenovirus in utero. Gene Ther 2004;11:70–78.

12 Boucher RC: New concepts of the pathogenesis of cystic fibrosis lung disease. Eur Respir J 2004;23:146–158.

13 Zhang L, Button B, Skiadopoulos MH, Dang Y, Bukreyev A, Gabriel S, Boucher RC, Collins PL, Pickles RJ: Restoration of periciliary liquid volume to normal levels in human CF airway epithelium after targeted expression of CFTR in ciliated cells with a parainfluenza virus type 3-CFTR gene transfer vector. Ped Pulm 2003;S25:256.

14 Boyd AC, Davidson HE, Doherty A, Davidson-Smith H, Sheppard DN, Cai Z, Webb S, Dorin JR, Porteous DJ: Construction and characterisation of genomic context vectors for CF gene therapy. Pediatr Pulm 1999; S19:237.

15 Koehler DR, Hannam V, Belcastro R, Steer B, Wen Y, Post M, Downey G, Tanswell AK, Hu J: Targeting transgene expression for cystic fibrosis gene therapy. Mol Ther 2001;4:58–65.

16 Ostrowski LE, Hutchins JR, Zakel K, O'Neal WK: Targeting expression of a transgene to the airway surface epithelium using a ciliated cell-specific promoter. Mol Ther 2003;8:637–645.

17 Ferrari S, Geddes DM, Alton EW: Barriers to and new approaches for gene therapy and gene delivery in cystic fibrosis. Adv Drug Deliv Rev 2002;54:1373–1393.

18 Ferrari S, Griesenbach U, Geddes DM, Alton E: Immunological hurdles to lung gene therapy. Clin Exp Immunol 2003;132:1–8.

19 Yonemitsu Y, Kitson C, Ferrari S, Farley R, Griesenbach U, Judd D, Steel R, Scheid P, Zhu J, Jeffery PK, Kato A, Hasan MK, Nagai Y, Masaki I, Fukumura M, Hasegawa M, Geddes DM, Alton EW: Efficient gene transfer to airway epithelium using recombinant Sendai virus. Nat Biotechnol 2000;18: 970–973.

20 Lukacs GL, Haggie P, Seksek O, Lechardeur D, Freedman N, Verkman AS: Size-dependent DNA mobility in cytoplasm and nucleus. J Biol Chem 2000;275:1625–1629.

21 Pickles RJ, McCarty D, Matsui H, Hart PJ, Randell SH, Boucher RC: Limited entry of adenovirus vectors into well-differentiated airway epithelium is responsible for inefficient gene transfer. J Virol 1998;72: 6014–6023.

22 Fasbender A, Zabner J, Zeiher BG, Welsh MJ: A low rate of cell proliferation and reduced DNA uptake limit cationic lipid-mediated gene transfer to primary cultures of ciliated human airway epithelia. Gene Ther 1997;4: 1173–1180.

23 Matsui H, Johnson LG, Randell SH, Boucher RC: Loss of binding and entry of liposome-DNA complexes decreases transfection efficiency in differentiated airway epithelial cells. J Biol Chem 1997;272:1117–1126.

24 Whittaker GR: Virus nuclear import. Adv Drug Deliv Rev 2003;55:733–747.

25 Gasiorowski JZ, Dean DA: Mechanisms of nuclear transport and interventions. Adv Drug Deliv Rev 2003;55:703–716.

26 Hebert E: Improvement of exogenous DNA nuclear importation by nuclear localization signal-bearing vectors: a promising way for non-viral gene therapy? Biol Cell 2003;95: 59–68.

27 Harvey BG, Maroni J, O'Donoghue KA, Chu KW, Muscat JC, Pippo AL, Wright CE, Hollmann C, Wisnivesky JP, Kessler PD,

Rasmussen HS, Rosengart TK, Crystal RG: Safety of local delivery of low- and intermediate-dose adenovirus gene transfer vectors to individuals with a spectrum of morbid conditions. Hum Gene Ther 2002;13:15–63.

28 Cao H, Koehler DR, Hu J: Adenoviral vectors for gene replacement therapy. Viral Immunol 2004;17:327–333.

29 Zabner J, Freimuth P, Puga A, Fabrega A, Welsh MJ: Lack of high affinity fiber receptor activity explains the resistance of ciliated airway epithelia to adenovirus infection. J Clin Invest 1997;100:1144–1149.

30 Walters RW, Grunst T, Bergelson JM, Finberg RW, Welsh MJ, Zabner J: Basolateral localization of fiber receptors limits adenovirus infection from the apical surface of airway epithelia. J Biol Chem 1999;274: 10219–10226.

31 Krasnykh VN, Douglas JT, Van Beusechem VW: Genetic targeting of adenoviral vectors. Mol Ther 2000;1:391–405.

32 Pickles RJ, Fahrner JA, Petrella JM, Boucher RC, Bergelson JM: Retargeting the coxsackievirus and adenovirus receptor to the apical surface of polarized epithelial cells reveals the glycocalyx as a barrier to adenovirus-mediated gene transfer. J Virol 2000;74: 6050–6057.

33 Davis B, Nguyen J, Stoltz D, Depping D, Excoffon KJ, Zabner J: Adenovirus-mediated erythropoietin production by airway epithelia is enhanced by apical localization of the coxsackie-adenovirus receptor in vivo. Mol Ther 2004;10:500–506.

34 Harvey BG, Leopold PL, Hackett NR, Grasso TM, Williams PM, Tucker AL, Kaner RJ, Ferris B, Gonda I, Sweeney TD, Ramalingam R, Kovesdi I, Shak S, Crystal RG: Airway epithelial CFTR mRNA expression in cystic fibrosis patients after repetitive administration of a recombinant adenovirus. J Clin Invest 1999;104:1245–1255.

35 Croyle MA, Chirmule N, Zhang Y, Wilson JM: 'Stealth' adenoviruses blunt cell-mediated and humoral immune responses against the virus and allow for significant gene expression upon readministration in the lung. J Virol 2001;75:4792–4801.

36 Jooss K, Turka LA, Wilson JM: Blunting of immune responses to adenoviral vectors in mouse liver and lung with CTLA4Ig. Gene Ther 1998;5:309–319.

37 Youil R, Toner TJ, Su Q, Chen M, Tang A, Bett AJ, Casimiro D: Hexon gene switch strategy for the generation of chimeric recombinant adenovirus. Hum Gene Ther 2002;13: 311–320.

38 Heemskerk B, Veltrop-Duits LA, Van Vreeswijk T, Ten Dam MM, Heidt S, Toes RE, Van Tol MJ, Schilham MW: Extensive cross-reactivity of CD4+ adenovirus-specific T cells: implications for immunotherapy and gene therapy. J Virol 2003;77:6562–6566.

39 Lu Y: Recombinant adeno-associated virus as delivery vector for gene therapy – A review. Stem Cells Dev 2004;13:133–145.

40 Aitken ML, Moss RB, Waltz DA, Dovey ME, Tonelli MR, McNamara SC, Gibson RL, Ramsey BW, Carter BJ, Reynolds TC:

41 Wagner JA, Nepomuceno IB, Messner AH, Moran ML, Batson EP, Dimiceli S, Brown BW, Desch JK, Norbash AM, Conrad CK, Guggino WB, Flotte TR, Wine JJ, Carter BJ, Reynolds TC, Moss RB, Gardner P: A phase II, double-blind, randomized, placebo-controlled clinical trial of tgAAVCF using maxillary sinus delivery in patients with cystic fibrosis with antrostomies. Hum Gene Ther 2002;13:1349–1359.

42 Flotte TR, Zeitlin PL, Reynolds TC, Heald AE, Pedersen P, Beck S, Conrad CK, Brass-Ernst L, Humphries M, Sullivan K, Wetzel R, Taylor G, Carter BJ, Guggino WB: Phase I trial of intranasal and endobronchial administration of a recombinant adeno-associated virus serotype 2 (rAAV2)-CFTR vector in adult cystic fibrosis patients: A two-part clinical study. Hum Gene Ther 2003;14: 1079–1088.

43 Moss RB, Rodman D, Spencer LT, Aitken ML, Zeitlin PL, Waltz D, Milla C, Brody AS, Clancy JP, Ramsey B, Hamblett N, Heald AE: Repeated adeno-associated virus serotype 2 aerosol-mediated cystic fibrosis transmembrane regulator gene transfer to the lungs of patients with cystic fibrosis: A multicenter, double-blind, placebo- controlled trial. Chest 2004;125:509–521.

44 Zabner J, Seiler M, Walters R, Kotin RM, Fulgeras W, Davidson BL, Chiorini JA: Adeno-associated virus type 5 (AAV5) but not AAV2 binds to the apical surfaces of airway epithelia and facilitates gene transfer. J Virol 2000;74:3852–3858.

45 Auricchio A, O'Connor E, Weiner D, Gao GP, Hildinger M, Wang L, Calcedo R, Wilson JM: Noninvasive gene transfer to the lung for systemic delivery of therapeutic proteins. J Clin Invest 2002;110:499–504.

46 Ding W, Yan Z, Zak R, Saavedra M, Rodman DM, Engelhardt JF: Second-strand genome conversion of adeno-associated virus type 2 (AAV-2) and AAV-5 is not rate limiting following apical infection of polarized human airway epithelia. J Virol 2003;77:7361–7366.

47 Ferrari S, Griesenbach U, Shiraki-Iida T, Shu T, Hironaka T, Hou X, Williams J, Zhu J, Jeffery PK, Geddes DM, Hasegawa M, Alton EW: A defective nontransmissible recombinant Sendai virus mediates efficient gene transfer to airway epithelium in vivo. Gene Ther 2004;11:1659–1664.

48 Zhang L, Peeples ME, Boucher RC, Collins PL, Pickles RJ: Respiratory syncytial virus infection of human airway epithelial cells is polarized, specific to ciliated cells, and without obvious cytopathology. J Virol 2002;76: 5654–5666.

49 Kobinger GP, Weiner DJ, Yu QC, Wilson JM: Filovirus-pseudotyped lentiviral vector can efficiently and stably transduce airway epithelia in vivo. Nat Biotechnol 2001;19: 225–230.

50 Sinn PL, Hickey MA, Staber PD, Dylla DE, Jeffers SA, Davidson BL, Sanders DA, B. MP: Lentivirus vectors pseudotyped with

filoviral envelope glycoproteins transduce airway epithelia from the apical surface independently of folate receptor alpha. J Virol 2003;77: 5902–5910.

51 Copreni E, Penzo M, Carrabino S, Conese M: Lentivirus-mediated gene transfer to the respiratory epithelium: a promising approach to gene therapy of cystic fibrosis. Gene Ther 2004;11(suppl 1):S67–S75.

52 Emerson M, Renwick L, Tate S, Rhind S, Milne E, Painter H, Boyd AC, McLachlan G, Griesenbach U, Cheng SH, Gill DR, Hyde SC, Baker A, Alton EW, Porteous DJ, Collie DDS: Transfection efficiency and toxicity following delivery of naked plasmid DNA and cationic lipid-DNA complexes to ovine lung segments. Mol Ther 2003;8: 646–653.

53 Parker AL, Newman C, Briggs S, Seymour L, Sheridan PJ: Nonviral gene delivery: techniques and implications for molecular medicine. Expert Rev Mol Med 2003;5:1–15.

54 Ziady AG, Davis PB, Konstan MW: Nonviral gene transfer therapy for cystic fibrosis. Expert Opin Biol Ther 2003;3:449–458.

55 Wiseman JW, Goddard CA, McLelland D, Colledge WH: A comparison of linear and branched polyethylenimine (PEI) with DCChol/DOPE liposomes for gene delivery to epithelial cells in vitro and in vivo. Gene Ther 2003;10:1654–1662.

56 Hyde SC, Southern KW, Gileadi U, Fitzjohn EM, Mofford KA, Waddell BE, Gooi HC, Goddard CA, Hannavy K, Smyth SE, Egan JJ, Sorgi FL, Huang L, Cuthbert AW, Evans MJ, Colledge WH, Higgins CF, Webb AK, Gill DR: Repeat administration of DNA/liposomes to the nasal epithelium of patients with cystic fibrosis. Gene Ther 2000;7: 1156–1165.

57 Montier T, Delepine P, Pichon C, Ferec C, Porteous DJ, Midoux P: Non-viral vectors in cystic fibrosis gene therapy: progress and challenges. Trends Biotechnol 2004;22: 586–592.

58 Koping-Hoggard M, Varum KM, Issa M, Danielsen S, Christensen BE, Stokke BT, Artursson P: Improved chitosan-mediated gene delivery based on easily dissociated chitosan polyplexes of highly defined chitosan oligomers. Gene Ther 2004;11:1441–1452.

59 Liu G, Li D, Pasumarthy MK, Kowalczyk TH, Gedeon CR, Hyatt SL, Payne JM, Miller TJ, Brunovskis P, Fink TL, Muhammad O, Moen RC, Hanson RW, Cooper MJ: Nanoparticles of compacted DNA transfect postmitotic cells. J Biol Chem 2003;278: 32578–32586.

60 Ziady AG, Gedeon CR, Miller T, Quan W, Payne JM, Hyatt SL, Fink TL, Muhammad O, Oette S, Kowalczyk T, Pasumarthy MK, Moen RC, Cooper MJ, Davis PB: Transfection of airway epithelium by stable PEGylated poly-L- lysine DNA nanoparticles in vivo. Mol Ther 2003;8:936–947.

61 Jenkins RG, Meng QH, Hodges RJ, Lee LK, Bottoms SE, Laurent GJ, Willis D, Ayazi Shamlou P, McAnulty RJ, Hart SL: Formation of LID vector complexes in water alters physicochemical properties and enhances pulmonary gene expression in vivo. Gene Ther 2003;10:1026–1034.

62 Ziady AG, Kelley TJ, Milliken E, Ferkol T, Davis PB: Functional evidence of CFTR gene transfer in nasal epithelium of cystic fibrosis mice in vivo following luminal application of DNA complexes targeted to the serpin-enzyme complex receptor. Mol Ther 2002;5:413–419.

63 Hyndman L, Lemoine JL, Huang L, Porteous DJ, Boyd AC, Nan X: HIV-1 Tat protein transduction domain peptide facilitates gene transfer in combination with cationic liposomes. J Control Release 2004;99:435–444.

64 Sangiuolo F, Bruscia E, Serafino A, Nardone AM, Bonifazi E, Lais M, Gruenert DC, Novelli G: In vitro correction of cystic fibrosis epithelial cell lines by small fragment homologous replacement (SFHR) technique. BMC Med Genet 2002;3:8.

65 Gruenert DC, Bruscia E, Novelli G, Colosimo A, Dallapiccola B, Sangiuolo F, Goncz KK: Sequence-specific modification of genomic DNA by small DNA fragments. J Clin Invest 2003;112:637–641.

66 Thorpe P, Stevenson BJ, Porteous DJ: Optimising gene repair strategies in cell culture. Gene Ther 2002;9:700–702.

67 De Semir D, Aran JM: Misleading gene conversion frequencies due to a PCR artifact using small fragment homologous replacement. Oligonucleotides 2003;13:261–269.

68 Krieg AM: Direct immunologic activities of CpG DNA and implications for gene therapy. J Gene Med 1999;1:56–63.

69 Yew NS, Wang KX, Przybylska M, Bagley RG, Stedman M, Marshall J, Scheule RK, Cheng SH: Contribution of plasmid DNA to inflammation in the lung after administration of cationic lipid:pDNA complexes. Hum Gene Ther 1999;10:223–234.

70 Reyes-Sandoval A, Ertl HC: CpG methylation of a plasmid vector results in extended transgene product expression by circumventing induction of immune responses. Mol Ther 2004;9:249–261.

71 Yew NS, Zhao H, Przybylska M, Wu IH, Tousignant JD, Scheule RK, Cheng SH: CpG-depleted plasmid DNA vectors with enhanced safety and long- term gene expression in vivo. Mol Ther 2002;5:731–738.

72 Paillard F: Promoter attenuation in gene therapy: causes and remedies. Hum Gene Ther 1997;8:2009–2010.

73 Gill DR, Smyth SE, Goddard CA, Pringle IA, Higgins CF, Colledge WH, Hyde SC: Increased persistence of lung gene expression using plasmids containing the ubiquitin C or elongation factor 1alpha promoter. Gene Ther 2001;8:1539–1546.

74 Walker WE, Porteous DJ, Boyd AC: The effects of plasmid copy number and sequence context upon transfection efficiency. J Control Release 2004;94:245–252.

75 Magin-Lachmann C, Kotzamanis G, D'Aiuto L, Cooke H, Huxley C, Wagner E: In vitro and in vivo delivery of intact BAC DNA – Comparison of different methods. J Gene Med 2004;6:195–209.

76 Borthwick DW, Shahbazian M, Krantz QT, Dorin JR, Randell SH: Evidence for stem-cell niches in the tracheal epithelium. Am J Respir Cell Mol Biol 2001;24:662–670.

77 Neuringer IP, Randell SH: Stem cells and repair of lung injuries. Respir Res 2004;5:6.

78 Bishop AE: Pulmonary epithelial stem cells. Cell Prolif 2004;37:89–96.

79 Schoch KG, Lori A, Burns KA, Eldred T, Olsen JC, Randell SH: A subset of mouse tracheal epithelial basal cells generates large colonies in vitro. Am J Physiol Lung Cell Mol Physiol 2004;286:L631–L642.

80 Hong KU, Reynolds SD, Watkins S, Fuchs E, Stripp BR: In vivo differentiation potential of tracheal basal cells: Evidence for multipotent and unipotent subpopulations. Am J Physiol Lung Cell Mol Physiol 2004;286:L643–L649.

81 Kotton DN, Summer R, Fine A: Lung stem cells: New paradigms. Exp Hematol 2004;32:340–343.

82 Hong KU, Reynolds SD, Giangreco A, Hurley CM, Stripp BR: Clara cell secretory protein-expressing cells of the airway neuroepithelial body microenvironment include a label-retaining subset and are critical for epithelial renewal after progenitor cell depletion. Am J Respir Cell Mol Biol 2001;24:671–681.

83 Engelhardt JF: Stem cell niches in the mouse airway. Am J Respir Cell Mol Biol 2001;24:649–652.

84 Vermeer PD, Einwalter LA, Moninger TO, Rokhlina T, Kern JA, Zabner J, Welsh MJ: Segregation of receptor and ligand regulates activation of epithelial growth factor receptor. Nature 2003;422:322–326.

85 Ferrari S, Kitson C, Farley R, Steel R, Marriott C, Parkins DA, Scarpa M, Wainwright B, Evans MJ, Colledge WH, Geddes DM, Alton EW: Mucus altering agents as adjuncts for nonviral gene transfer to airway epithelium. Gene Ther 2001;8: 1380–1386.

86 Glasspool-Malone J, Steenland PR, McDonald RJ, Sanchez RA, Watts TL, Zabner J, Malone RW: DNA transfection of macaque and murine respiratory tissue is greatly enhanced by use of a nuclease inhibitor. J Gene Med 2002;4:323–322.

87 Stern M, Caplen NJ, Browning JE, Griesenbach U, Sorgi F, Huang L, Gruenert DC, Marriot C, Crystal RG, Geddes DM, Alton EWFW: The effect of mucolytic agents on gene transfer across a CF sputum barrier in vitro. Gene Ther 1998;5:91–98.

88 Weiss DJ, Beckett T, Bonneau L, Young J, Kolls JK, Wang G: Transient increase in lung epithelial tight junction permeability: An additional mechanism for enhancement of lung transgene expression by perfluorochemical liquids. Mol Ther 2003;8:927–935.

89 Johnson LG, Vanhook MK, Coyne CB, Haykal-Coates N, Gavett SH: Safety and efficiency of modulating paracellular permeability to enhance airway epithelial gene transfer in vivo. Hum Gene Ther 2003;14:729–747.

90 Wells DJ: Gene therapy progress and prospects: Electroporation and other physical methods. Gene Ther 2004;11:1363–1369.

91 Dean DA, MacHado-Aranda D, Blair-Parks K, Yeldandi AV, Young JL: Electroporation as a method for high-level nonviral gene transfer to the lung. Gene Ther 2003;10:1608–1615.

92 Gersting SW, Schillinger U, Lausier J, Nicklaus P, Rudolph C, Plank C, Reinhardt D, Rosenecker J: Gene delivery to respiratory epithelial cells by magnetofection. J Gene Med 2004;6:913–922.

93 Zarnitsyn VG, Prausnitz MR: Physical parameters influencing optimization of ultrasound-mediated DNA transfection. Ultrasound Med Biol 2004;30:527–538.

94 Lechardeur D, Lukacs GL: Intracellular barriers to non-viral gene transfer. Curr Gene Ther 2002;2:183–194.

95 Fong S, Liu Y, Heath T, Fong P, Liggitt D, Debs RJ: Membrane-permeant, DNA-binding agents alter intracellular trafficking and increase the transfection efficiency of complexed plasmid DNA. Mol Ther 2004;10:706–718.

96 Dorin JR, Farley R, Webb S, Smith SN, Farini E, Delaney SJ, Wainwright BJ, Alton EWFW, Porteous DJ: A demonstration using mouse models that successful gene therapy for cystic fibrosis requires only partial gene correction. Gene Ther 1996;3:797–801.

97 Ramalho AS, Beck S, Meyer M, Penque D, Cutting GR, Amaral MD: Five percent of normal cystic fibrosis transmembrane conductance regulator mRNA ameliorates the severity of pulmonary disease in cystic fibrosis. Am J Respir Cell Mol Biol 2002;27:619–627.

98 Griesenbach U, Boyd AC: Pre-clinical and clinical endpoint assays for cystic fibrosis gene therapy. J Cystic Fibrosis 2005;4:88–100.

99 Rose AC, Goddard CA, Colledge WH, Cheng SH, Gill DR, Hyde SC: Optimisation of real-time quantitative RT-PCR for the evaluation of non-viral mediated gene transfer to the airways. Gene Ther 2002;9:1312–1320.

100 Kaji EH, Leiden JM: Gene and stem cell therapies. JAMA 2001;285:545–550.

101 Spencer H, Jaffe A: The potential for stem cell therapy in cystic fibrosis. J R Soc Med 2004;97 Suppl 44:52–56.

102 Grove JE, Lutzko C, Priller J, Henegariu O, Theise ND, Kohn DB, Krause DS: Marrow-derived cells as vehicles for delivery of gene therapy to pulmonary epithelium. Am J Respir Cell Mol Biol 2002;27:645–651.

103 Yant SR, Meuse L, Chiu W, Ivics Z, Izsvak Z, Kay MA: Somatic integration and long-term transgene expression in normal and haemophilic mice using a DNA transposon system. Nat Genet 2000;25:35–41.

104 Olivares EC, Hollis RP, Chalberg TW, Meuse L, Kay MA, Calos MP: Site-specific genomic integration produces therapeutic Factor IX levels in mice. Nat Biotechnol 2002;20:1124–1128.

105 Lipps HJ, Jenke AC, Nehlsen K, Scinteie MF, Stehle IM, Bode J: Chromosome-based vectors for gene therapy. Gene 2003;304:23–33.

106 Cavazzana-Calvo M, Hacein-Bey S, De Saint Basile G, Gross F, Yvon E, Nusbaum P, Selz F, Hue C, Certain S, Casanova JL, Bousso P, Deist FL, Fischer A: Gene therapy of human severe combined immunodeficiency (SCID)-X1 disease. Science 2000;288:669–672.

107 Abe S, Lauby G, Boyer C, Rennard SI, Sharp JG: Transplanted BM and BM side population cells contribute progeny to the lung and liver in irradiated mice. Cytotherapy 2003;5: 523–533.

108 Kleeberger W, Versmold A, Rothamel T, Glockner S, Bredt M, Haverich A, Lehmann U, Kreipe H: Increased chimerism of bronchial and alveolar epithelium in human lung allografts undergoing chronic injury. Am J Pathol 2003;162:1487–1494.

109 Davies JC, Potter M, Bush A, Rosenthal M, Geddes DM, Alton EW: Bone marrow stem cells do not repopulate the healthy upper respiratory tract. Pediatr Pulmonol 2002;34: 251–256.

110 Wagers AJ, Weissman IL: Plasticity of adult stem cells. Cell 2004;116:639–648.

111 Hacein-Bey-Abina S, Von Kalle C, Schmidt M, McCormack MP, Wulffraat N, Leboulch P, Lim A, Osborne CS, Pawliuk R, Morillon E, Sorensen R, Forster A, Fraser P, Cohen JI, De Saint Basile G, Alexander I, Wintergerst U, Frebourg T, Aurias A, Stoppa-Lyonnet D, Romana S, Radford-Weiss I, Gross F, Valensi F, Delabesse E, MacIntyre E, Sigaux F, Soulier J, Leiva LE, Wissler M, Prinz C, Rabbitts TH, Le Deist F, Fischer A, Cavazzana-Calvo M: LMO2-associated clonal T cell proliferation in two patients after gene therapy for SCID-X1. Science 2003;302: 415–419.

A. Christopher Boyd
Medical Sciences (Medical Genetics)
University of Edinburgh
Molecular Medicine Centre
Western General Hospital
Crewe Road
Edinburgh EH4 2XU (UK)
Tel. +44 131 651 1060
Fax +44 131 651 1059
E-Mail Chris.Boyd@ed.ac.uk

Digestive Tract

Bush A, Alton EWFW, Davies JC, Griesenbach U, Jaffe A (eds): Cystic Fibrosis in the 21st Century.
Prog Respir Res. Basel, Karger, 2006, vol 34, pp 232–241

Gut Disease: Clinical Manifestations, Pathophysiology, Current and New Treatments

C.J. Taylor[a]   J. Hardcastle[b]

[a]Sheffield Children's Hospital and [b]Department Biomedical Science, University of Sheffield, Sheffield, UK

Abstract

The intestinal epithelium regulates the transport of nutrients, electrolytes and water. CFTR, which is present in abundance at the luminal membrane, is central to these processes.
Dysfunction leads to impaired intestinal Cl^- and fluid secretion while Na^+ and Na^+-linked nutrient absorption are
enhanced. The resulting dehydration of luminal contents contributes to many of the gastrointestinal manifestations of the
disease, including meconium ileus and distal ileal obstruction
syndrome. Pancreatic exocrine insufficiency is the commonest
manifestation of gastrointestinal disease but malabsorption
may be compounded by inactivation of pancreatic enzymes by
hyperacidity and peptidases in the upper intestine, deranged
bile salt function, and associated enteropathies. Gastro-
oesophageal reflux disease is also common but underdiagnosed. Long-standing oesophageal hyperacidity may
predispose to complications including Barrett's metaplasia and
subsequent malignant change. An association between CF and
digestive tract tumours has been established and presents
new challenges to the long-term management of patients.

Gastrointestinal dysfunction remains the earliest and
most common manifestation of cystic fibrosis (CF).
Presentation is on prenatal sonography [1] or in the immediate postnatal period with intestinal obstruction secondary
to meconium ileus (MI) or jejunoileal atresia. There have
been major recent advances in the understanding of CF gastrointestinal disease.

Pathophysiology

The primary role of the intestine is the movement of
nutrients, electrolytes and water from the lumen to the
blood. Enterocytes also secrete electrolytes and water.
CFTR is localized at the luminal membrane [2], and in CF,
intestinal secretion is impaired [3] while Na^+ [4] and Na^+-
linked nutrient absorption are enhanced [5].

Intestinal fluid secretion plays an important role in
lubrication, providing Na^+ for Na^+-dependent nutrient
absorption and flushing immunoglobulins from the crypts.
The secretory process involves stimulation of electrogenic
Cl^- secretion coupled with an inhibition of electroneutral
NaCl cotransport (fig. 1a). Electrogenic Cl^- secretion is a
two-stage process [6]. Cl^- is taken up into the enterocyte at
the basolateral membrane via the NKCC (Na^+, K^+, $2Cl^-$)
cotransporter. This uses energy from the Na^+ gradient,
maintained by the Na^+ pump, to power Cl^- accumulation
within the cell. When the cell is stimulated to secrete, Cl^-
channels in the luminal membrane open, allowing the accu-
mulated Cl^- to move into the lumen down its electrochemical gradient. The resulting increase in luminal negativity
causes Na^+ to move into the lumen via the paracellular
pathway, with water following osmotically. At the basolateral membrane K^+ channels open causing a hyperpolariza-
tion that acts via the paracellular pathway to maintain the
luminal membrane potential and hence the driving force for
continued Cl^- secretion. Many endogenous and exogenous
agents can stimulate the secretory process and they act either
by increasing cyclic nucleotide production or by raising the

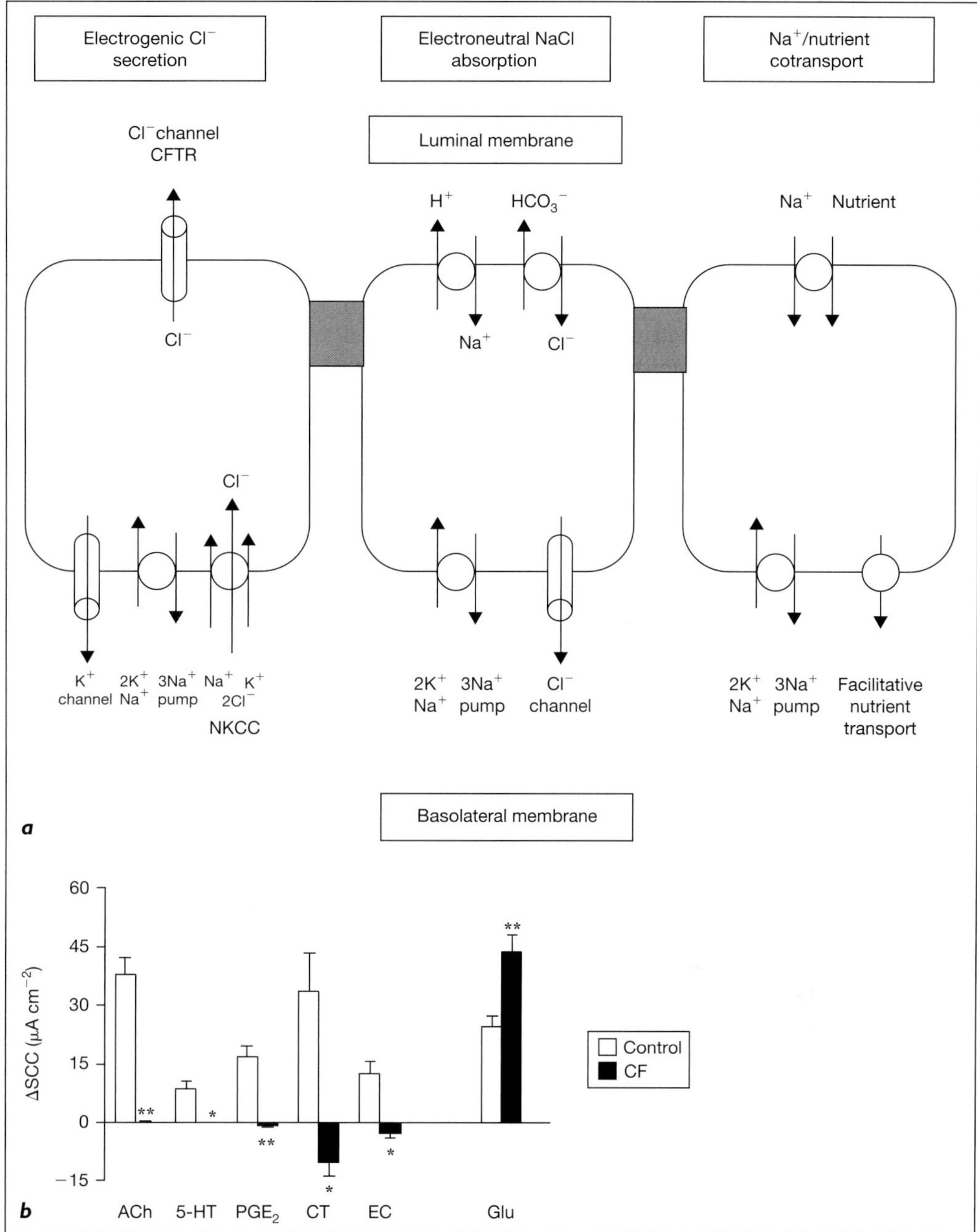

Fig. 1. *a* Secretory pathways and intracellular mediators in the enterocyte. *b* Responses of jejunal biopsies from control and CF patients to secretagogues and glucose. Acetylcholine [ACh; 10^{-3} M, n (control) = 53, n (CF) = 32], 5-hydroxytryptamine [5-HT; 2.6×10^{-5} M, n (control) = 7, n (CF) = 4] and prostaglandin E_2 [PGE$_2$; 10^{-3} M, n (control) = 13, n (CF) = 14] were added to the basolateral side of the tissues, while cholera toxin [CT; 5 µg/ml, n (control) = 4, n (CF) = 4], *E. coli* STa [EC; 50 U/ml, n (control) = 5, n (CF) = 4] and glucose [Glu; 10 mM, n (control) = 53, n (CF) = 32] were added luminally. Each bar represents the mean ± 1 SEM of the number of control [n (control), age range: 2 days to 14 years and 4 months] and CF [n (CF), age range: 2 months to 18 years and 6 months] patients. An unpaired t test was used to assess significance: * p < 0.01; ** p < 0.001.

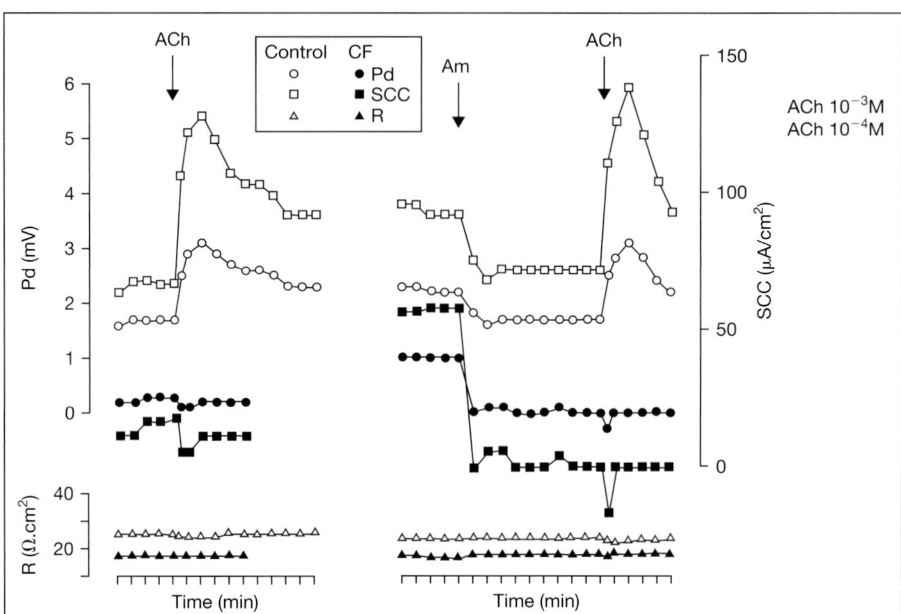

Fig. 2. Changes in potential difference (Pd) to secretagogue [acetylcholine (Ach), 10^{-3} *M*] and amiloride (Am, 10^{-4} *M*) stimulation in CF rectal biopsies compared to controls. Control tissues show response to cholinergic stimulation which is absent in CF biopsies. Pd falls in both tissues with amiloride.

cytosolic Ca^{2+} concentration within the enterocyte. As the mechanism of Cl^- secretion is electrogenic, its activity can be monitored by measuring transintestinal electrical activity. Using jejunal or duodenal [7] biopsy specimens or samples of tissue obtained surgically it has been shown that Cl^- secretion by the small intestine is absent in CF (fig. 1b). The endogenous secretagogues acetylcholine and 5-hydroxytryptamine (elevate cytosolic Ca^{2+} levels) and prostaglandin E_2 (stimulates cAMP production) increase the short-circuit current (SCC), a reflection of electrogenic Cl^- secretion, in control tissues, but fail to elicit a response in CF tissues. Moreover, the bacterial toxins cholera toxin (acts via cAMP) and the heat-stable enterotoxin of *Escherichia coli* (*E. coli* STa, acts via cGMP) are also ineffective in CF tissues [8]. In the colon and rectum the secretory response involves increases in the secretion of both Cl^- and K^+ [9]. In CF the Cl^- secretory component is absent [10], but K^+ secretion remains [11], so that a negative SCC change is often observed [12] (fig. 2). Experiments in conscious human subjects confirm that the failure of Cl^- secretion in isolated tissues is a reflection of events in vivo. The difference in the responses of control and CF intestine to secretagogue challenge is so marked that it can be used as an aid to diagnosis [11, 13].

The lack of a secretory response in CF intestine is not due to a failure of intracellular signalling pathways as the stimulation of both cAMP production by isolated enterocytes [14] and cytosolic Ca^{2+} levels in isolated crypts [15] are normal in CF. It is now clear that the luminal Cl^- channel that is the route for Cl^- exit from the enterocyte is CFTR [6] and hence defects in this protein explain the failure of Cl^- secretion in CF.

Stimulation of intestinal secretion involves not only an increase in electrogenic Cl^- secretion, but also an inhibition of electroneutral NaCl absorption (fig. 1a). NaCl uptake at the luminal face of the enterocyte occurs via parallel Na^+/H^+ and Cl^-/HCO_3^- exchangers. Na^+ is actively extruded from the cell at the basolateral membrane by the Na^+ pump while Cl^- leaves passively [9]. This process does not directly involve CFTR, yet its inhibition is impaired in CF [3, 16]. It is, therefore, likely that this is an example of an interaction of CFTR with other transport proteins. A list of known and likely interactions between CFTR and other channels/transport proteins is given in table 1.

CFTR can also influence the Na^+-dependent absorption of nutrients (fig. 1a). The active absorption of nutrients utilizes energy from the Na^+ gradient, harnessed by a cotransporter at the luminal membrane, to accumulate nutrients within the enterocyte. The nutrients leave the cell via facilitated diffusion across the basolateral membrane, while Na^+ is extruded by the Na^+ pump. As nutrient absorption is accompanied by increased Na^+ absorption the process is electrogenic and can be monitored by measuring the SCC. In jejunal biopsies the SCC associated with Na^+-dependent glucose absorption is enhanced in CF tissues (fig. 1b). Examination of the kinetics revealed that the maximum rate

	Transport protein/channel	Function	Reference
ENaC	Epithelial Na$^+$ channel	Amiloride-sensitive channel allowing Na$^+$ movement down its electrochemical gradient	Greger et al. [65]
ORCC	Outwardly rectifying Cl$^-$ channel	Channel allowing Cl$^-$ movement down its electrochemical gradient with current flow greater in outward direction	Greger et al. [65]
AQP3	Aquaporin 3	Water channel	Greger et al. [65]
DRA	Downregulated in adenoma	Cl$^-$/HCO$_3^-$ exchanger involved in electroneutral NaCl absorption by intestine	Wheat et al. [66]
NBC1	Sodium bicarbonate cotransporter 1	Cotransporter responsible for HCO$_3^-$ uptake at basolateral membrane that provides HCO$_3^-$ for basal and cAMP-stimulated secretion	Soleimani and Burnham [67]
SGLT1	Na$^+$/glucose cotransporter 1	Cotransporter responsible for Na$^+$-dependent glucose uptake by the small intestine	Baxter et al. [5].
IBAT	Ileal bile acid transporter	Cotransporter responsible for Na$^+$-dependent bile acid uptake by the ileum	Hardcastle et al. [68]

of glucose transport was increased in CF, with similar findings for the amino acid alanine [5]. A recent in vivo perfusion study failed to detect any difference in jejunal glucose absorption between control and CF groups [17]. This technique does not distinguish between active and passive glucose absorption, but did reveal that the glucose-dependent potential difference, an index of active sugar absorption, is increased in CF. It, therefore, appears that CFTR can also interact with Na$^+$-dependent nutrient cotransporters where its effects are similar to its actions on the epithelial Na$^+$ channel, ENaC.

Meconium Ileus

Pathophysiology

Bowel obstruction is secondary to the accumulation of inspissated intraluminal meconium. This differs in the composition and amount of protein, mucoprotein, mucopolysaccharides, and reducing sugars compared with meconium from healthy newborns. Lactase and β-*D*-fucosidase are present in the meconium of babies with CF but not of healthy infants. MI is usually seen in pancreatic-insufficient CF, but occurs in normals CF [18] and pancreatic-sufficient CF where immaturity of the myenteric plexus may be a factor [19].

More germane to the aetiology of MI is the demonstrable absence of cAMP-mediated Cl$^-$ transport in intestinal tract of both human and CF mouse models. Furthermore, the murine CF intestinal tract also shows defective cAMP regulation of electroneutral NaCl absorption, an inability to secrete HCO$_3^-$ and elevated electrogenic Na$^+$ transport in the distal colon [20]. These anomalies will combine to dehydrate the intestinal contents.

The tendency to MI may also reflect the action of modifier genes on CFTR function. Zielenski et al. [21] have detected a CF modifier locus for MI on human chromosome 19q13 and Norkina et al. [22] have shown upregulation of components of the innate immune system in the intestine of CF mice which exhibit a severe intestinal phenotype.

Clinical Outcome

Standard treatment is reviewed elsewhere. In summary, resection and primary anastomosis are usual [23]. This appears to be associated with less late operative intervention (20%) compared with other approaches (81%) [24] and gives an excellent overall survival (92.4%).

Expected late complications include a higher risk of distal ileal obstruction syndrome (DIOS; 20 vs. 6% for non-MI CF), and adhesive small bowel obstruction and blind loop syndrome (27%). Long-term follow-up suggests that the clinical status should not differ from those presenting later in childhood with pulmonary complications [25]. Lai et al. [26] have, however, documented poor growth and nutritional problems despite a good calorie intake (130% recommended dietary allowance), and Li et al. [27] reported worse lung function in children with MI compared to CF patients diagnosed through neonatal screening.

Gastro-Oesophageal Reflux Disease

A high incidence of pathologically increased gastro-oesophageal reflux has been reported in infants and children with CF [28]. Based on a fractional reflux time >10% with an oesophageal pH <4.0, almost 20% of CF infants <6 months of age will have gastro-oesophageal reflux disease (GERD) [29]. However, the normal range for reflux index during the first 12 months of life is about 10% (95 percentile), decreasing from 13% at birth to 8% at 12 months [30]. Incidence figures for reflux in older children with CF vary between 25 and 100%, depending on patient selection [29]. Similar high rates have been reported in CF adolescents and adults with suggestive symptoms.

Chest physiotherapy in postural drainage positions is believed to exacerbate gastro-oesophageal reflux in CF infants [31]. Button et al. [32] found that the number of reflux episodes per hour was increased during physiotherapy with head down tilt (MPT) compared with a modified technique omitting this. However, Phillips et al. [33] were unable to demonstrate a significant change in median oesophageal pH during chest physiotherapy using gravity-assisted positioning with head downward tip. Nevertheless, the Melbourne group were able to show better clinical progress and improved lung function in children treated from infancy with MPT [34].

GERD may exacerbate lung disease through aspiration and reflex bronchospasm. A combination of oesophageal manometry, acid perfusion and reflux provocation tests showed a strong correlation between reflux and chest radiograph and lung function scores. Long-standing acid reflux may predispose to Barrett's metaplasia [35] and, potentially, adenocarcinoma of the oesophagus.

Malabsorption, Diarrhoea and Associated Enteropathies

Although steatorrhoea and protein loss (azotorrhoea) in the stool principally reflect pancreatic exocrine failure, absorption may also be compromised by associated intolerances, enteropathies and adverse intraluminal conditions (table 2).

Duodenal hyperacidity and gastric pepsin may inhibit both endogenous and exogenous pancreatic enzyme activity. Duodenal pH is regulated by bicarbonate-rich pancreatic secretions and intestinal bicarbonate secretion. Not only is the former defective in CF but CFTR is also known to play an important role in cAMP-stimulated bicarbonate secretion in the duodenum. Duodenal HCO_3^- secretion involves both an electrogenic secretion via a CFTR HCO_3^- conductance and also an electroneutral secretion via a CFTR-dependent Cl^-/HCO_3^- exchange process that is closely associated with the carbonic anhydrase activity of the epithelium [36]. In the presence of a favourable cell-to-lumen HCO_3^- gradient, the CFTR-mediated HCO_3^- current accounts for about 80% of stimulated HCO_3^- secretion. Exposure of the duodenal mucosa to acidic pH triggers HCO_3^- secretion via an electroneutral, 4,4′-diisothiocyanato-stilbene-2,2′-disulfonic acid (DIDS)-sensitive Cl^-/HCO_3^- exchange process. Basal HCO_3^- secretion is significantly lower in the CF duodenal mucosa, and, in contrast to normal controls, db-cAMP-stimulated HCO_3^- secretion is absent in CF tissues [37]. In the CF duodenum, other apical membrane acid-base transporters retain function, thereby affording limited control of transepithelial pH. Activity of a Cl^--dependent anion exchanger provides near-constant HCO_3^- secretion in CF intestine, but under basal conditions the magnitude of secretion is lessened by simultaneous activity of a Na^+/H^+ exchanger. During cAMP stimulation of CF duodenum, a small increase in net base secretion is measured but the change results from cAMP inhibition of Na^+/H^+ exchanger activity rather than increased HCO_3^- secretion [38].

Reduction in duodenal hyperacidity and hence steatorrhoea can be achieved by adding either H_2 receptor antagonists or proton pump inhibitors to an adequate amount of lipolytic activity [39].

A large number of associated enteropathies have been reported, which are discussed in standard texts. These include coeliac disease, Crohn's disease, cow's milk protein and lactose intolerance. Malabsorption should never be assumed to be simply a reflection of inadequate pancreatic enzyme replacement therapy.

Table 2. Factors compounding malabsorption in CF

Primary defect in pancreatic ductular HCO_3^- and water secretion
Reduction in enzyme secretion secondary to acinar cell destruction
Inactivation of enzyme by hyperacidity and peptidases in upper intestine
Deranged bile salt function

Table 3. Differential diagnosis of distal ileal obstruction syndrome

Acute appendicitis
Appendix abscess
Intussusception
Mucocoele of the appendix
Constipation
Fibrosing colonopathy
Crohn's disease

Peptic Ulceration

Abdominal pain is common in CF, but the risk of peptic ulceration appears relatively low, despite the limited ability to neutralize acid as discussed above. The proximal duodenal mucosa is exposed to frequent pulses of gastric acid which, recent studies suggest, can lower the intracellular pH, pH(i), of duodenal epithelial cells via an epithelium-sensing mechanism including capsazepine-inhibitable vanilloid receptors, presumably similar to the vanilloid receptor TPVR-1 [40]. The intestinal mucosa is functionally 'leaky', increasing the importance of defence mechanisms such as the mucus gel layer, cellular acid/base transporters, bicarbonate secretion, and mucosal blood flow. The absence of peptic ulceration in CF supports the view that duodenal epithelial cell protection occurs as the result of pH(i) regulation rather than by neutralization of acid by HCO_3^- in the pre-epithelial mucus [41].

Distal Ileal Obstruction Syndrome

The variation in the reported prevalence of DIOS probably reflects the age range of the study population. DIOS presents either acutely with signs of abdominal obstruction or more commonly subacutely with crampy abdominal pain and relative constipation. The diagnosis is usually clinical and confirmed with a plain abdominal film. This characteristically shows a speckled faecal gas pattern in the right lower quadrant. Examination may reveal a tender mass in the right iliac fossa, the main differential diagnosis being appendicitis, constipation or intussusception. In cases of diagnostic uncertainty ultrasound examination or CT scan may provide additional information [42]. The differential diagnosis is given in table 3. The management is reviewed elsewhere. This section chiefly focuses on pathophysiology.

Pathogenesis of DIOS

Multiple factors may influence the development of DIOS; however, the actual cause remains unclear. DIOS is most often associated with pancreatic exocrine insufficiency but cases have been reported in pancreatic sufficient patients [43]. Koletzko et al. [44] showed that faecal fat losses were increased in pancreatic-insufficient patients with DIOS compared with pancreatic-insufficient patient controls (DIOS patients: faecal fat 31.2 ± 20.6 vs. 16.2 ± 17.6; faecal fat losses expressed as % of oral fat intake). DIOS has also been reported in patients who adhere poorly to pancreatin therapy or where too rapid changes in pancreatin dose have been made.

More relevant to the development of distal ileal obstruction is likely to be CF transmembrane conductance regulator (CFTR) dysfunction which leads to dehydration of luminal contents (see section on GI physiology). A similar failure in anion secretory function has also been observed in the large bowel (fig. 2). Additional factors operating within the CF ileum which may further dehydrate luminal contents include increased permeability and enhanced glucose and amino acid uptake.

Intestinal obstruction in CF may also reflect the effect of various modifying genes on CFTR function. The recent cloning of the human basolateral membrane K^+ (IK) channel gene ($KCNN_4$) has enabled detailed mapping of tissue gene expression. IK channel expression is found in gastrointestinal epithelia where it acts to set a negative membrane potential and facilitate chloride efflux thereby facilitating chloride, sodium and water secretion across the epithelium into the gut lumen. However, the effect of polymorphisms in the $KCNN_4$ gene on secretory dysfunction in CF remains speculative [45].

The intestine is also responsible for the enterohepatic circulation of bile salts. Bile acids are actively reabsorbed in the ileum by a sodium-dependent mechanism. This is significantly reduced in CF. Absorption is inhibited by a low sodium concentration and by competitive inhibition by other bile acids [46]. Bile acids, however, also provide a secretory stimulus throughout the intestinal tract. This process contributes to the fluidity of the intestinal contents and is disturbed in CF [47].

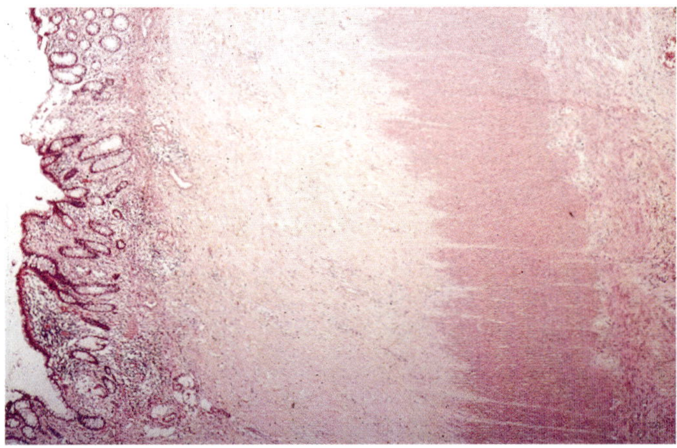

Fig. 3. FC histology: Cross section of stenotic segment of colon. Note dense submucosal fibrosis with widening of the submucosa causing severe narrowing of the lumen. The main muscle layers are undamaged and the mucosa is intact.

Abnormalities in gut motility may also predispose to intestinal obstruction. Pooled data suggest that gastric emptying may be normal, accelerated or delayed depending upon whether a solid or liquid test meal is used and upon the fat content of the meal [48]. Moreover, anaerobic fermentation of mucus and undigested carbohydrate results in excessive quantities of short chain fatty acids within the gut lumen which inhibit colonic peristaltic activity and may stimulate tonic activity [49]. Small bowel transit is approximately twice that seen in normal patients (small bowel transit: CF 50–390 min, mean 136 min; controls 30–50 min, mean 88 min). Colonic transit is also prolonged in CF; studies suggest a mouth to anus transit of between 25 and 55 h (normal 24.2 ± 6.8 h) [50]. Transit is slowest in the caecum and right colon and influenced by the fat content of the stool and presence or absence of bile salts.

Motility and transit may also be influenced by intestinal wall thickening. Ultrasound studies have shown thickening in the stomach, duodenum and ileum and more recently an ultrasound study of 33 children found that the CF colon was 50% thicker than in controls. This abnormality was found to be age related [51]. It is unclear whether these changes are primary or secondary events.

Liver disease may be important in the aetiology of DIOS. Patients with liver disease were more likely to suffer either MI or DIOS compared with those who have experienced neither complication (33.5 vs. 12.3%, $p < 0.01$). Unfortunately the net contribution from DIOS alone was not made clear [52]. Other predisposing factors include the use of anticholinergic and opiate drugs and dehydration, particularly due to diabetes mellitus.

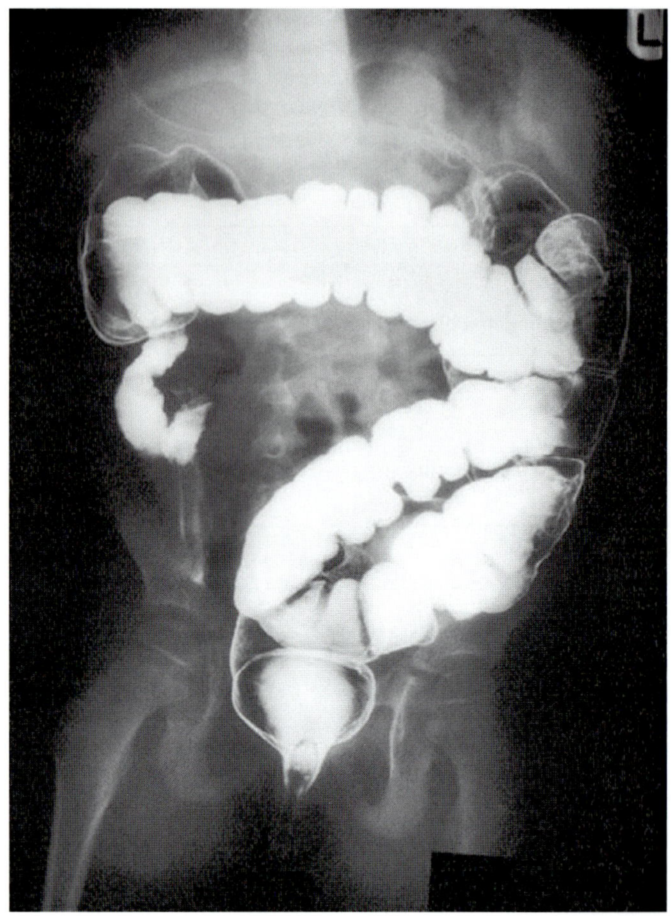

Fig. 4. FC contrast study: There is a segmental stricture in the ascending colon, and there is mucosal irregularity and speculation. Luminal narrowing affected long segments of the colon and longitudinal shortening of the colon is also a prominent feature in many cases.

Gastrointestinal inflammation has also been reported in patients with DIOS. Croft et al. [53] found significantly increased levels of IL-1β and TNF-α in gastrointestinal lavage fluid from 12 patients with DIOS, although the elevated cytokine levels may be from swallowed sputum [54]. However, Raia et al. [55] reported chronic inflammation in CF duodenal biopsy. Changes included an increased number of lamina propria mononuclear cells expressing ICAM-1, CD25, IL-2 and IFN-γ and villus enterocytes highly expressing transferrin receptor. Immunologic markers were reduced. Bruzzese et al. [56] found high levels of faecal calprotectin, a marker of acute inflammation, and rectal nitric oxide production in CF patients. Both markers were reduced after probiotic administration suggesting that intestinal microflora plays a major role in intestinal inflammation in CF children [57].

Fibrosing Colonopathy

Fibrosing colonopathy (FC) is a rare condition seen almost exclusively in children with CF. The characteristic histopathology comprises submucosal fibrosis with thickening of the muscularis propria and chronic mucosal inflammation chiefly affecting the caecum and ascending colon (fig. 3). This leads to a fusiform narrowing and shortening of the bowel with subsequent stricture (fig. 4). An association between FC and high-dose pancreatic supplementation was shown by case-control studies [58, 59].

The Medicines Control Agency (MCA) recommended in 1994 that 'unless special reasons exist, patients with CF should not use high-potency pancreatins' [60] and that the total dose used 'should not usually exceed 10,000 iu lipase/kg body weight daily', following which there was a dramatic fall in notifications of FC in the UK. The actual aetiology of the disease remains elusive. Recently, Serban et al. [61] reported typical FC in a 3-week-old infant prior to exposure to exogenous pancreatin. Histology revealed markers of a recent severe injury, and submucosal vasculitis, findings previously reported in FC associated with the use of high-strength pancreatins. Waters [62] described submucosal fibrosis on full-thickness rectal biopsy in an untreated neonate who presented with MI.

It appears that FC can evolve in the absence of, although undoubtedly exacerbated by, exogenous pancreatin; the mechanism is unclear. Alternative mechanisms considered include increased intestinal permeability and the co-existence of a modifier gene which may predispose a subgroup of infants with CF to develop FC following exposure to endogenous pancreatin.

Whether exposure of the abnormally permeable gut mucosa to viscid intestinal contents and pancreatic proteases can act as a trigger for FC remains speculative.

CF and Cancer

An association between CF and cancer has been suggested; however, the increased risk of cancer appears to relate chiefly to digestive tract tumours. The CF and Cancer Study Group identified 37 cancers ascertained between 1985 and 1992 from the US and Canadian CF registries [63] compared to an expected total of 45.6; thus there was no overall increased risk. Nevertheless, there was a significantly increased predisposition to digestive tract malignancy with an observed to expected ratio of 6.5 (95% CI, 3.5–11.1); tumours were identified in the oesophagus, stomach, small and large intestine, and biliary tract. Pancreatic cancer may also be found in association with CF, although the risk appears to relate to inherited pancreatic cancers, which represent approximately 5–10% of all pancreatic cancers [64]. An association with leukaemias was suggested but not confirmed.

References

1 Muller F, Simon-Bouy B, Girodon E, Monnier N, Malinge MC, Serra JL; French Collaborative Group: Predicting the risk of cystic fibrosis with abnormal ultrasound signs of fetal bowel: Results of a French molecular genetic study based on 641 prospective cases. Am J Med Genet 2002;110:109–115.

2 O'Loughlin EV, Hunt DM, Bostrom TE, et al: X-ray microanalysis of cell elements in normal and cystic fibrosis jejunum: Evidence for chloride secretion in villi. Gastroenterology 1996;110:411–418.

3 O'Loughlin EV, Hunt DM, Gaskin KJ, et al: Abnormal epithelial transport in cystic fibrosis jejunum. Am J Physiol 1991;260:G709–G716.

4 Mall M, Bleich M, Kuehr J, Brandis M, Greger R, Kunzelmann K: CFTR-mediated inhibition of epithelial Na$^+$ conductance is defective in cystic fibrosis. Am J Physiol 1999;277:G709–G716.

5 Baxter PS, Golghill J, Hardcastle J, Hardcastle PT, Taylor CJ: Enhanced intestinal glucose and alanine transport in cystic fibrosis. Gut 1990;31:817–820.

6 Barrett KE, Keely SJ: Chloride secretion by the intestinal epithelium: molecular basis and regulatory aspects. Annu Rev Physiol 2000;62:535–557.

7 Hallberg K, Reims A, Strandvik B: Electrogenic ion transport in duodenum: an aid to cystic fibrosis diagnosis. Scand J Gastroenterol 2000;35:1106–1109.

8 Baxter PS, Goldhill J, Hardcastle J, Hardcastle PT, Taylor CJ: Accounting for cystic fibrosis. Nature 1988;335:211.

9 Kunzelmann K, Mall M: Electrolyte transport in the mammalian colon: mechanisms and implications for disease. Physiol Rev 2001;82:245–289.

10 Hardcastle J, Hardcastle PT, Taylor CJ, Goldhill J: Failure of cholinergic stimulation to induce a secretory response from the rectal mucosa in cystic fibrosis. Gut 1991;32:1035–1039.

11 Goldstein JL, Shapiro AB, Rao MC, Layden TJ: In vivo evidence of altered chloride, but not potassium secretion in cystic fibrosis rectal mucosa. Gastroenterology 1991;101:1012–1019.

12 Veeze HJ, Sinaasappel M, Bijman J, Bouquet J, de Jonge HR: Ion transport abnormalities in rectal suction biopsies from children with cystic fibrosis. Gastroenterology 1991;101:398–403.

13 Taylor CJ, Dalton A, Hardcastle J, Hardcastle PT, Evans S: Diagnosing cystic fibrosis. Arch Dis Child 1997;77:454.

14 Hitchin BW, Dobson PRM, Brown BL, Hardcastle J, Hardcastle PT, Taylor CJ: Measurement of intracellular mediators in enterocytes isolated from jejunal biopsy specimens of control and cystic fibrosis patients. Gut 1990;31:893–899.

15 Klaren PHM, Hardcastle J, Evans S, et al: Acetylcholine induces Ca^{2+} mobilization in isolated distal colonic crypts from normal and cystic fibrosis mice. J Pharmacol 2001;53:371–377.

16 Clarke LL, Harline MC: CFTR is required for cAMP inhibition of intestinal Na$^+$ absorption in a cystic fibrosis mouse model. Am J Physiol 1996;270:G259–G267.

17 Russo MA, Hogenauer C, Coates SW, et al: Abnormal passive chloride absorption in

cystic fibrosis functionally opposes the classic chloride secretory defect. J Clin Invest 2003; 112:116–125.

18 Fakhoury K, Durie PR, Levison H, Canny GJ: Meconium ileus in the absence of cystic fibrosis. Arch Dis Child 1992;67:1204–1206.

19 Stringer MD, Brereton RJ, Drake DP, et al: Meconium ileus due to extensive intestinal aganglionosis. J Pediatr Surg 1994;29: 501–503.

20 Grubb BR, Gabriel SE: Intestinal physiology and pathology in gene-targeted mouse models of cystic fibrosis. Am J Physiol 1997;273: G258–G266.

21 Zielenski J, Corey M, Rozmahel R, et al: Detection of a cystic fibrosis modifier locus for meconium ileus on human chromosome 19q13. Nat Genet 1999;22:128–129.

22 Norkina O, Kaur S, Ziemer D, de Lisle R: Inflammation of the cystic fibrosis mouse small intestine. Am J Physiol Gastrointest Liver Physiol 2004;286:G1032–G1041.

23 Wit J, Sellin S, Degenhardt P, Scholz M, Mau H: Is the Bishop-Koop anastomosis in treatment of neonatal ileus still current? Chirug 2000;71:307–310.

24 Del Pin CA, Czyrko C, Ziegler MM, Scanlin TF, Bishop HC: Management and survival of meconium ileus. A 30 year review. Ann Surg 1992;215:179–185.

25 Fuchs JR, Langer JC: Long-term outcome after neonatal meconium obstruction. Pediatrics 1998;101:E7.

26 Lai HC, Korosok MR, Laxova A, Davis LA, FitzSimmon PC, Farrel PM: Nutritional status of patients with cystic fibrosis with meconium ileus: A comparison with patients without meconium ileus and diagnosed early through screening. Pediatrics 2000;105:53–61.

27 Li Z, Lai HC, Korosok MR, Laxova A, Rock MJ, Splaingard ML, Farrel PM: Longitudinal pulmonary status of cystic fibrosis children with meconium ileus. Pediatr Pulmonol 2004; 38:277–284.

28 Malfroot A, Dab I: New insights on gastro-oesophageal reflux in cystic fibrosis by longitudinal follow up. Arch Dis Child 1991;66: 1339–1345.

29 Heine RG, Button BM, Olinsky A, Phelan P, Catto-Smith AG: Gastro-oesophageal reflux in infants under 6 months with cystic fibrosis. Arch Dis Child 1998;78:44–48.

30 Vandenplas Y, Goyvaerts H, Helven R, et al: Gastro-oesophageal reflux, as measured by 24-hour pH monitoring, in 509 healthy infants screened for risk of sudden infant death syndrome. Pediatrics 1991;88:834–840.

31 Pencharz PB, Durie PR: Pathogenesis of malnutrition in cystic fibrosis, and its treatment. Clin Nutr 2000;19:387–394.

32 Button BM, Heine RG, Catto-Smith AG, et al: Postural drainage and gastro-oesophageal reflux in infants with cystic fibrosis. Arch Dis Child 1997;76:148–150.

33 Phillips GE, Pike SE, Rosenthal M, Bush A: Holding the baby: head downwards positioning for physiotherapy does not cause gastro-oesophageal reflux. Eur Respir J 1998;12: 954–957.

34 Button BM, Heine RG, Catto-Smith AG, et al: Chest physiotherapy in infants: To tip or not to tip? A five year study. Pediatr Pulmonol 2003; 35:208–213.

35 Hassal E, Israel DM, Davidson AG, et al: Barrett's esophagus in children with cystic fibrosis: Not a coincidental association. Am J Gastroenterol 1993;88:1934–1938.

36 Clarke LL, Harline MC: Dual role of CFTR in cAMP-stimulated HCO_3^- secretion across murine duodenum. Am J Physiol 1998;274: G718–G726.

37 Pratha VS, Hogan DL, Martensson BA, Bernard J, Zhou R, Isenberg JI: Identification of transport abnormalities in duodenal mucosa and duodenal enterocytes from patients with cystic fibrosis. Gastroenterology 2000;118: 1051–1060.

38 Clarke LL, Stien X, Walker NM: Intestinal bicarbonate secretion in cystic fibrosis mice. JOP 2001;2(suppl):263–267.

39 DiMagno EP: Gastric acid suppression and treatment of severe exocrine pancreatic insufficiency. Best Pract Res Clin Gastroenterol 2001;15:477–486.

40 Kaunitz JD, Akiba Y: Acid-sensing protective mechanisms of duodenum. J Physiol Pharmacol 2003;54(suppl 4):19–26.

41 Kaunitz JD, Akiba Y: Duodenal intracellular bicarbonate and the 'CF paradox'. JOP 2001; 2(suppl):268–273.

42 Dik H, Nicolai JJ, Schipper J, Heijerman HG, Bakker W: Erroneous diagnosis of distal intestinal obstruction syndrome in cystic fibrosis: Clinical impact of abdominal ultrasonography. Eur J Gastroenterol Hepatol 1995;7:279–281.

43 Millar-Jones L, Goodchild MC: Cystic fibrosis, pancreatic sufficiency and distal intestinal obstruction syndrome: A report of four cases. Acta Paediatr 1995;84:577–578.

44 Koletzko S, Corey M, Ellis L, et al: Effects of cisapride in patients with cystic fibrosis and distal intestinal obstruction syndrome. J Pediatr 1990;117:815–822.

45 Jensen BS, Strobeek D, Olesen SP, Christophersen P: The Ca^{2+}-activated K^+ channel of intermediate conductance: A molecular target for novel treatments. Curr Drug Targets 2001;2:401–422.

46 Hardcastle J, Hardcastle PT, Chapman J, Taylor CJ: Ursodeoxycholic acid action on transport function of the small intestine in normal and cystic fibrosis mice. J Pharmacol 2001;53:1457–1467.

47 Hardcastle J, Hardcastle PT, Chapman J, Taylor CJ: Taurocholic acid-induced secretion in normal and cystic fibrosis mouse ileum. J Pharmacol 2001;53:711–719.

48 Gregory PC: Gastrointestinal pH, motility/transit and permeability in cystic fibrosis. J Pediatr Gastroenterol Nutr 1996;23:613–623.

49 Cherbut C: Motor effects of short-chain fatty acids and lactate in the gastrointestinal tract. Proc Nutr Soc 2003;62:95–99.

50 Pai CG, Kurian G: A modified radiographic method for estimating segmental colinic transit time in subjects with rapid gut transit. Indian J Med Res 1999;110:22–26.

51 Connett GJ, Lucas JS, Atchley JT, Fairhurst JJ, Rolles CJ: Colonic wall thickening is related to age and not dose of high strength pancreatin microspheres in children with cystic fibrosis. Eur J Gastroenterol Hepatol 1999;11: 181–183.

52 Colombo C, Apostolo MG, Ferrari M, et al: Analysis of risk factors for the development of liver disease associated with cystic fibrosis. J Pediatr 1994;124:393–399.

53 Croft NM, Marshall TG, Ferguson A: Direct assessment of gastrointestinal inflammation and mucosal immunity in children with cystic fibrosis. Postgrad Med J 1996;72(suppl 2): S32–36.

54 Doull I, Langton-Hewer S: Gut inflammation in children with cystic fibrosis on high-dose enzyme supplementation. Lancet 1996;347: 327.

55 Raia V, Maiuri L, de Ritis G, et al: Evidence of chronic inflammation in morphologically normal small intestine of cystic fibrosis patients. Pediatr Res 2000;47:344–350.

56 Bruzzese R, Raia V, Gaudiello G, et al: Intestinal inflammation is a frequent feature of cystic fibrosis and is reduced by probiotic administration. Aliment Pharmacol Ther 2004;20:813–819.

57 Bruzzese E, Canari RB, De Marco G, Guarino A: Microflora in inflammatory bowel diseases: a paediatric perspective. J Clin Gastroenterol 2004;38(6 suppl):S91–S93.

58 Smyth RL, Ashby D, O'Hea U, et al: Fibrosing colonopathy in cystic fibrosis: Results of a case control study. Lancet 1995;345:1247–1251.

59 FitzSimmonds SC, Burkhardt GA, Borowitz D, et al: High dose pancreatic enzyme supplements and fibrosing colonopathy in children with cystic fibrosis. N Engl J Med 1997;336: 1283–1289.

60 Medicines Control Agency: Committee on Safety of Medicines. Fibrosing colonopathy associated with pancreatic enzyme. Curr Probl Pharmacovigilance 1994;21.

61 Serban DE, Florescu P, Mui N: Fibrosing colonopathy revealing cystic fibrosis before any pancreatic enzyme supplementation. J Pediatr Gastroenterol Nutr 2002;35:356–359.

62 Waters BL: Cystic fibrosis with fibrosing colonopathy in the absence of pancreatic enzymes. Pediatr Dev Pathol 1998;1:74–78.

63 Neglia JP, FitzSimmons SC, Maisonneuve P, et al: The risk of cancer among patients with cystic fibrosis. Cystic Fibrosis and Cancer Study Group. N Engl J Med 1995;332: 494–499.

64 Rulyak SJ, Brentnall TA: Inherited pancreatic cancer: Improvements in our understanding of genetics and screening. Int J Biochem Cell Biol 2004;36:1386–1392.

65 Greger R, Schreiber R, Mall M, Wissner A, Hopf A, Briel M, Bleich M, Warth R, Kunzelmann K: Cystic fibrosis and CFTR. Pflügers Arch 2001;443:S3–S7.

66 Wheat VJ, Shumaker H, Burnham C, Shull G, Yankaskas JR, Soleimani M: CFTR induces the expression of DRA along with Cl^-/HCO_3^- exchange activity in tracheal epithelial cells. Am J Physiol 2000;279:C62–C71.

67 Soleimani M, Burnham CE: Na$^+$ HCO$_{(3-)}$ cotransporters (NBC): cloning and characterization. J Membr Biol 2001;183:71–84.

68 Hardcastle J, Harwood MD, Taylor CJ: Absorption of taurocholic acid by the ileum of normal and transgenic DeltaF508 cystic fibrosis mice. J Pharm Pharmacol 2004;56: 445–452.

C.J. Taylor
Academic Unit of Child Health
Sheffield Children's Hospital
Western Bank
Sheffield S10 2TH (UK)
Tel. +44 114 271 7304
Fax +44 114 275 5364
E-Mail c.j.taylor1@sheffield.ac.uk

Bush A, Alton EWFW, Davies JC, Griesenbach U, Jaffe A (eds): Cystic Fibrosis in the 21st Century.
Prog Respir Res. Basel, Karger, 2006, vol 34, pp 242–250

Pancreatic Involvement: Clinical Manifestations, Pathophysiology and New Treatments

Keith J. Lindley

London Centre for Pancreatic Disease in Childhood,
Institute of Child Health and Great Ormond Street Hospital for Children NHS Trust, London, UK

Abstract

Cystic fibrosis affects the epithelia of multiple organs including the gastrointestinal and respiratory tracts and can result in suppurative lung disease and severe pancreatic exocrine insufficiency. The pancreatic phenotype is variable with differing manifestations in pancreatic-sufficient (PS) and pancreatic-insufficient (PI) individuals and the possibility of progression from PS to PI phenotype with the passage of time. This chapter focuses on the spectrum of pancreatic disease seen in association with mutations in ABCC7.

Spectrum of Pancreatic Disease in Cystic Fibrosis

The functional unit of the exocrine pancreas is composed of the acinus and its draining ductule. Ductular cells secrete bicarbonate under predominantly neural (vagal) and humoral (secretin) control. The acinar cells are specialized to synthesize, store and secrete digestive enzymes. Zymogen granules (the store of digestive enzyme precursor molecules) are concentrated in the apical pole of the cell the contents of which are secreted under neuro-humoral control. Zymogen granules contain a wealth of proteolytic, lipolytic, amylolytic and nuclease precursor molecules which are secreted in an inactive form. These inactive zymogens are washed along the pancreatic duct in an alkalinized fluid secreted by the pancreatic ductular cells. Activation of these pro-enzymes takes place predominantly in the intestinal lumen where enterokinase, a brush border glycoprotein peptidase, activates trypsinogen by hydrolysis

of an N-terminal fragment of the molecule (Val-Asp-Asp-Asp-Asp-Lys). The active form, trypsin, catalyzes the activation of the other zymogens. Small amounts of trypsin are also formed autocatalytically within the pancreas and are inactivated by a trypsin inhibitor which is secreted by the acinar cells – PSTI (pancreatic secretory trypsin inhibitor or SPINK1). SPINK1 forms a relatively stable complex with trypsin near its catalytic site hence inactivating the enzyme.

Signs of pancreatic involvement in cystic fibrosis (CF) appear to be correlated to a greater or lesser extent with the gestational age and pancreatic trypsinogen expression [1].

Pancreatic Sufficiency and Insufficiency

In the early descriptions of CF the demonstration of pancreatic exocrine insufficiency was a necessary criterion for diagnosis [2]. The subsequent demonstration of abnormal concentrations of sweat electrolytes in CF allowed earlier suggestions that exocrine function could be variable in CF with a spectrum of affection from pancreatic insufficiency (PI) to pancreatic sufficiency (PS) to be confirmed [3, 4]. CFTR is expressed at high levels in the pancreatic duct as in other epithelia [5, 6]. Most PS patients with CF have evidence of impaired pancreatic ductular function, with reduced secretion of anions and water [7, 8] and many also have reduced enzyme secretion [9].

With the subsequent identification of the CF gene, ABCC7, and of its gene product, CFTR, a cAMP-regulated chloride channel, it has been possible to start to dissect the pathophysiology of pancreatic exocrine dysfunction in CF and to establish genotype/phenotype relationships (see below) [9–12]. Moreover, it is now appreciated that CFTR and its gene have many other functions, regulating and/or interacting directly/indirectly with a host of other genes and proteins within and without the cell [13–16]. Clinical phenotype may be further influenced by modifier genes [17–21]. With this increasing insight has come the recognition that CFTR dysfunction can result in a spectrum of disease ranging from classical CF through to mild disease referred to as CFTR-related disease [22].

In infants with PI, pancreatic injury may begin in utero manifest as mucus inspissation within the pancreatic ductules, and a reduction in acinar volume [1, 23, 24]. In severe cases there may be acinar replacement by fat and/or fibrous tissue with preservation of endocrine tissue until the late stages of disease. In such cases there is complete loss of exocrine pancreatic tissue and islets of Langerhans are seen embedded in fibrous tissue and fatty stroma [25, 26]. In mild cases normal pancreatic morphology may be relatively preserved [25, 26]. Studies in a CFTR knockout mouse model have supported the notion that CF pancreatic disease is, at least in part, a consequence of chronic and progressive obstruction of small ducts and acini [27]. Studies of infants who are PS confirm that pancreatic disease can progress from PS to PI [28] although overall about 10–15% of individuals with classical CF will remain PS and do not require pancreatic enzyme replacement therapy [29, 30].

The high levels of CFTR in the pancreatic duct epithelium are instrumental in the generation of a high-volume alkaline secretion by the pancreas which both maintains the solubility of acinar derived enzymes and pro-enzymes and flushes them out of the pancreatic duct [31]. However, diminished ductal bicarbonate secretion in CF also leads to a primary defect in membrane trafficking at the apical plasma membrane of pancreatic acinar cells [32]. CFTR is required for secretion of bicarbonate by the pancreatic duct and duodenal epithelium. The classical dogma is that CFTR is primarily a cAMP-regulated Cl^- channel with secreted Cl^- being exchanged for HCO_3^- by the apical membrane Cl^--HCO_3^- exchanger. There is, however, increasing evidence for direct secretion of HCO_3^- through CFTR [33–35] (fig. 1). Defective alkalinization of the duodenum as a consequence of both reduced pancreatic ductal HCO_3^- secretion and reduced duodenal HCO_3^- secretion has been suggested to result in hyperstimulation of feedback mechanisms controlling pancreatic HCO_3^- secretion and in induction of a stress response in the pancreas (see later) [36, 37]. The development of functional PI status in formerly PS individuals with CF is, therefore, likely to be a multifactorial process which is more complex than simple plugging of pancreatic ductules with hyperviscous secretions.

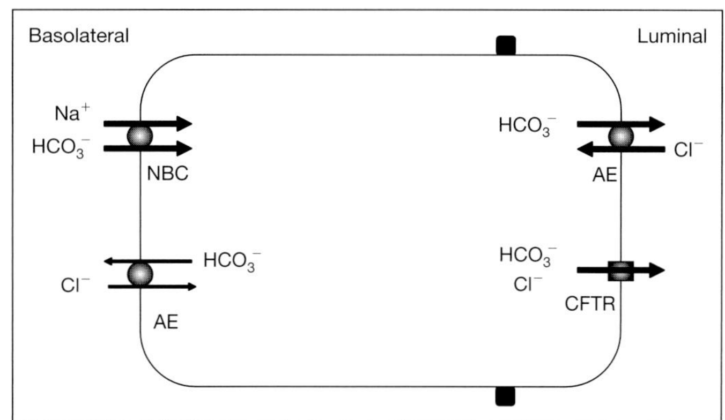

Fig. 1. Mechanisms of chloride and bicarbonate secretion by the pancreatic duct cell. NBC = Sodium-bicarbonate channel; AE = anion exchanger; CFTR = cystic fibrosis transmembrane regulator.

Genotype-Phenotype Correlations in PI

It is conventional wisdom that the pancreatic phenotype in CF has the best correlation with the genotype of any of the numerous manifestations seen in CF [12]. This view arose in part out of a publication by the Cystic Fibrosis Genotype Phenotype Consortium – a study of 800 homozygotes and age-matched compound heterozygotes for the ΔF508 mutation [38]. In this study compound heterozygotes having the genotype R117H/ΔF508 clearly differed from age- and sex-matched ΔF508 homozygotes being more often PS, older at first diagnosis and having lower sweat chloride concentrations (80 ±18 vs. 108 ±14 mmol/l; p < 0.001). No statistically significant differences between ΔF508 homozygotes and other compound heterozygotes were found in any other variable including nutritional indices, pulmonary function, pseudomonas colonization, chest radiology and complications of CF [38]. Contemporary studies have used an improved approach, correlating the predicted functional consequences of specific mutant alleles with the severity of PI [9]. The approach uses a five-class system to clarify the molecular consequences of ABCC7 mutation: class I: no CFTR protein formed, class II: CFTR mRNA formed but little functioning CFTR reaches the apical membrane due to misfolding/defective processing, class III: defective regulation of CFTR at the cell surface with no channel activity, class IV: reduced single channel conductance, and class V: reduced channel abundance. Mutations belonging to classes I–III are predicted to have severe functional consequences on CFTR function whilst mutations belonging to classes IV and V are expected to have some residual CFTR channel function [39, 40]. This approach has demonstrated

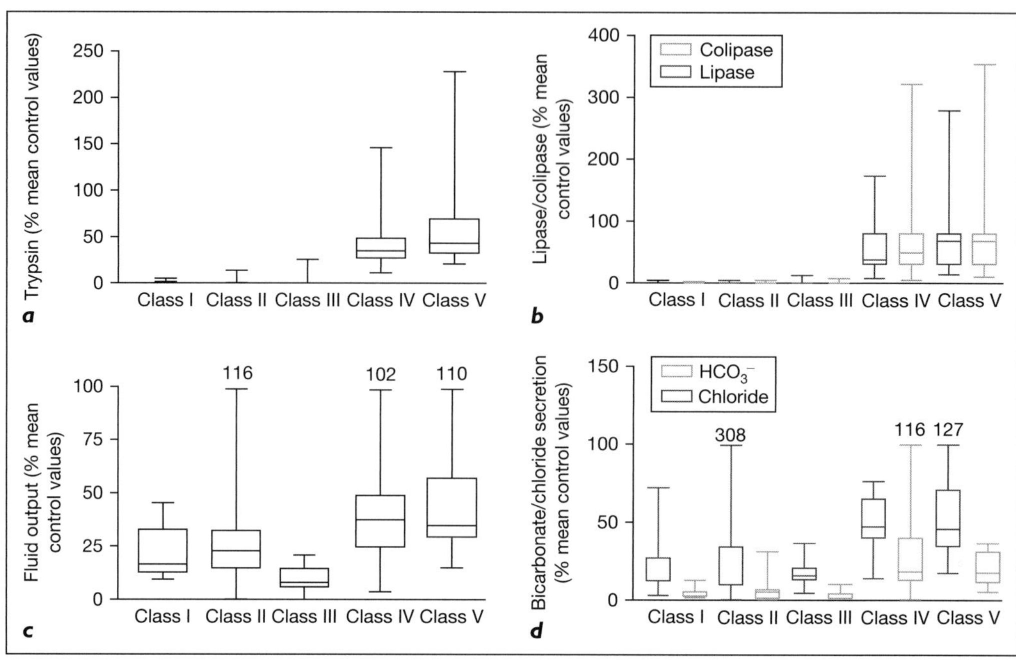

Fig. 2. Quantitative pancreatic acinar and ductular secretion following intravenous cholecystokinin/ secretin infusion in 70 patients, expressed as a percentage of mean control values. Trypsin (**a**), lipase and colipase (**b**), fluid output (**c**) and bicarbonate (HCO_3^-) and chloride secretion (**d**) are shown. Box margins represent the 25th and 75th percentiles, with the horizontal bar representing the median. Whiskers indicate the range of data. Comparison of classes I, II, and III with class IV and V mutations shows significant differences: p = 0.01 (fluid), p = 0.001 (bicarbonate and chloride), p < 0.0001 (trypsin, colipase and lipase) by Kruskal-Wallis, χ^2, with 4 d.f. [9].

a degree of preservation of both ductular and acinar function in individuals heterozygous for ΔF508 (a class II mutation) and a number of class IV and V mutations including ΔF508/R117H, ΔF508/R347H, ΔF508/S1251N, ΔF508/ V232D, ΔF508/3849 +10 kb C→T, ΔF508/3272-26A→G, ΔF508/A445E, ΔF508/P574H and ΔF508/5T [9] (fig. 2). Amongst patients with class IV and V mutations acinar function varied from 18% of control values to within the normal range. Three hundred and eighty-one patients were ΔF508/ΔF508. Sixteen of 381 had been PS at diagnosis although 15/16 had become PI. There were no significant differences in sweat chloride, FEV_1, weight for height, age at diagnosis nor direct indices of acinar function between PI and PS→PI groups [9]. Hence homozygosity or compound heterozygosity for class I, II and III mutations is strongly associated with PI status whereas compound heterozygosity for a class IV or V mutation is predictive of PS status.

Recurrent Acute and Idiopathic Chronic Pancreatitis
Some PS individuals with classical CF suffer from recurrent episodes of pancreatitis [41, 42] and a relationship

between mutations of ABCC7 and idiopathic chronic pancreatitis (ICP) is now well established [43, 44]. In the initial description of Sharer et al. [43] of 234 adults with chronic pancreatitis, 13% were heterozygous for an ABCC7 mutation and 10% had the 5T allele in intron 8 although only the most common mutations were screened for. A contemporaneous publication by Cohn et al. [44] concluded that the frequency of mutations in ABCC7 in ICP was 6 times higher than expected and the 5T allele twice as frequent as expected. Subsequent studies of patients with chronic pancreatitis in whom the entire ABCC7 coding sequence has been screened have demonstrated that 25–30% of individuals carry at least one mutation encoding CFTR [45, 46]. In these studies several patients were either compound heterozygous for mutations in ABCC7 or trans-heterozygous for an ABCC7 mutation and a mutation in SPINK1 (serine protease inhibitor, Kazal type 1) or PRSS1 (cationic trypsinogen, serine protease-1) [47–51]. Mutations in PRSS1 are associated with incomplete penetrance, with hereditary pancreatitis and SPINK1 is a modifier gene associated with ICP and with tropical calcific pancreatitis [19, 52–55].

There is increasing evidence that recurrent acute pancreatitis in individuals with classical CF is also associated with dysfunction of CFTR. In a heterogeneous series of children with acute pancreatitis Corpino et al. [56] found evidence of CFTR dysfunction in children with recurrent disease without an anatomical cause. Subsequent functional studies in individuals with pancreas divisum (a disorder found in up to 10% of the population and which is a bona fide risk factor for recurrent acute pancreatitis) [57] have strongly suggested a link between CFTR dysfunction and recurrent acute pancreatitis in patients with pancreas divisum [58]. Increasingly it is becoming apparent that CFTR dysfunction is an important risk factor in a 'multi-hit' hypothesis of the genesis of acute pancreatitis.

Pathophysiology of PI and Fat Maldigestion/Malabsorption

Exocrine pancreatic secretion in response to ingestion of a meal is divided into cephalic, gastric and intestinal phases. Vagal efferent nerves mediate the cephalic phase of secretion (principally VIPergic and GRPergic) promoting enzyme secretion rather than bicarbonate secretion. The gastric phase follows gastric distension, again involving vagal afferents and efferents and principally driving pancreatic enzyme secretion. The intestinal phase starts when gastric contents (gastric juice and meal) enter the duodenum as chyme. Secretin is released when duodenal pH falls below 4.5 resulting in ductular bicarbonate secretion. Cholinergic neural inputs potentiate this response. Intralumenal fatty acids, amino acids and to a lesser extent glucose result in pancreatic enzyme secretion via neurohumoral mechanisms. CCK is the major meal-related secretogogue which is released by duodenal mucosal enteroendocrine cells which continuously sample intestinal lumenal contents and act to promote secretion via vagal afferents [59]. Feedback inhibition of exocrine secretion is brought about by monitor peptide, secreted by the pancreas and CCK releasing peptide, secreted by the duodenum, which, if not digested by trypsin, will result in CCK release by enteroendocrine cells [60–62].

Functional PI can arise either as a consequence of failure of pancreatic acinar (and ductal) function (primary insufficiency) or because of inadequate/impaired neuroendocrine signalling to the exocrine pancreas (secondary insufficiency). Both mechanisms can be impaired in CF.

Primary PI

CFTR is necessary for bicarbonate secretion by the pancreatic duct where it functions as an anion channel that is permeable to both bicarbonate and chloride with experimentally measured HCO_3^- to Cl^- permeability ratio of between 0.5:1 and 0.02:1 [63–66] (fig.1). Mutated human CFTR may retain substantial chloride channel activity in the presence of reduced bicarbonate permeability [33]. CFTR mutations associated with PI often exhibit reduced bicarbonate permeability to levels <10% of normal but may have preserved chloride permeability; mutations associated with PS have bicarbonate transport ratios of 31 and 46% of the normal ratio [33]. Whilst residual HCO_3^- permeability has been explained by the presence of other CFTR-regulated bicarbonate transporters (e.g., SLC26) [34], there is now firm evidence in humans that CFTR is permeable to both Cl and HCO_3^- and that epithelial cells can selectively alter HCO_3^- permeability [67].

Impaired pancreatic ductular secretion of bicarbonate and chloride is the hallmark of CFTR dysfunction and is prevalent in individuals with ABCC7 mutations of all classes (I–V) when compared with healthy controls [9]. In association with impaired ductular function individuals with class IV and V mutations have some preservation of acinar function but those with class I mutations have severely impaired acinar function [9]. CFTR has been localized within the pancreas to both centroacinar and proximal intralobular duct cells and to the apical membranes of acinar cells [5, 6]. In rats, ductal release of bicarbonate into the pancreatic duct lumen couples endocytosis to exocytosis at the apical plasma membrane of the pancreatic acinar cell [68]. Recycling of the acinar apical membrane after exocytosis of the zymogen granule is influenced by lumenal pH and is impaired at the acidic ductal pH seen in association with impaired HCO_3^- secretion in CF. Hence pancreatic dysfunction in CF may also be in part due to progressive acidification of the acinar and duct lumen which impairs apical trafficking of zymogen granule membrane and solubilization of secretory enzymes/proenzymes [31]. Acidification of the acinar lumen of healthy cholecystokinin-stimulated acini results in histological findings similar to those observed in CF, including massive dilatation of the acinar lumen, decreased appearance of zymogen granules, loss of the apical pole of the acinar cell, and persistent aggregation of secretory (pro)enzymes released into the lumenal space [31]. It has also been suggested that pancreatic ductal dilatation in CF is not only a result of ductal obstruction by inspissated secretions but arises as a result of defective membrane recycling at the apical acinar plasma membrane [32].

It is tantalizing that severe (class I–III) mutations which commonly result in early onset atrophy of the exocrine pancreas and PI can occasionally (~1% of patients) be associated with clinical preservation of PS status (direct testing of pancreatic function may confirm exocrine impairment) [9, 69]. This variability has not, to date, been fully explained although other influences including modifier genes seem likely to be implicated [17, 18, 20, 21].

Secondary PI

Inflammatory conditions of the proximal small intestine are commonly associated with functional impairment of pancreatic exocrine secretion [70–74]. Morphometric analysis in such cases has demonstrated that even subtle inflammation within the duodenum is able to impair pancreatic secretion [73]. CF is associated with intestinal inflammation [75–79]. Emerging evidence suggests that, as in other tissues, this is a primary event in CF [80, 81]. The magnitude of any contribution of such secondary factors to the clinical impairment of pancreatic exocrine secretion in CF is an area of ongoing evaluation.

Other Factors Influencing Intralumenal Digestion in CF

The impaired secretion of digestive enzymes by the pancreas in CF represents only one part of the failure of intralumenal digestion and malabsorption encountered in CF [82]. Dietary fats are insoluble in water so that digestion by pancreatic lipase, a water-soluble enzyme, occurs at the oil-water interface. Mechanical mixing in the foregut emulsifies the fats into small droplets to facilitate intralumenal lipolysis. Emulsification is maintained through the combined contributions of polar lipids and bile salts such that stable particles approximately 0.5–1.0 μm in diameter are formed and maintained. In addition conjugated bile acids alter the pH optimum of pancreatic lipase from 8.9 to 6.7. Hence efficient duodenal hydrolysis of fat requires a number of events at the oil-water interface of duodenal chyme involving pancreatic lipase, colipase, phospholipase A_2, and bile salts in an environment where the pH must be close to neutrality and mechanical factors maintain mixing and emulsification. Lipolytic products (monoglycerides and free fatty acids) must then be shuttled through the unstirred water layer to the surface of the microvillus membrane by ionized bile salts, which must be present in sufficient concentrations to form micelles [reviewed in 82]. Failure of alkalinization of the duodenum arises in CF because of impaired pancreatic and duodenal HCO_3^- secretion and also gastric acid hypersecretion [83–85]. Acidic duodenal pH will impair lipase activity and precipitate bile acids so that bile acid concentrations will fall below the critical micellar concentration. Bacterial overgrowth of the small intestine in CF may result in deconjugation of bile acids which will both lower intralumenal bile acid concentrations and adversely alter the pH below the optimum for pancreatic lipase [86, 87]. Intrahepatic cholestasis and CF liver disease may further compromise the secretion of bile acids [88–91]. Deranged intestinal motility in CF has the potential to adversely influence emulsification and to further encourage bacterial overgrowth [79, 92]. There is emerging evidence that the absorptive phase which follows intralumenal digestion might also be directly impaired in CF [82, 93; Butt, unpubl. data]. Hence inefficient or ineffective intralumenal digestion by pancreatic enzymes in CF is not only the result of impaired secretion.

Pathophysiology of Recurrent Acute and Idiopathic Pancreatitis

The important observations that mutations of ABCC7 in PS patients are associated with ICP and recurrent acute pancreatitis are widely accepted [43, 44, 56, 58]. The mechanism whereby CFTR dysfunction is associated with pancreatic inflammation is less clear. The notion that poor intralumenal hydration of secreted proteins within the pancreatic ducts leads to inspissation of proteins within these small ducts resulting in episodes of autodigestion by entrapped autoactivated pancreatic enzymes is commonly cited as one factor. Given that the signs of pancreatic involvement in CF appear to be correlated with the gestational ontogeny of pancreatic trypsinogen expression it is perfectly logical to assume that the chronic pancreatitis of CF can be explained in the context of trypsinogen expression coupled with the inability to clear digestive enzymes from the pancreatic duct [1].

Patients with impaired, although not absent CFTR function are at risk of pancreatitis, which may be increased up to 80-fold [94]. Most people with this degree of impairment will not, however, develop pancreatitis. Hence a 'double hit' hypothesis of CFTR dysfunction unrelated to chloride channel dysfunction seems more likely. A variety of possibilities are being explored to explain this. It has been suggested that *CFTR* is one element of a polygenic disorder (for example involving *CFTR-SPINK1* heterozygosity and/or other genes) which predispose to pancreatitis [19, 53, 95]. Many other candidate genes have not however, at least in isolation, been associated with pancreatitis [96, 97]. It has also been suggested that mutations in ABCC7 usually involving exons 9 and 10 of the CFTR gene product which result in a CFTR molecule in which bicarbonate conductance is selectively altered might result in a disease phenotype which affects the pancreas more than the respiratory epithelium [33, 34, 67, 95]. Perhaps most attractive of all is

the notion that some CFTR variants may increase susceptibility to environmental factors including alcohol [98].
Polymorphisms in UDP-glucuronosyltransferase
(UGT1A7) are associated with alcoholic pancreatitis [99].
Downregulation of other genes associated with detoxifica-
tion of environmental toxins is described in gene array
analysis of animal models of CFTR dysfunction [16]. This
remains an area of ongoing investigation. It is an attractive
hypothesis that factors other than simple inspissation of
inadequately hydrated proteins within the pancreatic duct
are important.

Animal models of pancreatitis not infrequently use
'stress', for example cerulein hyperstimulation, to precipitate pancreatitits [100]. Hyperstimulation induces early
upregulation of P38MAPK and thereafter pancreatitis
[101, 102]. Recent studies have demonstrated that failure
to alkalinize the duodenum in animal models of CFTR dysfunction is associated with stress responses within the pancreatic acinus in which the inflammatory machinery of the
cell is activated [36, 37]. CFTR can activate such pathways
in respiratory epithelium independently of effects upon
chloride channel function [16, 81, 103–105]. There is a
strong teleological argument, therefore, to combine inhibition of P38MAPK activation with strategies to alkalinize
the duodenum to reduce the risk of pancreatitis in individuals with CFTR-related pancreatitis [102, 106–109].

Future Perspectives

There remain a number of challenges for the CF physician which constitute an exciting and fertile area for
research and clinical development with regard to pancreatic
disease. An understanding of the extent to which CF represents a primary inflammatory disease of the pancreatic acinus and the relationship of this to recurrent acute and
chronic pancreatitis will constitute a fruitful and clinically
important area for research effort.

References

1 Imrie JR, Fagan DG, Sturgess JM: Quantitative evaluation of the development of the
exocrine pancreas in cystic fibrosis and control infants. Am J Pathol 1979;95:697–707.
2 Fanconi G, Uehlinger E, Knauer C: Das
Coeliaksyndrom bei angeborener zystischer
Pankreasfibromatose und Bronchiektasien.
Wien Med Wochenschr 1936;86:753–756.
3 di Sant'Agnese PA, Darling RC, Perera GA,
Shea E: Abnormal electrolyte composition of
sweat in cystic fibrosis of the pancreas; clinical significance and relationship to the disease. Pediatrics 1953;12:549–563.
4 Gibbs GE, Bostick WL, Smith PM:
Incomplete pancreatic deficiency in cystic
fibrosis of the pancreas. J Pediatr 1950;37:
320–325.
5 Marino CR, Matovcik LM, Gorelick FS,
Cohn JA: Localization of the cystic fibrosis
transmembrane conductance regulator in
pancreas. J Clin Invest 1991;88:712–716.
6 Zeng W, Lee MG, Yan M, Diaz J, Benjamin I,
Marino CR, Kopito R, Freedman S, Cotton C,
Muallem S, Thomas P: Immuno and functional characterization of CFTR in submandibular and pancreatic acinar and duct
cells. Am J Physiol 1997;273:C442–C455.
7 Kopelman H, Durie P, Gaskin K, Weizman Z,
Forstner G: Pancreatic fluid secretion and
protein hyperconcentration in cystic fibrosis.
N Engl J Med 1985;312:329–334.
8 Kopelman H, Corey M, Gaskin K, Durie P,
Weizman Z, Forstner G: Impaired chloride
secretion, as well as bicarbonate secretion,
underlies the fluid secretory defect in the cys-

tic fibrosis pancreas. Gastroenterology
1988;95: 349–355.
9 Ahmed N, Corey M, Forstner G, Zielenski J,
Tsui LC, Ellis L, Tullis E, Durie P: Molecular
consequences of cystic fibrosis transmembrane regulator (CFTR) gene mutations in the
exocrine pancreas. Gut 2003;52:1159–1164.
10 Riordan JR, Rommens JM, Kerem B, Alon N,
Rozmahel R, Grzelczak Z, Zielenski J, Lok S,
Plavsic N, Chou JL: Identification of the cystic fibrosis gene: Cloning and characterization of complementary DNA. Science
1989;245: 1066–1073.
11 Rommens JM, Iannuzzi MC, Kerem B,
Drumm ML, Melmer G, Dean M, Rozmahel
R, Cole JL, Kennedy D, Hidaka N: Identification of the cystic fibrosis gene: Chromosome walking and jumping. Science 1989;
245:1059–1065.
12 Kerem E, Corey M, Kerem BS, Rommens J,
Markiewicz D, Levison H, Tsui LC, Durie P:
The relation between genotype and phenotype in cystic fibrosis– Analysis of the most
common mutation (delta F508). N Engl J
Med 1990;323:1517–1522.
13 Reddy MM, Light MJ, Quinton PM:
Activation of the epithelial Na$^+$ channel
(ENaC) requires CFTR Cl$^-$ channel function.
Nature 1999;402:301–304.
14 Konig J, Schreiber R, Voelcker T, Mall M,
Kunzelmann K: The cystic fibrosis transmembrane conductance regulator (CFTR)
inhibits ENaC through an increase in the
intracellular Cl$^-$ concentration. EMBO Rep
2001;2: 1047–1051.

15 Kunzelmann K: CFTR: Interacting with
everything? News Physiol Sci 2001;16:
167–170.
16 Norkina O, Kaur S, Ziemer D, De Lisle RC:
Inflammation of the cystic fibrosis mouse
small intestine. Am J Physiol Gastrointest
Liver Physiol 2004;286:G1032–G1041.
17 Acton JD, Wilmott RW: Phenotype of CF and
the effects of possible modifier genes.
Paediatr Respir Rev 2001;2:332–339.
18 Drumm ML: Modifier genes and variation in
cystic fibrosis. Respir Res 2001;2:125–128.
19 Pfutzer RH, Barmada MM, Brunskill AP,
Finch R, Hart PS, Neoptolemos J, Furey WF,
Whitcomb DC: SPINK1/PSTI polymorphisms
act as disease modifiers in familial and idiopathic chronic pancreatitis. Gastroenterology
2000;119:615–623.
20 Pirzada O, Taylor C: Modifier genes and cystic fibrosis liver disease. Hepatology 2003;
37: 714.
21 Salvatore F, Scudiero O, Castaldo G: Genotype-phenotype correlation in cystic fibrosis:
The role of modifier genes. Am J Med Genet
2002;111:88–95.
22 Knowles MR, Durie PR: What is cystic fibrosis? N Engl J Med 2002;347:439–442.
23 Harris A: The duct cell in cystic fibrosis. Ann
NY Acad Sci 1999;880:17–30.
24 Reid CJ, Hyde K, Ho SB, Harris A: Cystic
fibrosis of the pancreas: Involvement of
MUC6 mucin in obstruction of pancreatic
ducts. Mol Med 1997;3:403–411.
25 Kopito LE, Shwachman H: The pancreas in
cystic fibrosis: Chemical composition and

comparative morphology. Pediatr Res 1976;
10:742–749.

26 Oppenheimer EH, Esterly JR: Cystic fibrosis of the pancreas. Morphologic findings in infants with and without diagnostic pancreatic lesions. Arch Pathol 1973;96: 149–154.

27 Durie PR, Kent G, Phillips MJ, Ackerley CA: Characteristic multiorgan pathology of cystic fibrosis in a long-living cystic fibrosis transmembrane regulator knockout murine model. Am J Pathol 2004;164:1481–1493.

28 Waters DL, Dorney SF, Gaskin KJ, Gruca MA, O'Halloran M, Wilcken B: Pancreatic function in infants identified as having cystic fibrosis in a neonatal screening program. N Engl J Med 1990;322:303–308.

29 Gaskin K, Gurwitz D, Durie P, Corey M, Levison H, Forstner G: Improved respiratory prognosis in patients with cystic fibrosis with normal fat absorption. J Pediatr 1982;100: 857–862.

30 Couper RT, Corey M, Moore DJ, Fisher LJ, Forstner GG, Durie PR: Decline of exocrine pancreatic function in cystic fibrosis patients with pancreatic sufficiency. Pediatr Res 1992; 32:179–182.

31 Scheele GA, Fukuoka SI, Kern HF, Freedman SD: Pancreatic dysfunction in cystic fibrosis occurs as a result of impairments in luminal pH, apical trafficking of zymogen granule membranes, and solubilization of secretory enzymes. Pancreas 1996;12:1–9.

32 Freedman SD, Kern HF, Scheele GA: Pancreatic acinar cell dysfunction in CFTR($-$/$-$) mice is associated with impairments in luminal pH and endocytosis. Gastroenterology 2001;121:950–957.

33 Choi JY, Muallem D, Kiselyov K, Lee MG, Thomas PJ, Muallem S: Aberrant CFTR-dependent HCO_3^- transport in mutations associated with cystic fibrosis. Nature 2001; 410:94–97.

34 Ko SB, Shcheynikov N, Choi JY, Luo X, Ishibashi K, Thomas PJ, Kim JY, Kim KH, Lee MG, Naruse S, Muallem S: A molecular mechanism for aberrant CFTR-dependent HCO_3^- transport in cystic fibrosis. EMBO J 2002;21:5662–5672.

35 Clarke LL, Harline MC: Dual role of CFTR in cAMP-stimulated. Am J Physiol 1998;274: G718–G726.

36 De Lisle RC, Isom KS, Ziemer D, Cotton CU: Changes in the exocrine pancreas secondary to altered small intestinal function in the CF mouse. Am J Physiol Gastrointest Liver Physiol 2001;281:G899–G906.

37 Kaur S, Norkina O, Ziemer D, Samuelson LC, De Lisle RC: Acidic duodenal pH alters gene expression in the cystic fibrosis mouse pancreas. Am J Physiol Gastrointest Liver Physiol 2004;287:G480–G490.

38 Correlation between genotype and phenotype in patients with cystic fibrosis. The Cystic Fibrosis Genotype-Phenotype Consortium. N Engl J Med 1993;329:1308–1313.

39 Welsh MJ, Smith AE: Molecular mechanisms of CFTR chloride channel dysfunction in cystic fibrosis. Cell 1993;73:1251–1254.

40 Wilschanski M, Zielenski J, Markiewicz D, Tsui LC, Corey M, Levison H, Durie PR: Correlation of sweat chloride concentration with classes of the cystic fibrosis transmembrane conductance regulator gene mutations. J Pediatr 1995;127:705–710.

41 Shwachman H, Lebenthal E, Khaw KT: Recurrent acute pancreatitis in patients with cystic fibrosis with normal pancreatic enzymes. Pediatrics 1975;55:86–95.

42 Atlas AB, Orenstein SR, Orenstein DM: Pancreatitis in young children with cystic fibrosis. J Pediatr 1992;120:756–759.

43 Sharer N, Schwarz M, Malone G, Howarth A, Painter J, Super M, Braganza J: Mutations of the cystic fibrosis gene in patients with chronic pancreatitis. N Engl J Med 1998;339: 645–652.

44 Cohn JA, Friedman KJ, Noone PG, Knowles MR, Silverman LM, Jowell PS: Relation between mutations of the cystic fibrosis gene and idiopathic pancreatitis. N Engl J Med 1998;339:653–658.

45 Audrezet MP, Chen JM, Le Marechal C, Ruszniewski P, Robaszkiewicz M, Raguenes O, Quere I, Scotet V, Ferec C: Determination of the relative contribution of three genes – the cystic fibrosis transmembrane conductance regulator gene, the cationic trypsinogen gene, and the pancreatic secretory trypsin inhibitor gene – to the etiology of idiopathic chronic pancreatitis. Eur J Hum Genet 2002;10: 100–106.

46 Noone PG, Zhou Z, Silverman LM, Jowell PS, Knowles MR, Cohn JA: Cystic fibrosis gene mutations and pancreatitis risk: Relation to epithelial ion transport and trypsin inhibitor gene mutations. Gastroenterology 2001;121: 1310–1319.

47 Pfutzer R, Myers E, Applebaum-Shapiro S, Finch R, Ellis I, Neoptolemos J, Kant JA, Whitcomb DC: Novel cationic trypsinogen (PRSS1) N29T and R122C mutations cause autosomal dominant hereditary pancreatitis. Gut 2002;50:271–272.

48 Pfutzer RH, Whitcomb DC: Trypsinogen mutations in chronic pancreatitis. Gastroenterology 1999;117:1507–1508.

49 Whitcomb DC: Hereditary pancreatitis: A model for understanding the genetic basis of acute and chronic pancreatitis. Pancreatology 2001;1:565–570.

50 Witt H, Luck W, Hennies HC, Classen M, Kage A, Lass U, Landt O, Becker M: Mutations in the gene encoding the serine protease inhibitor, Kazal type 1 are associated with chronic pancreatitis. Nat Genet 2000;25: 213–216.

51 Witt H, Luck W, Becker M, Bohmig M, Kage A, Truninger K, Ammann RW, O'Reilly D, Kingsnorth A, Schulz HU, Halangk W, Kielstein V, Knoefel WT, Teich N, Keim V: Mutation in the SPINK1 trypsin inhibitor gene, alcohol use, and chronic pancreatitis. JAMA 2001;285:2716–2717.

52 Bhatia E, Choudhuri G, Sikora SS, Landt O, Kage A, Becker M, Witt H: Tropical calcific pancreatitis: Strong association with SPINK1 trypsin inhibitor mutations. Gastroenterology 2002;123:1020–1025.

53 Hassan Z, Mohan V, Ali L, Allotey R, Barakat K, Faruque MO, Deepa R, McDermott MF, Jackson AE, Cassell P, Curtis D, Gelding SV, Vijayaravaghan S, Gyr N, Whitcomb DC, Khan AK, Hitman GA: SPINK1 is a susceptibility gene for fibrocalculous pancreatic diabetes in subjects from the Indian subcontinent. Am J Hum Genet 2002;71:964–968.

54 Schneider A, Suman A, Rossi L, Barmada MM, Beglinger C, Parvin S, Sattar S, Ali L, Khan AK, Gyr N, Whitcomb DC: SPINK1/PSTI mutations are associated with tropical pancreatitis and type II diabetes mellitus in Bangladesh. Gastroenterology 2002;123: 1026–1030.

55 Threadgold J, Greenhalf W, Ellis I, Howes N, Lerch MM, Simon P, Jansen J, Charnley R, Laugier R, Frulloni L, Olah A, Delhaye M, Ihse I, Schaffalitzky de Muckadell OB, Andren-Sandberg A, Imrie CW, Martinek J, Gress TM, Mountford R, Whitcomb D, Neoptolemos JP: The N34S mutation of SPINK1 (PSTI) is associated with a familial pattern of idiopathic chronic pancreatitis but does not cause the disease. Gut 2002;50: 675–681.

56 Corpino M, Thompson M, Milla P, Jaffe A, Lindley KJ: Evidence of CFTR dysfunction in children with recurrent pancreatitis. Arch Dis Child 2003;88(suppl 1):A15.

57 DiMagno EP: Toward understanding (and management) of painful chronic pancreatitis. Gastroenterology 1999;116:1252–1257.

58 Gelrud A, Sheth S, Banerjee S, Weed D, Shea J, Chuttani R, Howell DA, Telford JJ, Carr-Locke DL, Regan MM, Ellis L, Durie PR, Freedman SD: Analysis of cystic fibrosis gene product (CFTR) function in patients with pancreas divisum and recurrent acute pancreatitis. Am J Gastroenterol 2004;99:1557–1562.

59 Owyang C, Logsdon CD: New insights into neurohormonal regulation of pancreatic secretion. Gastroenterology 2004;127:957–969.

60 Liddle RA: Regulation of cholecystokinin secretion in humans. J Gastroenterol 2000;35: 181–187.

61 McVey DC, Romac JM, Clay WC, Kost TA, Liddle RA, Vigna SR: Monitor peptide binding sites are expressed in the rat liver and small intestine. Peptides 1999;20:457–464.

62 Wang Y, Prpic V, Green GM, Reeve JR Jr, Liddle RA: Luminal CCK-releasing factor stimulates CCK release from human intestinal endocrine and STC-1 cells. Am J Physiol Gastrointest Liver Physiol 2002;282: G16–G22.

63 Drumm ML, Pope HA, Cliff WH, Rommens JM, Marvin SA, Tsui LC, Collins FS, Frizzell RA, Wilson JM: Correction of the cystic fibrosis defect in vitro by retrovirus-mediated gene transfer. Cell 1990;62:1227–1233.

64 Linsdell P, Tabcharani JA, Rommens JM, Hou YX, Chang XB, Tsui LC, Riordan JR, Hanrahan JW: Permeability of wild-type and mutant cystic fibrosis transmembrane conductance regulator chloride channels to polyatomic anions. J Gen Physiol 1997;110: 355–364.

65 O'Reilly CM, Winpenny JP, Argent BE, Gray MA: Cystic fibrosis transmembrane

conductance regulator currents in guinea pig pancreatic duct cells: Inhibition by bicarbonate ions. Gastroenterology 2000;118: 1187–1196.

66 Poulsen JH, Fischer H, Illek B, Machen TE: Bicarbonate conductance and pH regulatory capability of cystic fibrosis transmembrane conductance regulator. Proc Natl Acad Sci USA 1994;91:5340–5344.

67 Reddy MM, Quinton PM: Control of dynamic CFTR selectivity by glutamate and ATP in epithelial cells. Nature 2003;423:756–760.

68 Freedman SD, Kern HF, Scheele GA: Acinar lumen pH regulates endocytosis, but not exocytosis, at the apical plasma membrane of pancreatic acinar cells. Eur J Cell Biol 1998; 75:153–162.

69 Gaskin KJ, Durie PR, Lee L, Hill R, Forstner GG: Colipase and lipase secretion in childhood-onset pancreatic insufficiency. Delineation of patients with steatorrhea secondary to relative colipase deficiency. Gastroenterology 1984;86:1–7.

70 Carroccio A, Verghi F, Santini B, Lucidi V, Iacono G, Cavataio F, Soresi M, Ansaldi N, Castro M, Montalto G: Diagnostic accuracy of fecal elastase 1 assay in patients with pancreatic maldigestion or intestinal malabsorption: A collaborative study of the Italian Society of Pediatric Gastroenterology and Hepatology. Dig Dis Sci 2001;46: 1335–1342.

71 Gomez JC, Moran CE, Maurino EC, Bai JC: Exocrine pancreatic insufficiency in celiac disease. Gastroenterology 1998;114:621–623.

72 Salvatore S, Finazzi S, Barassi A, Verzelletti M, Tosi A, Melzi d'Eril GV, Nespoli L: Low fecal elastase: Potentially related to transient small bowel damage resulting from enteric pathogens. J Pediatr Gastroenterol Nutr 2003; 36:392–396.

73 Schappi MG, Smith VV, Cubitt D, Milla PJ, Lindley KJ: Faecal elastase 1 concentration is a marker of duodenal enteropathy. Arch Dis Child 2002;86:50–53.

74 Walkowiak J, Herzig KH: Fecal elastase-1 is decreased in villous atrophy regardless of the underlying disease. Eur J Clin Invest 2001;31: 425–430.

75 Smyth RL, Croft NM, O'Hea U, Marshall TG, Ferguson A: Intestinal inflammation in cystic fibrosis. Arch Dis Child 2000;82: 394–399.

76 Bruzzese E, Raia V, Gaudiello G, Polito G, Buccigrossi V, Formicola V, Guarino A: Intestinal inflammation is a frequent feature of cystic fibrosis and is reduced by probiotic administration. Aliment Pharmacol Ther 2004;20:813–819.

77 Maiuri L, Raia V, De Marco G, Coletta S, de Ritis G, Londei M, Auricchio S: DNA fragmentation is a feature of cystic fibrosis epithelial cells: A disease with inappropriate apoptosis? FEBS Lett 1997;408: 225–231.

78 Raia V, Maiuri L, de Ritis G, de Vizia B, Vacca L, Conte R, Auricchio S, Londei M: Evidence of chronic inflammation in morphologically normal small intestine of cystic fibrosis patients. Pediatr Res 2000;47: 344–350.

79 Smith VV, Schappi MG, Milla PJ, Lindley KJ: Distal ileal obstruction in cystic fibrosis is associated with lymphocytic leiomyositis and myenteric ganglionitis. Submitted.

80 Kiparissi F, Smith VV, Trimarco G, Jaffe A, Milla PJ, Lindley KJ: Meconium ileus is associated with transmural ileal leiomyositis and ileitis. J Pediatr Gastroenterol Nutr 2004; 39(suppl 1):S166.

81 Weber AJ, Soong G, Bryan R, Saba S, Prince A: Activation of NF-kappaB in airway epithelial cells is dependent on CFTR trafficking and Cl⁻ channel function. Am J Physiol Lung Cell Mol Physiol 2001;281: L71–L78.

82 Roy CC, Weber AM, Lepage G, Smith L, Levy E: Digestive and absorptive phase anomalies associated with the exocrine pancreatic insufficiency of cystic fibrosis. J Pediatr Gastroenterol Nutr 1988;7(suppl 1): S1–S7.

83 Robinson PJ, Smith AL, Sly PD: Duodenal pH in cystic fibrosis and its relationship to fat malabsorption. Dig Dis Sci 1990;35: 1299–1304.

84 Cox KL, Isenberg JN, Ament ME: Gastric acid hypersecretion in cystic fibrosis. J Pediatr Gastroenterol Nutr 1982;1:559–565.

85 Barraclough M, Taylor CJ: Twenty-four hour ambulatory gastric and duodenal pH profiles in cystic fibrosis: Effect of duodenal hyperacidity on pancreatic enzyme function and fat absorption. J Pediatr Gastroenterol Nutr 1996; 23:45–50.

86 O'Brien S, Mulcahy H, Fenlon H, O'Broin A, Casey M, Burke A, FitzGerald MX, Hegarty JE: Intestinal bile acid malabsorption in cystic fibrosis. Gut 1993;34:1137–1141.

87 Lewindon PJ, Robb TA, Moore DJ, Davidson GP, Martin AJ: Bowel dysfunction in cystic fibrosis: Importance of breath testing. J Paediatr Child Health 1998;34:79–82.

88 O'Brien SM, Campbell GR, Burke AF, Maguire OC, Rowlands BJ, FitzGerald MX, Hegarty JE: Serum bile acids and ursodeoxycholic acid treatment in cystic fibrosis-related liver disease. Eur J Gastroenterol Hepatol 1996;8:477–483.

89 Feranchak AP, Sokol RJ: Cholangiocyte biology and cystic fibrosis liver disease. Semin Liver Dis 2001;21:471–488.

90 Molmenti EP, Squires RH, Nagata D, Roden JS, Molmenti H, Fasola CG, Prestidge C, D'Amico L, Casey D, Sanchez EQ, Goldstein RM, Levy MF, Benser M, McPhail W, Andrews W, Andersen JA, Klintmalm GB: Liver transplantation for cholestasis associated with cystic fibrosis in the pediatric population. Pediatr Transplant 2003;7:93–97.

91 Shapira R, Hadzic N, Francavilla R, Koukulis G, Price JF, Mieli-Vergani G: Retrospective review of cystic fibrosis presenting as infantile liver disease. Arch Dis Child 1999;81: 125–128.

92 Schappi MG, Roulet M, Rochat T, Belli DC: Electrogastrography reveals post-prandial gastric dysmotility in children with cystic fibrosis. J Pediatr Gastroenterol Nutr 2004;39:253–256.

93 Kalivianakis M, Minich DM, Bijleveld CM, van Aalderen WM, Stellaard F, Laseur M, Vonk RJ, Verkade HJ: Fat malabsorption in

cystic fibrosis patients receiving enzyme replacement therapy is due to impaired intestinal uptake of long-chain fatty acids. Am J Clin Nutr 1999;69:127–134.

94 Cohn JA, Bornstein JD, Jowell PS: Cystic fibrosis mutations and genetic predisposition to idiopathic chronic pancreatitis. Med Clin North Am 2000;84:621–631, ix.

95 Whitcomb DC: Value of genetic testing in the management of pancreatitis. Gut 2004;53: 1710–1717.

96 Schneider A, Pogue-Geile K, Barmada MM, Myers-Fong E, Thompson BS, Whitcomb DC: Hereditary, familial, and idiopathic chronic pancreatitis are not associated with polymorphisms in the tumor necrosis factor alpha (TNF-alpha) promoter region or the TNF receptor 1 (TNFR1) gene. Genet Med 2003;5: 120–125.

97 Schneider A, Togel S, Barmada MM, Whitcomb DC: Genetic analysis of the glutathione s-transferase genes MGST1, GSTM3, GSTT1, and GSTM1 in patients with hereditary pancreatitis. J Gastroenterol 2004; 39:783–787.

98 Casals T, Aparisi L, Martinez-Costa C, Gimenez J, Ramos MD, Mora J, Diaz J, Boadas J, Estivill X, Farre A: Different CFTR mutational spectrum in alcoholic and idiopathic chronic pancreatitis? Pancreas 2004;28: 374–379.

99 Ockenga J, Vogel A, Teich N, Keim V, Manns MP, Strassburg CP: UDP glucuronosyltransferase (UGT1A7) gene polymorphisms increase the risk of chronic pancreatitis and pancreatic cancer. Gastroenterology 2003; 124:1802–1808.

100 Kern HF, Adler G, Scheele GA: Structural and biochemical characterization of maximal and supramaximal hormonal stimulation of rat exocrine pancreas. Scand J Gastroenterol Suppl 1985;112:20–29.

101 Hofken T, Keller N, Fleischer F, Goke B, Wagner AC: Map kinase phosphatases (MKP's) are early responsive genes during induction of cerulein hyperstimulation pancreatitis. Biochem Biophys Res Commun 2000; 276:680–685.

102 Wagner AC, Metzler W, Hofken T, Weber H, Goke B: p38 map kinase is expressed in the pancreas and is immediately activated following cerulein hyperstimulation. Digestion 1999;60:41–47.

103 Blackwell TS, Stecenko AA, Christman JW: Dysregulated NF-kappaB activation in cystic fibrosis: Evidence for a primary inflammatory disorder. Am J Physiol Lung Cell Mol Physiol 2001;281:L69–L70.

104 Tabary O, Escotte S, Couetil JP, Hubert D, Dusser D, Puchelle E, Jacquot J: Genistein inhibits constitutive and inducible NFkappaB activation and decreases IL-8 production by human cystic fibrosis bronchial gland cells. Am J Pathol 1999;155:473–481.

105 Tabary O, Escotte S, Couetil JP, Hubert D, Dusser D, Puchelle E, Jacquot J: Relationship between IkappaBalpha deficiency, NFkappaB activity and interleukin-8 production in CF human airway epithelial cells. Pflügers Arch 2001;443(suppl 1):S40–S44.

106 Kumar S, Boehm J, Lee JC: p38 MAP kinases: Key signalling molecules as therapeutic targets for inflammatory diseases. Nat Rev Drug Discov 2003;2:717–726.

107 Lee JC, Kumar S, Griswold DE, Underwood DC, Votta BJ, Adams JL: Inhibition of p38 MAP kinase as a therapeutic strategy. Immunopharmacology 2000;47:185–201.

108 Nick JA, Young SK, Arndt PG, Lieber JG, Suratt BT, Poch KR, Avdi NJ, Malcolm KC, Taube C, Henson PM, Worthen GS: Selective suppression of neutrophil accumulation in ongoing pulmonary inflammation by systemic inhibition of p38 mitogen-activated protein kinase. J Immunol 2002;169: 5260–5269.

109 Underwood DC, Osborn RR, Bochnowicz S, Webb EF, Rieman DJ, Lee JC, Romanic AM, Adams JL, Hay DW, Griswold DE: SB 239063, a p38 MAPK inhibitor, reduces neutrophilia, inflammatory cytokines, MMP-9, and fibrosis in lung. Am J Physiol Lung Cell Mol Physiol 2000;279:L895–L902.

Dr. Keith J. Lindley
Pancreatic and Gastrointestinal Services
Institute of Child Health
University College London
30 Guilford Street
London WC1N 1EH (UK)
Tel. +44 20 7905 2111
Fax +44 20 7404 6181
E-Mail K.Lindley@ich.ucl.ac.uk

Bush A, Alton EWFW, Davies JC, Griesenbach U, Jaffe A (eds): Cystic Fibrosis in the 21st Century.
Prog Respir Res. Basel, Karger, 2006, vol 34, pp 251–261

Cystic Fibrosis: Liver Disease

D. Westaby

Chelsea and Westminster Hospital, and Imperial College Medical School, London, UK

Abstract

Evidence of chronic liver disease is found in 25% of patients with cystic fibrosis (CF) and is the cause of liver decompensation in 2–3%. Liver injury is secondary to bile duct plugging and secondary bile-acid-related toxicity; almost all cases present in the first two decades of life. The marked variation in the presence and severity of disease may be due to modifier genes. Most cases are detected on routine screening and only a small proportion present with variceal bleeding, ascites or persistent jaundice. Abnormalities of liver function tests have a low sensitivity and specificity and the presence of established cirrhosis will be diagnosed on imaging. There is some evidence that the biliary liver disease of CF responds to ursodeoxycholic acid, although the degree of benefit remains uncertain. Liver transplantation has been successfully undertaken in the presence of isolated liver decompensation with maintained pulmonary function. Specific complications of cirrhosis including variceal haemorrhage, ascites and encephalopathy are managed by standard techniques applicable to all types of cirrhosis. There is accumulating evidence that established compensated cirrhosis does not adversely affect the outcome from lung transplantation.

Liver involvement is well recognized in cystic fibrosis (CF) and may also be an occasional dominant manifestation. Estimates of the prevalence of liver disease depend upon the tools and definitions used, the earliest from postmortems suggesting that in excess of 70% of adults in the third decade had some evidence of focal biliary cirrhosis (the pathognomic lesion of CF liver disease) [1]. Recent prospective studies suggest that approximately 20–25% of CF patients will develop liver disease but only 6–8% of these will have established cirrhosis [2, 3], the majority of which will present in the first 20 years of life. With the vast improvements in the care of pulmonary complications of CF as well as the availability of lung transplantation, it might have been anticipated that the prevalence of liver disease in a more elderly adult population would increase, but this does not appear to be the case.

Pathogenesis

The pathogenesis of chronic liver disease in CF is illustrated in figure 1. The characteristic hepatic lesion in CF is focal biliary cirrhosis consistent with that seen in partial biliary obstruction (fig. 2). The plugging of intra-hepatic bile ducts is similar to that seen in the pancreatic ducts of CF patients [4]. CF transmembrane conductance regulator (CFTR) has been localized to the apical membrane of the cells lining the intra-hepatic bile ducts [5]. The abnormalities of chloride transport inhibit the hydration of the canalicular-produced bile, resulting in increased viscosity. In addition, intra-hepatic biliary epithelial cells produce excessive mucus composed of proteoglycans, which increases the viscosity of CF bile [6]. The initial focal distribution of cirrhotic changes in CF can be explained by early patchy plugging of the intra-hepatic ducts; with increasing ductular involvement, the process becomes more diffuse, producing a fully established biliary cirrhosis with pan liver involvement. Whether biliary duct obstruction is alone sufficient to account for this process remains controversial.

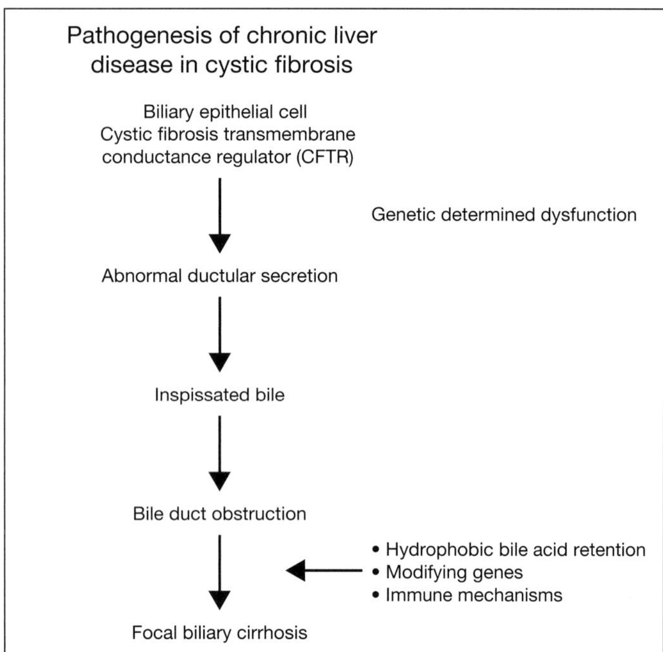

Fig. 1. Pathogenesis of chronic liver disease in cystic fibrosis.

Fig. 2. Post mortem liver showing the typical multifocal nodularity of cystic fibrosis related cirrhosis.

A study of CF liver disease based on both light and electron microscopy demonstrated features more in keeping with a destructive bile duct lesion than obstruction alone [7]; the authors suggested a bile-related toxin as the most likely explanation for these findings. Changes in the composition of the bile acid pool have also been suggested as possible causes [8]. Although studies have shown no significant difference in the serum bile acid profile between those with and without evidence of liver disease, the crude measures used may not accurately reflect the exposure of the hepatocyte to potentially hepatotoxic bile acids. Bile salt output remains normal or modestly reduced in CF, although the total volume of bile is significantly decreased [8, 9]. Thus a high concentration of bile acid is generated within the intrahepatic bile ducts. If there is partial or complete ductal obstruction, bile acid reflux could occur, exposing the hepatocyte to a high concentration of potentially toxic lipophilic bile acids, either primary (chenodeoxycholic acid) or secondary (deoxycholic and lithocholic acids).

Although these hypotheses provide a possible aetiological basis for chronic CF liver disease, they fail to account for the absence of liver involvement in the majority of patients or the wide spectrum of severity in those in whom this does occur. Speculation that with time the vast majority of patients with CF will develop liver disease has not been substantiated [10] and attempts to correlate the *CFTR* genotype with development of liver disease have been unsuccessful

[11]. It has been proposed that other factors such as environmental, nutritional or non-*CFTR* genetic influences may be important. Mutations in the alpha-1-antitrypsin, mannose-binding lectin (MBL) [12] and glutathione-S-transferase [13] genes may act as independent risk factors for CF liver disease (chapter 10). It has also been suggested that obstruction of the common bile duct as it passes through the diseased pancreas may contribute to liver disease as biliary cirrhosis is a recognized complication of long-term bile duct obstruction in chronic pancreatitis of other causes [14] although some studies have not confirmed this [15]. The presence of a sub-population of lymphocytes, cytotoxic to hepatocytes and directed towards the liver-specific lipoprotein, suggests that immune mechanisms might also be involved in the pathogenesis of CF liver disease [16].

Clinical Features

Deep cholestasis secondary to common bile duct obstruction with inspissated bile may be the earliest manifestation of CF [17, 18]. Fatty infiltration of the liver may sometimes produce massive hepatomegaly and abdominal distension, complicated by hypoglycaemia [19, 20]. Evidence of underlying cirrhosis may occur at any time, but new diagnoses are most frequently made during the first two decades of life. Historically, many cases of established chronic liver disease were detected as part of routine follow-up in patients with an established diagnosis of CF [21], hepato-splenomegaly being the commonest presentation. Abnormal liver function tests are common in CF; they may be of no significance, but might also be the only indicator of underlying chronic liver disease. Many large centres have established routine surveillance including sequential ultrasound scanning (see below), which has identified a small proportion of patients with no other clinical or laboratory evidence of liver disease. Variceal bleeding may be the presenting feature of established portal hypertension and may occur in the absence of any other signs of decompensation. As in other types of biliary cirrhosis, portal hypertension may occur in a pre-cirrhotic phase because of the pre-sinusoidal component to portal vascular resistance. Signs of decompensated biliary cirrhosis (jaundice, ascites or encephalopathy) are very unusual presenting features. In general, the clinical picture is one of very slowly progressive liver disease (as seen in other biliary cirrhotic disease such as primary biliary cirrhosis or primary sclerosing cholangitis [22, 23]). The natural history is, in fact, usually interrupted by mortality related to pulmonary disease.

Controversy remains as to whether the adverse effect of liver disease in CF is restricted to the 2–3% of patients with overt liver decompensation and the 1–2% with variceal bleeding, or it confers an adverse prognosis per se. One large study showed a decline in the prevalence of liver disease in the third decade, raising the possibility of premature mortality in those with the complication [10]. A large time-dependent, multivariate analysis reported liver disease as an independent risk factor for mortality in addition to pulmonary function and nutrition [24], although the mechanism has not been fully explained. The systemic and pulmonary haemodynamic abnormalities seen in all types of cirrhosis are also present in patients with CF liver disease [25]; the low peripheral vascular resistance, high cardiac out-put and increased pulmonary shunting might adversely affect patients with advanced pulmonary disease, although prospective studies will be required to determine whether this is the case.

Investigations

Liver Function Tests
Standard liver function tests have reasonable sensitivity but poor specificity, which is not surprising as they may be influenced by many factors such as infection, hypoxemia and medications. This is particularly the case with amino

Table 1. The ultrasound scoring system

Score	1	2	3
Hepatic parenchyma	Normal	Coarse	Irregular
Liver edge	Smooth	–	Nodular
Periportal fibrosis	None	Moderate	Severe

transferase levels. Markers of a biliary component such as alkaline phosphatase and γ-glutamine transpeptidase may be more helpful, particularly when they are elevated by a factor of 3–4, sustained over a period of months [26]. However, it must be remembered that a small proportion of patients with established cirrhosis will have entirely normal liver function tests [10, 26]. The most important role for standard liver function tests is to initiate a search for possible underlying liver disease, specifically, liver imaging.

Ultrasound Scanning
The availability of high quality transabdominal ultrasound scanning has provided a means of detecting liver disease cheaply and readily. In experienced hands ultrasound can both diagnose diffuse cirrhosis and detect focal disease [27]. Splenomegaly, a dilated portal vein and collateral vessels are all important markers of portal hypertension [27]. The use of doppler studies allows the detection of portal or splenic vein thrombosis, the incidence of which is increased in CF, most commonly as a consequence of associated chronic pancreatitis. An ultrasound scoring system has been developed for the detection of chronic CF liver disease in adults (table 1) [28] based on irregularity of the parenchyma and liver edge and periportal fibrosis (fig. 3).

Magnetic Resonance Imaging
A recent study has investigated the use of magnetic resonance imaging (MRI) and magnetic resonance cholangiopancreatography (MRCP) in the documentation of CF liver disease [29]. These techniques have produced excellent definition of the cirrhotic liver and the collateral circulation associated with portal hypertension (fig. 4). The MRCP technique has allowed visualization of the biliary tree required in the detection and management of common bile duct stones. With improving resolution, MRCP may also define the intra-hepatic biliary tree and the abnormalities of calibre characteristic of CF-related liver disease [30].

Radionuclide Imaging
Derivatives of iminodiacetic acid (IDA) labelled with technetium-99^m provide an alternative means of assessing

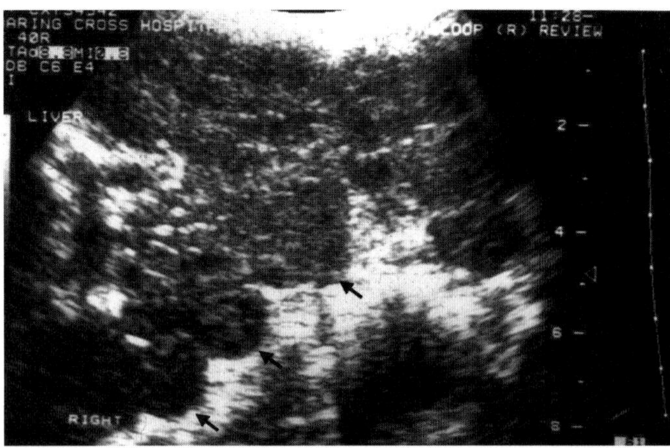

Fig. 3. Hepatic ultrasound showing marked nodularity of the liver surface characteristic of established cirrhosis.

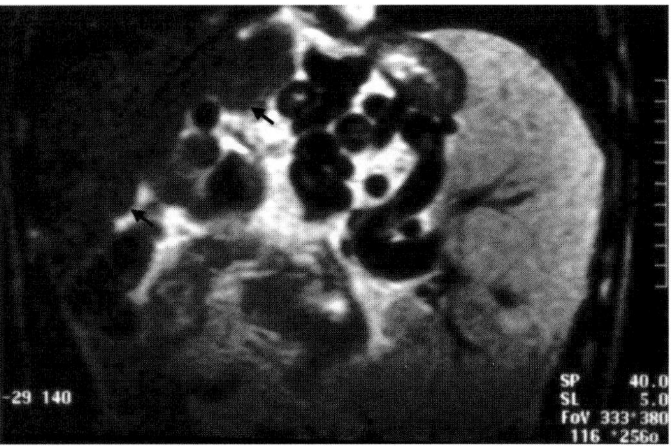

Fig. 4. A TW2 abdominal MRI scan with fast spin echo. The nodular surface of the cirrhotic liver is arrowed. Enlarged spleen (SPL) and large portal venous collateral vessels (col).

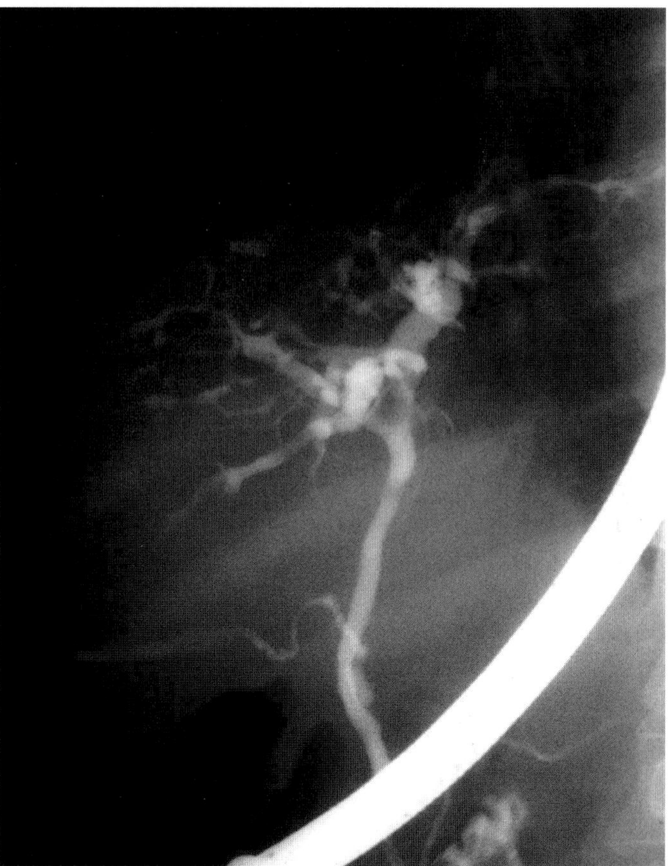

Fig. 5. A cholangiogram obtained at ERCP showing the calibre irregularities of the intrahepatic biliary tree (arrowed).

the biliary tree [31]. IDA, when injected systemically, is taken up by hepatocytes and then cleared rapidly into bile. In established cirrhosis there are documented delays in hepatocyte uptake and excretion, at the levels of both the intra- and extra-hepatic biliary trees, which may be one of the earliest abnormalities seen in those susceptible to the development of the biliary cirrhosis [32]. The advent of quantitative IDA imaging raises the possibility of monitoring objectively the degree of hepatocyte and biliary impairment and response to therapy [33].

Invasive Techniques

The characteristic intra-hepatic bile ductular change associated with established CF cirrhosis is irregularity of calibre caused by areas of stricture and dilatation and is similar to the picture in primary sclerosing cholangitis (fig. 5) [15]. These changes were first detected by trans gall bladder or endoscopic contrast cholangiography [15, 32] and are highly specific for established chronic liver disease, not being described in patients without this on the basis of ultrasound evidence [15]. Endoscopic retrograde cholangiography (ERCP) is an invasive procedure which is no longer considered an appropriate investigative technique for evaluating CF-related liver disease. With the increasing resolution of MRI/MRCP there is every expectation that these intra-hepatic ductular changes will be adequately delineated by this non-invasive technique. Endoscopy has a role in the management of common bile duct stones and a small proportion of patients in whom there is evidence of common bile duct obstruction at the level of the head of the pancreas. Histological assessment forms a fundamental basis for most aspects of hepatology. However, in CF liver disease the initial focal nature of the changes may result in considerable sampling error. Ultrasound-guided biopsy may reduce this risk if focal nodularity can be detected.

Westaby

There has been understandable reluctance to carry out liver biopsy in patients with CF on the basis that management was seldom changed and it carries the risk of pneumothorax. The imaging techniques described above, carried out in experienced hands may well provide sufficient diagnostic information for the vast majority of patients. Liver biopsy can then be reserved for a very small proportion of patients in whom other possible causes of liver damage need to be excluded.

Management

Bile Acid Therapy

Ursodeoxycholic acid (UDCA) is a hydrophilic bile acid, comprising 3% of the total bile acid pool in humans, which has been used extensively in cholestatic disorders [34], although the mechanism of action is unclear and is probably multi-factorial. Evidence points towards protection of cholangiocytes against the cytotoxic influence of hydrophobic bile acid [35]. In patients with primary biliary cirrhosis and primary sclerosing cholangitis there has also been evidence of reduced inflammatory reaction around the intra-hepatic bile ducts with UDCA therapy [36, 37]. UDCA has been shown to have a stimulatory effect upon biliary secretion by increasing the number and activity of carrier proteins in the apical cell membrane [38] and it may protect the hepatocyte against hydrophobic bile-acid-induced apoptosis [39]. A possible immuno-regulatory effect has been proposed related to the reversal of aberrant expression of HLA class-1 molecules on hepatocytes [40]. A number of studies have evaluated UDCA in patients with CF liver disease. Initial uncontrolled data suggested both bio-chemical and clinical improvement [41, 42] and the optimum dose appeared to be between 15 and 20 mg/kg [43, 44]. Attention has been paid to the need for taurine supplementation as part of UDCA therapy. Taurine deficiency is frequently observed in patients with CF secondary to malabsorption and faecal loss of taurine-conjugated bile salts and increases the proportion of glycol-conjugated bile acids, which are potentially hepatotoxic. However, a single study showed no additional affect upon liver function, but there was a degree of nutritional benefit [42]. A small unblinded controlled trial was the first to report both benefits in liver biochemistry and improvement in biliary excretion of IDA derivatives with UDCA [45]. A further placebo-controlled trial has also confirmed improvement in biochemistry and a general illness score [46]. However, neither of these trials has been of sufficient power or duration to allow assessment of the risk of decompensated liver disease, need for liver transplantation or associated mortality. A small study has evaluated the effect of UDCA on liver histology [47]. Using a scoring system based on bile duct proliferation, fibrosis, inspissation of bile and inflammatory changes, a significant histological benefit was confirmed after 1 or 2 years. There are clearly insufficient data upon which to base clear management guidelines [48]. It is highly unlikely that this drug is capable of reversing advanced liver disease and as such there is considerable justification for focussing further studies on the introduction of this drug in patients with early imaging evidence of liver disease [26]. More objective therapy may evolve as risk factors predicting the development liver disease are identified. In the meantime it is likely that treatment with UDCA will continue as the drug is well tolerated with very few adverse effects.

Liver Transplantation

Liver transplantation has been an important strategy for advanced chronic liver disease with survival rates in excess of 80% at 1 year and as high as 60% at 10 years. The criteria for inclusion of patients for liver transplantation have gradually expanded with increased experience. As for other types of cirrhosis, features of advanced liver decompensation are standard including encephalopathy, poorly controlled ascites and progressive jaundice. The use of liver transplantation in CF was initially discounted based on fears that the required immunosuppression would increase the risk of overwhelming pulmonary sepsis. However, beneficial results in several small series of patients undergoing liver transplantation have encouraged wider application [49, 50] and in fact, many patients demonstrated significant improvements in pulmonary function after this procedure, possibly related to resolution of splenomegaly or ascites, which impair diaphragmatic function, reduction of intra-pulmonary shunting or the immunosuppressive agents used. However, a number of pulmonary contra-indications to liver transplantation remain, including severely compromised lung function or frequent exacerbations of pulmonary infection. Infection with organisms such as *Burkholderia cepacia* or other multi-resistant bacteria may be considered relative contra-indications. A persistently raised arterial CO_2 indicating underlying ventilatory failure would also represent an absolute contra-indication to single organ liver transplantation. In appropriately selected CF patients, survival following isolated liver transplantation in the short and medium term (up to 5 years) appears to be similar to that in other types of cirrhosis [51]. The importance

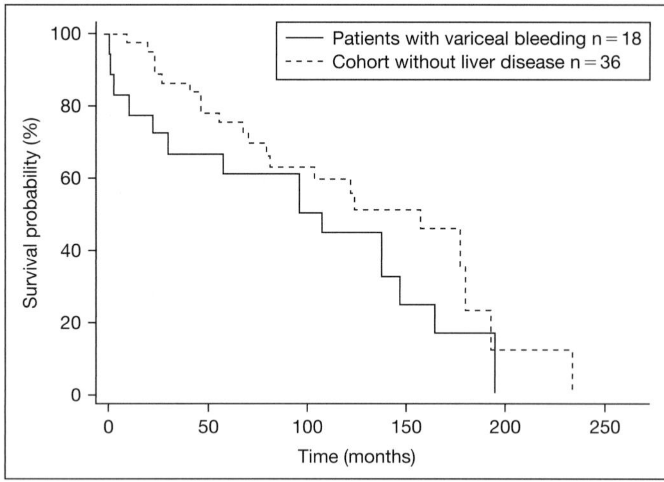

Fig. 6. A Kaplan-Meier survival analysis for patients from the time of first variceal bleed compared to a cohort without liver disease.

of portal hypertension and variceal bleeding as indicators for liver transplantation is controversial. There are a number of groups who use the presence of portal hypertension and the history of variceal bleeding as specific risk factors incorporated in scoring systems to identify those suitable for isolated liver transplantation [52]. However, in our large personal experience, we observed long-term survival following variceal bleeding to be comparable to that of the general CF population [author's unpubl. data]. This likely reflects the success of new endoscopic techniques for managing variceal bleeding (fig. 6) and the absence of other serious complications such as persistent ascites and encephalopathy. There remains a small but important group of patients with advanced liver disease and pulmonary disease of such severity that liver transplantation alone would not be feasible, in whom there are a number of reports of heart, lung, liver or lung/liver transplantation [51, 53, 54], one demonstrating 1-year survival of up to 70% [53], although medium- and long-term data are not available. With a shortage of donor organs there has been some reluctance to use those available in such a high-risk undertaking.

Management of Complications

Most CF patients with cirrhosis never develop specific complications and with the exception of UDCA, no specific therapy needs to be considered. However, some care should be taken with respect to nutritional requirements as patients with established cirrhosis have an increased resting energy expenditure [55] and may have deficiencies of micronutrients, fat-soluble vitamins and clotting factors [56], which may be further compounded by the contraindication to gastrostomy in patients with established portal hypertension and ascites.

Jaundice

Jaundice is occasionally seen as a consequence of bile duct obstruction in infancy and resolves when the plugging of the common bile duct is relieved. As a complication of cirrhosis it is unusual [15], is considered a poor prognostic feature and requires investigation to exclude other possible treatable causes, such as sepsis, drug toxicity or haemolysis. Trans-abdominal ultrasound scanning is essential to exclude bile duct obstruction, from either stones or a distal common bile duct stricture (see above). For those patients in whom jaundice represents the advanced stage of chronic liver disease UDCA has been shown to improve liver function and may lead to at least transient resolution of the jaundice.

Variceal Bleeding

This represents the commonest serious complication of chronic liver disease, but occurs infrequently, with a prevalence of under 2%. There is a well-recognized relationship between the severity of the underlying liver disease and the risk of first haemorrhage. Variceal bleeding should be managed using the same principles as in any other cirrhotic group of patients [57]. Initial resuscitation is critical. Replacement of blood loss is essential to maintain systemic haemodynamics and protect against renal impairment. There is a high risk of sepsis during an episode of variceal bleeding and prophylactic antibiotics covering a broad cross-section of gastro-intestinal-related bacteria are beneficial [58]. Injection sclerotherapy via fibre optic endoscopy was established as the first therapeutic technique to improve survival but has now been replaced by banding ligation (fig. 7) [59]. Success in controlling bleeding has been reported in 85–90% and in experienced hands the complication rate associated with banding ligation has been small [59]. Proton pump inhibitors [60] and sucralfate [61] are beneficial in the management of the mucosal ulceration associated with endoscopic therapy and also reduce the risk of early re-bleeding. There is a significant risk of early rebleeding (approximately 30%) and repeated banding ligation may be required. Both vasopressin and somatostatin analogues modulate portal blood flow and pressure by reducing splanchic inflow and, in the case of somatostatin, by a direct effect on the portal circulation itself. The vasopressin analogue tri-glycyl-lysine vasopressin (glypressin) was the first pharmacological agent to be reported to produce a survival benefit in active variceal

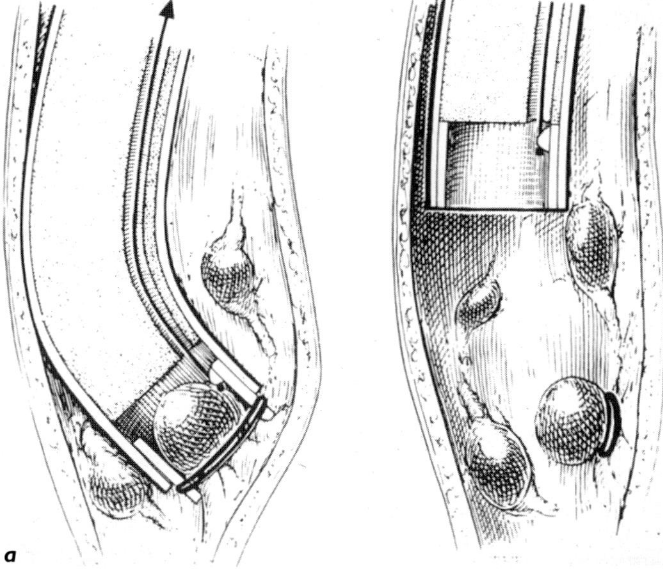

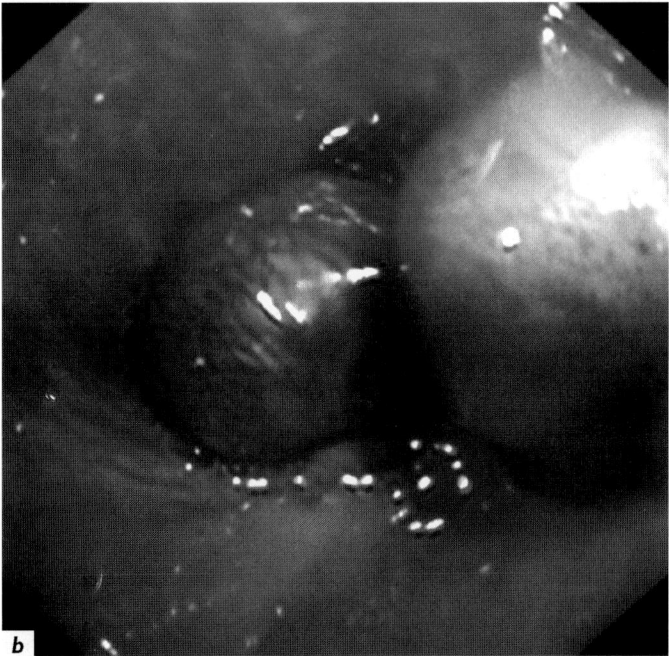

Fig. 7. (*a*) A diagrammatic representation of banding ligation for oesophageal varices. (*b*) endoscopic view of strangulated oesophageal varices after banding ligation.

bleeding [62]. This drug is easy to administer as an intravenous bolus and appears to have few cardiovascular side effects (the major drawback of vasopressin). Somatostatin and its analogue, octreotide, have benefit in active bleeding with few associated side effects although comparative studies suggest that glypressin is the most effective agent [63]. The major role for pharmacological agents is to buy time for endoscopic therapy. There is also some evidence that continuing these agents for 4–5 days after endoscopic therapy reduces the risk of early rebleeding [64]. In the presence of life-threatening bleeding, balloon tamponade [65] has been shown to be effective. It is at best a very uncomfortable and unpleasant experience for the patient and in the presence of significant pulmonary disease is associated with a high risk of complications, particularly aspiration. The most commonly used rescue procedure for persistent bleeding is the creation of a portal systemic shunt. Initial surgical shunts were highly effective at the expense of a marked reduction in liver blood flow and high morbidity and mortality from liver failure [66]. Over the last decade it has been possible to create a portal systemic shunt by a radiological technique termed a 'trans-jugular intra-hepatic portal systemic shunt (TIPPS)'. Because this is a much less invasive procedure, operative morbidity and mortality are small and the benefits with respect to arresting bleeding and preventing re-bleeding were maintained [67]. However, hopes that this intra-hepatic shunt might maintain a higher level of blood flow to the liver as compared to shunt surgery have not been realized. Post-procedural liver decompensation, specifically encephalopathy, is well recognized and represents the major drawback. There is however a small but important rescue role for this approach in patients in whom there has been a failure to control bleeding following the endoscopic technique and this has been successfully applied in CF patients [68]. Nonselective β-adrenoreceptor blockade has been shown to be effective for preventing recurrent variceal bleeding as well as reducing the risk of the first variceal haemorrhage in those patients who are known to have high risk oesophageal varices [69]. Drugs such as propranolol and nadolol have been widely used in this setting. However, in the presence of pulmonary complications of CF there are concerns that these drugs might precipitate bronchoconstriction. Endoscopic techniques have also been evaluated to prevent the first variceal bleed in high risk patients and in the case of banding ligation there is some accumulating evidence of benefit [70].

Ascites

The accumulation of a transudate within the peritoneal cavity is a well-recognized complication of cirrhosis and portal hypertension and is a poor prognostic factor. In CF liver disease it is a feature of advanced disease and decompensation (which may not be the case for variceal bleeding). For those cases in whom clinically significant and persisting ascites develops, the management is that applied in any case of underlying cirrhosis [71].

Encephalopathy

Hepatic encephalopathy is an extremely infrequent complication in patients with CF. It has only been described in patients with very advanced liver disease, often complicating some other adverse event such as ventilatory failure, sepsis, gastro-intestinal haemorrhage or persistent constipation. Where there is a clearly defined precipitating factor there may be expectation that the disturbed cerebration may revert promptly when the acute situation is resolved. More prolonged encephalopathy in the absence of a specific precipitant is an extremely poor prognostic factor. As with other complications of cirrhosis management is not different to that used more widely in the hepatological field [72].

Splenomegaly and Hypersplenism

Gross splenomegaly may occur in patients with CF related-cirrhosis. This may be the cause of abdominal pain, which on occasions may be severe (usually as a consequence of a splenic infarct). In most circumstances there is no indication for specific treatment, and simple non-opiate analgesia (avoiding non-steroidal inflammatory drugs) is sufficient. Pain alone is a very unusual indication for splenectomy. In the small proportion of patients with very large spleens there may be impairment of diaphragmatic function. Low platelet counts due to hypersplenism are frequently encountered, and in the rare instance in which spontaneous bleeding occurs is an indication for splenectomy [73] or partial splenectomy [74]. Alternative approaches have been to embolize the splenic artery to reduce splenic size. TIPPS has also been used as a means of reducing portal hypertension in this setting.

Influence of Liver Disease upon Organ Transplantation

There is now evidence from a small series that patients with well compensated cirrhosis tolerate lung transplantation without difficulty and have not presented problems with decompensation. Furthermore there is no evidence that variceal bleeding has been precipitated [75]. In our own series, 5 patients with established cirrhosis and portal hypertension have undergone heart/lung or lung transplantation without specific liver related complications. For those patients in whom there is evidence of liver disease but not fully established cirrhosis there should be no contraindication to lung transplantation.

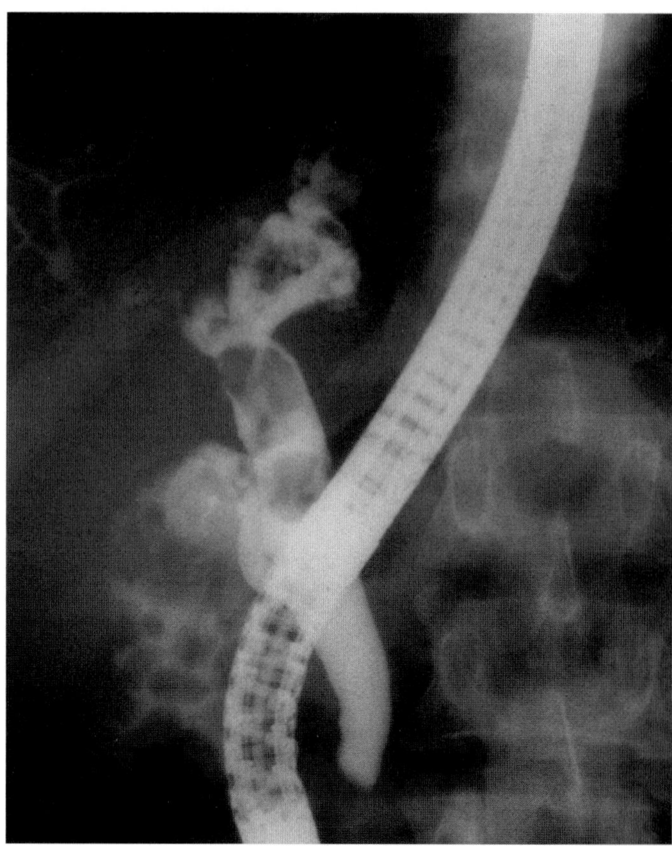

Fig. 8. A cholangiogram obtained at ERCP showing multiple stones in the biliary tree as well as in the gallbladder.

Extra-Hepatic Biliary Disease in Cystic Fibrosis

Abnormalities of the extra-hepatic biliary system, the pathogenesis of which is similar to that of intra-hepatic biliary disease, are commonly observed in CF. Approximately 25% of patients have non-functioning gallbladders and at post mortem 30% have micro-gallbladders (<1.5 cm in length and <0.5 cm in width) [27]. Stenosis or atresia of the cystic duct is also a common finding. At post mortem 24% of adult CF patients were found to have gallstones [27] (fig. 8). A reduced prevalence in younger patients suggests increasing risk with age which is typical of the pattern seen in the general population. These stones are almost always radiolucent and the majority are of cholesterol origin [76]. Analysis of stones removed from patients with CF has shown a composition including calcium bilirubinate as well as proteinaceous material. The most likely stimulus for stone formation is the low volume, high viscosity bile that follows on from the failure to hydrate the canalicular-produced bile. Complications of gallstones are commonly seen in CF including biliary colic, cholecystitis and extra-hepatic bile duct obstruction. The management of gall-stones in CF is similar to that in the general population [77]. Laparoscopic cholecystectomy is generally well tolerated in CF even in the presence of quite severe pulmonary disease. The endoscopic approach offers a minimally invasive means of managing common bile duct stones and in our own experience this has been extremely well tolerated with a very low risk of complications. The use of UDCA to dissolve gallbladder stones has not been specifically reported in CF and it is likely that in the presence of a diseased gallbladder this would not be successful. However, in selected cases in which gallbladder function is still maintained this may represent an alternative approach.

Hepatobiliary Malignancy

It is now well recognized that patients with CF have an increased risk of gut-related malignancy [78]. We have observed a single case of a gallbladder cancer in a patient presenting at the age of 40 with biliary colic. There has recently been another single case report of a hepatocellular carcinoma in a patient with CF-related cirrhosis [79].

References

1 Vawter GF, Shwachman H: Cystic fibrosis in adults: An autopsy study. Pathol Annu 1979; 14(pt 2):357–382.
2 Corbett K, Kelleher S, Rowland M, Daly L, Drumm B, Canny G, Greally P, Hayes R, Rourke B: Cystic fibrosis-associated liver disease: A population-based study. J Pediatr 2004;145:327–332.
3 Lamireau T, Monnereau S, Martin S, Marcotte JE, Winnock M, Alvarez F: Epidemiology of liver disease in cystic fibrosis: A longitudinal study. J Hepatol 2004;41:920–925.
4 Marino CR, Gorelick FS: Scientific advances in cystic fibrosis. Gastroenterology 1992;103: 681–693.
5 Cohn JA, Strong TV, Picciotto MR, Nairn AC, Collins FS, Fitz JG: Localization of the cystic fibrosis transmembrane conductance regulator in human bile duct epithelial cells. Gastroenterology 1993;105:1857–1864.
6 Bhaskar KR, Turner BS, Grubman SA, Jefferson DM, LaMont JT: Dysregulation of proteoglycans production by intrahepatic biliary epithelial cells bearing defective (delta-F508) cystic fibrosis transmembrane conductance regulator. Hepatology 1998;27: 7–14.
7 Lindblad A, Hultcrantz R, Strandvik B: Bile-duct destruction and collagen deposition: A prominent ultrastructural feature of the liver in cystic fibrosis. Hepatology 1992;16: 372–381.
8 Robb TA, Davidson GP, Kirubakaran C: Conjugated bile acids in serum and secretions in response to cholecystokinin/secretin stimulation in children with cystic fibrosis. Gut 1985;26:1246–1256.
9 Weizman Z, Durie PR, Kopelman HR, Vessely SM, Forstner GG: Bile acid secretion in cystic fibrosis: Evidence for a defect unrelated to fat malabsorption. Gut 1986;27:1043–1048.
10 Scott-Jupp R, Lama M, Tanner MS: Prevalence of liver disease in cystic fibrosis. Arch Dis Child 1991;66:698–701.
11 Duthie A, Doherty DG, Williams C, Scott-Jupp R, Warner JO, Tanner MS, Williamson R, Mowat AP: Genotype analysis for delta F508, G551D and R553X mutations in children and young adults with cystic fibrosis with and without chronic liver disease. Hepatology 1992;15:660–664.
12 Salvatore F, Scudiero O, Castalso G: Genotype-phenotype correlation in cystic fibrosis: The role of modifier genes. Am J Med Genet 2002;111:88–95.
13 Henrion-Caude A, Flamant C, Roussey M, Housset C, Flahault A, Fryer AA, Chadelat K, Strange RC, Clement A: Liver disease in paediatric patients with cystic fibrosis is associated with glutathione S-transferase P1 polymorphism. Hepatology 2002;36:913–917.
14 Gaskin KJ, Waters DL, Howman-Giles R, de Silva M, Earl JW, Martin HC, Kan AE, Brown JM, Dorney SF: Liver disease and common-bile-duct stenosis in cystic fibrosis. N Engl J Med 1988;318:340–346.
15 Nagel RA, Westaby D, Javaid A, Kavani J, Meire HB, Lombard MG, Wise A, Williams R, Hodson ME: Liver disease and bile duct abnormalities in adults with cystic fibrosis. Lancet 1989;ii:1422–1425.
16 Mieli-Vergani G, Psacharopoulos HT, Nicholson AM, Eddleston AL, Mowat AP, Williams R: Immune responses to liver membrane antigens in patients with cystic fibrosis and liver disease. Arch Dis Child 1980;55: 696–701.
17 Vlaman HB, France NE, Wallis PG: Prolonged neonatal jaundice in cystic fibrosis. Arch Dis Child 1971;46:805–809.
18 Furuya KN, Roberts EA, Canny GJ, Phillips MJ: Neonatal hepatitis syndrome with paucity of interlobular bile ducts in cystic fibrosis. J Pediatr Gastroenterol Nutr 1991;12:127–130.
19 Roy CC, Weber AM, Morin CL, Lepage G, Brisson G, Yousef I, Lasalle R: Hepatobiliary disease in cystic fibrosis: A survey of current issues and concepts. J Pediatr Gastroenterol Nutr 1982;1:469–478.
20 Treem WR, Stanley CA: Massive hepatomegaly, steatosis, and secondary plasma carnitine deficiency in an infant with cystic fibrosis. Pediatrics 1989;83:993–997.
21 Colombo C, Battezzati PM, Crosignani A, Morbito A, Costantini D, Padoan R, Glunta A: Liver disease in cystic fibrosis: A prospective study on incidence, risk factors, and outcome. Hepatology 2002;36:1374–1382.
22 Pares A, Rodes J: Natural history of primary biliary cirrhosis. Clin Liver Dis 2003;7: 779–794.
23 Farrant JM, Hayllar KM, Wilkinson ML, Karani J, Portman BC, Westaby D, Williams R: Natural history and prognostic variables in primary sclerosing cholangitis. Gastroenterology 1991;100:1710–1717.
24 Hayllar KM, Williams SG, Wise AE, Pouria S, Lombard M, Hodson ME, Westaby D: A prognostic model for the prediction of survival in cystic fibrosis. Thorax 1997;52:313–317.
25 Williams SG, Samways J, Innes JA, Guz A, Geddes DM, Hodson ME, Westaby D: Systemic haemodynamic changes in patients with cystic fibrosis with and without chronic liver disease. J Hepatol 1996;25:900–908.
26 Lenaerts C, Lapierre C, Patriquin H, Bureau N, Lepage G, Harel F, Marcotte J, Roy CC: Surveillance for cystic fibrosis-associated hepatobiliary disease: Early ultrasound changes and predisposing factors. J Pediatr 2003;143:343–350.
27 McHugo JM, McKeown C, Brown MT, Weller P, Shah KJ: Ultrasound findings in children with cystic fibrosis. Br J Radiol 1987;60: 137–141.
28 Williams SG, Evanson JE, Barrett N, Hodson ME, Boulthee JE, Westaby D: An ultrasound scoring system for the diagnosis of liver disease in cystic fibrosis. J Hepatol 1995;22:513–521.
29 King LJ, Scurr ED, Murugan N, Williams SG, Westaby D, Healy JC: Hepatobiliary and pancreatic manifestations of cystic fibrosis: MR imaging appearances. Radiographics 2000;20: 767–777.
30 Durieau I, Pellet O, Simonot L, Durupt S, Bellon G, Durand DV, Minh VA: Sclerosing cholangitis in adults with cystic fibrosis: A magnetic resonance cholangiographic prospective study. J Hepatol 1999;30:1052–1056.
31 Krishnamurthy S, Krishnamurthy GT: Technetium-99m-iminodiacetic acid organic anions: Review of biokinetics and clinical application in hepatology. Hepatology 1989;9: 139–153.

32 O'Brien S, Keogan M, Casey M, Duffy G, McErlean D, Fitzgerald MX, Hegarty JE: Biliary complications of cystic fibrosis. Gut 1992;33:387–391.

33 Colombo C, Castellani MR, Balistreri WF, Seregni E, Assaisso ML, Giunta A: Scintigraphic documentation of an improvement in hepatobiliary excretory function after treatment with ursodeoxycholic acid in patients with cystic fibrosis and associated liver disease. Hepatology 1992;15:677–684.

34 Paumgartner G, Beuers U: Ursodeoxycholic acid in cholestatic liver disease: Mechanisms of action and therapeutic use revisited. Hepatology 2002;36:525–531.

35 Van Nieuwkerk CM, Elferink RP, Groen AK, Ottenhoff R, Tytgat GN, Dingemans KP, Van Den Bergh Weerman MA, Offerhaus GJ: Effects of urodeoxycholate and cholate feeding on liver disease in FVB with a distrupted mdr2 P-glycoprotein gene. Gastroenterology 1996;111:165–171.

36 Poupon RE, Balkau B, Eschwege E, Poupon R: A multicenter, controlled trial of ursodiol for the treatment of primary bi-cirrhosis. UDCA-PBC study group. N Engl J Med 1991; 324:1548–1554.

37 Beuers U, Spengler U, Kruis W, Aydemir U, Wiebecke B, Heldwein W, Weinzierl M, Pape GR, Sauerbruch T, Paumgartner G: Ursodeoxycholic acid for treatment of primary sclerosing cholangitis: A controlled trial. Hepatology 1992;16:707–714.

38 Fickert P, Zollner G, Fuchsbichler A, Stumptner C, Pojer C, Zenz R, Lammert F, Stieger B, Meier PJ, Zatloukal K, Denk H, Trauner M: Effects of ursodeoxycholic and cholic acid feeding on hepatocellular tranexpression in mouse liver. Gastroenterology 2001;121:170–183.

39 Qiao L, Yacoub A, Studer E, Gupta S, Pei XY, Grant S, Hylemon PB, Dent P: Inhibition of the MAPK and P13K pathways enhances UDCA-induced apoptosis in primary rodent hepatocytes. Hepatology 2002;35:779–789.

40 Calmus Y, Gane P, Rouger P, Poupon R: Hepatic expression of class I and class II major histocompatibility complex molecules in primary biliary cirrhosis: Effect of ursodeoxycholic acid. Hepatology 1990;11: 12–15.

41 Colombo C, Setchell KD, Podda M, Crosignani A, Roda A, Curcio L, Ronchi M, Giunta A: Effects of ursodeoxycholic acid therapy for liver disease associated with cystic fibrosis. J Pediatr 1990;117:482–489.

42 Cotting J, lentze MJ, Reichen J: Effects of ursodeoxycholic acid treatment on nutrition and liver function in patients with cystic fibrosis and longstanding cholestasis. Gut 1990;31: 918–921.

43 Colombo C, Crosignani A, Assaisso M, Battezzati PM, Podda M, Giunta A, Zimmer-Nechemias L, Setchell KD: Ursodeoxycholic acid therapy in cystic fibrosis-associated liver disease: A dose-response study. Hepatology 1992;16:924–930.

44 Van de Meeberg PC, Houwen RH, Sinaasappel M, Heijerman HG, Bijleveld CM, Vanberge-Henegouwen GP: Low-dose venous high-dose ursodeoxycholic acid in cystic fibrosis-related cholestatic liver disease: Results of a randomized study with 10-year follow-up. Scand J Gastroenterol 1997;32: 369–373.

45 O'Brien S, Fitzgerald MX, Hegarty JE: A controlled trial of ursodeoxycholic acid treatment in cystic fibrosis related liver disease. Eur J Gastroenterol 1992;4:857–863.

46 Colombo C, Battezzati PM, Podda M, Bettinardi N, Giunta A: Ursodeoxycholic acid for liver disease associated with cystic fibrosis: A double-blind multicenter trial. The Italian group for the Study of Ursodeoxycholic Acid in cystic fibrosis. Hepatology 1996;23:1484–1490.

47 Lindblad A, Glaumann H, Strandvik B: A two year prospective study of the effect of ursodeoxycholic acid on urinary bile acid excretion and liver morphology in cystic fibrosis-associated liver disease. Hepatology 1998;27:166–174.

48 Cheng K, Ashby D, Smyth R: Ursodeoxycholic acid for cystic fibrosis related liver disease. Cichrane Database Syst Rev 2000;(2): CD00222.

49 Noble-Jamieson G, Barnes N, Jamieson N, Friend P, Calne R: Liver transplantation for hepatic cirrhosis in cystic fibrosis. J R Soc Med 1996;89(suppl 27):31–37.

50 Mack DR, Traystman MD, Colombo JL, Sammut PH, Kaufman SS, Vanderhoof JA, Antonson DL, Markin RS, Shaw BW Jr, Langnas AN: Clinical denouement and mutation analysis of patients with cystic fibrosis undergoing liver transplantation for biliary cirrhosis. J Pediatr 1995;127:881–887.

51 Milkiewicz P, Skiba G, Kelly D, Weller P, Bonser R, Gur U, Mirza D, Buckels J, Stableforth D, Elias E: Transplantation for cystic fibrosis outcome following early liver transplantation. J Gastroenterol Hepatol 2002; 17:208–213.

52 Noble-Jamieson G, Barnes ND: Liver transplantation for cirrhosis in cystic fibrosis. J Pediatr 1996;129:314.

53 Couetil JP, Soubrane O, Houssin DP, Dousset BE, Chevalier PG, Guinvarch A, Loulmet D, Achkar A, Carpentier AF: Combined heart-lung-liver, double lung-liver, and isolated liver transplantation for cystic fibrosis in children. Transpl Int 1997;10:33–39.

54 Dennis CM, McNeil KD, Dunning J, Stewart S, Friend PJ, Alexander G, Higenbottam TW, Calne RY, Wallwork J: Heart-lung-liver transplantation. J Heart Lung Transplant 1996;15: 536–538.

55 Madden AM, Morgan MY: Patterns of energy intake in patients with cirrhosis and healthy volunteers. Br J Nutr 1999;82:41–48.

56 Sokol RJ, Durie PR: Recommendations for management of liver and biliary tract disease in cystic fibrosis. Cystic Fibrosis Foundation Hepatobiliary Disease Consensus Group. J Pediatr Gastroenterol Nutr 1999;28(suppl 1): S1–S13.

57 Jalan R, Hayes PC: UK guidelines on the management of variceal haemorrhage in cirrhotic patients. British Society of gastroenterology. GUT 2000;46 S3–4:III1–III15.

58 Garcia-Tsao G: Bacterial infections in cirrhosis. Can J Gastroenterol 2004;18:405–406.

59 Vlavianos P, Westaby D: Management of acute variceal haemorrhage. Eur J Gastroenterol Hepatol 2001;13:335–342.

60 Gimson A, Polson R, Westaby D, Williams R: Omeprazole in the management of intractable esophageal ulceration following injection sclerotherapy. Gastroenterology 1990;99: 1829–1831.

61 Polson RJ, Westaby D, Gimson AE, hayes PC, Stellon AJ, Hayllar K, Williams R: Sucralfate for the prevention of early rebleeding following injection sclerotherapy for oesophageal varices. Hepatology 1989;10:79–82.

62 Ioannou G, Doust J, Rockey DC: Terlipressin for acute oesophageal variceal haemorrhage. Cochrane Database Syst Rev 2003;(1): CD002147.

63 Nevens F: Review article: A critical comparison of drug therapies in currently used therapeutic strategies for variceal haemorrhage. Aliment Pharmacol Ther 2004;20(suppl 3): 18–22.

64 de Franchis R: Somatostatin, somatostatin analogues and other vasoactive drugs in the treatment of bleeding oesophageal varices. Dig Liver Dis 2004;36(suppl 1):S93–S100.

65 Vlavianos P, Gimson AE, Westaby D, Williams R: Balloon tamponade in variceal bleeding: Use and misuse. BMJ 1989;298:1158.

66 Stern RC, Stevens DP, Boat TF, Doershuk CF, Izant RJ Jr, Matthews LW: Symptomatic hepatic disease in cystic fibrosis: Incidence, course, and outcome of portal systemic hunting. Gastroenterology 1976;70:645–649.

67 Luketic VA, Sanyal AJ: Esophageal varices. II TIPS (transjugular intrahepatic portosystemic shunt) and surgical therapy. Gastroenterol Clin North Am 2000;29:387–421.

68 Pozler O, Krajina A, Vanicek H, Hulek P, Zizka J, Michl A, Elias P: Transjugular intrahepatic portosystemic shunt in five children with cystic fibrosis: Long term results. Hepatogastroenterology 2003;50:1111–1114.

69 Abraczinskas DR, Ookubo R, Grace ND, Groszmann RJ, Bosch J, Garcia-Tsao G, Richardson CR, Matloff DS, Rodes J, Conn HO: Propranolol for the prevention of first esophageal variceal haemorrhage: A lifetime commitment? Hepatology 2001;34:1096–1102.

70 Schepke M, Kleber G, Nurnberg D, Willert J, Koch L, Veltzke-Schlieker W, Hellerbrand C, Kuth J, Schanz S, Kahl S, Fleig WE, Sauerbruch T: German study group for the primary prophylaxis of variceal bleeding. Ligation versus Propranolol for the primary prophylaxis of variceal bleeding cirrhosis. Hepatology 2004;40:65–72.

71 Garcia-Tsao G: Portal hypertension. Curr Opin Gastroenterol 2003;19:250–258.

72 Shawcross D, Jalan R: Dispelling myths in the treatment of hepatic encephalopathy. Lancet 2005;365:431–433.

73 Westwood AT, Millar AJ, Ireland JD, Swart A: Splenectomy in cystic fibrosis patients. Arch Dis Child 2004;89:1078.

74 Zach MS, Thalhammer GH, Eber E: Partial splenectomy in CF patients with hypersplenism. Arch Dis Child 2003;88:649.

75 Gilljam M, Chaparro C, Tullis E, Chan C, Keshavjee S, Hutcheon M: GI complications after lung transplantation in patients with cystic fibrosis. Chest 2003;123:37–41.

76 Angelico M, Gandin C, Canuzzi P, Bertasi S, Cantafora A, De Santis A, Quattrucci S, Antonelli M: Gallstones in cystic fibrosis: A critical reappraisal. Hepatology 1991;14: 768–775.

77 Lahmann BE, Adrales G, Schwartz RW: Choledocholithiasis-principles of diagnosis and management. Curr Surg 2004;61: 290–293.

78 Maisonneuve P, FitzSimmons SC, Neglia JP, Campbell PW 3rd, Lowenfels AB: Cancer risk of nontransplanted and transplanted cystic fibrosis patients: Year study. J Natl Cancer Inst 2003;95:381–387.

79 McKeon D, Day A, Parmar J, Alexander G, Bilton D: Hepatocellular carcinoma in association with cirrhosis in a patient with fibrosis. J Cyst Fibros 2004;3:193–195.

D. Westaby
Chelsea and Westminster Hospital
369 Fulham Road
London SW10 9NH (UK)
Tel. +44 208 846 1076
Fax +44 208 237 5007
E-Mail westaby.david@chelwest.nhs.uk

Other Organs

Bush A, Alton EWFW, Davies JC, Griesenbach U, Jaffe A (eds): Cystic Fibrosis in the 21st Century.
Prog Respir Res. Basel, Karger, 2006, vol 34, pp 264–269

Fertility, Contraception, Incontinence and Pregnancy

J.G. Thorpe-Beeston

Chelsea and Westminster Hospital, London, UK

Abstract

The advances in the management of cystic fibrosis have meant that parenthood is a realistic aim for many affected women and increasingly for affected men. However, such pregnancies are not without significant risk for both mother and fetus. The consequences of an unplanned pregnancy or a pregnancy in a woman with very poor lung function who fails to understand the significance of her disease may be disastrous. The issues regarding fertility and contraception should be addressed in childhood, and a responsible attitude to contraception generated. A multidisciplinary approach is essential, pre-pregnancy counselling strongly advocated and meticulous antenatal care in pregnancy is recommended.

Male Fertility

Ninety-five percent of the males with cystic fibrosis (CF) are infertile, but it is important to stress that sexual function is not affected. Infertility results from structural abnormalities of the reproductive tract giving rise to an obstructive azoospermia [1, 2]. A variety of abnormalities have been described, but the most common is congenital absence of the vas deferens. Histologically the testes are normal and active spermatogenesis occurs. Some studies have suggested an increased incidence of abnormal and immature sperm. Abnormal secretions within the reproductive organs may lead to impaction with inspissated secretions with resultant atrophy. A small percentage of males with CF will be fertile [1, 2]. It is essential to assume that prepubertal males are fertile until proven otherwise. The difficult issues of counselling adolescent boys about their fertility may in the past have been neglected, but most centres will now initiate open discussion of these issues. Semen analysis showing complete absence of sperm in the ejaculate confirms infertility; other features suggestive of infertility would include a low pH and a reduced ejaculate volume.

A number of techniques have more recently been developed that permit sperm retrieval in men with congenital bilateral absence of the vas deferens. These techniques include microsurgical epididymal sperm aspiration, percutaneous epididymal sperm aspiration and testicular biopsy. The assisted conception technique of intracytoplasmic sperm injection allows possible fertilization of an ovum using a single aspirated sperm. A recent study analyzed the outcome of intracytoplasmic sperm injection with fresh and frozen-thawed surgically retrieved spermatozoa from men diagnosed with congenital bilateral absence of the vas deferens. The presence of CF mutations was noted. In 27 cases spermatozoa were aspirated by microsurgical epididymal sperm aspiration, percutaneous epididymal sperm aspiration or open testis biopsy. A 17% embryo implantation rate was achieved and of 18 pregnancies which occurred, there were 14 live births without congenital anomaly. The study concluded that the presence of CF mutations in the male partner does not compromise in vitro fertilization treatment outcomes or the opportunity for healthy live births [3].

Female Fertility

For many years a 20% reduction in female fertility was incorrectly accepted, although the investigators did not provide details as to how fertility was assessed [4, 5]. Women with CF have a normal reproductive tract. Although there is some evidence that unspecified factors may influence menarche, in those women with good lung function and a normal BMI and percentage fat, menstruation and ovulation are likely to occur normally [6]. Those with a reduced BMI are more likely to suffer anovulatory cycles and secondary amenorrhoea. In women luteinizing hormone changes at puberty are indistinguishable from the normal population. However, follicular-stimulating hormone changes occur at lower levels and pubertal levels are reached 2 years later than by the normal population resulting in delayed menarche. The underlying mechanisms for these changes are not understood. Similarly, progesterone and oestrogen peaks are delayed. Recent studies have shown that the cystic fibrosis transmembrane conductance regulator (CFTR) gene has been isolated in the hypothalamus and it has been hypothesized that this may influence the release of gonadotrophin-releasing hormone [7].

The cervix, although structurally normal in women with CF, may be functionally compromised. CFTR has been documented in the epithelium of the cervix [8]. Hormonally mediated CFTR expression may affect the water content of the cervical mucus [4]. The normally observed increase in water content in mid-cycle fails to occur resulting in a thick tenacious mucus plug, present throughout the menstrual cycle [9].

Contraception

The majority of women with CF develop normal menstrual cycles. Those with amenorrhoea typically have a lower percentage of body fat and worse lung function [10]. Women with CF should be assumed to be fertile and contraception must be addressed in early adolescence. The awareness of issues relating to fertility and contraception in women and men is increasing. Although this is welcomed, in many cases it is clear that these issues are being addressed by parents rather than directly by CF centres [11–13].

In a study from Aberdeen, UK, 59% of women were found to be using contraception as compared to the national rate of 65% [13]. The choice of contraception is potentially difficult and needs to be individualized. Barrier methods remain a very effective and in the younger patient they minimize the risks of sexually transmitted disease and human papilloma virus infection. Couples should be reminded of the availability of emergency postcoital contraception. The use of oral contraception has some theoretical drawbacks. Concern has been raised that they may worsen diabetes, malabsorption and liver dysfunction. The progesterone component may adversely affect mucus production and viscosity. In addition the use of broad-spectrum antibiotics may adversely affect absorption and efficacy. Despite these concerns oral contraception has been widely used. Reassuringly a study of 10 women, using a combined contraceptive preparation, with moderate to severe lung disease found no significant deterioration in clinical status or pulmonary lung function [13]. In another study of 6 women the pharmacokinetics of the commonly used contraceptive steroids ethinyloestradiol and levonorgestrel were investigated. An increase in total body clearance of ethinlyoestradiol was compensated for by increased oral bioavailability. The pharmacokinetics of levonorgestrel did no significantly alter. These studies suggest that women with CF using oral contraception are likely to receive a similar contraceptive effect as healthy women [14].

The intrauterine contraceptive device is a common and highly acceptable form of contraception for many women. The traditional copper devices may be associated with heavier menstrual loss for the first 2–3 cycles and if this does not resolve may be unsuitable for women with borderline iron reserve. The risk of infection has generally been overstated. Recently the introduction of the progestogen-only intrauterine system releasing levonorgestrel directly in to the uterine cavity has become the treatment of choice for many women with heavy menses and is an excellent contraception. Since the progestogen is released close to the site of action, progestogenic side effects and interactions are minimal. Patients should be informed, however, to expect some irregular bleeding for up to 6 months following insertion.

Urinary Incontinence

Urinary incontinence is a common distressing problem which may have a significant impact on the quality of life of an affected individual. Chronic cough is thought to predispose to urodynamic stress incontinence (USI) by raising the intra-abdominal pressure and the pressure on the pelvic floor. The incidence of incontinence in a study of adult males and females was 4.8 and 37.9%, respectively [15]. This is in general agreement with other studies which suggest a prevalence in a female CF population of 30–59% [16, 17].

The frequency of incontinence is double that seen in a normal population and represents a significant problem which may need to be addressed for the first time in childhood [18]. The management of incontinence relies on initially confirming the diagnosis. A mid-stream urine sample is essential to exclude a simple treatable condition. In the majority of cases the underlying cause will be USI; however, because of the wide overlap of symptoms experienced by sufferers it may be appropriate to refer the patient for urodynamic assessment to make a formal diagnosis. Initially the management will be conservative including pelvic floor exercises, biofeedback, electrical stimulation or vaginal cones. These techniques are supervised by a specialist physiotherapist or incontinence advisor. Pelvic floor exercises are effective at improving and reducing leakage in the short term, but women may be reluctant to discuss their problem and poor at reliably undertaking the treatment programmes [19]. A more open approach to the problem such as routinely inquiring about possible symptoms at regular reviews may make it easier for affected women to ask for help.

In the normal population if conservative measures fail, a variety of surgical procedures have been developed. These are unlikely to provide long-term benefit for a CF population due to the chronic nature of their cough. Recently duloxetine [20], a selective serotonin and noradrenaline reuptake inhibitor, which increases the contractility of the urethral sphincter, has been introduced for USI, but it has yet to be tested in CF.

Pregnancy

The physiological changes of normal pregnancy are summarized in tables 1 and 2. They may impose a heavy burden on women with CF whose respiratory function may be compromised. The normal weight gain in pregnancy is approximately 10–12 kg. The total energy requirement has been calculated to be between 80,000 and 124,000 kcal. The three major components of energy expenditure are growth of the fetus and reproductive tissues, new maternal fat stores and increased maternal metabolism. Even in a healthy population it is recognized that there is enormous individual variation. Poorly nourished women try to maintain fetal growth at the expense of laying down body fat and by reducing their metabolic rate. This adaptation may compromise fetal well-being [21]. The hormonal and physical changes that occur in pregnancy increase the likelihood of nausea and vomiting in early pregnancy and dyspepsia and reflux at a more advanced gestation. The additional calorific requirements in women with CF may be very difficult to achieve and supplemental feeding may be required [22]. In a study reporting 26 term and 22 preterm live deliveries in women with CF the mean maternal weight gain of those pregnancies reaching term were 8.9 kg and only 2.6 kg in those delivering prematurely [23].

A degree of impaired glucose tolerance due to changes in oestrogens, progesterone and human placental lactogen is normal in pregnancy. There is a reduced renal threshold for glucose and an increased glomerular filtration rate in pregnancy resulting in glycosuria in 5–50% of all pregnant women. Gestational diabetes is more common in CF, particularly in the second half of pregnancy. In a study of 92 pregnancies in women with CF the prevalence of gestational diabetes was 14% [24], considerably higher than in the normal population. Screening for diabetes includes a random blood glucose at booking followed at 26 weeks' gestation by a 50-gram oral glucose load with blood glucose estimation 1 h later (see chapter 35). Although the

Table 1. Changes in respiratory function in normal pregnancy

Lung function	Change by term	Amount
Total lung capacity	Decreased	4%
Functional residual capacity	Decreased	1–20%
Vital capacity	Probable increase	200 ml
Tidal volume	Increased	40%
Inspiratory capacity	Increased	300 ml
Expiratory reserve	Decreased	200 ml
Residual volume	Decreased	20%
Oxygen consumption	Increased	20–33%
Minute volume	Increased	30%
FEV_1	Unchanged	
Peak expiratory flow	Unchanged	

Table 2. Cardiovascular changes in normal pregnancy

Cardiac function	Change by term	Amount
Heart rate	Increased	11–16 bpm
Stroke volume	Increased	20%
Cardiac output	Increased	30–40%
Systolic blood pressure	Increased in third trimester	
Diastolic blood pressure	Decreased to third trimester, then increased	
Blood volume	Increased	50%
Peripheral resistance	Decreased	

consequences of impaired glucose tolerance on pregnancy are debated, insulin-requiring diabetes is associated with a worse pregnancy outcome for both mother and baby.

Effect of Pregnancy on Clinical Status

Successful term pregnancies have been achieved in women with poor lung function and it has been suggested that stable pre-pregnancy lung function may be the key [25], although some advocate termination of pregnancy in women with poor lung function ($FEV_1 < 50\%$ predicted) or progressive pulmonary deterioration [26].These very difficult issues reinforce the crucial importance of pre-pregnancy counselling by clinicians directly involved in the management of such cases.

Edenborough et al. [26] were the first to report the outcome of such pregnancies together with serial lung function measurements. The data confirmed the earlier conclusions that patients with milder CF, reasonable lung function and a good nutritional status tended to tolerate pregnancy well. Those with poor lung function, reduced BMI, hepatic or pancreatic involvement or secondary pulmonary hypertension are at a significantly increased risk from pregnancy. This work suggested a decline of 13% in FEV_1 and 11% in FVC during pregnancy, although most of this was regained following delivery. In those with moderate to severe lung disease ($FEV_1 < 60\%$), there was an increased risk of preterm delivery and an increased loss of lung function as compared to those with more mild disease. It is clear that in this group of women there is a substantial risk of irretrievable loss of lung function and death, which may occur in the immediate postpartum period or over the first 2–3 years following delivery. Finally, of four cases of CF in pregnancy where the pre-pregnancy $FEV_1 < 50\%$ predicted, 3 died, all of whom were infected with *Burkholderia cepacia* [27].

Long-Term Effect of Pregnancy

Six studies have attempted to address the issue of the long-term effects of pregnancy on CF. Five of the studies reach similar conclusions. They provide little direct evidence for a statistically significant increased loss of lung function due to pregnancy as compared to non-pregnant matched controls [26, 28–31]. The most recent study concludes that pregnancy probably has a slight adverse effect on the health of CF women [31]. In the group of women with poor pre-pregnancy lung function it was clear that there was a much higher probability that the mother would

Table 3. Issues to consider in pre-pregnancy counselling

Issue	Discussion
Partner CF carrier status	Prenatal diagnosis
Pre-pregnancy lung function	Prognosis
Pre-pregnancy diabetes	Optimize control
Pre-pregnancy nutrition	Folic acid supplementation
Current medication	Review of therapy
Increased risk of premature delivery	Neonatal consequences
Increased risk of fetal growth restriction	Ultrasound assessment
Possible deterioration of lung function	Long-term prognosis
Increased hospital visits	
Risk of hospital admission	
Increased risk of Caesarean section	
Breast-feeding	

deliver prematurely or that she would suffer a significant loss of lung function. Soberingly, of 80 French women with CF who had 90 pregnancies in total, 12 deaths were recorded of which 3 were in the year following the pregnancy. All 3 women had an FEV_1 of $<50\%$ before pregnancy [31]. In contrast in a large cohort study of 680 pregnant women with CF, matched to over 3,000 control women with CF, it was concluded that pregnancy was not harmful in any group of CF patients including a subgroup with FEV_1 40% of predicted or diabetes [32].

Pre-Pregnancy Counselling

Individualized counselling prior to pregnancy is ideal (table 3). Each pregnancy will provide a different challenge and for some women with poor pre-pregnancy lung function, the issue of their own mortality needs to be frankly discussed. In this group, the women also need to have an understanding of the possible implications of preterm delivery for babies including the possible risk of long-term handicap. It is essential that the woman's partner is also involved at this stage, for in a worse case scenario he may potentially be faced with the prospect of losing his partner and then having the responsibility of looking after a disabled child. Whilst clearly there is no right or wrong answer under these circumstances, honest prenatal counselling will help focus the thoughts of couples considering pregnancy, making them aware not only of the risks but also the huge commitment in terms of time, monitoring and hospital visits pregnancy is likely to bring about. Whatever decisions prospective parents may make, it is important that they are aware that they will always be supported.

Obstetric Management

The role of the obstetrician is to become one part of the multidisciplinary team providing care for the individual. In addition to the obstetrician and primary chest physician, input from physiotherapists, dieticians, midwives, neonatologists and specialist anaesthetists will be required. Coordinating these services and ensuring good interprofessional communication are essential. Ideally, the obstetrician will build up a special expertize in CF.

It is important that once pregnancy is diagnosed consideration is given to ensuring all aspects of maternity care are catered for, including normal antenatal services to screen for the maternal blood group and possible congenital infections such as rubella, hepatitis, syphilis and HIV. In Europe and the USA toxoplasmosis is also screened routinely. The options for screening for chromosomal anomalies should be discussed and serial ultrasound scans instituted to look for possible evidence of fetal growth restriction. In addition assessment of the uterine and fetal circulation by Doppler ultrasound may provide useful information about the maternal and fetal well-being.

Careful liaison throughout the pregnancy with the lead chest physician permits appropriate outpatient or inpatient management. Infective exacerbations should be aggressively treated with intravenous antibiotics. It is helpful in early pregnancy for the mother to meet the obstetric anaesthetists. The level of monitoring at the time of delivery can then be agreed and an explanation given regarding epidural anaesthesia. Depending on the maternal lung function consideration may be given to elective admission in the third trimester for a course of antibiotics and intense physiotherapy to ensure the lung condition is optimal at delivery. If delivery is likely to occur before 34 completed weeks of gestation, maternal steroids should be administered to aid fetal lung maturation and help reduce the incidence of respiratory distress syndrome and associated complications. Steroids may safely be given even in diabetic mothers even though it is likely to destabilize the maternal blood sugars for a few days. An insulin pump or sliding scale dose regimen will be necessary at this time.

In general terms for most mothers it is reasonable in the absence of obstetric indications to aim for a vaginal delivery. Women should be encouraged to attend antenatal classes. Epidural anaesthesia allows for a relatively stress-free labour and consideration should be given to limiting the length of the second stage, depending on the maternal condition [33]. Elective assistance by forceps or ventouse in the second stage is likely to reduce the risk of overburdening the respiratory system, although if the mother is well and coping there is every chance she will deliver the baby spontaneously. Caesarean section may be indicated for obstetric considerations such as breech presentation. In women with particularly poor lung function, although the ideal remains vaginal delivery, often the mothers are clearly unable to cope with the demands of labour and operative delivery may be required [34]. The skill in management is in trying to judge the optimal time for delivery in the mother with poor lung function, before there is a rapid decline in reserve.

Care should be taken, in particular if an assisted delivery is undertaken, to minimize the risk of blood loss. Postnatal anaemia will contribute to a slow recovery. If intravenous fluids are administered at any stage, normal saline should be avoided. In the postnatal period, contrary to an early report [35], breast-feeding is a reasonable option and for preterm babies expressed beast milk is undoubtedly their best form of nutrition. A more careful study of breast milk concluded that CF breast milk does not contain elevated concentrations of sodium and is safe for infant feeds [36]. However, breast-feeding is potentially an exhausting process and in those women who are struggling to cope with everything in the postnatal period, it may be sensible to suggest a change to bottle feeding.

References

1 Taussig LM, Lobeck CC, di Sant'Annese PA, Ackerman DR, Kattwinkel J: Fertility in males with cystic fibrosis. N Engl J Med 1972;287:586–689.

2 Barreto C, Marques Pinto L, Duarte A: A fertile male with cystic fibrosis: Molecular genetic analysis. J Med Genet 1991;28:420–421.

3 Phillipson GT, Petrucco OM, Matthews CD: Congenital bilateral absence of the vas deferens, cystic fibrosis mutation analysis and intracytoplasmic sperm injection. Hum Reprod 2000;15:431–435.

4 Kopito LE, Kosasky HJ, Shwachman H: Water and electrolytes in cervical mucus from patients with cystic fibrosis. Fertil Steril 1973;24:512–516.

5 Kotloff RM, FitzSimmons SC, Fiel SB: Fertility and pregnancy in patients with cystic fibrosis. Clin Chest Med 1992;13:623–635.

6 Moshang T, Holsclaw DS: Menarchal determinants in cystic fibrosis. Am J Dis Child 1980;134:1139–1142.

7 Weyler RT, Altschuler SM, Reedstra WW: Regulation of neurosecretion by the cystic fibrosis transmembrane conductance regulator (CFTR). Pediatr Pulmonol 1998;17(suppl): A76.

8 Tizzano EF, Buchwald M: CFTR expression and organ damage in cystic fibrosis. Ann Intern Med 1995;123:305–308.

Thorpe-Beeston

9 Oppenheimer EA, Case AL, Esterly JR, Rothberg RM: Cervical mucus in cystic fibrosis: A possible cause of infertility. Am J Obstet Gynecol 1970;108:673–674.

10 Stead RJ, Hodson ME, Batten JC, Adams J, Jacobs HS: Amenorrhoea in cystic fibrosis. Clin Endocrinol 1987;26:187–195.

11 Hames A, Beesley J, Nelson R: Cystic fibrosis: What do patients know, and what else would they like to know? Respir Med 1991; 85:389–392.

12 Sawyer SM: Reproductive and sexual health in adolescents with cystic fibrosis. BMJ 1996; 313:1095–1096.

13 Fair A, Griffiths K, Osman LM: Attitudes to fertility issues among adults with cystic fibrosis in Scotland. Thorax 2000;55:672–677.

14 Stead RJ, Grimmer SF, Rogers SM, Back DJ, Orme ML, Hodson ME, Batten JC: Pharmacokinetics of contraceptive steroids in patients with cystic fibrosis. Thorax 1987;42:59–64.

15 White D, Stiller K, Roney F: The prevalence and severity of symptoms of incontinence in adult cystic fibrosis patients. Physiother Theor Pract 2000;16:35–42.

16 Cornacchia M, Zenorini A, Perobelli S, Zanolla L, Mastella G, Bragg C: Prevalence of urinary incontinence in women with cystic fibrosis. Br J Urol 2001;88:44–48.

17 Moran F, Bradley JM, Boyle L, Elborn JS: Incontinence in adult females with cystic fibrosis: A Northern Ireland survey. Int J Clin Pract 2003;57:182–183.

18 Prasad SA, Tannenbaum EL, Mikelsons C: Physiotherapy in cystic fibrosis. J R Soc Med 2000;93(suppl 38):27–36.

19 McVean RJ, Orr A, Webb AK, Bradbury A, Kay L, Philips E, Dodd ME: Treatment of urinary incontinence in cystic fibrosis. J Cystic Fibrosis 2003;2:171–176.

20 McCormack PL, Keating GM: Duloxetine: In stress incontinence. Drugs 2004;64: 2567–2573.

21 Williams DJ: Nutrition in pregnancy; in Warrell DA, Cox T, Firth JD, Benz E (eds): Oxford Textbook of Medicine. Oxford, Oxford University Press, 2003, pp 386–389.

22 Valenzuela GJ, Comunale FL, Davidson BH, Dooley RR, Foster TC: Clinical management of patients with cystic fibrosis and pulmonary insufficiency. Am J Obstet Gynecol 1988;159: 1181–1182.

23 Edenborough FP, Mackenzie WE, Stableforth DE: The outcome of 72 pregnancies in 55 women with cystic fibrosis in the United Kingdom 1977–1996. Br J Obstet Gynaecol 2000;107:254–261.

24 Gilljam M, Antoniou M, Shin J, Dupuis A, Corey M, Tullis D: Pregnancy in cystic fibrosis. Fetal and maternal outcome. Chest 2000; 118:85–91.

25 Canney GJ, Corey M, Livingstone RA, Carpenter S, Green L, Levison H: Pregnancy and cystic fibrosis. Obstet Gynecol 1991;77: 850–853.

26 Edenborough FP, Mackenzie WE, Conway SP: The effect of pregnancy on maternal cystic fibrosis vs nulliparous severity matched controls. Thorax 1996;51(suppl 3):A50.

27 Tanser SJ, Hodson ME, Geddes DM: Case reports of death during pregnancy in patients with cystic fibrosis – three out of four patients were colonized with *Burkholderia cepacia*. Respir Med 2000;94:1004–1006.

28 Frangolias DD, Nakeilna EM, Wilcox PG: Pregnancy in cystic fibrosis: A case-controlled study. Chest 1997;111:963–969.

29 Ahmed R, Wielinski CL, Warwick WJ: Effect of pregnancy on CF. Pediatr Pulmonol 1995; 12(suppl):289.

30 Fiel SB, Fitzsimmons SC: Pregnancy in patients with cystic fibrosis. Pediatr Pulmonol 1995;12(suppl):93–94.

31 Gillet D, de Braekeleer M, Bellis G, Durieu I: French Cystic Fibrosis Registry: Cystic fibrosis and pregnancy. Report from French data (1980–1999). Br J Obstet Gynaecol 2002;109: 912–918.

32 Goss CH, Rubenfeld GD, Otto K, Aitken ML: The effect of pregnancy on survival in women with cystic fibrosis. Chest 2003;124: 1460–1468.

33 Howell PR, Kent N, Douglas MJ: Anaesthesia for the parturient with cystic fibrosis. Int J Obstet Gynaecol 1993;2:152–158.

34 Bose D, Yentis SM, Fauvel NJ: Caesarean section in a parturient with respiratory failure caused by cystic fibrosis. Anaesthesia 1997; 52:578–582.

35 Whitelaw A, Butterfield A: High breast milk sodium in cystic fibrosis. Lancet 1977;ii: 1288.

36 Shiffman ML, Seale TW, Flux M, Rennert OR, Swender PT: Breast-milk composition in women with cystic fibrosis: Report of two cases and a review of the literature. Am J Clin Nutr 1989;49:612–617.

J.G. Thorpe-Beeston, MA, MD, FRCOG
Consultant Obstetrician and Gynaecologist
Chelsea and Westminster Hospital
Fulham Road
London SW10 9NH (UK)
Tel. +44 208 846 7902
E-Mail guytb@btopenworld.com

Bush A, Alton EWFW, Davies JC, Griesenbach U, Jaffe A (eds): Cystic Fibrosis in the 21st Century.
Prog Respir Res. Basel, Karger, 2006, vol 34, pp 270–277

Arthritis, Vasculitis and Bone Disease

Sarah Elkin

Royal Brompton Hospital, London, UK

Abstract

It is evident that with increasing survival, patients with cystic fibrosis (CF) develop joint and bone disease. This chapter discusses CF-related arthropathy, vasculitis and bone disease. The central principles for preventing bone disease are heightened surveillance, particularly during puberty, physical exercise, and supplementation with calcium and vitamins D and K. There have been recent reports of the use of oral or intravenous bisphosphonates for treating established disease. Future options for treating CF bone disease may include anabolic agents and human recombinant growth hormone.

Arthropathies

Symptoms relating to joint pathology occur in approximately 5–10% of individuals with cystic fibrosis (CF). Joint disease may be directly related to CF, a complication of drug treatment or due to an unrelated joint disease. The most commonly diagnosed conditions are episodic arthritis and hypertrophic pulmonary osteoarthropathy. These may occur at any age but more commonly affect adults (table 1). Factors, such as rashes, recent viral contacts, diarrhoea, sexually acquired diseases and drug history can help elucidate other aetiology.

Vasculitis

Vasculitis is a recognized but unusual complication of CF occurring in approximately 2–3% of patients [1–3]. The literature mainly comprises of case reports and case series. It appears that whilst vasculitis may present in childhood, the majority of cases occur in patients over 20 years of age. Most patients have severe pulmonary disease and the occurrence of vasculitis is often associated with a poor prognosis. The rash is purpuric and usually located distally on the legs around the ankles and on the dorsum of the feet. Onset can be associated with constitutional symptoms such as fever, malaise, arthralgia or myalgia. Histological examination of the rash is consistent with a leukocytoclastic vasculitis [4]. Although the aetiology of the vasculitis is not confirmed, immune complexes have been reported in CF patients with vasculitis and purpura due to cryoglobulinaemia has been described in one CF patient [5]. It has been postulated that autoantibodies against bactericidal/permeability-increasing protein antibodies (BPI) might play a role. These are found frequently in CF patients with vasculitis and levels appear to correlate with pseudomonal load, reduction in lung function and the presence of vasculitis [6]. BPI expresses a protective activity against lipopolysaccharide-induced injury on vascular endothelial cells [7]. Thus, if anti-BPI antibodies interfere with this function they may facilitate vasculitic inflammation. Whether this particular mechanism contributes to vasculitis in CF is unknown.

Low Bone Mineral Density in CF

Low bone mineral density (BMD) is almost universal in patients with severe disease [8, 9]. The risk factors for developing low BMD are numerous and are outlined in table 2.

Table 1. Episodic arthritis and hypertrophic osteoarthropathy in CF

	Incidence	Clinical presentation	Age	Joint distribution	Proposed pathogenesis	Correlates with lung disease	Treatment
Episodic arthritis	2–8% adults	Arthralgia Swelling Erythema	Teenage onwards	Large joints, asymmetrical	Circulating immune complexes (60-kDa heat shock protein autoantibodies)	Not closely linked	NSAIDS Glucocorticoids Methotrexate Azathioprine
HPOA	7% adults >males	Digital clubbing Arthralgia Pain over long bones	Mean age onset 20 years	Symmetrical: wrists, knees, ankles Dista: tibia, fibula femur, radius	Release of platelet-derived growth factor in small vessels	Yes	NSAIDS Glucocorticoids Intravenous pamidronate

HPOA = Hypertrophic pulmonary osteoarthropathy.

Table 2. Risk factors for the development of low BMD

Delayed puberty (girls >12 years – Tanner breast scale, boys >14 years – gonad size)
Secondary hypogonadism
Nutritional failure (ideal body weight < 90%)
Moderate/severe lung disease (FEV_1 < 60% predicted)
Pulmonary sepsis (high cytokine load)
Organ transplant candidate/posttransplant
Significant systemic glucocorticoid use (>90 days/year)
CF-related diabetes
Hypovitaminosis D (<30 ng/ml)
Hypovitaminosis K
Depot medroxyprogesterone acetate (DepoProvera), heparin and other drugs that may cause bone loss
Physical inactivity

Clinical Manifestations of Bone Disease in CF

Several cross-sectional studies have reported a higher rate of fracture in patients with CF as compared to controls. Aris et al. [8] reported a significant increase in fractures in 70 patients awaiting transplantation. Fracture rates were approximately 2-fold increased in females aged 16–34 years and males age 25–35 years.. These results were corroborated by both Elkin et al. [10] and more recently by Rossini [35]. These fractures can lead to pain and deformities that can hinder airway clearance and reduce lung function.

Prevalence of CF Bone Disease

Studies from around the world have found low BMD is common in children and adults with CF, with a greater prevalence found in adults (table 3).

Some believe the prevalence of low BMD is overestimated in CF studies as body and bone size have not been taken into account. Indeed, the importance of recruiting controls matched for height is demonstrated in several studies [11, 13–15], in which no difference in BMD was demonstrated once adjustments for height were made. Although reduced bone size probably does overestimate the prevalence of low BMD in smaller individuals with CF there is little doubt that the problem exists. Studies assessing regional BMD with quantified computerized tomography, which measures volumetric (3–dimensional) lumbar spine bone density, have demonstrated significant reductions in both children and adults with CF [16, 17]. This finding is now given more credence by the finding of reduced total bone volume in CF bone biopsies [18] (fig. 1).

Bone Histomorphometry in CF

Bone histomorphometry studies help to characterize CF bone disease by describing bone remodelling and structure [19]. They also confirm or refute the presence of defective mineralization in patient groups that have low serum levels

Table 3. Summary of selection of studies investigating CF bone disease

Study	Number of patients	Age years	FEV₁	BMI	Methods	Results	Correlates
Gibbens et al., 1988 [17]	57 (local controls)	3–21 (mean 12)			QCT	BMD 10% lower than controls	BMI, male sex, Shwachman score
Bhudhikanok et al., 1996 [34]	49 (local controls)	8–48		Z score: females −0.8; males −0.7	DEXA (BMAD calculated)	53% osteopenia	Age, BMI, glucocorticoid use, disease severity
Aris et al., 1998 [8]	70	18–39	Females: 34%; males: 33%	Females 17; males 19	DEXA	39% T <-1 57% T $<-$ 2.5	Age, glucocorticoid use, BMI
Haworth et al., 1999 [16]	151	15–52	58.8	21.2	DEXA QCT	34% Z score <-2	BMI, % FEV₁, physical activity, CRP, PTH
Conway et al., 2000 [31]	114	16–34 (approximately)	47% (13–101)	Females 18.7/20; males 19.5/20.6	DEXA	55% men and 43% women T score <-1	Male sex, Shwachman-Kulczycki score, BMI, Northern score, diabetes
Elkin et al., 2001 [10]	107	18–60	50.8	20.8		38% Z score <-1	% FEV₁, i.v. antibiotic usage, mean energy expenditure, steroid use, BMI
Buntain et al., 2004 [15]	153 (local controls)	5.3–55.8	Children: 83.5 Adolescents: 77.5 Adults: 56.6			5–10: no difference; adolescents: reduced TB; adults: reduced at all sites	Children: BMI, FEV₁, BCM, number of days in hospital Adults: BCM, number of days in hospital, activity

FEV₁ = Forced expiratory volume in 1 second; BMI = body mass index; QCT = quantitative computed tomography; DEXA = dual-energy x-ray absorptiometry; BMAD = bone mineral apparent density; CRP = C-reactive protein; BCM = body cell mass.

of 25-OH vitamin D (25-OHD), such as in CF [10, 16, 20–22]. There is, for obvious ethical reasons, a paucity of data from bone biopsy specimens in CF patients.

In 1985 Friedman et al. [23] reported the finding of osteomalacia from analysis of an iliac crest bone biopsy from a 25-year-old man with CF. This patient had low serum 25-OHD concentrations and secondary hyperparathyroidism. Repeat bone biopsy after 1 year of parenteral vitamin D showed healing of the osteomalacia. Haworth et al. [24] found no characteristics of osteomalacia in an analysis of autopsy bone samples from 15 patients with CF, 11 of whom had undergone lung transplantation. The study reported that there was severe osteopenia in both trabecular and cortical bone. They hypothesized that an uncoupled increase in osteoclastic activity resulted in an increase in resorptive surfaces. However, in the absence of tetracycline labelling, dynamic indices of bone formation and resorption could not be assessed. It is important to note that the majority of the samples were from autopsy implying that the patients had been severely ill, with chronic infection and raised cytokine levels. All of these factors would predispose to increased resorption [25, 26]. Data from a more recent histomorphometry study of clinically stable CF adults and low BMD demonstrated lower cancellous bone area and a trend towards a decrease in cancellous bone connectivity [18]. Bone formation rate at tissue level was significantly lower in CF patients and wall width, representing the amount of bone formed within individual

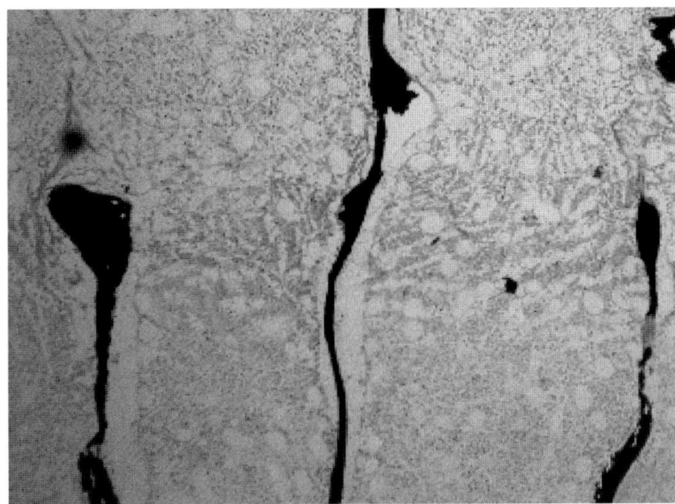

Fig. 1. Thin bone trabeculae in adult CF patient.

remodelling units, was decreased. Surprisingly, analysis of resorption cavities revealed a decrease in resorption in the CF patients when compared to controls. However, there was considerable heterogeneity in resorption measurements in the patients studied, at least 2 of the subjects having larger resorption cavities than the control mean value. In addition, the patient with the greatest mean cavity depth also had the highest resorption biomarker values. Although there was evidence of a mineralization deficit in the CF group as a whole, only one biopsy revealed osteomalacia. This patient had deficient levels of 25-OHD and raised serum PTH.

Bone Accrual and Loss in CF

In individuals with CF it appears that good health in childhood may lead to normal BMD. Puberty appears to be a crucial time with one longitudinal study reporting inadequate bone mineral accrual. Interestingly late gains in BMD were obtained in some individuals with associated weight gain [27]. Haworth et al. [28] followed 107 adults (aged 15–49) over 1 year and reported mean losses of 0.5, 2.1 and 1.8% at the lumbar spine, femoral neck and total hip, respectively. These rates are higher than those experienced by women following the menopause.

Bone marker studies indicate that both accelerated bone breakdown and inadequate bone formation are important in the pathogenesis of low BMD in CF [29, 30]. There appears to be a 'background' of low bone formation, particularly prominent at puberty, with episodes of increased resorption leading to uncoupling of bone turnover.

Clinical Correlates

There is little doubt that the incidence of low BMD is related to pulmonary disease severity. Many studies, both longitudinal [27, 28] and cross-sectional [10, 16, 31], have shown positive correlation with BMD and FEV_1 (forced expiratory volume in 1 s) and body mass index. Patients listed for transplantation invariably have low BMD [8]. Worsening pulmonary function will lead to reduced exercise and consequently reduced mechanical stress on bone. Infection will also lead to increased levels of inflammatory cytokines, IL-1, IL-6 and tumour necrosis factor (TNF-α) and increased bone resorption via an increase in osteoclast number. The strong correlation between BMD and disease severity is supported by studies that report normal BMD in well-nourished patients with minimal pulmonary disease. Salamoni et al. [11] reported BMD and body composition to be normal in well-nourished CF children. This finding was supported by a UK study which found no difference in regional and whole-body BMD [32]. Risk factors for low BMD are discussed further below.

Delayed Puberty and Early Gonadal Failure

There is increasing evidence that a normal puberty is crucial for bone health in individuals with CF [15, 27]. However, in-depth comprehensive studies have yet to be performed. It is well known that pubertal delay can occur in CF and is related to disease status [33]. Inadequate sex steroids during such a time in bone development may prevent peak bone mass being obtained. Hypogonadism is seen in adults with CF and is probably underrecognized. In studies to date, Bhudhikanok et al. [34] found pubertal stage to be a predictor of bone mineral status. They also found low testosterone levels in young CF adults but an effect on BMD was not demonstrated. The Brompton group [10] found 31 of 58 males investigated to have low total serum testosterone, 18% having low free testosterone. The latter significantly correlated with total body BMD. This finding was corroborated by Rossini et al. [35] who found significantly lower free testosterone and serum oestradiol levels in CF males with vertebral fractures. Hypogonadism is likely to be detrimental to bone health, especially if cytokine levels are high, as androgens have a protective effect on the skeleton.

Physical Inactivity

Physical activity and weight-bearing exercise are essential for the maintenance of skeletal mass. It is well known that immobilization leads to bone loss due to increased resorption and decreased formation [36]. Therefore, CF patients will be at risk of increased bone loss during periods of ill health. Decreased pulmonary function leads to a decrease in exercise level and numerous studies have shown a decrease in daily exercise level to predict low BMD [10, 16, 34]. Although this is most likely to be a reflection of the decreased pulmonary reserve of these patients it is provocative to hypothesize that weight-bearing activity might be an important determinant of BMD in CF patients [37].

Hypovitaminosis D

Low serum 25-OHD levels have been widely reported in CF patients [10, 16, 20–22]. To date only 2 studies have found a relationship between low serum 25-OHD and low BMD [38, 39]. Recently, Buntain et al. [15] reported a significant BMD deficit in the presence of vitamin D sufficiency in a large Australian cohort giving further credence that low BMD in CF is multifactorial. However, the failure to find an association between low 25-OHD and BMD does not equate to low vitamin D playing no role in the pathogenesis of low BMD in certain CF populations. Vitamin D levels may fluctuate from season to season and BMD reflects bone health over the lifetime of the patient; therefore, low serum levels probably contribute to the decreased bone mass found in some patients via intermittent episodes of secondary hyperparathyroidism. This hypothesis is supported by Haworth et al. [16], who reported a negative relationship between mean BMD Z score and serum PTH in 139 adults with CF. It should be noted that glucocorticoids are likely to compound this problem as they increase PTH further by decreasing calcium absorption.

It has become apparent that the degree of vitamin D absorption varies considerably amongst individuals with CF. Lark et al. [40] reported that CF adults absorbed less than one half of the amount of oral vitamin D_2 given as a dose in comparison to non-CF controls. The 25-OHD levels did not rise in response to the vitamin D_2 contrary to the control group who showed a doubling of serum levels, suggesting impairment in hepatic hydroxylation. Recent data show that the dose of vitamin D required to obtain adequate serum levels varies considerably and interestingly some patients do not reach sufficient levels despite very high oral

doses (Boyle NACF 2003). More research is required to determine the best preparation and route of administration in these patients.

Vitamin K

Vitamin K plays an essential role in haemostasis and also in bone metabolism. Osteocalcin undergoes a vitamin K-dependent, posttranscriptional γ-carboxylation of the glutamic acid residues, which results in a higher mineral binding coefficient with the calcium ions in the matrix at the mineralization front of new bone. It is feasible that vitamin K deficiency may in part be causing the abnormal bone formation reported in histomorphometry studies [41].

In 1999 Rashid et al. [45] performed a prospective study investigating 98 patients and 62 healthy controls. They reported that 78% of pancreatic-insufficient patients and all those with CF liver disease had raised PIVKA-II levels. A further study found 68 of 102 to have low serum K, raised PIVKA-II or both [47]. These results suggest a link between subclinical vitamin K deficiency and an uncoupling of bone formation and resorption. More recently Mosler et al. [46] concluded that vitamin K deficiency affecting bone or liver could occur independently and diagnosis required sophisticated expensive tests that could not be recommended as routine. They suggested all pancreatic-insufficient patients should receive routine supplementation. It is still unclear what dosage of supplementation will suffice. Five milligrams per week was found to raise serum vitamin K level and improve PIVKA-II and carboxylation of osteocalcin in 18 CF patients but PIVKA-II levels still remained high in 73% of patients indicating this dose to be inadequate [48]. Recommendations vary from 1 mg/day to 10 mg/week to 10 mg/day. Further studies are needed to decide upon the optimal effective dose and the effects of such dosage upon bone turnover and osteocalcin.

Glucocorticoids

Patients with CF are likely to be prescribed oral glucocorticoids as their pulmonary disease progresses or if they are diagnosed with allergic bronchopulmonary aspergillosis. This may lead to an initial rapid loss of bone followed by a slower phase of bone loss of about 2–5% annually. Some of the larger studies have detected an association between glucocorticoid use and low BMD [10, 31]. Recent studies provide evidence that the main mechanism of steroid-induced bone loss is by a decrease in bone formation. This

Elkin

appears to be due to suppression of osteoblastogenesis and promotion of the apoptosis of osteoblasts and osteocytes [49]. There is also an early increase in bone resorption. The cardinal histological features of glucocorticoid bone loss are decreased bone formation rate and a decreased trabecular wall thickness, findings not dissimilar to the CF bone histomorphometry results. Indeed, a subgroup analysis of the glucocorticoid-prescribed CF patients by the author (unpubl. data) revealed a greater reduction in bone formation at tissue level when compared to non-glucocorticoid users; however, osteoblastic activity was not affected and was depressed independently of steroid use. It is, therefore, apparent that oral glucocorticoid use contributes, but is not the sole cause, of decreased BMD in adults with CF.

Chronic Infection

Indirect evidence supports the role for inflammation associated with pulmonary infections playing a part in bone loss in CF patients. Elkin et al. [10] found the number of intravenous antibiotic courses in the previous 5 years to negatively correlate with BMD. Acute pulmonary infection in CF is associated with increases in circulating IL-6, IL-1 and TNF-α, which can both increase osteoclast formation and activity and inhibit osteoblast function. Levels drop significantly after treatment with antibiotics. This drop is concomitant with a decrease in bone resorption markers [26] suggesting increased inflammatory cytokines may be linked to increased bone resorption during acute exacerbations.

Genetics

The question of whether CFTR might be directly responsible for the low BMD has long been debated. Most cross-sectional studies have found no association between CF genotype and low BMD. However, recently King et al. [49a] reported a direct link with ΔF508 and low BMD. Haworth et al. [16] previously reported significant differences in bone turnover between ΔF508 homozygotes and non-homozygotes suggesting that there may be a genetic component influencing bone and calcium homeostasis in patients with CF. Clearly the important issue to clarify is whether chloride channels are linked to a pathophysiological process that may alter bone density. Interestingly, cortical and trabecular bone formation is significantly reduced in CF knockout mice and these mice do not have CF pulmonary disease, which suggests the low BMD is arising despite normal lungs. Future work should aim to determine if CFTR is expressed in osteoblasts or osteoclasts and if the CF mutations alter the biological effects.

Screening for Bone Disease

A recently published consensus document from the CF Foundation [50] recommends dual-energy x-ray absorptiometry (DEXA) scanning as the preferred method of measuring BMD in CF patients as it has a low radiation exposure and is sufficiently precise for serial monitoring. Measurements should be of the posteroanterior lumbar spine and proximal femur. Screening is not indicated under the age of 8 years, as there is no easily available normative data. Over the age of 8 children should only have a DEXA test if they have risk factors for low BMD (table 2) or a previous fragility fracture. At the age of 18 years or on entry to the adult clinic, all patients should have a baseline DEXA test.

Follow-up DEXA scans are recommended as follows: (1) Z score –1.0 and above: repeat every 5 years (sooner if risk factors develop) and (2) between –1.0 and –2.0: repeat every 2–4 years (sooner if risk factors develop).

The following patients should have yearly scans until stable or improved: Z score –2.0 or less, Z score above –2.0 (if there is significant loss on serial DEXA tests), chronic glucocorticoid use, organ transplantation, and treatment with bisphosphonate. If possible, repeat measurements should be performed by the same technologist using the same machine. The change in absolute BMD should be monitored, not Z or T scores.

Intervention should be considered if the patient has: (1) a fragility fracture and a BMD Z score of –1.0 or below, (2) a Z score –2.0 and below (T score –2.0 or below if over age 30) and (3) a significant loss of BMD (typically >3% at the spine, >5% at the hip).

Treatments

All patients should receive vitamin D supplementation aiming for serum levels of above 60 nmol/l (30–60 ng/ml). These levels may be difficult to obtain in more northern latitudes although doses of up to 50,000 IU weekly can be used. It does appear that response to vitamin D is highly individual and it has been suggested that phototherapy could be considered in these patients. Calcium and vitamin K supplementation should be given according to local and national guidelines.

Recently there have been several reports of the use of oral or intravenous bisphosphonates in patients with bone disease associated with CF [52]. Although the main mechanism of action of these agents is to inhibit the recruitment and function of osteoclasts there is increasing evidence that they also enhance the bone-forming activities of osteoblasts [51, 53] and prevent osteoblast apoptosis [54]. Therefore, the finding of decreased bone formation does not preclude a therapeutic effect of bisphosphonates in CF bone disease. Indeed, two studies (both using intravenous pamidronate) in adults with CF have shown significant increases in the BMD of the lumbar spine in the treatment group compared to controls, both in non-transplant [55] and post-lung transplant patients [56]. Unfortunately, significant adverse events occurred with pamidronate use in the pretransplant group, most notably severe bone pain. None of the patients taking oral corticosteroids developed bone pain suggesting a protective effect of this therapy. A 1-year study with oral alendronate in CF adults showed a significant increase in BMD in the treated group as compared to control in both the lumbar spine and hip [57]. Most bisphosphonates are licensed for use in steroid-induced osteoporosis and post-menopausal osteoporosis. They are not licensed for women of child-bearing age. It is, therefore, prudent to obtain consent from patients before commencing the medication. It is unclear how long patients should be treated to obtain a benefit. Long-term data on bone density and fracture incidence are needed.

Future options for treating CF bone disease might include anabolic agents in view of the evidence that CF bone disease is characterized by decreased bone formation. Preliminary data from the States show promise for the use of human recombinant growth hormone in children with CF [58].

References

1 Bourke S, Rooney M, Fitzgerald M, Bresnihan B: Episodic arthropathy in adult cystic fibrosis. Q J Med 1987;64:651–659.
2 Hodson ME: Vasculitis and arthropathy in cystic fibrosis. J R Soc Med 1992;85(suppl 19):38–40.
3 Schidlow DV, Goldsmith DP, Palmer J, Huang NN: Arthritis in cystic fibrosis. Arch Dis Child 1984;59:377–379.
4 Finnegan MJ, Hinchcliffe J, Russell-Jones D, Neill S, Sheffield E, Jayne D, Wise A, Hodson ME: Vasculitis complicating cystic fibrosis. Q J Med 1989;72:609–621.
5 Garty BZ, Scanlin T, Goldsmith DP, Grunstein M: Cutaneous manifestations of cystic fibrosis: Possible role of cryoglobulins. Br J Dermatol 1989;121:655–658.
6 Zhao MH, Jayne DR, Ardiles LG, Culley F, Hodson ME, Lockwood CM: Autoantibodies against bactericidal/permeability-increasing protein in patients with cystic fibrosis. QJM 1996;89:259–265.
7 Arditi M, Zhou J, Huang SH, Luckett PM, Marra MN, Kim KS: Bactericidal/permeability-increasing protein protects vascular endothelial cells from lipopolysaccharide-induced activation and injury. Infect Immun 1994;62:3930–3936.
8 Aris RM, Renner JB, Winders AD, Buell HE, Riggs DB, Lester GE, Ontjes DA: Increased rate of fractures and severe kyphosis: Sequelae of living into adulthood with cystic fibrosis. Ann Intern Med 1998;128:186–193.
9 Shane E, Silverberg SJ, Donovan D, Papadopoulos A, Staron RB, Addesso V, Jorgesen B, McGregor C, Schulman L: Osteoporosis in lung transplantation candidates with end-stage pulmonary disease. Am J Med 1996;101:262–269.
10 Elkin SL, Fairney A, Burnett S, Kemp M, Kyd P, Burgess J, Compston JE, Hodson ME: Vertebral deformities and low bone mineral density in adults with cystic fibrosis: A cross-sectional study. Osteoporos Int 2001;12:366–372.
11 Salamoni F, Roulet M, Gudinchet F, Pilet M, Thiebaud D, Burckhardt P: Bone mineral content in cystic fibrosis patients: Correlation with fat-free mass. Arch Dis Child 1996;74:314–318.
12 Laursen EM, Molgaard C, Michaelsen KF, Koch C, Miller J: Bone mineral status in 134 patients with cystic fibrosis. Arch Dis Child 1999;81:235–240.
13 Hardin DS, Arumugam R, Seilheimer DK, LeBlanc A, Ellis KJ: Normal bone mineral density in cystic fibrosis. Arch Dis Child 2001;84:363–368.
14 Elkin SL, Williams L, Moore M, Hodson ME, Rutherford OM: Relationship of skeletal muscle mass, muscle strength and bone mineral density in adults with cystic fibrosis. Clin Sci (Lond) 2000;99:309–314.
15 Buntain HM, Greer RM, Schluter PJ, Wong JC, Batch JA, Potter JM, Lewindon PJ, Powell E, Wainwright CE, Bell SC: Bone mineral density in Australian children, adolescents and adults with cystic fibrosis: A controlled cross-sectional study. Thorax 2004;59:149–155.
16 Haworth CS, Selby PL, Webb AK, Dodd ME, Musson H, McL Niven R, Economou G, Horrocks AW, Freemont AJ, Mawer EB, et al: Low bone mineral density in adults with cystic fibrosis. Thorax 1999;54:961–967.
17 Gibbens DT, Gilsanz V, Boechat MI, Dufer D, Carlson ME, Wang CI: Osteoporosis in cystic fibrosis. J Pediatr 1988;113:295–300.
18 Elkin SL, Vedi S, Bord S, Garrahan NJ, Hodson ME, Compston JE: Histomorphometric analysis of bone biopsies from the iliac crest of adults with cystic fibrosis. Am J Respir Crit Care Med 2002;166:1470–1474.
19 Compston J: Bone histomorphometry; in Feldman D, Glorieux FH, Pike MC (eds): Vitamin D. New York, Academic Press, 1987, vol 37, pp 573–586.
20 Hanly JG, McKenna MJ, Quigley C, Freaney R, Muldowney FP, FitzGerald MX: Hypovitaminosis D and response to supplementation in older patients with cystic fibrosis. Q J Med 1985;56:377–385.
21 Donovan DSJ, Papadopoulos A, Staron RB, Addesso V, Schulman L, McGregor C, Cosman F, Lindsay RL, Shane E: Bone mass and vitamin D deficiency in adults with advanced cystic fibrosis lung disease. Am J Respir Crit Care Med 1998;157:1892–1899.
22 Ott SM, Aitken ML: Osteoporosis in patients with cystic fibrosis. Clin Chest Med 1998;19:555–567.
23 Friedman HZ, Langman CB, Favus MJ: Vitamin D metabolism and osteomalacia in cystic fibrosis. Gastroenterology 1985;88:808–813.
24 Haworth CS, Webb AK, Egan JJ, Selby PL, Hasleton PS, Bishop PW, Freemont TJ: Bone histomorphometry in adult patients with cystic fibrosis. Chest 2000;118:434–439.
25 Ionescu AA, Nixon LS, Evans WD, Stone MD, Lewis-Jenkins V, Chatham K, Shale DJ: Bone density, body composition, and inflammatory status in cystic fibrosis. Am J Respir Crit Care Med 2000;162:789–794.
26 Aris RM, Stephens AR, Ontjes DA, Denene BA, Lark RK, Hensler MB, Neuringer IP, Lester GE: Adverse alterations in bone metabolism are associated with lung infection in adults with cystic fibrosis. Am J Respir Crit Care Med 2000;162:1674–1678.
27 Bhudhikanok GS, Wang MC, Marcus R, Harkins A, Moss RB, Bachrach LK: Bone acquisition and loss in children and adults with cystic fibrosis: A longitudinal study. J Pediatr 1998;133:18–27.
28 Haworth CS, Selby PL, Horrocks AW, Mawer EB, Adams JE, Webb AK: A prospective study

of change in bone mineral density over one year in adults with cystic fibrosis. Thorax 2002;57:719–723.

29 Baroncelli GI, De Luca F, Magazzu G, Arrigo T, Sferlazzas C, Catena C, Bertelloni S, Saggese G: Bone demineralization in cystic fibrosis: Evidence of imbalance between bone formation and degradation. Pediatr Res 1997; 41:397–403.

30 Aris RM, Ontjes DA, Buell HE, Blackwood AD, Lark RK, Caminiti M, Brown SA, Renner JB, Chalermskulrat W, Lester GE: Abnormal bone turnover in cystic fibrosis adults. Osteoporos Int 2002;13:151–157.

31 Conway SP, Morton AM, Oldroyd B, Truscott JG, White H, Smith AH, Haigh I: Osteoporosis and osteopenia in adults and adolescents with cystic fibrosis: Prevalence and associated factors. Thorax 2000;55:798–804.

32 Sood M, Hambleton G, Super M, Fraser WD, Adams JE, Mughal MZ: Bone status in cystic fibrosis. Arch Dis Child 2001;84: 516–520.

33 Landon C, Rosenfeld RG: Short stature and pubertal delay in male adolescents with cystic fibrosis. Androgen treatment. Am J Dis Child 1984;138:388–391.

34 Bhudhikanok GS, Lim J, Marcus R, Harkins A, Moss RB, Bachrach LK: Correlates of osteopenia in patients with cystic fibrosis. Pediatrics 1996;97:103–111.

35 Rossini M, Del Marco A, Dal Santo F, Gatti D, Braggion C, James G, Adami S: Prevalence and correlates of vertebral fractures in adults with cystic fibrosis. Bone 2004;35:771–776.

36 Donaldson CL, Hulley SB, Vogel JM: Effect of prolonged bed rest on bone mineral. Metabolism 1970;19:1071–1084.

37 McKay HA, Petit MA, Schutz RW, Prior JC, Barr SI, Khan KM: Augmented trochanteric bone mineral density after modified physical education classes: A randomized school-based exercise intervention study in prepubescent and early pubescent children. J Pediatr 2000; 136:156–162.

38 Hahn TJ, Squires AE, Halstead LR, Strominger DB: Reduced serum 25-hydroxyvitamin D concentration and disordered mineral metabolism in patients with cystic fibrosis. J Pediatr 1979;94:38–42.

39 Henderson RC, Madsen CD: Bone density in children and adolescents with cystic fibrosis. J Paediatr 1996;128:28–34.

40 Lark RK, Lester GE, Ontjes DA, Blackwood AD, Hollis BW, Hensler MM, Aris RM: Diminished and erratic absorption of ergocalciferol in adult cystic fibrosis patients. Am J Clin Nutr 2001;73:602–606.

41 Conway SP, Wolfe SP, Brownlee KG, White H, Oldroyd B, Truscott JG, Harvey JM, Shearer MJ: Vitamin K status among children with cystic fibrosis and its relationship to bone mineral density and bone turnover. Pediatrics 2005;115: 1325–1331.

42 de Montalembert M, Lenoir G, Saint-Raymond A, Rey J, Lefrere JJ: Increased PIVKA-II concentrations in patients with cystic fibrosis. J Clin Pathol 1992;45:180–181.

43 Choonara IA, Winn MJ, Park BK, Littlewood JM: Plasma vitamin K1 concentrations in cystic fibrosis. Arch Dis Child 1989;64: 732–734.

44 Durie PR: Vitamin K and the management of patients with cystic fibrosis. CMAJ 1994;151: 933–936.

45 Rashid M, Durie P, Andrew M, Kalnins D, Shin J, Corey M, Tullis E, Pencharz PB: Prevalence of vitamin K deficiency in cystic fibrosis. Am J Clin Nutr 1999;70:378–382.

46 Mosler K, von Kries R, Vermeer C, Saupe J, Schmitz T, Schuster A: Assessment of vitamin K deficiency in CF – how much sophistication is useful? J Cyst Fibros 2003;2:91–96.

47 Conway SP: Vitamin K in cystic fibrosis. J R Soc Med 2004;97(suppl 44):48–51.

48 Beker LT, Ahrens RA, Fink RJ, et al: Effect of vitamin K supplementation on vitamin K status in cystic fibrosis patients. J Pediatr Gastroenterol Nutr 1997;24:512–517.

49 Manolagas SC, Weinstein RS: New developments in the pathogenesis and treatment of steroid-induced osteoporosis. J Bone Miner Res 1999;14:1061–1066.

49a King SJ, Topliss DJ, Kotsimbos T, Nyulasi IB, Bailey M, Ebeling PR, Wilson JW: Reduced bone density in cystic fibrosis: DeltaF508 mutation is an independent risk factor. Eur Respir J 2005;25:54–61.

50 Aris RM, Merkel PA, Bachrach LK, Borowitz DS, Boyle MP, Elkin SL, Guise TA, Hardin DS, Haworth CS, Holick MF, et al: Guide to bone health and disease in cystic fibrosis. J Clin Endocrinol Metab 2005;90:1888–1896.

51 Bukowski JF, Dascher CC, Das H: Alternative bisphosphonate targets and mechanisms of action. Biochem Biophys Res Commun 2005; 328:746–750.

52 Brenckmann C, Papaioannou A: Bisphosphonates for osteoporosis in people with cystic fibrosis. Cochrane Database Syst Rev 2001;4:CD002010.

53 Reinholz GG, Getz B, Pederson L, Sanders ES, Subramaniam M, Ingle JN, Spelsberg TC: Bisphosphonates directly regulate cell proliferation, differentiation, and gene expression in human osteoblasts. Cancer Res 2000;60: 6001–6007.

54 Plotkin LI, Weinstein RS, Parfitt AM, Roberson PK, Manolagas SC, Bellido T: Prevention of osteocyte and osteoblast apoptosis by bisphosphonates and calcitonin. J Clin Invest 1999;104:1363–1374.

55 Haworth CS, Selby PL, Adams JE, Mawer EB, Horrocks AW, Webb AK: Effect of intravenous pamidronate on bone mineral density in adults with cystic fibrosis. Thorax 2001;56:314–316.

56 Aris RM, Lester GE, Renner JB, Winders A, Denene BA, Lark RK, Ontjes DA: Efficacy of pamidronate for osteoporosis in patients with cystic fibrosis following lung transplantation. Am J Respir Crit Care Med 2000;162: 941–946.

57 Aris RM, Lester GE, Caminiti M, Blackwood AD, Hensler M, Lark RK, Hecker TM, Renner JB, Guillen U, Brown SA, Neuringer IP, Chalermskulrat W, Ontjes DA: Efficacy of alendronate in adults with cystic fibrosis with low bone density. Am J Respir Crit Care Med 2004;169:77–82.

58 Hardin DS, Rice J, Ahn C, Ferkol T, Howenstine M, Spears S, Prestidge C, Seilheimer DK, Shepherd R: Growth hormone treatment enhances nutrition and growth in children with cystic fibrosis receiving enteral nutrition. J Pediatr 2005;146:324–328.

Dr. Sarah Elkin
Royal Brompton Hospital
Sydney St, London SW3 6NP (UK)
Tel. +44 207 351 8232
Fax +44 207 351 8473
E-Mail s.elkin@rbh.nthames.nhs.uk

Bush A, Alton EWFW, Davies JC, Griesenbach U, Jaffe A (eds): Cystic Fibrosis in the 21st Century.
Prog Respir Res. Basel, Karger, 2006, vol 34, pp 278–283

Diabetes in Cystic Fibrosis

Nicola Bridges[a]   Karen Spowart[b]

[a]Consultant Paediatric Endocrinologist, and [b]Paediatric Diabetes Nurse Specialist,
Chelsea and Westminster Hospital, London, UK

Abstract

Pancreatic-insufficient individuals with cystic fibrosis (CF) develop abnormalities in insulin secretion which can be detected in childhood. The prevalence of impaired glucose tolerance and diabetes increases with age. Lung function and clinical status deteriorate in the years prior to a diagnosis of diabetes and survival is reduced in individuals with diabetes. Regular screening with oral glucose tolerance tests combined with clinical vigilance allows for early intervention with insulin to prevent clinical decline. Diabetes management must take into account the complex medical and psychological issues for individuals with CF, aiming to maintain nutritional and clinical status and prevent long-term complications of diabetes.

The increased prevalence of diabetes in cystic fibrosis (CF) has been recognized for many years, but the clinical significance of diabetes has increased as those with CF live longer. Most children with CF can now expect to live to their 30s or 40s – and unfortunately they can also expect to have a more than 50% chance of developing diabetes [1]. CF-related diabetes (CFRD) is different from type 1 and type 2 diabetes, and different treatment strategies are needed. The goals of managing CFRD are also different: in addition to preventing long-term diabetic complications, management is aimed at preventing the negative effects of diabetes on nutrition and lung function.

The Pathology of Cystic-Fibrosis-Related Diabetes

The basic defect underlying CFRD appears to be a reduction in beta cell number and loss of insulin secretion. CFRD is only seen in those with pancreatic insufficiency. Autopsy studies have demonstrated reduced numbers of beta cells in the pancreas in CF, with a greater reduction in CFRD [2].

Standard Definitions of Impaired Glucose Tolerance and Diabetes

Criteria for the diagnosis of diabetes were revised by the World Health Organization in 1999 (table 1) [3] – categories of impaired glucose tolerance and impaired fasting glucose were introduced. The cut-off values are based on an assessment of the increased cardiovascular risk associated with elevated glucose levels.

The Physiology of Cystic-Fibrosis-Related Diabetes

The nature of the defect in insulin secretion in CF has been studied using a number of different methods:
- measuring glucose and insulin during a standard oral glucose tolerance test (OGTT);

Table 1. Criteria for the diagnosis of diabetes [3]

Definition	Clinical features

Diabetes	Symptoms plus
	  random glucose concentration >11.1 mmol/l
	  or
	  Fasting glucose of >7.0 mmol/l
	  or
	  2-hour plasma glucose over >11.1 mmol/l after OGTT

Impaired glucose tolerance	Fasting plasma glucose <7.0 mmol/l, and
	  2-hour plasma glucose >7.8 mmol/l to <11.1 mmol/l
	  after OGTT

Impaired fasting glucose	Fasting glucose of >6.1 to <7.0 mmol/l

- studies of insulin sensitivity using insulin clamp (insulin is infused at a fixed rate and insulin sensitivity estimated from the glucose volume required to maintain normoglycaemia), and
- continuous glucose-monitoring systems (CGMS), which measure the glucose content of interstitial fluid via a disposable subcutaneous glucose electrode.

Insulin Secretion

Impaired glucose tolerance and diabetes in CF is caused by defects in the timing and quantity of insulin secretion. As glucose tolerance worsens, the timing of the insulin peak after a glucose load is increasingly delayed and there is a decline in total insulin secretion [3–5]. Because of the defect in insulin timing, glucose can be abnormally elevated during an OGTT despite a normal glucose concentration at 2 h (fig. 1) [5, 6]. Fasting glucose concentrations may be normal when OGTT is diabetic [7]. CGMS has demonstrated that subjects can still be significantly hyperglycaemic despite having normal fasting glucose, HbA1c and OGTT [7].

Insulin Sensitivity

Normal or increased insulin sensitivity has been reported in some studies [4, 5], and reduced insulin sensitivity in others [6–8]. The differences are probably explained by differences in the clinical status of the patients, with decreased insulin sensitivity related to worse clinical status [8, 9].

The Prevalence of Impaired Glucose Tolerance and Diabetes

European and US surveys of CF patients in the 1990s found that 4.9 and 6.1%, respectively, had a diagnosis of

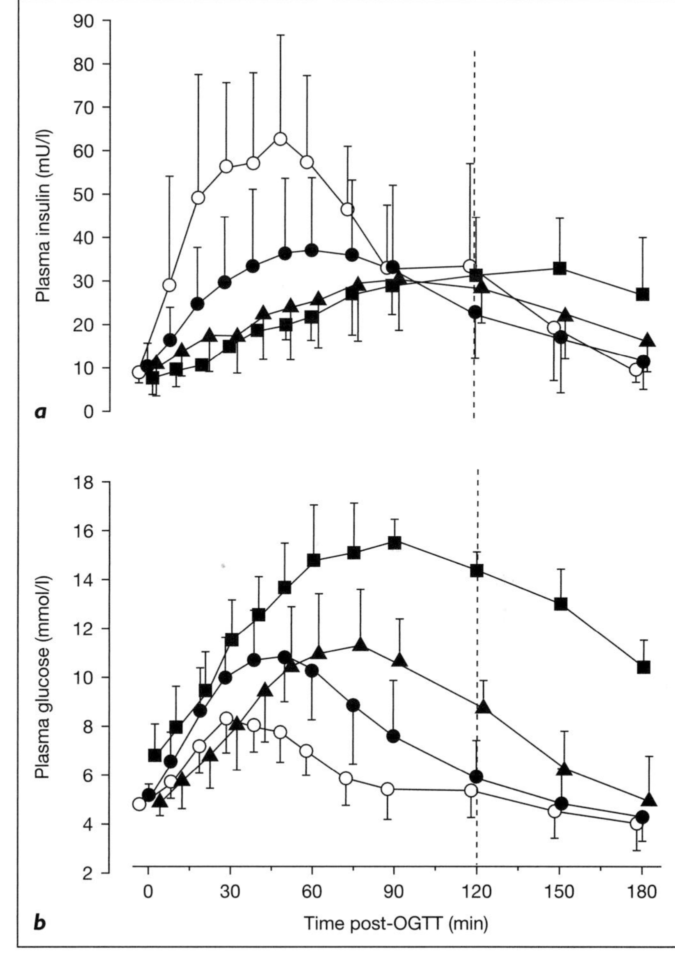

Fig. 1. Study demonstrating changes in insulin secretion as glycaemic status declines in CF. If the glucose is measured fasting and then only at 2 h during an OGTT, this may miss significant abnormalities in glucose handling. Mean plasma insulin (**a**) and glucose profiles (**b**) following extended oral glucose load in healthy controls (○; n = 8) and CF patients with differing glycaemic status (● = normal glucose tolerance, n = 16; ▲ = impaired glucose tolerance, n = 6; ■ = diabetic glucose tolerance, n = 2). The 120-min cut-off time point for determining glycaemic status is also shown. Reproduced from Yung et al. [4].

diabetes [10, 11]. Neither group was screened for diabetes, and a study screening patients with OGTT found that 17% of patients over the age of 5 years had CFRD without fasting hyperglycaemia and 11% had CFRD with fasting hyperglycaemia [4]. Reduced insulin secretion has been demonstrated in children from 5 years old and the prevalence of impaired glucose tolerance and CFRD increases with age [12, 13]. CFRD is unusual under 10 years old, but screening with OGTT found 24% of patients aged 20 had CFRD and 76% of those over 30 years [1].

The Relationship between Clinical State and Cystic-Fibrosis-Related Diabetes

Individuals with CFRD have worse lung function and reduced survival rates compared with non-diabetic subjects of the same age [14]. This relationship remains if allowance is made for the fact that only those with more severe CF mutations develop diabetes [15]. There is an insidious decline in lung function and nutritional parameters as glucose tolerance deteriorates, with significant differences up to 4 years before a diagnosis of diabetes [15]. Abnormal glucose tolerance predicts a greater decline in lung function over the next 4 years [16]. Nutrition is tightly linked to lung function in CF [17], and the gradual loss of the anabolic effect of insulin is likely to be a factor in the clinical decline. In children with CF, insulin secretion correlates with growth velocity [18]. Increased blood glucose concentrations result in an increase in the glucose concentration in nasal and airway secretions [19] (fig. 2). Increased infection risk, associated with higher glucose concentrations in the airways may contribute to clinical deterioration.

Diabetic Complications in Cystic-Fibrosis-Related Diabetes

Acute Complications
Diabetic ketoacidosis is rare in CFRD but has been described [20]. Symptomatic hyperglycaemia (polyuria and polydipsia) can occur.

Long-Term Complications of Diabetes in Cystic-Fibrosis-Related Diabetes
In type 1 and type 2 diabetes, the risk of developing microvascular complications (retinopathy, nephropathy, microalbuminuria and neuropathy) is related to level of glycaemic control and to the time since onset of diabetes [21, 22]. Ten to twenty-three percent of patients with CFRD

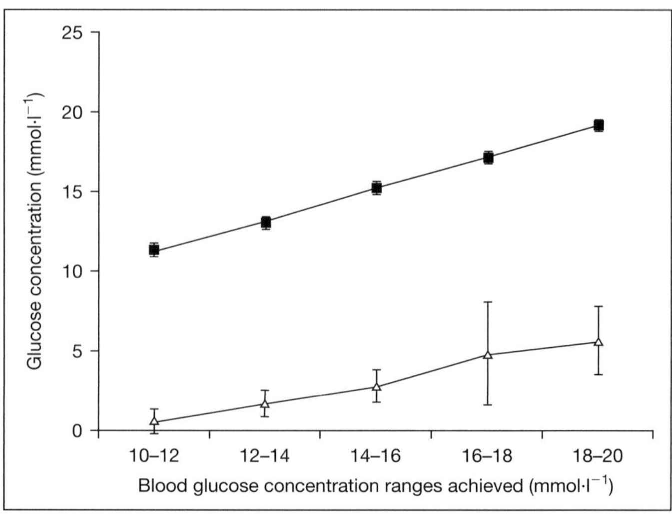

Fig. 2. Rise in nasal glucose concentrations as the blood glucose concentration is raised. Increasing airway glucose concentrations may have a role in the deterioration in clinical status with the development of CFRD. Reproduced from Wood et al. [20]. Blood glucose concentrations in 6 normal individuals (■) were raised to 20 mmol/l and nasal glucose concentrations (Δ) were measured. Values are means of six experiments and error bars show SD.

have been found to have microvascular complications [23, 24]. Diabetic retinopathy has been reported in 16% of patients after 5 years of CFRD and 23% after 10 years [25]. It is likely that the risk of microvascular complications in CFRD is similar to that seen in type 1 or type 2 diabetes. Macrovascular complications have not been described in CFRD, probably because cholesterol levels are low in CF and hypertension is unusual unless the patient has been transplanted.

Screening and Diagnosis of Cystic-Fibrosis-Related Diabetes
Significant hyperglycaemia can be present for years with no symptoms. Fasting glucose levels and HbA1c are frequently normal in CFRD and are not reliable as screening tests [1, 26]. Recent guidelines for the UK suggest yearly OGTT after 12 years of age [27]. Most clinicians use standard criteria for defining glucose tolerance in CF (table 1) [3]. These definitions are based on the increased risk of cardiovascular disease associated with different levels of glucose tolerance. In CFRD the decision to treat may also be based on the effect of glucose tolerance on lung function and clinical state. The results of OGTT in CF can vary with clinical status and an abnormal OGTT may revert to normal over time [1]. A report of early insulin treatment in 4 paediatric patients with normal OGTT but random blood glucose

Table 2. Analogue insulin preparations

Name	Approximate duration of action	Uses
Short-acting analogues	4 h	Very rapid peak action – can be given
Aspart (Novorapid©)	4 h	immediately before or after food
Lispro (Humalog©)		
Long-acting analogues	24 h	'basal' insulins with no peak, less risk of
Glargine (Lantus©)	20 h	hypoglycaemia compared with
Detemir (Levemir©)		conventional long-acting insulins

levels over 11.1 mmol/l describes improvements in weight and lung function [28]. This suggests that insulin deficiency may be of clinical significance without an abnormal OGTT. Insulin treatment in CFRD results in improvements in BMI, lung function and clinical condition [16, 17, 29], and because of this insulin should be the first-line treatment in CFRD. Although hypoglycaemic agents are widely used, there are few studies examining the administration of oral hypoglycaemic agents. Sulphonylureas [30] and repaglinide [31] have been demonstrated to control blood glucose in CFRD. Early intervention with insulin treatment is likely to reduce the decline in clinical status which occurs with worsening glucose tolerance. Regular screening with OGTT (with reassessment between screening tests if the clinical state changes) is currently the best way to detect individuals who might benefit from insulin treatment.

Treatment of Cystic-Fibrosis-Related Diabetes with Insulin

A period of glucose monitoring is helpful in designing an insulin regimen. If patients have erratic eating habits, short- and long-acting insulin analogues can provide a flexible insulin regimen (table 2). Twice daily mixed insulin is simpler, but may not give good control if food intake is not regular. Targets for glycaemic control and for glucose monitoring will vary according to the clinical condition of the patient. Management often involves a compromise between ideal diabetes management and what the patient can actually manage. Insulin delivery by dry powdered inhalation is being studied for patients with diabetes; it is unlikely that patients with CF will benefit from this, because lung disease would result in lack of absorption.

Dietary Management

The dietary management of CFRD is dealt with in detail in chapter 37. The primary aim of treatment in CFRD is to maintain nutrition, and the maintenance of a high energy content diet is important. Individuals should be advised not to limit their intake of refined carbohydrates or high fat foods [25]. Maintaining an adequate intake of carbohydrate is helpful in maintaining steady blood glucose levels. Glucose- or sucrose-containing drinks can sometimes increase the blood glucose and it may help to omit them, or only take them with food. Many people are aware of the dietary restrictions that are part of the treatment of type 1 and type 2 diabetes and it must be made clear that these do not apply to CFRD. Information should be given about appropriate management of hypoglycaemia. Many individuals with CF have erratic eating habits – it is better to try to tailor their insulin regimen to their pattern of eating than to try and change life style. Nasogastric or PEG feeds can be covered with soluble insulin if they run over a few hours, or intermediate- or long-acting insulin if they run overnight. Relatively large doses of insulin may be required and patients and carers must be made aware of the risk of severe hypoglycaemia if the insulin is given and not followed by the feed, or the feed is stopped early.

Psychological Issues

The diagnosis of CFRD may have a great psychological impact on the patient and her or his family. Many are not aware of the risk of diabetes until they are diagnosed [32]. The adverse side effects of diabetes are widely known and frequently cause concern. As chronic disorders where nutrition is central to management, increased prevalence of abnormal eating behaviour and eating disorders have been reported in type 1 diabetes [33] and cystic fibrosis [34] (see also chapter 39). In type 1 diabetes missing or reducing insulin doses to control weight gain is common; up to 30% of women with type 1 diabetes have done this [35]. Children and young women with CF are less likely than control subjects to wish to be thinner, but this observation is limited to those who are thin already [36, 37]. Manipulation of insulin dose in order to control weight may be less common in CFRD than type 1 diabetes but is not impossible.

Children and Adolescents

In younger children the insulin regimen may need to take account of how much self-care the child can manage. Because of the importance of avoiding hypoglycaemia in children, blood glucose targets will be less tight than in adults. During adolescence there is increasing independence, and the peer group becomes more important. Risk-taking behaviour is common. In adolescents with CF, psychological issues may result in reduced adherence to their treatment [38, 39]. Insulin doses may be missed and blood glucose testing may be erratic. Paying attention to their feelings and being non – judgemental will enable the clinician to develop a rapport with the young person [40]. It is essential to explore the extent to which the adolescent discusses issues with parents and to support the adolescent-parental relationship.

Special Situations – Pregnancy and Heart Lung Transplants

A diagnosis of diabetes prior to pregnancy does not appear to increase the risk of complications in pregnancy for women with CF [41]. The prevalence of gestational diabetes is greatly increased in women with CF [42, 43] (see chapter 33). Diabetes is not a contraindication to lung transplantation – there is no reduction in long-term survival or increase in the risk of peri-operative death related to sepsis [44]. Many patients require short-term insulin treatment [45].

Conclusions

There is a relatively small resource of research and clinical experience on which to base management of CFRD. Our current knowledge suggests that prospective screening with OGTT and early intervention with insulin treatment is likely to be the best approach. Screening with OGTT identifies those with CFRD and also some who are likely to need insulin in the near future. Discussion of the problem in advance is likely to reduce the psychological impact of having to start insulin. If over 50% of individuals with CF are going to develop CFRD, there is a case for increasing knowledge about diabetes and its management for all those with CF and their families. Clinics for type 1 and type 2 diabetes are not likely to be the best solution for follow-up of CFRD. In this setting it is more difficult to look at the other aspects of CF, and clinicians may not have significant experience in CFRD. Integrating diabetes care into CF management is likely to be a better solution, ideally from a team experienced in CFRD. Effective management of CFRD is important because of the impact of diabetes on nutrition and lung function, and because of the risk of long-term diabetic complications which would add to the morbidity of CF.

References

1 Lanng S, Hansen A, Thorsteinsson B, Nerup J, Koch C: Glucose tolerance in patients with cystic fibrosis: Five year prospective study. BMJ 1995;311:655–659.

2 Iannucci A, Mukai K, Johnson D, Burke B: Endocrine pancreas in cystic fibrosis: An immunohistochemical study. Hum Pathol 1984;15:278–284.

3 World Health Organisation 1999: Definition, diagnosis and classification of Diabetes Mellitus and its complications. World Health Organisation, Department of Non Communicable Disease Surveillance, Geneva.

4 Moran A, Doherty L, Wang X, Thomas W: Abnormal glucose metabolism in cystic fibrosis. J Pediatr 1998;133:10–17.

5 Yung B, Noormohamed FH, Kemp M, Hooper J, Lant AF, Hodson ME: Cystic Fibrosis-related diabetes: The role of peripheral insulin resistance and beta-cell dysfunction. Diabet Med 2002;19:221–226.

6 Holl RW, Wolf A, Bernhard M, Buck C, Missel M, Heinze E, Von der Hardt H, Teller WM: Insulin resistance with altered secretory kinetics and reduced proinsulin in cystic fibrosis. J Pediatr Gastroenterol Nutr 1997;25: 188–193.

7 Dobson L, Sheldon CD, Hattersley AT: Conventional measures underestimate glycaemia in cystic fibrosis patients. Diabet Med 2004; 21:691–696.

8 Hardin DS, LeBlanc A, Lukenbaugh S, Seilheimer DK: Insulin resistance is associated with decreased clinical status in cystic fibrosis. J Pediatr 1997;6:948–956.

9 Austin A, Kalhan SC, Orenstein D, Nixon P, Arslanian S: Roles of insulin resistance and beta-cell dysfunction in the pathogenesis of glucose intolerance in cystic fibrosis. J Clin Endocrinol Metab 1994;79:80–85.

10 Hardin DS, LeBlanc A, Marshall G, Seilheimer DK: Mechanisms of insulin resistance in cystic fibrosis. Am J Physiol Endocrinol Metab 2001;281:E1022–E1028.

11 Rosenecker J, Eichler I, Kuhn L, Harms H, von der Hardt H: Genetic determination of diabetes mellitus in patients with cystic fibrosis. Multicenter Cystic Fibrosis Study Group. J Pediatr 1995;127:441–443.

12 Konstan MW, Butler SM, Schidlow DV, Morgan WJ, Julius JR, Johnson CA: Patterns of Medical Practice in Cystic Fibrosis. II. Use of Therapies. Pediatr Pulmonol 1999;28:248–254.

13 Solomon MP, Wilson DC, Corey M, Kalnins D, Zielenski J, Tsui L-C, Pencharz P, Durie P, Sweezey NB: Glucose intolerance in children with cystic fibrosis. J Pediatr 2003;142: 128–132.

14 Taylor AM, Thomson A, Bruce-Morgan C, Ahmed ML, Watts A, Harris D, Holly JMP, Dunger DB: The relationship between insulin, IGF-I and weight gain in cystic fibrosis. Clin Endocrinol 1999;51:659–665.

15 Koch C, Rainisio M, Madessani U, Harms HK, Hodson ME, Mastella G, McKenzie SG, Navarro J, Strandvik B: Presence of cystic fibrosis-related diabetes mellitus is tightly linked to poor lung function in patients with cystic fibrosis: Data from the European Epidemiologic registry of cystic fibrosis. Pediatr Pulmonol 2001;32:343–350.

16 Nousia-Arvanitakis S, Galli-Tsinopoulou A, Karamouzis M: Insulin improves clinical status of patients with cystic-fibrosis-related diabetes mellitus. Acta Paediatr 2001;90: 515–519.

17 Milla CE, Warwick WJ, Moran A: Trends in pulmonary function in patients with cystic fibrosis correlate with the degree of glucose intolerance at baseline. Am J Respir Crit Care Med 2000;162:891–895.

18 Laursen EM, Koch C, Petersen JH, Muller J: Secular changes in anthropometric data in cystic fibrosis patients. Acta Paediatr 1999;88: 169–174.

19 Ripa P, Robertson I, Cowley D, Harris M, Brent Masters I, Cotterill AM: The relationship between insulin secretion, the insulin-like growth factor axis and growth in children with cystic fibrosis. Clin Endocrinol 2002;56: 383–389.

20 Wood DM, Brennan AL, Philips BJ, Baker EH: Effect of hyperglycaemia on glucose concentration of human nasal secretions. Clin Sci 2004;106:527–533.

21 Atlas AB, Finegold DN, Becker D, Trucco M, Kurland G: Diabetic ketoacidosis in cystic fibrosis. Am J Dis Child 1992;146: 1457–1458.

22 UK Prospective Diabetes Study (UKPDS) Group: Intensive blood-glucose control with sulphonylureas or insulin compared with conventional treatment and risk of complications in patients with type 2 diabetes (UKPDS 33). Lancet 1998;352:837–853.

23 DCCT Research Group: The effect of intensive treatment of diabetes on the development and progression of long-term complications in insulin dependent diabetes mellitus. N Engl J Med 1993;329:977–986.

24 Sullivan MM, Denning CR: Diabetic microangiopathy in patients with cystic fibrosis. Pediatrics 1989;84:642–647.

25 Wilson DC, Kalnins D, Stewart C, Hamilton N, Hanna AK, Durie PR, Tullis E, Pencharz PB: Challenges in the dietary treatment of cystic fibrosis related diabetes. Clin Nutr 2000;19:87–93.

26 Yung B, Landers A, Mathalone B, Gyi KM, Hodson ME: Diabetic retinopathy in adult patients with cystic fibrosis-related diabetes. Respir Med 1998;92:871–872.

27 Garagorri JM, Rodriguez G, Ros L, Sanchez A: Early detection of impaired glucose tolerance in patients with cystic fibrosis and predisposition factors. J Pediatr Endocrinol Metab 2001;14:53–60.

28 Report of the UK Cystic fibrosis trust diabetes working group: Management of cystic fibrosis related diabetes mellitus. Cystic Fibrosis Trust 2004.

29 Dobson L, Hattersley AT, Tiley S, Elworthy S, Oades PJ, Sheldon CD: Clinical improvement in cystic fibrosis with early treatment. Arch Dis Child 2002;87:430–431.

30 Rolon MA, Benali K, Munck A, Navarro J, Clement A, Tubiana-Ru N, Czernichow P, Polak M: Cystic fibrosis-related diabetes mellitus: Clinical impact of prediabetes and effects of insulin therapy. Acta Paediatr 2001; 90:860–867.

31 Rosenecker J, Eichler I, Barmeier H, von der Hardt H: Diabetes mellitus and cystic fibrosis: Comparison of clinical parameters in patients treated with insulin versus oral glucose-lowering agents. Pediatr Pulmonol 2001;32: 351–355.

32 Moran A, Phillips J, Milla C: Insulin and glucose excursion following premeal insulin lispro or repaglinide in cystic fibrosis-related diabetes. Diabetes Care 2001;24:1706–1710.

33 Connor S, Cowperthwaite C, Clegg C, Casson I, Walshaw M: The psychosocial impact of its diagnosis on adult cystic fibrosis (CF) patients. Diabetes Res Clin Pract 2000; 50(suppl 2):236.

34 Rydall AC, Rodin GM, Olmsted MP, Devenyi RG, Daneman D: Disordered eating behavior and microvascular complications in young women with insulin-dependent diabetes mellitus. New Engl J Med 1997;336: 1849–1854.

35 Raymond NC, Chang PN, Crow SJ, Mitchell JE, Dieperink BS, Beck MM, Crosby RD, Clawson CC, Warwick WJ: Eating disorders in patients with cystic fibrosis. J Adolesc 2000;23:359–363.

36 Bryden KS, Neil A, Mayou RA, Peveler RC, Fairburn CG, Dunger DB: Eating habits, body weight, and insulin misuse. A longitudinal study of teenagers and young adults with type 1 diabetes. Diabetes Care 1999;22: 1956–1960.

37 Truby H, Paxton SJ: Body image and dieting behavior in cystic fibrosis. Pediatrics 2001; 107:92–99.

38 Walters S: Sex differences in weight perception and nutritional behaviour in adults with cystic fibrosis. J Hum Nutr Dietet 2001;14: 83–91.

39 DiGirolamo AM, Quittner AL, Ackerman V, Stevens J: Identification and assessment of ongoing stressors in adolescents with a chronic illness: An application of the behavior-analytic model. J Clin Child Psychol 1997; 26:53–66.

40 DeLambo KE, Ievers-Landis CE, Drotar D, Quittner AL: Association of observed family relationship quality and problem-solving skills with treatment adherence in older children and adolescents with cystic fibrosis. J Pediatr Psychol 2004;29:343–353.

41 Azzopardi K, Lowes L: Management of cystic fibrosis related diabetes in adolescence. Br J Nurs 2003;12:359–363.

42 Goss CH, Rubenfeld GD, Otto K, Aitken ML: The effect of pregnancy on survival in women with cystic fibrosis. Chest 2003;124: 1460–1468.

43 Gilljam M, Antoniou M, Shin J, Dupuis A, Corey M, Tullis DE: Pregnancy in cystic fibrosis. Fetal and maternal outcome. Chest 2000;118:85–91.

44 Ødegaard I, Stray-Pedersen B, Hallberg K, Haanaes OC, Trond Storrøsten O, Johannesson M: Maternal and fetal morbidity in pregnancies of Norwegian and Swedish women with cystic fibrosis. Acta Obstet Gynecol Scand 2002;81:698–705.

45 De Soyza A, Archer L, Wardle J, Parry G, Dark JH, Gould K, Corris PA: Pulmonary transplantation for cystic fibrosis: Pre-transplant recipient characteristics in patients dying of peri-operative sepsis. J Heart Lung Transplant 2003;22:764–769.

46 Wiebea K, Wahlersa T, Harringera W, Hardt HVD, Fabel H, Haverich A: Lung transplantation for cystic fibrosis – a single center experience over 8 years. Eur J Cardiothorac Surg 1998;14:191–196.

N. Bridges
Chelsea and Westminster Hospital
369 Fulham Road
London SW10 9NH (UK)
Tel. +44 20 8746 8687, Fax +44 20 8746 8644
E-Mail n.bridges@imperial.ac.uk

Multidisciplinary Care

Bush A, Alton EWFW, Davies JC, Griesenbach U, Jaffe A (eds): Cystic Fibrosis in the 21st Century.
Prog Respir Res. Basel, Karger, 2006, vol 34, pp 286–292

Challenges for Nurses

Susan Madge

Royal Brompton and Harefield NHS Trust, London, UK

Abstract

The role of the Cystic Fibrosis Clinical Nurse Specialist (CF CNS) has changed dramatically over the last decade. Children nowadays are generally well and the adult population is increasing; however, as young adults live longer new problems are being revealed. Common challenges facing the CF CNS today include an increasing demand for support in the home, workplace or college, and guidance with travel, starting families, financial problems such as insurance and pensions, transplantation and terminal care. The CF CNS has had to adapt and learn new skills to meet these challenges acknowledging that this will continue to be an ongoing development.

Cystic Fibrosis Care in the UK

Cystic fibrosis (CF) care is ideally managed at a Specialist CF Centre. Often the Specialist Centre is the local hospital, although geography often means that the local hospital acts as a shared care resource with the Specialist Centre. When organized well, shared care can contribute significantly to a reduction in the total treatment-related burden. The child or adult with CF is required to attend hospital appointments on an outpatient basis 2–3 monthly throughout his or her life so that all aspects of the disease can be monitored and changes in management planned and discussed. Where good communication pathways are set up between the specialist multidisciplinary team and the local team, patients will benefit.

Gradually, as the individual's condition deteriorates, inpatient admissions for 2–3 weeks at a time are required, often several times a year. Supported and supervised home-care is an option offered by many Specialist CF Centres and may help to reduce the stress associated with repeated hospitalization. However, taking the hospital into the home may have longer-term implications for the individual and their families; therefore, the Cystic Fibrosis Clinical Nurse Specialist (CF CNS) must be continually aware of changing needs.

Challenges for the Nurse

Looking after people with CF is complex and demanding. Nurses who care for children or adults in hospital or in the community must ensure that they liaise closely with the nearest Specialist CF Centre. Nursing is a partnership between the patients, carers and other professionals. Advances in medical management over the last few years have improved both quality of life and longevity. However, due to the complexity of the disease it is widely recognized that people with CF should continue to be cared for using a multidisciplinary team approach. CF teams in the UK include doctors, specialist nurses, dietitians, physiotherapists, psychologists, social workers and administrators. In addition it is recommended that these teams be supported by staff such as respiratory function technicians and ward nurses skilled in caring for people with CF [1]. The multidisciplinary team works very closely together to ensure a holistic approach to the care of their patients. Good communication is paramount in managing this patient group; therefore, the team must allow time to regularly meet.

All Specialist CF Centres have a clinical specialist nurse (CF CNS) with expert knowledge of CF although their exact roles may vary. In some centres the CF CNS will co-ordinate and facilitate community care in liaison with local or generic teams. Other centres will have a dedicated CF CNS who is able to offer direct care to patients in the community and some will have both. The CF CNS, whether community or hospital based, can provide skilled support, advice and care directly to the patient and family wherever it is needed. Nurses working with patients with CF in areas where there is no opportunity to work within a CF Team are strongly advised to make contact with the nearest Specialist CF Centre, both for their own support and to assure optimum care for their patients [2].

The role of the nurse caring for individuals with CF, whether children or adults, incorporates key elements. These elements form the basis of good clinical care and include advocacy, clinical management, support, advice, education, research and management [3, 4]. There are many demands made on both the patient and the CF Team; therefore, patient and family advocacy is one of the most important roles for the nurse. Patient well-being and satisfaction with care are paramount and successful advocacy can ensure this [5]. Nurses must also take part in decision making and monitoring of care. In addition to the practical, day-to-day care, nurses need to be aware of all treatment modalities used in the management of CF and ensure that each patient receives optimum care for their individual needs. Patients, families and professional colleagues will benefit from support, liaison and advice co-ordinated by the nurse [6]. Problems are often resolved more easily if pathways of communication are well established for all parties.

There are many treatment regimens that have to be learnt throughout the patient's life, for example nebulizer therapy, enteral feeding and intravenous antibiotic administration. Successful teaching of the patient and all concerned (parent, carer, school, work colleagues), and their understanding of the disease process will ensure that treatment is carried out safely and effectively. Ongoing support and instruction as well as educational sessions designed to meet the needs of individual issues surrounding adherence to treatment can be successfully dealt with as they occur.

Participating in research either as a collaborator or as a lead investigator allows nurses to develop the most appropriate methods of treatment and care for their patient group. All nurses are responsible for developing their own professional practice and maintaining evidence of professional development is a responsibility that must be sustained. By ensuring that they keep up to date with new advances in treatment and new developments in the world of CF research the CF CNS can optimize the service that they offer to their patient group [7, 8].

Recommendations for Best Nursing Practice

There are particular areas of CF care that present the nurse with challenges; these include diagnosis, inpatient and outpatient care, care in the community, adolescence and young adulthood, family planning, transition from paediatric to adult care, and end of life issues. The treatment of CF has been reported by parents and patients alike as monotonous and repetitive. Daily treatment regimens mainly revolve around careful diet planning, administration of pancreatic enzyme supplements, chest physiotherapy, and the administration of both inhaled and oral medication. Patients suffering from exacerbations of their chest condition may also require intravenous antibiotic therapy, with this need increasing as their condition deteriorates. Those severely affected may eventually require continuous nutritional support via a gastrostomy, oxygen therapy, some ventilatory support and become wheelchair bound due to breathlessness and decreasing exercise ability.

Specialist nursing care commonly overlaps with the contribution from other members of the CF Team, physiotherapy, dietetics and psychology, often with nurses working in close partnership with colleagues. However, with good communication the involvement of nurses in these areas ensures co-ordination and support for patients and a point of contact for all concerned. For clarity, the roles of the CF CNS are described here at the various stages of life and disease progression, from diagnosis, through the early stages of disease, becoming an adult and end-stage disease. Clearly, many of these roles will be applicable at several times throughout life.

Diagnosis
Although neonatal screening exists in some areas of the UK, diagnosis of CF is usually made on clinical grounds and can be made at any age. While most patients are diagnosed within the first year of life it is not unusual for adults, especially males, to be diagnosed later on in life often due to infertility problems. Diagnosis of a life-limiting disease can be likened to bereavement and needs to be handled with honesty and sensitivity by skilled personnel. Ideally, diagnosis should be carried out in a Specialist CF Centre experienced in the techniques being used. Where CF is

confirmed, immediate referral must be made to a CF Team [9]. The Team will ensure that support and counselling is available, with a nurse present when the diagnosis is being given [10, 11]. Giving a new diagnosis often means presenting the parent and patient with new and potentially frightening information. Involving a nurse in the process will help to determine the appropriate timing for the introduction of information [11]. Providing ongoing support is often a role that the nurse offers to the patient and their family, especially around areas such as continuing education about the disease, the genetic implications to the immediate and extended family and the expectations of care [12].

Routine Care in Childhood and Adulthood

The majority of children with CF will be well most of the time, with short periods of exacerbation. Most of their contact with hospital staff will come from outpatient visits, with a minority of children requiring inpatient care. The majority of Specialist CF Centres provide a homecare service for individuals where there is a need. Even in adulthood, many people with CF with mild or moderate disease will have a similar pattern of care and the roles of the CF CNS will also apply.

Outpatient Care

At the time of diagnosis parents are taught an often complicated regimen of treatment to maintain the health of their child at home. Essentially, this regimen will involve chest physiotherapy, the administration of medicines (oral, inhaled, nebulized and intravenous), and attention to diet. The treatment regimen will vary between individuals and change as the child grows up; therefore, regular attendance at a Specialist CF Centre is essential to ensure that they are receiving the most appropriate treatment. During the visit the patient will have key aspects of their treatment reviewed; however, all patients with CF receive a comprehensive annual assessment [13, 14]. This assessment is an opportunity to evaluate all aspects of CF care and plans for the following year are made between members of the CF Team, the patients and their carers. CF CNSs are familiar with many aspects of treatment and can offer help and advice on indwelling venous access devices, gastrostomies, the use of nebulizers and monitoring CF-related diabetes.

Of particular concern in CF is the issue of cross-infection; many Specialist CF Centres hold different clinics for patient groups depending on the microbiology of their sputum (*Burkholderia cepacia* complex, MRSA, multiresistant *Pseudomonas aeruginosa*). It is important for all staff to be aware of the cross-infection policies for the Specialist CF

Table 1. Areas of CF nursing that require specialist knowledge

Issues surrounding diagnosis	Osteoporosis
	Pre- and posttransplant management
Nutritional requirements	End-stage and symptom management
Psychological issues	Nebulizer therapy
Infection control	Enteral feeding
CF-related diabetes	Intravenous therapy
Adherence	Care of indwelling venous access devices
Adolescents/young adults	
Transition to adult care	Oxygen therapy
Antenatal and postnatal care	Respiratory support

Centre they work at and be able to liaise with patient and family on issues surrounding cross-infection and infection control [14].

Inpatient Care

Although CF is generally managed at home, it is inevitable that there will be times of hospitalization. Often patients are admitted to a general paediatric ward or general respiratory medicine ward; it is helpful, therefore, for ward nurses to have access to a CF CNS, who has knowledge and experience of CF, the disease process and the clinical and psychological outcomes [15] (table 1). The CF CNS will also be available to offer advice, education and support to the patient, their families/carers and all staff involved [16]. In liaison with ward staff, the CF CNS will ensure that discharge-planning needs are met with special reference to the family doctor, the primary health care team, the shared care hospital and school or workplace (where appropriate).

Community Care

Whenever safe and effective care can be delivered in the community; patients and their families should be offered the choice. Care supported by the CF CNS in the community includes supporting the newly diagnosed, intravenous antibiotic therapy, follow-up after admission, teaching new treatments, liaising with schools, colleges and the workplace, support for pregnant women, support for those on a transplant waiting list, terminal care and support for the bereaved. As families manage the majority of CF care in the home it is important that they receive a level of support that is equal to that offered in hospital. Recommendations suggest that each patient should have this support offered by a CF CNS [9]. However, this is not always possible; therefore, nurses caring for patients with CF in the community should at least be supported by a CF CNS who has knowledge and experience of CF. This will include knowledge of the disease process and the clinical and psychological outcomes of the treatment modalities to ensure safe and effective care meeting the minimum nursing standards [17]. The CF CNS will also ensure that there is close liaison between the CF Team, the GP, the primary health care team, the shared care hospital, work or school and other relevant agencies [17].

Adolescence and Adulthood

Adolescence and young adulthood are particularly difficult times for those having to deal with both the psychological and practical burden of CF. Many treatment-related issues can present difficulties and the nurse must be alert to problems, especially those surrounding adherence and conflict between teenager and parents [18]. It is important, therefore, that the nurse plays an instrumental role in promoting self-care and responsibility in the young adult and offering support and advice to the parents [19].

It has been acknowledged more recently that encouraging independence with treatment regimens can be difficult for both the child to accept responsibility and for the parents to 'let go'. Professionals working in CF now advocate a concept of interdependence between parents and child with support being given and received from either side. Anecdotal clinical reports indicate that adherence to rigorous and complex treatment regimens appears to be improved where this philosophy has been adopted. Treatment for CF, therefore, is time consuming, intrusive, complicated, differentiating and restrictive. No immediate benefits are felt from most of the treatments as they are largely preventative. These regimens can be far too complex for the young person to carry out alone; therefore, some parental/carer involvement is advocated.

Growing up with CF does not preclude the young person from experiencing life as their peers do; unfortunately some behaviours may have a detrimental effect on medical management. Parents and young adults often find that the responsibility for administering correct doses at the right times of the day overwhelming. Anecdotally, older children and adults report poor adherence to both treatment regimens and taking medication as it is often seen as time wasting, boring, and different from the normal life of their peers. These young adults are determined to live as normal a life as possible and acknowledge that they are often making an informed decision not to carry out their treatment. Literature reports non-adherence to prescribed medical treatment in both children and adults as an ongoing concern, despite supervision and support from parents or other family members [20]. The consequences of poor adherence to treatment regimens can become devastating causing often irreparable damage to pulmonary and nutritional status. It is, therefore, important to understand factors influencing adherence in promoting good health status. It appears that despite understanding the reasons for carrying out treatment, there is an unspoken acknowledgement between patients and families that normal daily life needs to continue.

The role of the CF CNS, therefore, is crucial during these difficult times. Nurses working with these groups of patients need to ensure that they receive appropriate knowledge regarding issues such as fertility, pregnancy, contraception and safe sex, cross-infection, smoking and substance abuse, further education and employment [21]. Academic examinations can be a particularly important issue and the CF CNS must remember to liaise with schools and colleges in supporting continuing education [6]. Being a constant and reliable advocate and source of support can often help adolescents and young adults pass through this stage in their lives and move onto a more responsible adulthood.

Transition from Paediatric to Adult Care

Transition from paediatric care to adult care is an issue not only for the young person but also for their family. Nurses should use experience and knowledge to advise on the appropriate time for transition to adult care for each patient and the most appropriate programme to follow. This will ideally include a programme incorporating liaison and communication between the paediatric and adult Specialist CF Centres about all aspects of care, the co-ordination of joint transition clinics, parallel care and visits to the Specialist Adult CF Centre where appropriate and the provision of adequate information and ongoing support for patients and parents during the transition period [22, 23].

Sadly, despite an ever-increasing adult population, the transition from paediatric care to adult care continues to remain a challenge. Many authors have identified major problems with transition, either with an existing programme or more frequently, with a lack of any formal programme [24, 25]. Problems described have included a reluctance to 'let go' even when there are trained physicians available, parental resistance, lack of information, adolescent and parental beliefs about adult care, poor communication between the paediatric and adult teams and adolescent concerns not being met [26, 27].

Adolescents are going through many transitions in their lives. Early preparation and the availability of information will encourage the adolescent, their parents and professionals caring for them to prepare for the move. Successful elements of a transition programme include responding to concerns such as finance, independence, fertility and sexuality, optimism about the future, meeting the 'adult team', the presence of family members, and the provision of information [23, 28, 29]. The CF CNS is crucial in the organization and co-ordination of a transition programme involving both paediatric and adult teams.

Pregnancy and Fertility

Having CF does not preclude women wanting to start a family. They have grown up with plans and dreams about their future and in many cases, as with their peers, this includes having a baby. Fertility is not a problem in well-nourished women with CF and even those with poorer lung function ($FEV_1 < 50\%$ predicted) occasionally manage to conceive. The CF Team have a responsibility to offer advice and information regarding the risks of pregnancy to women; however, women who go ahead with pregnancy against advice must feel able to approach the Team for support. Pregnancy management in CF requires the input of many professionals and the CF CNS plays a key role in co-ordinating care between members of the CF Team, the obstetric team and the primary healthcare team. Support for the mother must be offered both before and after the birth, in hospital and in the home. Often the mother becomes unwell during the final stages of pregnancy and has to hospitalized until delivery. This need for hospitalization may continue after the birth and preparation must be made in the medical ward to allow this to happen (see also chapter 33).

Most men with CF are infertile, most commonly due to bilateral absence of the vas deferens. Fostering, adoption and artificial insemination by donor are all options that have been explored; however, more recently men are favouring the option of epididymal sperm aspiration followed by intracytoplasmic sperm injection. Assisted reproduction involves both parents; therefore, the CF CNS should be in a position to offer counselling and provide appropriate information.

For both men and women with CF having a baby (pregnancy or going through assisted reproduction) is exhausting, both emotionally and physically. The CF CNS will need to be aware of these demands and support the individual in maintaining their health throughout the process. Balancing treatment time with looking after a child can be very difficult and many parents with CF admit that treatment suffers as their child always comes first. As a parent with CF gets older, more time and energy will have to be devoted to maintaining their health.

All parents make plans for their child; however, the CF CNS will need to support the parent and their partner in thinking about the future from the time that they start planning a baby. The parent with CF needs to think realistically about the unlikelihood of seeing their child grow up; therefore, the CF CNS working with colleagues in the CF Team can help with preparation and planning for the future.

Advanced Disease

Organ Transplantation

Recognizing the most appropriate time to discuss the option of organ transplant (lungs, heart/lungs or liver) can be very difficult. Patients and their families will either prompt discussion or the CF Team will approach patients and their families following a joint decision to proceed. Introducing the idea of transplantation, especially lungs or heart/lung, acknowledges an irreversible deterioration in the patient's health and can become a difficult issue. The CF CNS should be actively involved in supporting the patient and their family throughout this time by allowing opportunities to ask questions and providing up-to-date information. Assessment for an organ transplant can be a time-consuming process; where this happens away from the Specialist CF Centre the CF CNS should be familiar with the investigations involved so that appropriate assistance can be offered. Acceptance onto a waiting list brings responsibilities and anxieties. Health status must be optimized and patients should be aware of the need to be contacted any time day or night. The transplant centre needs to be kept up to date with any deterioration in health and holidays or trips planned, and it is often the CF CNS who co-ordinates between the patient and the transplant centre. Waiting for organs can create concern and stress in the family and the CF CNS can be a helpful point of contact for any member of the family.

Transplantation is not an option for everyone; for some it may be their own decision to decline transplant and for others it may be a decision made following the assessment. Where it is patient choice researchers have found that individuals often question their decision and can swing between refusal and acceptance [30]. The CF CNS must pay special attention to this group of patients because whether by choice or clinical decision they have no further options for treatment and this can become difficult to accept and live with. This topic is discussed in further detail in chapter 23.

End-Stage and Symptom Management

CF remains a life-limiting disease and preterminal grief at diagnosis onwards must be acknowledged in both the patient and their family. Treatment choice, the dilemma of aggressive management versus palliative care, can become controversial for the patient, their family and the CF Team. The preterminal and terminal stages must be handled with sensitivity and compassion together with sound clinical judgement and involve the patient, their family and staff. Nurses involved in end of life management must be responsive to the patient's complexity of care and changing needs. This will enable them to offer support to the patient and their family to help them come to terms with and adapt to changes on a more personal level [31]. It is important to both the patient and their family that there is someone they know and trust who will advocate on their behalf. In playing the role of an advocate the nurse helps them accept new ways of coping, recognizes denial, respects their wishes and decisions about treatment, and allows discussion around issues of dying [32]. By providing support in this way the nurse ensures that the patient and family receive sufficient knowledge to make informed decisions including the flexibility of treatment choice. Treatment decisions that are often discussed at this time include terminal care – hospital or home, continuation of enteral feeding and intravenous therapy, options for respiratory support, pain management and symptom control and dealing with complications such as haemoptysis, and pneumothorax, transplantation, and the management of cardiorespiratory failure [33].

The extended role of the nurse during this time includes involving other support and specialist agencies where appropriate and supporting the healthcare team – both hospital and community based – in accepting the outcome [31]. Many families have had a long-term relationship with the CF Team and, in liaison with colleagues, it is important to acknowledge this by ensuring that appropriate bereavement support is offered to the family both in the terminal stages and after death. Caring for the dying patient is complex and stressful; however, the CF CNS working with the CF Team and the available support network can ensure that each individual is assured a dignified and peaceful death.

The Future

The number of adults with CF is increasing in the UK. In the future this population will include many healthy young adults who demand support while living a full and active life at college, in the workplace and at home with their own families. Professionals will also be faced with the challenges of an older, much sicker population presenting the CF Team with new clinical concerns. To meet the changing needs of patients nurses should be receptive to the many new and exciting developments becoming available. Technology is improving methods of communication, from using mobile phones more innovatively to telemedicine and a link with personal computers in each patient's home to their Specialist CF Centre. People with CF are part of this technological world and welcome a partnership in developing inventive ways of offering an individualized package of care.

The lifelong and multisystem nature of CF poses many challenges to those involved in the care. Equally, however, professionals working with this patient group – children and adults – learn a great deal about living with a life-limiting condition. Increasing life expectancy with improved quality of care are aims for all, although the day-to-day management issues should remain the nurse's priority. CF is a multisystem disease; the care we provide, as nurses, should, therefore, reflect this through a holistic, individualized, multidisciplinary team approach.

References

1 Clinical Standards Advisory Group: Cystic Fibrosis: Access to and Availability of Specialist Services. London, HMSO, 1993.
2 Madge S, Khair K: Multi-disciplinary teams in the United Kingdom: Problems and solutions. J Pediatr Nurs 2000;15:131–134.
3 Dyer J: Cystic fibrosis nurse specialist: A key role. J R Soc Med 1997;90(suppl 31):21–25.
4 Brenner P: From Novice to Expert. Excellence and Power in Clinical Nursing Practice. Addison Wesley, 1984.
5 Sutor JA: Can nurses be effective advocates? Nurs Stand 1993;7:30–32.
6 Dyer J, Morais A: Supporting children with cystic fibrosis in school. Prof Nurse 1996;11:518–520.
7 Bamford O, Gibson F: The clinical nurse specialist role: Key components identified. Managing Clin Nurs 1998;2:105–109.
8 Borbasi SA: Advanced practice/expert nurses: Hospitals can't live without them. Aust J Adv Nurs 1999;16:21–28.
9 Clinical Standards and Accreditation Group Standards of Care. London, Cystic Fibrosis Trust, 2001.
10 Madge S, Khair K: Genetic counselling: A new set of skills for primary care? Nurs Pract 2002;27:10–12.
11 Mitchie S, Marteau TM, Bobrow M: Genetic counselling; the psychological impact. J Med Genet 1997;34:237–241.
12 Madge S: The challenges of cystic fibrosis for nurses. Ir Nurse 2004;6:34–36.
13 Carr SB, Dinwiddie R: Annual review or continuous assessment? J R Soc Med 1996;89(suppl 27):3–7.
14 Littlewood J, Dodd M, Elborn S, Goven J, Hart CA, Heaf D, Madge S, Webb K: A Statement on *Burkholderia cepacia*. London, Cystic Fibrosis Trust, 1999.
15 The Cystic Fibrosis Trust: The Care of Patients with Cystic Fibrosis – A Patient's Charter. London, Cystic Fibrosis Trust, 1994.
16 Chuk PK: Clinical nurse specialists and quality patient care. J Adv Nurs 1997;26:501–506.
17 Cottrell J, Burrows E: Community-based care in cystic fibrosis: Role of the cystic fibrosis nurse specialist and implications for patients and families. Disabil Rehabil 1998;20:254–261.
18 Shepherd SL, Hovell MF, Harwood IR, et al: A comparative study of the psychosocial assets of adults with cystic fibrosis and their healthy peers. Chest 1990;97:1310–1316.
19 Coe L, Baker K: Growing up with a chronic condition: Transition to young adulthood for the individual with cystic fibrosis. Holistic Nurse Pract 1993;8:8–15.

20 Czajkowski DR, Koocher GP: Medical compliance and coping with cystic fibrosis. J Child Psychol Psychiatry 1987;28:311–319.
21 Bryon M, Madge S: Transition from paediatric to adult care: Psychological principles. J R Soc Med 2001;94(suppl 40):5–7.
22 Pownceby J: The Coming of Age Project. London, Cystic Fibrosis Trust, 1996.
23 Madge S, Bryon M: A model for transition of care in cystic fibrosis. J Pediatr Nurs 2002;17: 283–288.
24 Zack J, Jacobs CP, Keenan PM, Harney K, Woods ER, Collins AA, Emans SJ: Perspectives of patients with cystic fibrosis on preventive counseling and transition to adult care. Pediatr Pulmonol 2003;36:363–365.
25 Flume PA, Taylor LA, Anderson DL, Gray S, Turner D: Transition programs in cystic fibrosis centers: Perceptions of members. Pediatr Pulmonol 2004;37:4–7.
26 Anderson DL, Flume PA, Hardy KK, Gray S: Transition programs in cystic fibrosis centers: Perceptions of patients. Pediatr Pulmonol 2002;33:327–331.
27 Boyle MP, Farukhi Z, Nosky ML: Strategies for improving transition to adult cystic fibrosis care based on patient and parent views. Pediatr Pulmonol 2001;32:428–436.
28 Palmer ML, Boisen LS: Cystic fibrosis and the transition to adulthood. Soc Work Health Care 2003;36:45–58.
29 Brumfield K, Lansbury G: Experiences of adolescents with cystic fibrosis during their transition from paediatric to adult health care: A qualitative study of young Australian adults. Disabil Rehabil 2004;26:223–234.
30 Gotz I: Survival without transplant. J Cyst Fibros 2003;2:55–57.
31 Tonelli MR: End of life care in cystic fibrosis. Curr Opin Pulm Med 1998;4:332–336.
32 Thornes R: Care of Dying Children and Their Families. National Association of Health Authorities Report, 1988.
33 Robinson WM, Ravilly S, Berde C, Wohl M: End-of-life care in cystic fibrosis. Pediatrics 1997;100:205–208.

Susan Madge, SRN, RSCN, MSc, MCGI, PhD
Consultant Nurse, Royal Brompton Hospital
Sydney Street
London SW3 6NP (UK)
Tel. +44 020 7352 8121, ext. 4053
E-Mail s.madge@rbh.nthames.nhs.uk

Bush A, Alton EWFW, Davies JC, Griesenbach U, Jaffe A (eds): Cystic Fibrosis in the 21st Century.
Prog Respir Res. Basel, Karger, 2006, vol 34, pp 293–300

Dietetics

Sue Wolfe[a]   Alison Morton[b]

[a]Regional Paediatric, and [b]Regional Adult Cystic Fibrosis Unit, Leeds, UK

Abstract

With present-day treatment, the majority of people with cystic fibrosis (CF) should have a normal rate of growth, nutritional status and body composition. Diagnosis through neonatal screening enables early nutritional intervention and helps to achieve this. Regular dietetic counselling, initially focussing on optimizing energy, nutrient and pancreatic enzyme intakes should be available to all patients throughout their lives. With increased life expectancy patients face new nutritional challenges including those related to CF-related diabetes (CFRD), osteoporosis, pregnancy and transplantation.

Growth and Nutritional Assessment

All patients should be seen regularly by a dietitian experienced in the management of cystic fibrosis (CF), who will monitor growth and nutritional status to enable dietary advice to meet individual needs. A dietary and enzyme diary or a structured one day dietary/enzyme recall and food frequency questionnaire will give an estimate of the adequacy of the diet and pancreatic enzyme replacement therapy (PERT). Ideally, this should be recorded every 1 or 2 years or more frequently if clinically indicated.

Maximizing Dietary Intakes

Energy requirements vary widely depending on the degree of fat maldigestion and malabsorption [1], the extent of respiratory disease [2] and the presence of glycosuria associated with glucose intolerance [3]. Treatment-related factors may also increase energy expenditure. Consequently, energy requirements generally increase with age as the disease progresses. An infant may have normal energy needs but patients with severe respiratory disease may need over 150% [4] of the age-appropriate recommendations.

Studies report that patients fail to achieve their recommended energy intakes [5]. Chronically poor appetites, infection-related anorexia, abdominal pain, depression and feeding behaviour problems all contribute to reduced food intake.

Encouraging a high-fat diet, with appropriate PERT will maximize energy intakes. Oral dietary supplements may also be beneficial for some patients [6]. The type and amount are recommended on an individual basis depending on age, taste preference and energy requirements. They are best taken after or between meals to avoid reducing the appetite, and altering the type and flavour periodically will help to prevent taste fatigue. However, there are some concerns that they may replace food intake. Therefore the effectiveness of oral supplements in CF requires further clarification from clinical trials [7]. A randomized controlled trial examining the efficacy of oral supplements has recently been completed and the results should be published in 2005 [8].

Feeding Infants, Toddlers and School Age Children

Neonatal screening for CF enables early diagnosis and the opportunity to prevent nutritional problems early in life.

Diagnostic age is critical in contributing to nutritional status, with the most consistently observed benefits of neonatal screening being nutritional in nature [9].

Most infants thrive if given breast milk or a normal infant formula [10, 11]. Breast milk contains lipase and nutritional and growth factors which could be beneficial to the infant. If malabsorption is controlled, 150–200 ml formula/kg/day or demand breast feeding will support normal growth. Approximately 15% of infants are born with meconium ileus. These infants are nutritionally compromised [12] and may have special nutritional needs. If surgical resection of meconium ileus is required, cows' milk protein intolerance may occur and a hydrolyzed protein, medium-chain triglyceride feed may be required. Gastro-oesophageal reflux is also relatively common [13] and may improve by thickening the infant's feed, giving a pre-thickened formula or by reducing gastric acid [14].

Weaning foods should be introduced at 4–6 months. Methods to achieve a higher energy intake should always be advised on an individual basis, taking into account the infant's growth, hunger, fluid intake and absorption. If weight gain is slow, the use of a high-energy infant formula and/or energy supplementation to food should be considered. A normal weaning diet including full-fat cow's milk should be introduced by one year of age. As the diet becomes more varied, the need for enzyme variation according to the fat content of the food becomes greater (see below). As the child gets older, food is often used as an effective tool to obtain parental attention and behavioural food refusal can become a problem [15]. Unless carefully handled, this behaviour can persist for a number of years resulting in a poor dietary intake and growth. Attention to the behavioural aspects of feeding has been shown to improve energy intakes [16] and feeding behaviour management strategies are now inherent in nutritional management guidelines [17–19] (table 1).

Feeding Adolescents and Adults

Adolescence is a challenging time physically and emotionally and this may impact on the patient's health. Malnourished adolescents have a marked decline in lung function [20], therefore the nutritional goal is to maximize weight gain. Teenagers have a profound desire to become independent, make their own decisions and engage in risk-taking behaviour. Food choice is a target for testing this independence, with many existing solely on junk food. Poor adherence to enzyme, vitamin and nutritional supplements is also common and compromise is essential to minimize the consequences of this behaviour.

Table 1. Management of feeding behaviour problems

Encourage family meals, so that the child is seated with other children/adults and will learn correct feeding behaviour
Avoid other distractions e.g. having the television on at mealtimes
Make food as attractive as possible
If your child is slow at eating, gentle encouragement will help
Never lose your temper when food is refused. In the child's eyes any reaction is attention. This will encourage the bad behaviour
Encourage 'good' or 'positive' behaviour and give lots of praise for it
Limit meal times to 30 min. Research has shown that allowing mealtimes to drag on rarely results in any more food being eaten. After 30 min remove the food without comment, and wait for the next meal/snack time
Never produce a second meal if the first is refused
Make sure there is a consistent approach from all who are involved with feeding

The main growth spurt occurs during this period, although timing will vary depending on gender, genetics and pubertal delay. Energy and protein requirements are highest at the time of peak growth and adequate intakes of all nutrients are essential. Iron deficiency is particularly common at this time [21].

Patients are living longer and the adult CF population is predicted to continue increasing [22]. Studies suggest that the prevalence of malnutrition increases with age [20] and is most common in those with a severe CFTR genotype [23]. Even adults with a normal BMI have depletion of fat-free mass related to disease severity [24].

Adolescents with chronic illness are at risk for disordered eating [25]. Reports in people with CF are conflicting [26]. Formal eating disorders are not prevalent but eating disturbance is reported [26]. However, eating behaviour and attitudes are similar to those of healthy peers [27] but nutritional support [28] and CFRD may alter these findings [29].

Pancreatic Enzyme Replacement Therapy

Infants with overt signs of malabsorption should start PERT (see chapters 30 and 31) as soon as CF is diagnosed; pancreatic status can still be confirmed by measurement of faecal pancreatic elastase. Fat provides over 50% of the energy content of breast milk or formula, therefore poor control of malabsorption has a devastating effect on nutritional status. If there is no evidence of malabsorption, pancreatic status should be assessed before PERT is commenced.

The dose of PERT required varies and is dependent on the degree of residual pancreatic function, the enzyme

Wolfe/Morton

preparation and pathophysiological factors such as intestinal pH [1].

There are various types of PERT available, the most effective being the enteric-coated, acid-resistant preparations. During infancy, all infants with pancreatic insufficiency (PI) are given a standard strength preparation. The capsules are opened and the mini/microspheres mixed with milk or fruit puree and given by spoon at the beginning of the feed. Initially, approximately 2,500 IU lipase per feed is used. The dose for a newly diagnosed older patient is 1–2 capsules per meal and a 0.5–1 capsule with fat-containing snacks. Enzymes must not be crushed/chewed, as this will reduce their effectiveness. The dose is gradually increased according to clinical symptoms, appearance of the stools and objective assessment of weight gain, growth and absorption. All food and drinks containing fat require PERT and the dose is titrated against the fat content of the food. The patient should be aware of foods that do not need enzymes, e.g. non-fat items such as fruit gums. Multiple factors affect enzyme efficacy and dose requirements can vary between 500 and 4,000 IU lipase/g fat. Enzymes are best taken at the beginning, middle and end of the meal, especially if it takes longer than 30 min to eat. Dividing the dose will also reduce the chance of overdosing if the meal is refused half way through.

In the UK the Committee on Safety of Medicines has advised a maximum daily intake of 10,000 IU lipase/kg/day, regardless of the preparation used [30]. If higher intakes are necessary, enzyme efficacy may be improved by reducing the gastric acid with H_2 receptor antagonists [31] or proton pump inhibitors [32]. If symptoms still persist, a full gastro-intestinal investigation should be performed to rule out other conditions [1].

Pancreatic Enzyme Replacement in the Ventilated Patient

Enteric coated pancreatic enzymes cannot be put down standard enteral feeding tubes. If a patient is unconscious and unable to take enzymes orally a powdered preparation may be used. Approximately 0.5 g should be flushed down the tube every 3–4 h. The enzymatic activity of powdered enzymes is largely destroyed in the acid environment of the stomach; however, a proton pump inhibitor will help to preserve some of the activity.

Pancreatic Enzymes with Overnight Feeds

All fat-containing elemental and polymeric feeds require PERT. The dose must be worked out on an individual basis taking into consideration the type, fat content and rate of infusion of the feed. If the feed is infused over a long

Table 2. Vitamin supplement doses [17]

Age	Vitamin A, IU	Vitamin D, IU	Vitamin E, IU
<1 year	4,000	400	10–50
>1 year	4,000–10,000	400–800	50–100
Adults	4,000–10,000	800–2,000	100–200

period only small doses of PERT may be required. This may be due to stimulation of gastric lipase [33]. The enzymes are usually given at the beginning of the feed or at the beginning and at the end of the feed.

Vitamin Supplementation

Malabsorption of fat-soluble vitamins is common in PI patients. Deficiency of these vitamins has been found by 2 months of age in untreated screened infants with CF [34]. All PI patients should receive supplements of the fat-soluble vitamins [17–19] (table 2) and annual monitoring of serum vitamin levels is essential. The need for supplementation in pancreatic-sufficient patients must also be assessed annually. Ideally fasting plasma levels of vitamins A, D, E, total cholesterol and vitamin E:cholesterol ratio should be monitored. Patients should also be advised to omit vitamin supplements for 12 h before measurements are taken.

Vitamin A. Deficiency of vitamin A causes night blindness and can progress to severe xerophthalmia [35]. Low vitamin A levels are associated with increasing age, lower lung function and weight Z score [36]. High levels have been reported following transplantation [37]. Low levels are difficult to interpret and to avoid hypervitaminosis A supplements should not be increased indiscriminately. Vitamin A and retinol-binding protein (RBP) fall transiently during the acute phase response to infection due to decreased hepatic release of RBP [38]. Levels should therefore be measured when clinically stable or alongside a marker of infection. There is a positive correlation between RBP and zinc status, therefore zinc supplements may be required if levels are low [39]. Consideration must be given to vitamin A supplementation in pregnancy (see below).

Vitamin D. Overt vitamin D deficiency is rare in CF. However, subclinical deficiency may contribute to bone disease, which is relatively common (see chapter 34). Low levels may be due to dietary inadequacy, fat and vitamin D malabsorption [40], low vitamin D binding protein levels and inadequate sunlight exposure.

Plasma 25-hydroxy vitamin D (25-OHD) is used to measure status. Plasma 25-OHD levels are largely dependent on sunlight exposure with studies reporting a wide seasonal variation despite supplementation with 400–2,000 IU/day [41, 42]. To achieve the recommended levels of >30 ng/ml (>75 nmol/l) [43] an intake of 50 µg (2,000 IU) per day has been suggested for the normal population [44], the amount required in CF may be even higher. Vitamin D is usually given in combination with vitamin A, therefore an increase in dose may lead to potentially toxic intakes of vitamin A and a separate vitamin D preparation may be required. A high vitamin A intake may also contribute poor bone mineralization [45]. Supplementation with the more potent vitamin D analogues e.g. calcitriol [46] has also been considered but monitoring for hypercalcaemia is important.

Vitamin E. Severe vitamin E deficiency causes neurological problems and contributes to anaemia. Vitamin E is an important antioxidant protecting cell membranes from oxidative damage [47] by reducing the effect of free radicals produced by chronic infection.

Vitamin K. Malabsorption, bile salt deficiency, liver disease and antibiotic therapy commonly contribute to vitamin K deficiency in CF. Prothrombin time is often used to assess vitamin K status but its accuracy is poor. Prothrombin induced in vitamin K absence (PIVKA II) levels suggest that the majority of patients are at risk of subclinical deficiency [48]. However, PIVKA II assesses vitamin K status of the liver and not bone [49]. The adequacy of vitamin K status for bone metabolism is assessed by undercarboxylated osteocalcin. The need for routine vitamin K supplements and the most appropriate dosing schedule have yet to be established.

Water-Soluble Vitamins. Routine supplementation of water-soluble vitamins is not required [17, 18]. Case reports of water-soluble vitamin deficiencies do occur [50] and supplementation may be needed on an individual basis. Parenteral vitamin B_{12} may be required following terminal ileal resection for meconium ileus [17, 18]. Supplementation with the antioxidants vitamin C and β-carotene is an area of current research.

Enteral Tube Feeding

Nutritional failure in CF is multifactorial (see fig. 1) and is reviewed in detail elsewhere [51]. Consensus reports recommend constant surveillance of nutritional status with staged nutritional intervention for those who have or are at risk of nutritional failure [17–19]. Enteral tube feeding

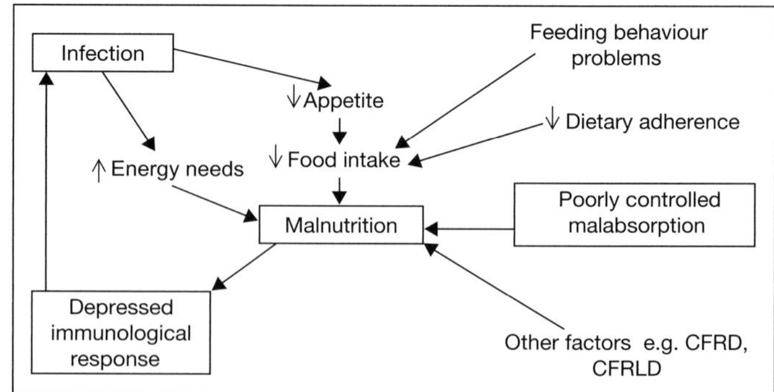

Fig. 1. Factors affecting nutritional status.

improves weight gain and nutritional status [16] and stabilizes or slows the rate of decline in respiratory function [52]. Early intervention is associated with improved outcome [52, 53]; however, there are no randomized controlled trials assessing the efficacy of enteral tube feeding in CF [54].

In the United Kingdom 10% of all people with CF receive nasogastric or gastrostomy feeding [55]. Feeds are usually administered overnight and patients are encouraged to eat through the day. Most tolerate whole protein, polymeric feeds with a high energy density; however, some may benefit from an elemental or semi-elemental formula. Hyperglycaemia requiring insulin therapy may be precipitated by enteral tube feeding, therefore the introduction of feeds should be closely monitored to assess the effect on glycaemia.

Parenteral nutrition is only used when enteral feeding is impractical, e.g. after major gastro-intestinal surgery. Although parenteral nutrition has beneficial effects, because of the cost and associated risks, it is not routinely recommended.

Appetite-Enhancing and Anabolic Agents

Megestrol acetate has been used as an appetite stimulant in CF [56, 57]. Although improvements in weight and respiratory function are reported, it is recommended that due to its potential side effects larger clinical trials are required before it is routinely used [18]. The use of cyproheptadine hydrochloride has also been shown to stimulate the appetite in patients with CF [58].

The anabolic agents insulin like growth factor I [59] and early insulin therapy [60] have also been studied in CF but evidence of their value is poor. In two small studies growth hormone improved weight gain and growth [61, 62].

Creatine supplementation has also resulted in improved muscle strength, patient well-being and weight [63].

Cystic-Fibrosis Related Diabetes

On diagnosis of CFRD (see chapter 35), individual review of energy requirements and consideration of the nutritional and clinical status should be made to determine the most appropriate dietary regimen. Most people will need to maintain a diet high in energy, fat, and protein with no restriction of refined carbohydrate or salt [64–66] although some may need to restrict the intake of sugary drinks. Unless taught carbohydrate counting, regular meals with a similar carbohydrate content should be encouraged. If tube feeding is required its introduction should be monitored to assess the need for adjustment of insulin therapy.

Pregnancy

Pregnancy in CF is dealt with in more detail in chapters 33 and 36. Women who are planning to become pregnant should optimize their nutritional status. Those who receive preconceptional nutritional advice achieve greater weight gain during pregnancy and have heavier babies [67]. At this time food safety and nutritional advice regarding folic acid supplementation should be given.

Maternal health post-partum is related to nutritional status during pregnancy, therefore constant nutritional surveillance is essential. Tube feeding may be required to achieve adequate weight gain. Changing gastrostomy buttons to tubes will help to reduce the risk of stoma closure should the tube come out. Essential fatty acid (EFA) deficiency has been reported in healthy women during pregnancy and lactation [68] and this may further exacerbate the already deranged EFA status of women with CF [69]. Adequate intakes of all nutrients are essential with particular attention given to protein, calcium, and vitamin D.

High intakes of vitamin A are teratogenic and supplements of 10,000 IU or more have been associated with an increased risk of birth defects [70]. The recommended maximum supplement dose is 10,000 IU (3,000 μg) daily [17, 71]. Pregnant women who are not receiving supplemental vitamin D should be given 10 μg/day [72].

Successful breast feeding in mothers with CF has been achieved [73]. Breast feeding increases maternal nutritional requirements and so individual advice about the method of infant feeding should be given according to the mothers' clinical condition.

Bone Health

The foundation for good bone health (see chapter 34) is established during infancy, childhood and adolescence, therefore there should be a focus on minimizing risk factors during these years. The main nutritional risk factors include poor nutritional status, vitamin D, calcium and vitamin K deficiencies.

Nutritional Status. Studies report a significant correlation with BMI in adults [74] and children [75] and reduction in bone mineral density (BMD). Poor nutrition can result in delayed puberty, which may also compromise bone health.

Vitamin D. Vitamin D is responsible for maintaining calcium homeostasis by increasing intestinal calcium absorption and stimulating bone mineral accretion. Low levels also increase the production of parathyroid hormone, which stimulates bone resorption hence contributing to osteoporosis. In the USA, the consensus group on osteoporosis has recommended that levels of 25-OHD of >30 ng/ml (>75 nmol/l) should be achieved for optimal bone health [43]. The methods for achieving this level are discussed above.

Vitamin K. Vitamin K is a co-factor in the λ carboxylation of the proteins osteocalcin and matrix GLA involved in bone metabolism [76], with under carboxylated proteins being functionally defective. The most accurate way of assessing vitamin K adequacy for bone metabolism is by measuring undercarboxylated osteocalcin. Consensus reports vary in their recommendations for vitamin K supplementation [17–19] and therefore the need for, and dose of, routine supplementation are yet to be determined.

Calcium. Approximately 30% of bone mineral is comprised of calcium, therefore the importance of regular assessment of calcium intake, especially during the pubertal years cannot be over-emphasized. In CF risk factors for deficiency include poor dietary intake, malabsorption [77] and vitamin D deficiency. In addition, oral corticosteroids and a high dietary sodium intake may increase requirements. In the UK dietary reference values for calcium are relatively low, in contrast the American consensus group on osteoporosis in CF has recommended a dietary intake of 1,300–1,500 mg for children over 9 years old [43].

Transplantation

A poor nutritional status is linked with increased mortality in patients awaiting lung transplantation (see chapter 23) [78]. Nutritional status should be optimized by intensifying nutritional support during the pre-transplantation period.

In the early post-operative period a 'clean' food diet is usually recommended. Patients should also be advised about foods and medication that affect immunosuppressive drugs. Distal intestinal obstruction syndrome is common in the early post-operative period [79]. Other nutritional complications include gastro-oesophageal reflux [80], diabetes mellitus [81], osteoporosis [82], hyperlipidaemia and abnormally high plasma vitamin A and E levels [37].

Essential Fatty Acids (EFA's)

Deficiency of EFA characterized by low plasma linoleic and docosahexaenoic acid levels and raised arachidonic acid levels have long been recognized in CF. Poor absorption of dietary fat and abnormal fatty acid metabolism related to the CFTR defect are believed to be involved [83]. Deficiency may contribute to poor weight gain [12] and increased susceptibility to staphylococcus and pseudomonas infections. It has also been suggested that EFA deficiency may be responsible for the phenotypic expression of CF [84]. Despite the research activity in this area, at the present time further clinical trials are required before supplementation with high doses of EFAs are recommended [85].

Conclusion

A normal nutritional status and rate of growth are achievable goals for the majority of patients. Nutritional intervention should begin as soon as the diagnosis is made in order to prevent or resolve malnutrition. With increasing life expectancy, the demands of nutritional management are becoming greater. Attention to detail is essential to prevent the consequences of subclinical nutrient deficiencies. The recent publication of nutritional management guidelines in the UK [17], Europe [18] and the USA [19] will help to improve dietary treatment, standardize practice and hopefully lead to further improvements in morbidity and mortality.

References

1 Littlewood JM, Wolfe SP: Control of malabsorption in cystic fibrosis. Paediatr Drugs 2000;2:205–222.
2 Bell SC, Bowerman AM, Nixon LE, Macdonald IA, Elborn JS, Shale DJ: Metabolic and inflammatory responses to pulmonary exacerbation in adults with cystic fibrosis. Eur J Clin Invest 2000;30:553–559.
3 Moran A, Doherty L, Wang X, Thomas W: Abnormal glucose metabolism in cystic fibrosis. J Pediatr 1998;133:10–17.
4 Wolfe S: The nutritional management of cystic fibrosis. CML – Clinical nutrition 2003;Issue 11.4.
5 White H, Morton AM, Peckham DG, Conway SP: Dietary intakes in adult patients with cystic fibrosis-do they achieve guidelines? J Cyst Fibros 2004;3:1–7.
6 Skypala IJ, Ashworth FA, Hodson ME, Leonard CH, Knox A, Hiller EJ, Wolfe SP, Littlewood JM, Morton A, Conway S, Patchell C, Weller P, McCarthy H, Redmond A, Dodge J: Oral nutritional supplements promote significant weight gain in cystic fibrosis patients. J Hum Nutr Diet 1998;11:95–104.
7 Smyth R, Walters S: Oral calorie supplements for cystic fibrosis. Cochrane Database Syst Rev 2000;2:CD000406.
8 Poustie VJ, Russell JE, Watling RM, Ashby D, Smyth RL: The CALICO multicentre randomised controlled trial of oral calorie supplements for children with cystic fibrosis. Pediatr Pulmonol 2004:(suppl 27):Abs 414.
9 Farrell PM, Kosorok MR, Rock MJ, Laxova A, Zeng L, Lai HC, Hoffman G, Laessig RH, Splaingard ML: Early diagnosis of cystic fibrosis through neonatal screening prevents severe malnutrition and improves long term growth. Pediatrics 2001;107:1–13.
10 Holliday K, Allen J, Waters DL, Gruca MA, Thompson SM, Gaskin KJ: Growth of human milk fed and formula fed infants with cystic fibrosis. J Pediatr 1991;118:77–79.
11 Ellis L, Kalnins D, Corey M, Brennan J, Pencharz P, Durie P: Do infants with cystic fibrosis need a protein hydrolysate formula? A prospective, randomised comparative study. J Pediatr 1998;132:270–276.
12 Lai HC, Kosorok MR, Laxova A, Davis LA, FitzSimmon SC, Farrell PM: Nutritional status of patients with cystic fibrosis with meconium ileus: A comparison with patients without meconium ileus and diagnosed early through neonatal screening. Pediatrics 2000;105:53–61.
13 Heine RG, Button BM, Olinsky A, Phelan PD, Catto-Smith AG: Gastro-oesophageal reflux in infants under 6 months with cystic fibrosis. Arch Dis Child 1998;78:44–48.
14 Brodzicki J, Trawinska-Bartnicka M, Korzon M: Frequency, consequences and pharmacological treatment of gastroesophageal reflux in children with cystic fibrosis. Med Sci Monit: 2002;8:CR529–CR537.
15 Duff AJA, Wolfe SP, Dickson C, Conway SP, Brownlee KG: Feeding behaviour problems in children with cystic fibrosis in the UK; prevalence and comparison with healthy controls. J Pediatr Gastroenterol Nutr 2003;36:443–447.
16 Jelalian E, Stark LJ, Reynolds L, Seifer R: Nutrition intervention for weight gain in cystic fibrosis: A meta analysis. J Pediatr 1998;132:486–492.
17 UK Cystic Fibrosis Trust Nutrition Working Group. Nutritional management of cystic fibrosis. Cystic Fibrosis Trust, 2002.
18 Sinaasappel M, Stern M, Littlewood J, Wolfe S, Steinkamp G, Heijerman HGM, Robberecht E, Doring G: Nutrition in patients with cystic fibrosis: A European consensus. J Cyst Fibros 2002;2:51–75.
19 Borowitz D, Baker RD, Stallings V: Consensus report on nutrition for pediatric patients with cystic fibrosis. J Pediatr Gastroenterol Nutr 2002;35:246–259.
20 Steinkamp G, Wiedemann B: On behalf of the German CFQA Group. Relationship between nutritional status and lung function in cystic fibrosis: Cross sectional and longitudinal analyses from the German Quality Assurance Project. Thorax 2002;57:596–601.
21 Pond M, Morton A, Conway S: Functional iron deficiency in adults with cystic fibrosis. Respir Med 1996;90:409–413.
22 Yankaskas JR, Marshall BC, Sufian B, Simon RH, Rodman D: Cystic fibrosis adult care: Consensus Conference Report. Chest 2004;125:1S–39S.
23 Dray X, Kanaan R, Bienvenu T, Desmazes-Dufeu N, Dusser D, Marteau P, Hubert D: Malnutrition in adults with cystic fibrosis. Eur J Clin Nutr 2005;59:152–154.
24 Ionescu AA, Evans WD, Pettit RJ, Nixon LS, Stone MD, Shale DJ: Hidden depletion of fat-free mass and bone mineral density in adults with cystic fibrosis. Chest 2003;124:2220–2228.
25 Neumark-Sztainer D, Story M, Falkner NH, Beuhring T, Resnick MD: Disordered eating among adolescents with chronic illness and

disability. Arch Pediatr Adolesc Med 1998; 152:871–878.

26 Shearer JE, Bryon M: The nature and prevalence of eating disorders and eating disturbance in adolescents with cystic fibrosis. J R Soc Med 2004;97(suppl 44):36–42.

27 Abbott J, Conway SP, Etherington C, Fitzjohn J, Gee L, Morton A, Musson H, Webb AK: Perceived body image and eating behavior in young adults with cystic fibrosis and their healthy peers. J Behav Med 2000;23: 501–517.

28 Abbott J, Conway S, Etherington C, Fitzjohn J, Gee L, Morton A, Musson H, Webb AK: Enteral nutrition and body satisfaction in CF adults. Neth J Med 1999;54(suppl):abstract 202.

29 Abbott J, Conway SP, Etherington C, Fitzjohn J, Gee L, Morton A, Musson H, Webb AK: Cystic fibrosis related diabetes, eating behaviours and body satisfaction. Pediatr Pulmonol 1998;17(suppl):395.

30 Committee on Safety of Medicines. Report of the Pancreatic Enzymes Working Party. London, 1995.

31 Francisco MP, Wagner MH, Sherman JM, Theriaque D, Bowser E, Novak DA: Ranitidine and omeprazole as adjuvant therapy to pancrealipase to improve fat absorption in patients with cystic fibrosis. J Pediatr Gastroenterol Nutr 2002;35:79–83.

32 Proesmans M, De Boeck K: Omeprazole, a proton pump inhibitor, improves residual steatorrhoea in cystic fibrosis patients treated with high dose pancreatic enzymes. Eur J Pediatr 2003;162:760–763.

33 Armand M, Hamosh M, Philpott JR, Resnick AK, Rosentein BJ, Hamosh A, Perman JA, Hamosh P: Gastric function in children with cystic fibrosis: Effect of diet on gastric lipase levels and fat digestion. J Pediatr Res 2004;55: 457–465.

34 Feranchak AP, Sontag MK, Wagener JS, Hammond KB, Accurso FJ, Sokol RJ: Perspective long term study of fat soluble vitamin status in children with cystic fibrosis identified by newborn screening. J Pediatr 1999;135:601–610.

35 Campbell DC, Tole DM, Doran RML, Conway SP: Vitamin A deficiency in cystic fibrosis resulting in severe xerophthalmia. J Hum Nutr Diet 1998;11:529–532.

36 Greer RM, Buntain HM, Lewindon PJ, Wainwright CE, Potter JM, Wong JC, Francis PW, Batch JA, Bell SC: Vitamin A levels in patients with CF are influenced by the inflammatory response. J Cyst Fibros 2004;3: 143–149.

37 Brotherwood M, Stephenon A, Roberts R, Durie P, Verjee Z, Chaparro C, Corey M, Tullis E: Abnormally high vitamin A & E levels following lung transplantation in adult cystic fibrosis patients. Pediatr Pulmonol 2004(suppl 27):P372.

38 Thurnham DI, McCabe GP, Northrop-Clewes CA, Nestel P: Effects of subclinical infection on plasma retinol concentrations and assessment of prevalence of vitamin A deficiency: Meta-analysis. Lancet 2003;362:2052–2058.

39 Navarro J, Desquilbet N: Depressed vitamin A and retinol binding protein in cystic fibrosis.

Effects of zinc supplementation. Am J Clin Nutr 1998;34:1439–1440.

40 Lark RK, Lester GE, Ontjes DA, Blackwood AD, Hollis BW, Hensler MM, Aris RM: Diminished and erratic absorption of ergocalciferol in adult cystic fibrosis patients. Am J Clin Nutr 2001;73:602–606.

41 Elkin SL, Burgess J, Hodson ME: Hypovitaminosis D is common in adult patients with cystic fibrosis. Pediatr Pulmonol 1999(suppl 19):P469.

42 Wolfe SP, Conway SP, Brownlee KG: Seasonal variation in vitamin D levels in children with cystic fibrosis in the United Kingdom. Abstracts 24th European CF Conference 2001; P115.

43 Joseph PM: Screening for CF bone disease. Pediatr Pulmonol 2002(suppl 24):178–179.

44 Vieth R: Vitamin D supplementation, 25-hydroxyvitamin D concentrations, and safety. Am J Clin Nutr 1999;69:842–856.

45 Promislow JH, Goodman-Gruen D, Slymen DJ, Barrett-Connor E: Retinol intake and bone mineral density in elderly: The Rancho Bernado Study. J Bone Miner Res 2002;17: 1349–1358.

46 Brown SA, Ontjes DA, Lester GE, Lark RK, Hensler MB, Blackwood AD, Caminiti MJ, Blacklund DC, Aris RM: Short term calcitriol administration improves calcium homeostasis in adults with cystic fibrosis. Osteoporos Int 2003;14:442–449.

47 Brown RK, Wyatt H, Price JF, Kelly FJ: Pulmonary dysfunction in cystic fibrosis is associated with oxidative stress. Eur Respir J 1996;9:334–339.

48 Rashid M, Durie PR, Andrew M, Kalnins D, Shin J, Corey M, Tullis E, Pencharz PB: Prevalence of vitamin K deficiency in cystic fibrosis. Am J Clin Nutr 1999;70:378–382.

49 Mosler K, von Kries R, Vermeer C, Saupe J, Schmitz T, Schuster A: Assessment of vitamin K deficiency in CF: How much sophistication is useful? J Cyst Fibros 2003;2:91–96.

50 McCabe H: Riboflavin deficiency in cystic fibrosis: Three case reports. J Hum Nutr Dietet 2001;14:365–370.

51 Pencharz PB, Durie PR: Pathogenesis of malnutrition in cystic fibrosis, and its treatment. Clin Nutr 2000;19:387–394.

52 Walker SA, Gozal D: Pulmonary function correlates in the prediction of long-term weight gain in cystic fibrosis patients with gastrostomy tube feedings. J Pediatr Gastroenterol Nutr 1998;27:53–56.

53 Oliver MR, Heine RG, Hang Ng C, Volders E, Olinsky A: Factors affecting clinical outcome in gastrostomy-fed children with cystic fibrosis. Pediatr Pulmonol 2004;37:324–329.

54 Conway SP, Morton A, Wolfe S: Enteral tube feeding for cystic fibrosis. Cochrane Database Syst Rev 1999;3:CD001198.

55 UK CF Database Annual Data Report 2002. www.cystic-fibrosis.org.uk

56 Marchand V, Baker SS, Stark TJ, Baker RD: Randomized, double-blind, placebo-controlled pilot trial of megestrol acetate in malnourished children with cystic fibrosis. J Pediatr Gastroenterol Nutr 2000;31:264–269.

57 Eubanks V, Koppersmith N, Wooldridge N, Clancy JP, Lyrene R, Arani RB, Lee J,

Modawer L, Atchison J, Sorcher EJ, Makris CM: Effects of megestrol acetate on weight gain, body composition, and pulmonary function in patients with cystic fibrosis. J Pediatr 2002;140:439–444.

58 Homnick DN, Homnick BD, Reeves AJ, Marks JH, Pimentel RS, Bonnema SK: Cyproheptadine is an effective appetite stimulant in cystic fibrosis. Pediatr Pulmonol 2004; 38:129–134.

59 Bucuvalas JC, Chernausek SD: Growth hormone and cystic fibrosis: Good for more than growth? J Pediatr 2001;139:616–618.

60 Ripa P, Robertson I, Cowley D, Harris M, Masters IB, Cotterill AM: The relationship between insulin secretion, the insulin-like growth factor axis and growth in children with cystic fibrosis. Clin Endocrinol 2002;56: 383–389.

61 Hardin DS, Ellis KJ, Dyson M, Rice J, McConnell R, Seilheimer DK: Growth hormone improves clinical status in prepubertal children with cystic fibrosis: Results of a randomized controlled trial. J Pediatr 2001;139: 636–642.

62 Darmaun D, Hayes V, Schaeffer D, Welch S, Mauras N: Effects of glutamine and recombinant human growth hormone on protein metabolism in prepubertal children with cystic fibrosis. J Clin Endocrinol Metab 2004;89: 1146–1152.

63 Braegger CP, Schlattner U, Wallimann T, Utiger A, Frank F, Schaefer B, Heizmann CW, Sennhauser FH: Effects of creatine supplementation in cystic fibrosis: Results of a pilot study. J Cyst Fibros 2003;2:177–182.

64 UK Cystic Fibrosis Trust Diabetes Working Group: Management of Cystic Fibrosis Related Diabetes Mellitus. Cystic Fibrosis Trust, 2004.

65 Wilson DC, Kalnins D, Stewart C, Hamilton N, Hanna AK, Durie PR, Tullis E, Pencharz PB: Challenges in the dietary management of cystic fibrosis related diabetes mellitus. Clin Nutr 2000;19:87–93.

66 Moran A, Hardin D, Rodman D, Allen HF, Beall RJ, Borowitz D, Brunzell C, Campbell III PW, Chesrown SE, Duchow C, Fink RJ, Fitzsimmons SC, Hamilton N, Hirsch I, Howenstine MS, Klein DJ, Madhun Z, Pencharz PB, Quittner AL, Robbins MK, Schindler T, Schissel K, Schwarzenberg SJ, Stallings VA, Tullis DE, Zipf WB: Diagnosis, screening and management of cystic fibrosis related diabetes mellitus A consensus conference report. Diabetes Res Clin Pract 1999;45:61–73.

67 Morton A, Wolfe S, Conway SP: Dietetic intervention in pregnancy in women with CF – the importance of pre-conceptional counselling. Isr J Med Sci 1996;32(suppl):S2.

68 Holman RT, Johnson SB, Ogburn PL: Deficiency of essential fatty acids and membrane fluidity during pregnancy and lactation. Proc Natl Acad Sci USA 1991;88:4835–4839.

69 Strandvik B: Fatty acid metabolism in cystic fibrosis. N Engl J Med 2004;350:605–607.

70 Rothman KJ, Moore LL, Singer MR, Nguyen UDT, Mannino S, Milunsky A: Teratogenicity of high vitamin A intake. N Engl J Med 1995; 333:1369–1373.

71 World Health Organization. Safe vitamin A dosage during pregnancy and lactation: Recommendations and report of a consultation. Document NUT/98.4. Geneva, WHO, 1998.

72 Department of Health: Report on health and social subjects 41. Dietary reference values for food energy and nutrients for the United Kingdom. London, HMSO, 1991.

73 Gilljam M, Antoniou M, Shin J, Dupuis A, Corey M, Tullis DE: Pregnancy in cystic fibrosis. Fetal and maternal outcome. Chest 2000;118:85–91.

74 Conway SP, Morton AM, Oldroyd B, Truscott JG, White H, Smith AH, Haigh I: Osteoporosis and osteopenia in adults and adolescents with cystic fibrosis: Prevalence and associated factors. Thorax 2000;55: 798–804.

75 Fok J, Brown NE, Zuberbuhler P, Tabak J, Tom M: Low bone mineral density in cystic fibrosis patients. Can J Diet Prac Res 2002;63: 192–197.

76 Weber P: Management of osteoporosis: Is there a role for vitamin K? Int J Vit Nutr Res 1997;67:350–356.

77 Schulze KJ, O'Brien KO, Germain-Lee EL, Baer DJ, Leonard AL, Rosenstein BJ: Endogenous faecal losses of calcium compromise calcium balance in pancreatic-insufficient girls with cystic fibrosis. J Pediatr 2003; 143:765–771.

78 Snell GI, Bennetts K, Bartolo J, Levvey B, Griffiths A, Williams T, Rabinov M: Body mass index as a predictor of survival in adults with cystic fibrosis referred for lung transplantation. J Heart Lung Transpl 1998;17: 1097–1103.

79 Gilljam M, Chaparro C, Tullis E, Chan C, Keshavjee S, Hutcheon M: GI complications after lung transplantation in patients with cystic fibrosis. Chest 2003;123:37–41.

80 Young LR, Hadjiliadis D, Davis RD, Palmer SM: Lung transplantation exacerbates gastroesophageal reflux disease, Chest 2003;124: 1689–1690.

81 Paolillo JA, Boyle GJ, Law YM, Lawrence K, Wagner K, Pigula FA, Griffith BP, Webber SA: Post transplant diabetes mellitus in pediatric thoracic organ recipients receiving tacrolimus-based immunosuppression. Transplantation 2001;71:252–256.

82 Aris RM, Neuringer IP, Weiner MA, Egan TM, Ontjes D: Severe osteoporosis before and after lung transplantation. Chest 1996;109: 1176–1183.

83 Strandvik B, Gronowitz E, Enlund F, Martinsson T, Wahlstrom J: Essential fatty acid deficiency in relation to genotype in patients with cystic fibrosis. J Pediatr 2001; 138:650–655.

84 Freedman SD, Katz MH, Parker EM, Laposata M, Urman MY, Alvarez JG: A membrane lipid imbalance plays a role in the phenotypic expression of cystic fibrosis in cftr −/− mice. Proc Natl Acad Sci USA 1999;96: 13995–14000.

85 Beckles Willson N, Elliott TM, Everard ML: Omega-3 fatty acids (from fish oils) for cystic fibrosis. Cochrane Database Syst Rev. 2002; 3:CD002201.

Sue Wolfe
Regional Paediatric Cystic Fibrosis Unit
Children's Day Hospital
St. James' University Hospital
Beckett Street
Leeds LS9 7TF (UK)
Tel. +44 113 206 4960
Fax +44 113 206 7011
E-Mail sue.wolfe@leedsth.nhs.uk

Bush A, Alton EWFW, Davies JC, Griesenbach U, Jaffe A (eds): Cystic Fibrosis in the 21st Century.
Prog Respir Res. Basel, Karger, 2006, vol 34, pp 301–308

Physiotherapy

Jennifer A. Pryor[a]   Eleanor Main[c]   Penny Agent[b]   Judy M. Bradley[d]

[a]Department of Cystic Fibrosis and [b]Physiotherapy Department, Royal Brompton Hospital and
[c]Portex Anaesthesia, Intensive Therapy & Respiratory Medicine Unit, Institute of Child Health, London, and
[d]Health and Rehabilitation Sciences Research Institute, University of Ulster and Regional Adult Cystic
Fibrosis Centre, Belfast City Hospital, Belfast, UK

Abstract

There has been steady progress in some areas of cystic fibrosis (CF) research, but there are still many aspects of physiotherapy that do not have a strong evidence base. This chapter focuses on current themes in research and practice related to physiotherapy for CF, in particular those associated with the child, nebulization, physical training and airway clearance.

Introduction

Physiotherapy is usually introduced soon after the diagnosis of cystic fibrosis (CF) has been made. Physiotherapy for CF has been synonymous with postural drainage and percussion, but today airway clearance is only a part of the physiotherapy management of people with CF. CF is no longer a disease of childhood but effective management of the child influences outcomes during adolescence and adulthood, and importantly, the increase in longevity is associated with improvements in quality of life. Problems of particular relevance to the therapist include not only excess bronchial secretions but also reduced exercise capacity, breathlessness, incontinence, arthropathy, osteoporosis and cardiac abnormalities.

Paediatric Physiotherapy

Uncertainty Surrounding Physiotherapy in Infancy
There is general agreement that the infant with CF passively receives physiotherapy treatment administered by an adult, while from the age of 2 years more active participation in physiotherapy is encouraged until suitable independent techniques are identified during adolescence [1]. There is also agreement that the effects of aerobic and anaerobic exercise are valuable for children during both stable disease and acute exacerbations [2]. However, there are two topics which currently cause deliberation and disagreement in the physiotherapy management of newborn children with CF. The first is the management of asymptomatic newborn babies and infants, and the second is whether it is harmful to use head-down postural drainage positions [3].

Management of the Asymptomatic Newborn Baby and Infant
Until recently, referral to physiotherapy following diagnosis of CF was accompanied by respiratory symptoms and thus the instigation of daily physiotherapy was justified. The introduction of newborn screening for CF has, however, generated a need to reassess the place of physiotherapy in newborn infants who are diagnosed by screening but are apparently free of respiratory symptoms.

Those who endorse immediate commencement of a daily physiotherapy regimen following CF diagnosis by screening do so for two reasons:
(1) The daily physiotherapy routine should be introduced as early as possible so that instigation of this practice, when children are older, is not resented. This view is not evidence based.
(2) There is consistent evidence that inflammation, impaired mucociliary transport and abnormalities of airway structure and lung function are present in

children with CF, even if there are no apparent respiratory symptoms [4–7]. The argument is that physiotherapy treatments have a place in compensating for impaired mucociliary clearance and preventing or retarding inflammatory processes [1]. However, there is little evidence to support this and the benefit of physiotherapy in the inflammatory, early manifestations of CF remains unclear. In support of this viewpoint, studies are cited which suggest that early diagnosis and instigation of a comprehensive care package (including physiotherapy) result in important beneficial effects on the outcome and clinical course of the condition [8–10]. It is, however, impossible to assess the proportional effect of physiotherapy on long-term outcomes and it appears that respiratory infections may be more important predictors of long-term pulmonary outcomes [3, 11].

The arguments against starting a daily physiotherapy regimen in asymptomatic screened infants are 3-fold:

(1) There is good evidence that physiotherapy is considered a significant burden for children with CF and their families and that adherence to daily physiotherapy is often suboptimal [12–14]. To introduce physiotherapy in the absence of either symptoms or tangible benefits of treatment would be to reduce the potential for adherence when benefits are obvious. This reasoning is not evidence based.

(2) While it can be demonstrated that physiotherapy is helpful for removing copious airway secretions and reversing atelectasis, there is no evidence that physiotherapy is helpful for inflammation in CF.

(3) The limited evidence suggesting that gastro-oesophageal reflux (GOR) is exacerbated by physiotherapy may suggest that traditional physiotherapy, involving postural drainage, may worsen inflammatory processes by the potential for aspiration of gastric contents.

There appears to be no consensus in opinion amongst physiotherapy experts and clinical judgement, based on experience and the needs of individual patients, will continue to determine physiotherapy treatments for the foreseeable future. It remains essential to discuss treatment options truthfully with families as well as the potential risks and benefits associated with them. Parents and their children at an appropriate age should be central to the process of decision making about their care, and receive appropriate information to enable them to make an informed choice [15].

Head-Down Postural Drainage Positions

Traditionally and currently, the majority of physiotherapists in Europe apply and teach head-down postural drainage positioning in young infants, with the rationale that it increases the movement of mucous in the trachea [16]. However, a growing body of research has challenged the efficacy and safety of traditional treatments in infants and suggests that postural drainage positions facilitate or exacerbate GOR, with the associated risks of lung and oesophageal damage and tracheal micro-aspiration [17, 18].

GOR is relatively common in both healthy infants and those with CF, with the suggestion that prevalence is a little higher in the CF infant population [17]. The controversy arises from research suggesting that GOR is exacerbated by postural drainage positions and that more episodes of GOR are measurable during head-down postural drainage treatments than during treatments which avoid such positions. This research in a small population cannot be considered definitive, but confirms other studies from the same research team in the last decade [17–19]. One study conducted in the United Kingdom could not reproduce these results, possibly because of less acute postural drainage angles and a slightly older infant population [20].

Is the evidence currently available of high enough quality to justify a change of current practice? Probably not, but then it is useful to question whether current practice itself is underpinned by evidence of high enough quality. In this particular debate, the answer is also probably not. The evidence to promote the benefits of postural drainage in any population remains insubstantial and indeed absent in the infant population, so there is no strong incentive to insist on its use. In addition, the potential damage to lung and oesophageal tissue from GOR is serious enough to warrant careful consideration by physiotherapists treating infants with CF. It is also serious enough to justify keeping this question in the spotlight until it is more satisfactorily answered.

The Challenge of Physiotherapy Research in Preschool and Young Children

There are wide international variations in the therapeutic options offered to children with CF and their families and these will often depend on therapist or institutional preference, geographical region or training. The large and growing variety of treatment techniques reflects the absence of international consensus or high quality evidence on optimal treatment. However, physiotherapists aiming to improve the research evidence in paediatric care face the challenge of selecting appropriate outcomes. Median survival in CF has increased progressively over the past few decades as a result of closer monitoring and more aggressive multidisciplinary treatment of early CF lung disease. The forced expiratory volume in 1 s (FEV_1) is commonly

Pryor/Main/Agent/Bradley

used as an outcome to assess the efficacy of therapeutic interventions. However, as a result of improved therapies, many school-aged children with CF now have an FEV_1 within the normal range and it is no longer sensitive enough to detect changes from physiotherapy interventions in childhood [21]. In addition, reliable FEV_1 manoeuvres are difficult to obtain in children under the age of 5 years, with testing in the infant and preschool age groups being largely confined to specialist laboratories [21]. There is thus an urgent need to find alternative and sensitive outcome measures of lung function in children of all ages for assessing the efficacy of therapeutic interventions.

Recent studies have suggested that multiple-breath washout detects abnormal lung function in children with CF more readily than plethysmography or spirometry [21]. Mucociliary clearance has also been proposed as a sensitive alternative outcome, but remains less attractive in paediatrics because of the radiation exposure [22]. The development of and accessibility to more sensitive outcome measures such as these will advance the ability to expand the evidence base for physiotherapy interventions in children with CF.

In the interim, a specialized and individualized approach to physiotherapy care in children with CF remains appealing. The National Service Framework (UK) recommends that children, young people and their families should have the opportunity to become 'expert patients'. To achieve this, they should have access to services that help them to develop the self-confidence and self-management skills needed to deal with the impact of the condition on the child and their family [15]. Physiotherapists should be highly specialized: trained in and with access to all available airway clearance techniques without bias towards geographical or personal preferences, know the child and family so well that they understand their abilities, commitment, likelihood of adherence and attitude to exercise. They should track clinical changes, interpret them correctly and modify treatment accordingly. They should assess systematically the suitability of airway clearance techniques for individuals, taking into account factors such as changes in lung function, frequency of exacerbations, personal preference and quality of life [1].

'Intelligent' Nebulizers

Effective and efficient inhalation therapy is an important component of CF care. There have been recent advances in nebulizer delivery systems, resulting in smaller, quicker [23, 24] and more efficient devices that give improved deposition of a range of medications [25], and some of which can adapt to an individual's breathing pattern or control their breathing pattern [24]. These intelligent nebulizers are not only important for delivering maintenance therapies such as inhaled antibiotics and recombinant human DNase, but also for future advances such as a possible route for administering gene therapy.

Many factors contribute to reducing the efficiency of inhalation such as patient adherence with a treatment regimen [26], aerosol particle size, the individual's breathing pattern and airway geometry. The new generation of nebulizer systems has advanced from traditional jet nebulizers to address some of these factors in order to improve efficient delivery of medications:

Adaptive Aerosol Delivery. These devices adapt to the individual's breathing pattern, only releasing aerosol into a predetermined proportion of their inspiration. Devices are programmed to deliver a pre-set dose with audible feedback signalling the successful completion of the dose.

Vibrating Mesh Technology. This technology couples the vibration ($\sim$116 kHz) of a piezoelectric crystal directly to a metallic membrane with thousands of tapered holes, through which liquid medication is forced. This creates an aerosol with a median particle size of 2.5 μm.

Controlled Inhalations. Devices incorporate a limited inhalation flow, to enable the individual to inhale slowly and deeply, resulting in high peripheral airway deposition [27]. Some devices can be used in combination with jet nebulizers, or are coexistent in combination with other inhalation technologies.

The evolution of these intelligent nebulizers has resulted in shorter treatment times in combination with improved efficiency and efficacy of deposition, and is an important development in inhalation therapy focusing on reducing the burden of care.

Physical Training

Assessment of Fitness

With limited availability of laboratory testing in many CF centres for the assessment of aerobic/anaerobic fitness, the focus of recent research has been in the use of field exercise tests such as the modified shuttle test, sprint tests or muscle strength tests for individual muscle groups [28–30]. The psychometric properties (validity, reliability, and repeatability) of field tests have been assessed, but there are still little data on normal ranges or on what represents a clinical important change in these tests in CF. New technologies such as the Life Shirt System can facilitate more

---

Fig. 1. Life Shirt System which can be used to facilitate in-depth assessment of the cardiorespiratory responses to activity.

Fig. 2. Computerized inspiratory muscle training system which can be used to facilitate assessment of inspiratory muscle strength and endurance.

in-depth assessment of the cardiorespiratory responses to field tests (fig. 1). The importance of flexibility testing is underemphasized in CF. Flexibility is assessed by determining the full range of motion for each muscle group. Simple tools such as a flexible ruler can be used to facilitate measurement of postural abnormalities in CF. Computerized systems are available to facilitate assessment of inspiratory muscle strength and endurance (fig. 2). Assessment of 'motivation to exercise' facilitates prescription of an appropriate individualized exercise programme [31].

Evidence for Exercise Training

A Cochrane systematic review provides a useful summary of randomized controlled trials on exercise training in

CF [32]. Aerobic or anaerobic physical training has a positive effect on exercise capacity, strength and lung function and quality of life; however, the improvements are not consistent among studies and are influenced by the type of training programme [32, 33]. Evidence for other suggested benefits of exercise training in CF are based on non-randomized controlled clinical trials or expert opinion. Recent randomized controlled trials provide evidence of the benefits of inspiratory muscle training in CF [34, 35]. The benefits are primarily focused on direct improvements in inspiratory muscle strength and endurance; the impact on dyspnoea and exercise capacity is less consistent.

Principles of Exercise Prescription

The CF Trust Guidelines 2002 provide information on exercise prescription: this section aims to provide supplementary material [36]. The prescribed exercise programme should include components of aerobic, anaerobic and flexibility training, individualized to patient need. Patients with mild disease should be encouraged to participate in aerobic activities with their peers and for these patients the role of the health professional may be primarily advisory. As disease progresses, aerobic exercise should be more individualized. An alternative option for guiding the intensity of aerobic exercise is to prescribe exercise at a speed equivalent to e.g. 70% of the maximum speed attained on a maximal field test. In more severe patients and for patients on the transplant waiting list, interval training including bouts of relatively intense work, separated by periods of rest or low intensity training may be more appropriate. Anaerobic training involves either strength training programmes (e.g. Delorme programme, Oxford programme, McQueen programme) or sprint training [28]. Flexibility training should focus on targeting muscle groups that have been identified as problem areas in the flexibility assessment. Sports that carry a medical risk for patients with CF and important considerations for exercise prescription in children have been previously reviewed [36]. The American College of Sports Medicine provides guidance on the content of inspiratory muscle strength and endurance training programmes [37]. It remains unclear as to how long after cessation of training the benefits are maintained; however, a general guide is that the benefits of training will be lost at a rate of 1% per day after 1 week of cessation of exercise [38].

Special Considerations for People with CF
Non-Invasive Ventilation

There is no research or published guidelines for the use of non-invasive ventilation during exercise training in CF although its use during exercise (fig. 3) has been

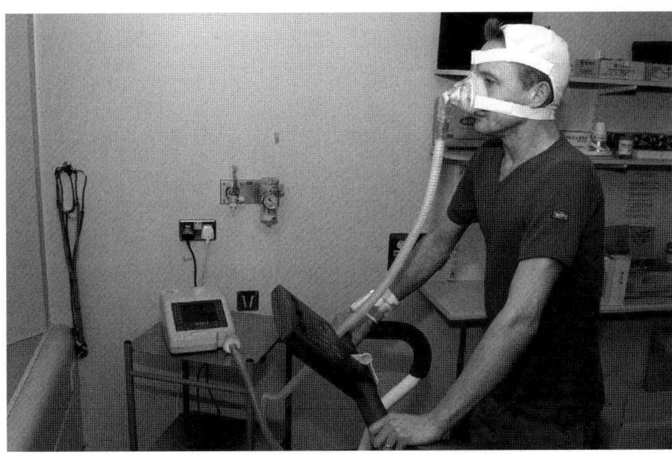

Fig. 3. Patient using non-invasive ventilation during exercise training.

demonstrated in other respiratory diseases. In patients with very severe disease or in those awaiting transplantation non-invasive ventilation may be considered to meet the increased demands during exercise [39, 40].

Acute Exacerbations

Patients may experience heightened respiratory symptoms and/or increased joint manifestations during an acute exacerbation; however, studies have reported beneficial effects of exercise training during an acute exacerbation of CF respiratory disease without any side effects [32].

Arthropathy

There is still limited evidence of the effects of exercise on CF-related arthropathies, but there is evidence to support the safety and efficacy of exercise in other inflammatory arthropathies. Although without a substantial evidence base, exercise performed in water to utilize the buoyancy, assistance and resistance of warm water for pain relief and muscle relaxation may be beneficial. The effects of immersion in patients with severe disease should be considered [41].

Osteoporosis

Decreased activity is a possible contributory factor to low bone mineral density in CF [42, 43]. A recent consensus report suggests that physical activity including appropriately prescribed weight bearing exercise can be used in prevention and management of CF bone disease [43] (see also chapter 34).

Nutrition

Accurate assessment of energy use and fluid loss, as well as nutritional advice on adjustment of daily intake to compensate for additional requirements of exercise, is important [43] (see also chapters 30 and 37).

CF-Related Diabetes

Self-glucose monitoring is an important component of exercise prescription, as exercise can cause hypoglycaemia either during or immediately after activity or up to 12 h later. Patients may be advised to consume food containing more carbohydrate than normal before activity and also after activity to prevent late hypoglycaemia (see also chapter 35).

Infection Control

It is important to assume that all patients with CF have transmissible pathogens, even if not identified by culture, and patients should exercise independently from other patients with CF [44]. Patients who use swimming pools should be aware that recreational amenities involving the use of water may be a potential source of *Pseudomonas aeruginosa* in CF [45].

Oxygen

There is some evidence from short-term studies to rationalize the use of supplemental oxygen during exercise in patients with moderate to severe CF [43, 46, 47]. Patients with CF and chronic hypoxaemia may require titration of oxygen during exercise and patients without chronic hypoxaemia, who show evidence of exercise-induced desaturation, require assessment for ambulatory oxygen [47, 48]. Considerable amounts of supplemental oxygen or high flows for correction of exercise-induced desaturation may be required. Availability of new systems for ambulatory oxygen and the use of oxygen-conserving devices will facilitate the use of oxygen for longer periods of exercise [48]. However, it should be noted that there is no evidence for the utility of oxygen therapy in CF, unlike that for chronic obstructive pulmonary disease.

Airway Clearance

'Forced expiratory manoeuvres are probably the most effective part of chest physiotherapy' [49]. It is the addition of the forced expiratory manoeuvre of the huff combined with breathing control [50] which has increased the effectiveness of many of the airway clearance regimens [51, 52]. Positive expiratory pressure (PEP) [53], high PEP [54], and oscillating PEP (acapella [55], RC-Cornet [56] and flutter [57]) now include the forced expiration technique from the active cycle of breathing techniques [58]. Autogenic drainage [59] and modified autogenic drainage [60], in their pure forms, do not include forced expiratory manoeuvres, but the techniques utilize breathing at different lung volumes to influence the movement of secretions from different parts of the airways and a huff or cough is used to clear secretions mobilized to the upper airways. The above regimens are all relatively inexpensive, easily portable and enable the user to be independent of an assistant. High frequency chest wall oscillation [61] and intrapulmonary percussive ventilation [62] are available and widely used in some countries, but are not as portable and are considerably more expensive.

Choice of Technique

It is probable that all the airway clearance regimens, by slightly different mechanical means, enhance airflow and reduce, in the short term, mucus viscosity [63–65]. Airflow is essential for airway clearance [66]. Figure 4 demonstrates the different flow and pressure characteristics of the oscillating systems of the acapella, RC-Cornet and flutter. The effects of these characteristics on the viscosity and mobilization of mucus in vivo require further study. The choice of technique may be related to personal preference rather than physiological principles, as each regimen can be adapted to suit an individual and adapted by the individual within a treatment session. Adherence to treatment is likely to be higher if the individual is involved in the selection process. There is increasing evidence that an airway clearance regimen combined with physical exercise may lead to additional improvements in pulmonary function, compared with airway clearance techniques alone [33, 67].

Gravity-Assisted Positioning

Improvements in the medical management of people with CF, improved nutrition and the advent of more effective airway clearance techniques have led to increased physical health and reduced need for the use of the head-down tipped position [68] or even the side-lying position in airway clearance [53, 57, 61]. The increase in ventilation to the dependent and more compliant parts of the lung may be more effective in the loosening and mobilization of excess mucus than the 'drainage' effects on more distended uppermost lung [69].

The Evidence

There is a dearth of evidence to support any long-term benefits from the use of airway clearance techniques [70] but few, if any, CF centres would support a randomized controlled trial with 'no airway clearance' as the control arm. A recent Cochrane systematic review compared

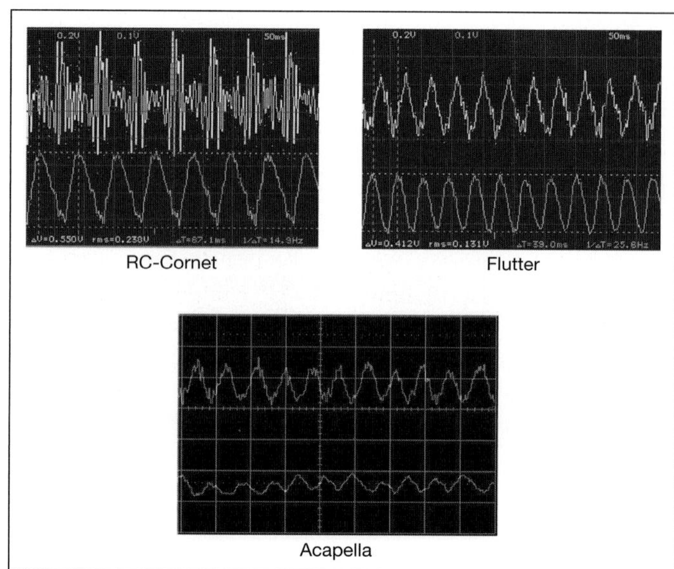

Fig. 4. Graphs showing the different pressure and flow characteristics for the oscillating PEP devices of the RC-Cornet (position 3–4), flutter and acapella. The upper line represents flow and the lower line represents pressure. Pressure was produced by a compressor with a flow of 38 litres per minute incorporating a 'whistle' with a capacity of 12 litres. Each square represents 50 ms on the horizontal axis and either 45 litres per minute (flow) or 20 mbars (pressure) on the vertical axis [reproduced with permission from work by Professor Ulrich Cegla, Montabaur, Germany].

'conventional' chest physiotherapy (postural drainage and percussion) with other contemporary airway clearance techniques. There appeared to be no advantage of 'conventional' chest physiotherapy over other airway clearance techniques in the primary outcome measure of lung function. There was an indication that individuals prefer self-administered regimens [71]. Pryor et al. (unpubl. data) studied 75 patients with CF over a period of 1 year. Each subject was randomized to one of the regimens of the active cycle of breathing techniques, autogenic drainage, RC-Cornet, flutter or PEP. There were no differences among the regimens in lung function, exercise capacity or quality of life. This concurs with work by Accurso et al. [72] in their long-term (3-year) multicentre airway secretion clearance study comparing postural drainage and percussion, flutter and high frequency chest wall oscillation.

The Definitive Physiotherapy Treatment

There may be no such thing as the perfect physiotherapy treatment, currently pursued by conventional study designs.

There is growing support for the hypothesis that individuals with CF are unique in terms of their genetic and physiological expression of the disease, and that individuals also have unique social, cultural, financial and psychological experiences that influence how they will adhere to, identify with or find benefit from different physiotherapy regimens. Specific treatments may be perfectly suited to some and unfavourable to others. Specific treatments may be better suited to stable disease while others are better during acute exacerbations. Since any or all of these factors are changeable over each individual's lifetime in terms of age, severity of disease and changes in life circumstances, the challenge is that of optimizing individual treatment supported by the evidence. It would be impractical for individuals with CF to undertake long-term tryouts of multiple individual airway clearance techniques in order to find their own optimum solution, especially if some treatments are ultimately deleterious. However, short-term outcomes do not necessarily reflect long-term outcomes and multicentre trials should be considered. In the interim, research should aim to provide guidance on which therapies are useful in particular circumstances and which risk factors (for example wheeze, clinical scores, psychosocial circumstances or GOR) preclude others, so that choices for individuals are limited to those most likely to succeed.

References

1 Lannefors L, Button BM, McIlwaine M: Physiotherapy in infants and young children with cystic fibrosis: Current practice and future developments. J R Soc Med 2004; 97(suppl 44):8–25.

2 Selvadurai HC, Blimkie CJ, Meyers N, Mellis CM, Cooper PJ, Van Asperen PP: Randomized controlled study of in-hospital exercise training programs in children with cystic fibrosis. Pediatr Pulmonol 2002;33:194–200.

3 Orenstein DM: Heads up! clear those airways! Pediatr Pulmonol 2003;35:160–161.

4 Sly PD, Brennan S: Detecting early lung disease in cystic fibrosis: Are current techniques sufficient? Thorax 2004;59:1008–1010.

5 Armstrong DS, Grimwood K, Carlin JB, Carzino R, Gutierrez JP, Hull J, et al: Lower airway inflammation in infants and young children with cystic fibrosis. Am J Respir Crit Care Med 1997;156:1197–1204.

6 Nixon GM, Armstrong DS, Carzino R, Carlin JB, Olinsky A, Robertson CF, et al: Early airway infection, inflammation, and lung function in cystic fibrosis. Arch Dis Child 2002; 87:306–311.

7 Khan TZ, Wagener JS, Bost T, Martinez J, Accurso FJ, Riches DW: Early pulmonary inflammation in infants with cystic fibrosis. Am J Respir Crit Care Med 1995;151: 1075–1082.

8 Orenstein DM, Boat TF, Stern RC, Tucker AS, Charnock EL, Matthews LW, et al: The effect of early diagnosis and treatment in cystic fibrosis: A seven-year study of 16 sibling pairs. Am J Dis Child 1977;131:973–975.

9 Dankert-Roelse JE, te Meerman GJ: Long term prognosis of patients with cystic fibrosis in relation to early detection by neonatal screening and treatment in a cystic fibrosis centre. Thorax 1995;50:712–718.

10 Farrell PM, Kosorok MR, Rock MJ, Laxova A, Zeng L, Lai HC, et al: Early diagnosis of cystic fibrosis through neonatal screening prevents severe malnutrition and improves long-term growth. Wisconsin Cystic Fibrosis Neonatal Screening Study Group. Pediatrics 2001;107:1–13.

11 Farrell PM, Li Z, Kosorok MR, Laxova A, Green CG, Collins J, et al: Bronchopulmonary disease in children with cystic fibrosis after early or delayed diagnosis. Am J Respir Crit Care Med 2003;168:1100–1108.

12 DeLambo KE, Ievers-Landis CE, Drotar D, Quittner AL: Association of observed family relationship quality and problem-solving skills with treatment adherence in older children and adolescents with cystic fibrosis. J Pediatr Psychol 2004;29:343–353.

13 Glasscoe CA, Quittner AL: Psychological interventions for cystic fibrosis. Cochrane Database Syst Rev 2003;3:CD003148.

14 Prasad SA, Cerny FJ: Factors that influence adherence to exercise and their effectiveness: Application to cystic fibrosis. Pediatr Pulmonol 2002;34:66–72.

15 The National Service Framework for Children, Young People and Maternity Services Department of Health, 2004; www.dh.gov.uk/healthtopics.

16 Button BM, Oberwaldner B, Carr M, Story I: The development of international guidelines for the physiotherapy management of infants with cystic fibrosis; results of a worldwide survey forming the foundations of the project – A European Respiratory Society (ERS) taskforce project endorsed by the IPG/CF. Pediatr Pulmonol 2001(suppl 22):A399.

17 Button BM, Heine RG, Catto-Smith AG, Phelan PD, Olinsky A: Chest physiotherapy, gastro-oesophageal reflux, and arousal in infants with cystic fibrosis. Arch Dis Child 2004;89:435–439.

18 Button BM, Heine RG, Catto-Smith AG, Olinsky A, Phelan PD, Ditchfield MR, et al: Chest physiotherapy in infants with cystic fibrosis: To tip or not? A five-year study. Pediatr Pulmonol 2003;35:208–213.

19 Button BM: Postural drainage techniques and gastro-oesophageal reflux in infants with cystic fibrosis. Eur Respir J 1999;14: 1456–1457.

20 Phillips GE, Pike SE, Rosenthal M, Bush A: Holding the baby: Head downwards positioning for physiotherapy does not cause gastro-oesophageal reflux. Eur Respir J 1998;12: 954–957.

21 Aurora P, Gustafsson P, Bush A, Lindblad A, Oliver C, Wallis CE, et al: Multiple breath inert gas washout as a measure of ventilation distribution in children with cystic fibrosis. Thorax 2004;59:1068–1073.

22 Robinson M, Eberl S, Tomlinson C, Daviskas E, Regnis JA, Bailey DL, et al: Regional mucociliary clearance in patients with cystic fibrosis. J Aerosol Med 2000;13:73–86.

23 Langman H, Orr A, McVean R, Riley D, Redfern J, Webb AK, Jones A, Dodd ME: Delivering a concentrated dose of colistin using AAD reduces treatment time. J Cyst Fibros 2004;3:S57.

24 Prince I, Dixon E, Agent P, Pryor JA, Hodson ME: Evaluation of a guided breathing manoeuvre for nebulised therapy in cystic fibrosis patients. Pediatr Pulmonol 2004;S27: 313.

25 Mullinger B, Sommerer K, Herpich C, Meyer G, Scheuch G: Inhalation therapy can be improved in CF patients by controlling the breathing pattern during inspiration. J Cyst Fibros 2004;3:S65.

26 Conway S, Dodd M, Marsden R, Paul E, Weller P: Comparison of compliance in cystic fibrosis patients using either a Halolite adaptive aerosol delivery (AAD) system or a conventional high output nebuliser. J Cyst Fibros 2002;1:S67.

27 Scheuch G, Meyer T, Mullinger B, Sommerer K, Brand P, Haeussinger K, Heyder J: Optimising aerosol lung deposition in CF patients. J Cyst Fibros 2002;1:S67.

28 Klijn PHC, Oudshoorn A, van der Ent CK, van der Net J, Kimpen JL, Helders PJM: Effects of anaerobic training in children with cystic fibrosis. Chest 2004;125:1299–1305.

29 Bradley JM, Howard J, Wallace E, Elborn S: Validity of the modified shuttle test in adult cystic fibrosis. Thorax 1999;54:437–439.

30 Bradley JM, Howard J, Wallace E, Elborn S: Reliability, repeatability, and sensitivity of the modified shuttle test in adult cystic fibrosis. Chest 2000;117:1666–1671.

31 Keele-Smith R, Leon T: Evaluation of individually tailored interventions on exercise adherence. West J Nurs Res 2003;25:623–640.

32 Bradley JM, Moran F: Physical training in cystic fibrosis (Cochrane Review); in The Cochrane Library, Issue 2. Oxford Update Software, 2003.

33 Moorcroft AJ, Dodd ME, Morris J, Webb AK: Individualised unsupervised exercise training in adults with cystic fibrosis: A 1 year controlled trial. Thorax 2004;59:1074–1080.

34 Albini S, Rath R, Renner S, Eichler I: Additional inspiratory muscle training intensifies the beneficial effects of cycle ergometer training in patients with cystic fibrosis. J Cyst Fibros 2004;3(suppl 1):63.

35 Enright S, Chatham K, Ionescu AA, Unnithan VB, Shale DJ: Inspiratory muscle training improves lung function and exercise capacity in adults with cystic fibrosis. Chest 2004;126: 405–411.

36 Association of Chartered Physiotherapists in Cystic Fibrosis (ACPCF): Clinical Guidelines for the Physiotherapy Management of Cystic Fibrosis. Recommendations of a Working Group. Bromley, Cystic Fibrosis Trust, 2002.

37 Balady GJ, Berra KA, Golding LA, Gordon NF, Mahler JN, et al: Exercise prescription for pulmonary patients; in Franklin BA, Whaley MH, Howley ET (eds): American College of Sports Medicine's (ACSM) Guidelines for Exercise Testing and Prescription, ed 6. Philadelphia, Lippincott Williams & Wilkins, 1999.

38 McArdle WD, Katch F, Katch V: Exercise Physiology: Energy, Nutrition, and Human Performance. Philadelphia, Lippincott Williams & Wilkins, 2001.

39 Ambrosino N: Exercise and non-invasive ventilatory support. Monaldi Arch Chest Dis 2000;55:242–246.

40 Fauroux B, Hart N, Lofasa F: Non-invasive mechanical ventilation in cystic fibrosis: Physiological effects and monitoring. Monaldi Arch Chest Dis 2002;57:268–272.

41 Anstey KH, Roskell C: Hydrotherapy: Detrimental or beneficial to the respiratory system? Physiotherapy 2000;86:2–56.

42 Hecker TM, Aris RM: Management of osteoporosis in adults with cystic fibrosis. Ther Pract 2004;64:133–147.

43 Yankaskas JR, Marshall BC, Sufian B, Simon RH, Rodman D: Cystic fibrosis adult care consensus conference report. Chest 2004;125: 1s–39s.

44 Saiman L, Siegel J; CF Foundation Consensus Conference on Infection Control Participants: Infection control recommendations for patients with cystic fibrosis: Microbiology important pathogens and infection control practices to prevent patient-to-patient transmission. Am J Infect Control 2003;31: S6–S62.

45 Moore JE, Heaney N, Millar BC, Crowe M, Elborn JS: Incidence of Pseudomonas in recreational and hydrotherapy pools. Commun Dis Public Health 2002;5:23–26.

46 McKone EF, Barry SC, Fitzgerald MX, Gallagher CG: The role of supplemental oxygen during submaximal exercise in patients with cystic fibrosis. Eur Respir J 2002;20: 134–142.

47 Royal College of Physicians: Domiciliary oxygen therapy services: Clinical guidelines and advice for prescribers. A Report of the Royal College of Physicians, 1999.

48 British Thoracic Society Working Group on Home Oxygen Services: Clinical component for the home oxygen service in England and Wales, 2004; www.brit-thoracic.org.uk/ public_content.asp.

49 van der Schans CP: Forced expiratory manoeuvres to increase transport of bronchial mucus: A mechanistic approach. Monaldi Arch Chest Dis 1997;52:367–370.

50 Thompson B, Thompson HT: Forced expiration exercises in asthma and their effect on FEV_1. N Z J Physiother 1968;3:19–21.

51 Pryor JA, Webber BA, Hodson ME, Warner JO: The Flutter VRP1 as an adjunct to chest physiotherapy in cystic fibrosis. Respir Med 1994;88:677–681.

52 Pike SE, Machin AC, Dix KJ, Pryor JA, Hodson ME: Comparison of flutter VRP1 and forced expirations (FE) with active cycle of breathing techniques (ACBT) in subjects with cystic fibrosis (CF). Neth J Med 1999;54: S55–S56.

53 Falk M, Kelstrup M, Andersen JB, Kinoshita T, Falk P, Støvring S, Gøthgen I: Improving the ketchup bottle method with positive expiratory pressure, PEP, in cystic fibrosis. Eur J Respir Dis 1984;65:423–432.

54 Oberwaldner B, Evans JC, Zach MS: Forced expirations against a variable resistance: A new chest physiotherapy method in cystic fibrosis. Pediatr Pulmonol 1986;2:358–367.

55 Volsko TA, DiFiore J, Chatburn RL: Performance comparison of two oscillating positive expiratory pressure devices: Acapella versus Flutter. Respir Care 2003;48:124–130.

56 Cegla UH, Bautz M, Fröde G, Werner T: Physiotherapy in patients with COPD and tracheobronchial instability – A comparison of two oscillating PEP systems (RC-Cornet®, VRP1 Desitin). Pneumologie 1997;51: 129–136.

57 Konstan MW, Stern RC, Doershuk CF: Efficacy of the Flutter device for airway mucus clearance in patients with cystic fibrosis. J Pediatr 1994;124:689–693.

58 Pryor JA, Webber BA, Enright S, Howarth A, Potter H, Tagg L: Physiotherapy techniques; in Pryor JA, Prasad SA (eds): Physiotherapy for Respiratory and Cardiac Problems, ed 3. Edinburgh, Churchill Livingstone, 2002, pp 161–242.

59 Chevaillier J: Autogenic drainage (AD); in Physiotherapy in the Treatment of Cystic Fibrosis (CF), ed 3. International Physiotherapy Group for Cystic Fibrosis (IPG/CF), 2002, pp 12–15. www.cfww.org/pub/physiotherapy. pdf

60 Kieselmann R: Modified AD (M AD); in Physiotherapy in the Treatment of Cystic Fibrosis (CF), ed 3. International Physiotherapy Group for Cystic Fibrosis (IPG/CF), 2002, pp 16–17. www.cfww.org/pub/physiotherapy.pdf.

61 Kluft J, Beker L, Castagnino M, Gaiser J, Chaney H, Fink RJ: A comparison of bronchial drainage treatments in cystic fibrosis. Pediatr Pulmonol 1996;22:271–274.

62 Scherer TA, Barandun J, Martinez E, Wanner A, Rubin EM: Effect of high-frequency oral airway and chest wall oscillation and conventional chest physical therapy on expectoration in patients with stable cystic fibrosis. Chest 1998;113:1019–1027.

63 App EM, Kieselmann R, Reinhardt D, Lindemann H, Dasgupta B, King M, et al: Sputum rheology changes in cystic fibrosis lung disease following two different types of physiotherapy: Flutter vs autogenic drainage. Chest 1998;114:171–177.

64 Feng W, Deng WW, Huang SG, Cheng QJ, Cegla UH, King M: Short-term efficacy of RC-Cornet in improving pulmonary function and decreasing cohesiveness of sputum in bronchiectasis patients. Chest 1998;114(suppl 4):320S.

65 Selsby D, Jones JG: Some physiological and clinical aspects of chest physiotherapy. Br J Anaesth 1990;64:621–631.

66 Lapin CD: Airway physiology, autogenic drainage, and active cycle of breathing. Respir Care 2002;47:778–785.

67 Thomas J, Cook DJ, Brooks D: Chest physical therapy management of patients with cystic fibrosis: A meta-analysis. Am J Respir Crit Care Med 1995;151:846–850.

68 Cecins NM, Jenkins SC, Pengelley J, Ryan G: The active cycle of breathing techniques – To tip or not to tip? Respir Med 1999;93: 660–665.

69 Lannefors L, Wollmer P: Mucus clearance with three chest physiotherapy regimes in cystic fibrosis: A comparison between postural drainage, PEP and physical exercise. Eur Respir J 1992;5:748–753.

70 van der Schans C, Prasad A, Main E: Chest physiotherapy compared to no chest physiotherapy for cystic fibrosis (Cochrane Review); in The Cochrane Library, Issue 4. Chichester, Wiley, 2004.

71 Main E, Prasad A, van der Schans C: Conventional chest physiotherapy compared to other airway clearance techniques for cystic fibrosis (Systematic Review). Cochrane Cystic Fibrosis and Genetic Disorders Group. Cochrane Database Syst Rev 2005;CD002011.

72 Accurso FJ, Sontag MK, Koenig JM, Quittner AL: Multi-center airway secretion clearance study in cystic fibrosis. Pediatr Pulmonol 2004(suppl 27):314.

Jennifer A. Pryor
Royal Brompton Hospital
Sydney Street
London SW3 6NP (UK)
Tel. +44 20 7352 8121, ext. 4925
Fax +44 20 7351 8052
E-Mail j.pryor@rbht.nhs.uk

Bush A, Alton EWFW, Davies JC, Griesenbach U, Jaffe A (eds): Cystic Fibrosis in the 21st Century.
Prog Respir Res. Basel, Karger, 2006, vol 34, pp 309–313

Psychological Interventions

Mandy Bryon

Great Ormond Street Hospital for Children NHS Trust, London, UK

Abstract

Psychological interventions are now generally accepted as a necessary component of the multidisciplinary care of cystic fibrosis. This chapter reviews recent developments in psychological knowledge to inform best clinical practice. These are ordered across the lifespan from diagnosis to end of life with the most salient psychological difficulties of each stage discussed and the current recommended psychological interventions described.

Introduction

The emotional impact of a diagnosis of cystic fibrosis (CF) is significant and without appropriate psychological guidance, children and families suffer in respect to their interpersonal relationships, health status and quality of life. Psychological input as an integral component of the multidisciplinary team is well established and most children and families now receive routine preventative support. The following chapter reviews some of the most recent areas of psychological understanding of the impact of CF on the developing child and family which can be used to guide psychological treatment interventions.

Impact of the Diagnosis on the Family

In the majority of cases the diagnosis of CF is given within the first few months of life. At a time when parents are feeling intense emotions towards their newborn infant they are asked to assimilate the information that their baby has a life-limiting condition. Families now have to expand their network to include a group of uninvited strangers comprising the CF team. Decisions about their child's future are no longer under their complete control; many of these will be taken by a stranger. The delivery of the diagnosis of CF marks a significant first step in the potentially long-term relationship between the parents, the doctor and a range of other health care professionals. Good communication is associated with better doctor-patient relationships [1], more patient satisfaction and increased likelihood that treatment advice will be followed [2]. It would not be difficult to imagine that the diagnosis of CF is followed by sorrow, depression, loss and disempowerment for the parents [3–5].

Having CF does not interfere with the child's drives to form an attachment, nor necessarily with the major carer's drive to respond to the infant's needs [6]. There have been some recent concerns that the impact of neonatal screening for CF may harm the mother-child attachment by disrupting the relationship with the disclosure of a diagnosis in an infant that had not displayed any symptoms [7]. Studies that have begun to examine that question have found no evidence to support this hypothesis [8]. Although mothers of children diagnosed by neonatal screening have higher frequencies of 'at-risk' scores for parenting stress than mothers of traditionally diagnosed children [9], it is suggested that the way in which the diagnosis is communicated and subsequent parental support are key factors in maternal well-being.

Postdiagnosis counselling must continue beyond the disclosure interview and further contact with families must not solely focus on the clinical features of the condition but need to include the family's emotional well-being [10]. Sawyer and Glazner [11] have evaluated a support programme to delineate the essential components for continuous service delivery.

Parents are at risk of overprotecting of their children with CF. A fear of infection can result in parents restricting the physical movements, toys and social contacts with other children. Studies of the play interactions of parents of preschool children with CF indicate that mothers are much more interfering and less supportive of their children and the child correspondingly shows less persistence and compliance in play activities [12]. The child-rearing practices of parents of children with CF may by affected by the overwhelming need to protect the child. All daily interactions from dressing to feeding to bedtimes may be indirectly altered by the diagnosis. Mealtimes are the area of most reported difficulty.

Managing Mealtime Behaviour
Parents are under tremendous pressure to achieve a high daily calorific intake for their young children and as a result they are less likely to allow their child to take increasing responsibility for self-feeding. Stark et al. [13] conducted a large observational study of the mealtime behaviours of preschool children with and without CF. She found that the children with CF ate more than their healthy peers though they failed to reach the set dietetic targets. The same type of mealtime behaviour problems were seen in the two groups although children with CF stayed at the table an average of 6 min longer and showed almost twice the frequency of food refusals, non-compliance to commands to eat and leaving the table. Parents of the children with CF were significantly more likely than the control groups to report these behaviours as problematic and engage in behaviours to get their child to eat using strategies such as coaxing, commanding, physical prompts and feeding the child themselves. None of these behaviours are found to be effective in increasing the amount consumed [14–16].

Behavioural interventions to modify parental behaviour at mealtimes have been found to be effective at increasing calorie intake and weight gain in young children with CF [17–19]. Observations of parental behaviour at mealtimes demonstrate that parents of children with CF focus attention on non-eating behaviours, reinforcing undesired rather than desired behaviours. Interventions are therefore aimed at training parents to shift their attention to eating behaviour.

The Impact of CF on School-Age Children

Many parents report that once they have established a routine for treatment, the threat of CF does not impact too greatly on the family's life. The school-age child's social world is expanding and with it their psychosocial development. These children are much more aware of 'social rules', how their behaviour affects others and how they expect others to behave towards them. Consequently, they start to realize the differences CF brings, in particular the daily treatments. They will no longer accept that they have to take tablets, have physiotherapy and/or nebulized drugs without questioning why it is not the same for their siblings or friends. The challenge is to enable the school-age child with CF to traverse the complexities of normal development with a positive self-concept intact.

Talking to Children about Their CF
The child's understanding of illness has been the subject of debate amongst developmental theorists for some time. Previously, clinicians were advised that the child's understanding of their own illness followed a stage approach mirroring the stages of development proposed by Piaget [20]. However, anecdotally it has been noted that young children with chronic illnesses develop a precocious comprehension of their own condition and have a much more competent understanding of illness than can be explained by the stage theory. It is suggested that children have much more elaborate concepts for some events [21–23] and for sick children this comes about by virtue of their frequent experience with hospital and illness [24]. Studies have found that young children can understand and operate the concept of contagion as originating from invisible sources; they can accept germs causing illness [23, 25] though they do not fully understand the causal processes [26]. Raman and Winer [26] propose that children and adults use multiple types of reasoning to explain illness from scientific reasoning to personal, subjective naïve beliefs. Though the level of understanding would vary according to age and cognitive development, contextual variables could alter the belief at any one time.

Children, therefore, are capable of more sophisticated understanding of CF than previously thought. Whilst it should be possible to offer rational explanations for the condition without stigma and clarify queries about peer comparisons without damaging self-concept, there exists a barrier to effective communication with children. On interview about their condition, children with CF tend to give stock responses to questions, they know that CF affects the respiratory (85%) and digestive (80%) systems, though do not know why or how and the majority do not know the importance of nutrition (70%) [27]. Children tend to separate their understanding and definitions of CF from how it affects them personally. Though they are quite capable of a decent knowledge of their own health and can engage in patient-clinician discussions about treatment, they tend not to. Children with CF define themselves as 'healthy' and this phenomenon is also found in other chronic health conditions, such as asthma [28] and heart disease [29].

This state of partial knowledge may relate to reluctance for open discussion of CF in the family and the CF health professionals' failure to include the child in consultations. Parents are often used by children as envoys and information brokers; they act as buffers from unpleasant information [30]. However, accurate knowledge of one's own medical condition correlates with less distress, less confusion, improved relationships with the medical team, better adherence to medication and an improved emotional well-being [29, 31, 32]. Parents often need guidance on how to adopt a more relaxed method of communication about CF with their children. Advising parents on effective communication with their children can be complex and requires ongoing support [33].

Managing CF in Adolescence

Adolescents with CF face the same tasks as their healthy peers: Physical growth and sexual development, personal individuation, intimate relationships, finding a comfortable social group, educational goals and preparation for an occupation. The medical progress and continued optimism for treatments for CF mean that it is a different disease than it was a few years ago; the emotional and behavioural preparation of adolescents and young adults has not kept pace [34]. There is a paucity of guidance for psychosocial support of adolescents with CF relative to that available for younger age groups [35].

Chronic illness will interfere with the normal developmental tasks of adolescence dependent on the course of the condition. Adequate nutrition becomes a greater problem as their growth demands even more calories than can be consumed, resulting in a notably underweight body shape. Correspondingly, there is increasing evidence that adolescents are at greater risk than their healthy peers for developing eating disturbance [36]. Other physical distortions such as an overinflated chest, clubbed finger tips, chronic cough and external treatment supplementations such as a subcutaneous venous access device or a gastrostomy tube all contribute to a negative body image and poor self-esteem [36, 37]. Concomitant with a delayed puberty are reported delays in developing a sexual identity and forming intimate relationships especially in girls [38, 39].

Adherence to Treatment

Adherence to a treatment regimen is perhaps the most well-documented area of adolescent rebellion in CF [40]. The adolescent years have been described as the period of the 'imaginary audience' [41] whereby the teenager considers him/herself to be constantly under public scrutiny. Consideration of how one measures up to one's peers is an overwhelming preoccupation. Adolescents with CF would be acutely aware of the impact CF and its intrusive treatment have as a visible marker of their difference. The fact that adolescents have the intellectual capacity to master the self-administration of a skilled treatment technique does not mean that they will carry out the task competently. Responsibility for treatment does not correlate with good rates of adherence [42]. Clearly, the prescribers of treatment for the CF adolescent must accept that a degree of poor adherence will be the norm. A traditional proscriptive approach will fail to uncover any incompatibility between medical criteria and the adolescent's criteria for treatment success. Although adult patients are encouraged to become active collaborators in treatment decision making, adolescents are not often awarded the same status. A medical relationship with a teenager is more complicated than with an adult above the legal age for consent. There are ethical and legal considerations of dealing with a teenager's refusal to assent to treatment and respecting the decisions of both adolescent patient and parents [43]. Generally, the paediatrician's relationship in treating adolescents with CF is to manage a balance between the patient's autonomy, the rights of the parents and the law [44].

Behavioural interventions, such as behaviour contracting, have been found effective in increasing targeted symptoms [45] but not as successful in maintenance of effect nor generalization to other symptoms. Psychoeducational approaches my improve understanding of treatment but do not result in increased adherence [46]. Therapeutic approaches to facilitate improved adherence to treatment must incorporate strategies aimed at changing behaviour of the adolescent, and strategies to improve the negotiation skills of parents and health professionals. A behavioural family systems approach [47] lends itself to this sort of problem as it combines the behavioural techniques of skills training with a systemic focus on structural problems in the family such as weak parental coalitions and negative belief systems. The intervention has been evaluated for families with an adolescent with CF [40].

Improved medical technologies mean that death is no longer seen as the inevitable outcome of life-threatening diseases, rather death is perceived as medical failure [48]. The medical management of end of life in CF has become problematic and more difficult to predict [49], discussion and decisions around it almost taboo. Grief has been pathologized, and expression of intense emotion seen as an abnormality that must be expertly dealt with by a trained professional [50]. In medicine, a crisis recovery model has been adopted, the bereaved must be 'cured' of their grief. Whilst understanding and support are called for, health professionals must accept that people do not recover from their loss. Bereaved people enter a process, albeit immeasurably difficult, in confronting the reality of the loss and an acceptance of the changed way of life without the loved one [51].

Grief reactions are normal, expected and take on a huge variety of courses according to a range of variables including personality, gender, culture, religious beliefs and support networks [52–54]. This means that the most appropriate support that can be given by the CF team is to acknowledge the wide diversity of normal grief reactions. There is no right or wrong way to grieve.

The most important factor in the effective intervention of health care professionals at end of life is communication [55]. Discussions about dying and death are aversive for the health professional and so tend to be avoided. However, studies have demonstrated that discussions about death are, of course, distressing but not damagingly stressful for the patient or relatives [56]. A booklet produced by the Cystic Fibrosis Trust in the UK [57] is intended to provide information on the processes of dying, death and bereavement. The booklet can be used directly by patients and their family or by the CF team members to help them communicate more effectively with families at the time of loss.

Bereaved Children

Increasingly, people with CF are having children of their own; the pregnancy itself can result in a physical deterioration in a woman with CF in relatively good pre-conception health [58]. Inevitably this means that children will be bereaved and the CF team may be called on for advice. Bereaved children may show more depression, withdrawal, anxiety, lower self-esteem and less hope for the future than non-bereaved children. Adults who lost parents as children are at risk of panic and anxiety disorders if they were excluded from the funeral and explanation of the loss; though perhaps well-meaning this results in what has been documented as a 'disenfranchised grief' [59]. A key debilitating factor for grieving children is a sense of loss of control; early interventions with children at the time of death are highly recommended to help reduce future anxieties [60].

References

1   Maguire P, Pitceathly C: Key communication skills and how to acquire them. BMJ 2002; 325:697–700.
2   Roter DL, Hall JA, Barker LR, Cole KA, Roca RP: Improving physician's interviewing skills and reducing patients' emotional distress. Arch Intern Med 1995;155:1877–1884.
3   Lemons PM, Weavers DD: Beyond the birth of a defective child. Neonatal Netw 1987;5: 13–20.
4   Myer PA: Parental adaptation to cystic fibrosis. J Pediatr Health Care 1998;2:20–28.
5   Jedlicka-Kohler I, Gotz M, Eicher I: Parents' recollection of the initial communication of the diagnosis of cystic fibrosis. Pediatrics 1996;97:204–209.
6   Simmons RJ, Goldberg S, Washington J, Fischer-Fay A, Maclusky I: Nutrition and attachment in children with cystic fibrosis. J Dev Behav Pediatr 1995;16:183–186.
7   Young SS, Kharrazi M, Pearl M, Cunningham G: Cystic fibrosis screening in newborns: Results from existing programs. Curr Opin Pulm Med 2001;7:427–433.
8   Baroni MA, Anderson YE, Mischler E: Cystic fibrosis newborn screening: Impact of early screening on parental stress. Pediatr Nurs 1997;23:143–151.
9   Boland C, Thompson NL: Effects of newborn screening of cystic fibrosis on reported maternal behaviour. Arch Dis Child 1990;65: 1240–1244.
10   Mitchell W, Sloper P: Information that informs rather than alienates families with disabled children: Developing a model of good practice. Health Soc Care Community 2002;10:74–81.
11   Sawyer SM, Glazner JA: What follows neonatal screening? An evaluation of a residential education program for parents of infants with newly diagnosed cystic fibrosis. Pediatrics 2004;114:411–416.
12   Goldberg S, Gotowiec A, Simmons RJ: Behaviour problems in chronically ill children. J Dev Psychopathol 1995;7:267–282.
13   Stark LJ, Jelalian E, Mulvihill MM, Powers SW, Bowen AM, Spieth LE, Keating K, Evans S, Creveling S, Harwood I, Passero MA, Hovell MF: Eating in preschool children with cystic fibrosis and healthy peers: Behavioural analysis. Pediatrics 1995;95:210–215.
14   Stark LJ, Mulvihill MA, Jelalian E, Bowen AM, Powers SW, Tao S, Creveling S, Passero MA, Harwood I, Light M, Lapey A, Hovell MF: Descriptive analysis of eating behaviour in school age children with cystic fibrosis and healthy controls. Pediatrics 1997;99:665–671.
15   Stark LJ, Jelalian E, Powers SW, Mulvihill MM, Opipari LC, Bowen A, Harwood I, Passero MA, Lapey A, Light M, Hovell MF: Parent and child mealtime behaviours in families of children with cystic fibrosis. J Pediatr 2000;136:195–200.
16   Spieth LE, Stark LJ, Mitchell MM, Schiller M, Cohen LL, Mulvihill MM, Hovell MF: Observational assessment of family functioning at mealtime in preschool children with cystic fibrosis. J Pediatr Psychol 2001;26: 215–222.
17   Stark LJ, Bowen AM, Tye VL, Evans SJ, Passero MA: A behavioural approach to increasing calorie consumption in children with cystic fibrosis. J Pediatr Psychol 1990; 15:309–326.
18   Stark LJ, Knapp LG, Bowen AM, Powers SW, Jelalian E, Evans S, Passero MA, Mulvihill MM, Hovell M: Increasing calorie consumption of children with cystic fibrosis: Replication with two-year follow-up. J Appl Behav Anal 1993;26:435–450.
19   Stark LJ, Powers SW, Jelalian E, Rape RN, Miller DL: Modifying problematic mealtime

interactions of children with cystic fibrosis and their parents via behavioural parent training. J Pediatr Psychol 1994;19:751–768.

20 Piaget J: The Origins of Intelligence in Children. New York, International Universities Press, 1952.

21 Kalish C: Causes and symptoms in preschoolers' concepts of illness. Child Dev 1996;67:1646–1670.

22 Springer K: Beliefs about causality among preschoolers with cancer: Evidence against immanent justice. J Pediatr Psychol 1994;19:91–101.

23 Springer K, Ruckel J: Early beliefs about the cause of illness. Cogn Dev 1992;7:429–443.

24 Eiser C: Children's understanding of illness: A critique of the 'stage' approach. Psychol Health 1989;3:93–101.

25 Rosen AB, Rozin P: Now you see it … now you don't: The preschool child's conception of invisible particles in the context of dissolving. Dev Psychol 1993;29:300–311.

26 Raman L, Winer GA: Children's and adults' understanding of illness: Evidence in support of a coexistence model. Genet Soc Gen Psychol Monogr 2002;128:325–356.

27 Frey M: Behavioural correlates of health and illness in youths with chronic illness. Appl Nurs Res 1996;9:167–176.

28 Veldtman GR, Matley SL, Kendall L, Quirk J, Gibbs JL, Parsons JM, Hewison J: Illness understanding in children and adolescents with heart disease. West J Med 2001;174: 171–174.

29 Young B, Dixon-Woods M, Windridge KC, Heney D: Managing communication with young people who have a potentially life-threatening chronic illness: Qualitative study of patients and parents. BMJ 2003;326: 305–314.

30 Kuther T: Medical decision-making and minors: Issues of consent and assent. Adolescence 2003;38:343–351.

31 Stewart M: Effective physician-patient communication and health outcomes: A review. Can Med Assoc J 1995;152:1423–1433.

32 Rushforth H: Communicating with hospitalized children: Review and application of research pertaining to children's understanding of health and illness. J Child Psychol Psychiatry 1999;40:683–691.

33 Towle A, Godolphin W: Framework for teaching and learning informed shared decision-making. BMJ 1999;319:766–771.

34 Mullins LL, Pace TM, Keller J: Cystic fibrosis: Psychological issues; in Olsen RA, Mullins LL, Gillman JB, Chaney JM (eds): The Sourcebook of Pediatric Psychology. Boston, Allyn & Bacon, 1994, pp 204–217.

35 Drotar D: Commentary: Cystic fibrosis. J Pediatr Psychol 1995;20:413–416.

36 Shearer JE, Bryon M: The nature and prevalence of eating disorders and eating disturbance in adolescents with cystic fibrosis. J R Soc Med 2004;97(suppl):36–42.

37 Bywater EM: Adolescents with cystic fibrosis: Psychological adjustment. Arch Dis Child 1981;56:538–543.

38 Sawyer SM, Rosier MJ, Phelan PD, Bowes G: The self-image of adolescents with cystic fibrosis. J Adolesc Health 1995;16:204–208.

39 Orr DP, Weller SC, Satterwaite B, Pless IB: Psychosocial implications of chronic illness in adolescence. J Pediatr 1984;104:152–157.

40 Quittner AL, Drotar D, Iveres-Landis C, Slocum N, Seidner D, Jacobsen J: Adherence to medical treatments in adolescents with cystic fibrosis: The development and evaluation of family-based interventions; in Drotar D (ed): Promoting Adherence to Medical Treatment in Chronic Childhood Illness. Mahwah, Earlbaum, 2000, pp 383–407.

41 Elkind D: Egocentrism in adolescence. Child Dev 1967;38:1025–1034.

42 Gudas LJ, Koocher GO, Wypij D: Perceptions of medical compliance in children and adolescents with cystic fibrosis. J Dev Behav Pediatr 1991;12:236–242.

43 Committee on Bioethics: Informed consent, parental permission and assent in pediatric practice. Pediatrics 1995;95:314–317.

44 Kuther T: Medical decision-making and minors: Issues of consent and assent. Adolescence 2003;38:343–351.

45 Wysocki T, Harris MA, Greco P, Bubb J, Danda CE, Harvey LM, Mc Donell K, Taylor A, White NH: Randomised controlled trial of behaviour therapy for families of adolescents with insulin-dependent diabetes mellitus. J Pediatr Psychol 2000;25:23–33.

46 Rubin DH, Bauman LJ, Lauby JL: The relationship between knowledge and reported behaviour in childhood asthma. J Dev Behav Pediatr 1989;10:307–312.

47 Robin AL, Foster SL: Negotiating Parent Adolescent Conflict: A Behavioural-Family Systems Approach. New York, Guilford, 1989.

48 McCue JD: The naturalness of dying. JAMA 1995;273:1039–1043.

49 Robinson WM, Ravilly S, Berde C, Wohl ME: End-of-life care in cystic fibrosis. Pediatrics 1997;100:205–209.

50 Wolfelt A: Healing the Bereaved Child: Grief Gardening, Growth through Grief and Other Touchstones for Caregivers. Colorado, Companion Press, 1996.

51 Stroebe MS: New directions in bereavement research: Exploration of gender differences. Palliat Med 1998;12:5–12.

52 Sheldon F: Bereavement; in Fallon M, O'Neill B (eds): ABC of Palliative Care. London, BMJ Books, 1998, pp 63–65.

53 Walter T: On Bereavement: The Culture of Grief. Buckingham, Open University Press, 1999.

54 Martin TL, Doka KJ: Men Don't cry … Women Do: Transcending Gender Stereotypes of Grief. Philadelphia, Brunner/Mazel, 2000.

55 Faulkner A: ABC of palliative care: Communication with patients, families and other professionals. BMJ 1998;316:130–132.

56 Emanuel EJ, Fairclough DL, Wolfe P, Emanuel LL: Talking with terminally ill patients and their caregivers about death, dying and bereavement: Is it stressful? Is it helpful? Arch Intern Med 2004;164:1999–2004.

57 Cystic Fibrosis Trust: Dying, Death and Bereavement. Help for All the Family When Someone Dies or Is Dying from Cystic Fibrosis. Bromley, CF Trust, 2004.

58 Edenborough FP, Stableforth DE, Webb AK, Mackenzie WE, Smith DL: Outcome of pregnancy in women with cystic fibrosis. Thorax 1995;50:170–174.

59 Doka KJ: Disenfranchised Grief: Recognizing Hidden Sorrow. New York, Lexington Books, 1989.

60 Goldman LE: Life and Loss: A Guide to Help Grieving Children, ed 2. New York, Taylor & Francis, 2000.

Mandy Bryon
Consultant Clinical Psychologist
Department of Psychological Medicine
Great Ormond Street Hospital for
Children NHS Trust
London WC1N 3JH (UK)
Tel. +44 20 7829 8679
Fax +44 20 7829 8657
E-Mail bryona@gosh.nhs.uk

The Future

Bush A, Alton EWFW, Davies JC, Griesenbach U, Jaffe A (eds): Cystic Fibrosis in the 21st Century.
Prog Respir Res. Basel, Karger, 2006, vol 34, pp 316–322

The Future

Andrew Bush[a]   Uta Griesenbach[b]   Jane C. Davies[a,b]   Eric W.F.W. Alton[b]   Adam Jaffe[c]

[a]Department of Paediatric Respiratory Medicine, Royal Brompton Hospital,
[b]Department of Gene Therapy, Imperial College,
[c]Department of Paediatric Respiratory Medicine, Great Ormond Street Children's Hospital,
 London, UK

Abstract

The concept of cystic fibrosis (CF) has progressed from a lethal diarrhoeal and respiratory disease of children to a multisystem disease, with survival into adult life being common. Concurrently, the rapidly fatal prognosis of the 'CF child' has changed to the well adult or child who happens to have CF. New knowledge will lead to a further broadening of diagnostic criteria and diagnostic testing. New multisystem complications are being detected, and strategies for prevention must be put in place. Fundamental knowledge of molecular and cell biology has led to the transition from symptom-directed treatment to the start of an era of specific treatments directed against the underlying defect, a trend which will continue until a specific and hopefully curative treatment is developed.

The individual chapters of this book have reviewed the new developments in cystic fibrosis (CF). The purpose of this chapter is to speculate provocatively about where we are going in the next 5 years, based on the current position and a look back into the past. At the least, it is hoped that reading this chapter in 5 years time will be a source of innocent merriment if some or all of the prophecies turn out to have been radically wrong!

Cystic Fibrosis: An Expanded Diagnosis

A commonly used opening statement to any communication about CF is to describe it as 'a disease caused by a mutation in the cystic fibrosis transmembrane regulator (CFTR) gene on the long arm of chromosome 7'. Increasingly, it is becoming clear that, although this is true for the vast majority of patients with CF, for a small number it is just untrue. Clinical CF has been described with a completely normal CFTR gene sequence [1]. How do we account for this?

The complexities of processing of CFTR, from transcription, to post-transcriptional modification in the journey to the apical cell membrane, and the recycling and reprocessing that goes on at the membrane, have been reviewed in detail. CFTR interacts with at least 20 proteins on the journey, and with at least 10 more at the cell surface. Is it conceivable that mutations in one or more of these proteins could prevent CFTR reaching its destination and functioning when it gets there? Genotype-phenotype correlations were described in chapter 8, and the effects of modifier genes in chapter 10, from which it is clear that the end stage disease phenotype is the product of complex interactions between the CFTR locus, gene products from other multiple loci, and the environment. Could dysfunction of a powerful modifier gene, even in the presence of normal CFTR, actually produce CF? In diffuse panbronchiolitis, a disease mostly described in the Japanese, the clinical phenotype is remarkably similar to CF, in that patients have recurrent sinusitis, bronchiectasis and chronic infection with bacteria including *Pseudomonas aeruginosa*. Although an increased prevalence of CFTR mutations in these patients has been reported, most have a normal CFTR gene and ion transport, suggesting that genes other than CFTR are important. Recently, cases of a CF-like disease have been presented in abstract, apparently caused by mutations in the epithelial sodium channel (ENaC) gene, not however severe enough to cause pseudohyperaldosteronism [2]. It seems likely that rare cases of clinical CF disease with normal CFTR will continue to be discovered.

CFTR: Expanded Functions

Once thought to be a simple chloride channel, it has become clear that CFTR has a multiplicity of functions. We do not know which functions are important in the production of particular manifestations of CF, but simple observation reveals that there is no clear correlation between 'CFTR function' and organ system disease. There are pancreatic-sufficient patients with severe lung disease, and pancreatic-insufficient with no discernable respiratory problems. Males may be fertile but have severe lung disease. It may be that in the future we will resolve this paradox by discerning which CFTR functions are important for particular organs, or discover mutations in genes which are responsible for the organ-specific expression of CFTR which may modify the disease. A hint of this came in a recent paper, in which nasal potential differences were measured in a large group of patients [3]. Residual CFTR function, as measured by the (reduced) response to low chloride/isoprenaline solution correlated best with pancreatic disease, and the size of the deflection to amiloride, reflecting ENaC, with lung disease. We need more understanding of which of the many functions of CFTR (or indeed, as yet undetermined functions) are important in which organ. This has implications both for diagnosis, and for the new therapies which are so eagerly anticipated.

Cystic Fibrosis: New Diagnostic Tests

The sweat test has long been the bedrock of CF diagnosis and is likely to be so for the foreseeable future. In one respect it has done us a disservice – just because measurement of chloride concentration is an excellent and discriminatory diagnostic test should not be taken to mean that any manifestation of the disease, with the probable exceptions of pseudo-Bartter's syndrome and heat exhaustion, have anything to do with chloride function. This challenging possibility needs to be at least borne in mind, despite the currently ever more compelling evidence linking ion channel dysfunction to CF lung disease. The implications of the multifunctional nature of CFTR include the possibility that a mutation might only affect some functions but not all;

speculatively and hypothetically, chloride channel function might be preserved, whereas glutathione transport could be lost, giving a very unusual CF phenotype, with normal sweat electrolytes. Diagnosis might be made by gene sequencing, or through the development of newer biochemical tests for the disease. These many other functions may be regulated differently. For example, chloride transfer requires ATP, glutathione ADP [4], raising the possibility of abnormalities in some functions of a normal molecule, due to disorder of a regulatory pathway. Diagnostic issues will be further complicated when newborn screening is universal. The common clinical triggers to initiate tests for CF will virtually disappear. Instead, clinical astuteness will be needed, to be sensitive to what is now atypical, but will become 'typical', i.e. the common late presentations of CF in a screened population. The panel of tests utilized will change, with the sweat test, although still being used, being increasingly complemented by electrical tests (chapter 13) and sophisticated genetic analyses. In the age of really exciting science, clinical acumen is alive, well and has a good prognosis.

Cystic Fibrosis: New Attitudes

The improved prognosis of CF, combined with earlier diagnosis through screening, has already started to result in a change in attitude. The hitherto scrawny, ill, infected CF infant who will die before adult life is increasingly being replaced by a fit individual, who has only ever had minimal if any symptoms, who happens to have a problem called CF. Clearly, the lessons of history for the older CF physician include absolutely to beware of complacency. How will the new generation of CF patients react to the demands of a complex treatment regimen?

It is quite clear that we need to work to minimize the demands made on the well patient, while continuing to maintain vigilance. The completely well child will simply not do 30 minutes of postural drainage twice daily. Physiotherapy techniques, reviewed in chapter 38, will need to become more creative, and fit more easily into a normal life style. Airway drug delivery must become more speedy and efficient, using powerful intelligent nebulizers, or more likely, dry powder devices to deliver antibiotics and mucolytics. We need better agents to assist in mucus clearance; recombinant human DNase (rhDNase) gives worthwhile clinical improvement in up to half of those who trial it, and hypertonic saline may help some rhDNase failures, but there remain a hard core who are not helped by current agents. We need to know more about the components of airway secretions, and determine new ways of preventing their accumulation.

Recent work, summarized in chapters 16, and 25–27, have painted a picture that is not all rosy. What is a 'well' CF patient? Even young patients, apparently without any overt respiratory disease, have evidence of airflow obstruction, and ongoing airway inflammation. We must beware of over-interpreting small changes, but be aware they may have long-term effects. From the world of asthma, middle-aged adults who had 'wheezy bronchitis' as children, presumed to be on the basis of developmental reductions in airway calibre, and who 'recovered' completely in later childhood, are now showing an accelerated decline in lung function [5].

The concept of the ageing CF patient is becoming more important; will we need to switch our focus to preventing accelerated lung ageing in this population also?

Cystic Fibrosis: How to Assess in the Modern Era?

The annual assessment of the CF patient is a ritual hallowed by tradition, which has provided the means to achieve the worthy aim of taking stock of how the CF patient is progressing. With the increasing appreciation that CF is a multisystem disease, the length and complexity of the process may be getting out of hand, and there becomes a risk that we obsess on process without considering outcome. Tests which have been included are review by at least six professionals (physician, nurse specialist, physiotherapist, dietician, psychologist, social worker); full blood work, including an ever-increasing range of vitamins and other levels; chest radiograph, high resolution CT scan, exercise testing, bone densitometry, oral glucose tolerance test, abdominal ultrasound; full lung function, including lung volumes; stool collection for fat estimation; tests of 8th nerve function; and the list is by no means exhaustive. There is a real danger that we may all become so exhausted by this information-accumulating exercise that we may forget to interact with the family, and eventually reach the point where the next assessment is due to start as the first one finishes. This is not to decry the usefulness of any of these activities, when applied appropriately. Nor should it be denied that the approach of ever more intensive investigation and treatment, when compared in particular to a casual, *laissez-faire* attitude, has brought dividends in terms of prolonging survival. However, perhaps the time has come when the 'everyone gets everything I can possibly think of as often as possible' is replaced by a more thoughtful consideration of who needs what, so the benefits of the old approach can be retained while minimizing what we ask of our patients.

Cystic Fibrosis: Implications of Longevity – New Complications

The need to re-evaluate strategies which have in the past been excellent, but may now be producing new problems, is nowhere more evident than in the treatment of infection. In the early days of CF, *S. aureus* was the major scourge; as anti-staphylococcal agents became more effective, *P. aeruginosa* dominated, and currently chronically infects around 80% of all CF patients. Aggressive strategies to treat initial isolates with combinations of oral, intravenous and, above all, nebulized antibiotics have been dramatically effective. The use of suppressive nebulized antibiotics, and regular intravenous treatments, has been advocated by many. However, there is no doubt that the benefits of antibiotic therapy come with two prices. The first is the emergence of new organisms. The widespread use of a broad-spectrum antibiotic to prevent infection with *S. aureus* was linked to an increased acquisition of *P. aeruginosa* [6]. Another example is that *P. aeruginosa* secretes exo-products which are toxic to *Aspergillus fumigatus*, and perhaps the laudable aim of eliminating the bacterium has resulted in more problems with the fungus. These problems can be overcome, but the principle that all good things come with a price remains valid. Nebulized gentamicin and repeated courses of intravenous antibiotics have been identified as risk factors for the emergence of *Stenotrophomonas maltophilia*. As we get better at treating the current crop of gram-negative rods, it is inevitable that new resistant pathogens will emerge. The other problem is antibiotic side-effects; subtle disturbances of renal tubular function, 8th cranial nerve dysfunction, and antibiotic allergy. One group reported a correlation between renal function and the number of courses of intravenous aminoglycosides [7]. Some patients had received more than 100 courses. We have encouraged the liberal use of antibiotics in the past, with excellent benefit; perhaps in the next 5 years we need to reign back on their use, and be more targeted and focused. There is precedent for reigning in on the 'more is better' approach; the morbidity of fibrosing colonopathy could have been avoided by more cautious use of pancreatic enzymes.

CF was previously thought to be a disease of the lungs and pancreas; now the multisystem nature has been appreciated, and this is reflected in chapters 33–35 in particular. The next two future challenges are to institute preventive programs where possible (difficult in the context of the 'well person who happens to have CF', see above) and also to detect and manage early manifestations of the problem.

It is clear that bone disease is a major problem in CF adults, with at least one third having osteopenia. How should this alter paediatric practice? Overspill of pro-inflammatory cytokines from the airway is one factor causing bone disease, but irrespective of any systemic consequences, we would treat lung disease energetically. However, strategies that have been shown to improve bone mineral density, such as extra milk (in an asthmatic population) [8] – which is of course already encouraged in CF – and weight bearing exercise, should be advocated. Furthermore, delayed puberty needs to be detected and prevented, not just for the psychological effects on the child, but also to preserve bone health. However, a word of caution: the recommendation of additional vitamins such as vitamin K, in the absence of satisfactory clinical therapeutic data has led to a wide discrepancy with respect to dosages prescribed. With the introduction of new therapies, it is essential that we assess their efficacy in a scientific controlled manner, otherwise the opportunity will be lost to evaluate them critically, and we may perhaps subject patients to a lifetime of therapies without any evidence base, just because it was intuitively felt to be a 'good thing'. One such example is that of inhaled steroids, currently widely used, but which the large CF WISE study has shown to be greatly over-prescribed. However, there is another tension: large trials are expensive and time-consuming, and there are only limited numbers of suitable patients. Thus it is unlikely that every single aspect of CF treatment will be subject to a randomized controlled trial, and selectivity is inevitable when deciding where the time, money and effort need to be expended in trial work.

There are at least some preventive strategies for bone disease, but the serious problems of diabetes and incontinence seem at the moment to be unpreventable. Could some manipulation of the destroyed exocrine pancreas prevent the progression to endocrine disease? Would the early institution of pelvic floor exercises prevent the later development of stress incontinence? There is a real need to make progress with these two complications. The close interactions between adult physicians and paediatricians will mean that as longevity throws up new complications for the adult physician to treat, so preventive strategies will need to be initiated in the paediatric clinic.

Cystic Fibrosis: How Will We Know Whether We Are Doing Better?

This question can be answered in two ways. The first is for an individual clinic, using National and International databases to compare their results with the best available, and compare the different clinic procedures to ensure that best practice is followed. The second more difficult aspect of this question is how, if we discover a new specific molecular or cell-based treatment, will we know that it is beneficial? Mortality is no longer a feasible end-point, so surrogates must be obtained. Each has a unique problem. In the main, pulmonary function is so stable that rate of change of spirometry as an end-point requires large numbers studied for a long time [9]. The use of molecular end-points might be thought to be the best; but mRNA expression may not equate to functional protein at the apical cell membrane, or clinical outcome. Even protein detection may not be the answer; the protein may be present, but not function effectively. Functional assays may overcome this in part, for example potential difference measured in upper or lower airway (chapter 13); however, these beg the question as to what functions are important (above). There is clearly a hierarchy of potential usefulness; the CF mouse has brown not white teeth, but tooth colour is unlikely to be of importance. However, reducing the increased adherence of *P. aeruginosa* to respiratory epithelial cells would likely be highly beneficial, if this led to the reduction in the prevalence of chronic infection. Disease end-points might also be of use; would the reduction of an inflammatory mediator be accepted as beneficial? Again, caution is needed; which mediator is truly important? A trial of a leukotriene B_4 receptor antagonist was stopped because of *more* infections in the treated group. There is no doubt that as our patients are in better health, the detection of benefit will become harder, and this is an ongoing research area.

Cystic Fibrosis: New Clinical Research and Treatments

This arm of the clinical research effort is unlikely to cure CF, but could increase the length and quality of life. New aids to airway clearance, more efficient drug delivery to the airway, and prevention and treatment of new complications have been discussed above. New knowledge of airway defences (chapter 15) will hopefully allow the development of new, potent antimicrobials. A potentially fruitful class of compounds are the macrolides. The beneficial effects of azithromycin are one of the major recent therapeutic advances (chapter 25), although the mechanisms of benefit are still unknown. There are more than 1,700 macrolides in nature; can we find a designer macrolide, more potent for whatever the beneficial mechanism may be, but without the potential problems of gastrointestinal disease or induction of antibiotic resistance?

Another challenge is to find more options for the patient dying of CF. Many who opt for transplantation die waiting for an organ. Living-related donation has proved to be part of the answer, despite the formidable ethical and logistic issues; perhaps transgenic pig lungs will be used to address the current organ shortage, Although re-transplantation is performed, the results are inferior to those of first transplant, and one must question the ethics of this procedure in an era of organ shortage. An unlimited supply of transgenic pig lungs would make the widespread use of re-transplantation eminently practical.

More futuristically, will we ever build an artificial lung? The heart is a much simpler organ, but the artificial heart, initially requiring the patient to be permanently attached to apparatus the size of a domestic freezer, is now fully implantable. Can we make a mechanical lung? Or grow a lung in a sophisticated tissue culture system? Or even use retinoids or some other compound to regenerate the lung, with therapy recapitulating normal lung development?

Cystic Fibrosis: New Specific Therapies

The major aim of CF search is to find a cure, probably for lung disease initially, then for other organ disease. One encouraging aspect of the current research position is the multiplicity of different approaches.

With the advancement of new scientific technologies, molecular techniques and mass spectroscopy are becoming more accessible and cheaper, and are being increasingly applied to dissect out the molecular pathways and develop novel therapies. One major accomplishment has been the completion of sequencing the human genome in recent years. The next scientific goal is comparative genomics and gene function which will allow us to survey the whole genome to identify single nucleotide polymorphisms in an effort to identify modifier genes as discussed in chapter 10. Structural genomics determines CFTR three-dimensional structure for use in future structure-based drug design as discussed in chapter 4. These studies, aimed at identifying candidate modifier proteins may identify key genes and proteins that will be potential therapeutic targets [10].

The detailed knowledge of CFTR processing, and the delineation of different classes of mutations, have led to the concept of pharmacological therapies, either to move mutant CFTR to the apical cell membrane, increase its survival time when it reaches the membrane, or stimulate one or more of its functions. Another approach is the insertion of the missing normal CFTR gene, or even repairing the mutant gene. This challenging task will call for the skills of aerosol scientists, as well as molecular biology, and it should be recalled that every day, the airway is exposed to more than 7,000 litres of air contaminated with bacteria, allergens, viruses and pollutants, and is uniquely efficiently designed to ensure that all these are removed while causing minimal damage. Thus the tempting and superficially easily accessible airway has formidable natural barriers, which will not distinguish between malign environmental influences and the potentially beneficial interventions of the gene therapist. This is particularly likely to be true for viral vectors. Finally, will cellular-based therapy with stem cells be the answer? Curing CF with a bone marrow transplant is attractive, in particular if marrow was harvested from the CF patient, corrected with in vitro gene therapy, and an autologous transplant performed. However, the lack of increased engraftment over time in the transplanted lung would seem to militate against significant traffic between the bone marrow and the airway under normal conditions [11]. This may change if the airway is damaged; it would be a brave investigator who in the context of CF airway inflammation and infection would perform an autologous marrow transplant and inflict further damage on the lung to promote cell trafficking from the marrow. But bravery may pay dividends! Yet another alternative approach has been to target not CFTR itself, but other molecules which may replace CFTR missing function(s).

Another possibility is that the 'cure' for CF will come not from directed research, but as the result of 'blue skies' studies, possibly in a completely different field. We will need to combine focus on the areas of CF science advancement with an awareness of results in other fields, which might advantageously be applied to CF.

Which is the likely 'winner' of the race? It may be that different approaches may work for different classes of mutations. We need to hold the balance between the necessary passionate belief in the approach one is oneself working on, and the need to be aware that multiple different approaches may be needed to help the whole range of CF patients. It would be foolhardy to discard all but one approach, whichever it might be, in the current state of knowledge. What we do know, is that multicentre collaborative efforts, such as the therapeutics development

Bush/Griesenbach/Davies/Alton/Jaffe

programme supported and coordinated through the Cystic Fibrosis Foundation are more likely to achieve results. Other collaborative models have been the UK CF Gene Therapy Consortium, and the funding of multicentre studies such as TOPIC and CALICO. Such approaches are expensive, and the challenge will be to find adequate funding to do the trials efficiently.

Research in Children?

Two factors have made this subject pertinent: the introduction of uniform newborn screening, and the realization that many therapies probably will work best in the early stages of disease, before irreversible damage has supervened. However, there are many ethical and practical difficulties [12].

Many aspects of early treatment of CF are anecdote- not evidence-based. We need to follow the lead of the Paediatric Oncologists, and ensure every screened baby is entered into a clinical trial. This would mean that over a 5-year trial period we could find out what really is the best prophylactic antibiotic, whether early institution of ursodeoxycholic acid therapy prevents liver disease, and sort out many other important questions. Funding is again an important issue; therapeutic research in children has been traditionally underfunded, and funding bodies are also reluctant to put money into studies which will take years to complete.

The selection of the timing of the introduction of new treatments is a difficult issue. The risks must be considered; the cases of lymphoproliferative disease in children treated with gene therapy for severe immunodeficiency have highlighted the potential problems [13]. In this case, the increased longevity of CF patients militates against therapeutic research. An early death in a baby who would have been dead within months is a tragedy, but much worse if instead the median survival is 40 years. So we need to find ways of delineating a higher risk young population in whom therapeutic trials are more justified; and we need to address the ethical issues around how early a child can take part in trials of a novel approach like gene therapy. At what point can we say that a treatment is so efficacious that it would be unfair discrimination to deny it to a 6-year-old, particularly if as is likely, invasive techniques such as bronchoscopy are needed to determine efficacy? The ideal is that therapeutic efficacy is demonstrated in adults, and non-invasive end-points discovered, before application to children. But what if the nature of the disease, or the changes of normal lung development, mean that a treatment is only efficacious if applied early. How do we avoid missing detecting such a therapy, while remaining ethical?

Summary and Conclusions

At the start of the 20th century, CF had not even been identified as a separate disease; half way through the century, it was a lethal respiratory and digestive disease of children, cause unknown; by the start of the 21st century, a huge amount of basic airway and CFTR biology has been discovered; there are nearly as many adults as children alive with the disease; and the multi-organ implications are being grasped. Treatment has gone from fire-fighting (let us treat an infection which is causing a problem now) to prevention of complications of current problems (nebulized antibiotics for the well CF patient who has isolated *P. aeruginosa*) and now to strategies to prevent future complications such as bone disease, and to the development of specific cell- and molecular-based therapies aimed at the fundamental disease process.

What will be being written half way through the 21st century? How will CF have changed? Possibly to a gastrointestinal disease which is a mild nuisance to middle aged people? Or even confined to the museums with smallpox, of historical interest only? Whatever the outcome, there is no reason to suppose that the rate of gain of knowledge of the disease will slacken in the near future, and a follow up volume in a few years is likely to be unrecognizable from this one, thanks to the ongoing efforts of the multidisciplinary team and basic scientists from many different areas.

References

1   Groman JD, Meyer ME, Wilmott RW, Zeitlin PL, Cutting GR: Variant cystic fibrosis phenotypes in the absence of CFTR mutations. N Engl J Med 2002;347:401–407.
2   Sheridan MB, Groman JD, Fong P, Conrad C, Flume P, Diaz R, et al: Mutations in the epithelial sodium channel cause a novel phenotype resembling non-classic cystic fibrosis (abstract). Pediatr Pulmonol 2004(suppl 27): 222–223.
3   Fajac I, Hubert D, Guillemot D, Honore I, Bienvenu T, Volter F, et al: Nasal airway ion transport is linked to the cystic fibrosis phenotype in adult patients. Thorax 2004;59: 971–976.
4   Kogan I, Ramjeesingh M, Li C, Kidd JF, Wang Y, Leslie EM, Cole SP, Bear CE: CFTR directly mediates nucleotide-regulated glutathione flux. EMBO J 2003;22:1981–1989.

5 Edwards CA, Osman LM, Godden DJ, Douglas JG: Wheezy bronchitis in childhood: A distinct clinical entity with lifelong significance? Chest 2003;124:18–24.

6 Ratjen F, Comes G, Paul K, Posselt HG, Wagner TO, Harms K: Effect of continuous antistaphylococcal therapy on the rate of *P. aeruginosa* acquisition in patients with cystic fibrosis. Pediatr Pulmonol 2001;31:13–16.

7 Al Aloul M, Miller H, Alapati S, Stockton PA, Ledson MJ, Walshaw MJ: Renal impairment in cystic fibrosis patients due to repeated intravenous aminoglycoside use. Pediatr Pulmonol 2005;39:15–20.

8 Bonjour JP, Chevalley T, Ammann P, Slosman D, Rizzoli R: Gain in bone mineral mass in prepubertal girls 3.5 years after discontinuation of calcium supplementation: A follow-up study. Lancet 2001;358:1208–1212.

9 Davis PB, Byard PJ, Konstan MW: Identifying treatments that halt progression of pulmonary disease in cystic fibrosis. Pediatr Res 1997;41:161–165.

10 Boucher RC: New concepts of the pathogenesis of cystic fibrosis lung disease. Eur Respir J 2004;23:146–158.

11 Spencer H, Rampling D, Aurora P, Bonnet D, Hart SL, Jaffe A: Transbronchial biopsies provide longitudinal evidence for epithelial chimerism in children following sex mismatched lung transplantation. Thorax 2005;60:60–62.

12 McIntosh N, Bates P, Brykczynska G, Dunstan G, Goldman A, Harvey D, Larcher V, McCrae D, McKinnon A, Patton M, Saunders J, Shelley P: Guidelines for the ethical conduct of medical research involving children. Royal College of Paediatrics, Child Health: Ethics Advisory Committee. Arch Dis Child 2000;82:177–182.

13 Hacein-Bey-Abina S, Von Kalle C, Schmidt M, McCormack MP, Wulffraat N, Leboulch P, Lim A, Osborne CS, Pawliuk R, Morillon E, Sorensen R, Forster A, Fraser P, Cohen JI, de Saint Basile G, Alexander I, Wintergerst U, Frebourg T, Aurias A, Stoppa-Lyonnet D, Romana S, Radford-Weiss I, Gross F, Valensi F, Delabesse E, Macintyre E, Sigaux F, Soulier J, Leiva LE, Wissler M, Prinz C, Rabbitts TH, Le Deist F, Fischer A, Cavazzana-Calvo M: LMO2-associated clonal T cell proliferation in two patients after gene therapy for SCID-X1. Science 2003;302:415–419.

Prof. A. Bush
Department of Pediatric Respiratory Medicine
Royal Brompton Hospital
Sidney Street
London SW3 6NP (UK)
Tel. +44 20 7352 8121
Fax +44 20 7351 8473
E-Mail a.bush@rbh.nthames.nhs.uk

Subject Index

eradication of early infection 181
genome sequence comparisons 140
hypermutation 142, 143
pathogenicity islands 141, 142
proteomics studies 112, 142
strain transmission studies 133–135
Psychological issues, cystic fibrosis
adolescents and adherence to treatment
311, 312
children 310, 311
diabetes 281, 282
diagnosis impact on family 309, 310
mealtime behavior management 310
terminal care 312
Puberty, delay in cystic fibrosis 273
Pulmonary function testing
infants 56, 57, 195–198
low-frequency forced oscillation
technique 203
multiple-breath inert gas washout
201–203
preschool children 200, 201
raised volume rapid thoracoabdominal
technique 196, 197, 199, 201
resistance interrupter technique 200

Radionuclide imaging, liver disease in
cystic fibrosis 253, 254
Raised volume rapid thoracoabdominal
technique, pulmonary function testing
196, 197, 199, 201
R domain
mechanism of action 42
phosphorylation 41, 42
structure 41
Resistance interrupter technique, pulmonary
function testing 200
Respiratory failure

progression 173, 174
terminal care 175
ventilation management 174

Single-nucleotide polymorphisms (SNPs)
abundance in human genome 78
cystic fibrosis modifier genes, *see also*
specific genes
classification 78–81
gastrointestinal and liver disease 81
study design 81, 82
definition 77
high-throughput analysis 77, 78
Sinusitis, cystic fibrosis related disease 72
Sirolimus, immunosuppression in lung
transplantation 178
Sodium/potassium-ATPase, respiratory
epithelial ion transport 102, 103
Sodium/proton exchanger, cystic fibrosis
transmembrane conductance regulator
interactions 48
Splenomegaly, management 258
Staphylococcus aureus, see also Methicillin-
resistant *Staphylococcus aureus*
antibiotic therapy 155, 156, 180, 181
chronic infection in cystic fibrosis 153,
154
epidemiology 154
lung disease exacerbation in cystic
fibrosis 155
microbiology 154
Stem cell therapy, cystic fibrosis 225, 226
Stenotrophomonas maltophilia
antibiotic therapy 149, 150
epidemiology of infection in cystic
fibrosis 149
identification 149
opportunistic infection 149

reservoirs 148, 149
taxonomy 150
Sweat chloride test, cystic fibrosis diagnosis
70

Thapsigargin, ΔF508 cystic fibrosis
transmembrane conductance regulator
mutant effects 214
Transforming growth factor-β (TGF-β),
cystic fibrosis modifier gene 80, 81
Tumor necrosis factor-α (TNF-α), cystic
fibrosis modifier gene 79
Twin studies, overview 78

Ultrasonography, liver disease in cystic
fibrosis 253
Urinary continence, cystic fibrosis patients
265

Varices, management 256, 257
Vasculitis, cystic fibrosis 270
Vitamin A, supplementation in cystic
fibrosis 295
Vitamin D
deficiency in cystic fibrosis 274,
297
supplementation in cystic fibrosis 275,
295, 296
Vitamin E, supplementation in cystic
fibrosis 296
Vitamin K
deficiency in cystic fibrosis 274, 297
supplementation in cystic fibrosis 275,
296

Zafirlukast, anti-inflammatory therapy
189